PHYSICAL SCIENCE
for Christian Schools

PHYSICAL SCIENCE for Christian Schools

by
Emmett L. Williams, Ph. D.
and
George Mulfinger, Jr., M. S.

Illustrated by Robert W. Martin

BOB JONES UNIVERSITY PRESS
GREENVILLE, SOUTH CAROLINA 29614

Library of Congress Cataloging in Publication Data

Williams, Emmett L
Physical science for Christian schools.

Includes index.
1. Science. 2. Religion and science.
3. Scientists. I. Mulfinger, George, 1932—
joint author. II. Title.
Q161.2.W53 500.2 74-22073
ISBN 0-89084-000-8

PHYSICAL SCIENCE
for Christian Schools

by Emmett L. Williams
and George Mulfinger, Jr.

Printed in the United States of America

Dedication

"Thine, O Lord, is the greatness, and the power, and the glory, and the victory, and the majesty: for all that is in the heaven and in the earth is thine; thine is the kingdom, O Lord, and thou art exalted as head above all" (I Chronicles 29:11).

"Thou art worthy, O Lord, to receive glory and honour and power: for thou hast created all things, and for thy pleasure they are and were created" (Revelation 4:11).

"Even so every good tree bringeth forth good fruit; but a corrupt tree bringeth forth evil fruit" (Matthew 7:17).

The two great opposing world-views, evolutionism and creationism, can be thought of as two trees, each producing its own fruit. But how different are the fruits!

Evolutionism spawns a disrespect for authority, for moral values, and for God Himself. Evolutionism basically destroys man by convincing him he is a mere accident of nature, a clever animal at best.

Creationism engenders in its followers a respect for the Creator. It keeps one mindful that he will someday face God. Creationism is the foundation stone that underlies all other doctrines of Scripture. Historically, creationism has had the effect of holding back the influence of evil on society. Creationism gives man hope as God's highest creation—hope for this life, and hope for the life to come. It is to the propagation of this world-view as it is presented in the Word of God that this book is dedicated.

Acknowledgments

Special mention should be made of the tireless dedication of two members of the authors' families—Mary Williams and Linda Mulfinger, a wife and a daughter respectively, who not only typed seemingly endless pages of original manuscript, but also wrote many hundreds of letters to obtain photographs for the book. As you can readily observe by leafing through the following pages, their correspondence bore rich fruit.

We are also particularly indebted to our colleague and division chairman, Dr. Joseph Henson, whose encouragement and promotional efforts initiated an extensive science-textbook-writing project in which we are still engaged. Our close association and collaboration with Dr. Henson in countless Bible-science seminars across the country has brought about a mutual enrichment and a fusion of our ideas. As a result, many of his perceptive insights appear in the book, especially in the first three chapters.

We are also greatly indebted to our colleague, Paul Brown of the Bob Jones University chemistry department, whose indefatigable efforts in proofreading manuscript at an early stage of the writing process did much to avert potential problems later on. Mr. Brown assisted not only in the capacity of one having scientific expertise, but as a grammarian as well.

Several persons helped out in an editorial capacity. Of these we would especially like to acknowledge the very extensive and capable labors of Richard Adams and Barbara Rumminger.

Special thanks are due to Evert Fruitman of Phoenix, Arizona, who undertook a number of difficult photographic assignments for us on request. His beautiful shots greatly enhance the appearance of the book. Also, the following individuals provided an almost endless supply of usable high quality photos: Robert A. Carlisle, head, Still Photo Branch, Department of the Navy; and James E. Westcott, photo officer, United States Atomic Energy Commission. The many governmental agencies and industrial concerns that furnished photos are far too numerous to mention here, but the appropriate credit is given on pages 626-627. Substantial help with the photographic illustrations was also received from Pete Steveson, Dayton Walker, and Roy E. Waite.

We would like to express our sincere appreciation for a variety of valuable suggestions to the following: Isabel Lambert, Bill Pinkston, Tom Hill, Janet Flower, Ed Birkenstock, Walter Fremont, and Rachel Larson. The following were kind enough to trial-teach portions of the book when the manuscript was still in rough form: Brian Jenkins, Emil Walker, and Judy Lanier.

The anonymous benefactors who made this textbook possible by providing support for the authors deserve the heartfelt commendation and thanks of all concerned. It was truly an answer to prayer that Christians who were in a position to do something felt the burden of this undertaking and realized the importance of this ministry of reaching young people with the truth.

The authors share a special feeling of pride in the artwork in this textbook. In a sense, Robert W. Martin, whose imaginative drawings adorn its pages, is our "discovery." We first made his acquaintance at a Bible-science seminar; we can perceive the very definite leading of the Lord in this meeting. Much to our satisfaction, Bob has since joined the staff at the Bob Jones University Press, not only as part of our textbook project, but also as an illustrator for the magazine, *FAITH for the Family*.

Finally, a word is in order concerning the setting in which the book was written. By its very nature, such an undertaking requires not only an auspicious intellectual climate for its successful effectuation, but also a healthy spiritual atmosphere. The incomparable campus of the "World's Most Unusual University®" was found to fulfill both requirements in most generous measure.

While we wish to give all the credit that is due to those who assisted in so many ways, the authors accept the full responsibility for any errors in factual content or judgment. We would be most happy to hear from any readers who wish to offer comments or suggestions.

Greenville, South Carolina
July 1974

Emmett L. Williams
George Mulfinger, Jr.

Preface

Physical science needs to be put into a proper perspective for the Christian high school student. The limitations as well as the accomplishments of scientists should be pointed out.

Science is generally overglorified in existing physical science textbooks. The many limitations inherent in science are usually glossed over in the standard textbooks. But the discipline of science is a human activity, and therefore subject to all the limitations of human nature. Adhering to a Biblical framework, we point out that man is a fallible, fallen creature, severely restricted in what he can ascertain and accomplish (Isaiah 55:9, I Corinthians 1:18-21, 3:19-21).

The failure of scientists to explain the nature of such basic things as matter, energy, and gravity, the tentativeness of scientific ideas, and the inability of scientists to make legitimate pronouncements on the subject of origins, are but a few of the many shortcomings.

Observation is the key ingredient of the scientific method. If a phenomenon cannot be observed by man's senses, either directly or with the aid of instruments, then it cannot be dealt with scientifically. In accordance with such restrictions, all speculations concerning the remote past, the distant future, and such conjectured unobservables as extraterrestrial life are eliminated from the lawful domain of science.

Considered collectively these limitations will help the student to realize that science is not a fit object of worship. The great prevalence of science worship today can be traced in part to a very specific defect in science courses and textbooks: the failure of teachers and authors to present a realistic picture of science.

Historical Approach The human element in physical science is important in yet another sense. The use of a historical approach to the subject can give the student an appreciation for the personalities who had a part in structuring the various laws and theories, and for the arbitrariness of the choices made. Blunders and blind alleys in the history of chemistry and physics, often deliberately omitted from textbooks, are here reexplored and assessed.

At certain key points, biographical sketches of outstanding Christian men of science are included. In each case a brief description of the scientist's work is presented, together with his testimony, beliefs, and insights.

All of the scientists who have been selected for biographical sketches were fundamental, Bible-believing Christians. It is hoped that this feature will serve to dispel the erroneous idea that the greatest scientists have been atheists or agnostics.

When scientific concepts such as the atomic theory are presented from a historical standpoint, the student immediately develops an appreciation of the changeableness of scientific ideas. The fact that scientific truth is not absolute is illustrated clearly. Hopefully, then, when furnished with enough insight from the history of science, the student will learn to evaluate scientific statements for himself. We are concerned that the Christian student have the necessary scientific facts at his disposal, but he should also be able to use his mind to develop a proper Christian philosophy of science.

Many scientific ideas are left rather mysterious. Present-day science is not presented as the final word. It is hoped that this will encourage many young people to spend their lives in scientific activity. Some might even be able to change the structure of scientific knowledge.

Creation and Evolution One might expect that evolution and a general anti-God philosophy would not constitute a major problem in a course which deals only with chemistry and physics. A close study of existing texts, however, reveals a number of objectionable statements that reflect an evolutionary bias of the authors. Evolution is presented and promoted, whether or not it is directly related to the subject.

For example, in one book, in a section on the lever, a highly imaginative description was given concerning a primitive subhuman and his use of a lever to pry rocks from the entrance of a cave. Such allusions are thoroughly unscientific in that they are based on unenlightened guesswork rather than actual observations.

We have endeavored scrupulously in every instance to trace to Biblical origins rather than to hypothetical beginnings conjured up by man's imaginations. Metal working, for example, is traced to Tubal-cain in Genesis 4:22 rather than to some speculative "neolithic man."

Our textbook is not only a Christian book; it is a creationist book. Whenever possible, creationist applications are developed. For example, cave formations and the formation of fossil fuel are discussed within a creation-worldwide flood framework. A young earth model is offered when the scientific dating methods are discussed. Dr. Thomas Barnes' information on the decaying magnetic field of the earth is covered in the chapter on magnetism.

In what is probably a unique chapter for a ninth-grade physical science book, the first and second laws of thermodynamics are discussed and then related to the concept of evolution.

Why do authors of most ninth-grade physical science texts avoid a discussion of these laws to any great depth despite the fact that the laws are the best science we have today? Is it possible that these laws offer such a devastating denial to supposed molecules-to-man evolution that most authors choose to ignore this crowning achievement of science?

Scientists may construct models that are in general accord with these laws. But the impact of thermodynamics on evolutionary thinking is often overlooked or shunned. Sections on organic chemistry and biochemistry can be used by a Christian teacher to discuss and debunk chemical evolution.

Teleology When creation is mentioned, teleology is inferred. The opposing world view, evolution, when consistently applied, implies *dys*teleology. When a student sees these two world views sharply contrasted he will be more inclined to think and examine, rather than to accept pronouncements by scientists in blind faith. For too long the idea of purposiveness or design in nature has been rejected as old-fashioned and unscientific.

Teleology is presented clearly in a number of areas, such as: (a) Some of the intricate mechanisms of the human eye are discussed in the section on optics. (b) The amazing details of the ear and the wonders of the human voice are described in the chapter on sound. (c) The protection afforded by the ozone layer of the stratosphere is covered under the chemistry of oxygen. (d) The function of the earth's magnetic field in shielding the earth's inhabitants from cosmic rays is discussed in the chapter on magnetism. (e) The unique properties of water and their significance for living organisms are taken up in a chapter on water. Other teleological arguments are included at appropriate points; therefore, such examples of God's handiwork should carry a strong message to the mind that is not blinded completely by sin.

Other Distinctives The effect of drugs, alcohol, and tobacco on the human body is presented as the primary physical pollution problem. The authors are in favor of reasonable environmental protection but are opposed to hysterical and hasty measures. Because present-day industrial technology is so highly advanced, engineering and industrial applications of scientific principles as well as research laboratory examples are given.

Many different scientific ideas are pictured through the use of cartoons. The use of humor can do much to increase student interest and aid in the grasp of nebulous models. Sterile scientific discussions may be satisfactory for graduate classes, but the ninth-grader needs to be stimulated before he will attempt to grapple with an abstract idea.

We pray the Lord's richest blessing on every teacher and student who uses this book. May it serve to draw Christians closer to the Lord and to demonstrate to believers and unbelievers alike that true science supports the claims of Scripture.

Contents

UNIT VIII MECHANICS

UNIT IX HEAT

UNIT X ELECTRICITY AND MAGNETISM

UNIT XI WAVE PHENOMENA

UNIT XII TWENTIETH CENTURY PHYSICS

CHRISTIAN MEN OF SCIENCE: BIOGRAPHICAL SKETCHES

To the Reader

Certain material in this book is starred. This material exceeds the scope generally accepted for high school physical science courses. It is included, however, for study by the advanced or highly motivated student and for use in special projects, as well as for the edification of the general reader. When a star appears at the top of a chapter title page (for example, Chapter 15), this indicates that the entire chapter is advanced material. In other places, the star applies to only the question or experiment indicated. Stars that appear within a subsection are applicable to the remainder of that subsection.

A number of experiments are included in this book. The would-be experimenter is cautioned that virtually any experiment can be hazardous when done carelessly or without adequate safety equipment. For this reason, and because the potency of chemical reagents sometimes varies, it is recommended that experiments be conducted only in a laboratory while under qualified supervision.

Photo next page
An awesome sight such as the Great Orion Nebula should inspire humility and reverence in the researcher.

Introduction

Suggestions on How to Study This book was written to be read. You will be putting yourself at a disadvantage if you fail to read any section that is assigned to you. Be careful, however, that you do not merely go through the motions of studying without any true learning taking place. Here are some suggestions to help you learn as much as possible in a minimum of time:

1. Be rested and alert; give this course your best mental effort.
2. Have a notebook and pen or pencil ready to jot down important facts.
3. Study actively by grouping facts together in your mind as you go along.
4. Ask yourself questions; write them down and try to answer them. If you are unable to answer your own questions after reading the assignment carefully, bring them to class the next day.

0:1 *Proper study conditions make for efficient learning.*

You will improve your understanding if you can correlate what you learn with what you read in newspapers and magazines. Perhaps you can relate some of what you are learning to current happenings in science and technology. The authors hope you will develop a measure of insight as you go through this course so that you learn to distinguish true science from "science falsely so-called" (I Timothy 6:20). There is a vast difference between that which is solidly based on observation and that which is only guesswork. However, the dividing line between the two is often cleverly concealed.

0:2 *Reading a journal to learn of current developments.*

Try to master new words as quickly as possible so that your understanding will not be hindered the second or third time a word occurs. Each unit is arranged cumulatively, and it is essential that you learn one part well before proceeding to the next. Sometimes, you may wish to refer to the glossary in the back of the book. It would also be wise to keep a good dictionary handy.

Avoiding Distractions It is important to provide yourself with physical conditions that contribute to learning. You should be comfortable when studying, but not *too* comfortable. Many students do their best learning at a desk, seated in a straight-back chair. The area where you work should be well lighted; inadequate lighting can cause eyestrain. You should have as few distractions as possible; sounds from a radio, record player, television, or from other people can waste your time by interrupting your thoughts.

There can also be distractions from within. Everyone has random thoughts that cross his mind as he reads. You should discipline your mind to stay on the subject. You can greatly increase your ability to learn by developing the habit of concentration.

Study Often You will find that your most effective learning in science or any other subject takes place when you study a little each day. Research has shown that far more can be accomplished by studying a few minutes on a daily basis than by studying the same length of time in a lump sum every few days. The more often you think about something, the more quickly it becomes a part of you. The students who get the most out of this course, therefore, will be the ones who are faithful in their daily work.

UNIT I SCIENCE AND THE CHRISTIAN

Chapter 1 Science and You

1:1 Defining Science Before you start your study of physical science, you ought to know what science is. Some possibilities for a definition of science are:

1. *the sum total of man's experiences*
2. *knowledge gained and confirmed by (exact) observation*
3. *knowledge gained through observation and classification*
4. *knowledge gained through man's investigation*
5. *man's efforts to understand God's creation*

From the above suggested definitions, develop a satisfactory definition of science that is realistic and truthful.

★An extended definition of science might include the following:

★6. *an exact and systematic statement or classification of knowledge on some subject; the knowledge acquired by scientific investigation.*

★7. *the knowledge obtained in scientific investigation systematized into hypotheses, theories, and laws.*

Scientists must communicate their findings (observations) to others in such a manner that the findings will be understood. Hypotheses, theories, and laws (these will be discussed in Chapter 3) are useful, because in these forms scientific knowledge can be subjected to continual testing to determine if that knowledge is true in every case. Scientific ideas are often revised and even discarded on occasion. This may sound strange to you, because you may think of scientific findings as

1:1 *What is science?*

unchangeable, like Biblical Truth. This is not the case: Biblical Truth is forever settled in Heaven, whereas scientific concepts are as changeable as clothing fashions.

1:2 Limitations of Science People sometimes get exaggerated ideas of what science is and what it is capable of doing. We do not wish to curb anyone's enthusiasm, but it is necessary to point out some of the limitations of science.

For one thing, *science deals only with the physical universe.* Science seems unable to cope with the spiritual domain. Perhaps the instruments of science are too crude, and the phenomena too subtle to detect. Suppose, for example, you were to attempt a research project on the subject of prayer. You are interested in finding out how prayer is transmitted. Are there "prayer waves" of some kind? You would find yourself wondering where to begin. You cannot build a detector for such supposed waves if you do not know what they are, and you cannot find out what they are until you can detect them. Such a problem may discourage you before your research project ever gets under way. Many informed individuals are of the opinion that science will never be able to give us answers to questions such as these.

Even in areas where the objects of study are somewhat better understood there can be difficulties. For example, an experimental process may work perfectly in the laboratory, but fail dismally when put to an actual test. The laboratory situation is, of necessity, artificial. Forming hasty conclusions from the results of experiments can be especially hazardous. Medical research is often carried out using experimental animals. A new drug may look promising with guinea pigs, but miss the mark with humans. The manner in which an experimental animal learns to go through a maze does not necessarily have anything to do with learning processes in humans. Yet researchers often attempt to relate the behavior of animals to man, using evolutionary "logic."

Another limitation of science is its inability to prove a **universal negative.** A universal negative is *a blanket statement of*

1:2 *Can experiments with laboratory animals yield information concerning human behavior?*

denial. Suppose someone makes the statement, "There is no such thing as a sea monster." How would you go about proving or disproving this claim? You would probably investigate reports of large, unclassified marine creatures—carcasses that have been washed ashore, footprints on beaches, and similar clues. Suppose that you are able to discredit every clue you are given. In each instance you are able to demonstrate that the report is either a misidentification of some commonplace creature or an outright hoax. You still have not proved what you set out to prove. To prove that sea monsters do not exist would require that you study every bit of sea water, at all depths, over the entire globe. Not only that; you would have to observe each region at the same time because, while you were looking in one place, the slippery sea monster could have gone somewhere else. *Science cannot establish the universal nonexistence of anything.* Only a very limited, localized nonexistence can be proved, and that only in certain situations. For instance, a limited statement, such as, "There are no sea monsters within 1000 feet of Monhegan Island," could be investigated and proved, if you had enough equipment to make the necessary observations.

1:3 *Do sea monsters exist? Is it possible to disprove their existence?*

The same line of reasoning applies to miracles. Many people take the easy way out by choosing not to believe in anything they do not personally experience. That is their privilege. But to say that science has disproved the existence of miracles is completely untrue. Remember, science cannot establish the universal nonexistence of anything. A statement such as "There is no God," therefore, can never be proved or disproved by science. A person who makes such statements is simply telling you his beliefs; science in no way lends support to his anti-God philosophy. Moreover, God's Word labels such an individual as a fool (see Psalm 14:1, 53:1).

1:4 *The floating axe head: "And he cut down a stick, and cast it in thither; and the iron did swim" (II Kings 6:6). Can science analyze miracles such as this?*

1:5 *Science unleashed the power of nuclear energy but offered no moral guidelines for its use.*

Another major shortcoming of science is its inability to make value judgments. Surprising as it may sound, *science cannot tell right from wrong.* Through science we obtain atomic energy, but we are not given any guidelines for its use. Science suggests ways of improving the human race genetically, but offers no plan for deciding who should select which couples will have children. You might think that a computer could pass judgment on moral issues for us, but a computer is nothing more than a complex piece of machinery. It has no way to produce answers of any kind until it is programmed with specific guidelines. If it is given Christian guidelines, the computer will make Christian judgments. Atheistic guidelines will result in atheistic answers.

Science also has the drawback of never producing final or absolute answers. The findings of science must be taken as merely temporary answers. If science were absolute, science textbooks could be written once and left unchanged, with no need for revised editions. *But science is changeable.* In view of this uncertainty, is it not strange that some people choose to worship science rather than God? The Christian puts his faith and trust in the One Who is "the same yesterday, and today, and forever" (Hebrews 13:8). With God there is "no variableness, neither shadow of turning" (James 1:17). His Word represents Absolute Truth, which stands unchanged from one generation to the next. How much better for a person to have his faith grounded on the Solid Rock than on shifting sand!

1:6 *Which is more reliable and enduring, the Bible or science?*

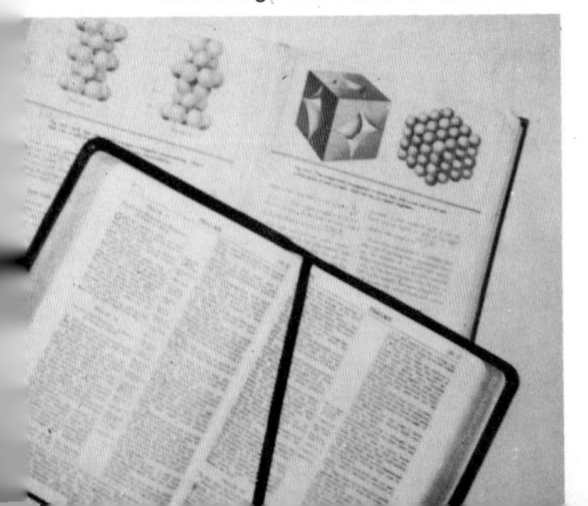

1:3 Need for Discipline in Science A great deal of mental alertness and control is required in science. Details are extremely important. A researcher or student who is careless and too impatient to concern himself with details can easily cause an experiment to end in failure. Also, there is danger connected with some experiments—acid burns, fires, explosions, and electrical shocks are but a few of the accidents that can occur. The successful experimenter stays alert at all times.

The side of science which deals with theory takes its share of discipline as well. The student must be willing to memorize many fundamentals. It is not practical to keep looking up each fact from previous chapters as you proceed through the book. Your teacher will guide you concerning which things should be memorized.

1:7 *A student wearing safety glasses while performing an experiment—a precaution that might save his eyes in case of an accident.*

1:4 Use of Mathematics in Science It will be necessary on occasion to think **quantitatively**—that is, *using numbers.* Quite possibly, you have previously studied science and mathematics as two separate subjects. In some parts of this course you will be required to draw these two fields of study together. Using numbers to describe ideas allows us to develop a more exact science than would be possible by just using words.

When a scientist sits down to examine the facts he has gathered, he will almost always use mathematics in one form or another. Mathematics plays a key role in science. Many scientists will show their final results in the form of a mathematical equation or on a curve (graph).

The more highly refined science becomes, the more mathematics is used. Nearly all of chemistry and physics involves mathematics. Modern concepts of the atom, such as the quantum theory, are highly mathematical. Anyone who earns a doctor's degree in chemistry or physics will need a thorough knowledge of mathematics. More and more biologists are beginning to use mathematics, too. Therefore, if you are thinking seriously about becoming a scientist, you will need to study mathematics.

1:5 Use of Composition in Science When a scientist completes an experiment or develops a new theory, he normally does not travel around the world to tell other scientists about his work. He simply publishes his observations and conclusions in a **scientific journal** (magazine), so that any interested person can read about his work. **Creationist** scientists (scientists who believe the Biblical description of creation) can report their findings in the quarterly journal of the Creation Research Society (CRS), or the Bible-Science Association newsletter.

It is helpful, then, for the scientist to have a good command of the English language and to know proper grammar, so that he is able to express his ideas both orally and in writing. If you are planning to become a scientist, study hard in your English courses. Develop the tools of good composition and grammar; they will be very important to you when you publish your scientific work and will also help you to use proper language when you speak.

1:8 *(a) A scientist using a slide rule. (b) A scientist using a calculator.*

1:9 *Journals in which scientists report the findings of their research.*

1:6 Unanswered Questions One of the first things we must realize is that science, by its very nature, can never answer all of our questions. In fact, one scientist has estimated that for each question science is able to answer, 10 new questions are raised. This might be somewhat discouraging for the young student of science until he realizes that this situation is according to the plan of God.

As we study God's handiwork and seek to find out the ordinances that govern our physical world, we should develop a deep sense of humility. We should realize that we cannot understand everything, for man is totally incapable of thinking on God's level. God declares, "For as the heavens are higher than the earth, so are my ways higher than your ways, and my thoughts than your thoughts" (Isaiah 55:9). Job repeats this theme: "God thundereth marvellously with his voice; great things doeth he, which we cannot comprehend" (Job 37:5). Thus, humbleness and reverence are proper attitudes for the serious scientific investigator.

Actually, we do not even understand some of the most basic things that we deal with in science. What is **matter**? We usually define it as *anything which occupies space and has mass.* But this definition, rather than stating what matter really is, merely describes two of its properties.

For as the heavens are higher than the earth, so are my ways higher than your ways, and my thoughts than your thoughts.
Isaiah 55:9

1:10 *Man tries to think God's thoughts, but his efforts never bring him to absolute truth. All of our successes are still far beneath God.*

1:11 *An awesome sight such as the Great Orion Nebula should inspire humility and reverence in the researcher. (See introduction page)*

What is **energy**? Energy is usually defined as *the ability to do work.* But again, this only tells us what it does, not what it is. We can calculate the amount of energy, and we can follow it through various transformations. But the true "essence" of energy is not known.

What is **gravity**? It is *a force that pulls objects toward the earth.* How does it work? Again, we must plead ignorance. We can calculate the strength of the force for a given object, how fast it will be falling at any given instant, and how long it will take to hit the ground. But to explain how the earth "reaches up" and pulls the object down is beyond the knowledge of present-day science. In fact, no more is understood about exactly "how" gravity works than at the time Sir Isaac Newton set forth his law of gravitation 300 years ago. The same is true of magnetism and electrostatic forces (attraction and repulsion of electrical charges). Their strength can be calculated, but the details of their operation are not yet understood. Keep in mind Job's declaration: "Great things doeth he, which we cannot comprehend."

1:12 *What is the mysterious force called gravity that pulls a falling pencil to the floor? Science does not yet have the answer.*

List of Terms

creationist	quantitative
energy	scientific journal
gravity	universal negative
matter	

Questions

1. In what book of the Bible is the atheist labeled a fool?
2. Science cannot determine right from wrong. True or False?
3. Name four accidents that could happen to the careless laboratory worker.
4. How does a scientist normally make the results of his research known to other scientists?
5. "CRS" stands for "Christian Research Society." True or False?
6. Science is a collection of unchangeable truths. True or False?
7. How is matter usually defined?
8. How is energy usually defined?
9. How magnetism works is well understood today. True or False?
10. List five limitations of science.
11. Why is memorization necessary in science?
12. Scientists have recently disproved the existence of "prayer waves." True or False?
13. Science can never prove a universal negative. True or False?
14. Will science ever have "all the answers"?
15. Why are reverence and humility appropriate attitudes for a scientist?
16. Science has disproved the existence of "sea monsters." True or False?
17. Discuss some recent development in science which has appeared in a magazine or newspaper article.
18. In what respect is a laboratory situation artificial?
19. A computer is nothing more than a very complex machine. True or False?
20. What is science?
21. Why is mathematics necessary in science?
22. Quantitative reasoning involves the use of mathematics. True or False?
23. Why are composition and grammar necessary in science?
24. Newton set forth his law of gravitation in the 1600's. True or False?
25. Can science prove that there is no God? Explain.

Photo next page

When Galileo observed the heavens with his telescope in 1609, he became the first man to verify the truth of Jeremiah 33:22: "The host of heaven cannot be numbered."

UNIT I SCIENCE AND THE CHRISTIAN

Chapter 2 Science and the Bible

2:1 Relation of Science to Scripture Science and Scripture are not in the same category, but they do overlap at times. Table 2-1 shows some comparisons and contrasts between the two fields.

Table 2-1 *Comparison of Scripture and Science*

Scripture	Science
a. unchangeable	a. changeable
b. Absolute Truth	b. workability (a concept works but it is not necessarily true)
c. God's revelation	c. man's physical observation
d. given by a perfect, holy God	d. developed by imperfect, fallible man

From this comparison, it should be apparent that God's Word should never be viewed in the light of modern science; modern science should be viewed in the light of God's Word. Science can never change Scripture; that is why such theories as **theistic evolution** (the belief that evolution was God's method of creation) are worthless. When a Christian studies any subject, including science, he must constantly check to see if what he is reading agrees with Scripture. The Bible has the final word on every subject. If an idea or method is unscriptural, we cannot adopt it without incurring the disfavor of our Lord Jesus Christ.

There are many statements in the Bible that can be subjected to scientific investigation.

2:1 *Science did not become aware that air has weight (Job 28:25) until the seventeenth century. We now use the weight of air to support lighter-than-air craft.*

2:2 *Men rejected the idea that the earth is hung upon nothing (Job 26:7) until the time of Copernicus.*

2:3 *When Galileo observed the heavens with his telescope in 1609, he became the first man to verify the truth of Jeremiah 33:22: "The host of heaven cannot be numbered."*

(See chapter title page)

Table 2-2 *Some Scientific Statements in Scripture*

Air has weight.	Job 28:25
The earth hangs upon nothing.	Job 26:7
The host of heaven (stars) cannot be numbered.	Jeremiah 33:22

The Christian should not be afraid to subject the Bible to honest investigation. The Scriptures have withstood all criticism (honest and dishonest). Investigation shows that the Bible is scientifically accurate. Consequently, science will agree with the Bible when considering scientific *facts* (observed phenomena). However, the Bible will not agree with many scientific *interpretations* (philosophical speculations), such as **evolution** (the theory that complex things develop from simple things over a period of time). No Christian should try to twist Biblical Truth to fit a framework of evolutionary speculation. When any interpretation, scientific or otherwise, is not Biblical, the Christian should denounce it.

Can a Christian, therefore, adopt an evolutionary philosophy? He can, but we believe that he is wrong. The first two chapters of the book of Genesis present creation as the *direct acts* of God, not as processes taking millions of years. God said, "Let there be light, and there was light" (Genesis 1:3). Creation occurred in six days; God spoke and it was finished. No evolutionary processes are possible, because there was not enough time for evolution. If a Christian accepts evolution, he must reject the plain teaching of the Old and New Testaments. The Lord Jesus Christ, the Creator (John 1:1-3), said that "from the beginning of creation God made them (humans) male and female" (Mark 10:6). When a Christian embraces evolution, he willfully or unknowingly calls God a liar. Normally, such a person believes that there are errors in the original manuscripts of the Scriptures.

God gave us a mind to use; we can use our good sense to check science against the Scripture. We discover that the Scriptural viewpoint is a superior system of interpretation. The facts of science line up more logically with a Creation-Flood framework than with an evolutionary framework. We will cover many of these areas in this book.

The scientists in the Creation Research Society have been working to show that the Creation-Flood interpretation is superior. Their publications, some of the finest on this topic, assert that the Christian position is the most reasonable position on origins. Many excellent Bible-science books have been written to demonstrate the creationist thesis. Some of them are listed at the end of this chapter. It would be helpful for you to read several of them.

2:2 Science and Origins Why make such an issue over the origin of the universe, the origin of the solar system, and the origin of life? Simply because there are certain things that are beyond scientific investigation. Before man appeared on the earth, no scientific activity was possible. **Evolutionists** believe that the earth is about five *billion* years old (this estimate has been increasing through the years). However, many evolutionists say that man has been on the earth only two million years. Whatever took place on the earth during the *imagined* 4.998 billion years before man is beyond scientific investigation, because there was no man there to observe what happened.

As **creationists,** we believe that God created the universe (including the earth) in six days. However, God did not create man until the sixth day. Whatever happened before man was created cannot be subjected to scientific investigation.

Thus, the subject of origins must be beyond the field of science. This is an obvious limitation of science which many evolutionary scientists refuse to acknowledge. They are supposedly carrying on experiments to determine the origin of life. That is impossible! If some scientist could generate life in a test tube, this would be no guarantee that it happened in nature exactly the way the scientist caused it to occur in his laboratory.

As creationists, we have a superior position on the subject of origins. Who was there in the beginning to observe how things were created? God, the Creator, was there. He tells us what He did in the first two chapters of the book of Genesis. God *spoke* everything into existence; this is the nearest thing to a scientific account that you will ever read on origins.

Since evolution and creation cannot be subjected to scientific study, you must decide which one you prefer to believe. The choice is yours. Science cannot prove either. God will not force you to believe in creation; you have a free will. Before making such a decision, you should desire to know if scientific facts fit better into an evolutionary world view or into a creationist world view. We will compare the two views as we study physical science throughout this book.

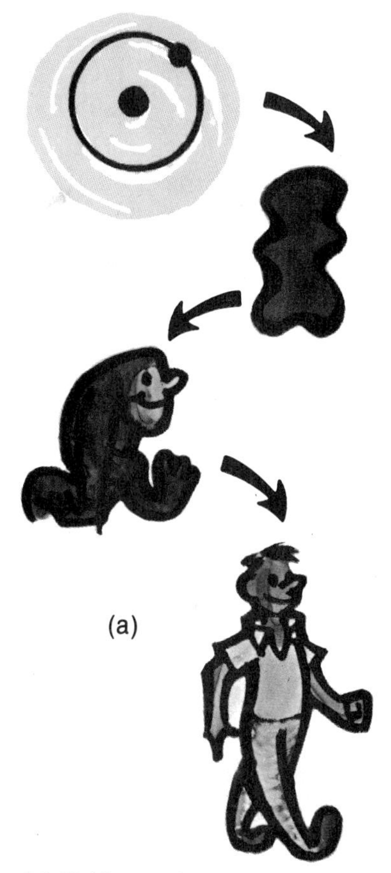

2:4 *Neither evolution from molecules to man (a) nor creation by direct acts of God (b and c) can be demonstrated scientifically. Creation, however, can be shown to be a more reasonable world view.*

2:3 Historical Blunders of Science Every age has been beset with its false ideas. Very often these errors masquerade as science. For 15 centuries it was generally believed that the sun and other members of the solar system revolved around the earth. The **geocentric theory,** as it was called, was undoubtedly one of the most long-lived theories in the history of science. But it was wrong.

In the 17th and 18th centuries, the **phlogiston theory** was widely believed by the leading scientists of the day and taught almost as fact at the world's major universities. The theory said that every substance which burned contained the magic ingredient "phlogiston," which gave it its combustibility. After it had burned, the ashes were said to be "dephlogisticated," rendering them unable to burn further. The phlogiston theory was finally refuted by French chemist Antoine Lavoisier, who showed that combustion involved the element oxygen. Yet prior to the overthrow of the phlogiston theory, a person risked being banished from the scientific community for failing to believe it!

At about the same time, the **caloric theory** was extremely popular in Europe. It stated that heat was a form of matter which was capable of flowing from a hot object to a cold object. However, through the careful work of England's Count Rumford and others, heat came to be recognized as a form of energy rather than a material substance.

2:5 *Our budding young scientist realizes that many scientific concepts are no longer valid; so he has painted over them. He wonders if and when the modern atomic theory will be ready for the scientific scrap heap.*

No age thus far has been immune to such errors. Science is a human undertaking and man is fallible. In I Corinthians 3:19 we are admonished, "For the wisdom of this world is foolishness with God."

In 1796, Pierre Simon de Laplace, a brilliant French mathematician but an atheist, put forth his **nebular hypothesis,** a scheme to explain the origin of the solar system apart from God. Laplace asserted that the sun and all of the astronomical bodies revolving around it condensed from a vast cloud solely by natural processes over a long period of time. Although he failed to explain how the cloud got there in the first place, Laplace assured his followers that no Creator was needed to account for it. In the mid-1800's, however, the nebular hypothesis was carefully inspected by James Clerk Maxwell, a Christian physicist from Scotland; he found it to be impossible. Maxwell showed that the cloud could never condense into planets as Laplace had said. What Laplace had attempted to establish using mathematics, Maxwell tore down using better mathematics. The Bible says God "disappointeth the devices of the crafty .... He taketh the wise in their own craftiness" (Job 5:12-13).

These are only a few examples of the historical blunders of science. Every age has been beset with them. We would be most foolish if we were to insist that our present-day science is free from error. The best way to avoid such blunders, of course, is to build our science upon a solid, Scriptural framework. Such an approach avoids the fickleness and changeability of ideas so prevalent in the world today. Once we possess the truth concerning a particular phenomenon, we should not have to keep changing our description of it.

2:4 Teleology Creationist scientists often speak about **teleology,** but teleology is a word that is unfamiliar to most people. It means *purposiveness* or *design in nature.* The Christian man of science, in studying an object or an event in nature, will try to understand God's creative purpose behind it. Sometimes this is easy; in other cases it is very difficult, if not impossible.

The Christian researcher will readily recognize the human body as something that is "fearfully and wonderfully made" (Psalm 139:14). The unbeliever, because of his spiritual blindness, can study the same facts and miss the whole point. The Christian sees each part of the body as a marvelously engineered device especially suited to performing its assigned function. In many cases the functions are obvious: the heart is designed for pumping blood, the eye for seeing, the ear for hearing. In other cases our research is not advanced enough to enable us to state the specific function of an organ. However, our knowledge is continually expanding, and science is determining the functions of organs formerly thought to be useless. At one time there were 180 organs classified as vestigial (useless). This number has now been reduced to a mere handful. The list may eventually be eliminated altogether. In the meantime, we should follow the

2:6 *Does nature evidence design (top) or chaos (bottom)?*

principle stated by evangelist Dr. Bob Jones, Sr., of "giving the Lord, rather than the devil, the benefit of the doubt," trusting fully that God understood what He was doing when He created the human body.

We also see clear evidence of design when looking at the earth and its near surroundings. The earth is just the right distance from the sun, which in turn shines with exactly the right amount of energy to give the proper temperatures for living things on the earth. The atmosphere contains precisely the right mixture of gases. The earth's magnetic field protects us from cosmic rays. The ozone layer in the stratosphere protects us from ultraviolet rays. The earth's rotation speed provides ideal periods of day and night. All these things, to the mind that is not blinded by sin, speak of God's creative design.

2:7 *An example of teleology. Our earth is at the optimum distance from the sun for life as we know it; Venus is too close; Mars is too far.*

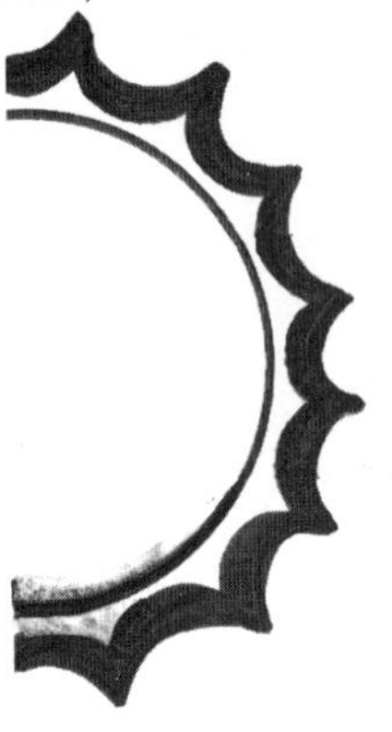

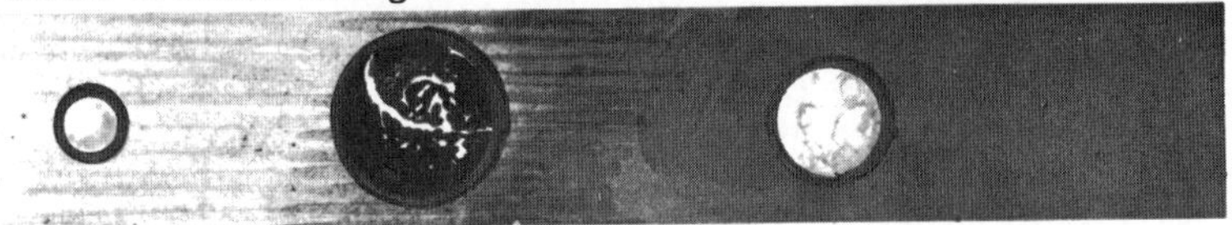

On the other hand, there are other observations that seem unexplainable to us. With further study, however, these too may fit into the general picture of creative design. Whether or not man is capable of understanding its significance, every fact of nature exists and operates in accordance with some part of God's overall plan. There are no "brute facts"; that is, facts that lack meaning until man gives them an interpretation. Every fact in the universe is already known to God.

2:5 Command to Subdue the Earth In the first chapter of Genesis we find God's command for man to do scientific work:

> And God blessed them, and God said unto them, Be fruitful, and multiply, and fill the earth and subdue it; and have dominion over the fish of the sea, and over the fowl of the air, and over every living thing that moveth upon the earth (Genesis 1:28).

Them refers to the first man and woman, Adam and Eve. God gave man authority over all life on the earth. He told man to "subdue" the earth. Man must study and experiment to know how to exercise his authority and how to properly subdue the earth. How would you subdue an automobile? You would observe someone else drive, or study a book to learn how to drive. If a person teaches you, you watch him closely and listen carefully. When you have learned to drive, you can subdue the car.

2:8 *Science is not in a position to declare the divine purpose behind each object encountered in the heavens.*

Genesis 1:28 is one commandment in Scripture that man has obeyed diligently. Scientific activity is not wrong, but it can be pursued in a wrongful manner. Attempting to produce an evolutionary history of the universe is not a proper aim or purpose in science. Even if evolution were true (which it is not), it could never be proved scientifically.

2:6 Worshiping Science In the preceding sections we have tried to point out the strengths and weaknesses of science. It is beyond question that scientific activity has been successful. But success in the physical realm does not imply success in the spiritual realm. Our life-span may have been lengthened and our lives made more comfortable by scientific discoveries, but science can do nothing for our souls.

Many people actually worship scientists or the scientific method because of this physical success. Some people feel that anything scientists say must be true simply because scientists say it. To these people, science or the scientist is a god. This is the most unnecessary kind of idolatry. We hope many of you will become scientists in the future. But always keep things in their proper perspective; put the Lord and His Word first, and science will be much more enjoyable for you.

The authors of this book are trained in science. We have done scientific research, and we have taught science. We enjoy scientific work, but we would not stake our souls' eternal destiny on some finding of science or on a scientist's word.

We have put our trust in the Lord Jesus Christ as personal Saviour. He is the only One to Whom you should entrust your soul. The Lord Jesus stated, "I am the way, the truth, and the life: no man cometh unto the Father but by me" (John 14:6). If you do not know the Lord Jesus Christ as your personal Saviour, we hope you will trust Him now. He will give you the peace that science cannot offer.

2:9 *God should be worshiped, not science.*

Creationist Societies

Bible-Science Association
Box 1016
Caldwell, Idaho 83605

Creation Research Society
2717 Cranbrook Road
Ann Arbor, Michigan 48104

Selected Creationist Literature

Clark, Harold W. *Fossils, Flood and Fire.* Escondido, California: Outdoor Pictures, 1968.

Clark, Robert E. D. *Darwin: Before and After.* Chicago, Illinois: Moody Press, 1966.

Cook, Melvin A. *Prehistory and Earth Models.* London: Max Parish, 1966.

Davidheiser, Bolton. *Evolution and Christian Faith.* Philadelphia, Pennsylvania: Presbyterian and Reformed Publishing Company, 1969.

Davidheiser, Bolton. *Science and the Bible.* Grand Rapids, Michigan: Baker Book House, 1971.

Frair, Wayne and William P. Davis. *The Case for Creation.* Chicago, Illinois: Moody Press, 1967.

Klotz, John W. *Genes, Genesis and Evolution.* St. Louis, Missouri: Concordia Publishing House, 1961.

Lammerts, Walter, ed. *Scientific Studies in Special Creation.* Philadelphia, Pennsylvania: Presbyterian and Reformed Publishing Company, 1971.

Lammerts, Walter, ed. *Why Not Creation?* Philadelphia, Pennsylvania: Presbyterian and Reformed Publishing Company, 1970.

Marsh, Frank L. *Evolution, Creation and Science.* Washington, D. C.: Review and Herald Publishing Association, 1947.

Moore, John N. and Harold S. Slusher, eds. *Biology: A Search for Order in Complexity.* Grand Rapids, Michigan: Zondervan Publishing House, 1970.

Morris, Henry M. *The Bible and Modern Science.* Chicago, Illinois: Moody Press, 1951.

Morris, Henry M. *Biblical Cosmology and Modern Science.* Grand Rapids, Michigan: Baker Book House, 1970.

Morris, Henry M. *Evolution and the Modern Christian.* Philadelphia, Pennsylvania: Presbyterian and Reformed Publishing Company, 1967.

Morris, Henry M., William W. Boardman, Jr., and Robert F. Koontz. *Science and Creation: A Handbook for Teachers.* San Diego, California: Creation-Science Research Center, 1971.

Morris, Henry M. *Science, Scripture and Salvation.* Denver, Colorado: Baptist Publications, 1965.

Morris, Henry M. *Studies in the Bible and Science.* Philadelphia, Pennsylvania: Presbyterian and Reformed Publishing Company, 1966.

Morris, Henry M. *The Twilight of Evolution.* Grand Rapids, Michigan: Baker Book House, 1964.

Mulfinger, George L. *The Flood and the Fossils.* Greenville, South Carolina: Bob Jones University Press, 1969.

Mulfinger, George L. *How Did the Earth Get Here?* Greenville, South Carolina: Bob Jones University Press, 1972.

Nelson, Byron C. *After Its Kind.* Minneapolis, Minnesota: Bethany Fellowship, 1932.

Nelson, Byron C. *The Deluge Story in Stone.* Minneapolis, Minnesota: Bethany Fellowship, 1935.

Rehwinkel, Alfred M. *The Flood.* St. Louis, Missouri: Concordia Publishing House, 1951.

Rushdoony, R. J. *The Mythology of Science.* Nutley, New Jersey: Craig Press, 1967.

Smith, A. E. Wilder. *Man's Origin, Man's Destiny.* Wheaton, Illinois: Harold Shaw Publishers, 1968.

Tinkle, William J. *Heredity—A Study in Science and the Bible.* Houston, Texas: St. Thomas Press, 1967.

Whitcomb, John C., Jr.,and Henry M. Morris.*The Genesis Flood.* Philadelphia, Pennsylvania: Presbyterian and Reformed Publishing Company, 1961.

Whitcomb, John C., Jr. *The Origin of the Solar System.* Philadelphia, Pennsylvania: Presbyterian and Reformed Publishing Company, 1964.

Zimmerman, Paul A. *Creation, Evolution and God's Word.* St. Louis, Missouri: Concordia Publishing House, 1972.

Zimmerman, Paul A. *Darwin, Evolution and Creation.* St. Louis, Missouri: Concordia Publishing House, 1959.

List of Terms

caloric theory	nebular hypothesis
creationist	phlogiston theory
evolution	teleology
evolutionist	theistic evolution
geocentric theory	

Questions

1. In what verse of the Bible does God command man to do scientific work?
2. Why is the practice of attempting to produce an evolutionary history of the universe not a proper aim of science?
3. The Bible states that air has weight. True or False?
4. The scientist takes "brute facts" and attempts to impart meaning to them. True or False?
5. Compare Scripture with science.
6. Name some scientific facts mentioned in the Bible; give verse references.
7. Science worship is one kind of idolatry. True or False?
8. Why should a Christian be uncomfortable if he believes in evolution?
9. The subject of origins is outside the realm of true science. Explain why this is true.
10. Science is fallible. True or False?
11. Who was the Christian man of science who disproved Laplace's *nebular hypothesis*?
12. Can science answer the question of "why" something happens? Explain.
13. Theistic evolution is not a logical position. True or False?
14. Define teleology.
15. What are the two opposing views of origins?
16. What is the most authoritative source on origins? Why?
17. The Bible states that the total number of stars in the heavens cannot be numbered (by man). True or False?
18. If life could ever be created in a test tube, it would disprove the Bible. True or False?
19. What are "brute facts"?
20. Name four historic blunders of science.
21. Genesis gives us a picture of a rapid, miraculous creation. True or False?
22. Discuss the relationship of teleology to the creation-evolution issue.
23. The earth's magnetic field protects us from harmful ultraviolet rays. True or False?
24. Name four ways in which the earth illustrates teleology.

Student Activity

Read one Bible-science book listed at the end of the chapter and write a report on it.

Photo next page
Microbiological research laboratory of a large pharmaceutical company.

UNIT I SCIENCE AND THE CHRISTIAN

Chapter 3 The Scientific Method

3:1 The Senses We Use By now your class should have settled on a satisfactory definition of science. Somewhere in that definition should be the mention of *human* knowledge obtained by *human* investigation. In other words, scientific work requires the presence of human beings to sense what is happening. Science is man-made. If a human was not around to observe an event, it cannot come under the category of science.

Since science demands human **perception**, what senses do we have to perceive our surroundings? Taste, touch, smell, sight, and hearing are the fundamental tools a scientist has with which to do his work.

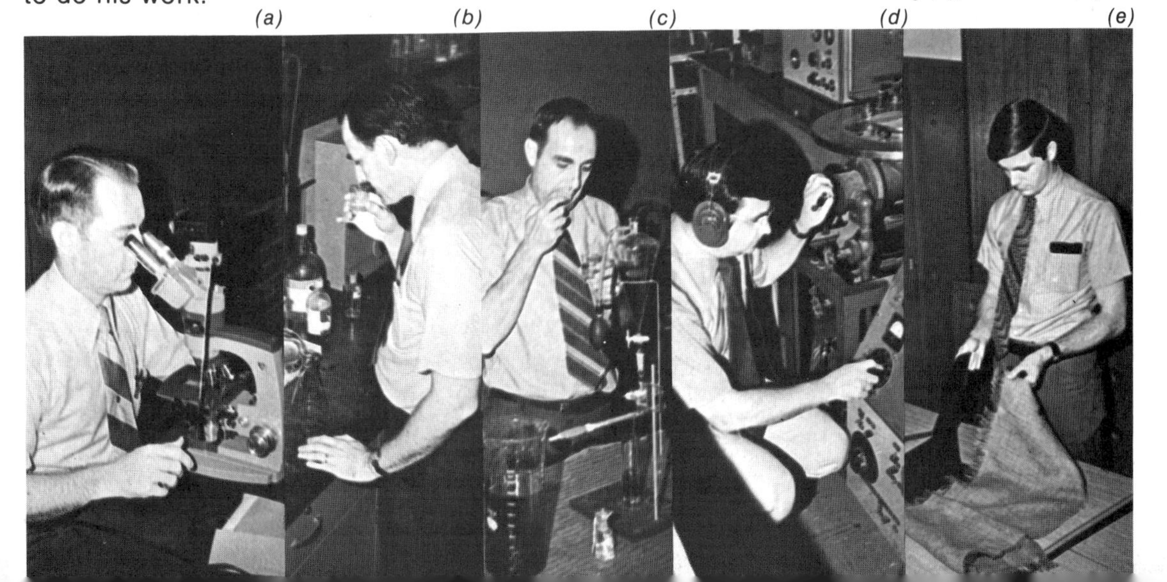

3:1 *Scientists using different senses in performing their work: sight (a); smell (b); taste (c); hearing (d); and touch (e).*

One of the unfortunate facts of life is that our senses are fallible; we are far from perfect. You can make serious mistakes by relying only on your senses. For example, in Figure 3:2 are clever drawings that will fool you if you are not careful.

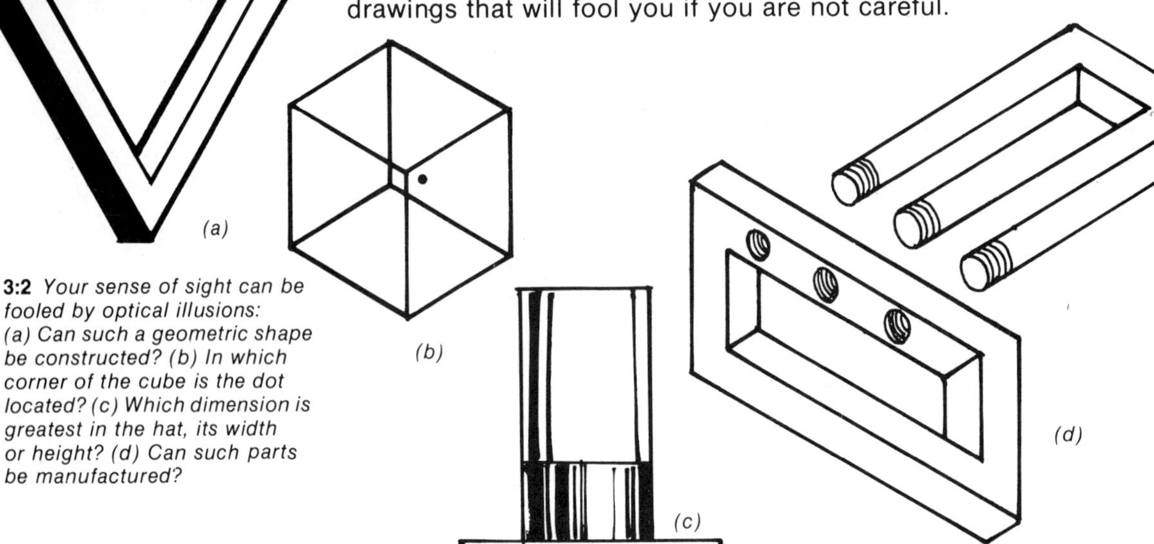

3:2 *Your sense of sight can be fooled by optical illusions: (a) Can such a geometric shape be constructed? (b) In which corner of the cube is the dot located? (c) Which dimension is greatest in the hat, its width or height? (d) Can such parts be manufactured?*

Many automobile accidents are caused by people who misjudge distance while driving. Often when you have heard a voice but have not seen the person, you have guessed incorrectly as to who was speaking; you may have said your ears were "playing tricks" on you. Our senses are fallible; therefore, fallible information will be gathered in scientific work. To be blunt, science is fallible. This will be the case in any activity where men are involved. Science is man-made, man-controlled work.

Probably the sense that man uses most in scientific investigation is his sight. Biologists and metallurgists look at cells or metals through a microscope. A chemist watches as a chemical reaction takes place in a beaker. Physicists, chemists, and engineers read dials on instruments and record what they indicate; the dial might show temperature, electrical current, or voltage, to name a few possibilities. Thus, observation is basic to scientific work. If someone asked you to give a one-word definition of the scientific method, your answer should be **observation**. This definition would not be completely correct, but it would be as simple an answer as could be given.

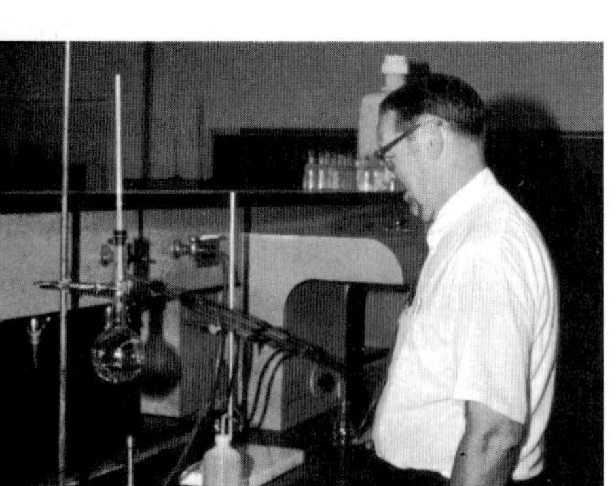

3:3 *A chemist watches the progress of a chemical reaction.*

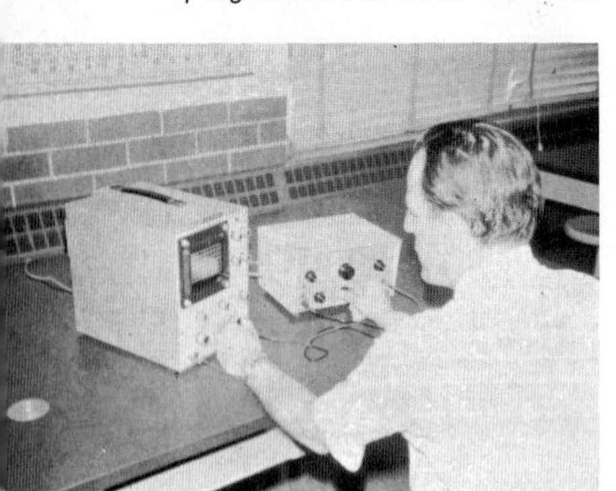

3:4 *A physicist observes experimental results on an oscilloscope.*

3:2 Gathering Scientific Data How often should you do an experiment? Suppose you were timing the swings of a pendulum. A good scientist would time the swings many, many times to make sure that he had enough readings. One test would not be enough to give an accurate answer; something could have been wrong with the pendulum. A good scientist does the experiment two, three, or even five times to check his results. The scientist will feel increasingly confident about his work the more times he performs the experiment.

Once all of the observations are recorded, there is still a lot of work to be done. Next, the scientist *collects* his **data,** the information from his experiment. Collecting data is somewhat of a **classification** operation. It is easy to read data that have been collected and put into a table. But if a person cannot sensibly collect his data, the information is worthless. Here is how a table of his data might look if he did five trials:

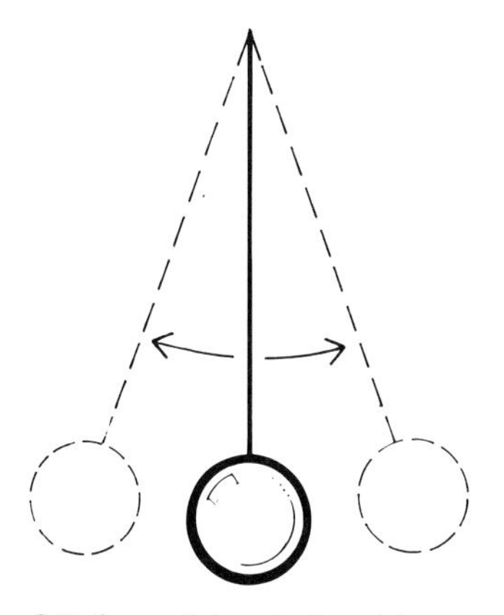

3:5 *A pendulum bob, set in motion, swings from side to side.*

Table 3-1 *Time of 50 Swings of Pendulum*

Trial	**Time** (seconds)
1	101
2	98
3	100
4	99
5	103

But what is the time of one swing? Now the scientist must analyze his data. This becomes a matter of choice, and here is where the scientist *chooses* the answer. He might decide to divide each of his experimental runs by 50 and get the time per swing as 2.02, 1.96, 2.00, 1.98, and 2.06 seconds, respectively. He then averages these and gets a final answer of 2.004 seconds.

The experimenter could also choose the **median** (the middle number in an arranged set of numbers) of his results by collecting his data in the following manner:

Table 3-2 *Time of 50 Swings of Pendulum*

Trial	**Time** (seconds)
2	98
4	99
3	100 median
1	101
5	103

In the middle of these data is the trial of 50 swings in 100 seconds, which would yield 2 seconds per swing. The scientist then selects the answer he thinks is best, either the average or the median. In this case it would not make much difference (2.004 or 2), but in many cases it would make a drastic difference.

Suppose the data from the same experiment, but with a different pendulum, are as follows:

Table 3-3 *Time of 50 Swings of Pendulum*

Trial	**Time** (seconds)
1	70
2	80
3	90 median
4	120
5	140

The median is 1.8 seconds per swing (90 seconds divided by 50 swings); however, the average is 2.0 seconds per swing (500 total seconds divided by 250 total swings).

If you want to keep accurate time with the pendulum, the result you choose will make a great deal of difference. If one person chooses 1.8 seconds and another prefers 2.0 seconds, there would be a 10 percent difference in their answers.

Remember that no scientist is ever forced to accept the result of a scientific experiment, nor does he *discover* an answer; he *chooses* an answer. The scientist can use several methods to work out a final result, but in the end he selects the one he feels is best. He may try to be as unbiased as possible, but he cannot be neutral toward his own experiment.

3:6 *(a) A scientist spends much of his time thinking about proper ways to do his work, and (b) he spends much of his time doing actual experimental work.*

3:3 Which Comes First—Experiment or Hypothesis? Most scientific writings lead the reader to believe that scientists collect literal mountains of experimental results *before* they invent a theory or hypothesis. This can happen. A scientist may examine his findings and build up a scientific argument to "explain why" he got the results he did. This scientific argument is called **interpretation** of experimental results. It may contain very logical scientific **mechanisms** and mathematical precision. However, you should never confuse the experimental results with the interpretation. The *observed* results are true science; the interpretation is simply the scientist's personal view of what the results mean. This speculation is fruitful in that it may suggest other experiments (to check on the interpretation or aid in prediction). Interpretation deals with scientific models, not with reality.

Probably the approach used most in science is (1) the scientist thinks up a hypothesis, (2) he develops an experiment to prove or disprove his idea, and (3) he performs the experiment. The hypothesis comes first, then the experiment. The results of the experiment may cause the scientist to revise his idea several times until he comes up with a workable theory. Notice that he is interested primarily in **workability.** If it works, it is useful; if it does not work, it is not useful.

3:7 *A hypothesis is considered the "roughest" of scientific concepts; a theory is more refined, and a scientific law is the most reliable.*

3:4 Hypotheses and Theories Although the word **"hypothesis"** (plural—hypotheses) sounds very impressive, a hypothesis is nothing more than a guess. We could say that it is a temporary explanation based on very limited knowledge. As new knowledge accumulates, a hypothesis will often require some changes or even replacement.

Occasionally a **working hypothesis** will be adopted to give direction to a series of experiments. Madame Curie observed that there was too much radioactivity in pitchblende ore to be explainable by the uranium it contained. Her "working hypothesis" that there was some other substance present in the mixture eventually proved to be correct and led to the discovery of the element radium.

There is a fine line of distinction between a hypothesis and a **theory.** A theory is usually framed only after a large body of information has been assembled. It, too, is speculative and may eventually be proved wrong. A theory is an idea that appears to be backed by some observed facts. It helps us to understand individual observations as part of a larger picture. A good theory should tie together a number of separate observations. James Clerk Maxwell's electromagnetic wave theory, for example, gave us a framework for understanding, at least in part, radio, radar, television, light, infrared, ultraviolet, gamma rays, and X-rays.

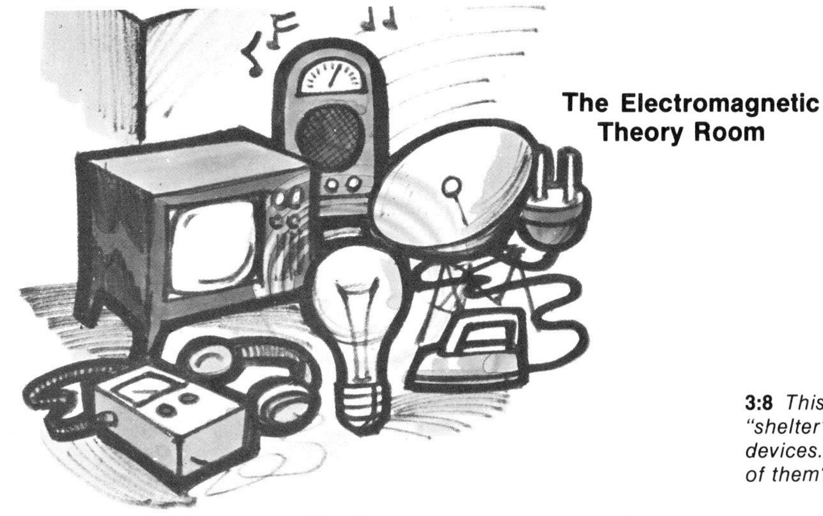

The Electromagnetic Theory Room

3:8 *This room provides a "shelter" for many modern devices. Can you identify any of them?*

Ironically, it is far more difficult to prove a correct theory than to disprove a false theory. Often a theory will predict certain things which can be tested. If it fails such tests consistently, the theory should be discarded. On the other hand, even if the theory passes every test with flying colors, we have still not completely proved it to be true. Thus, the theory may be giving sufficiently correct results to "pass" the test, even though it contains some errors.

Scientists should be willing to study new facts as they turn up to see if their theories agree with the facts. Any theory that does not agree with the facts should be discarded. Occasionally, however, theories have become elevated to a place where they are believed to be almost sacred and totally immune to new facts.

Evolution is one such theory. In 1968, a human shoeprint was discovered in rock that evolutionists estimate to be between 500 and 600 million years old. Something is clearly wrong with this theory, because evolutionists say that man is only about 2 million years old. How could the shoeprint of 2 million-year-old man be found in rock that is 500 or 600 million years old? This shoeprint was taken to a large state university where, sad to say, members of the geology department flatly refused even to look at it. They have become so sure that the theory of evolution is correct that any evidence against it is automatically disregarded.

(a)

(b)

3:9 *(a) Antelope Springs looking southwest at the site where the trilobite fossils were found embedded in (b) the impression of a human shoeprint.*

This is not science; it is thoroughly unscientific and shamefully dishonest. (You may be interested in reading the account of this remarkable shoeprint discovery in the December 1968 issue of the *Creation Research Society Quarterly.*) This is by no means the only evidence that has turned up against evolution, but each new find has been automatically rejected so that the accumulated body of evidence against the theory of evolution never has to be confronted.

Imagine a courtroom trial in which an innocent man is about to be convicted on the basis of circumstantial evidence and incorrect testimony. At the last minute, new facts are discovered which could prove his innocence, but the judge refuses to admit this information into the trial record. So convinced is the judge of the defendant's guilt that he assumes the new evidence to be faulty. Whether it be science or law, all the facts should be considered.

3:5 Scientific Laws A scientific **law** is a statement of a uniform pattern of behavior. If, under the same conditions, a sequence of events in nature is observed to occur the same way every time, a law can be formulated to describe this behavior. (Note that the law does not *explain* the behavior; it merely *describes* the behavior.)

Some laws are **qualitative**—that is, they do not involve numbers. For example, the law of electrical charges states that like charges repel and unlike charges attract. It does not tell us how strongly they attract or repel.

3:10 *Many evolutionists are like this judge; he is so convinced of the defendant's guilt that he will not admit any evidence to the contrary.*

If we wish to find the strength of the force, we can use Coulomb's law. This is a **quantitative** relationship in the form of a mathematical equation that allows us to calculate the force if we know the charge on each body and the distance between the bodies.

It may come as a surprise to most students to learn that scientific laws are *chosen* rather than "discovered." The scientist who formulates a law does so by choosing a verbal statement or mathematical equation from a vast number of possibilities. He will pick the one which, in his opinion, is most suitable.

Frequently, a graph will be made from experimental data. Then a mathematical equation will be chosen which seems best to follow the experimental points. The problem is complicated, however, by the fact that every measurement contains a certain amount of experimental error. If there were no error, the experimental data would give us distinct points on a graph, such as shown in Figure 3:12. A single curve can be drawn through these points, and it would yield only one possible law. But since there *is* some error in the measurements, there is some uncertainty as to where each point should be placed. Therefore, we show *boxes* on the graph, rather than *points,* to represent this uncertainty. There are any number of mathematical curves that can be drawn through the three boxes; it is the duty of the scientist to decide which curve fits the best. It will not be a perfect fit, but it should give reasonably accurate predictions for future experiments.

This, in principle, is how many of our scientific laws have been set up. They do not represent absolute truth, because the measurements upon which they are based are imperfect. And the selection process used by the scientist may be imperfect as well. Occasionally, with better experimental data (more points and smaller boxes), a law will be modified. In more drastic cases it may have to be completely rewritten.

3:6 Verification and Predictability Once scientific data are collected and generalizations are made so that hypotheses, theories, and laws are formed, the scientist will conduct experiments to test or check on these ideas. This is a continuing process. The hypotheses, theories, and laws are subject to change and are restructured according to what is revealed by further testing. Many times scientific ideas are discarded completely. Scientists the world over are re-evaluating the very structure of science each day! What is considered science today may be error tomorrow. This is quite different from God's Word, which is true yesterday, today, and forever. God does not change His mind, but men change their minds many times.

Once scientists develop their ideas they try to use them in productive speculation. This is called **prediction** (just as a weatherman tries to forecast the weather). A scientist collects some data and works these into a mathematical equation. From this he tries to predict what will happen under conditions that have never been observed. For example, suppose an engineer is testing the gas mileage on a new type of engine in an automobile. The table of his experimental data is as follows:

Table 3-4 *Gas Usage by Test Automobile*

Speed *(miles per hour)*	**Gas mileage** *(miles per gallon)*
20	15
30	14
40	13
50	12
60	11
70	10

3:11 *A scientist mathematically evaluates his experimental data, carefully choosing the proper equation to fit the data.*

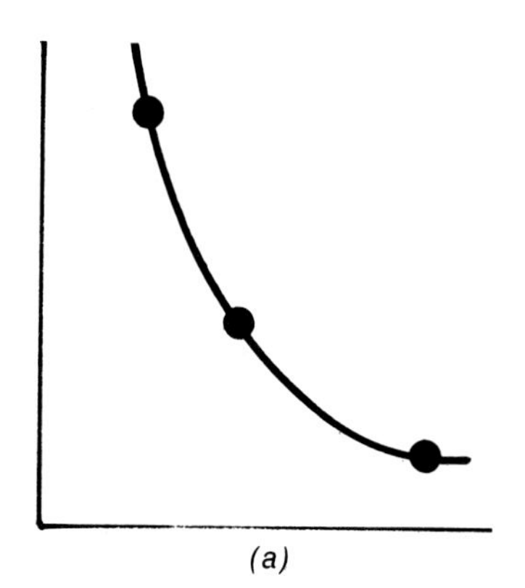

(a)

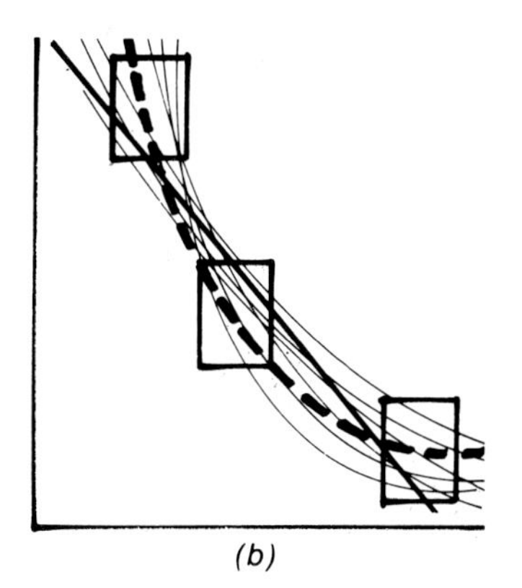

(b)

3:12 *(a) No error in experimental measurements—an impossible situation. (b) Many different curves can be drawn using the same experimental measurements—an actual situation.*

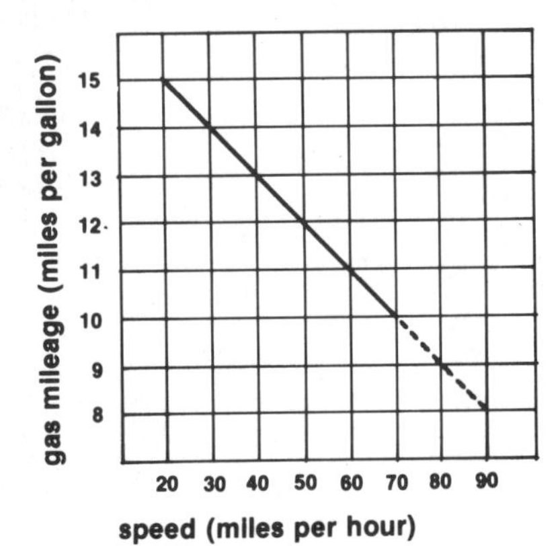

3:13 *(a) Prediction from experimental results (dotted line)*
(b) Actual performance above 70 miles per hour (dotted line)

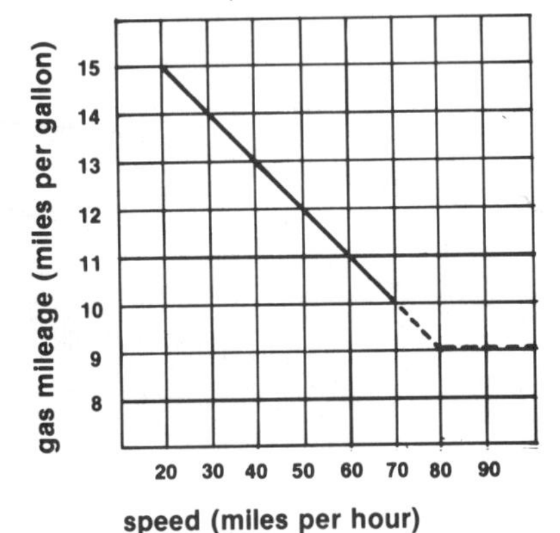

Based on these results the engineer could predict that at 80 miles per hour the car will get 9 miles per gallon of gas, and at 90 miles per hour, 8 miles per gallon of gas. This information could be plotted in the form of a graph (see Figure 3:13a). The engineer is making an extension of the line into an unknown region (one that has never been observed before) as shown by the dotted line. This is a reasonable prediction. However, you can never be sure that the line will continue in the same manner.

Many times predictions fail. Suppose the engineer decided to run the car at 90 miles per hour to check the gas mileage. He found that the gas mileage was 9 miles per gallon at 90 miles per hour instead of the expected 8 miles per gallon. Something obviously happened to the engine between 70 and 90 miles per hour that the engineer did not expect.

Prediction is a necessary and helpful part of science, because tests cannot be run for every imaginable circumstance. It also gives scientists an idea of what could happen under different conditions or over long periods of time. Say, for instance, that a company manufactures a dishwasher which they want to last 20 years. However, they cannot wait for 20 years of testing before they try to sell it. So the company tests as much as possible under all expected conditions, and then "extends" its information 20 years: that is, they predict how well the machine will operate in 20 years based on the information collected during their tests.

Many products, of course, do not last as long as manufacturers predict they will. Often parts have to be replaced quickly with redesigned new parts because the original parts could not withstand the service conditions. Airplanes have crashed because of unusual conditions in service that were never present during testing. A crash of an Electra was an alarming example of this (see Figures 3:14a and 3:14b).

(a)

(b)

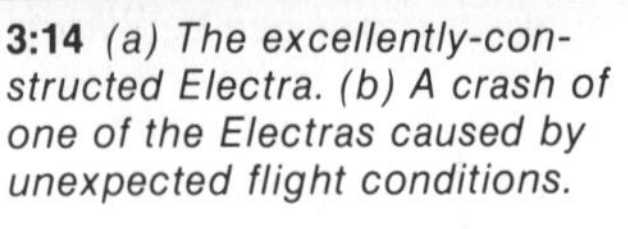

3:14 *(a) The excellently-constructed Electra. (b) A crash of one of the Electras caused by unexpected flight conditions.*

Most engineers are quite conservative when they make predictions, so that what they say is reasonably reliable. An engineer will specify the conditions of his experiment to make certain that any predictions are understood within this scope.

On the other hand, newspaper and magazine articles on science often throw caution to the winds and make unreasonable statements and unreliable predictions. These are usually in the form of the wildest speculation (especially in newspapers) and are written mostly to be spectacular. The scientific accuracy of such accounts is usually questionable.

You should be especially skeptical when reading articles detailing evolutionary predictions. A notable failure of evolutionary predictions concerned the surface features of the moon. Scientists once firmly believed that the dust covering the moon's surface was hundreds of feet deep. Because of the Apollo moon landings, we now know that the surface dust is only several feet deep at the most. Scientists had even suggested that disease-carrying microorganisms had evolved on the moon's surface. That is why the astronauts had to endure an elaborate quarantine procedure after their return to earth. The quarantine is no longer used because it now appears that there are no such microorganisms on the moon. Many evolutionary predictions about the moon have already been proved wrong; undoubtedly most, if not all, of them will be disproved conclusively before very long. When you start with the wrong assumptions, you can only get wrong answers.

3:7 Science, Models, and Reality It is difficult for many people to realize that science does not deal with reality. What a scientist dreams up in his mind is not what really exists. Remember, our senses are fallible and we cannot understand the creation fully or correctly. We see everything through a clouded glass (I Corinthians 13:12). We do experiments under carefully controlled conditions that will never exist anywhere except in a laboratory. We do not study matter; we study our ideas of matter (**models**). For instance, the atomic theory is a model. The gravitational theory has also been developed to account for certain things. A good model is a workable model; a bad model is discarded.

3:8 Workability in Science As you can see, scientific work is fallible and prone to error. But this should not make us say, "Well, if this is the case, science is worthless." Absolutely not! Science is very useful. Look at all of the modern conveniences and medical advances wrought by scientists. Next time you listen to your transistor radio or take a dose of some "miracle" drug, remember that this is a result of scientific effort. If something works in science, we use it; if it does not work, we do not use it. Scientists admit that sometimes it is hard to make things work.

3:9 Prejudice of Scientists Many times a scientist's **prejudices** (likes and dislikes) will determine what type of model he will choose or develop. A creationist will choose a different model from one an evolutionist would choose to "explain" origins. The authors of this book are Bible-believing Christians who accept the creation account in the book of Genesis. We have "prejudiced" this book toward creation and God's Word.

3:15 *The Moon. Many evolutionary predictions about the moon were proved wrong by the Apollo program.*

3:16 *Penicillin mold. This symmetrical colony of green mold is* Penicillium chrysogenum, *a mutant form which now produces almost all of the world's commercial penicillin.*

3:17 *A transistorized radio.*

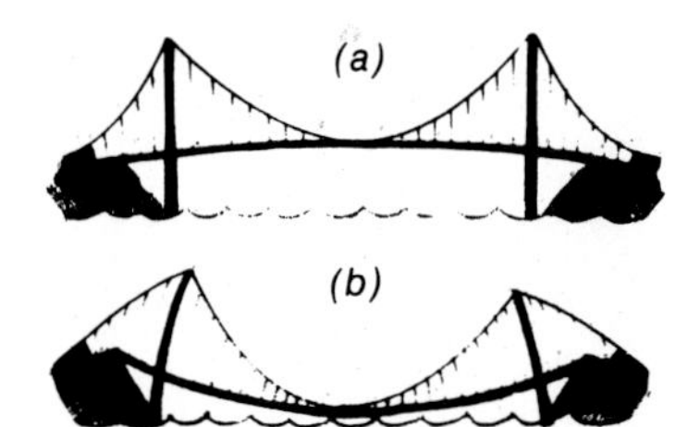

3:18 *(a) A bridge properly constructed using steel and concrete as building materials. (b) A bridge improperly constructed using lead as a building material.*

Liberal journalists, professors, and politicians have created an atmosphere in America today that makes prejudice a dirty word. Actually, prejudices are necessary traits to have. Education is a process of prejudicing. The most highly educated person is a very prejudiced person. A civil engineer, for example, has been prejudiced in his education against using lead to build a bridge. He has learned that lead "creeps" under its own weight and, therefore, is not a good structural material. A prejudiced engineer would not use lead to build a bridge; an unbiased (uneducated) engineer might do so, however, and be rather surprised at the finished product, a sagging bridge.

You do not play with poisonous snakes; why? Because you have been educated (prejudiced) against doing so. You may never have been bitten by one, but you have been told that poisonous snakes will hurt you and you do not wish to find out firsthand. Your prejudice can save your life many times; it is not a trait to be avoided.

3:19 *A prejudiced person jumps a safe distance away from a coiled snake because he knows that the snake may bite him.*

When you read scientific literature, you must realize that it is the work of a prejudiced person. A scientist is not forced into any conviction; he invents his theories. His theories come from a mind that has been influenced by his training and personal philosophy.

List of Terms

classification	observation
data	perception
hypothesis	prediction
interpretation	prejudice
law	qualitative
mechanism	theory
median	workability
model	working hypothesis

Questions

1. Four scientists have obtained different results from an experiment conducted to determine the viscosity (how a fluid resists flow) of a liquid. Each ran his test by dropping a small steel ball through a column of the liquid to see how long it would take the ball to fall to the bottom of the column. All four give you the information they have obtained and ask you to calculate the speed of the ball in falling through the liquid. The data are as follows:

Scientist	*Length of Liquid in Column (feet)*	*Time of Fall (seconds)*
A	3	1.0
B	2	0.6
C	5	1.4
D	6	1.8

   (a) Find the speed of the ball in each case. Speed = distance divided by time.
   (b) How do you decide what will be the final result of all of the work, assuming each scientist did the work under the same conditions?
   (c) Is this average exactly the same as any of the actual results?
   (d) Is scientific work absolute? Why?
2. Why is continual testing necessary in science?
3. Why is prediction a helpful part of science?
4. What is dangerous about prediction?
5. What is a model in science?
6. Why are prejudices necessary?
7. How is a Christian prejudiced?
8. How can smell be used in scientific work?
9. Why is the sense of sight used more than the other senses in scientific work?
10. Why do scientists choose answers?
11. What is a scientific hypothesis?
12. What is a scientific theory?
13. What is a scientific law?
14. Why is workability the chief aim of science?

Student Activity

Investigate in detail a disproved scientific law or theory. State what the theory proposes; develop the arguments of the theory; then show how and why it failed.

Photo next page

Apollo 15 mission commander David R. Scott salutes shortly after the United States flag was unfurled at the landing site in the Hadley-Apennine region. The lunar module, Falcon, is seen at upper right sitting slightly atilt. The astronauts used many items made possible by chemical research.

UNITED
STATES

UNIT II THE STRUCTURE OF MATTER

Chapter 4 The Composition of Matter

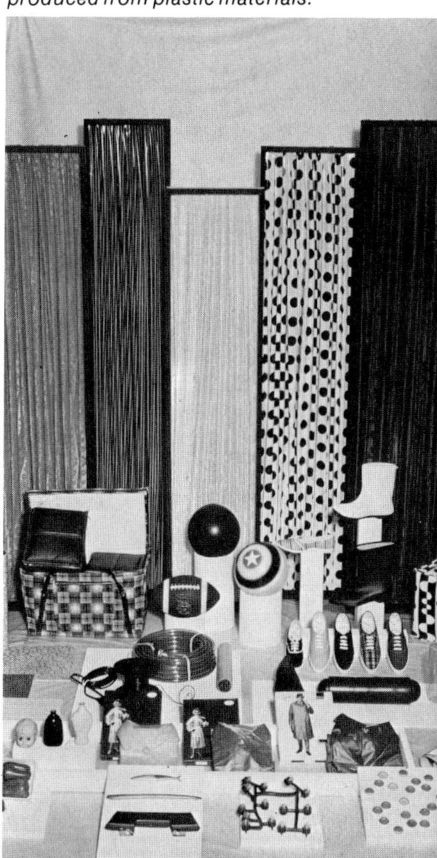

4:1 *Thousands of products for home and commercial use are produced from plastic materials.*

4:1 Defining Chemistry The first half of this book is devoted to the branch of science called **chemistry**. Chemistry is *the study of matter*. We could define it more completely this way: Chemistry is the science that deals with what **matter** is made of and the **transformations** (changes) it undergoes. What is matter? Matter is *anything that occupies space and has mass*. You can see then that chemistry is a very far-reaching subject.

Of all the sciences, chemistry has probably made the greatest number of contributions to our modern way of life. Drugs to fight disease, fuels to heat our homes and power our automobiles, synthetic fabrics and dyes for our clothing, preservatives and flavorings for our food, paints, glass, cosmetics, inks, bleaches, soaps, detergents, plastics, and glues are just a few of the benefits chemistry has helped to bring about.

How old is chemistry? Its roots can be traced back almost to the beginning of man himself. Because chemical processes are needed to separate metals from their ores, chemistry dates back at least as far as Tubal-cain, the father of metalworking (Genesis 4:22).

The Biblical record makes it clear that man was highly intelligent from the very outset; he was able to develop quickly the technology (know-how) needed for a number of remarkably sophisticated processes. When the Tower of Babel was begun (Genesis 11:3), brick and mortar were already being successfully produced. Pottery, which requires both technology and artistic

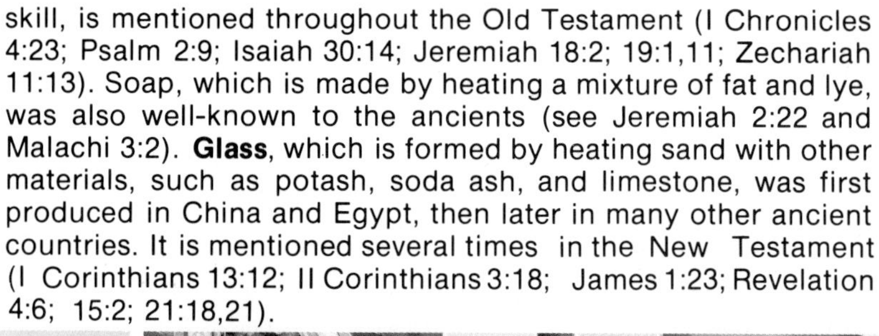
skill, is mentioned throughout the Old Testament (I Chronicles 4:23; Psalm 2:9; Isaiah 30:14; Jeremiah 18:2; 19:1,11; Zechariah 11:13). Soap, which is made by heating a mixture of fat and lye, was also well-known to the ancients (see Jeremiah 2:22 and Malachi 3:2). **Glass**, which is formed by heating sand with other materials, such as potash, soda ash, and limestone, was first produced in China and Egypt, then later in many other ancient countries. It is mentioned several times in the New Testament (I Corinthians 13:12; II Corinthians 3:18; James 1:23; Revelation 4:6; 15:2; 21:18,21).

4:2 *Anhydrous ammonia being applied to a field. Much chemical technology is involved in the production of fertilizers.*

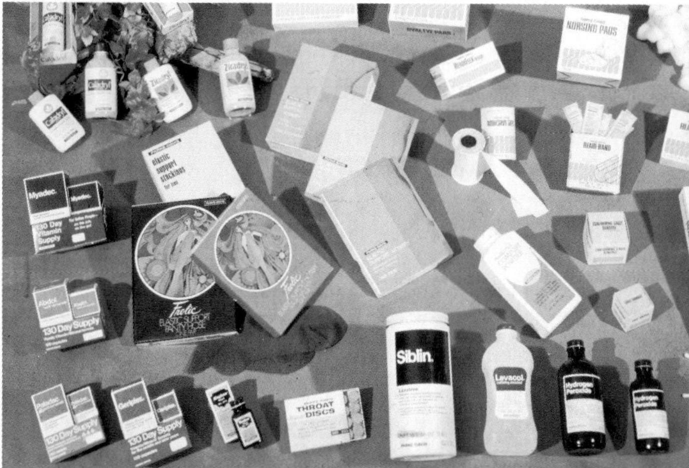

4:3 *Assorted drugs and medical supplies. The drug industry depends greatly on chemical breakthroughs for development of new products.*

4:4 *A researcher at work in an analytical chemistry laboratory.*

There are many specialized branches of chemistry. *Analytical chemistry* is concerned with the analysis (taking apart) of materials to see what they are made of. Analytical chemists try to find out which elements are present in a substance, and how these elements are combined with one another into mixtures and compounds. The study of *which* elements are present is called *qualitative analysis*. The study of *how much* of each element is present is *quantitative analysis*.

Chemistry plays an important role in *geology*, the study of the earth. *Geochemistry* is the chemistry of materials which make up the earth, such as rocks, minerals, and soil. *Mineralogy* is a specialized branch of geochemistry dealing with the composition of minerals. In many ways mineralogy is of great economic importance—for example, the study of coal, oil, natural gas, and metal ores. *Chemical metallurgy* is the division of chemistry that seeks methods for extracting metals from their ores.

Chemistry also plays a part in *astronomy*. Stars, planets, and other heavenly objects, judging by information obtained from an analysis of their light, are made up of many of the same elements as the earth. The principles developed in our earth-bound chemical laboratories, therefore, can be used to further our knowledge of the rest of the universe.

Organic chemistry deals with compounds that contain the element *carbon*. Because hundreds of thousands of compounds contain carbon, organic chemistry is an extensive field. Important subdivisions of organic chemistry are *biochemistry* and *molecular biology*; both are concerned with the chemistry of living systems.

Biologists (scientists who investigate living things) have come to depend more and more upon chemistry to understand the basic substances in living organisms—proteins, fats, carbohydrates, enzymes, and hormones—and the processes by which these are formed, used, and changed into other substances. As a result of detailed study of the structure of organic molecules scientists are now able to synthesize (build up from component parts) many new compounds. By means of organic synthesis, chemists have enriched our daily lives in many ways.

The latter half of this book is devoted to **physics**. Whereas chemistry is the study of matter, physics is *the study of energy.*

Chemistry and physics overlap in certain areas. Every chemical reaction involves an exchange of energy. Consequently, the principles of physics must be used to help us understand what is happening when a chemical change takes place. Areas of mutual interest to both chemists and physicists are usually studied in college-level courses called *physical chemistry* or *chemical physics*.

4:5 *Apollo 15 mission commander David R. Scott salutes shortly after the United States flag was unfurled at the landing site in the Hadley-Apennine region. The lunar module,* Falcon, *is seen at upper right. The astronauts used many items made possible by chemical research.*

(See chapter title page)

4:6 *Chemistry has a role in producing rubber, gasoline, alloys, paints, and plastics that go into automobiles.*

4:2 States of Matter Why bother to classify matter? Is it necessary? Absolutely! Suppose you get up in the morning to dress for school and go to a chest of drawers to pick out a shirt, socks, and trousers to wear. If the drawer in which these items can be found is in a chaotic mess, you will have to spend considerable time selecting your clothing for the day. If the clothing is organized (classified) into piles of shirts, socks, and trousers, your job of selection is much easier. The **classification** of the articles of clothing in the drawer enables you to "study" the various items quickly and arrive at a more intelligent decision. Scientists classify matter because it helps them in their future studies.

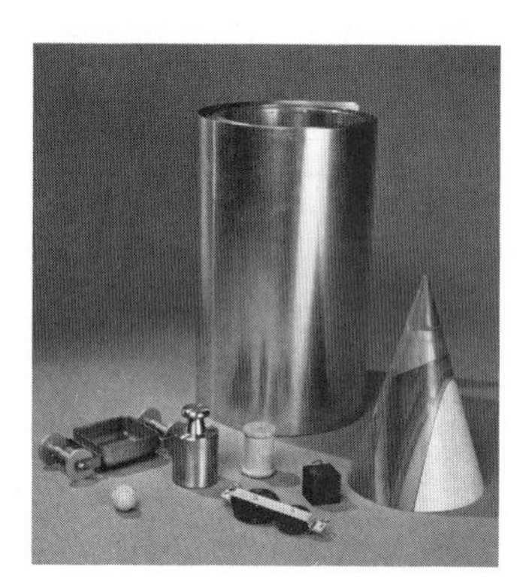

4:7 *Assorted solids.*

Here is another example. Pretend someone gives you a coin collection in a shoe box. One day you get the coins out and find that you have many from all over the world, including one copy of every dime and penny ever minted in the United States. A friend comes over to look at your collection, and you boast about having a complete collection of U.S. dimes and pennies. He looks at the jumbled mess of coins and promptly loses interest.

However, suppose you had carefully piled the U.S. dimes and pennies into separate stacks. Then you had arranged them in order of date minted from the oldest to the most recent. Further, you had mounted them on a board making a very pleasing display. Your classification work enhanced your collection, and it now interests your friend greatly. He can "make sense" out of your collection. Scientists also "make sense" out of matter by classifying it.

Everything in the world may be classified into four broad "states of matter": solid, liquid, gas, or plasma. These states are sometimes called *phases*.

A **solid** has definite shape and volume. It tends to retain its shape and resists deformation under load.

A **liquid** has a definite volume but no shape of its own. It assumes the shape of the container into which it is poured.

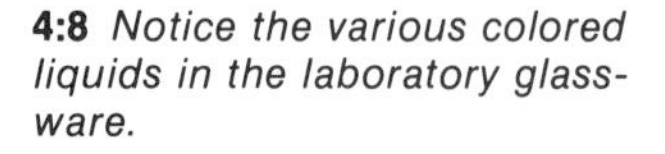

4:8 *Notice the various colored liquids in the laboratory glassware.*

A **gas** has no definite volume or shape. It is very compressible and can expand indefinitely in all directions.

Plasma is an extremely hot, ionized gas. It is present on the surface of stars.

4:9 *A plasma arc is used in cutting metals.*

Glass looks solid but is not. It is considered a *supercooled liquid.*

4:10 *Various sizes and shapes of glass products.*

Other subclassifications of matter, which we will define and discuss later, are *solutions, colloids, sols,* and *gels.*

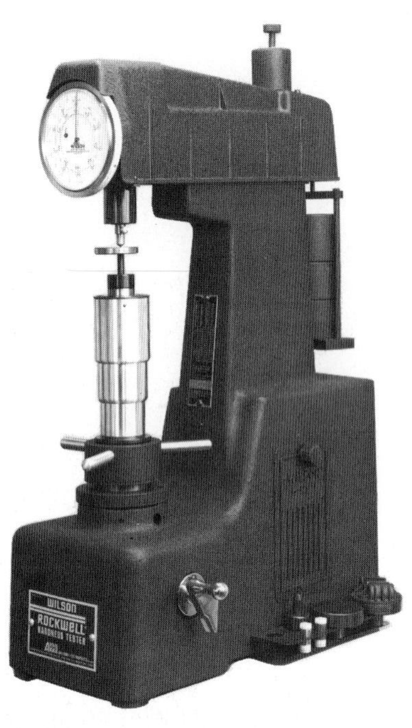

4:11 *Rockwell Hardness Tester.*

4:3 Properties of Matter In classifying matter, we consider both its *physical properties*—such as color, hardness, strength—and its *chemical properties*—how does it react when it comes into contact with other substances? A physical property is considered so that we can distinguish between different substances without changing them into some other kind of matter.

Suppose your class has to decide on the hardness of several metals. The only tools you have are a knife, a hammer, and your fingernails. After everyone has cut and beat on the metals (and probably torn a few fingernails), the metals are classified as follows:

Table 4-1 *Classification of Metal Hardness*

Metal	**Hardness**
A	very hard
B	hard
C	medium hard
D	not so hard
E	soft
F	very soft

Then someone brings in metal G, and the class tests it. You decide it is not as hard as metal B but harder than metal C. You must now make up a new classification—maybe harder than medium hard.

Suppose someone brings in 10 more metals. To classify these new metals would be an almost hopeless task, and the classification would become unbelievably complicated.

The job would be much easier if we could assign a number to the hardness rather than some vague expression. Let us pretend that we have a hardness tester and test the metals using this.

We can now classify the metals as follows:

Table 4-2 *Quantitative Classification of Metal Hardness*

Metal	**Hardness** (Rockwell C)
A	62
B	58
C	50
D	46
E	42
F	38

Now if someone brings in metal G and we test it, we will find that it has a hardness of 53. The classification is simple; no matter how many metals we test, we can easily classify them according to hardness. Using numbers (mathematics) is the best way to classify physical properties.

Considering *chemical properties* is also a good way to classify substances. If you find something that you think is gold, but someone tells you it is fool's gold (iron sulfide), a very simple chemical test will prove who is right. Drop the material into some nitric acid; gold will not dissolve, but fool's gold will. You are using a chemical property (solubility, or ability to dissolve, in nitric acid) to distinguish between two substances.

Zinc will **react** (combine) with sulfur to form zinc sulfide. This is a chemical reaction. The product of the reaction (zinc sulfide) has different chemical and physical properties from the zinc and sulfur.

Suppose you wanted to show someone else how these two elements would react. You would get tired of writing if you wrote it the following way:

Zinc reacted with sulfur yields zinc sulfide. This is as unnecessary as writing

six plus four equals ten

When we can, we like to use labor-saving symbols. For instance:

$$6 + 4 = 10$$

is much easier to write than spelling out the numbers. Similarly,

zinc + sulfur = zinc sulfide

is a little better way to write out our reaction, but the best way to express it is

$$Zn + S \longrightarrow ZnS$$

using chemical symbols. Just as 6 is the symbol for six, so Zn is the symbol for zinc and S the symbol for sulfur. A Japanese chemist, a German chemist, a Russian chemist—*all* chemists—would recognize that

$$Zn + S \longrightarrow ZnS$$

means "zinc reacted with sulfur yields zinc sulfide" without our having to write it out in Japanese, German, or Russian. These chemical symbols are used all over the world. This is a chemical shorthand you must learn. There is a different symbol for each of the 105 elements, and all scientists know the meaning of these symbols.

(a)

4:12 *(a) Pyrite (fool's gold)— FeS_2.*
(b) Gold bars.

(b)

Table 4-3 *Examples of Chemical Symbols*

Element	Symbol
Hydrogen	H
Oxygen	O
Iron	Fe
Copper	Cu
Gold	Au
Silver	Ag
Nitrogen	N
Argon	Ar

Returning to the chemical reaction,

$$Zn + S \longrightarrow ZnS$$

the arrow (→) indicates the direction of the transformation. Zinc and sulfur are *combining* to form the product zinc sulfide. If you write

$$ZnS \longrightarrow Zn + S$$

this would indicate a chemical reaction for the *breakdown* of zinc sulfide into zinc and sulfur. It says "zinc sulfide decomposes into zinc and sulfur."

Another way to separate different kinds of matter is to use the idea of *purity*. We can draw a spiritual parallel here. Christians are supposed to remain spiritually pure by keeping themselves separated from the corrupting influences of the world. The more separated you are, the purer your life will be.

Practically all matter in the physical world is in an impure state (mixture).

(a)

(b)

4:13 *(a) Chemically pure substances. (b) Mixtures.*

Seawater, for example, is a mixture of salt and water (plus many impurities). Suppose we separated the salt and the water—it would take a considerable amount of **energy** to do this. In the same manner, to separate yourself, as a Christian, from the world will cost you something, such as unsaved friends and worldly entertainment. If we get rid of all of the other impurities in seawater, we will have two pure substances, salt and water. A **pure substance** is a kind of matter with a definite, fixed **chemical composition.** Pure salt will always be sodium chloride; you can recognize it chemically. Similarly, a separated Christian can always be recognized by the way he acts, talks, and dresses.

A **mixture** is a material containing different substances in varying proportions in which the substances keep their individual properties.

4:14 *Add salt to pure water to form a homogeneous mixture of salt and water.*

Pure water+pure salt=Homogeneous mixture

A **solution** is a **homogeneous mixture**. When salt dissolves in water, it forms a solution of salt in water. The salt is called the *solute*; the water is called the *solvent*. A homogeneous solution looks as if it is a pure substance but it is not. There are also **heterogeneous mixtures**, such as granite and steel. With the naked eye you can see that granite is made up of different substances, and with a microscope, you can see the varied make-up of steel (see Figure 4:15). One type of heterogeneous mixture is called a **suspension**, which is undissolved particles held in a fluid, such as muddy water. If a suspension is left to sit for a short time,

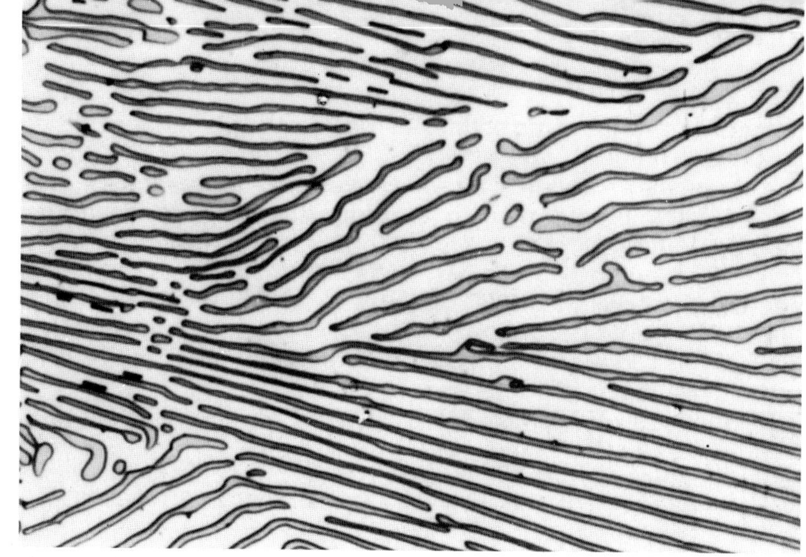

4:15 *A photomicrograph of a steel at 2500 magnifications. The structure shown is pearlite, a heterogeneous mixture of iron carbide and alpha iron.*

the particles will settle to the bottom. Stirring or shaking will again suspend the particles for a time. Other examples of a suspension are certain types of medicines, salad dressings such as Italian and old-fashioned French, and other products whose labels advise "Shake well before using."

A homogeneous mixture is like a Christian with hidden sin in his life; you cannot tell it outwardly. A heterogeneous mixture is like a Christian who is living in open sin; it is evident to everyone.

We can take salt and break it down into sodium (a metal) and chlorine (a gas). However, we cannot chemically break down sodium or chlorine any further. Sodium and chlorine, because they cannot be broken down further chemically, are called elements. Sodium chloride (salt), because it can be broken down further chemically, is called a compound. An **element**, then, is *a substance that cannot be decomposed by ordinary chemical means.* A **compound** is *a substance that is made up of two or more elements chemically combined.*

Purity of matter can be illustrated by a classification table:

Table 4-4 *Chemical Classification of Matter*

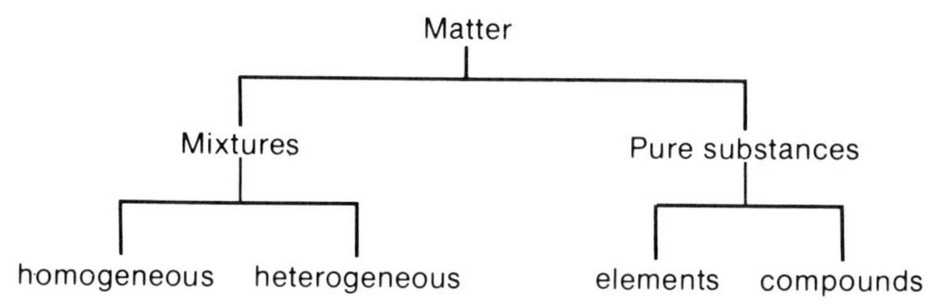

4:4 Changes in Matter Also mentioned in the complete definition of chemistry is transformation. The form and properties of matter can be changed either physically or chemically.

Physical change in a substance is a change which alters its properties without changing it into a new substance. Examples of this are the freezing or boiling of water and the condensing of steam to a liquid.

Most matter tends to be solid at very low temperatures. As a pure substance is heated to a certain temperature, called its melting point, it will change to its liquid state. If the substance is heated to higher temperatures, it will become a gas at its boiling point. If you could heat the gas to the temperature of the surface of stars, it would become a plasma.

As you cool a gas, it *condenses* into a liquid at the *boiling point*. The boiling point and condensation point of a pure substance are the same. If you continue to cool the liquid, it freezes into a solid at the *melting point.*

Sublimation is an interesting physical change. This happens when a solid goes directly to a gas and skips the liquid state. It can also go back to the solid state without passing through the liquid state. Ask your teacher to demonstrate this change with iodine crystals. Camphor and naphthalene also can be used to illustrate the process of sublimation.

Table 4-5 *Relation of Phase and Temperature of a Pure Substance*

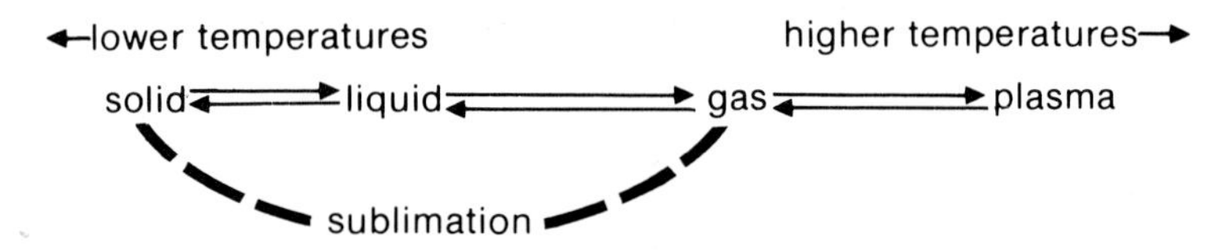

Chemical change is one in which a substance loses its characteristics and changes into one or more new substances with new properties. An example of this is the reaction between hydrogen and oxygen (do not try it) to form a liquid, water. Burning is also a chemical change. The products of most burning or combustion operations are carbon dioxide and water.

List of Terms

chemical change	matter
chemistry	mixture
classification	physical change
composition (chemical)	physics
compound	plasma
element	pure substance
energy	react
gas	solid
glass	solution
heterogeneous	sublimation
homogeneous	suspension
liquid	transformation

Questions

1. Define (a) chemistry, (b) transformation, (c) matter, (d) solid, (e) liquid, and (f) gas.
2. Is glass a solid?

3. The freezing point and melting point of a pure substance are two different temperatures. True or False?
4. Define (a) physical change and (b) chemical change.
5. Name a physical change that could happen to gasoline. Name a chemical change that could happen to gasoline.
6. What is the importance of classification in science?
7. How would you classify gases according to density? Explain.
8. Someone gives you 10 metals with the following tensile strengths (the force per square inch required to pull a metal apart).

metal	*strength* (pounds per square inch)
A	70,000
B	10,000
C	11,000
D	18,000
E	75,000
F	50,000
G	42,000
H	31,000
I	68,000
J	22,000

(a) Arrange them in a sensible classification table.
(b) Suppose someone wants you to tell him what metals could be used on a job that required a metal with a strength of at least 35,000 psi. What metals could be used? (Refer to your classification table.)

9. Define (a) pure substance, (b) mixture, (c) element, and (d) compound.
10. Name five homogeneous mixtures.
11. Name five heterogeneous mixtures.
12. Name (a) five compounds and (b) five elements.
13. Name (a) five metallic elements and (b) five gaseous elements. (In those states at room temperature.)
14. Classify the following substances according to Table 4-4: air, gasoline, water, dirt, paper, gold, and glass.
15. (a) Write a chemical reaction for iron plus oxygen yielding iron oxide.
    (b) Write a chemical reaction for the decomposition of copper oxide into copper and oxygen.
16. Write a chemical reaction for sodium and potassium reacting with chlorine to form sodium chloride and potassium chloride.
17. Why must the principles of physics be studied at the time you are studying chemistry?
18. Why must man have had a knowledge of chemistry back in the days of Tubal-cain?
19. List three Scripture references where glass is mentioned.
20. Name three physical changes.
21. Name three chemical changes.
22. Your body is giving off heat. Is this the result of chemical or physical change?

Student Activities

1. Collect several rocks and minerals from your area and identify them according to Table 4-4.
2. Make a mixture; then separate it back into the original substances by some means.

Photo next page
If an astronaut failed to wear a pressurized space suit on the moon, his blood would boil because of the absence of pressure on his body.

UNIT II THE STRUCTURE OF MATTER

Chapter 5 Molecules in Motion

5:1 Some Evidences to Consider The results of many different kinds of experiments have led scientists to believe that

1. matter is made up of tiny particles, and
2. the tiny particles are continually moving.

We will look at some evidences of this very briefly.

We are all familiar with the fact that a lump of sugar can be crushed into smaller pieces. We can further pulverize these smaller pieces into a very fine powder with the use of a **mortar and pestle** (see Figure 5:1). Finally, we can subdivide the powder into much smaller units by dissolving it in water. The particles of sugar in the water will be so small that they cannot be seen even with the most powerful microscope. Yet they must be there, because the sugar can be tasted in every part of the water.

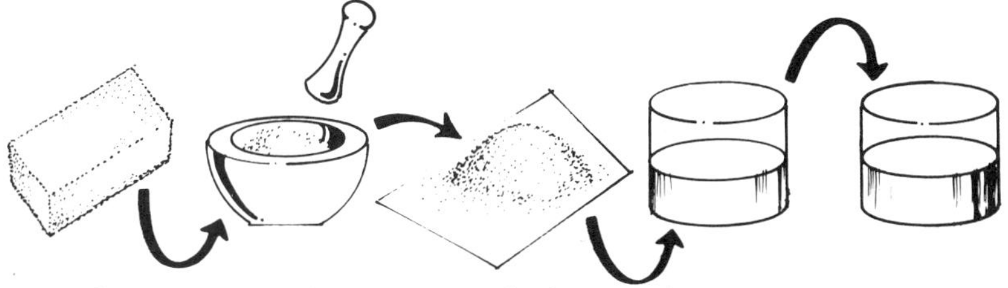

5:1 *Division of matter. A lump of sugar ground to powder in a mortar by a pestle and stirred into water can be dissolved completely. The particles of sugar are now too small to be seen even with the most powerful microscope. We know they must be present, however, because the sugar can be tasted in every part of the water.*

Suppose we place a crystal of potassium permanganate (a purple chemical) in a beaker of water and stir it until it is completely dissolved. We observe that all the water has been colored. The crystal has somehow divided itself and spread throughout the water in the beaker. If we now pour the contents of the beaker into a much larger container of water and stir the solution again, the color becomes fainter, but it is still there. The material from the crystal seems to be able to keep dividing almost indefinitely. For this to be the case there must be a very large number of particles.

The fact that matter can be divided in this way is just one kind of evidence that it is made up of particles. We can reverse the procedure by boiling off the water. The potassium permanganate particles apparently come back together again when we do this, because we are again left with the same dark-colored material that we had at the beginning.

5:2 *Dissolving a crystal of potassium permanganate and diluting the solution thus formed. The purple color becomes fainter with increasing dilution.*

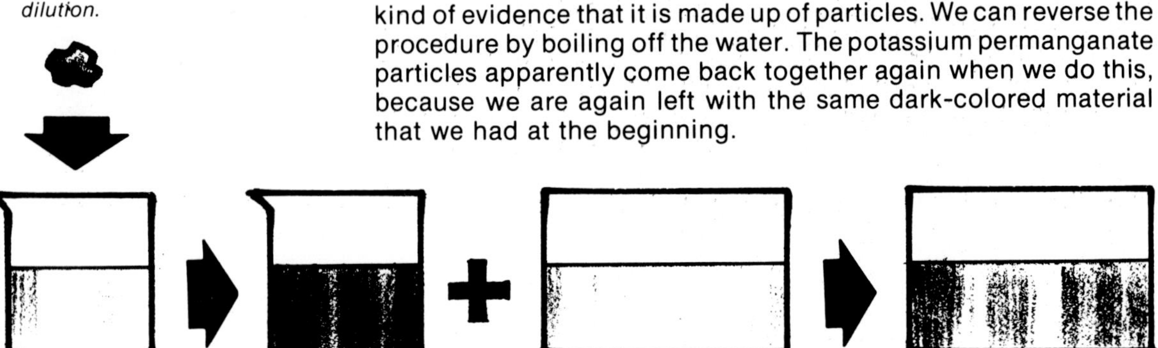

Now let us change the procedure slightly by simply dropping a crystal of potassium permanganate into a beaker of water *without stirring*. If we patiently observe, we will see that the crystal dissolves (but much more slowly) and that eventually the potassium permanganate and the water are completely mixed. One way of accounting for this mixing without stirring is to assume that the particles that make up the water are in *constant motion*. This process of mixing is called **diffusion**. The constant motion mixes the particles of potassium permanganate through the water.

Diffusion happens far more rapidly with gases, leading us to believe that the particles move more rapidly in gases than they do in liquids. For instance, a few drops of perfume spilled from a bottle on one side of a room can soon be smelled on the far side of the room. Perfume vaporizes quickly. Presumably, the particles of air are in very rapid motion and soon scatter the particles of perfume vapor about the room.

Diffusion in the gaseous state can be demonstrated visually with bromine vapor and air. Bromine vapor is brown and very dense; it is about 5½ times as heavy as air. Suppose that we place a container of air upside down over a similar container of bromine vapor having its open end facing up but separated by a glass plate. When the glass plate is removed, the bromine will start diffusing upward into the air. Air also will move downward into the bromine vapor. This cannot be explained by gravity, because the air—which is lighter—goes *downward*, while the heavier bromine vapor moves *upward*. In a few minutes the two gases will be evenly mixed. The particles of the air and the bromine must each have motion of their own to mix themselves so thoroughly and quickly (see Figure 5:3).

5:3 *Diffusion in gases. The fact that bromine and air mix spontaneously when the partition between them is removed cannot be explained by gravity because bromine is heavier than air.*

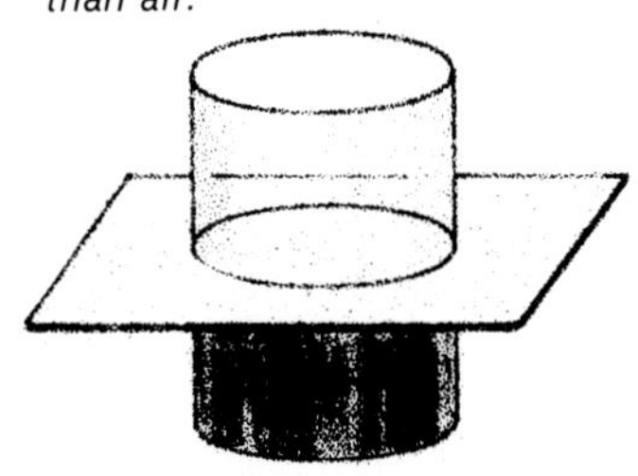

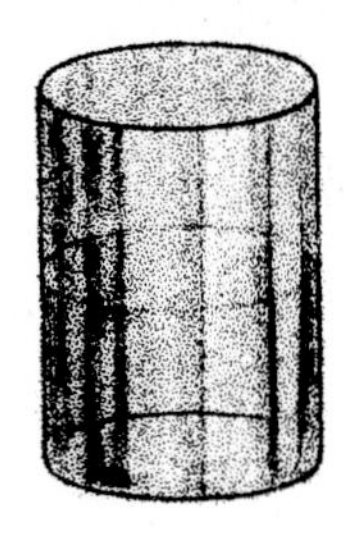

Experiment 5:1 *Diffusion*

1. Fill 3 identical beakers with water.
2. Heat 1 beaker of water to the boiling point. Add a drop of food coloring to the water and time how long it takes to become completely mixed. (Do not stir or shake the beaker.) Record the time required.
3. Heat the second beaker of water to 60° C. Again add a drop of food coloring and record the time required for complete mixing.
4. Repeat the procedure using the beaker of water at room temperature. Record the time required.
5. What relationship do you find between temperature and the speed of mixing?

5:4 *Brownian motion. An individual plant spore suspended in water might be observed with a microscope to follow a path such as the one shown.*

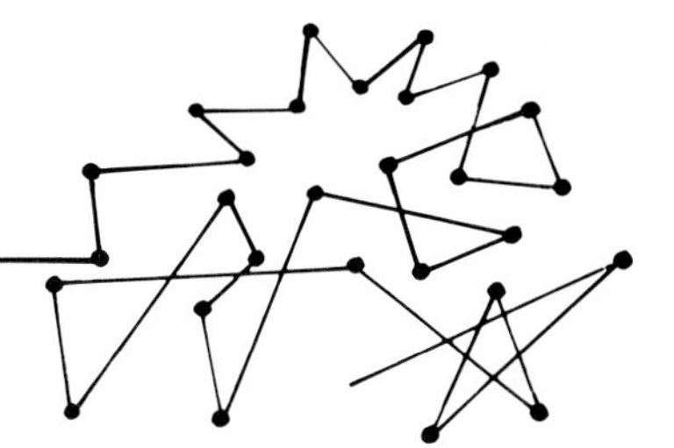

There is a more direct type of evidence that matter is in constant motion. In 1827, Robert Brown (1773-1858), an English botanist, examined through a microscope plant spores (tiny seeds) floating in water. Much to his amazement he saw the spores jostling back and forth as if they were being struck repeatedly from different sides. Yet nothing was touching them except the water in which they were floating. Today we account for this by assuming that the individual water particles are in constant, random motion. This effect is called **Brownian motion** or **Brownian movement** in honor of its discoverer (see Figure 5:4).

Experiment 5:2 *Brownian Motion—First Method*

1. Place a small amount of lycopodium powder (spores from club mosses) on a piece of paper. Place a drop of water on a microscope slide. Blow the powder onto the slide with the droplet as shown in Figure 5:5.
2. Place the slide on the microscope stage under the objective lens. Be sure the droplet is well lighted.
3. Focus the microscope so that the grains of powder may be seen moving about in the water. This is Brownian motion. Notice how the particles act when they collide.
4. Pick out 1 grain; follow its motion and sketch the motion as well as you can. Sketch what happens when grains collide. Do this 3 times.

5:5 *Experiment 5:1. Procedure used to add lycopodium powder to a drop of water on a slide.*

Experiment 5:3 *Brownian Motion—Second Method*

1. Place a drop of India ink on a microscope slide.
2. Blow some chalk dust onto the ink using the same technique as shown in Figure 5:5. Examine the slide under the microscope and follow the instructions in Experiment 5:2, steps 3 and 4.

5:6 *Power lines sag lower in the summer than in the winter, an example of the thermal expansion of a solid.*

These, then, are a few of the evidences that matter is made up of exceedingly tiny particles which are in constant, rapid motion. Amadeo Avogadro (1776-1856), an Italian physicist and chemist, chose to call these particles **molecules**. *A molecule is the smallest particle into which matter can be subdivided without changing its chemical nature.* Or we could say that it is the smallest particle of any substance which has all the properties of the whole substance. Molecules are unbelievably small. It would take 100 million water molecules placed side by side to make a row an inch long. If a drop of water could be magnified to the size of the earth, the molecules in it would still be only as large as baseballs! Or consider this "thought experiment": If each molecule in a thimbleful of air could be changed into a grain of sand, it would make enough sand to fill a container a mile wide, a mile long, and a mile deep.

5:2 The Kinetic Theory The *kinetic theory* states that matter is made up of moving molecules. As we have seen, there is excellent *experimental evidence* that there are "little somethings" that make up matter, and that these "molecules" are moving. The overall idea is still called a "theory," and rightly so, even though some parts of it have been quite firmly established.

Solids are thought of as consisting of molecules that are rather tightly bound together. It is difficult to compress a solid. The molecules must be held together by some kind of "glue" or *intermolecular attraction* that allows the material to keep a definite size and shape. The molecules of a solid seem to be free to vibrate within their own little region of space, but limited as to how far they can move by the presence of neighboring molecules. If the molecules are already touching one another, there is practically nowhere for them to go when we try to squeeze them together. At low temperatures there is little molecular motion. As temperature increases, the molecules move more and more energetically against their neighbors, pushing them farther apart in every direction. The effect is that the material expands. When the temperature rises to the **melting point**, the intermolecular forces (forces betweer
molecules are no longe
liquid.

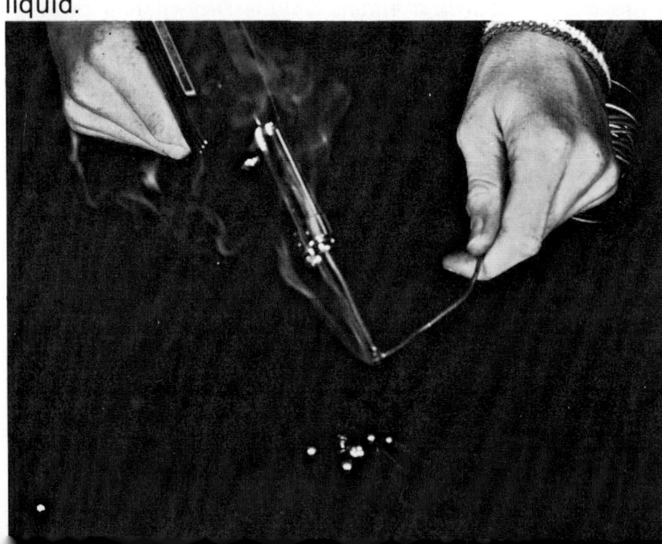

5:7 *Melting a solid. Solder, an alloy of tin and lead, is easily melted. The solder gun supplies heat energy which breaks down the intermolecular forces in the solder.*

Liquids are pictured as having molecules that are in contact but free to roll over one another. BB's poured from one beaker to another give us a crude picture of this. Using this concept, we can account for the fact that *liquids have a definite volume but no definite shape*. At low temperatures the molecules crawl over one another rather lazily. A few at the surface may be able to escape into the air. For example, a glass of water evaporates slowly at room temperature. At higher temperatures the motion becomes more vigorous, and a great many more molecules escape from the surface. As in the case of solids, liquids expand as the temperature increases. This is caused by the increased forces of the molecules against each other in all directions. The rising of the liquid in a thermometer is a common example of this. If enough heat is supplied, a liquid will reach its **boiling point**, the temperature at which the upward pressure of the escaping molecules equals the downward pressure of the atmosphere. The liquid then rapidly changes to the gaseous state or **vapor** state.

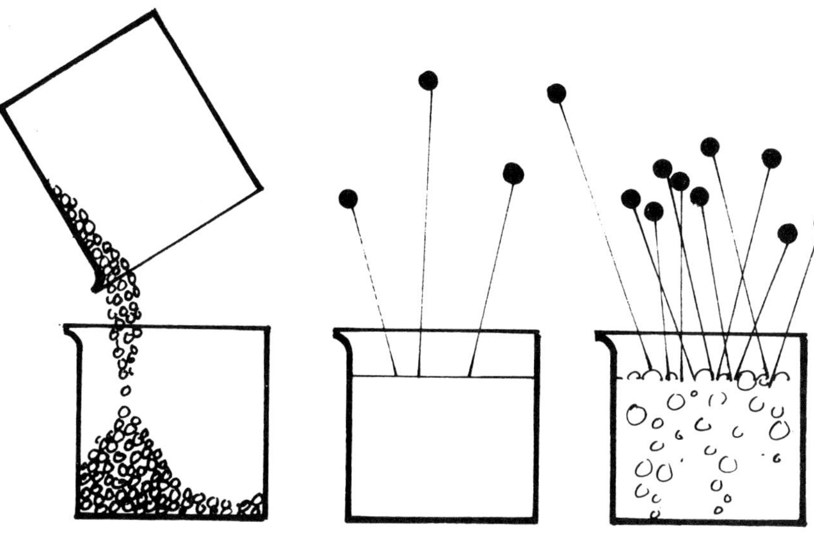

5:8 *Lead shot poured from one beaker to another (left) furnishes a crude model of a liquid. At low temperatures only a few molecules escape from the surface of a liquid (center). At higher temperatures many molecules leave the surface (right).*

5:9 *A liquid reaches the boiling point when its upward pressure equals the downward pressure of the atmosphere.*

Gases are quite different from solids and liquids. At ordinary temperatures and pressures gas molecules are very far apart. You will recall from the previous chapter that gases have no definite size *or* shape, and they expand to fill any container they occupy.

As you may know, air can be liquefied if it is cooled enough and put under sufficient pressure. There is a tremendous amount of space between the molecules of ordinary air. The cubes in Figure 5:11 have been drawn to help us compare the size of an air molecule to the space between the molecules. The larger cube (regular air) has a side roughly 9 times as long as the side of the smaller cube (liquid air). We can even go one step further: When air is compressed into a liquid, 9 units of length in any direction compress to *one* unit of length. Therefore, we can say that, on the average, the distance between two molecules of ordinary air is about 9 times the diameter of an air molecule.

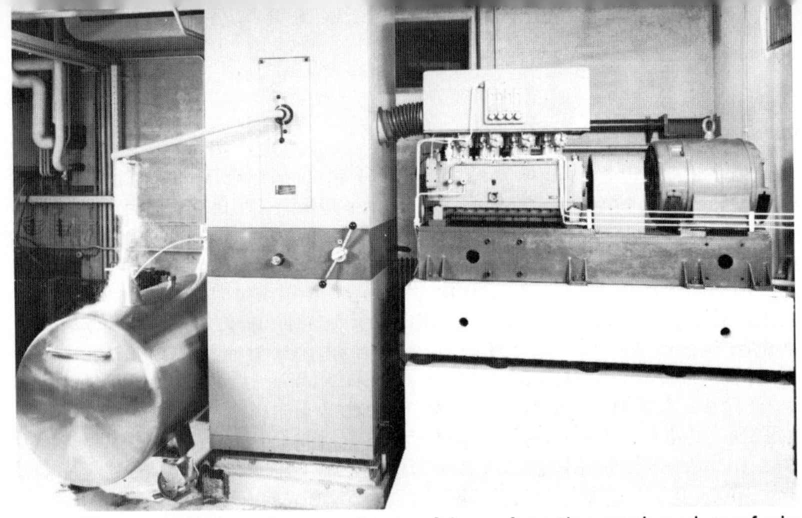

5:10 *Cryogenic equipment. Liquid air can be seen at the left.*

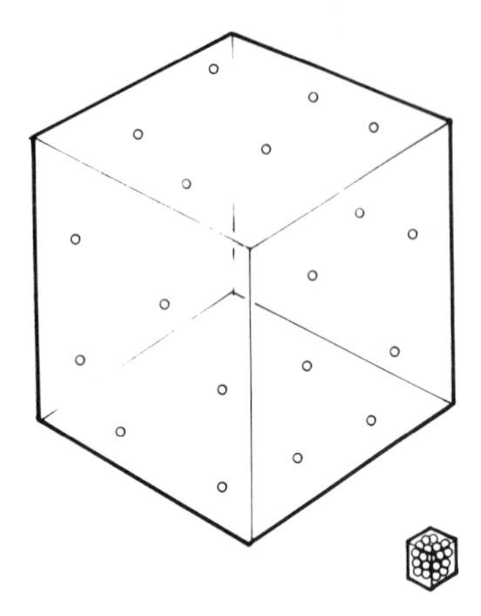

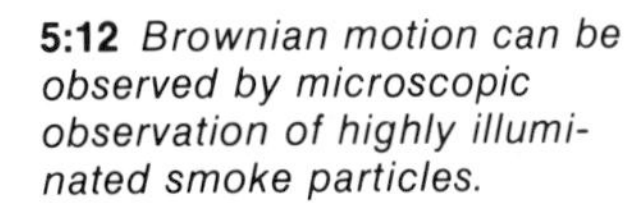

5:11 *767 cubic units of ordinary air (left) liquefy to give one cubic unit of liquid air (right). The larger cube has a side roughly 9 times the side of the smaller cube.*

We can also get some idea of *how fast* the molecules of air are moving. The figures are too involved to include here, but the result would show that at room temperature a typical air molecule is traveling at more than 1100 miles per hour! This sounds fast, and it is *extremely* fast. Each individual air molecule does not get very far, though, because there are always other molecules in the way. Usually they travel only a tiny fraction of an inch before bumping into another molecule. After this collision both molecules bounce away in directions that are different from their original paths.

These ideas are supported by experimental evidence. Smoke particles suspended in air can be observed with a microscope if they are illuminated with a bright enough light. The smoke particles can be seen knocking about in a vibratory kind of motion, apparently caused by the high-speed impacts (bumping together) of the air molecules. This is another example of Brownian motion.

5:12 *Brownian motion can be observed by microscopic observation of highly illuminated smoke particles.*

5:3 Understanding Pressure Molecules of a gas enclosed in a container experience *two* kinds of collisions—collisions with one another and collisions with the walls of the container. Try to imagine the pressure exerted by an enclosed gas as being produced by the molecules bombarding the walls of the container—millions of them hitting each square inch every second! With such a vast number of molecules, the effect of an individual molecule is never noticed. Collectively, however, their effect can be very great.

A pumped-up football or basketball has an inside pressure greater than atmospheric pressure. The molecules have been packed together with the pump more closely than in ordinary air. Therefore, during any unit of time, more molecules are bouncing off every unit of area on the inside than on the outside, keeping the ball tightly pushed out in all directions.

5:13 *The molecules are being packed more closely together in the ball than in ordinary air.*

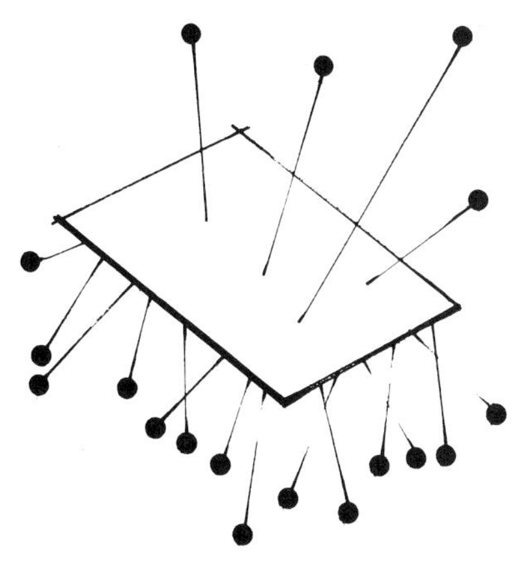

5:14 *Visualizing molecules bombarding the surface of the ball shown in Figure 5:13. The pressure on the inside (lower surface) is greater than that on the outside (upper surface). Hence there are more molecular impacts on the inside than on the outside in any given interval of time.*

5:15 *Automobile tires are pumped to a pressure of approximately two atmospheres, maintaining a continuous outward pressure on the rubber.*

5:16 *If an astronaut failed to wear a pressurized space suit on the moon, his blood would boil because of the absence of pressure on his body.*

(See chapter title page)

5:4 The Effect of Heat It is now understood that, in any gas, different molecules have different speeds. Some are speeded up by their collisions; others are slowed down. If we make a graph of the number of molecules in a container of gas traveling at various speeds, we will obtain a bellshaped curve. Such a curve is shown in Figure 5:17. This curve is called a **"Maxwellian distribution"** in honor of James Clerk Maxwell (1831-1879), who developed the mathematical groundwork for the kinetic theory of gases. (See the biographical sketch of James Clerk Maxwell in Chapter 34.)

At any given moment most molecules will have average speeds. There will be a few exceptionally fast ones and a few very slow ones, but the greatest number will always be found toward the middle of the distribution. The dotted line on the graph shows what happens when the gas is heated. The average is shifted to the right, toward higher speeds. The very fastest moving molecules will be faster than *any* of the molecules at the lower temperature, and there will be fewer of the slowest ones. Notice that the *area* under the curve remains the same; this is because the total number of molecules remains unchanged.

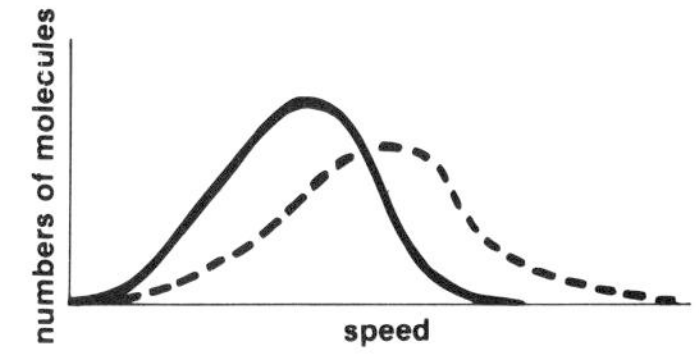

5:17 *Maxwellian distribution for gas molecules at two different temperatures. The higher temperature is indicated by the dashed curve. In both cases we can readily see that most of the molecules have average speeds; only a few are very fast or very slow.*

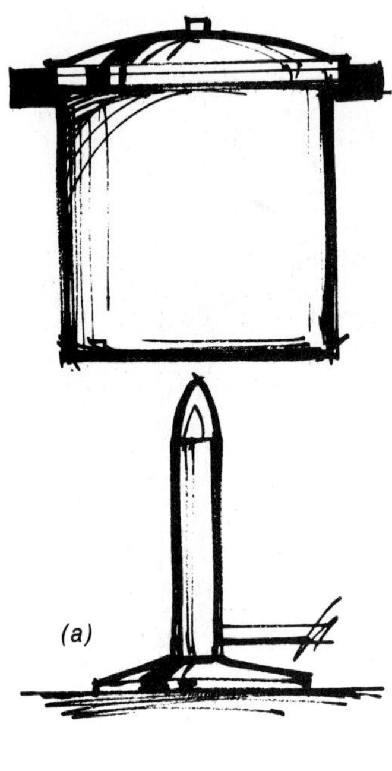

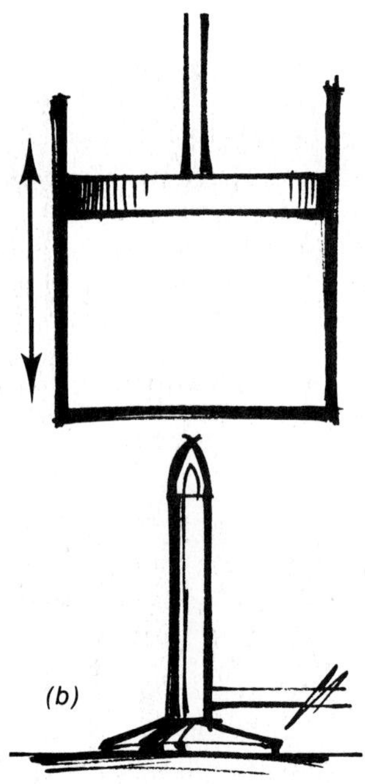

5:18 *(a) Heating a gas at constant volume and (b) at constant pressure.*

A gas may be heated in at least two ways: at *constant volume* and at *constant pressure*. When a gas is heated at constant volume (see Figure 5:18a), the pressure is observed to go up. This is what happens in a pressure cooker. The heat makes the molecules of air and water vapor in the pressure cooker travel faster. This increased rate of bombardment shows up on a pressure gauge as an increase in pressure. Thus, we again have a means of picturing what is happening on the basis of the individual molecules.

When a gas is heated at constant pressure (see Figure 5:18b), the container is allowed to increase in size in some way to accommodate the expanding gas. Thus, the volume increases but the pressure remains the same. In the case of Figure 5:18b, some of the energy that is supplied by the heat is used in raising the piston. The rest of the energy remains in the molecules as increased energy of motion (higher speeds). After the container has increased in size, the molecules are spread through a greater volume. They now have to bombard a greater surface than before. How can they still give the same pressure as before when they are more spread out? The answer is that they are moving faster and are capable of covering much more distance in a unit of time. Once more the molecular approach "works."

We have seen that the idea of molecules is very helpful. It fits the observed facts and helps us to comprehend what is going on behind the scenes. In some cases it even allows us to predict what is *going* to happen.

But the picture is still not complete. Some things that occur *cannot* be accounted for on the basis of molecules alone; so it is necessary to go one step further. Molecules themselves *must* be made up of something still smaller. We call these smaller somethings **atoms**. *An atom is the smallest part of an element that can enter into chemical combination with other elements.* Some molecules are made up of hundreds and even thousands of atoms. The hemoglobin molecule, for example, contains 9520 atoms of six different kinds. The water molecule, on the other hand, contains only three atoms of two different kinds. Molecules of the rare gases such as helium and neon contain only a single atom. In other words, in these special cases, the atom and the molecule are the same thing. But for most substances the molecules are made up of two or more atoms. In the next two chapters we shall be considering the very intriguing questions of what atoms are and how they are held together to form molecules.

List of Terms

atom
boiling point
Brownian movement (motion)
diffusion
Maxwellian distribution
melting point
molecule
mortar and pestle
vapor

Questions

1. The moon could not retain an atmosphere even if one could be placed around it artificially. How can this be understood on the basis of the motion of the molecules?
2. What is the difference between an atom and a molecule? Give definitions of each.
3. What is the kinetic theory of matter?
4. How can we explain gas pressure on the basis of the kinetic theory?
5. On the basis of the kinetic theory, explain why reducing the volume of a gas by half should double its pressure. (Assume constant temperature.)
6. What is the boiling point of a liquid?
7. (a) How could the boiling point of a liquid be lowered?
   (b) How could it be raised?
   (c) What device makes use of a raised boiling point?
8. Explain the expansion of a solid on the basis of the kinetic theory.
9. What happens on the molecular level when the temperature of a solid is raised to its boiling point?
10. What is diffusion?
11. (a) Is diffusion more rapid in a liquid or in a gas?
    (b) Why?
12. (a) How many water molecules placed side by side would be required to make a row an inch long?
    (b) Calculate how many it would take to make a row a centimeter long. (1 inch = 2.54 centimeters.)
13. Why does a glass of water evaporate, even at room temperature?
14. How is air liquefied?
15. Give an example of a substance in which the atom and the molecule are the same thing.
16. Smoke particles suspended in the air are observed to be knocked about in a random vibratory motion when strongly illuminated and viewed through a microscope.
    (a) What name is given to this phenomenon?
    (b) How can it be explained?
17. How do the distances between molecules in ordinary air compare with the diameters of the molecules?
18. On the basis of the kinetic theory, explain why the pressure on your body decreases as you climb a high mountain.
19. Who first proposed the use of the word "molecule"?
20. A gas is heated at constant volume. On the basis of the kinetic theory, explain why the pressure increases.
21. Why is it difficult to compress solids and liquids but easy to compress gases?
22. Why do liquids have a definite volume but not a definite shape?

Student Activities

1. Perform the operations with potassium permanganate ($KMnO_4$) described in Section 5:1.
2. Carefully mix 50 milliliters of water and 50 milliliters of absolute ethyl alcohol. If this is done properly, you will observe that the combined volume is somewhat *less* than 100 milliliters. This is generally explained by saying that some of the water molecules occupy spaces between alcohol molecules. (Note—The milliliter is a small unit of volume commonly used for laboratory work. There are 946 milliliters in a quart.)

Photo next page
Atomic planes in silicon.

UNIT II THE STRUCTURE OF MATTER

Chapter 6 Atomic Structure

6:1 Background Ideas about the atom have changed throughout the years. We probably will never be able to see what an atom looks like. The atoms described in books are nothing more than models, and what we talk about in this section may be out of date in 10 years or less.

Atomic models have been formulated in men's imaginations to account for certain scientific experiments and experiences. Remember, no one has seen an atom, not even the world's most famous scientists. The idea of atoms must be taken by faith. Isn't it strange that people can believe in certain things that they have not seen, such as atoms, yet they claim they cannot believe in God because they have never seen Him!

6:1 *John Dalton.*

6:2 Dalton Model The first atomic model we will discuss was invented by John Dalton (1766-1844). Dalton did a lot of work on atmospheric gases. As a matter of fact, he made observations on the weather for 57 years.

hydrogen gas

nitric oxide

carbon dioxide gas

6:2 *Dalton's atomic models. Dalton thought of atoms as being hard particles with "atmospheres of heat" radiating from them. The lines around the particles represent the atmospheres of heat.*

Dalton's model of an atom was a solid core of the element itself surrounded by an envelope of heat. The atoms of different elements had different sizes. If the atoms were of the same kind, the heat forces radiating out from the core would be alike, and the atoms would form neat layers like BB's packed together on a table. If the atoms were not the same size, the heat forces would also be unequal. This would cause the atoms of different gases to move around each other, and the gases would mix and stay mixed.

Dalton's atoms look like many-legged insects. The heat envelope takes the form of a cloud with caloric force lines radiating out into the cloud like legs on an insect. The *caloric theory* was very popular when Dalton developed his atomic ideas. Scientists thought that heat (caloric) was much like matter and could be transferred from place to place just as you transport water from place to place. Today we consider heat as one form of energy and matter as a different form of energy.

It is interesting to note that when we talk about the "cloud" around an atom today, we refer to it as an **electron cloud**. But in Dalton's day it was known as a caloric cloud. The negative charge of the electron cloud on an atom repels the negative charge of the cloud on another atom. However, in the Dalton model the envelope of caloric served as the repellent.

Since heat travels from a hot body to a cold body, it could be imagined that heat would "repel" heat and "attract" cold.

Dalton's atoms of different elements could combine to form what we call compounds. His atoms did not account for electrical charge or magnetic force; these concepts had not yet been developed. Note also that Dalton's atom could not be divided. It had no inner structure; it was solid. Scientists had not yet made any observations that needed to be accounted for by an inner structure of the atom.

6:3 Electrical Nature of Matter By the late 1800's the caloric cloud in Dalton's atom had been replaced by a fog to account for electrical charge.

Electricity was invented to account for the fact that matter sometimes attracts and sometimes repels other matter. If you rub a hard rubber rod with flannel or silk, it will attract certain objects and repel others. To express these two different observations (attraction and repulsion) the names "positive charges" and "negative charges" were invented. For one substance to attract another, one must be positively charged and the other negatively charged. For two substances to repel each other, they both must have the same charge—both positively charged or both negatively charged.

The name **electron** has been given to the negative charge and **proton** to the positive charge. If an object is negatively charged, it has an excess of electrons; if it is positively charged, it has an excess of protons. When a positively charged body and a negatively charged body touch, there is a flow of charge; and the positive and negative charges neutralize each other. The electrical models of particles of matter (atoms) have proved to be very successful in picturing the structure of the atom.

6:4 Lenard Model In 1903 Philipp Lenard (1862-1947) showed that **cathode rays** (stream of electrons) could go through a sheet of aluminum. Therefore, matter—or what matter is made of (atoms)—is not so solid after all! If rays can pass through tightly packed atoms, atoms must not be so "solid." A lot of empty space must be in them and between them. To account for this unusual observation Lenard developed his model of an atom. He invented a shell for an atom with **"dynamids"** and empty space. A dynamid was made up of a positive charge combined with a negative charge. The number of dynamids was proportional to the atomic weight of the element. These dynamids formed an impenetrable barrier; but the rest of the atom was empty, and electrons could go through easily.

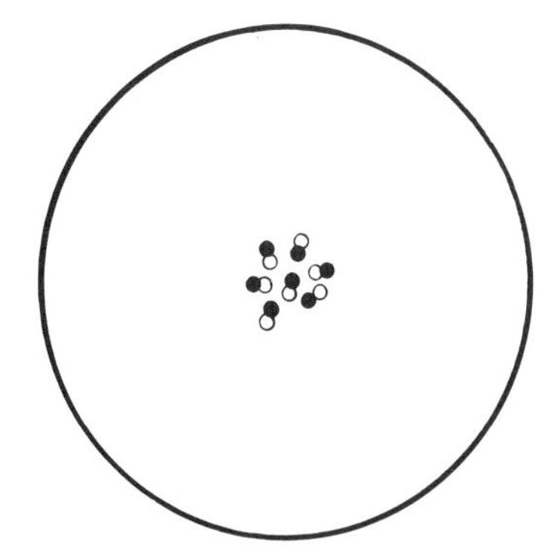

6:3 *Lenard model of the atom showing the "dynamids" and empty spaces. The positive charge (black) is combined with a negative charge (white) to form a dynamid.*

6:5 Nagaoka Model At about this same time Hantaro Nagaoka, a Japanese physicist (1865-1950), devised a different kind of model—one resembling the planet Saturn. In his representation a positive sphere was surrounded by rings of electrons.

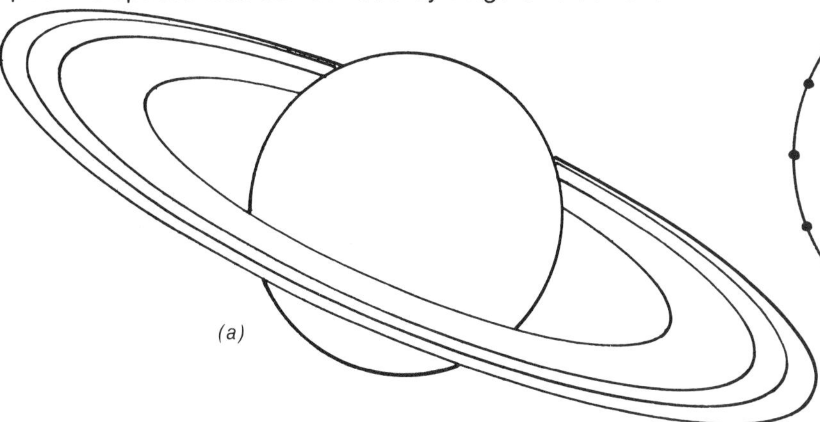

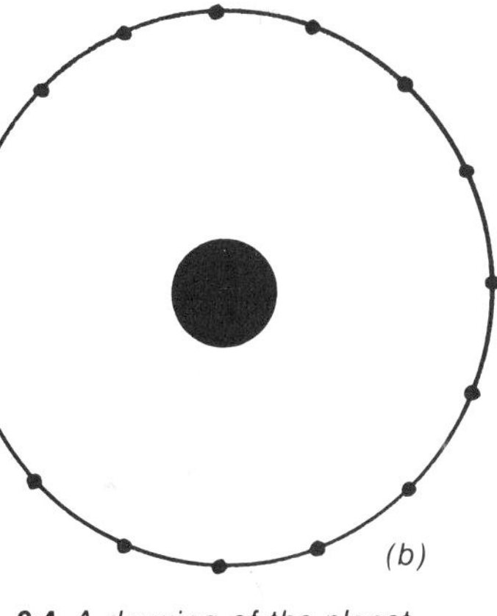

6:4 *A drawing of the planet Saturn (a) The Nagaoka or Saturnian model of the atom (b) The electrons lie in the same plane in the model, similar to the rings of Saturn.*

By the time Lenard and Nagaoka developed their atomic models, it was known that matter could give off electrons and form positive ions. Most scientists preferred Nagaoka's model because electrons could be imagined to be more easily "lost" from the "Saturn" model than from the Lenard "dynamid" model. The dynamids would not give up a negative charge so closely associated with a positive charge as would a loosely-bound halo of electrons. Why?

Also, it was known that if matter was heated to very high temperatures, light would be given off from the "atoms" of the substance. Scientists felt that this light came from vibrating electrons (from Maxwell's electromagnetic theory); it was easy to imagine electrons vibrating in the Saturn halo, whereas it was difficult to think of the closely held electrons in Lenard dynamids vibrating at all.

6:5 *(a) The negative charge is held closely by the positive charge in the Lenard model. (b) The negative charge, however, is not held so closely by the positive charge in the Nagaoka model. Thus, (b) allows for the vibration of the negative charges (electrons).*

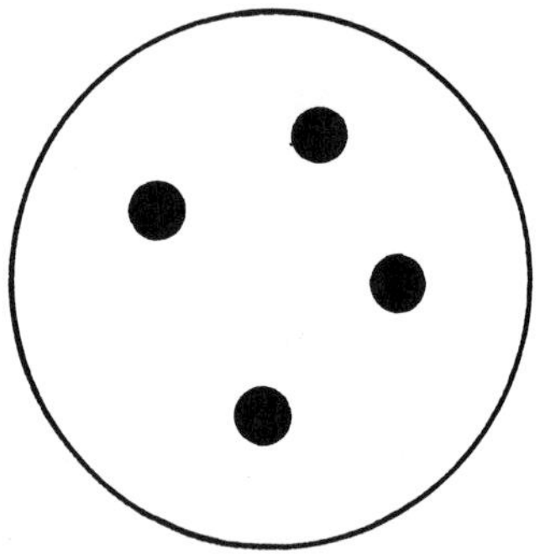

6:6 *Thomson model of the atom. Electrons embedded in a positive jelly.*

6:6 Thomson Model In 1904 the English physicist Joseph John Thomson (1856-1940) invented yet another atomic model. It is sometimes referred to as the "plum pudding" or "raisin muffin" atom. The electrons (plums or raisins) were embedded in a uniform positive core of matter (pudding or muffin). This was the most popular model among scientists at this time.

According to Thomson's model, if two atoms collided (caused by heating), the electrons would start vibrating in the positive pudding or muffin. This electronic vibration would set up the radiation of light from the atom. Each element would give off different wavelengths of light.

6:7 *A bright line spectrum of calcium (representation).*

At this point in the history of the atom the three models—Lenard's, Nagaoka's, and Thomson's—were not satisfactory. Lenard's model could account for the emptiness of matter but did not accurately explain the light given off by excited matter. Thomson's models could better account for light given off but not for the emptiness of matter. Nagaoka never tried to adjust his model to experimental findings. Later scientific experiments revealed that these models were indeed unsatisfactory, and even the popular Thomson model was eventually abandoned.

6:7 Rutherford Model In 1908 Ernest Rutherford of England (1871-1937) and Hans Geiger of Germany (1882-1947) conducted a very important scientific experiment. They bombarded a thin metal foil with **alpha particles** (*helium nuclei*). Some of the positively charged particles went through the metal foil almost in a straight line, and some went through slightly bent from their original path. But some particles were reflected almost straight backwards!

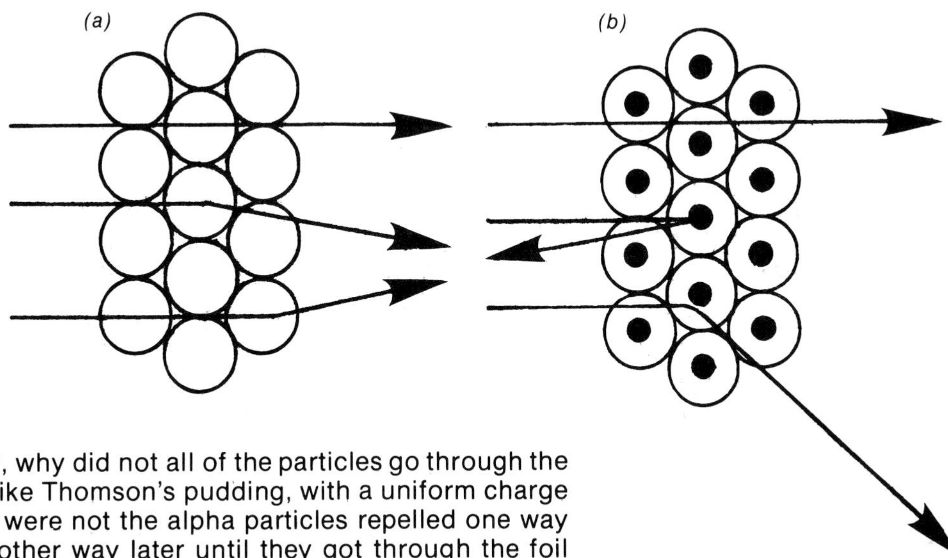

6:8 *(a) Alpha rays passing through Thomson-type atoms and being slightly deflected as they move through. (b) Alpha rays passing through Rutherford-type atoms. Any alphas that collide directly with a concentrated positive charge will be reflected directly backwards as shown.*

If matter is empty, why did not all of the particles go through the foil? If matter is like Thomson's pudding, with a uniform charge distribution, why were not the alpha particles repelled one way one time and another way later until they got through the foil only slightly affected?

A man hurrying through a standing crowd of people may be bumped to the left sometimes and to the right at other times, but eventually he will work his way through the crowd. The only way he can be forced to go back the way he came is if some very *repulsive* person knocks him backwards out of the crowd. What could knock an alpha particle back the way it came? Since it is a positively charged particle, a very strong, highly concentrated positive charge could repel it.

6:9 *(a) Positive man (alpha particle) moves through a Thomson-model crowd and is bumped around slightly but manages to get through the crowd in the direction he was moving. (b) Positive man (alpha particle) moves through a Rutherford-model crowd and makes headway until he meets a very repulsive positive man (concentrated positive charge) and is knocked completely backwards.*

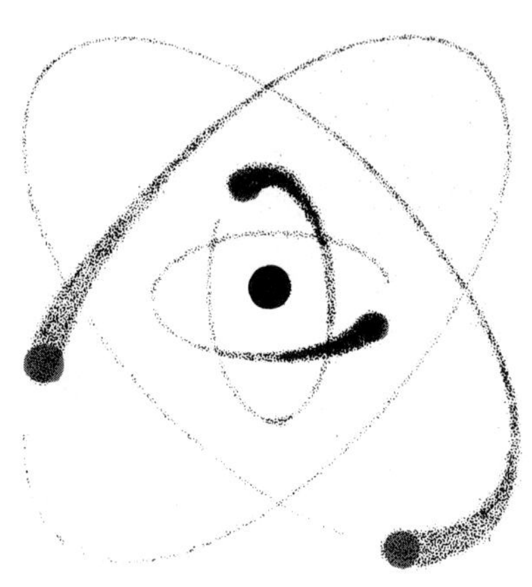

6:10 *Rutherford model of the atom.*

In 1911 Rutherford invented an atomic model that would account for this unusual repulsion. In his model the positive charge of an atom and most of its mass is contained in a central region called a **nucleus**. The protons in the nucleus provide all of the positive charge and some of the mass. The rest of the mass is provided by particles called **neutrons**. A neutron weighs about the same as a proton, but it has no charge. Any proton or neutron that exists inside the nucleus is called a **nucleon**. The nucleus is extremely small and we know very little about it. One or more electrons, depending upon the charge of the nucleus, revolve around the nucleus but at a great distance away from it. This model is mostly empty space, which allows most of the bombarding alpha particles to pass through unaffected; but any particles which come near the nucleus are deflected backwards. Electronic vibrations by excitation could be easily visualized also.

There is one major problem with the Rutherford atomic model. With all of the positive charge concentrated at the nucleus, the negatively charged electrons should fall into the nucleus because of the strong electrical attraction. The positive nucleus would attract the negative electrons into itself and the atom would collapse! Even if the electrons would circle the nucleus in the way the planets go around the sun in our solar system, they would still eventually fall into the nucleus. According to Maxwell's theory of electromagnetism, any accelerating charge (an electron changing its velocity) must radiate energy. Enough energy would leave the atom until the electrons would stop, and they would then be drawn into the nucleus. This is a very serious problem indeed.

Because of this objection, all atomic models that are like a planetary system, or have electrons moving in definite **orbits** (paths) like planets, are wrong. You probably have come to think of atoms in this way; but as you have seen, this idea cannot be true.

6:8 Bohr Model Niels Bohr of Denmark (1885-1962) invented a model of the atom to avoid this major problem in the Rutherford model. Bohr's model has a central nucleus like Rutherford's model. However, Bohr stated that as long as an electron travels in an orbit, it does not radiate energy. Only when an electron drops from one orbit to another will it emit energy. Bohr also developed his model so that the emitting electron would give off energy only in specific units called **quanta** (singular = quantum). This means that the electrons would have definite amounts of energy at all times. When these excited electrons drop back to lower energy levels and give off these specific amounts of energy, they produce definite spectral lines (brightly colored lines on a dark background). Bohr's model is very good in accounting for these spectral lines. It also works very well for the hydrogen atom but is not successful in picturing elements with higher atomic numbers.

Bohr restricted himself to circular electronic orbits. Arnold Sommerfeld of Germany (1868-1951) later introduced elliptical (oval) orbits into Bohr's model. However, even with all of the improvements of the Bohr model, it is still basically a planetary system with definite energy levels; and many observations will not fit into the model.

One of the problems with all planetary atomic models is that they try to explain too much. They claim to show the structure of something that cannot be seen. Other models could be developed that might account for experimental observations, but they would have to have a structure different from a planetary system. So what has been gained? We emphasize over and over again that **what cannot be observed is not science.** Planetary models of the atom go further beyond observed facts than it is safe to venture.

6:9 Quantum Model Physicists in the 1920's developed models of the atom based mainly on the **quantum theory** and **wave mechanics**. The subject is too involved for this book, but we will give you some general ideas about the quantum model of the atom. The theory was developed by men such as Heisenberg (1901-), de Broglie (1892-), and Schrödinger (1887-1961).

Since we cannot see an electron or follow its orbit, we have

6:11 *Bohr model of the atom. This shows a model of the hydrogen atom, an electron spinning about a proton.*

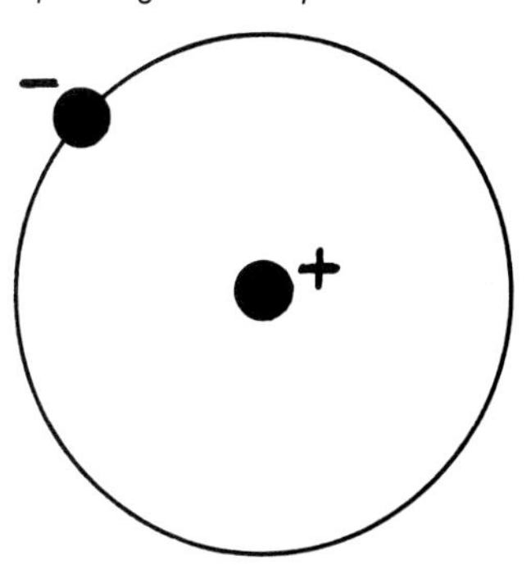

6:12 *Electron energy diagrams of hydrogen shown as concentric circles (a) and as levels (b). Electrons dropping from higher levels to lower levels give off energy in the form of electromagnetic radiation.*

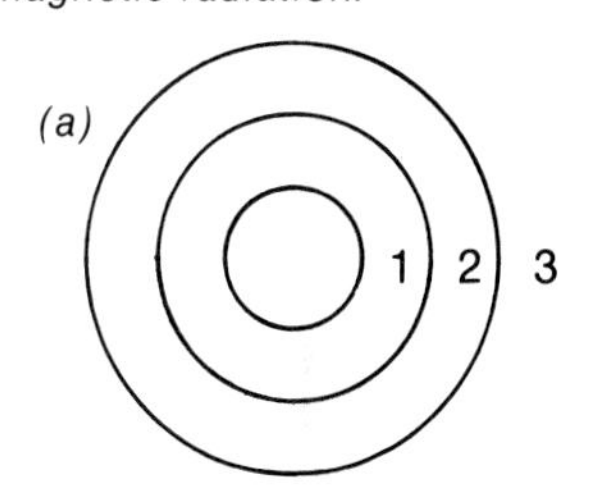

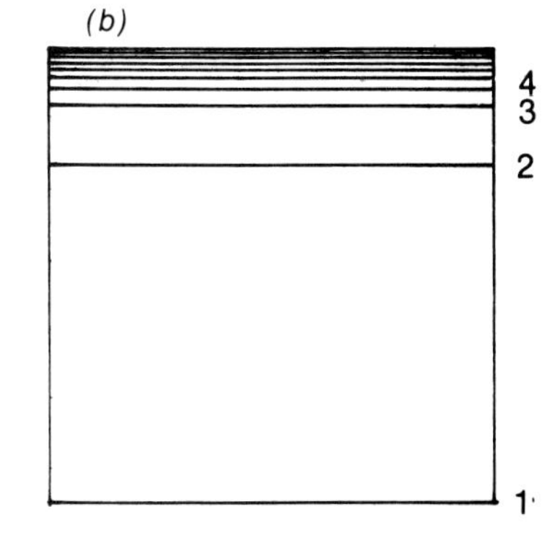

6:13 *One of the possible standing waves on a string.*

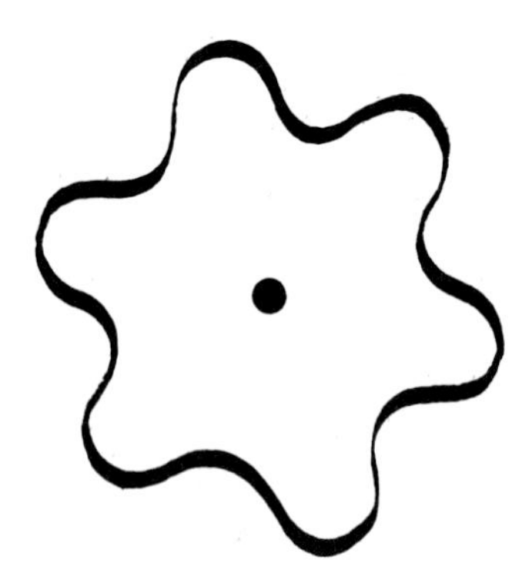

6:14 *An incorrect model, treating an electron as a standing wave in an orbit.*

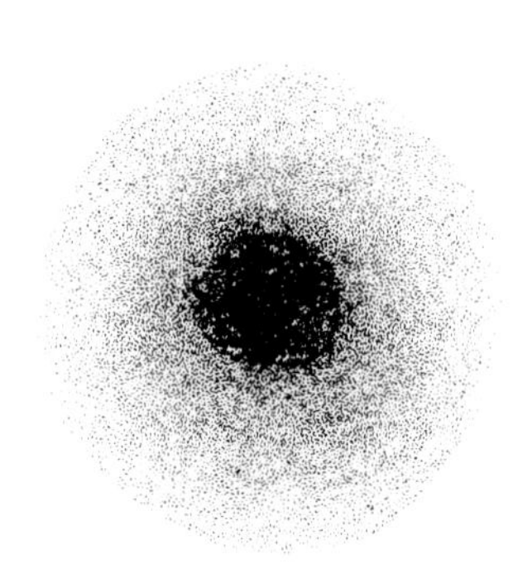

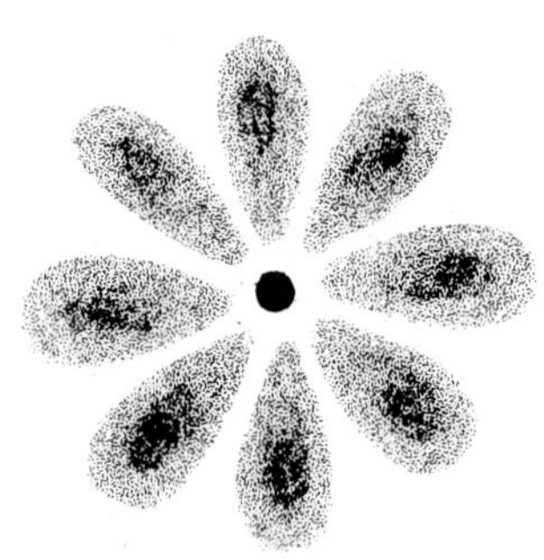

6:15 *Two possible states for electrons in hydrogen. The probability of finding the electron in the dark areas is very high.*

no right to assume electronic orbits exist. In the quantum model, the electron is viewed as a wave and not as a particle! In an atom, an electron is thought to act somewhat like a standing wave (see Chapter 33). The wave exists inside the atom. We cannot say exactly *where* the wave exists since we cannot see it inside the atom, but using wave mechanics we can tell you approximately where we might expect to find it at a given time. There is a great deal of mathematics involved in the quantum model. If you go to college and enroll in a good modern physics course, wave mechanics will be covered in that course. For example, some of these topics are discussed thoroughly in the junior year physics course at Bob Jones University.

After you have applied complicated mathematics, you might expect to find an electron in places similar to what you see in Figure 6:15. These are various "state functions" of the electron in hydrogen. The darker the area, the greater the chance of finding the electron in that area at any time. These "state functions" are the solutions to wave mechanical equations. Therefore, the atomic model of today is more a mathematical concept than an actual physical model. This is more truthful, because we have never seen an atom. So far, the quantum model is the most accurate model we have.

6:10 What Model Do You Use? The modern atomic concept is nothing more than mathematics! It is some very difficult math at that! However, many scientists continue to use pictures of electrons, orbits, atoms, and nuclei because they are useful in everyday work. The atom model that is used depends upon the problem you want to solve, even though the model may not be mathematically correct.

Remember, if something works, scientists will use it. Your teacher may use a model that is not quite correct, but he is using it so that you can get some idea of what we think the atom is. He cannot resort to complicated mathematics, for this would only confuse you.

Keep in mind that the atom is only a model. Charles H. Bachman, a physics professor, has stated, "An atomic model is any configuration which represents a conception of an atom. Consequently, there can be many atomic models. They can exist built up of materials, on paper or in the imagination. A model used in explaining one feature of an atom may not be useful for explaining any other feature; in fact, it may be contradictory."[1]

If you do not like any of the atomic models we have discussed, then you can enter some field of scientific work and possibly invent an atomic model of your own. If it is a good model and satisfies the scientific community, then it will be accepted. It is fitting to close this section with these words by the noted scientist Alfred Romer: "Do not think for a moment, though, that now you know the 'real' atom. The atom is an idea, a theory, a hypothesis; it is whatever you need to account for the facts of experience. . . . A good deal will happen in the future and the changes in the atom will continue. An idea in science, remember, lasts only as long as it is useful."[2]

[1]Charles H. Bachman, *Physics: Understanding Our Physical Environment* (Croton-on-Hudson, New York: Bogden & Quigley, Inc., 1970), p. 62.

[2]Alfred Romer, *The Restless Atom* (Garden City, New York: Doubleday & Co., Inc., 1960), p. 175.

List of Terms

alpha particle	nucleon
Bohr model	nucleus
cathode rays	orbit
Dalton model	proton
dynamid	quanta
electron	quantum model
electron cloud	quantum theory
Lenard model	Rutherford model
Nagaoka model	Thomson model
neutron	wave mechanics

Questions

1. What is the significance of the study of the atomic theory from a historical viewpoint?
2. Atoms are (a) actual physical entities, (b) models, or (c) elements?
3. Has any human ever seen an atom?
4. What physical phenomena did Dalton's model of the atom include?
5. What did his model lack?
6. What are cathode rays?
7. What is the importance of being able to send cathode rays through matter?
8. The term "dynamid" is connected with which atomic model?
9. What three things did Lenard's model of the atom account for?
10. The planet Saturn reminds you of what atomic model?
11. What interesting feature did the Nagaoka model have that none of the previous models had?
12. What is the Thomson model of the atom?
13. What is the Rutherford model like?
14. What evidence did Rutherford use to disprove the Thomson model and help gain acceptance for his own model?
15. The charge on a proton is (a) +1, (b) −1, or (c) 0?
16. The charge on a neutron is (a) +1, (b) −1, or (c) 0?
17. The charge on an electron is (a) +1, (b) −1, or (c) 0?
18. What are nucleons?
19. What are the major problems with the Bohr and Rutherford models of the atom?
20. What is the basic difference between the Bohr and Rutherford models?
21. What are electron clouds? How do they differ from electron orbits?

Student Activities

1. Make paper, cardboard, and wire replicas of some or all of the atomic models we have discussed.
2. Think up an experiment that will demonstrate the electrical nature of matter without using electrical current.

Photo next page

Catalytic cracking plant for petroleum. Large molecules are split apart and formed into smaller, more useful molecules. Chemical bonds are rearranged in the process.

UNIT II THE STRUCTURE OF MATTER

Chapter 7 Chemical Bonding

7:1 Ionic Bonding When we talk about **bonding** between elements to form compounds, it is sometimes helpful to use a combination of the Rutherford and Bohr models. Other times it is helpful to use the quantum model. In this book we shall always choose the model that works best. Scientists are concerned chiefly with workability.

Let us look at a typical chemical reaction.

$$2Na + Cl_2 \longrightarrow 2NaCl$$

To "explain" why sodium and chlorine combine, imagine the atoms to look like the models shown in Figure 7:1. The nucleus is positively charged and attracts the negatively charged electrons. The electrons closer to the nucleus are more tightly held by the nuclear charge; electrons which are farthest away from the nucleus are very loosely held.

Notice that sodium has a single electron in its outermost orbit. Besides being far away from the sodium nucleus, it is "screened off" from the nucleus by the closer electrons. This makes that outer electron very loosely held indeed.

Each atom of an element is considered electrically neutral. Counting the electrons in the sodium atom, we find that it has 11. Thus, there must be 11 protons in the nucleus to make the atom electrically neutral. In comparison, the chlorine atom has 17 electrons and 17 protons.

Suppose we add some sodium to chlorine gas. During the reaction, sodium atoms and chlorine atoms come very close

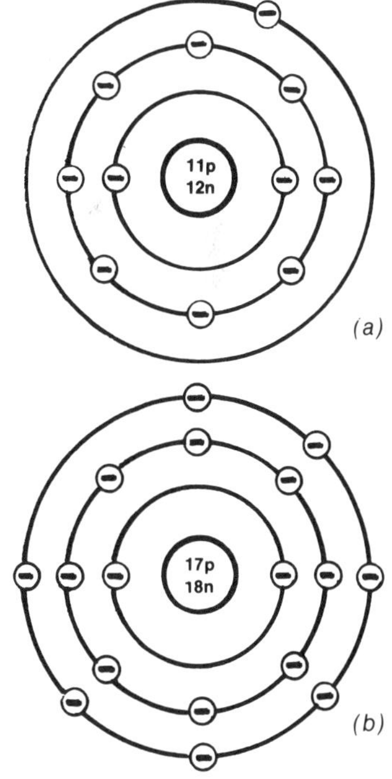

7:1 *(a) Model of a sodium atom. (b) Model of a chlorine atom.*

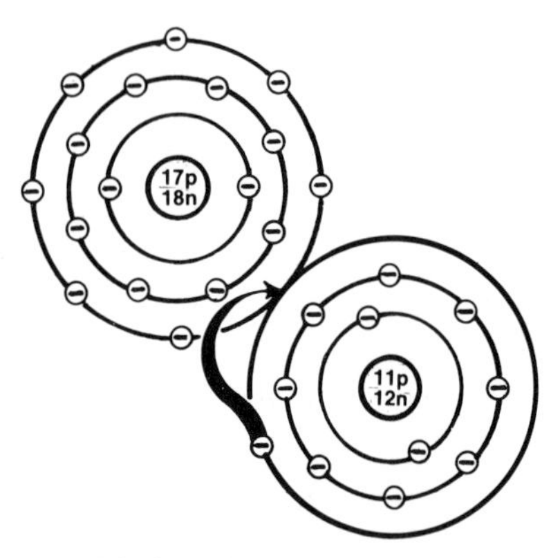

7:2 *A sodium atom donates an electron to a chlorine atom to form sodium and chloride ions.*

together. We imagine that they almost touch; this is represented in Figure 7:2.

The outermost electron of the sodium is now as close to the chlorine nucleus as it is to the sodium nucleus. However, the chlorine nucleus has a greater positive charge than the sodium nucleus (17 compared to 11). So the outside sodium electron becomes attracted to the greater charge of the chlorine nucleus.

7:3 *The hero, Mr. Chlorine, grabs Miss Electron away from the villain, Mr. Sodium. The villain will never be able to get her back from the strong hero.*

The electron leaves the sodium atom *permanently* and goes into the outermost electron shell of the chlorine atom. Let us discuss a spiritual analogy to this electron transfer. Before being saved, a person belongs to Satan's family; he is a child of the devil. Then he comes in contact with a child of God who witnesses to him. The Holy Spirit attracts him to accept Jesus Christ as his Saviour. When he confesses his sin and receives Him, he then is adopted into God's family. His transfer from one family to another is permanent.

When an atom loses or gains one or more electrons, the atom becomes an **ion**. An ion is no longer neutral, because its proton-electron balance has been upset. The sodium atom lost an electron, leaving it unbalanced; the sodium ion has 11 protons and 10 electrons. Thus, the sodium ion is positively charged. The chlorine atom gained an electron. Therefore, the chlorine ion has 17 protons and 18 electrons. This makes the chlorine ion negatively charged.

These two oppositely charged ions—the positively charged sodium ion and the negatively charged chloride ion—attract each other and come together to form the compound sodium chloride. This type of bond formation is called **ionic bonding**. Actually,

$NaCl$ is Na^+Cl^- because when they bond, sodium and chlorine are ions. Sodium lost an electron which chlorine gained. The number of electrons gained or lost by an atom in a chemical reaction is called its **valence**. Sodium has a valence of +1 (it donates one electron in a reaction). Chlorine has a valence of -1 (it accepts one electron in a reaction). The positive ion is called a **cation**, and the negative ion is called an **anion**. In this discussion we have considered only a single sodium ion combining with a single chloride ion. Actually ionic bonding involves vast numbers of ions combining to form crystals of the ionic compound.

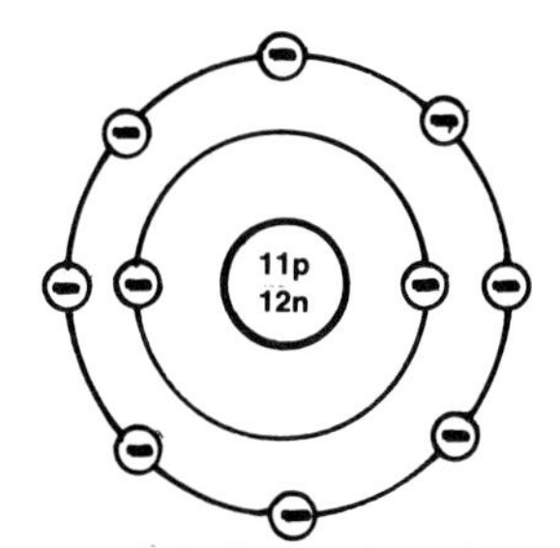

model of a sodium ion.

7:2 Stable Octet of Electrons Look at the illustration (Figure 7:4) of the model of the sodium and chloride ions. Notice the outer electron shell of each. They have the same number of electrons—eight. This is considered a very stable arrangement of electrons. Many chemists believe that elements react together so that they can obtain a stable **octet** of electrons in their outer electron shell.

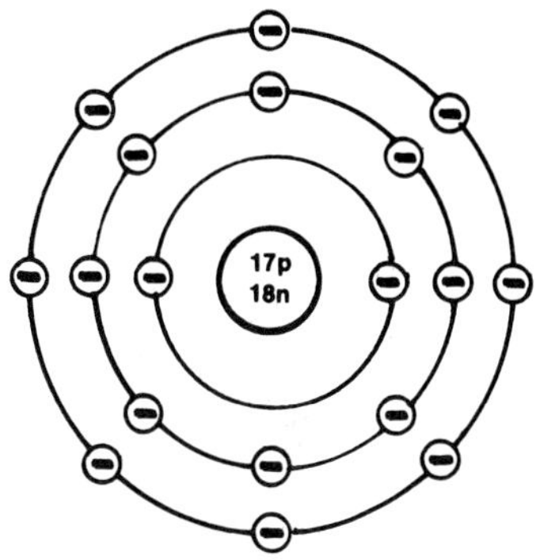

model of a chloride ion.

7:4 *Models of sodium ion and chloride ion. (Numbers in nucleus designate protons and neutrons.)*

7:3 Acids A model of the hydrogen atom is shown below. The nucleus consists of one proton, and there is one orbiting electron. The atom becomes an ion when it loses the electron (-1), and the proton (+1) is left. So the hydrogen ion is actually a proton. This ionization reaction is expressed in chemical symbols as

$$H \longrightarrow H^+ + e^-$$

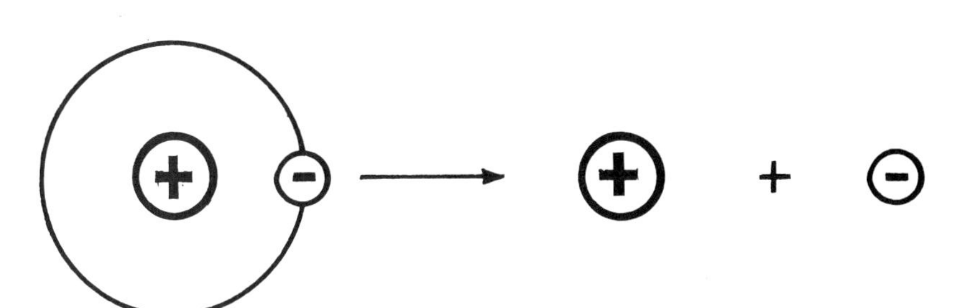

7:5 *A hydrogen atom loses its electron to form a hydrogen ion (a proton).*

Any solution with an excess number of protons in it is considered an acid solution. In 1923 J. N. Brönsted (1879-1947) of Denmark and T. M. Lowry (1874-1936) of England suggested that an **acid** be defined as a substance which can donate a proton to any other substance.

Experiment 7:1 *Properties of Acids*

1. Put a small amount of concentrated lemon juice and vinegar into separate beakers. Test the solutions with blue litmus paper. What does this indicate? Record the results. What substances in the lemon juice and vinegar are causing this?
2. Dip a finger into each solution and taste it. Describe the taste.

Experiment 7:2 *The Action of Acids on Metals*

1. Drop a new piece of aluminum foil into dilute HCl.
2. Remove the foil after 15 minutes and observe the surface of the foil. What reaction occurred in the solution? Write the chemical reaction equation.

Typical acids are hydrochloric (HCl), sulfuric (H_2SO_4), and nitric (HNO_3). In water, their ionization reactions are represented as follows:

$$HCl + H_2O \longrightarrow H_3O^+ + Cl^-$$
$$H_2SO_4 + 2H_2O \longrightarrow 2H_3O^+ + SO_4^{-2}$$
$$HNO_3 + H_2O \longrightarrow H_3O^+ + NO_3^-$$

If you add an acid to water, many hydrogen ions (protons) will result. These will react with water to form **hydronium ions** (H_3O^+). You can test for an acid solution by using **litmus paper**. Acids will turn blue litmus paper a reddish color. Some acids we work with every day in the kitchen are lemon juice and vinegar. And battery acid is essential to our automobiles.

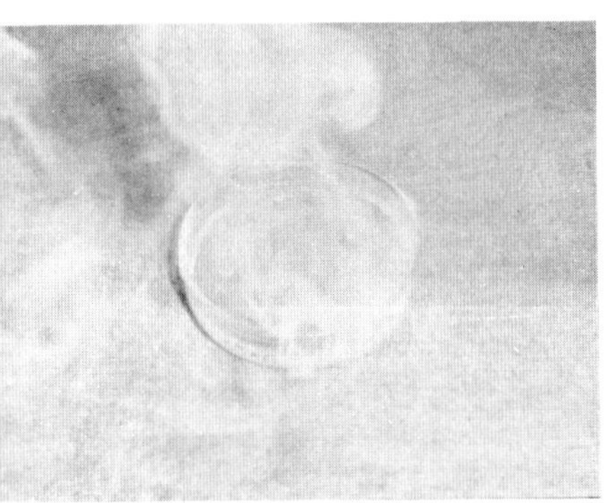

7:6 Sodium reacting with water to form sodium hydroxide and hydrogen. The reaction is so violent that it generates enough heat to boil the water. Steam can be seen coming out of the reaction vessel.

7:4 Bases According to the Bronsted-Lowry theory an acid is a substance which donates a proton to another substance. The substance which accepts the proton is called a **base**. The ion that gives the base its properties is the **hydroxide ion (OH^-)**. A common base (found in most soaps) is sodium hydroxide (NaOH). Its ions separate in water solution as follows:

$$NaOH \xrightarrow{H_2O} Na^+ + OH^-$$

Another base we find in the home is household ammonia.

A base can also be thought of as a substance which will donate a hydroxide ion to another substance. Normally, bases are formed when a metal reacts with water. For instance,

$$2Na + 2H_2O \longrightarrow 2NaOH + H_2\uparrow$$

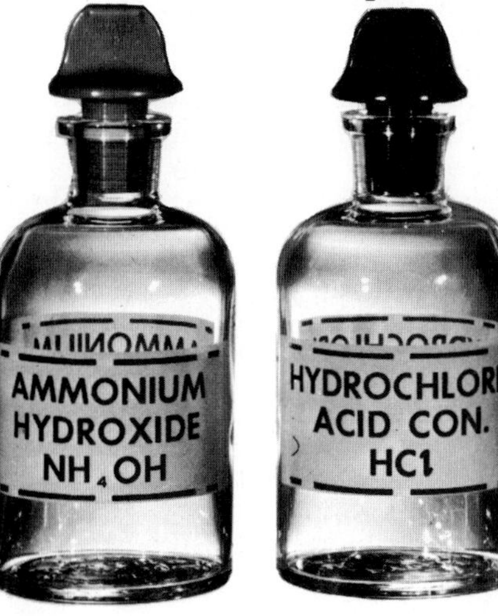

7:7 Laboratory bottles for storing acids and bases.

Hydrogen gas (H_2) is given off as a by-product of the reaction. The reaction of sodium with water is quite vigorous. The hydrogen ion takes the electron from the sodium and becomes molecular hydrogen gas.

$$2Na + 2H^+ \longrightarrow H_2\uparrow + 2Na^+$$
$$Na^+ + OH^- \longrightarrow NaOH, \text{ or}$$
$$2Na + 2H^+ + 2OH^- \longrightarrow 2Na^+ + 2OH^- + H_2\uparrow$$

A basic solution turns red litmus paper a bluish color.

Experiment 7:3 *Properties of Bases*

1. Obtain small pieces of sodium and calcium metals. Do not handle the metals with your bare hands!
2. Drop each piece of metal into different beakers containing a small amount of water. In each case note the speed of reaction.
3. Test the solutions with pink litmus paper. Record the results. What does this indicate?
4. Dip your fingers into the solutions and rub them together. How do the solutions feel on your fingers? (Wash your hands immediately afterwards.)
5. Write the equations for the chemical reactions that occurred. Save the reacted mixtures for Experiment 7:4.

Experiment 7:4 *The Action of Acids and Bases on Organic Materials*

1. Place a small amount of sugar in the bottom of a test tube and carefully add a small amount of concentrated H_2SO_4. The acid completely decomposes the sugar.
2. Now add concentrated HCl to a small piece of synthetic cloth placed in the bottom of a test tube. Again, the acid destroys the organic material.
3. Prepare a solution of 5 grams of NaOH in 100 milliliters of water. Heat the solution until it boils. Drop a piece of wool or cotton into the boiling basic solution. The cloth will fall apart. Now you know why you should not spill acids or bases on your clothing!

7:5 The pH Scale The acidity or basicity of a solution is rated by a numerical system called the **pH scale**. The *pH* of a solution can be measured by means of multi-colored indicator papers (see Figure 7:8) or read directly on a *pH* meter (see Figure 7:9).

A neutral solution, such as water, has a *pH* of 7. *Acids* have *pH* values *less than* 7. The lower the number, the more acidic the solution. *Bases* have *pH* values *greater than* 7. The higher the number, the more basic the solution. Some typical *pH's* are as follows:

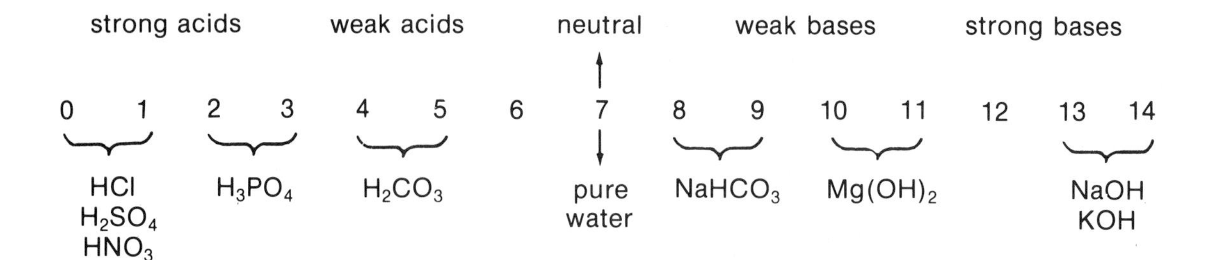

The chemist often needs to make sure that the proper range of *pH* is maintained during his experiment. In certain instances he will use a **buffer**—a chemical or mixture of chemicals capable of restoring the desired *pH* in a solution if the *pH* should start to change. Living organisms have very complex buffering systems which are capable of holding their *pH's* within extremely narrow limits.

7:8 *pH-sensitive paper. It will turn different colors when dipped into acid or base solutions, giving an indication of the pH of the solution.*

7:6 Salts Suppose we add an acid to a base, such as hydrochloric acid to sodium hydroxide. The acid has an excess of hydrogen ions (H^+). The base has an excess of hydroxide ions (OH^-). The ions will react to form water

$$H^+ + OH^- \longrightarrow H_2O$$

The total reaction is

$$HCl + NaOH \longrightarrow NaCl + H_2O$$

Actually, as the NaCl forms, it dissolves in the water and separates into individual sodium ions and chloride ions.

$$NaCl \xrightarrow{H_2O} Na^+ + Cl^-$$

If you get rid of the water by heating or evaporation, you will find NaCl (table salt) crystals left behind. A **salt** forms when we add an acid to a base.

$$\text{Acid} + \text{Base} \longrightarrow \text{Salt} + \text{Water}$$

This is called a **neutralization reaction**. The base neutralizes the acid, and the acid neutralizes the base.

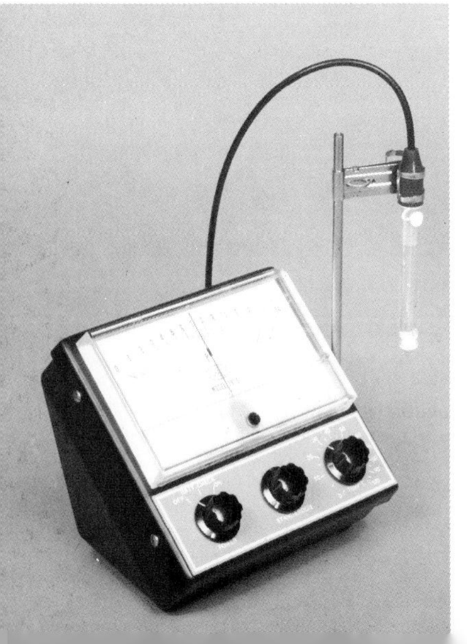

7:9 *Laboratory pH meter. Using the proper electrode, the pH of various solutions can be determined accurately.*

Experiment 7:5 *Neutralization Reaction*

1. Add 2 milliliters of concentrated HCl to 98 milliliters of water in a beaker. Mix thoroughly.
2. Add 1 gram of sodium hydroxide to 100 milliliters of water in another beaker. Stir until the NaOH has completely dissolved.
3. Fill 1 buret with acid and another with base.
4. Add 20 milliliters of acid to an empty beaker or flask.
5. Add 1 or 2 drops of phenolphthalein to the acid solution. Phenolphthalein is an indicator that is colorless in an acid solution but turns pink in a basic solution.
6. Add drop by drop the basic solution into the acid solution until the entire mixture turns slightly pink. Do not get in a hurry or you will overshoot the neutralization point by adding too much base. Continually stir the solution in the beaker to mix the incoming drops. Also, place a piece of white paper behind the beaker so that when the indicator changes to a very light pink you can recognize the change.
7. You may have to add acid drop by drop if you miss the neutralization point. You know that the mixture is neutral when 1 drop of acid will turn the indicator colorless and a drop of base will turn it back to pink.
8. After neutralization has been achieved, dip your finger in the solution and put it in your mouth. Do you recognize the taste?
9. Place some of the neutral solution in an evaporating dish and boil off the water. Taste the residue. Do you recognize the taste now? If so, write the chemical reactions of what you have accomplished during neutralization and during evaporation.

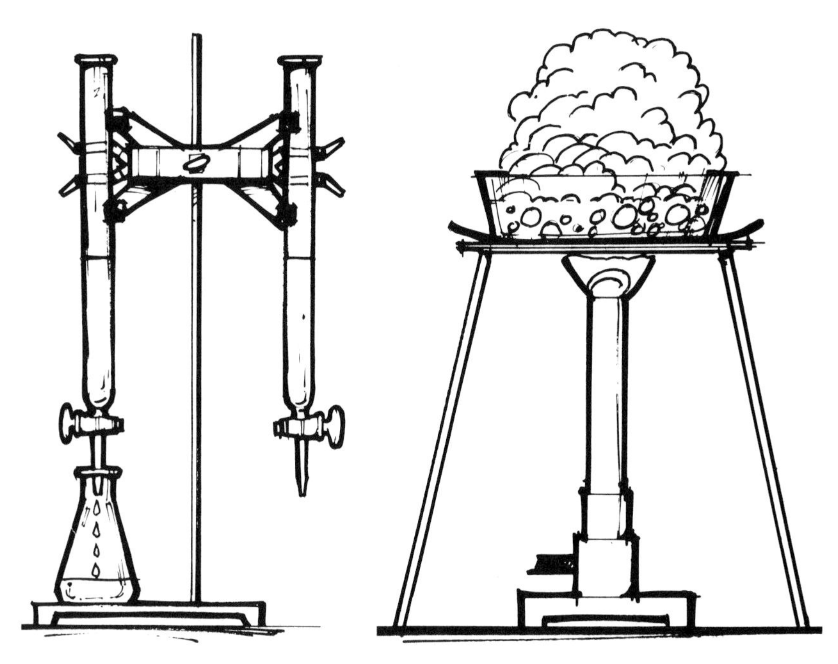

7:10 *Burets used to titrate acid and base in neutralization experiment.*

7:11 *Boiling off liquid in an evaporating dish.*

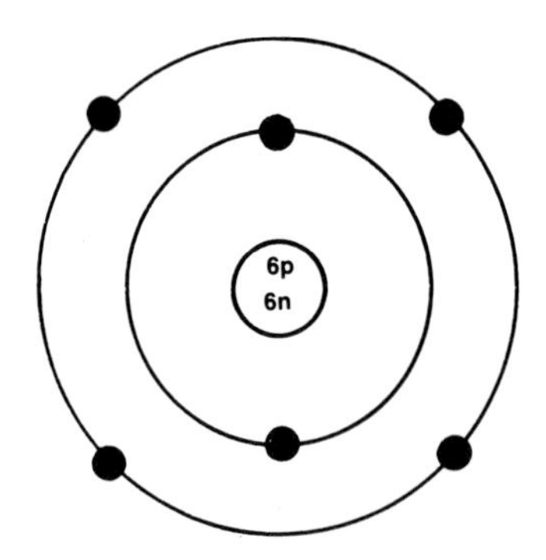

7:12 *Model of a carbon atom.*

7:7 Ionization in Water Even water can ionize by itself.

$$H_2O \longrightarrow H^+ + OH^-$$

However, as fast as it ionizes, the proton reacts with the hydroxide ion to re-form water.

$$H^+ + OH^- \longrightarrow H_2O$$

Both reactions can be written together as

$$H^+ + OH^- \rightleftharpoons H_2O$$

A forward (→) and a backward (←) reaction are going on at the same time. This is called a **reversible reaction**. Actually, in the case of water, the forward reaction is more likely to occur. We show this by

$$H^+ + OH^- \rightleftharpoons H_2O$$

The backward reaction occurs so seldom that we represent it with a small arrow (←). The principle of reversible reaction is important because most of the chemical reactions in your body are reversible.

7:8 Covalent Bonding—Bohr Model Many compounds in nature have elements which cannot achieve a stable octet of outer electrons by donating or accepting electrons to form ions. However, in keeping with the model of a stable octet, *sharing of electrons* was invented as a means to "explain" the formation of certain types of compounds. This sharing of electrons is called **covalent bonding** and the compounds formed as a result of this are called covalent compounds.

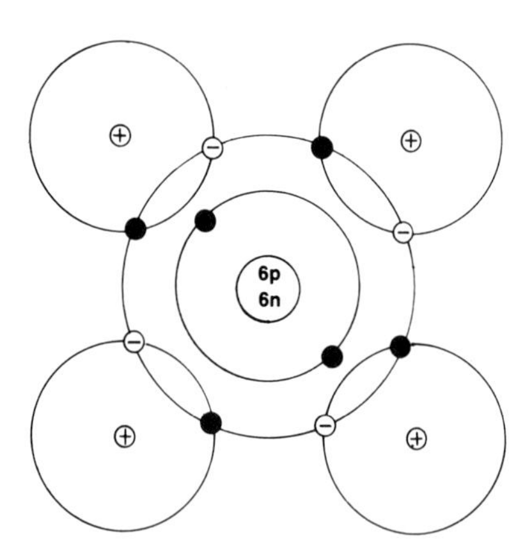

7:13 *Model of a methane molecule.*

Consider the compound methane (CH_4). Carbon has a valence of +4. Hydrogen has a valence of +1. Carbon and hydrogen cannot form an ionic bond because both would be positively charged. If the carbon atom could gain four more electrons, it would have a stable outer electronic shell of eight. This can be accomplished if it shares electrons with four hydrogen atoms. Also, the four hydrogen atoms will fill up their outermost shell by sharing the four carbon electrons. This will put a total of two electrons in the outer shell of each hydrogen atom.

The outer shell of the hydrogen atom will accept only one additional electron. This shell is stable when it contains two electrons. Hydrogen would have to share or accept nine electrons to get a stable octet in the next electron shell, and this is impossible. With the sharing arrangement, both the carbon and hydrogen are very stable (according to the model). The shared pair of electrons "spend most of their time" between the carbon and hydrogen nuclei. This sharing "holds" the atoms together.

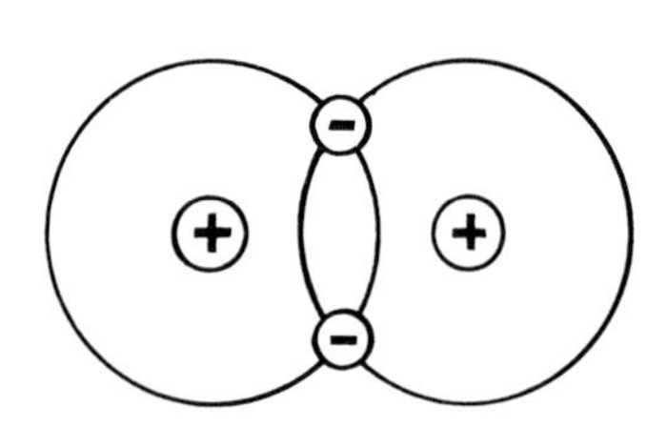

7:14 *Model of a hydrogen molecule.*

Other interesting examples of covalent bonding are gas molecules, such as hydrogen and chlorine. A single hydrogen atom does not have its outer **electron shell** filled. However, if it shares its electron with another hydrogen atom, each atom in the molecule will have two electrons in its outer shell part of the time. This configuration is more stable than a single hydrogen atom by itself.

As for chlorine, each atom has seven electrons in its outer shell. The stable condition for this shell is eight electrons. If two chlorine atoms share electrons, some of the time chlorine atom *A* will have eight electrons and at other times atom *B* will have eight electrons. This is shown in Figure 7:15. You might note that, according to the model, where one chlorine atom has eight electrons and the other has six, one would be positively charged and the other negatively charged. This means that, if the model is correct, covalent bonding has some ionic character to it.

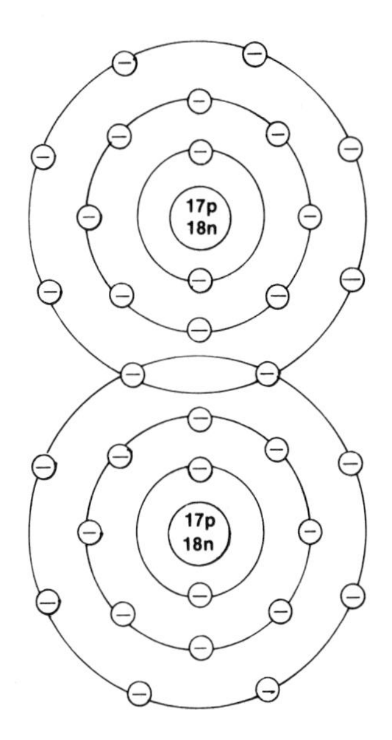

7:15 *Model of a chlorine molecule.*

★7:9 Covalent Bonding Using Electron Clouds As you can see, the Bohr-Rutherford model becomes rather unwieldy when we consider covalent bonding. Most chemists prefer to use models obtainable from quantum mechanics. Let us look at each molecule which was considered previously in covalent bonding from this more modern approach.

Do you remember what we said earlier about "state functions" for electrons (see Figure 6:15)? These "state functions" are sometimes called **electron orbitals** or electron clouds. These orbitals or clouds are places where you would most likely find the electrons. The orbitals exist (in theory) whether electrons are in them or not. They are darker on the inside and lighter as you move outward. Each orbital can hold no more than two electrons. Several different kinds of possible orbitals are shown in Figure 7:16.

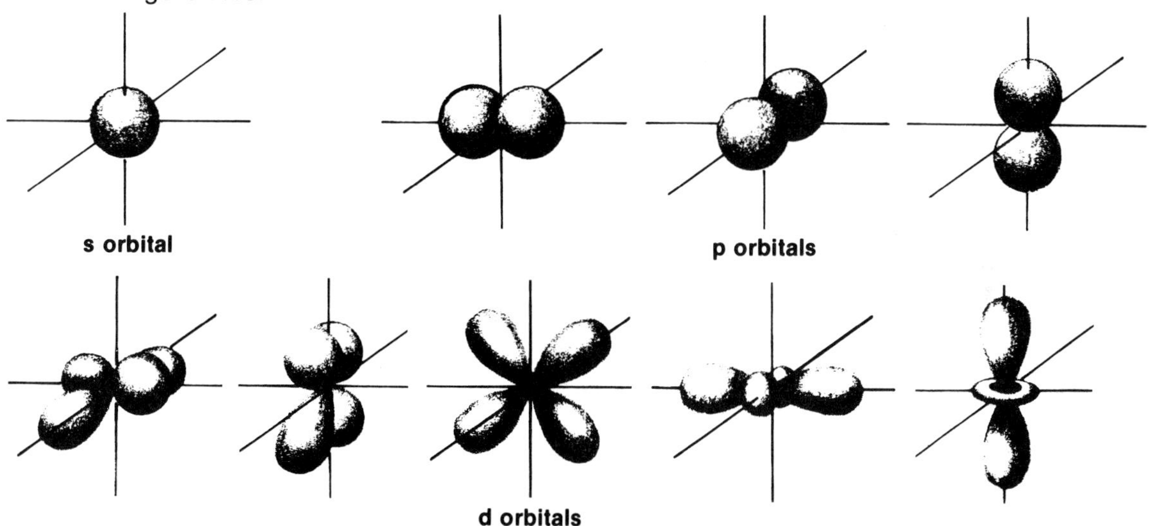

7:16 *Some possible electron orbitals.*

Notice that the *p* and *d* orbitals are completely directional in character. This very nice feature is the greatest advantage of the quantum model over the Bohr model.

Methane is a carbon atom sharing electrons with four hydrogen atoms. The electron in hydrogen is in an *s* orbital. The one carbon electron is in an *s* orbital and the other three are in *p* orbitals. When they react, the electrons are more likely to be found *between* the respective carbon and hydrogen atoms. The molecular orbitals appear enlarged between the atoms and "shrink" on opposite sides of the overlapped orbitals.

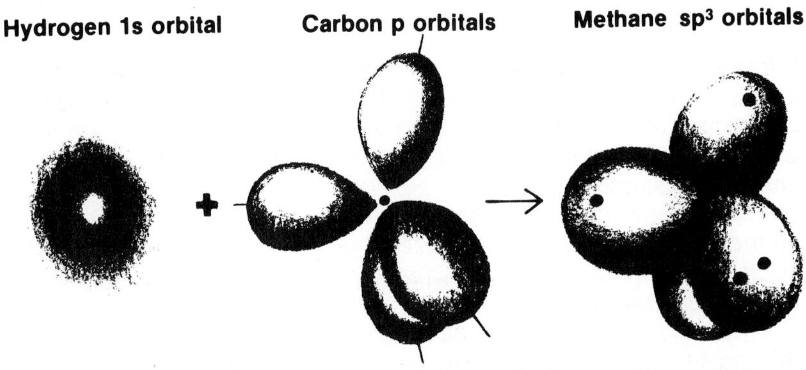

7:17 *Molecular-orbital-bonding model of carbon and hydrogen to form methane. The carbon nucleus is in the geometric center of the molecule and the hydrogen nuclei are in* sp^3 *orbitals.*

7:18 *Molecular orbital model of hydrogen* (H_2).

Bonding occurs when atomic orbitals overlap and form molecular orbitals. The hydrogen and chlorine molecules are illustrated in Figures 7:18 and 7:19b using the molecular orbital concept.

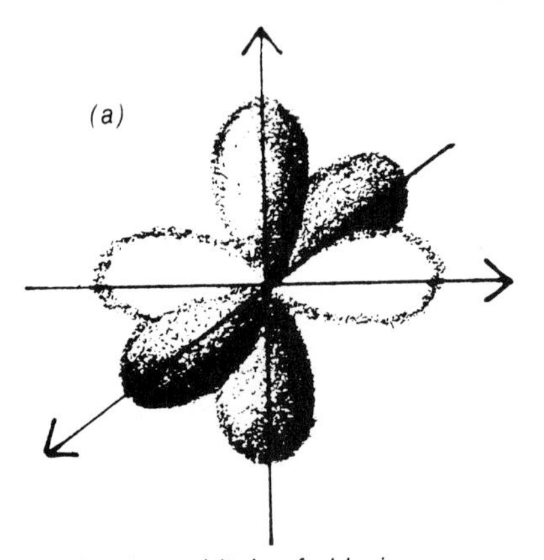

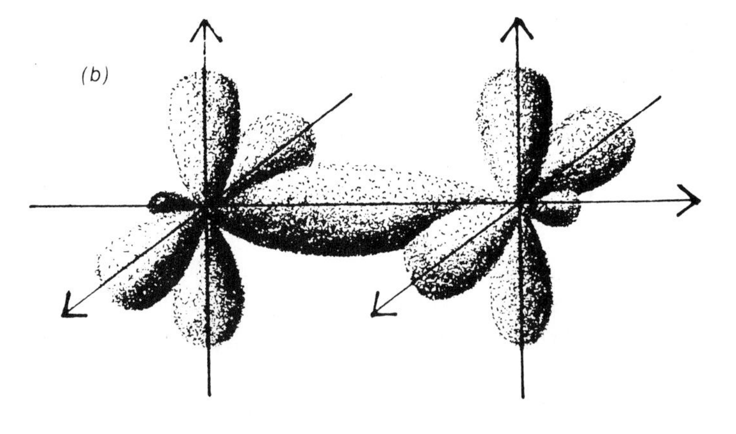

7:19 *(a) p orbitals of chlorine (b) Molecular orbital model of chlorine* (Cl_2). *When two chlorine atoms combine to form a chlorine molecule, two of their p orbitals overlap and enlarge between the two nuclei and shrink on opposite sides of the bond.*

We cannot go into any more detail on the various bonding models, although it is a fascinating study. If it interests you, we suggest you take more chemistry in high school. Perhaps the Lord will even lead you into a career in the field of chemistry.

★ **7:10 Metallic Bonding** How do metals "stick together"? That is an interesting question. Take a pure metal like aluminum: Is the bond ionic? No. Each aluminum atom has a valence of +3 (it wants to donate three outer electrons in a reaction). There is no simple way to imagine aluminum atoms forming ions with other aluminum atoms.

What about aluminum forming a covalent bond? Again, the answer is no. There is no simple way to imagine aluminum sharing electrons and ending up with stable octets in the outer shells.

Many theories have been developed to handle this problem. The one we will use is the **free electron theory**. Since there can be no localized sharing of electrons, it is imagined that all of the valence electrons in a metal are shared by all of the atoms. This is sharing on a grand scale rather than on a level of two, three, or

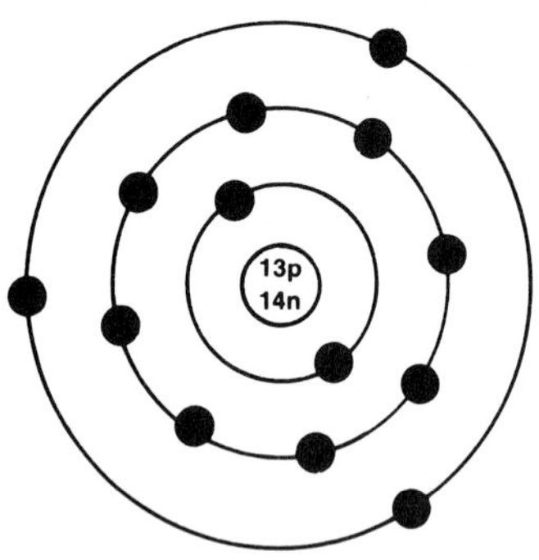

7:20 *Model of an aluminum atom.*

four atoms. A positive core of aluminum ions would be embedded in a sea of negative electrons. Many of the valence electrons would be free to "roam" throughout the entire metal structure. Some scientists think of this roaming as an extended covalent sharing of electrons.

The free electrons would "explain" why metals conduct electricity so readily. These electrons could drift easily with an applied electrical potential. There are many things wrong with this theory, but it is still useful.

It is agreed among scientists that metallic bonding is different from all others. If you are interested, this subject is dealt with in solid state physics and physical metallurgy.

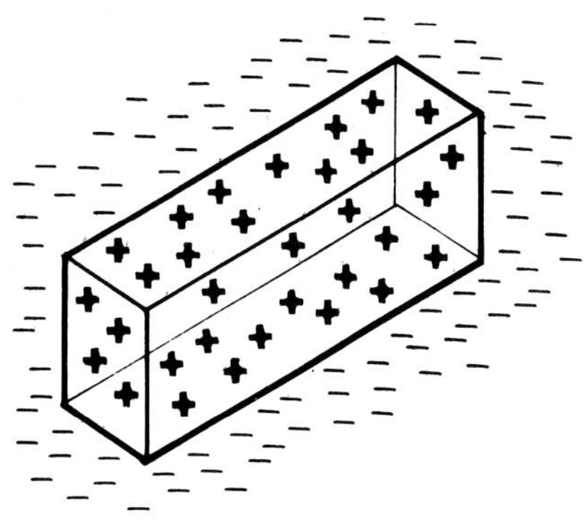

7:21 *Free electron model of a metal structure, a positive core of nuclei submerged in a sea of negative electrons.*

7:11 Chemical Formulas Knowing the valence of particular elements will help you in writing the chemical formulas of their compounds. For instance, the valence of magnesium is +2 and the valence of chlorine is −1. What, then, would be the chemical formula of magnesium chloride? Magnesium gives away two electrons in an ionic reaction. One chlorine atom can accept one electron. Therefore, it would take two chlorine atoms to accept two electrons from magnesium. Thus, the chemical formula for magnesium chloride is $MgCl_2$.

Here are the chemical formulas of various chlorides:

Table 7-1 *Valences of Some Common Elements*

metal	Na	Mg	Al	C
valence	+1	+2	+3	+4
chloride	NaCl	$MgCl_2$	$AlCl_3$	CCl_4

The chemical names of binary compounds (2-element compounds) end in the suffix *ide*: sodium chlor*ide* (NaCl), carbon tetrachlor*ide* (CCl_4), magnesium ox*ide* (MgO).

Suppose we react some elements with oxygen to form oxides. Oxygen has a valence of −2. Some typical formulas would be as follows:

Na_2O	CaO	Al_2O_3	CO_2
sodium oxide	calcium oxide	aluminum oxide	carbon dioxide

7:12 Radicals When two or more elements make up an ion, it is called a **radical**. The hydroxide ion (OH^-) is called a radical. Radicals react with substances as a unit. Some of the more common radicals are listed here:

SO_4^{-2} sulfate
NO_3^- nitrate
PO_4^{-3} phosphate
CO_3^{-2} carbonate

7:13 Chemical Reaction Equations Chemical equations have been previously introduced. However, more detail is necessary. Again, it is essential that you know valences to do the work correctly. Chemical reaction equations are like mathematical equations—the left- and right-hand sides must balance

$$Zn + S \longrightarrow ZnS$$

$$\text{not} \quad Zn + S \longrightarrow Zn_3S_4$$

There is one zinc atom on the left-hand side and one zinc atom on the right-hand side. The same is true for sulfur. In writing chemical equations there must always be a conservation of matter. *You cannot create or destroy any matter in a chemical reaction.* This is called **the law of conservation of matter**. It works for chemical equations, but it does not work when you consider nuclear reactions.

Suppose we give you the reaction for the synthesis (forming) of oxygen from mercuric oxide

$$\underset{\text{mercuric oxide}}{HgO} \longrightarrow \underset{\text{mercury}}{Hg} + \underset{\text{oxygen}}{O_2\uparrow}$$

What is wrong? (Check the left- and right-hand sides.) To balance, it should read

$$2HgO \longrightarrow 2Hg + O_2\uparrow$$

Let us classify reactions and give an example of each so that you can see the usefulness of chemical reactions. They are simply a quick way of showing you what is going on chemically. The types of reaction we will discuss are **synthesis**, **decomposition**, and **replacement**.

7:14 Synthesis and Decomposition To synthesize means to form something. In the previous reaction, oxygen was synthesized. If we heat a mixture of zinc (Zn) and sulfur (S), we get zinc sulfide.

$$Zn + S \xrightarrow{\Delta} ZnS$$

The triangle over the arrow represents heat. We have to heat zinc and sulfur before they will react.

Decomposition is the opposite of synthesis; it means to break something apart. Carbonic acid decomposes into carbon dioxide and water

$$H_2CO_3 \longrightarrow CO_2\uparrow + H_2O$$

The arrow upward means that CO_2 is a gas. Carbonic acid is in soft drinks. When you open a soft drink bottle, the "pop" sound and bubbles are caused by the carbon dioxide coming out of the solution (the carbonic acid in the soft drink is decomposing).

7:15 Replacement or Substitution Certain elements are able to replace other elements in compounds. In the reaction

$$2Na + 2H_2O \longrightarrow 2NaOH + H_2\uparrow$$

sodium replaces hydrogen in the compound HOH (water).

If we react zinc with copper sulfate,

$$Zn + CuSO_4 \longrightarrow ZnSO_4 + Cu$$

zinc replaces copper in the sulfate.

Potassium can replace sodium in its compounds, but sodium cannot replace potassium.

$$Na + K_2SO_4 \longrightarrow \text{no reaction}$$

Active metals react more vigorously than **passive metals** and can replace the latter in their compounds. Chemists have arranged elements in an **activity series** (another type of classification) that shows this relationship. The most reactive elements are listed first.

Table 7-2 *Activity Series of Some Elements*

Metals	Nonmetals
potassium	fluorine
sodium	chlorine
calcium	bromine
magnesium	iodine
aluminum	
zinc	
iron	
nickel	
tin	
lead	
hydrogen	
copper	
mercury	
silver	
platinum	
gold	

The element potassium can replace any metal below it. Nickel can replace any metal below it, but also can be replaced in its compounds by any metal above it. Gold is so *nonreactive* that it can be replaced in its compounds by *all* metals above it. Gold is considered a *noble metal* because of its lack of chemical activity; this property makes the metal very valuable.

Degree of activity can be written

$$Mg > Zn > Fe > Cu$$

which means magnesium is more active than zinc, which is more active than iron, which is more active than copper.

Experiment 7:6 *Metals Differ in Chemical Activity*

1. Place small amounts of Cu, Fe, Mg, and Zn into different tubes.
2. Pour a small amount of dilute HCl (hydrochloric acid) into the tube containing zinc. Hold your thumb over the top of the tube until the pressure of the gas given off is felt.
3. Place the mouth of the tube close to a flame and remove your thumb. The gas (hydrogen) burns with a mild explosion. The pop of the explosion will be heard. Notice how vigorously the zinc reacts.
4. Now add dilute HCl to each test tube and observe how vigorously each metal reacts. Arrange the metals in an activity series, listing the most reactive first.

7:16 Double Replacement Reactions The neutralization of hydrochloric acid with sodium hydroxide is an example of still another type of chemical reaction:

$$HCl + NaOH \longrightarrow NaCl + H_2O$$

In this case the ions "swap partners" to form new compounds. This general type of reaction is called **double replacement**.

List of Terms

acid
active metals
activity series
anion
base
bonding
buffer
cation
covalent bond
decomposition reaction
double replacement reaction
★ electron orbitals
electron shell
★ free electron theory of metals
hydronium ion
hydroxide ion
ion
ionic bond
law of conservation of matter
litmus paper
★ metallic bond
neutralization reaction
octet
passive metals
pH
radical
replacement reaction
reversible reaction
salt
synthesis reaction
valence

Questions

1. What do atoms look like?
2. Explain ionic bonding.
3. What is an anion?
4. What is a cation?
5. What is the valence of sodium? What does the valence of an element mean?
6. Is an octet of eight outermost electrons in an electron shell a stable situation?
7. What is an acid?
8. What is a hydrogen ion?
9. What is a hydronium ion?
10. What is a base?
11. What is the hydroxide ion?
12. What is a salt?
13. What is a reversible reaction?
★14. If you know what a reversible reaction is, what do you think an irreversible reaction is?
15. What is covalent bonding?
16. Why don't carbon and hydrogen form an ionic bond in methane?
★17. Why does hydrogen have a stable situation with only two electrons in its outermost electron shell in covalent bonding, rather than a stable octet of eight?
★18. What is the advantage of using electron orbitals to "explain" covalent bonding?
★19. Why do we need metallic bonding?

★ 20. What is the free electron theory of metallic bonding?
21. If calcium has a valence of +2, write the chemical formula for calcium chloride.
22. Draw a Bohr model of the calcium ion.
23. What is the purpose of a chemical formula?
24. (a) How many atoms are in a NaCl molecule?
(b) How many are in an $AlCl_3$ molecule?
(c) How many are in a CCl_4 molecule?
25. The chemical symbol for silicon is Si. Write out the name of this compound: $SiCl_4$.
26. What is a radical?
27. What is the chemical formula for the carbonate ion? What kind of drinks do you think have this ion in them?
28. Why are chemical equations similar to mathematical equations?
29. Balance the following equations.
(a) $Na + O_2 \longrightarrow Na_2O$
(b) $Na^+ + CO_3^{-2} \longrightarrow Na_2CO_3$
(c) $Fe + O_2 \longrightarrow Fe_2O_3$
★ 30. Why is conservation of matter so important in chemistry?
31. Identify the following reaction as synthesis, decomposition, or replacement.
(a) $2K + Cu_2O \longrightarrow K_2O + 2Cu$
(b) $4V + 5O_2 \longrightarrow 2V_2O_5$
(c) $CaCO_3 \longrightarrow CaO + CO_2\uparrow$
(d) $Ba + Cl_2 \longrightarrow BaCl_2$
(e) $2Li + Na_2CO_3 \longrightarrow Li_2CO_3 + 2Na$
32. Why is gold considered a noble metal?
33. (a) Can nickel replace copper out of its compounds?
(b) Can silver replace gold out of its compounds?
(c) Can nickel replace zinc out of its compounds?
34. Are the following reactions possible?
(a) $PbCl_2 + Hg \longrightarrow HgCl_2 + Pb$
(b) $PbCl_2 + Mg \longrightarrow MgCl_2 + Pb$
(c) $3KCl + Fe \longrightarrow FeCl_3 + 3K$
35. Name three Bible verses in which gold is mentioned.

Student Activities

1. Write a paper on gold discussing the unusual properties of the metal. Include chemical inertness, inactivity, malleability, form found in nature, physical and chemical properties, and Scripture references to the metal.
2. Write a paper on how a particular element was discovered.
3. If you have a chemistry set at home, perform several experiments with it, reacting various substances (be sure to follow directions carefully). Write out the balanced reaction equation for each of the reactions you performed. Note whether the reaction gave off heat or absorbed heat, the colors of the reactants, and the products.
4. React several different metals with HCl. Try to estimate the speed of reaction and arrange the metals in an activity series. Try to save some of the reaction products by boiling off the solution in a hood after it has reacted several days. Record the color and appearance of the residue (powdery, crystalline, hard, or soft). Make a display board of a piece of metal before test, a piece of the same metal after test, and the reaction product. Keep your observation on the speed of reaction in a notebook.

Photo next page
The element germanium, the existence of which was predicted by Mendeleev's original periodic table, is valuable to the semiconductor industry.

Al
Si
P
33
As
Ga
50
In
32
72.59
2830
4
937.4
Ge
5.32
[Ar]3d10 4s2 4p2
Germanium

UNIT II THE STRUCTURE OF MATTER

Chapter 8 The Periodic Table

8:1 Organizing the Elements Dmitri Mendeleev (1834-1907), a Russian chemist, and Lothar Meyer (1830-1895), a German chemist, are considered to have each developed the **periodic table** of the elements independently. However, the horizontal arrangement of Mendeleev's has become more popular than Meyer's vertical order. Also, Mendeleev's predictions, which his table enabled him to make concerning new elements and their properties, have enjoyed amazing success. Hence, we generally hear more about Mendeleev than about Meyer.

8:1 *Dmitri Mendeleev, originator of the horizontal form of the periodic table used today.*

As recently as the mid-1800's, chemistry was still in a state of serious disorder. There was, as yet, no systematic grouping or arrangement of elements. Sometime in the late 1860's Mendeleev made a card catalog of the 63 then-known elements. After experimenting with various arrangements of the cards, he decided to try putting them in order of increasing **atomic weight** (relative weight of each atom). When he did so, Mendeleev found that the properties of the elements repeated periodically. To illustrate this principle, we will look at the first 17 elements in his listing.

8:2

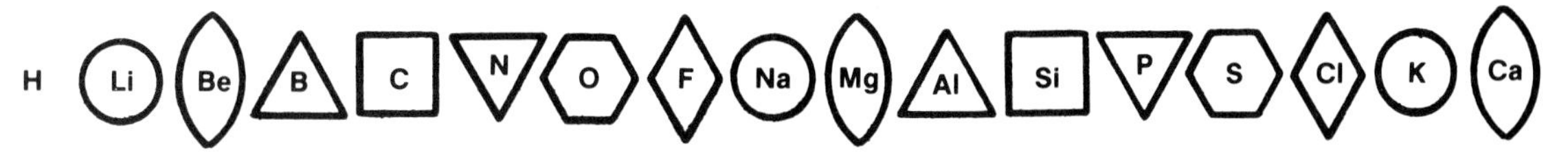

These are in order of *increasing* atomic weight, starting with the very lightest element. Let us consider the chemical properties of these elements: lithium, sodium, and potassium, all shown with circles, are very active *metals*. In each case, similar geometric figures have been drawn around elements having similar properties. It will be seen, however, that hydrogen does not exactly match up with anything. Hydrogen was a problem for Mendeleev and continues to be a problem with our present tables. Mendeleev experimented first with a vertical arrangement, published in 1869. By 1871 he had devised his horizontal table, which was the forerunner of our modern periodic table of elements. The 17 elements listed above were then organized as follows:

8:3

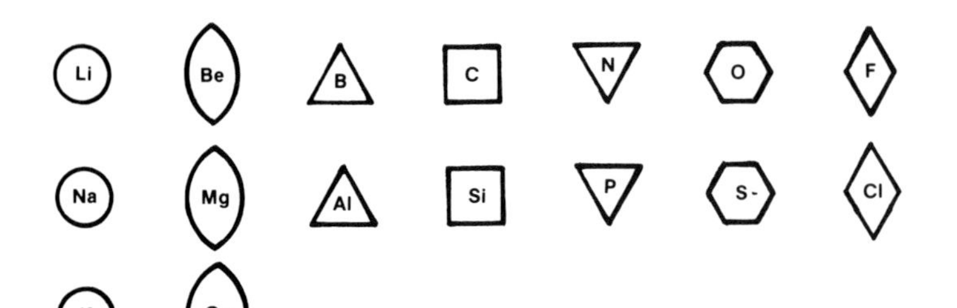

As you can readily see, matching elements line up in vertical columns. Any member of a column has properties in common with other members of the column. The elements are still in order of increasing atomic weight, proceeding from left to right across each row in order, starting from the top. Mendeleev's complete table was much more extensive than this, but this sampling shows the periodic principle.

There is some question, as mentioned above, about where hydrogen should be placed. In some compounds, it resembles the alkali metals of column 1. In other compounds, it is like the halogens of column 7. In some modern tables it is placed at the top of both columns. Mendeleev chose to place it in column 1, probably because it usually exhibits a **metallic character** (tendency to give up electrons).

If you count the columns in the above sample of Mendeleev's table and compare it with the number of main groups in a modern table, you will see that they are different. There is one more column in the modern table. The reason for this is simple: some of the elements in our present table had not yet been discovered in Mendeleev's time.

By the late 1890's, most of the rare gases (helium, neon, argon, and the like) had been discovered. It seemed logical to give these rare gases a column of their own—column 8. Helium has an atomic weight greater than that of hydrogen but less than that of lithium; therefore, it is clearly entitled to the first place in column 8. Neon has an atomic weight greater than that of fluorine but less than that of sodium; it fits nicely into the second space of column 8. But argon creates a problem; its atomic weight is greater than either of the elements it is supposed to go between (chlorine and potassium)! How should this problem be solved?

8:4

H							He
Li	Be	B	C	N	O	F	Ne
Na	Mg	Al	Si	P	S	Cl	Ar ?
K	Ca						

You will recall Mendeleev's original decision to arrange the elements in order of increasing atomic weights. This was a personal choice on the part of Mendeleev. Maybe there is a better way to do it. What we are looking for in science is *workability*; if an idea is unworkable, something more suitable should be put in its place.

Suppose that instead of using atomic weights, we arrange the elements in order of increasing **atomic numbers**—that is, increasing number of protons in the nucleus. X-ray research performed by English physicist Henry G. Moseley (1887-1915) suggested that such an arrangement would be workable. When this is done, the elements that were in the proper order stay in order, and the problem with argon is solved. Chlorine has 17 protons, argon 18, and potassium 19. Several other potential difficulties further along in the table are also avoided by this move.

8:5 *Henry G. Moseley, the English physicist whose X-ray research led to the arrangement of the elements by atomic number.*

Our modern periodic table, then, is based on atomic numbers. Going through the table in order, we see that each element has one more proton than the element preceding it. We can also state the periodic nature of the elements in the form of a law: *The chemical properties of the elements are periodic functions of their atomic numbers*. In other words, when the elements are arranged according to atomic numbers, the properties of the elements repeat. This is known as the **periodic law**.

8:2 A Simplified Periodic Table Below is a simplified form of the periodic table. It is substantially the same table we have just been discussing but is carried all the way through. However, we have left out several dozen elements! The X's mark the positions of the missing elements—10 positions between calcium and gallium, 10 between strontium and indium, 24 between barium and thallium, and 17 following radium. These missing elements are all metals. More specifically, they are the **transition elements**. All these metals have one or two electrons in their outer shells.

Most of them are *incomplete* in their next-to-last shell. You can see which elements have been omitted by looking at the complete table in Figure 8:8.

I	II	III	IV	V	VI	VII	VIII
H							He
Li	Be	B	C	N	O	F	Ne
Na	Mg	Al	Si	P	S	Cl	Ar
K	Ca X	Ga	Ge	As	Se	Br	Kr
Rb	Sr X	In	Sn	Sb	Te	I	Xe
Cs	Ba X	Tl	Pb	Bi	Po	At	Rn
Fr	Ra X						

8:6 *A simplified period table.*

The dark staircase-like dividing line separates metals on the left from nonmetals on the right. This divides the elements of the simplified table about equally—23 metals as compared to 21 nonmetals. When the transition metals are included in the tally, the metals "win" by a 4 to 1 ratio—84 metals to 21 nonmetals.

8:3 The Spiral Form In the simplified table we had a question concerning the location of hydrogen. One way of solving this problem is to use a spiral arrangement that places hydrogen at the center. As before, the transition metals are omitted and indicated by X's. Elements having similar properties are located along radii of the spiral, like the spokes of a bicycle wheel. In spite of the eye-catching appeal of the spiral form, it has not enjoyed widespread use. One drawback with this arrangement is that it is difficult to see the periods (see Section 8:6). The end of one period merges into the beginning of the next, whereas in table form it is easy to see where one line ends and the next begins.

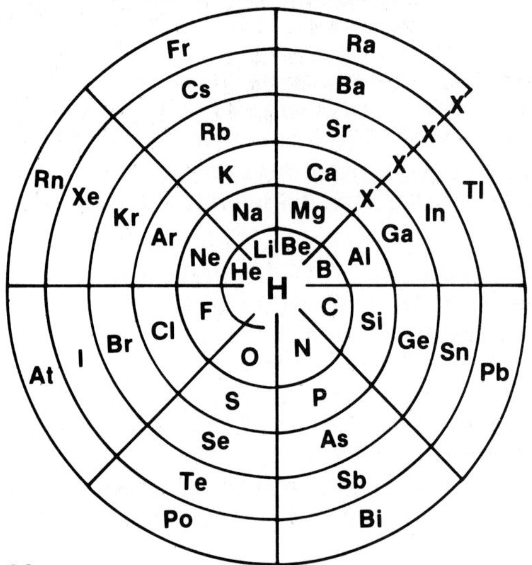

8:7 *Spiral form of periodic table.*

8:4 The Complete Periodic Table The complete table shown in Figure 8:8 is representative of those in general use today. This type is called the **"long-form" table** because it is 18 columns wide. Another variation, called the **"short-form" table**, has fewer columns because some of the transition elements are interspersed with the **main group elements**. Both forms place the **actinide** and **lanthanide** transition elements at the bottom. The most recent tables include elements 104, called *kurchatovium*, and 105, called *hahnium*. They are placed immediately after *actinium*, element 89.

Roman numerals are used to number the columns in both the "long-form" and "short-form" tables. In a "long-form" table, such as that shown in Figure 8:8, the letters "A" and "B" are used after the Roman numerals to differentiate the main groups from the transition elements, the letter "A" signifying a main group, the letter "B" a transition group.

8:5 Family Resemblances The elements in Group IA all have one electron in their outer shells. Because it is very easy to lose a single electron to form a stable configuration (see Chapter 7), these elements are all quite active and, with the exception of hydrogen, are metals. This is the **alkali metals** group. A valence of +1 in compounds is characteristic of this group.

It should be observed, however, that *all* elements have a valance of *zero* until they react with some other element. The valence of an atom or ion can be determined by figuring the sum of all the positive and negative charges within it. Until an atom combines with an atom of another element, its total number of negative charges exactly equals its total number of positive charges. After an electron transfer has taken place, there will be a charge imbalance one way or the other, and an ion will be formed.

The **alkaline-earth metals** (Group IIA) all have two electrons in their outer shells. Two electrons can be lost with reasonable ease, although not as easily as a single electron. Hence, these elements are somewhat less active than the Group IA elements. The alkaline-earth metals exhibit a valence of +2 in compounds.

Group IIIA, IVA, VA, and VIA elements have 3, 4, 5, and 6 electrons in their outer shells, respectively. Again, there are strong family resemblances within each group.

Group VIIA, called the **halogens**, consists of elements having 7 electrons in their outer shells. Because they accept an additional electron to achieve a stable octet, they have a valence of −1 in compounds. Fluorine, for example, starts off with 9 electrons and 9 protons. After completing its octet, it has 10 electrons and 9 protons. The sum of +9 and −10 is −1.

Group VIIIA, called the **rare gases** or **noble gases**, consists of elements having complete outer shells. Since they are already chemically "satisfied," they do not usually participate in chemical reactions to form compounds.

The principle of family resemblances can sometimes be used in predicting how an element will behave in a certain

8:8 *Periodic Table of the Elements (C-12 = 12.0000).*

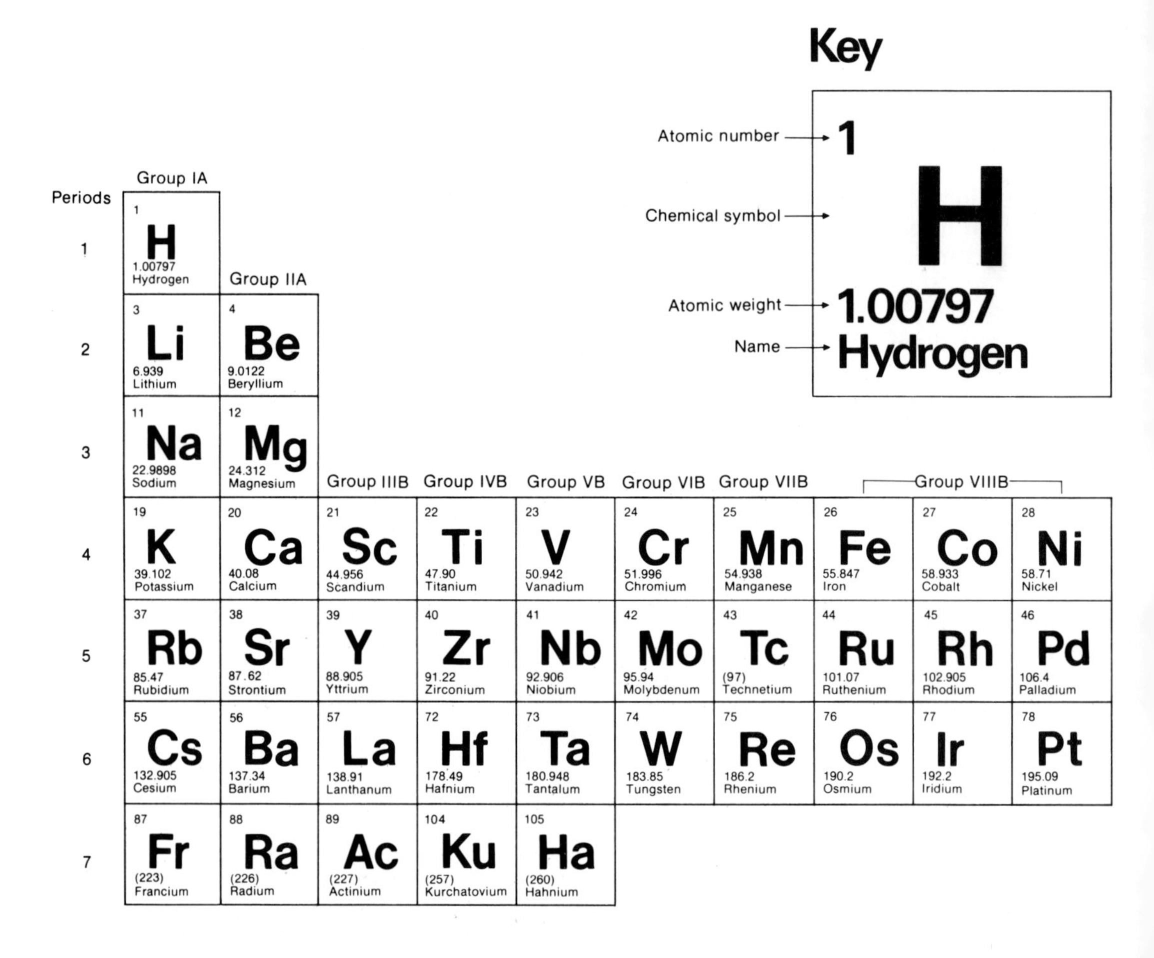

Rare Earths

Lanthanide Series	58 **Ce** 140.12 Cerium	59 **Pr** 140.907 Praseodymium	60 **Nd** 144.24 Neodymium	61 **Pm** (145) Promethium	62 **Sm** 150.35 Samarium	63 **Eu** 151.96 Europium
Actinide Series	90 **Th** 232.038 Thorium	91 **Pa** (231) Protactinium	92 **U** 238.03 Uranium	93 **Np** (237) Neptunium	94 **Pu** (244) Plutonium	95 **Am** (243) Americium

Group IB	Group IIB	Group IIIA	Group IVA	Group VA	Group VIA	Group VIIA	Group VIIIA
							2 **He** 4.0026 Helium
		5 **B** 10.811 Boron	6 **C** 12.01115 Carbon	7 **N** 14.0067 Nitrogen	8 **O** 15.9994 Oxygen	9 **F** 18.9984 Fluorine	10 **Ne** 20.183 Neon
		13 **Al** 26.9815 Aluminum	14 **Si** 28.086 Silicon	15 **P** 30.9738 Phosphorus	16 **S** 32.064 Sulfur	17 **Cl** 35.453 Chlorine	18 **Ar** 39.948 Argon
29 **Cu** 63.54 Copper	30 **Zn** 65.37 Zinc	31 **Ga** 69.72 Gallium	32 **Ge** 72.59 Germanium	33 **As** 74.922 Arsenic	34 **Se** 78.96 Selenium	35 **Br** 79.909 Bromine	36 **Kr** 83.80 Krypton
47 **Ag** 107.870 Silver	48 **Cd** 112.40 Cadmium	49 **In** 114.82 Indium	50 **Sn** 118.69 Tin	51 **Sb** 121.75 Antimony	52 **Te** 127.60 Tellurium	53 **I** 126.904 Iodine	54 **Xe** 131.30 Xenon
79 **Au** 196.967 Gold	80 **Hg** 200.59 Mercury	81 **Tl** 204.37 Thallium	82 **Pb** 207.19 Lead	83 **Bi** 208.980 Bismuth	84 **Po** (210) Polonium	85 **At** (210) Astatine	86 **Rn** (222) Radon

64 **Gd** 157.25 Gadolinium	65 **Tb** 158.924 Terbium	66 **Dy** 162.50 Dysprosium	67 **Ho** 164.930 Holmium	68 **Er** 167.26 Erbium	69 **Tm** 168.934 Thulium	70 **Yb** 173.04 Ytterbium	71 **Lu** 174.97 Lutetium
96 **Cm** (247) Curium	97 **Bk** (247) Berkelium	98 **Cf** (251) Californium	99 **Es** (254) Einsteinium	100 **Fm** (253) Fermium	101 **Md** (256) Mendelevium	102 **No** (254) Nobelium	103 **Lr** (257) Lawrencium

reaction. If the behavior of one member of the family is known, the others can be assumed to follow suit, unless contradictory information is obtained.

8:6 Periods A **period** or **series** is a horizontal row of elements in the periodic table. How do we know where a period begins and ends? The answer is found in the outer shell electrons of its elements.

Each period begins with an element having one electron in its outer shell. The first period consists only of hydrogen and helium because it takes only two electrons to fill the K shell (the K shell is the first shell, counting from the nucleus outward). Hydrogen has one electron in the outer shell; helium has two. We are using Bohr-type models of the atoms here (sometimes called **bull's-eye diagrams**), keeping in mind their many shortcomings. The Bohr model gives us a convenient way of keeping track of the outer electrons, but they in no way "look like" the phenomenon we are representing.

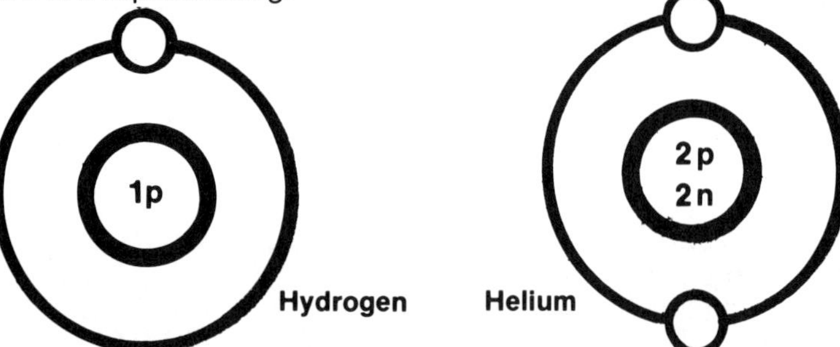

8:9 The first period of the periodic table, in which the K shell is filling with electrons.

The second period consists of eight elements because it takes eight electrons to fill the L shell (the L shell is the second shell, counting from the nucleus outward). These are shown in order here, adding one electron at a time:

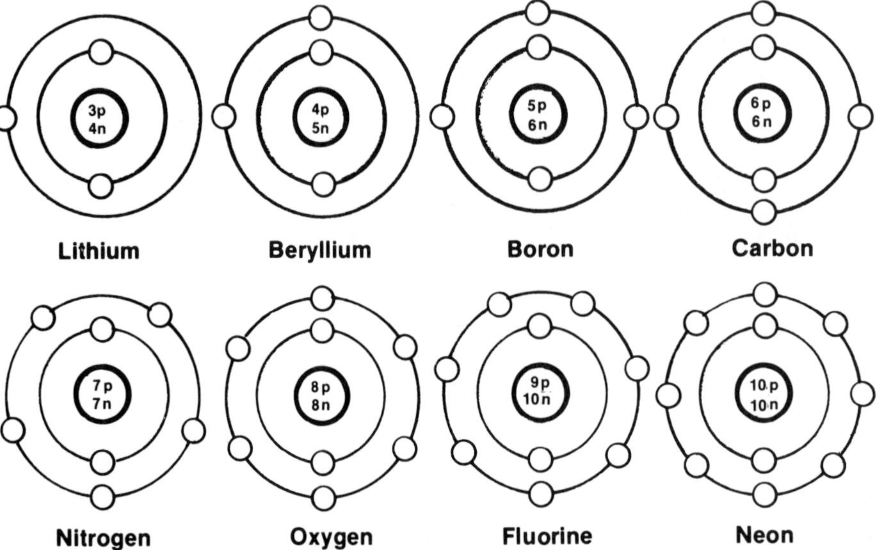

8:10 The second period of the periodic table, in which the L shell is filling with electrons.

The L shell in neon is full. Therefore in the next element, sodium, the last electron is forced to start a new shell, making sodium the first element of the third period. This is another eight-element period, ending with argon, which again has a full outer **octet** (see Section 7:2). In this period we illustrate only the first and last elements.

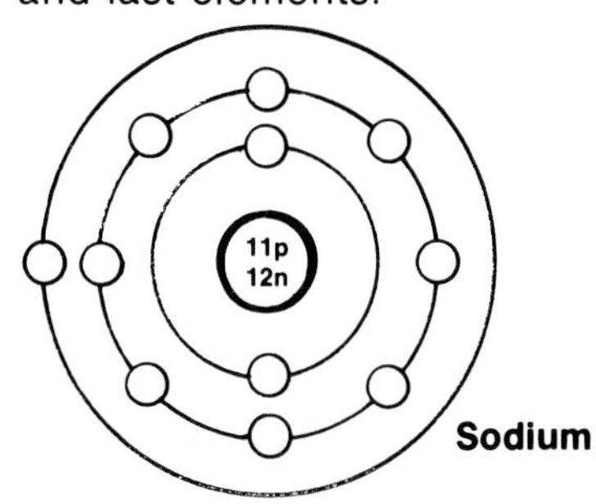

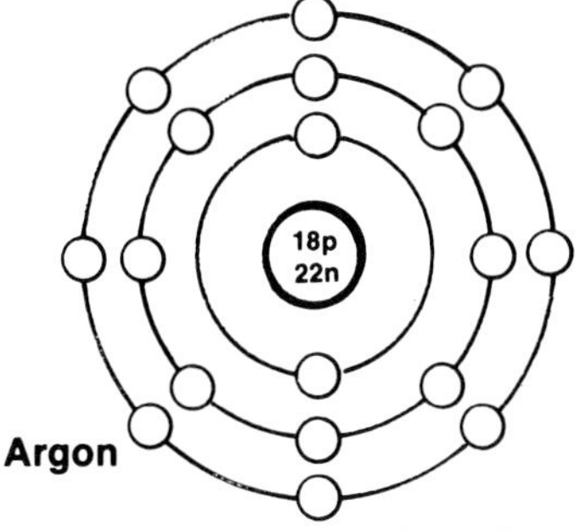

8:11 *The first and last elements of the third period of the periodic table.*

In like manner, electrons continue to add in each period. The main group elements add one each in their outer shell; the transition elements add in a less predictable way to the next-to-last shell.

8:7 Some Generalizations You will recall the fact that the metals are located on the left-hand side of the table, the nonmetals on the right. The staircase-like line is again the dividing line as in the simplified table. Some of the elements located along the borderline are called *metalloids*. They are neither definitely metals nor nonmetals. In this category are boron, silicon, germanium, arsenic, antimony, tellurium, and astatine.

The periodic table also gives us an approximate arrangement by *density*. Lightweight elements such as hydrogen and helium are located at the top of the table. Heavy elements such as lead, mercury, gold, uranium, osmium, and iridium are found toward the bottom of the table. This is true because an element's density depends upon the makeup of the nucleus, and the numbers of protons and neutrons in each element increase as we progress from the top to the bottom of the table.

It is interesting to note that the most active metals and the most active nonmetals are found clustered at *opposite* corners of the table—the most active metals at the lower left, the most active nonmetals at the upper right. Metals are made active by their ability to *lose* electrons readily. The fewer electrons there are in the outer shell, the more easily they can be lost.

Another important factor is the tenacity with which the nucleus holds the outer electrons. The heavier atoms have more electron shells between the nucleus and the outer electrons. These in-between shells make a "**screening effect**," tending to weaken the pull of the nucleus on the outer electrons. Considering both of the above factors, the most active metal could be expected to be francium (Fr). However, because francium is highly radioactive, it has not been possible to isolate enough of the metal to test this prediction. But the other metals in the lower left-hand corner of the table are highly active.

A nonmetal is made active by its ability to *capture* electrons readily. Fluorine, near the upper right-hand corner of the table, has seven electrons in its outer shell. It needs only one electron to complete its octet, and its nucleus has very little "screening." Therefore, it can effectively capture the needed electron if the electron is available. You can imagine that there would be an extremely vigorous reaction between francium and fluorine to form the ionic compound francium fluoride, if enough of the metal francium could be obtained to try the reaction.

8:8 Other Information The periodic table is a handy reference chart for chemists and physicists alike. It gives the atomic number (number of protons) and atomic weight for each element.

The atomic weights in the most recent tables are based on carbon-12. That is, the isotope of carbon, which has an atomic weight of 12, is set equal to 12.0000. Every other element's weight will be found to differ somewhat from a whole number. One reason for this is that the atomic weights given in the table represent an *average* of the naturally occurring isotopes. *Isotopes* are different forms of the same element that have the same atomic number but different atomic weights. Naturally occurring carbon has an atomic weight slightly greater than the whole number (12.01115) because of small percentages of heavier isotopes that contribute to the average.

Often the following additional information will be included in a periodic table: electronegativity, ionization potential, atomic radius, electron configuration, and characteristic valences for each element. Mendeleev's simple card file has grown into an extensive storehouse of useful information.

List of Terms

actinide elements
alkali metals
alkaline-earth metals
atomic number
atomic weight
bull's-eye diagram
halogens
lanthanide series
long-form table
main group elements
metallic character
noble gases
octet
period
periodic law
periodic table
screening effect
series
short-form table
transition elements

Questions

1. State the periodic law.
2. In the modern periodic table, the elements are arranged in order of increasing atomic __________ .
3. In what part of the periodic table are the least dense elements found? The most dense elements?

4. The elements in Group IA of the periodic table are called ________.
5. Elements which have one or two electrons in their outer shells and which usually possess an incomplete next-to-last shell are called ________.
6. Which elements are the most numerous—metals, nonmetals, or metalloids?
7. How were the elements organized prior to Mendeleev's time?
8. Name seven metalloids.
9. State the number of electrons in each shell of the following elements: (a) nitrogen (b) magnesium (c) potassium (d) argon.
10. Show how argon does not "fit" in the periodic table if the elements are arranged on the basis of atomic weights. What other elements would cause similar problems?
11. The elements in Group VII A are called ________.
12. How do the Group VII A elements vary in activity within the family?
13. Elements in Group VIII A are called ________.
14. The elements of Group I A all display-a valence of ________ in compounds.
15. How do the chemical properties of elements change as one progresses from left to right within a given period?
16. All unreacted elements have a valence of ________.
17. In the simplified periodic table the ________ are omitted.
18. The actinide and lanthanide series each contain ________ elements.
19. Elements in the same ________ have the same number of electrons in their outer shells.
20. All the elements in the same ________ have the same number of electron shells.
21. Why does potassium react more vigorously with water than sodium does?
22. Why is bromine a more active nonmetal than iodine?
23. Draw bull's-eye diagrams for (a) neon (b) boron (c) calcium (d) lithium.
24. (a) What considerations would favor the inclusion of the element hydrogen with the Group I A metals? (b) What considerations would favor its inclusion with the Group VII A nonmetals?
25. ★ Which transition elements of periods 4 and 5 have full inner shells?
26. All of the rare gases have octets except ________ which has only two electrons in its outer shell.
27. The elements called the ________ all have seven electrons in their outer shells.
28. Atomic weights in the present-day periodic table are based on an isotope of the element ________.
29. There are now ________ known chemical elements.
30. Where are the two most recently discovered elements located in the periodic table?

Student Activity

Prepare a report on Dmitri Mendeleev, giving special attention to his correct predictions concerning the properties of the elements scandium (Sc), gallium (Ga), and germanium (Ge).

Photo next page
Making measurements on cryogenic (low temperature) equipment. Meters are read to determine the operating conditions inside the unit.

CTi
MODEL 140

UNIT III USING NUMBERS IN SCIENCE

Chapter 9 Measurement

9:1 The Importance of Numbers and Units Lord Kelvin (1824-1907), a leading scientist of his day and an outstanding Christian, made this statement concerning the necessity for making measurements in science: "When you can measure what you are speaking about, and express it in numbers, you know something about it; but when you cannot express it in numbers, your knowledge is a meagre and unsatisfactory kind; it may be the beginning of knowledge, but you have scarcely, in your thoughts, advanced to the stage of science. . . ."

As we have seen in Chapter 3, measurements are needed to give us the data for formulating scientific laws. However, numbers must have **units** to have meaning. If someone tells you that the length of a piece of string is "10," is it 10 inches, 10 feet, 10 yards, 10 miles, 10 meters, or 10 centimeters? Once the units are specified, you have the information you need.

How do we define the different units of measurement? What systems of measurement have been accepted in the scientific world? These are questions we will answer in this chapter.

9:1 *Measuring a piece of wood. Accurate measurements are needed for good results in both carpentry and science.*

9:2 The Metric System The system of measurement with which you are probably most familiar is the **English system**. It uses such cumbersome units as inches, feet, yards, miles, quarts, pounds, tons, gallons, seconds, minutes, and hours. Figuring mathematical problems using the English system can sometimes be very tedious. For example, suppose you want to know the number of feet in 15 miles. There are 5280 feet in one mile; 5280 is not a very easy number to remember. Multiply 15 times 5280 and you have the answer—there are 79,200 feet in 15 miles.

The corresponding operation in the **metric system** would be much simpler. You would be determining the number of meters in 15 kilometers. Since there are 1000 meters in each kilometer, the answer is, simply, 15,000 meters.

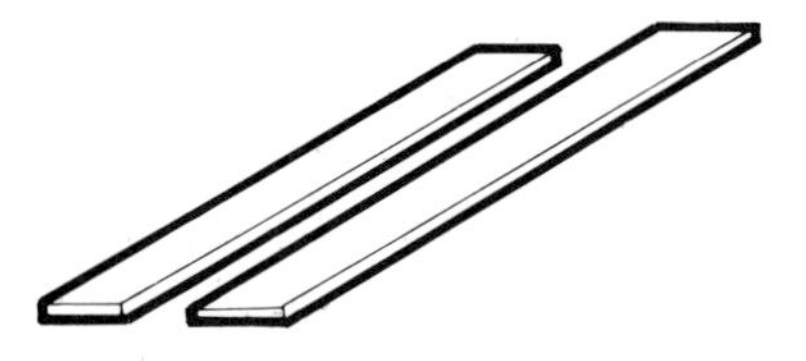

9:2 *Comparative lengths of the meter and the yard. The meter (right) is approximately 10 percent longer than the yard (left).*

For ease of calculation, scientists and mathematicians all over the world have agreed to use the metric system for making measurements and expressing quantitative information. In the metric system, all units of one kind are related to each other by the number 10. Therefore, it is usually only necessary to move the decimal point to make a calculation. In the above problem, to figure the number of meters in 15 kilometers, you merely move the decimal point three places to the right. Moving the decimal point three places to the right is the same as multiplying by 1000.

9:3 Length The metric system is based on the unit called the **meter**. The meter is approximately 10 percent longer than a yard—a yard is 36 inches, a meter is 39.37 inches. Originally defined as 1/10,000,000 of the distance from the equator to the North Pole, it is now (since 1960) specified in such a way that it can be duplicated in any laboratory that has the necessary equipment. The new definition of the meter is 1,650,763.73 wavelengths of the orange-red light from krypton-86. Krypton-86 is a heavy isotope of the rare gas krypton. A quantity of the gas is placed in a glass tube cooled with liquid nitrogen and caused to glow by passing an electric current through it. Through the use of a device called an interferometer, 1,650,763.73 wavelengths are measured out to give exactly one meter. The wavelength of this light is exceedingly small; there are about 42,000 waves to the inch! Using such a small unit gives us a very exact means of getting the right overall length for a meter. Any scientist can light a krypton lamp in his own laboratory and use its light as a standard against which measurements can be checked.

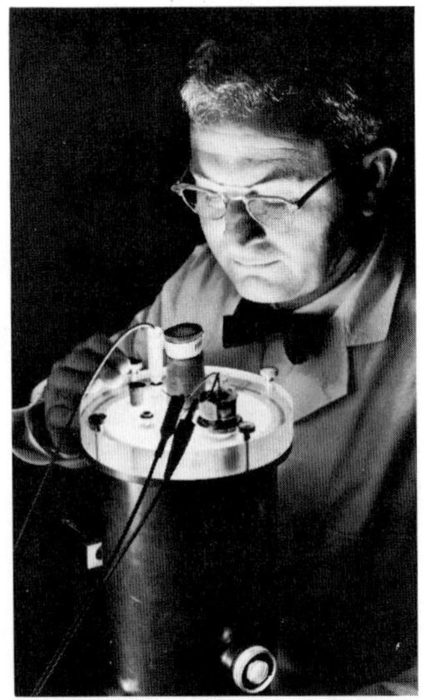

9:3 *Krypton-86 lamp. The wavelength of the orange light emitted by krypton-86 is used to define the standard meter.*

The meter is too long a unit for measuring, say, the length of a test tube or the thickness of a sheet of metal. Therefore, it has been subdivided into smaller units called **centimeters** and **millimeters**. *A centimeter is the hundredth part of a meter.* (There are 2.54 centimeters in an inch.) The prefix "centi-" means 1/100; this can be easily remembered by recalling that one cent is 1/100 of a dollar. *A millimeter is the thousandth part of a meter.* The prefix "milli-" means 1/1000. There are, therefore, 10 millimeters in each centimeter.

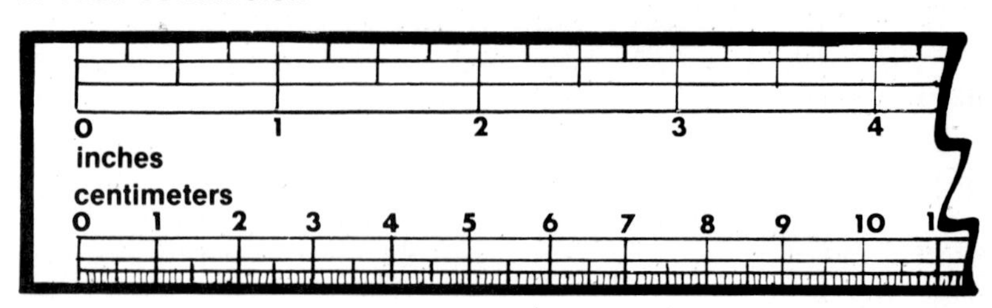

9:4 *Portion of a meter stick showing comparison between inches and centimeters.*

The meter is too short a unit for expressing the distance, say, between two cities. For this purpose we use a multiple of the meter called the **kilometer**. A kilometer is equal to 1000 meters. The kilometer is 0.621 miles; 1.61 kilometers equal a mile.

Some useful length relations are summarized in Table 9-1.

Table 9-1 *Length Relationships*

Metric—Metric	Metric—English
1000 millimeters = 1 meter	2.54 centimeters = 1 inch
100 centimeters = 1 meter	30.48 centimeters = 1 foot
1000 meters = 1 kilometer	1 meter = 39.37 inches
	1 meter = 3.28 feet
	1 kilometer = 0.621 miles

The following abbreviations are in common usage for length units:

centimeter: cm — meter: m

millimeter: mm — kilometer: km

(It is customary *not* to use a period after an abbreviated unit.)

Experiment 9:1 *Determining the Value of Pi by Measurement*

1. Find the circumference of (distance around) a softball by wrapping a string around the ball and measuring the string to the nearest 1/10 of a centimeter.
2. Find the diameter (distance through the middle) of the ball by placing the ball on a table and holding a book on top of it as you measure the distance from the table surface to the book. Be sure the book is held level. Make the reading to the nearest 1/10 of a centimeter.
3. Calculate your experimental value of pi (π) by dividing the diameter into the circumference. How close is your experimental value to the accepted value of pi (3.14)? The true value of pi is the ratio of the circumference to the diameter in a perfect circle.

9:4 Area In the English system, **area** is expressed in square inches or square feet. In the metric system, the most commonly used units are **square centimeters** and **square meters**. The area of a sheet of notebook paper, for example, would be given in square centimeters, whereas the area of a gymnasium floor would be expressed in square meters. These are abbreviated as follows:

square centimeters: cm^2

square meters: m^2

The area of a rectangle is found by multiplying its length by its width. Expressed as a formula:

$$A = LW$$

In words, this means "area equals length times width." (Note that the multiplication symbol need not be used between the L and the W; multiplication is indicated simply by placing the letters alongside each other.)

Using this formula, a slip of paper measuring 6 centimeters by 8 centimeters has an area of 48 square centimeters (6 cm x 8 cm = 48 cm^2). When centimeters are multiplied by centimeters, the result is square centimeters.

The area of a square is occasionally needed in working physics problems. The square may be considered as a special rectangle with its length equal to its width. Expressed as a formula,

$$A = s^2$$

The symbol s is the length of one side of the square. The number 2 in this formula is called an **exponent**. It indicates that the length of the side is to be multiplied by itself. A square whose side is 4 cm will have an area of 16 cm^2 (4 cm x 4 cm = 16 cm^2).

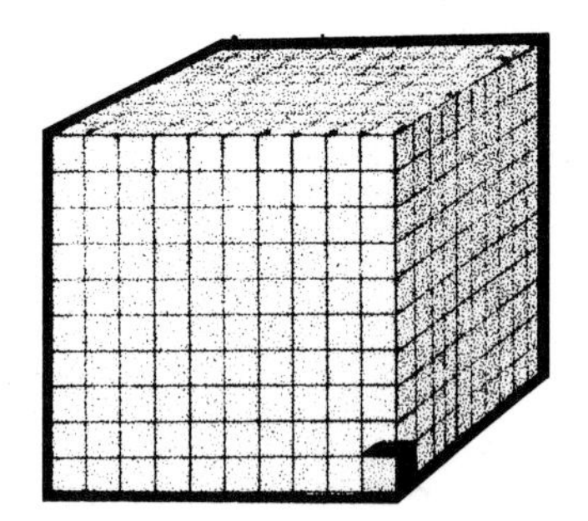

9:5 *A liter cube. A cube measuring 10 centimeters on a side has a volume of one liter. The small cube in the lower righthand corner has a volume of one cubic centimeter. There are 1000 cubic centimeters in a liter.*

9:5 Volume By the **volume** of an object we mean how much space it occupies. In the English system, volume is measured in units such as cubic inches, cubic feet, cubic yards, and cubic miles. In the metric system, we begin with the **cubic centimeter**. Picture a **cube** (a solid with six square sides) that measures one centimeter on each side. This is a tiny unit of volume equal to about 1/2 to 1/3 of a thimbleful. The cubic centimeter is used extensively in both chemistry and physics.

To form a larger unit of volume, let's make a cube that is exactly 10 centimeters on one side. The unit of volume thus formed is called the **liter** and is equal to 1000 cubic centimeters. It is slightly larger than a quart; one liter equals 1.06 quarts.

For measuring the volume of a very large object, such as a tall building, the **cubic meter** would be a good unit to use. The cubic meter is a large unit formed by making a cube that is one meter on each side. A cubic meter is equal to 1000 liters.

Some of the more useful volume relationships are included in Table 9-2.

Table 9-2 *Volume Relationships*

Metric—Metric	Metric—English
1000 cubic centimeters = 1 liter	1 liter = 1.06 quart
1000 liters = 1 cubic meter	946 cubic centimeters = 1 quart

The abbreviations for the volume units are:
cubic centimeter: cm^3
liter: l

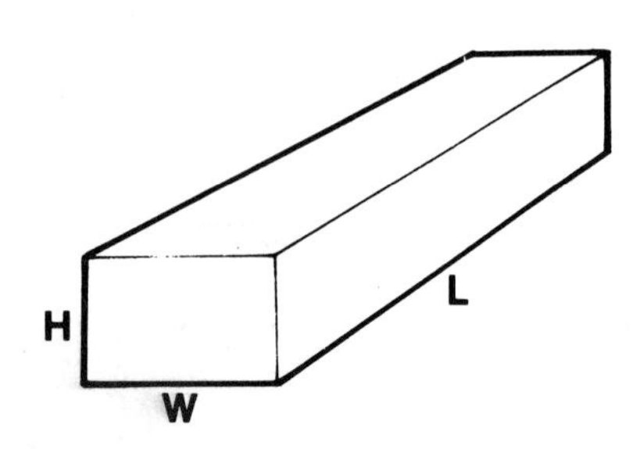

9:6 *The volume of a rectangular solid is found by multiplying length times width times height.*

Since the cubic centimeter is the thousandth part of a liter, it is sometimes called the **milliliter**, abbreviated ml.

To find the volume of a **regular solid** (a solid which has a standard geometric shape), we can simply measure and calculate. For example, the volume of a rectangular solid measuring 6 cm long, 3 cm wide, and 2 cm high (Figure 9:6) can be determined by the formula

$$V = LWH$$

In words, this formula means "volume equals length times width times height."

Multiplying the three numbers together, we obtain 36 cubic centimeters (6 cm x 3 cm x 2 cm = 36 cm^3).

A frequently encountered regular solid is the cube. The volume of a cube can be computed by the formula shown above. Since all three dimensions—length, width, and height—are the same, the formula reduces to

$$V = s^3$$

The symbol *s* is the length of one side. The exponent 3 indicates that the length of the side should be cubed (multiplied together three times). A cube whose side is 5 cm will have a volume of 125 cm^3 (5 cm x 5 cm x 5 cm = 125 cm^3).

For irregular solids (solids which do not have a standard geometric shape), volume can be determined by the water displacement method (see Figure 9:7). A quantity of water is placed in a **graduate cylinder**—enough to submerge the object whose volume is being determined. A reading is taken using the lowest part of the curved surface of the water. The object is then carefully placed in the water and another reading taken, again using the bottom of the curve. Let's say that our first reading was 33 cm^3 and and the final reading 78 cm^3. The volume of the object is the difference between these two numbers:

$$78\ cm^3 - 33\ cm^3 = 45\ cm^3$$

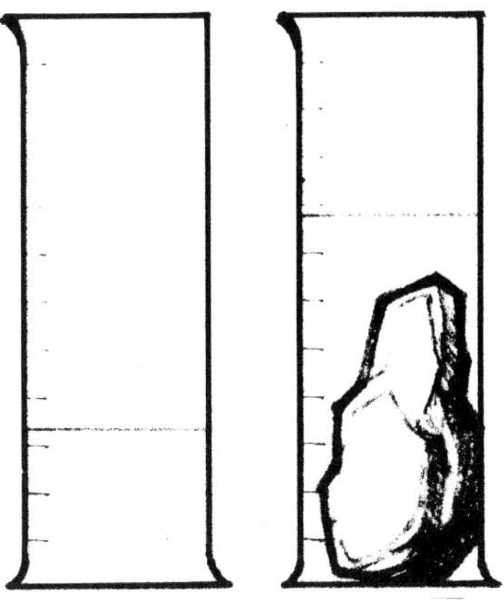

9:7 *The water displacement method for finding the volume of an irregular solid. The difference between the two readings on the graduate cylinder is equal to the volume of the object.*

This is a very handy way to find the volume of an object such as a stone. Attempting to calculate the volume of a stone from its shape would require an extremely complex mathematical formula, and even then the answer would only be approximate.

For larger irregular objects that will not fit into a graduate cylinder, an **overflow can** is sometimes used (see Figure 9:8). An overflow can is prepared for use by filling it with water to the point where it begins to flow out the spout. An empty beaker is then placed below the spout and the object is carefully placed in the overflow can. The water that runs out is transferred to a graduate cylinder to measure the volume of the water. This volume will be the same as the volume of the object.

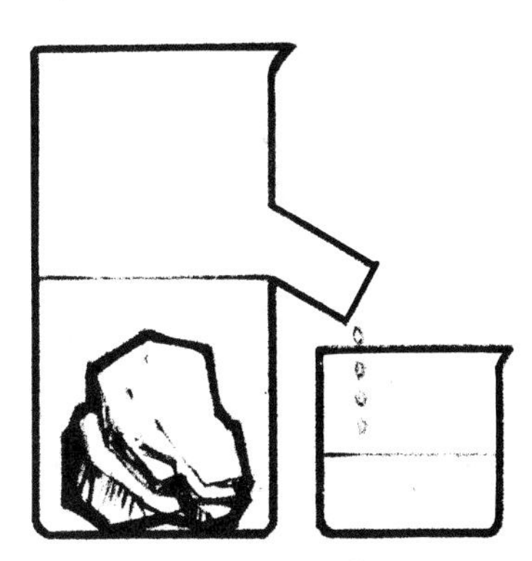

9:8 *Alternate method for finding the volume of an irregular solid, using an overflow can and beaker.*

9:6 Weight and Mass The downward force an object experiences as a result of the earth's gravitational pull is called its **weight**. The extent of this downward pull varies with location. On a high mountain the pull is less than at sea level.

Mass is a measure of the quantity of matter in a body. Mass is related to weight, but is a more basic and permanent property of an object than weight. An astronaut in orbit around the earth becomes weightless but he still retains his mass. An astronaut on the moon weighs only 1/6 as much as he does on the earth, but his mass stays the same.

The mass units in the metric system are defined by using water. The mass of one cubic centimeter of water at 4° C (four degrees Celsius temperature) is one **gram**. A medium-size paper clip has a mass approximately equal to one gram. The mass of one liter of water at 4° C is one **kilogram**. There are 1000 grams in a

9:9 *The standard kilogram, kept in the National Bureau of Standards, Washington, D.C., is a secondary standard, being an exact duplicate of the primary standard at Sèvres, France.*

kilogram. For very small mass measurements use is made of the **milligram**, the thousandth part of a gram. Abbreviations of the mass units are

gram: g
kilogram: kg
milligram: mg

The **primary standard** for the kilogram is a platinum-iridium cylinder located at Sèvres, France, near Paris. **Secondary standards** have been measured from this and are in use in various countries. The United States' secondary standard, designated as kilogram number 20, is kept at the Bureau of Standards in Washington, D. C. (see Figure 9:9). Comparisons can be made against this secondary standard if very accurate mass determinations must be made. For most purposes, however, the mass of one liter of water at 4°C is satisfactory.

The English unit of mass is called the **slug**. A slug is *a quantity of material that has a weight of 32 pounds at the earth's surface.* As long as we restrict ourselves to measurements on the earth, we can make the English to metric conversions between mass and weight as follows.

1 pound = 454 grams

1 kilogram = 2.205 pounds

We can determine the mass of an unknown object by comparing it with a set of known masses, usually called weights (see Figure 9:10). The traditional way to accomplish this would be to balance the unknown mass against known masses on an equal-arm balance. For many laboratory operations, however, it is easier and quicker to use one of the balances shown in Figures 9:11 and 9:12. These represent a very practical compromise between cost and accuracy, and do not require the use of loose weights. Measurements to the nearest 1/10 or 1/100 of a gram, such as these instruments give, are usually accurate enough for the purposes of a particular experiment. When greater accuracy is needed, an analytical balance, such as that shown in Figure 9:13, can be used. The better analytical balances give readings to the nearest 1/10 of a milligram.

9:10 *Typical laboratory weights.*

9:11 *Triple beam platform balance.*

9:12 *Quadruple-beam pan-type balance.*

9:13 *A high-accuracy, single-pan, analytical balance.*

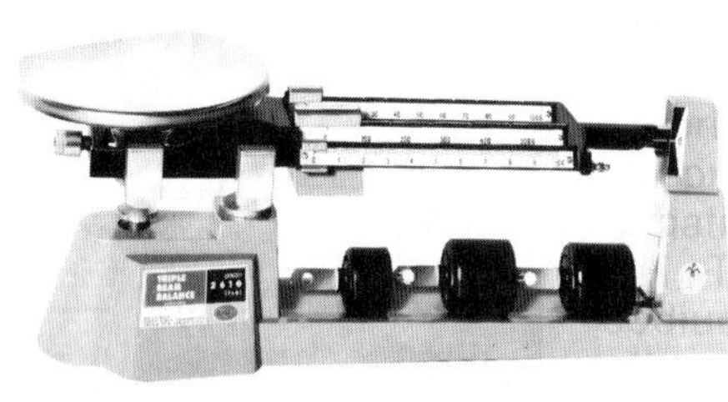

9:11

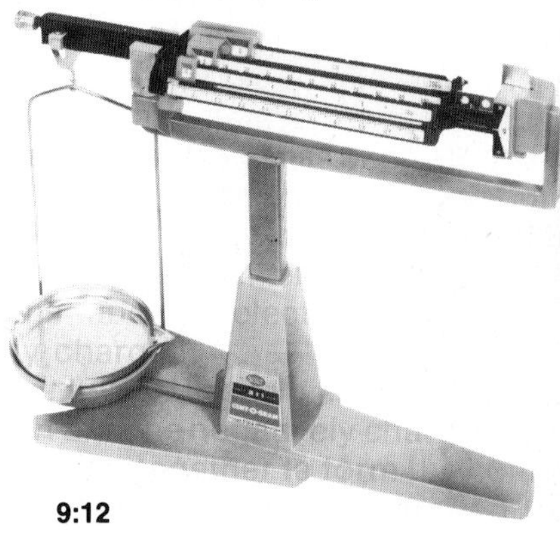

9:12

9:13

9:7 Time The **second** is the basic unit of time in both the English and metric systems. Prior to 1967, the second was defined as a fractional part of the year 1900. It is now defined as *the time required for 9,192,631,770 vibrations of an atom of cesium-133.* The device used for this standard is the cesium-beam atomic clock, which can measure time intervals with an error of less than one second in 1000 years! (see Figure 9:14).

9:14 *Cesium beam atomic clock. The standard second is now defined as the time required for 9,192,631,770 vibrations of an atom of cesium-133.*

For most purposes, cesium-beam atomic clock accuracy is not needed. In fact, for many experiments an ordinary stopwatch will be sufficient. If greater accuracy is desired, the time signals broadcast by radio stations WWV or CHU can be used. WWV is the National Bureau of Standards station located at Fort Collins, Colorado. It broadcasts 24 hours a day, 7 days a week, at frequencies of 2.5, 5, 10, 15, 20, and 25 megahertz, sending extremely accurate pulses (ticks) to mark each second. CHU is the Dominion Observatory station at Ottawa, Ontario, Canada. It broadcasts time signals continuously at 3.333, 7.335, and 14.67 megahertz. Both of these stations can usually be received on an ordinary shortwave receiver anywhere in North America.

9:15 *National Bureau of Standards station WWV, Fort Collins, Colorado. Time signals of extremely high accuracy are broadcast by the short wave transmitters at this location.*

9:16 *Control equipment used at National Bureau of Standards station WWV.*

Other common units of time are defined as multiples of the second:

1 minute = 60 seconds
1 hour = 3600 seconds
1 day = 86,400 seconds

For measuring very short intervals of time the second is subdivided:

1 millisecond = 0.001 second
1 microsecond = 0.000001 second
(The prefix "micro-" means "one millionth.")

9:8 Density The quantity which expresses how massive an object is for its size is called density. **Density** is defined as *the mass of an object divided by its volume*. The units most commonly used for density are grams per cubic centimeter (g/cm^3).

Let's think again about the rectangular solid described earlier in the chapter. Since its measurements were 6 cm, 3 cm, and 2 cm, its volume was 36 cm^3. Now suppose its mass is measured on a balance and found to be 407 grams. The density of the rectangular solid can then be calculated by dividing its mass by its volume:

$$\frac{407g}{36cm^3} = 11.3 \ g/cm^3$$

Note how the units are handled. Grams divided by cubic centimeters give grams per cubic centimeter in the answer. The general formula for this operation is

$$D = M/V$$

D, M, and V are density, mass, and volume, respectively.

From the table of densities (Table 9-3) you will see that 11.3 g/cm^3 is the density of lead. Let us assume that it is known beforehand that the rectangular solid you are measuring is, in fact, lead. We therefore know that our result, 11.3 g/cm^3, is correct.

Lead is a fairly dense substance, but it is by no means the *most* dense. The substance which has the highest known density is osmium (22.48 g/cm^3). Iridium (22.42 g/cm^3) is a close second. Some of the lighter substances include ice (0.92 g/cm^3), benzene (0.88 g/cm^3), ethyl alcohol (0.79 g/cm^3), and ethyl ether (0.71 g/cm^3).

The table gives only a sampling of some solids and liquids. Of those listed, water is the only substance having a density of 1

9:17 *Ice, having a density of only 0.92 g/cm³, floats in water.*

Table 9-3 *Densities of Some Solids and Liquids (in g/cm³)*

Solids		Liquids	
osmium	22.48	mercury	13.6
iridium	22.42	concentrated sulfuric acid	1.84
platinum	21.45	carbon tetrachloride	1.59
gold	19.3	carbon disulfide	1.26
uranium	18.7	glycerin	1.26
lead	11.3	water	1.00
silver	10.5	olive oil	0.92
copper	8.92	benzene	0.88
iron	7.86	ethyl alcohol	0.79
aluminum	2.70	ethyl ether	0.71
beryllium	1.85		
magnesium	1.74		
ice	0.92		

g/cm³. It is important to understand why the density of water should be equal to one. Since the gram is defined as the mass of 1 cm³ of water at 4°C, the density of water at any reasonable temperature must be very close to 1 g/cm³. Any solid having a density less than 1 g/cm³ should *float* in water. Solids having a density greater than 1 g/cm³ should *sink*. Ice floats because it has a density less than 1 g/cm³. Metals sink because their densities are greater than 1 g/cm³. (We are considering, of course, solid metal samples with no air spaces in them.)

Mercury has a density of 13.6 g/cm³, which means that every metal having a density less than 13.6 g/cm³ (even lead) will float in mercury. Brass, having a density between 8 and 9 g/cm³, floats fairly high in mercury, and aluminum (2.70 g/cm³) and magnesium (1.74 g/cm³) remain almost on the surface.

9:18 *Brass weight, floating in mercury.*

On one occasion the Bob Jones University science department was given the problem of determining the composition of a collection of ancient Roman coins. One requirement was that the coins could not be tested chemically or defaced in any way. The museum curator who sent them assured the science department that the coins were either copper, gold, or some mixture of the two metals. Fortunately, copper and gold have widely differing densities. Each coin was weighed and its volume determined by the water displacement method. The mass was then divided by the volume in each case. One coin, which was typical of the collection, had a volume of 1.2 cm³ and a mass of 10.5 grams. Its density was, therefore,

$$D = \frac{M}{V} = \frac{10.5\ g}{1.2\ cm^3} = 8.7\ g/cm^3$$

For 10 coins tested, the densities ranged from 7 to 10 g/cm^3. Copper has a density of 8.92 g/cm^3. Gold is more than twice as dense, 19.3 g/cm^3. The conclusion was that the coins were all predominantly copper, the copper varying in purity from one coin to the next. If the coins had contained very much gold, the densities would have been much higher.

Using the relationship $D = M/V$, we can find any of the three quantities—density, mass, or volume—if the other two are known. We have already calculated density. Let us now calculate mass.

Problem: Find the mass of 30 cm^3 of silver.

Solution: The formula $D = M/V$ should be solved for mass before the numbers are substituted. Multiplying both sides of the equation by V we obtain

$$DV = M$$

Turning the equation around, we arrive at the desired form of the relationship:

$$M = DV$$

From the table of densities we note that silver has a density of 10.5 g/cm^3. Putting in the numbers

$$M = (10.5\ g/cm^3)\ (30\ cm^3)$$

$$M = 315\ g$$

The answer is 315 grams. In a case such as this, the units are cancelled like fractions. The cubic centimeters drop out, leaving only grams in the answer.

Another type of problem remains—that of finding the volume.

Problem: What is the volume of a kilogram of gold?

Solution: A kilogram is 1000 grams. The density of gold is 19.3 g/cm^3. To solve the formula for volume requires two steps.

Starting with $D = M/V$, both sides are multiplied by V, as before, to give $DV = M$. Now both sides are divided by D to yield $V = M/D$.

The numbers are then put in:
$V = M/D = 1000g/19.3\ g/cm^3 = 51.8\ cm^3$

The grams cancel, and the cm^3, which was in the denominator *of* the denominator, appears in the answer. Here again, we see the close similarity between units and numbers.

Experiment 9:2 *Density of a Regular Solid*

1. Obtain a wooden or metal block. Measure the thickness, width, and length of the block in centimeters.
2. Compute the volume of the block in cubic centimeters (cm^3).
3. Find the mass of the block in grams (g) using a triple-beam or equal-arm balance.
4. Calculate the density of the block in grams per cubic centimeter (g/cm^3).

Experiment 9:3 *Density of an Irregular Solid*

1. Obtain a small, hard stone measuring an inch or 2 in its largest dimension.
2. Find the mass of the stone in grams (g) using a balance.
3. Fill a graduate cylinder approximately three-fourths full of water and read the lowest point of the surface of the liquid. Record this reading.
4. Carefully place the stone in the water and read the level of the water again. Record this second reading.
5. Subtract the smaller reading from the larger to determine the volume of the stone in cubic centimeters (cm^3).
6. Calculate the density of the stone in grams per cubic centimeter (g/cm^3).

★ **9:9 Problems of Measurement** No measurement is absolutely perfect. There is always some error. No matter how careful a researcher is, he is still limited by the capabilities of his instruments. An inexpensive balance will give a reading such as 14.1 grams for a sample of, say, sodium chloride. A better balance, such as a triple-beam unit, might give a reading of 14.12 grams for the same sample. This is better, but not accurate enough for some purposes. A very expensive analytical balance could give a reading of 14.1201. This is very good, but it still does not represent perfection in any sense. To achieve more accuracy than this would cost a tremendous amount of money. Unfortunately, perfect measurements cannot be bought at any price. The wise student will understand from this that science can never give us absolute truth.

Scientists have developed a code to tell each other how accurately a measurement has been made. In the following list the code appears at the left and its meaning at the right:

Measurement reported as:	**The actual value lies between:**
8	7.5 and 8.5
8.0	7.95 and 8.05
8.00	7.995 and 8.005
8.000	7.9995 and 8.0005

The four measurements on the left are said to have, respectively, one significant figure, two significant figures, three significant figures, and four significant figures. The more **significant figures** given, the greater is the accuracy being claimed for the measurement.

Regardless of the number of significant figures used, the last digit is always understood to be a **doubtful figure**; that is, there is a good chance that it is correct, but it still could be somewhat "off." With increasingly better instruments (and, therefore, readings of more significant figures) the doubtfulness becomes less important because the last figure becomes a smaller part of the total reading.

THE MODERNIZED METRIC SYSTEM

The International System of Units-SI is a modernized version of the metric system established by international agreement. It provides a logical and interconnected framework for all measurements in science, industry, and commerce. Officially abbreviated SI, the system is built upon a foundation of seven base units, plus two supplementary units, which appear on this chart along with their definitions. All other SI units are derived from these units. Multiples and sub-multiples are expressed in a decimal system. Use of metric weights and measures was legalized in the United States in 1866, and since 1893 the yard and pound have been defined in terms of the meter and the kilogram. The base units for time, electric current, amount of substance, and luminous intensity are the same in both the customary and metric systems.

COMMON CONVERSIONS
Accurate to Six Significant Figures

Symbol	When You Know	Multiply By	To Find	Symbol
in	inches	[1]25.4	[2]millimeters	mm
ft	feet	[1] 0.3048	meters	m
yd	yards	[1] 0.9144	meters	m
mi	miles	1.609 34	kilometers	km
yd^2	square yards	0.836 127	square meters	m^2
	acres	0.404 686	[3]hectares	ha
yd^3	cubic yards	0.764 555	cubic meters	m^3
qt	quarts (lq)	0.946 353	[4]liters	l
oz	ounces (avdp)	28.349 5	grams	g
lb	pounds (avdp)	0.453 592	kilograms	kg
°F	Fahrenheit temperature	[1]5/9 (after subtracting 32)	Celsius temperature	°C
mm	millimeters	0.039 370 1	inches	in
m	meters	3.280 84	feet	ft
m	meters	1.093 61	yards	yd
km	kilometers	0.621 371	miles	mi
m^2	square meters	1.195 99	square yards	yd^2
ha	[3]hectares	2.471 05	acres	
m^3	cubic meters	1.307 95	cubic yards	yd^3
l	[4]liters	1.056 69	quarts (lq)	qt
g	grams	0.035 274 0	ounces (avdp)	oz
kg	kilograms	2.204 62	pounds (avdp)	lb
°C	Celsius temperature	[1]9/5 (then add 32)	Fahrenheit temperature	°F

[1] exact

[2] for example, 1 in = 25.4 mm, so 3 inches would be (3 in)(25.4 mm/in) = 76.2 mm

[3] hectare is a common name for 10 000 square meters

[4] liter is a common name for fluid volume of 0.001 cubic meter

Note: Most symbols are written with lower case letters; exceptions are units named after persons for which the symbols are capitalized. Periods are not used with any symbols.

MULTIPLES AND PREFIXES
These Prefixes May Be Applied To All SI Units

Multiples and Submultiples	Prefixes	Symbols
1 000 000 000 000 = 10^{12}	tera	T
1 000 000 000 = 10^9	giga	G
1 000 000 = 10^6	mega	M
1 000 = 10^3	kilo	k
100 = 10^2	hecto	h
10 = 10^1	deka	da
Base Unit 1 = 10^0		
0.1 = 10^{-1}	deci	d
0.01 = 10^{-2}	centi	c
0.001 = 10^{-3}	milli	m
0.000 001 = 10^{-6}	micro	μ
0.000 000 001 = 10^{-9}	nano	n
0.000 000 000 001 = 10^{-12}	pico	p
0.000 000 000 000 001 = 10^{-15}	femto	f
0.000 000 000 000 000 001 = 10^{-18}	atto	a

9:21 *The metric system.*

SEVEN BASE UNITS

meter-m

Length

The meter (common international spelling, metre) is defined as 1,650,763.73 wavelengths in vacuum of the orange-red line of the spectrum of krypton-86.

The SI unit of area is the **square meter** (m^2).

The SI unit of volume is the **cubic meter** (m^3). The liter (0.001 cubic meter), although not an SI unit, is commonly used to measure fluid volume.

kilogram-kg

Mass

The standard for the unit of mass, the kilogram, is a cylinder of platinum-iridium alloy kept by the International Bureau of Weights and Measures at Paris. A duplicate in the custody of the National Bureau of Standards serves as the mass standard for the United States. This is the only base unit still defined by an artifact.

The SI unit of force is the **newton** (N). One newton is the force which, when applied to a 1 kilogram mass, will give the kilogram mass an acceleration of 1 (meter per second) per second.

$$1N = 1kg \cdot m/s^2$$

The SI unit for pressure is the **pascal** (Pa).

$$1Pa = 1N/m^2$$

The SI unit for work and energy of any kind is the **joule** (J).

$$1J = 1N \cdot m$$

The SI unit for power of any kind is the **watt** (W).

$$1W = 1J/s$$

second-s

Time

The second is defined as the duration of 9,192,631,770 cycles of the radiation associated with a specified transition of the cesium-133 atom. It is realized by tuning an oscillator to the resonance frequency of cesium-133 atoms as they pass through a system of magnets and a resonant cavity into a detector.

The number of periods or cycles per second is called frequency. The SI unit for frequency is the **hertz** (Hz). One hertz equals one cycle per second.

The SI unit for speed is the **meter per second** (m/s).

The SI unit for acceleration is the **meter per second per second** (m/s^2).

Standard frequencies and correct time are broadcast from WWV, WWVB, and WWVH, and stations of the U.S. Navy. Many short-wave receivers pick up WWV and WWVH, on frequencies of 2.5, 5, 10, 15, and 20 megahertz.

ampere-A

Electric current

The ampere is defined as that current which, if maintained in each of two long parallel wires separated by one meter in free space, would produce a force between the two wires (due to their magnetic fields) of 2×10^{-7} newtons for each meter of length.

The SI unit of voltage is the **volt** (V).

$$1V = 1W/A$$

The SI unit of electric resistance is the **ohm** (Ω).

$$1\Omega = 1V/A$$

Kelvin-K

Temperature

The **kelvin** is defined as the fraction 1/273.16 of the thermodynamic temperature of the triple point of water. The temperature 0° K is called "absolute zero".

On the commonly used Celsius temperature scale, water freezes at about 0° C and boils at about 100° C. The ° C is defined as an interval of 1° K, and the Celsius temperature 0° C is defined as an interval of 1° K, and the Celsius temperature 0° C is defined as 273.15° K.

The Fahrenheit degree is an interval of 5/9 ° C or 5/9 ° K; the Fahrenheit scale uses 32° F as a temperature corresponding to 0° C.

The standard temperature at the triple point of water is provided by a special cell, an evacuated glass cylinder containing pure water. When the cell is cooled until a mantle of ice forms around the reentrant well, the temperature at the interface of solid, liquid, and vapor is 273.16° K. Thermometers to be calibrated are placed in the reentrant well.

mole-mol

Amount of substance

The **mole** is the amount of substance of a system that contains as many elementary entities as there are atoms in 0.012 kilogram of carbon 12.

When the mole is used, the elementary entities must be specified and may be atoms, molecules, ions, electrons, other particles, or specified groups of such particles.

The SI unit of concentration (of amount of substance) is the **mole per cubic meter** (mol/m^3).

candela-cd

Luminous intensity

The candela is defined as the luminous intensity of 1/600,000 of a square meter of a blackbody at the temperature of freezing platinum (2045° K).

The SI unit of light flux is the **lumen** (lm). A source having an intensity of 1 candela in all directions radiates a light flux of 4π lumens.

A 100-watt light bulb emits about 1700 lumens.

TWO SUPPLEMENTARY UNITS

★ **radian-rad**

Plane angle

The radian is the plane angle with its vertex at the center of a circle that is subtended by an arc equal in length to the radius.

★ **steradian-sr**

Solid angle

The steradian is the solid angle with its vertex at the center of a sphere that is subtended by an area of the spherical surface equal to that of a square with sides equal in length to the radius.

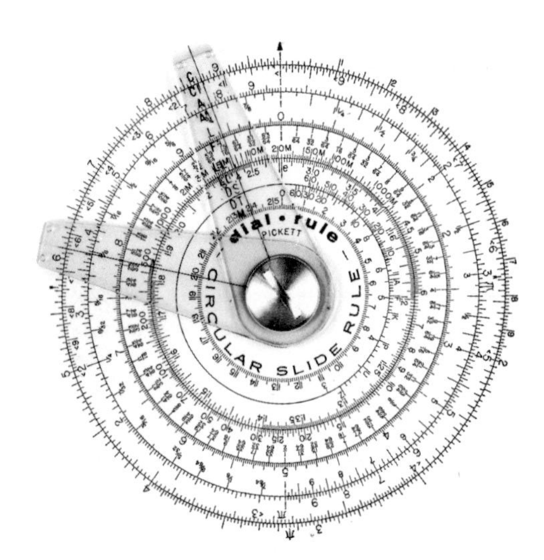

9:19 *Circular slide rule.*

An important rule to remember is that there can be no more significant figures in the answer than appear in the *least* accurate of the measurements that enter into a computation. Consider, for example, the calculation of the area of a room. The length of the room is measured very accurately to four significant figures and recorded as 5.824 meters. The width is measured with less accuracy to three figures and recorded as 4.20 meters. Substituting these numbers into the area formula, A = LW

$$A = (5.824\ \text{m})\ (4.20\ \text{m})$$
$$A = 24.4608\ \text{m}^2$$
$$\text{or, rounded off, } 24.5\ \text{m}^2$$

Only three figures should be given in the answer, the same as the *least* accurate of the two original measurements. The last digits are, therefore, worthless.

In many kinds of scientific work, it is common practice to use three significant figures for both measurements and computations. This is a good, workable compromise between accuracy and ease of computation. Such calculations can be carried out rapidly on a slide rule, a device which automatically rounds its answers to three significant figures.

9:20 *Student slide rule.*

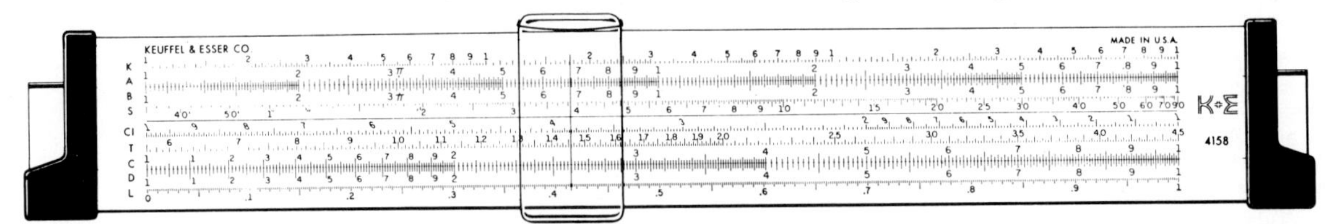

★ 9:10 English-Metric and Metric-English Conversions The great beauty of the metric system lies in the ease of computation if all measurements are made in metric units *and* all calculations based on the data are kept *within* the system. Occasionally, however, it is necessary to convert from metric to English or from English to metric. A few examples of such conversions are presented here.

In each case the answer is rounded to three significant figures.

Problem: Change 28.0 meters to feet.
Solution: There are 3.28 feet in each meter. Therefore, 28.0 should be multiplied by 3.28:

$$(28.0\ \cancel{m})\ (3.28\ \frac{ft}{\cancel{m}}) = 91.8\ ft.$$

The meters cancel, leaving only feet in the answer.

Problem: What is the height of a 6-foot, 2-inch man in centimeters?
Solution: 6-foot, 2-inches = 74.0 inches. There are 2.54 centimeters in one inch. Again, multiplication is the operation called for.

$$(74.0\ \cancel{in})\ (2.54\ \frac{cm}{\cancel{in}}) = 188\ cm$$

Problem: How many liters are there in 5.00 gallons?
Solution: 5.00 gallons = 20.0 quarts. There are 1.06 quarts to a liter.

Studying the units can help you decide whether to multiply or divide to get the answer. Multiplication would give qt^2/l, which is obviously wrong; we desire to have our answer in liters. Dividing the numbers gets rid of the quarts by cancellation, and the liters in the denominator *of* the denominator appear in the answer. The numbers must therefore be divided:

$$\frac{20.0\ \cancel{qt}}{1.06\ \cancel{qt}/l} = 18.9\ liters$$

Note: Since the liter is a slightly larger unit than the quart, the number of liters equivalent to 20.0 quarts had to be somewhat less than 20.0.

Problem: Express the mass of a 128-pound person in kilograms.
Solution: There are two ways to approach this problem. Since there are 454 grams in a pound, the two numbers can be multiplied to give the answer in grams, which can then easily be changed to kilograms.

$$(128\ \cancel{lb})\ (454\ \frac{g}{\cancel{lb}}) = 58{,}100\ grams = 58.1\ kg$$

Using the fact that there are 2.205 pounds in a kilogram, the answer can be obtained directly in kilograms by dividing.

$$\frac{128\ \cancel{lb}.}{2.205\ \cancel{lb}/kg} = 58.1\ kg$$

Of the two methods, the multiplication is preferable if the calculation is being performed longhand. With a slide rule the two are about equal in ease of computation.

LORD KELVIN (1824-1907)

Irish physicist

Lord Kelvin (William Thomson) is probably best known for the absolute scale of temperature that bears his name. He was also the mastermind behind the building of the first transatlantic telegraph cable and the first scientist to adopt the use of the term "energy," the most important of all quantities dealt with in physics. Kelvin was equally learned in both the theoretical and the practical branches of science. He contributed much to our knowledge about thermodynamics and pure mathematics. He also patented 70 inventions.

This illustrious giant of science was born in Belfast, Ireland, in 1824, the fourth of seven children of James Thomson, a Scottish mathematics professor. The Thomson children were brought up in the Established Church of Scotland (Presbyterian). The preaching they heard was fundamental and practical.

William responded favorably to the Gospel message and was saved at an early age. As a student at Cambridge University, he was a brilliant scholar and maintained a good Christian testimony before his professors and fellow students. Upon graduation he was awarded, among other academic honors, a special citation "in consideration of his great mind and exemplary conduct."

After receiving his bachelor's degree in 1845, young Thomson chose to broaden his horizons by working for a year at the Regnault Laboratories in Paris, France. He earned his master's degree, also from Cambridge University, in 1848. At the age of 22, Thomson accepted an appointment to the Chair of Natural Philosophy at the University of Glasgow, a post he filled until his retirement 53 years later.

His years at Glasgow proved to be both enjoyable and fruitful. He had few teaching responsibilities, and this left him ample time for research. Thomson was able to turn out hundreds of papers covering an unbelievably wide range of subjects. A few of the subjects included in his investigations were heat, light, sound, gravitation, electricity, magnetism, pure mathematics, hydrodynamics, viscosity, telegraphy, elasticity, astronomy, meteorology, geology, navigational aids, tides, mechanical calculators, osmosis, and radioactivity. His famous physics textbook, coauthored with Professor Peter Tait of Edinburgh, was the unchallenged leading work in its field for many years.

Five men—Sadi Carnot, Benoit Clapeyron, Rudolf Clausius, William Rankine, and Lord Kelvin—are generally credited with having made important contributions to the formulation of the second law of thermodynamics. (See Chapter 29 for a more complete discussion of this law.) Kelvin's definitive paper on the subject appeared in 1852. Although most people have probably never heard of this law, its meaning is extremely important. It indicates that the entire universe is degenerating (running down). The universe, therefore, cannot be eternal (infinitely old), or it would long since have run down completely. Thus, by the mid-1800's science had established the fact that there must have been, at some point in the past, a definite time of creation. Skeptics could still deny that there ever was a creation by God, but they would now do so in opposition to the clear findings of science.

The second law of thermodynamics also strongly contradicts any theory of evolution that has ever been proposed. If, indeed, destruction rather than improvement is the trend of nature, we could hardly expect increasing order and complexity to arise from less organized forms of matter. It was no secret that Lord Kelvin was strongly opposed to the evolutionary teachings of Charles Darwin.

Moreover, the Irish physicist stoutly rejected the idea that life on the earth could ever have arisen spontaneously. "Mathematics and dynamics fail us," he wrote, "when we contemplate the earth, fitted for life but lifeless, and try to imagine the commencement of life upon it. This certainly did not take place by any action of chemistry, or electricity, or crystalline grouping of molecules under the influence of force, or by any possible kind of fortuitous concourse (chance coming together) of atoms. We must pause, face to face with the mystery and miracle of the Creation of living creatures." We can readily see the sharp contrast between Kelvin's clear insight, derived from the Biblical account of creation, and the philosophy of today, which regards life as a mere collection of chemicals.

In 1856 Kelvin was elected a director of the Atlantic Telegraph Company, which was formed for the purpose of installing a transatlantic telegraph cable from Ireland to North America. By this time the use of Samuel Morse's telegraph had spread through much of the world. Wires were humming with messages on both the North American and European continents. Short lengths of submarine cables had been successfully laid, such as the cable connecting England with the Netherlands. But the feat of spanning the 2,500 miles between the European and American continents still posed many problems. The ocean was known to be three miles deep in some places, and there was no ship in existence that could carry the tremendous amount of cable needed.

The Atlantic Telegraph Company decided to use two ships. The British battleship *Agamemnon* and the United States frigate *Niagara* were acquired for the purpose. When the company's electrician fell ill and was unable to accompany the expedition, Kelvin volunteered his services and supervised the operations from aboard the *Agamemnon*. After several discouraging failures, the cable was finally completed from Valencia Bay, Ireland, to Heart's Content, Newfoundland, Canada, on August 5, 1858. There were wild celebrations on both sides of the Atlantic Ocean as Queen Victoria and President Buchanan exchanged congratulatory messages. Many newspaper headlines hailed it as the feat of the century.

For this and other accomplishments Kelvin was knighted. In 1892 he was again honored by the Queen of England; a barony was conferred on him, and his official title became Baron Kelvin of Largs. Although he had been born a commoner, William Thomson was now considered a nobleman. The total list of distinctions that he earned grew to an imposing length and included 21 honorary doctorates from universities throughout Europe and America.

Kelvin's was an unusually full life. He derived a wonderful sense of fulfillment from his life's work, counting the privilege of scientific investigation one of the Creator's greatest gifts to mankind. He lived to the age of 83, enjoying the blessings of good health and a keen mind until a month or two before his death. Living several years into the 20th century, Kelvin witnessed the advent of the automobile, the airplane, the theory of relativity, the quantum theory, and the first voice transmissions by radio. He is buried in Westminster Abbey in London, where he has been honored with a magnificent Gothic stained glass window bearing the inscription: *In memory of Baron Kelvin of Largs, Engineer, Natural Philosopher, B:1824 D:1907.*

List of Terms

area
centimeter
cube
cubic centimeter
cubic meter
density
doubtful figure
English system
exponent
graduate cylinder
gram
kilogram
kilometer
liter
mass
meter
metric system
milligram
milliliter
millimeter
overflow can
primary standard
regular solid
second
secondary standard
significant figure
slug
square centimeter
square meter
units
volume
weight

Questions

1. (a) How is the meter defined today? (b) How was it previously defined?
2. How does the length of a meter compare with the length of a yard?
3. What is the main advantage of the metric system?
4. How are the centimeter and the millimeter defined?
5. (a) How is the kilometer defined? (b) How does the kilometer compare with the mile?
6. Which holds more—a quart container or a liter container?
7. Explain the difference between mass and weight.
8. (a) Define density in words. (b) Define density mathematically.
9. (a) How is the second defined today? (b) How was it previously defined?
10. What are WWV and CHU?
11. Give the meaning of each prefix: (a) centi- (b) milli- (c) kilo- (d) micro-
12. What is meant by the volume of an object?
13. A certain liquid has a density of 1.54 g/cm^3. (a) Would a solid object whose density is 1.25 g/cm^3 float in it? (b) What would happen if the solid object had exactly the same density as the liquid?
14. What is the unit of mass in the English system?

★15. How many significant figures are there in each of these numbers? (a) 125 (b) 18.005 (c) 152.0 (d) 0.00025

Problems

1. How many cubic centimeters are there in 17.1 liters?
2. Change 1480 milligrams to grams.
3. What is the mass of 10.0 cubic centimeters of water?
4. What is the volume of 3.50 kilograms of water?
5. Change 9.60 kilograms to grams.
6. Find the density of a sample of rock which has a mass of 115 grams and a volume of 43.0 cubic centimeters.

7. Find the density of a liquid if 5.75 liters of it has a mass of 4.00 kilograms.
8. Find the mass of 100 cm^3 of ice.
9. Find the volume of 50.0 grams of iron.
10. Find the mass of 24.0 cm^3 of mercury.
11. How much space is occupied by 850 grams of magnesium?
★ 12. Change 184 feet to meters.
★ 13. How many quarts are there in 350 liters?
★ 14. Find the mass in kilograms of a 25.0-pound bag of flour.
★ 15. A ton equals 2000 pounds. Express this in kilograms.
★ 16. There are 28.35 grams in one ounce. A heavy letter weighs 76.8 grams. At the present postage rate, how much would it cost to mail the letter to another point in the United States?
★ 17. The speed limit on a certain European road is 80.0 kilometers per hour. What is the speed limit in miles per hour?
★ 18. One cubic foot of a certain type of wood weighs 48.2 pounds. Find the density of the wood in g/cm^3. If a cubic foot of water weighs 62.4 pounds, would the wood float?
19. How many milliseconds are there in 1/15 of a second?
20. How many microseconds are there in 85 milliseconds?
★ 21. The speed of light in the metric system is 300,000,000 meters per second. Find the corresponding speed in the English system. (Round to three significant figures and use zeros for the remaining digits.)
★ 22. Find the volume of a cube which measures 2.50 centimeters on a side. Round your answer to three significant figures.
★ 23. What is the total surface area of all the sides of the cube in Question 22? Round your answer to three significant figures.
24. A piece of modeling clay having a volume of 512 cm^3 is shaped into a cube. Find the length of one side of the cube.
★ 25. A weather report includes a barometer reading of 29.90 inches of mercury. (a) Express this in centimeters (b) Express this in millimeters of mercury.
★ 26. The distance between two cities is 180 miles. Calculate the distance in kilometers.
★ 27. Find the density of mercury in lb/ft^3 using the fact that water is 62.4 lb/ft^3.
★ 28. Find the value of a cubic foot of mercury if it is worth $10.00 per pound.

Student Activities

1. Determine the area of a page of this book in square centimeters.
2. Find the approximate volume of your classroom in cubic meters.
3. Study the labels on items in your kitchen cupboards to see if any metric units are used. Make a list of those you are able to find.
4. Weigh yourself and calculate your mass in kilograms.
5. Prepare a report on Biblical weights and measures.
6. Prepare a report on the work of the Sealer of Weights and Measures in your area.
7. Prepare a report on the National Bureau of Standards.

Photo next page
A weather balloon is a good illustration of Boyle's law. The balloon increases in size as it rises, due to the reduced pressure at higher altitudes.

UNIT III USING NUMBERS IN SCIENCE

Chapter 10 The Gas Laws

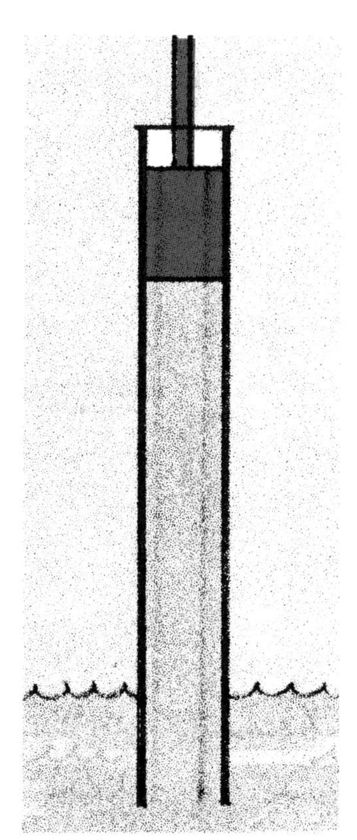

10:1 *Water can be drawn upward by suction to a height of 34 feet, but no higher. Atmospheric pressure pushing on the water below supports the water in the column.*

10:1 Historical Background The Bible contains many remarkable statements of scientific fact. Numerous examples can be found in Scripture where a fact was stated centuries before it became known to science. One such instance occurs in Job 28:25—"to make the weight for the winds." The fact that wind or air has *weight* was not realized by the scientific world until the seventeenth century A.D., when Evangelista Torricelli (1608-1647), a student of Galileo Galilei (1564-1642), built the first **barometer**. The book of Job was written at least a millennium and a half before the time of Christ, 32 centuries before Torricelli.

Galileo had observed that water could, by suction, be drawn up to a height of 34 feet in a glass tube, but no farther. Pondering this question one day, Torricelli was struck with an idea: Might not the water be held up *in* the tube by the weight of the air pushing down on the surface of the water *outside* the tube? Picture to yourself all of the water around the entire world being pressed down upon by the weight of the air. If a suction is created at one point, the water should rise at that point as a means of relieving some of the pressure.

Torricelli reasoned that if the weight of the air was actually supporting the column of water, mercury, being considerably more dense, should rise proportionately less. Since mercury is 13.6 times as dense as water, he predicted that, under suction, it should rise in a tube only 1/13.6 as high as water. When he experimented he found, much to his satisfaction, that the mercury rose only about 30 inches in the tube—almost exactly 1/13.6 of the height of the water. (The height of the water in inches was 34 feet times 12 inches per foot, or 408 inches.) Thus, in 1643, Torricelli invented the **mercurial barometer**, an extremely important device that is still in use today (see Figure 10:3 a & b).

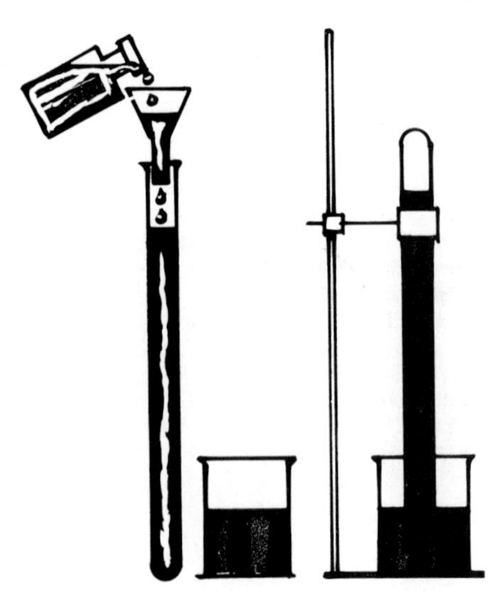

10:2 *Experiment 10:1. Making a barometer.*

Experiment 10:1 *Making a Barometer*

Fill a small beaker about half full of mercury. Using a funnel, fill a barometer tube completely with mercury. Hold your thumb over the open end of the tube; invert the tube and place the open end below the surface of the mercury in the beaker. Remove your thumb. The mercury should fall somewhat in the tube. Secure the tube with a clamp attached to a ringstand. Measure the height changes from one day to the next. Observe how the height changes from one day to the next.

10:2 The Pressure of a Gas Watching his barometer from day to day, Torricelli observed that the height of the mercury changed somewhat, depending on the weather conditions. When a high-pressure region of the atmosphere stood above the barometer, the mercury rose *above* 30 inches. A low-pressure system caused the mercury to fall *below* 30 inches. Thus, the barometer provides a means of keeping track of the atmospheric pressure. A typical reading given in a weather report might be "29.92 inches and falling."

Another classic experiment was conducted by Blaise Pascal (1623-1662), a noted French physicist and theologian. Pascal's brother-in-law, Florin Perrier, carried a mercurial barometer up a high mountain in central France and observed that the mercury fell several inches. This experiment gives strong support to the idea that we live at the bottom of a sea of air. A barometer at higher altitudes has less air pushing down on it from above, and therefore indicates less pressure. Air pressure is similar to liquid pressure—the deeper the liquid, the greater the pressure (see Figure 10:4).

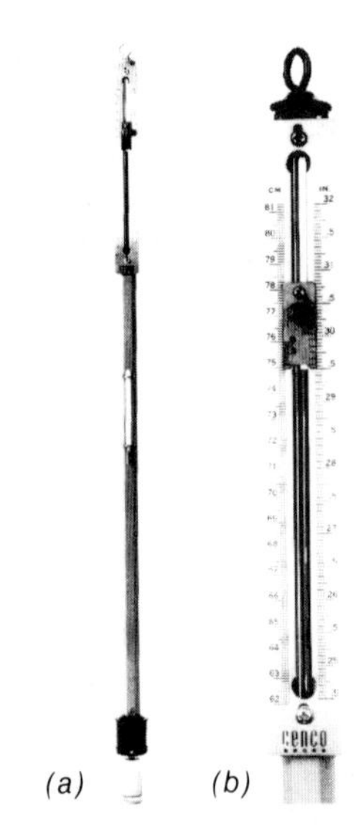

10:3 *(a) Wall-type mercurial barometer. (b) Close-up of top of mercurial barometer of Figure 10:3a, showing vernier scale.*

10:4 *Pressure in a fluid varies with depth. The greater the depth, the greater the pressure. The principle applies equally well to gases and liquids.*

10:5 *An altimeter is actually a pressure detector. The less the pressure on the instrument, the higher the altitude indicated.*

By **pressure** we mean how much force acts on a unit of area. By **force** we mean a push or a pull (see Figure 10:7). We can write the definition of pressure as a mathematical relationship:

$$P = \frac{F}{A}$$

P is pressure, F is force, and A is area. If 20 pounds of force act on an area of 4 square inches, the pressure is:

$$\frac{20 \text{ lb}}{4 \text{ in}^2} = 5 \text{ lb/in}^2$$

10:6 *Deep submergence vehicles must be specially designed because of the great pressures they are called upon to sustain.*

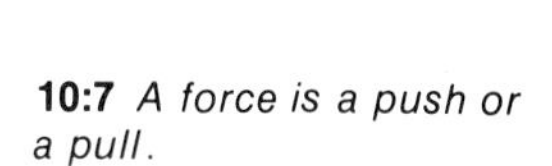

10:7 *A force is a push or a pull.*

Atmospheric pressure can be given in several different ways. All of the following are approximately equivalent and acceptable for expressing standard atmospheric pressure:

30 inches of mercury
76 centimeters of mercury
760 millimeters of mercury
14.7 pounds per square inch
1013 millibars
1 atmosphere

(a)

(b)

10:8 *Two types of barometers in common use. (a) An* **aneroid barometer** *for home use. (b) A recording aneroid barometer or barograph. A permanent record of pressure changes is traced on graph paper as the drum rotates.*

10:3 The Density of a Gas In the previous chapter we discussed the densities of solids and liquids. We can also compute densities for gases. At ordinary pressures such densities are exceedingly low compared to those of liquids and solids. The air you are breathing has a density of about 0.0012 g/cm^3. Since this is such a small number, it has been agreed to use *grams per liter* for densities of gases rather than grams per cubic centimeter.

If a cubic centimeter of air at the earth's surface has a mass of 0.0012 grams, a liter, which is 1000 cm^3, has a mass of 1.2 grams. Very roughly, a gram of air occupies a liter of space. There is more substance to air than you might think: the weight of the air in a typical classroom is several hundred pounds! It is this property of substance that provides support for planes and helicopters, and also permits a parachutist to descend safely to the ground. If, instead of possessing an atmosphere, the earth were surrounded by a vacuum, there would be nothing under the wings of a plane to hold it up. A parachutist, with nothing to slow him down, would plummet to the ground at ever-increasing speeds.

10:9 *Air has substance. It is this substantiality that provides support for planes and helicopters.*

10:10 *Further proof of the substantiality of air. Without the air to slow him down, a parachutist would plummet rapidly to the ground.*

10:11 *Hurricane damage provides yet additional proof of the fact that air has substance.*

Experiment 10:2 *The Weight of Air*

Place one end of a stopcock through a one-hole rubber stopper and fit the stopper into a Florence flask. Remove the air from the flask with a vacuum pump and shut the stopcock. Balance exactly with weights on an equal-arm balance. Open the stopcock. As the air rushes in, notice that there is an increase in the weight of the flask.

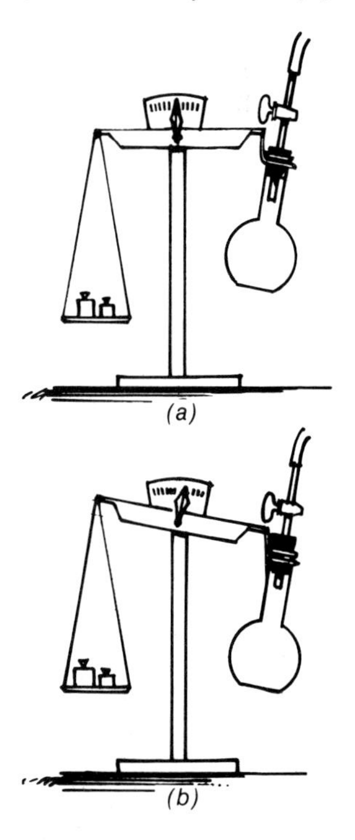

10:12 *Setup for Experiment 10:2. Evacuated flask is balanced as shown in (a). When air is allowed into the flask it becomes noticeably heavier (b).*

Air is, of course, more dense at lower altitudes. The atmosphere thins out so rapidly with increasing altitude that one-half of its weight lies below $3\frac{1}{2}$ miles. The other half is from $3\frac{1}{2}$ miles up, but there is no definite upper limit; it extends upward several hundred miles, gradually tapering off to practically nothing.

Table 10-1 gives the densities of some common gases at atmospheric pressure and 0° Celsius, the freezing point of water. This condition of low temperature accounts for the fact that the density of air is somewhat higher here than the 1.2 g/l mentioned above.

The densities in the table are given at standard conditions of temperature and pressure. Standard temperature is defined as 0° Celsius (Centigrade). Standard pressure is 760 millimeters of mercury. The letters **S.T.P.** (meaning standard temperature and pressure) are used to represent these conditions. The expression "a liter of hydrogen" has no meaning, because we have not specified the temperature and pressure at which the volume was measured. The expression "a liter of hydrogen at S.T.P.," on the other hand, leaves no room for doubt. It refers to a definite quantity of hydrogen.

Table 10-1 *Densities of Some Common Gases (in Grams Per Liter) Measured at 760 mm Pressure and 0° C*

chlorine	3.21
sulfur dioxide	2.93
carbon dioxide	1.98
hydrogen chloride	1.64
hydrogen sulfide	1.54
oxygen	1.43
air	1.29
nitrogen	1.25
carbon monoxide	1.25
ammonia	0.77
helium	0.18
hydrogen	0.09

From the table, it can be seen that there is quite a range of density from the lightest to the heaviest. Hydrogen and helium are both considerably lighter than air. Chlorine is an unusually heavy gas which can actually be observed to creep along the top of a table and fall over the edge.

10:4 Boyle's Law The density of a gas depends, as we have seen, on its pressure. Since gases are highly compressible, comparatively high densities can be obtained at increased pressures. The oxygen in a tank used for welding, for example, is compressed to 200 times its normal density (close to 300 grams per liter!), and can therefore deliver 200 times its own volume to the flame.

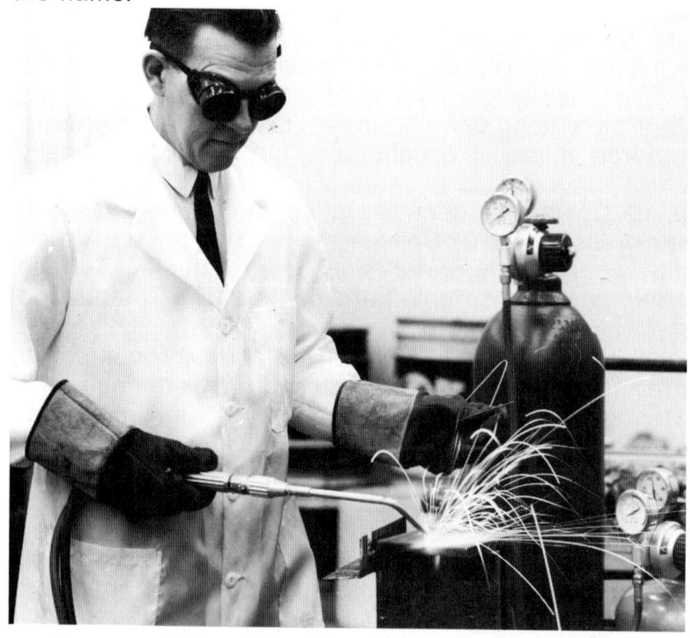

10:13. *Oxygen in tank used for welding is compressed to 200 times its normal density.*

Robert Boyle (1627-1691), one of the truly outstanding Christian men of science, made careful measurements on the compressibility of air. In carrying out the experiments that led to the law bearing his name, he developed the J-shaped tube that has since come to be called the "Boyle's law tube" (see Figure 10:14). The tube is sealed at the lower end but open at the top.

By pouring mercury in from the top, Boyle was able to trap a quantity of air in the lower portion of the tube. After leveling the mercury on the two sides, he began increasing the pressure on the trapped air by adding more and more mercury, taking care to let the air cool back to the temperature of its surroundings after each compression. Much to his delight he observed the volume of the air to be down to exactly half of its original value when the pressure had been doubled. And at four times the pressure, the volume was down to one-fourth of its original value.

After data had been collected on about two dozen in-between points, he was ready to formulate what he had been observing as a rule or law: The volume of the enclosed air varies inversely with the pressure, provided that the temperature is constant. To express this in another way, the product of the pressure and volume remains constant at constant temperature. The same experiment has been tried using many different gases. For every gas tried thus far, Boyle's law has given acceptable results at ordinary conditions of temperature and pressure. There *are* deviations from the law at high pressures and low temperatures, but these do not concern us here. We can state, then, for gases in general: *The volume of a certain amount of dry gas is inversely proportional to the pressure, provided the temperature remains constant*. Expressed as a mathematical equation,

$$PV = C$$

P is the pressure, *V* is the volume, and *C* is a constant. A **constant** is a number that stays the same throughout an experiment or calculation. PV = C simply tells us that when the pressure on an enclosed gas is changed, the volume will adjust itself in such a way that the product of pressure and volume remains the same.

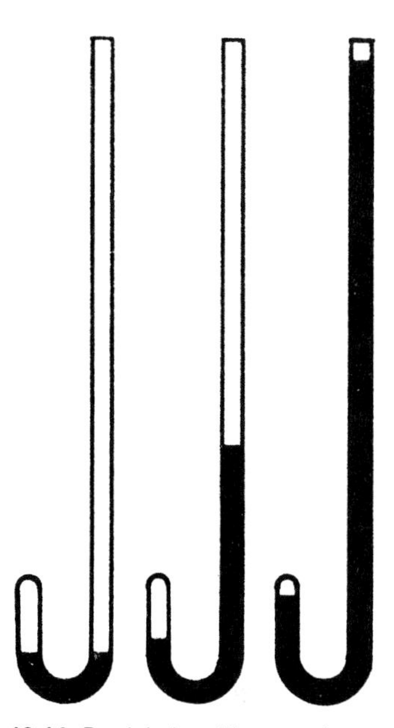

10:14 *Boyle's law illustrated. The volume of the air trapped in the tube decreases as the pressure on it is increased.*

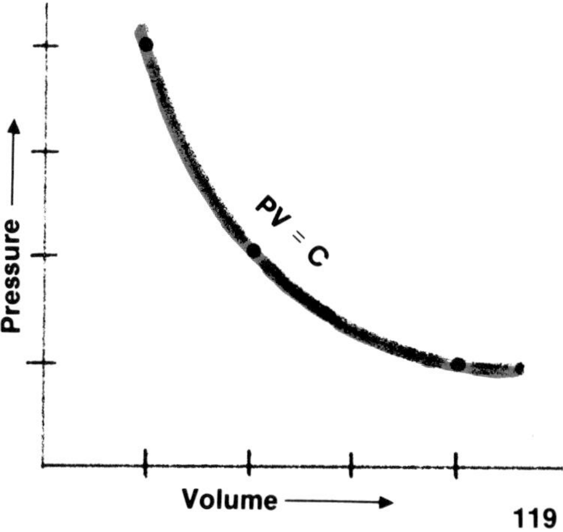

10:15 *Boyle's law plot is part of a hyperbola.*

★ **10:5 Use of Boyle's Law** A way to express Boyle's law that makes knowledge of the actual value of the constant unnecessary uses **subscripts**. Subscripts are small numbers written just after, and slightly below, an algebraic symbol. In any given problem or experimental situation we are generally concerned with two sets of conditions—the pressure and volume before and after some change is made. P_1 is used to represent the first pressure—that is, the pressure before it is changed. P_2 represents the second pressure, after it is changed. Likewise, V_1 and V_2 represent the volumes before and after the change.

Before any change is made, the product of the pressure and the volume, according to Boyle's law, must be equal to a constant:

$$P_1V_1 = C$$

After the change is made, the product of the new pressure and the new volume must still be equal to the same constant.

$$P_2V_2 = C$$

Because things equal to the same thing are equal to each other, we can write:

$$P_1V_1 = P_2V_2$$

We will use Boyle's law in this handy form. Let us illustrate its use with two problems.

Problem: 500 cm^3 of hydrogen is collected from a chemical generator when the pressure is 800 mm of mercury. What volume will the gas occupy when the pressure is 760 mm of mercury (normal atmospheric pressure)?

Solution: It is advisable to write down the given information and what is to be found in terms of our symbols.

$$P_1 = 800 \text{ mm}$$
$$V_1 = 500 \text{ cm}^3$$
$$P_2 = 760 \text{ mm}$$
$$V_2 = ?$$

Will the gas occupy more or less space at the new reduced pressure? It will expand somewhat and occupy *more* volume than before. Let us solve the equation for the term we want to find, V_2:

$$P_1V_1 = P_2V_2$$

To do this we divide both sides of the equation by P_2 and turn it around:

$$\frac{P_1V_1}{P_2} = V_2$$

$$V_2 = \frac{P_1V_1}{P_2}$$

Substituting the numbers,

$$V_2 = \frac{(800 \text{ mm})\ (500 \text{ cm}^3)}{760 \text{ mm}} = 526 \text{ cm}^3$$

The answer has been rounded to 3 figures. Notice that the millimeters cancel, leaving only cubic centimeters in the answer.

Problem: An oxygen tank of volume 2.00 ft^3 is filled to a pressure of 3000 lb/in^2. What was the original volume of the oxygen when it was at a pressure of 14.7 lb/in^2?

Solution: Will the original volume be more or less than the volume it has in the tank? Considerably more! Let us write down the given information and what is to be found, as before:

$$P_1 = 14.7\ lb/in^2$$
$$V_1 = ?$$
$$P_2 = 3000\ lb/in^2$$
$$V_2 = 2.00\ ft^3$$

Solve the equation for V_1 by dividing both sides by P_1:

$$P_1V_1 = P_2V_2$$
$$V_1 = \frac{P_2V_2}{P_1} = \frac{(3000\ lb/in^2)\ (2.00\ ft^3)}{14.7\ lb/in^2}$$
$$V_1 = 408\ ft^3$$

As before, the pressure units cancel out, leaving only the volume units in the answer.

Boyle's law is not absolute in any sense of the word. It does not dictate to a gas what it must do. It is merely a convenient way of describing the behavior of gases at ordinary conditions of temperature and pressure. As has been stated, deviations from Boyle's law have been observed at high pressures and low temperatures. But this principle has been retained as part of our general body of scientific knowledge for more than three centuries (since 1660), and it is still considered to be a highly useful relationship.

10:6 Effect of Temperature on a Gas Intuition or previous learning would probably tell us that gases expand when heated and contract when cooled. We can now go a step further and consider *how much* a gas expands with a given temperature increase. To do this, however, we need a suitable temperature scale. The **Fahrenheit scale** with which you are probably the most familiar has one major drawback; the numbers are not proportional to the "hotness" we are measuring. A day that is 40° F is not twice as hot as a day that is 20° F. The reason for this is that there are 460 degrees *below zero* on the Fahrenheit scale. So rather than comparing 40 and 20 degrees, we are really comparing 460 + 40 with 460 + 20. Seen in this light, the first day is not even close to being twice as hot as the second day. In fact, it is only 4 percent hotter!

Would the **Celsius (Centigrade) scale** be more suitable for our purposes, then? The Celsius scale, named in honor of its inventor, Anders Celsius (1701-1744), uses 0° instead of 32° for the freezing point of water, and 100° instead of 212° for the boiling point of water. There are 100 divisions between these two points instead of 180, hence the alternative name "Centigrade," meaning "100 divisions." While this scale gives us numbers that are easier to use for certain applications, there is still the same problem as before. We cannot compare things proportionately at two different temperatures. The reason, again, is that a large part of the scale (in this case, 273 degrees) lies below zero (see Figure 10:16).

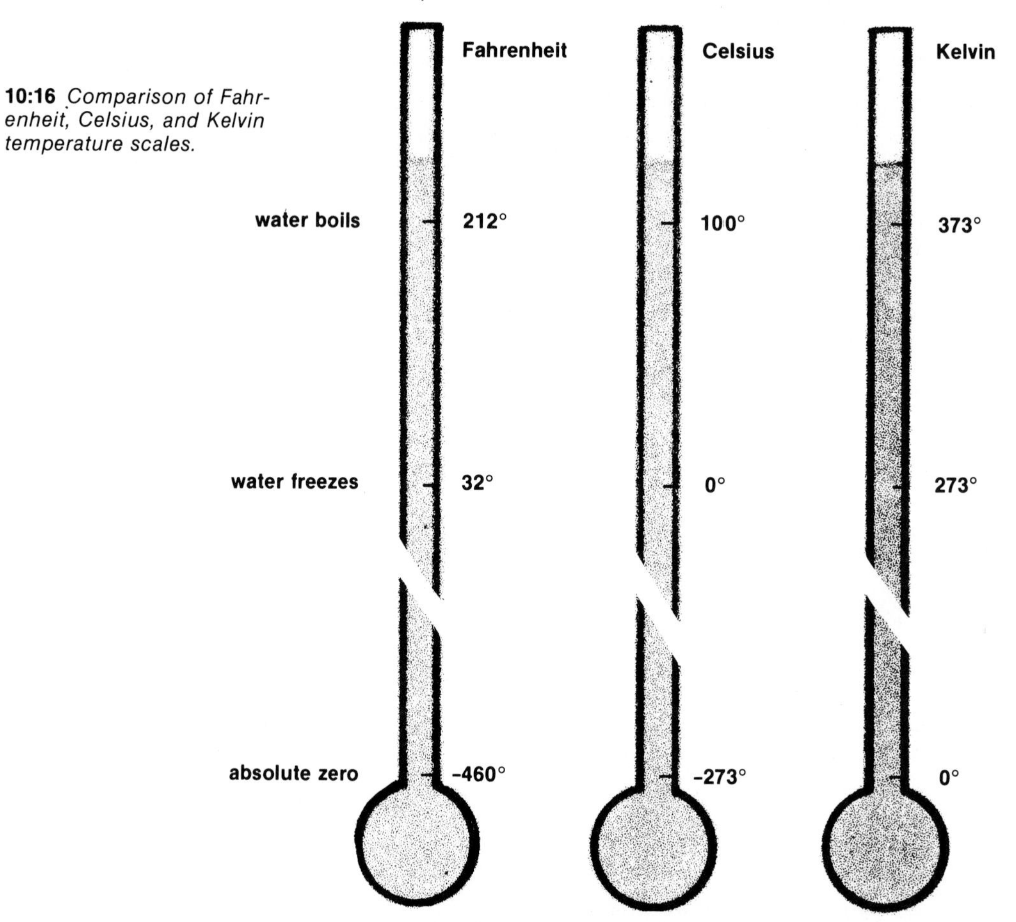

10:16 *Comparison of Fahrenheit, Celsius, and Kelvin temperature scales.*

In 1848 Lord Kelvin suggested that a satisfactory temperature scale for gas calculations could be devised by taking as the zero point the lowest temperature that is theoretically attainable—273 degrees below zero on the Celsius scale—and counting upward by Celsius degrees. This would eliminate the need for negative numbers and give us a scale where 40° really *is* twice as hot as 20°. This scale is called the **Kelvin (or Absolute) temperature scale**, and is abbreviated by °K. The comparison of the Fahrenheit, Celsius, and Kelvin scales is shown in Figure 10:16.

In gas problems involving temperatures we must be careful to change our readings to Kelvin degrees before calculating with them. To convert from Celsius to Kelvin is easy. Just add 273 to the number. Written as an equation,

$$K = C + 273$$

To change between Fahrenheit and Celsius requires a little more work. Since the zero points differ by 32 degrees and the size of a Fahrenheit degree is only 100/180 as large as a Celsius degree, we have to include these factors in our conversion equation. The 100/180 reduces to 5/9, giving us the relationship:

$$C = \frac{5}{9}(F - 32)$$

This equation is not difficult to use. It tells us simply that to change a reading from Fahrenheit to Celsius we subtract 32 from the Fahrenheit reading and multiply the answer thus obtained by 5/9.

To change from Celsius to Fahrenheit it is most convenient to use a different form of the relationship:

$$F = 1.8C + 32$$

Simply multiply the Celsius reading by 1.8 and add 32.

Here are three sample problems:

Problem: What Kelvin temperature corresponds to 53° C?
Solution:

$K = C + 273$
$K = 53 + 273$
$K = 326°$

Problem: The temperature in a certain pressure cooker is 113° F. Find the corresponding Celsius temperature.
Solution:

$C = 5/9\ (F - 32)$
$C = 5/9\ (113 - 32)$
$C = 5/9\ (81)$
$C = 45°$

Problem: The temperature of a certain room is 20° Celsius. What would be the reading on a Fahrenheit thermometer placed in the room?
Solution:

$F = 1.8C + 32$
$F = (1.8)\ (20) + 32$
$F = 36 + 32$
$F = 68°$

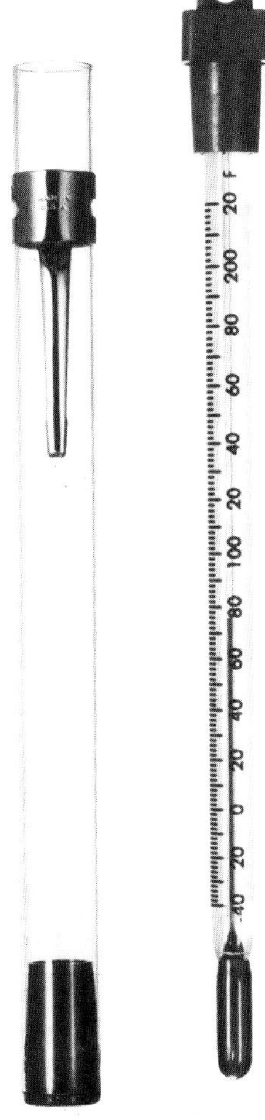

10:17 *Pocket thermometer calibrated in Fahrenheit degrees.*

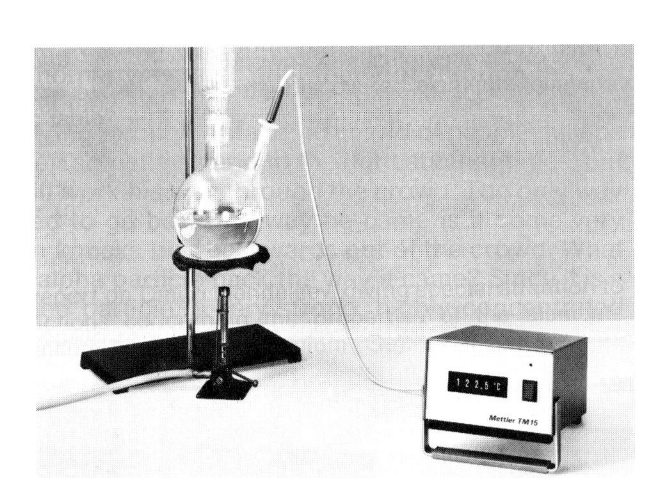

10:18 *Thermometer with digital readout. A temperature-sensing probe in the reaction vessel is electrically connected to the scaler at the right.*

10:7 Charles' Law Jacques Charles (1746-1823) and Joseph Gay-Lussac (1778-1850) independently studied the effect of temperature on the volume of a gas. The conclusion reached by both these men was that all gases expand at about the same rate—a gas at 0° C increases 1/273 of its volume for each Celsius degree increase in temperature. This information is summarized in the graph of Figure 10:19. Note that the pressure must remain constant for this relationship to be true.

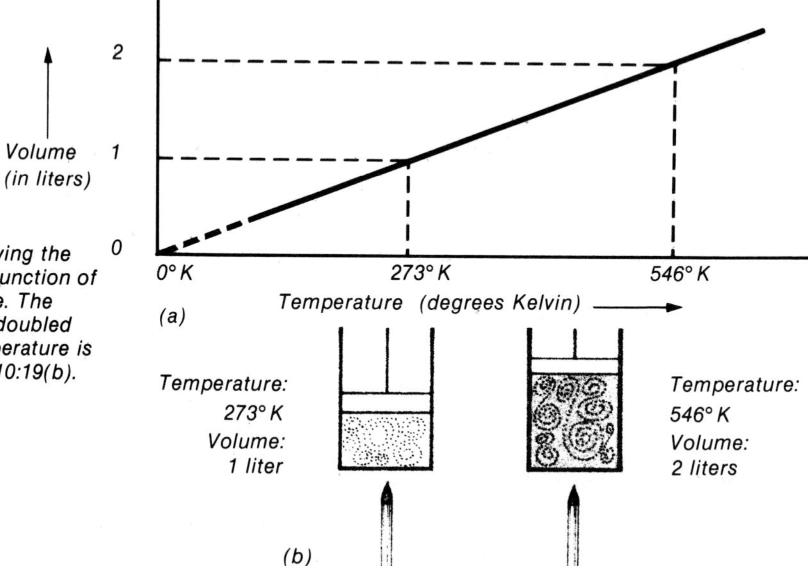

10:19 *(a) Graph showing the volume of a gas as a function of its Kelvin temperature. The volume of the gas is doubled when the Kelvin temperature is doubled. See Figure 10:19(b).*

Historians of science tell us that Charles arrived at this principle as early as 1787, but failed to publish his results. Gay-Lussac, who published his findings, did not carry out his definitive research until 1802. Contrary to the usual procedure in science, we call it Charles' law today, even though Charles never published his findings.

Now that we know the Kelvin temperature scale, we can restate this law in a simpler way. Studying Figure 10:19, we can see that the volume is twice as great at 546° K as it is at 273° K. Doubling the Kelvin temperature doubles the volume. The volume is therefore directly proportional to the absolute temperature. This can be expressed mathematically with the proportionality symbol,

$$V \propto T$$

or we can replace the symbol with an equal sign and a constant: (remember that the *pressure* must remain *constant* for the relationship to be true.)

$$V = CT$$

Dividing both sides by T we obtain

$$\frac{V}{T} = C$$

★ **10:8 Use of Charles' Law** Now we can use subscripts, as we did with Boyle's law, to distinguish between the volumes and temperatures before and after some change is made, V_1 and T_1 being the quantities *before* the change and V_2 and T_2 the corresponding quantities *after* the change. The initial conditions are described by

$$\frac{V_1}{T_1} = C$$

The final conditions are described by

$$\frac{V_2}{T_2} = C$$

Since things equal to the same thing are equal to each other, we can write

$$\frac{V_1}{T_1} = \frac{V_2}{T_2}$$

This is **Charles' law**. One way of stating this relationship in words is as follows: *The volume of a certain amount of dry gas varies directly with the Kelvin temperature, provided the pressure remains constant.*

It is not difficult to get correct answers from the above equation so long as you use Kelvin degrees for the temperatures. You will probably forget to do this the first time you attempt a Charles' law problem. Most people make this mistake at least once or twice before "catching on." Let us try a sample problem:

Problem: What volume will 50 cm^3 of a gas at 0°C occupy when heated to 27°C at constant pressure?

Solution:

$V_1 = 50\ cm^3$
$T_1 = 0°C$
$V_2 = ?$
$T_2 = 27°C$

Changing the temperatures to Kelvin degrees, we obtain

$T_1 = 273°K$
$T_2 = 300°K$

We now solve Charles' law for V_2 and substitute the numbers.

$$\frac{V_1}{T_1} = \frac{V_2}{T_2}$$

$$V_2 = \frac{V_1T_2}{T_1} = \frac{(50\ cm^3)\ (300°K)}{273°K} = 54.9\ cm^3$$

As in the case of Boyle's law, Charles' law does not describe perfectly what is happening. When we cool a gas to extremely low temperatures, its contraction does not follow a straight-line relationship. Charles' law predicts that the volume of a gas should keep decreasing as it is cooled, until finally the gas *vanishes* at **absolute zero**! Common sense tells us that something is wrong with that prediction. What actually happens is that the gas turns into a liquid before the lowest attainable temperatures are reached. This happens at different temperatures for different gases, and Charles' law in no way predicts what this temperature will be for any of them. But for the more commonly encountered temperatures Charles' law works very well.

★ **10:9 The Combined Gas Law** Suppose that the temperature and pressure of an enclosed gas *both* change. Can we still calculate the volume at the end of these changes? The answer is "yes," but there is more work involved.

Boyle's law and Charles' law can be combined into a single relationship:

$$\frac{P_1V_1}{T_1} = \frac{P_2V_2}{T_2}$$

This is known as the **combined gas law**. If any five quantities in it are known, the sixth can be calculated. Some students prefer to learn this one relationship rather than the two separate laws. If a problem is encountered in which the temperature remains constant ($T_1 = T_2$), the temperatures cancel out and the equation reduces to Boyle's law. In a problem where the pressure remains constant ($P_1 = P_2$), the pressures cancel, leaving Charles' law.

There is still another possibility. We may change the temperatures of a gas *at constant volume* and observe what happens to its pressure. In this case $V_1 = V_2$ and the equation reduces to $P_1/T_1 = P_2/T_2$, which is sometimes referred to as **Gay-Lussac's law**.

Here is an example in which both the pressure and the volume change:

Problem: 200 cm^3 of chlorine at 127°C and 800 mm pressure is cooled to 27°C as the pressure is reduced to 760 mm. Find the new volume.

Solution:

$V_1 = 200\ cm^3$
$T_1 = 127°C + 273 = 400°K$
$P_1 = 800\ mm$
$V_2 = ?$
$T_2 = 27°C + 273 = 300°K$
$P_2 = 760\ mm$

Solve the combined gas law for V_2 and substitute the numbers.

$$\frac{P_1V_1}{T_1} = \frac{P_2V_2}{T_2}$$

$$V_2 = \frac{P_1V_1T_2}{T_1P_2} = \frac{(800\ mm)\ (200\ cm^3)\ (300°K)}{(400°K)(760\ mm)}$$

$$V_2 = 158\ cm^3$$

The volume has decreased. The slight increase resulting from the reduction in pressure has been more than offset by the large temperature decrease.

In the above calculations some of the numbers can be cancelled rather than multiplying them all out. The slide rule is a good aid for this type of problem. When a slide rule is used, the consecutive operations can be carried out without writing down the partial products or quotients.

The combined gas law, like Boyle's law and Charles' law, has its imperfections. Deviations from the law become especially objectionable at high pressures and low temperatures. However, here again the law has been retained because it yields reasonably good answers under ordinary conditions of temperature and pressure. An **ideal gas** would be one which follows Boyle's law and Charles' law exactly. No such gas exists in reality, but a chemist will often pretend that a gas is ideal to simplify his calculations.

ROBERT BOYLE (1627-1691)

Irish physicist

Robert Boyle, one of the most outstanding Christians in the field of science, was born in Munster, Ireland, in 1627, the fourteenth in a family of 15 children. His father, the Earl of Cork, was reputed to be the wealthiest man in Great Britain. A devout Christian, the Earl attributed his great prosperity to the goodness and providence of God. Undoubtedly his godly influence played an important role in shaping the thoughts of the young lad who was to become the leading chemist of the seventeenth century.

Boyle studied the Scriptures in their original languages and was intimately familiar with all the important theological writings of his day. Having been converted in his early teens, he dedicated his scientific endeavors to be a witness to God's creation and control of the universe. Writing on a variety of scientific and religious subjects, Boyle became a powerful force for reproving evil and combatting heresy. His influence was felt even after his death. In his will he provided funds for the "Boyle Lectures"—a series of eight sermons to be delivered each year to demonstrate that Christianity is, in fact, intellectually defensible and far more reasonable than the various philosophies that seek to discredit it.

Although we usually associate the name Boyle with Boyle's law, a physics concept, his major work was in the field of chemistry. He has been called by many scientific historians "the father of chemistry." It was Boyle who guided the great transition from alchemy to true chemistry. Before his time, men spoke of elixirs and essences. After his work had made its impact on the scientific world, men spoke in terms of elements and compounds. The "spagyrists" and "hermetic philosophers" gave way to the chemists of the eighteenth century—men like Antoine Lavoisier—who were able to launch out in new directions, unhindered by the errors of previous generations. But it took a man of courage to oppose the traditions of the alchemists, and Boyle's spiritual make-up afforded him that courage in generous measure.

Boyle was strictly orthodox in his Christian beliefs. He did his utmost to defend and uphold the great doctrines of Scripture in both word and deed. Concerning the Lord Jesus Christ, he wrote of "His passion, His death, His resurrection and ascension, and all those wonderful works He did during His stay upon earth, in order to confirm mankind in the belief of His being God as well as man." (L. T. More, *The Life and Works of the Honorable Robert Boyle*, Oxford University Press, 1944, page 171.) He was thoroughly intolerant of preachers who spiritualized or allegorized important truths of the Bible rather than accepting them at face value.

Throughout his life Boyle read the Bible each morning, in spite of illness, eye trouble, and other adverse circumstances. As a result of his faithfulness and his clear-cut testimony before his fellowmen, he was repeatedly offered the highest positions in the Anglican Church. Each time he refused, stating that a layman's testimony for the Christian faith could have even more impact than that of a clergyman.

During his later years Boyle became intensely interested in world-wide evangelism. A man of considerable means, he supported missionary endeavors in Ireland, Scotland, Wales, India, and North America. In addition, he commissioned translations of the four Gospels and the book of Acts into Turkish, Arabic, and Malayan, and an Arabic translation of *The Truth of the Christian Religion* by Hugo Grotius (1583-1645).

Boyle's greatest burden was for his fellow Irishmen. In spite of great opposition from the Irish clergy, he financed a new Irish translation of the entire Bible, so that the people themselves could have access to the Word of God. Thousands of these Bibles were distributed throughout the British Isles at Boyle's expense.

When Boyle died in 1691, the loss was keenly felt by the scientific community around the world. His contributions in both physics and chemistry had been substantial. He was an outstanding researcher who used his science to exalt the name of the Lord.

List of Terms

absolute temperature scale
★ absolute zero
aneroid barometer
barometer
Boyle's law
Celsius temperature scale
Centigrade temperature scale
★ Charles' law
★ combined gas law
constant
Fahrenheit temperature scale
force
★ Gay-Lussac's law
★ ideal gas
inverse proportion
Kelvin temperature scale
mercurial barometer
pressure
S.T.P.
★ subscript

Questions

1. A soft-drink bottle is fitted with a one-hole stopper through which extends a glass tube. Explain why liquid cannot be withdrawn from the bottle by suction with your lips.
2. You have undoubtedly noticed the effect of atmospheric pressure on your eardrums when changing altitude rapidly in an elevator or airplane. Explain why swallowing helps to relieve the discomfort on your eardrums.
3. Why do astronauts use a pressurized space suit when exploring on the surface of the moon?
4. Why is the term "a cubic foot of oxygen" unsatisfactory?
5. Explain why a helium-filled balloon expands as it rises to higher altitudes. Which law is involved? (Remember that the law does not constitute an *explanation*, only a *description* of what is happening.)

6. Dented table tennis balls can sometimes be restored to normal use by placing them in hot water. What principle is involved here?
7. What happens to the size of an air bubble as it rises to the surface of a lake? What principle is involved?
8. Why is it dangerous to incinerate aerosol (spray) cans?
9. (a) What is standard temperature? (b) What is standard pressure?
10. Would there be more air in an "empty" bottle on a day when the atmospheric pressure is high than when the atmospheric pressure is low?
11. Define force.
12. Define pressure.
13. Explain how a medicine dropper draws up water from a beaker.

★14. At constant volume, how is the pressure of a gas related to its absolute temperature?

★15. (a) What is an ideal gas? (b) Does such a thing exist?

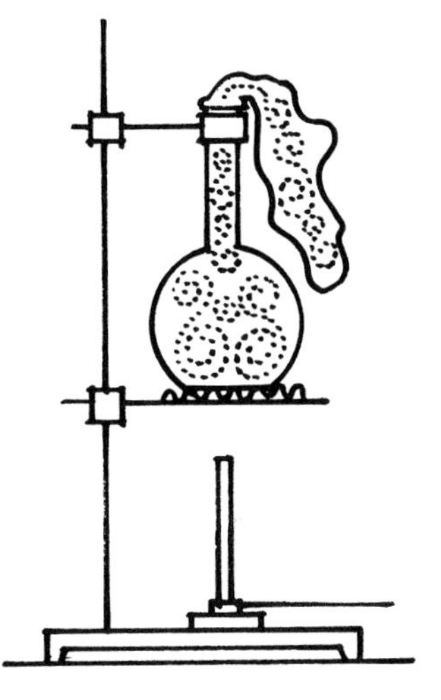

Problems

1. What is the total downward force in pounds due to atmospheric pressure on the top of a table 3 feet long and 2 feet wide?
2. Change to Kelvin: (a) 21°C (b) -78°C
3. Change to Celsius: (a) 500°K (b) 80°K
4. Change to Fahrenheit: (a) 15°C (b) 80°C
5. Change to Kelvin: (a) 54°F (b) -100°F
6. At what temperature are the Celsius and Fahrenheit readings the same?

★ 7. 700 cubic centimeters of ammonia gas are collected from a generator at a pressure of 1140 millimeters of mercury. What volume will the gas occupy when the pressure is 760 millimeters of mercury (normal atmospheric pressure)?

★ 8. What volume will 150 cm^3 of nitrogen occupy if it is initially at 27°C and is then heated to 527°C?

★ 9. A gas collected when the pressure is 900 mm has a volume of 425 cm^3. What volume will the gas occupy at standard pressure?

★10. A gas measures 500 cm^3 at a temperature of 25°C. Find its volume at 85°C.

11. What pressure in lb/in^2 would a barometer indicate at an altitude of $3\frac{1}{2}$ miles above sea level?

★12. One liter of hydrogen chloride gas is collected at 30°C and 900 mm pressure. What volume will the gas have at standard temperature and pressure? (Find to the nearest hundredth of a liter.)

Student Activities

1. Compute the weight of the air in your classroom by the following steps:
   a) Determine the volume of the room in cubic meters.
   b) Convert to liters.
   c) Find the mass of the air in grams, assuming a density of 1.2 grams per liter.
   d) Convert to pounds using 454 grams to the pound.
2. Place about 100 cm^3 of water in a one-gallon rectangular can and heat to boiling. When steam is issuing abundantly from the opening, discontinue heating, close the can with a tight-fitting stopper and place under cold running water. Explain the result.
3. Connect a balloon to a Florence flask as shown in Figure 10:20 and heat, taking care not to burn the rubber. What do you predict should happen? What do you observe? What principle is involved?

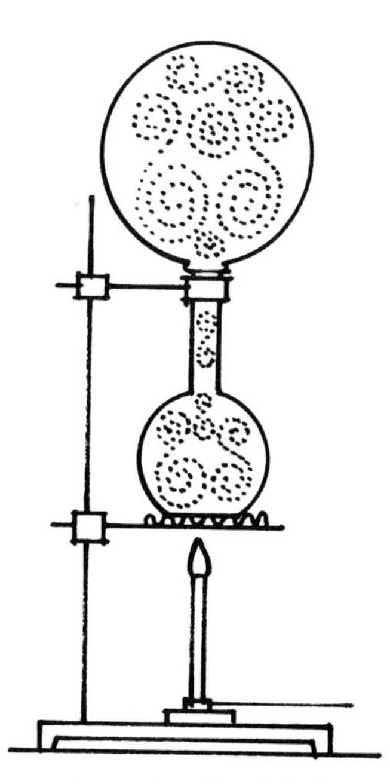

10:20 *Heating the air in the flask causes the balloon to expand.*

UNIT IV WATER AND ITS ELEMENTS

Chapter 11 Water

11:1 The Importance of Water Water is one of the many blessings that we are apt to take for granted. Water covers 71 percent of the earth's surface, in some places to a depth of more than six miles. There are vast quantities of water under the ground, in the atmosphere, and in the ice sheets of Antarctica and Greenland.

But when the entire universe is considered, water is a rare commodity. Our space probes have indicated that the rest of the solar system has very little water. Evolutionary theories of the origin of the solar system claim that all of the planets were formed by the same processes. But, in addition to their many other shortcomings, these theories cannot explain why, when all the other planets are barren wastelands, our earth has been so wondrously blessed with water.

11:1 *Space probes have failed to reveal liquid water on any other planet. Yet 71 percent of the earth's surface is covered with water.*

11:2 *Perhaps the most picturesque form of water in nature is the condensed moisture in the atmosphere. (See chapter title page)*

11:3 *A portion of the earth's water is in "cold storage" in glaciers. Coronation Glacier on Baffin Island is a typical valley glacier.*

Water is indispensable for life. All living organisms carry on their life processes in a watery environment. Our bodies require an almost continual supply of water, in spite of the fact that they are *already* about 5/7 water by weight. In addition, we use water for cooking, washing, irrigating crops, transportation, and recreational activities such as swimming and boating. To say that water is "important" would be a gross understatement.

Water has played a major role in the history of the earth. At the beginning of creation, before the dry land had appeared, the whole earth was covered with water (Genesis 1:2, 9-10). Some time later the entire world was again engulfed by water—the Flood of Noah's time recorded in Genesis chapters 6 through 9.

Many of the miracles recorded in the Bible had to do, either directly or indirectly, with water—the crossing of the Red Sea (Exodus 14:21-31), the crossing of the Jordan (Joshua 3:15-17), the ax head floating in water (II Kings 6:5-6), the curing of a leper in the Jordan River (II Kings 5:14), water turning to blood in Egypt (Exodus 7:20), and the Lord Jesus Christ's walking on water (Matthew 14:25), to name a few. Water is used symbolically to represent the Word of God (Ephesians 5:26) and the Holy Spirit (John 4:10-15, 7:38-39). Baptism in water is used to represent the death, burial, and resurrection of our Lord.

11:2 Physical Properties Pure water, at room temperature, is a transparent, odorless, tasteless liquid. Although water in small quantities is colorless, deep water is slightly blue because of its special way of absorbing sunlight. Water freezes at 0°C (32°F) and boils at 100°C (212°F) at standard pressure (760 mm of mercury).

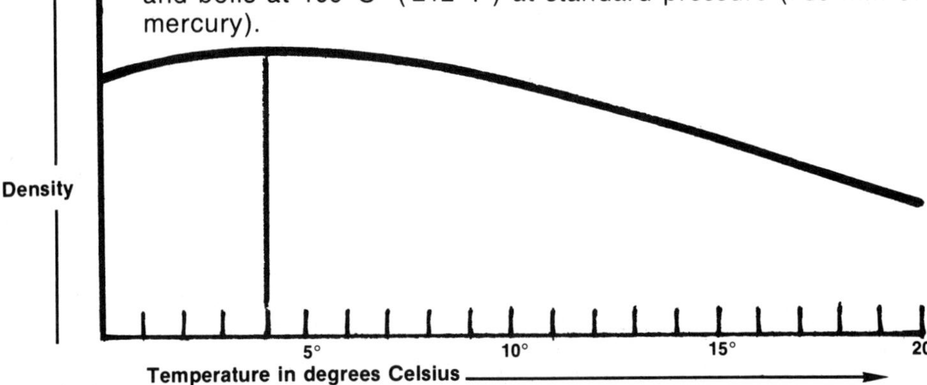

11:4 *Water exhibits its greatest density at 4° Celsius.*

You will recall that when we defined the gram as the mass of 1 cm^3 of water, we specified that the water had to be at 4°C. This is the temperature at which water exhibits its maximum density. If the temperature is hotter or colder than 4°C, the density of water decreases (see Figure 11:4). This is a unique property of water and an evidence of God's design in nature. Most liquids exhibit their greatest density at the freezing point. If this were the case with water, however, lakes and rivers would freeze from the bottom up in the winter and would destroy the plants and animals living in them.

When water freezes, it expands about 8 percent. The ice which is formed has a density of 0.92 g/cm^3. Again, this is evidence of God's design. Water is one of the very few substances which expand upon freezing. As a result, ice remains at the surface of a body of water rather than sinking to the bottom. Apparently, when water molecules form ice, the spacing between them increases.

Water heated to its boiling point turns to steam and, in so doing, expands about 1700 times at atmospheric pressure. True steam (**water vapor**) is invisible. You may think you have seen water vapor, but you were actually looking at droplets of liquid water as they condensed from the vapor. Although we cannot actually see it, we can detect the presence of water vapor in the air by means of a **hygrometer** (humidity meter), such as that shown in Figure 11:6.

You will recall that the *boiling point* is defined as the temperature at which the upward pressure of the liquid equals the downward pressure of the air above it. At reduced pressures, water can be boiled at temperatures far below 100°C. This procedure, called **vacuum distillation** or vacuum evaporation, is used in making syrup from sugar cane and in preparing evaporated milk. At increased pressures, water boils at temperatures above 100°C. A pressure cooker can easily cook food at 120°C, greatly reducing the time and fuel required.

Water can be purified by **distillation**. When the apparatus shown in Figure 11:8 is used, the impurities remain in the distilling flask while pure water collects in the receiving flask.

11:5 *Unlike most other liquids, water freezes from the top down.*

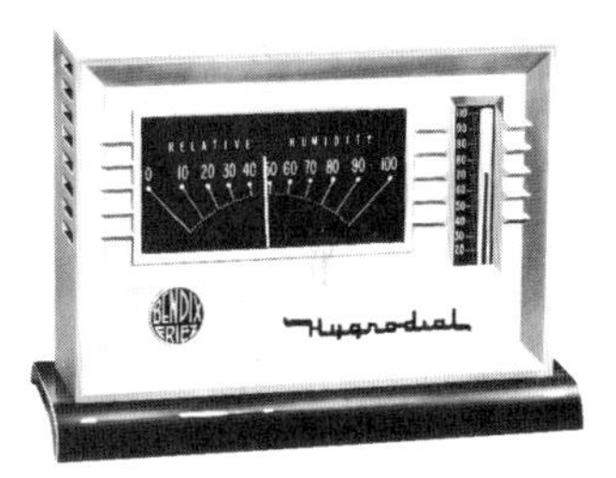

11:6 *Hair hygrometer (humidity meter) for home use.*

11:7 *Raising the pressure on a liquid increases its boiling point. The higher temperature of the water in a pressure cooker decreases the time required for cooking food.*

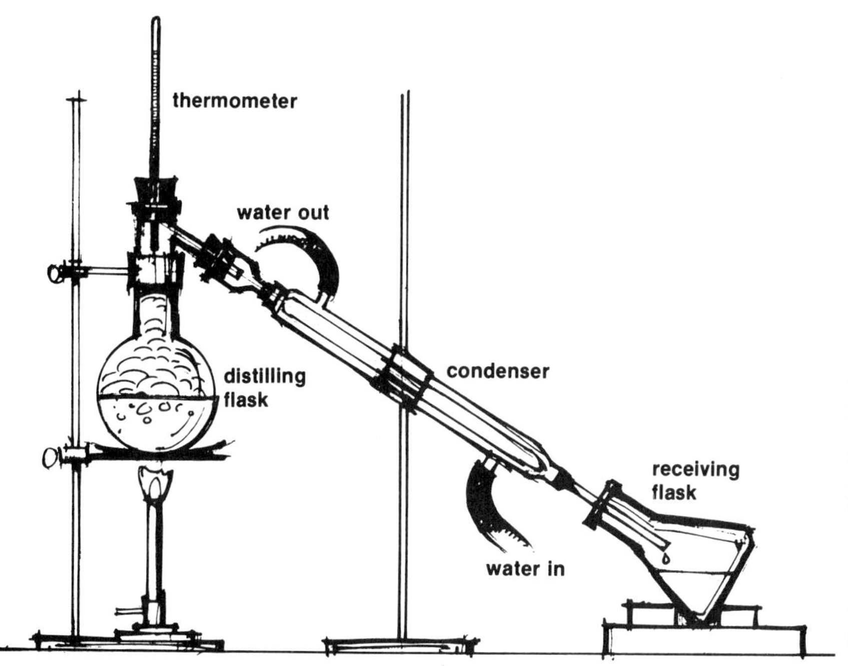

11:8 *Distillation apparatus—used to purify water from dissolved solids.*

Separating water from certain types of liquid and gaseous impurities is difficult. Often distillation is not the answer; some other method must be used.

Experiment 11:1 *Distillation*

1. Set up the apparatus shown in Figure 11:9 using a 250 ml or larger flask.
2. Place several crystals of potassium permanganate ($KMnO_4$) and 125 cm^3 of water in the flask.
3. Heat the flask until the test tube is half full of water. Is the water free from potassium permanganate?
4. Repeat the procedure using 100 cm^3 of water and 25 cm^3 of dilute sulfuric acid in the flask. (Caution: sulfuric acid, even when diluted, stains and disintegrates clothing. Be extremely careful not to spill any.)
5. Again, distill until the test tube is approximately half full.
6. Test the distillate with both red and blue litmus paper. Has the water in the test tube been completely separated from the acid?
7. Repeat the procedure using a mixture of 25 cm^3 of water and 25 cm^3 of denatured alcohol in the flask.
8. Distill until the test tube is half full.
9. Shut off the flame and open the distilling flask. Smell the liquid in the flask and the liquid in the test tube. Is either free from alcohol?

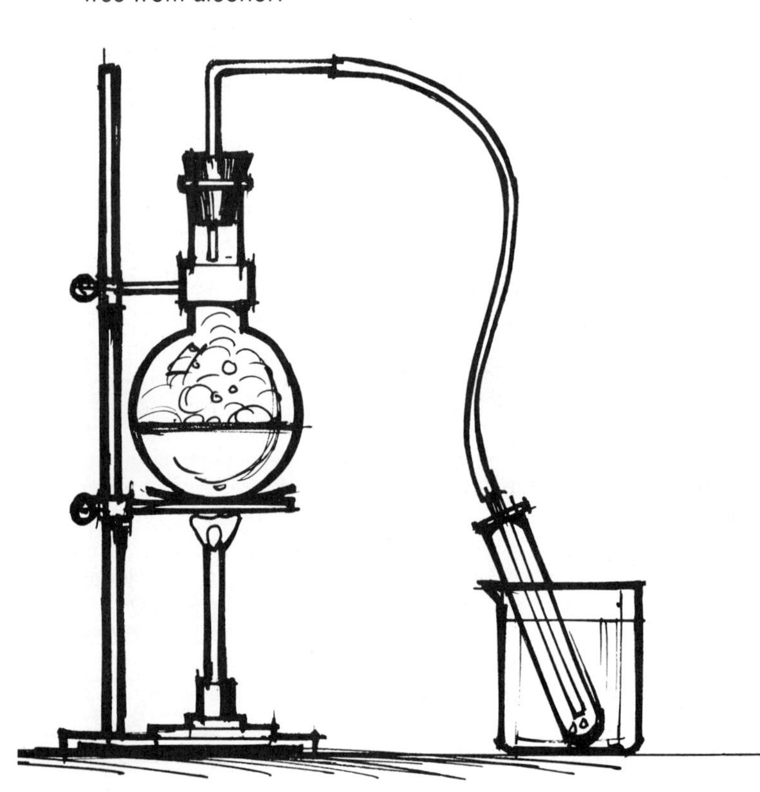

11:9 *Setup for Experiment 11:1.*

11:3 The Chemical Nature of Water Water is a compound of hydrogen and oxygen. A *compound*, as you know, is a chemical union of two or more elements. Its *formula*, H_2O, indicates that each water molecule consists of two atoms of hydrogen and one atom of oxygen.

Chemically, water is a very **stable** compound; the bonds that tie the atoms together in the molecule are so strong that a temperature of more than 2700° C is required to vibrate the atoms apart. Water can, however, be decomposed (separated into its elements) at room temperature by means of an electric current.

The angle between the two bonds in the water molecule has been determined to be about 105 degrees; so we can represent its structure very crudely as shown in Figure 11:10a. The bonds, though shown here as single lines, are actually *shared* electron pairs. They are *covalent bonds*. However, the electrons are not shared equally. The oxygen atom has a greater pull on the electrons; thus it acquires a partial negative charge. This leaves both hydrogen atoms somewhat positive. Thus, even though the bonds are covalent, this partial charge shows that they have some **ionic character**. If the electron pairs were shared equally, the bonds would have zero percent ionic character. If the oxygen atom completely dominated the electrons, keeping them away from the hydrogen atoms altogether, the bonds would have 100 percent ionic character. Figure 11:10b shows the eight outer electrons in the water molecule. The six electrons represented by *o's* were donated by the oxygen atom. The two designated by *x's* each came from the hydrogen atoms. Thus, a stable octet (see Section 7:2) is achieved.

The arrangement of charges in the water molecule can be represented as shown in Figure 11:10c.

Since it has a negative and two positive "poles," the water molecule is said to be **polar**. The polar nature of water makes it an excellent **solvent** for ionic substances, such as sodium chloride. Because of these positive and negative charges, water molecules interact with one another. Unlike charges attract from one molecule to the next. This attraction produces weak bonds called **hydrogen bonds** (11:10d). Two or more molecules can be tied together in this way. At lower temperatures there can be large groups of molecules extending in all directions. As the temperature is raised, more and more hydrogen bonds are broken. Finally, at the boiling point, the bonds are practically all broken, and the molecules go into the vapor phase separately.

Hydrogen bonds give water its unusual properties. For example, they play a key role in the formation of ice crystals (see Figure 11:11). Each oxygen atom in ice is connected to four hydrogen atoms—two by covalent bonds in its own molecule, and two by hydrogen bonds to hydrogen atoms in other molecules. This permits an open structure with "holes" between molecules, giving ice its characteristic low density. Also, if it were not for hydrogen bonds, water would not even be a liquid at room

11:10

(a)

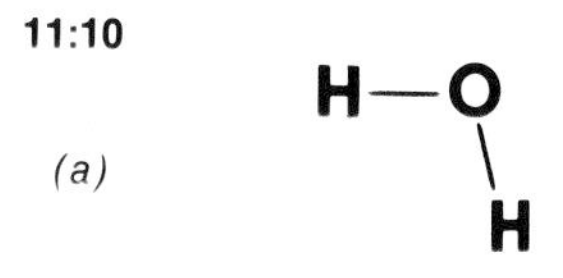

(b)

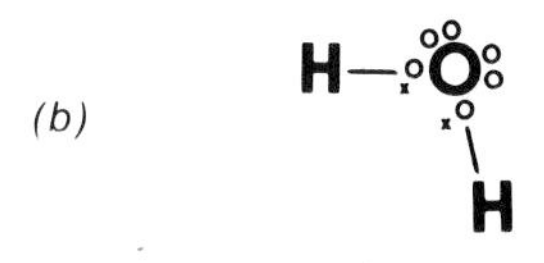

(c)

(d)

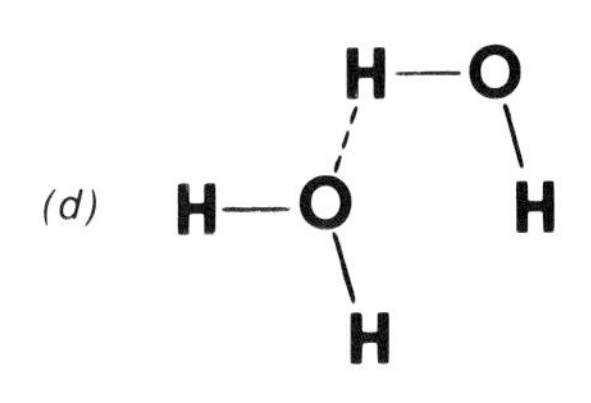

11:11 *Hydrogen bonds are thought to be responsible for the spaces between molecules in ice. The spaces give ice a lower density than liquid water. The black atoms are oxygen.*

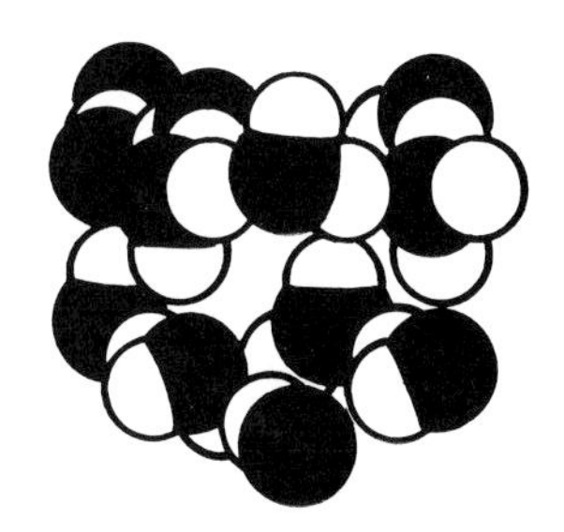

temperature. Other compounds having molecules of about the same weight, such as methane (CH_4) and ammonia (NH_3), are gases.

11:4 Reactions of Water with Metals Active metals, such as sodium, replace hydrogen from water at room temperature

$$2Na + 2H_2O \longrightarrow 2NaOH + H_2\uparrow$$

The NaOH solution produced is basic (turns red litmus blue) because of the hydroxide ion (OH^-). Similar reactions occur with calcium (Ca) and potassium (K).

$$Ca + 2H_2O \longrightarrow Ca(OH)_2 + H_2\uparrow$$
$$2K + 2H_2O \longrightarrow 2KOH + H_2\uparrow$$

Calcium (Ca) reacts more slowly than sodium (Na), whereas potassium (K) is more violent.

Magnesium (Mg) reacts only with *boiling* water.

$$Mg + 2H_2O \xrightarrow{\Delta} Mg(OH)_2 + H_2\uparrow$$

Red-hot iron reacts with steam to produce Fe_3O_4 (magnetic oxide of iron) and hydrogen.

11:5 Reactions of Water with Nonmetals Fluorine (F_2), chlorine (Cl_2), and bromine (Br_2) react with water at room temperature to form acids. Chlorine, for example, yields a mixture of hydrochloric (HCl) and hypochlorous (HClO) acids.

$$Cl_2 + H_2O \longrightarrow HCl + HClO$$

Hot carbon (coke) is reacted with steam to give a mixture of carbon monoxide (CO) and hydrogen.

$$C + H_2O \xrightarrow{\Delta} CO\uparrow + H_2\uparrow$$

The mixture of gases that is formed (called "water gas") has commercial importance as a fuel.

11:6 Reactions of Water with Metal Oxides Quicklime (CaO) reacts with water at ordinary temperatures to produce slaked lime ($Ca(OH)_2$).

$$CaO + H_2O \longrightarrow Ca(OH)_2$$

Oxides of sodium and potassium also form the corresponding bases:

$$Na_2O + H_2O \longrightarrow 2NaOH$$
$$K_2O + H_2O \longrightarrow 2KOH$$

In general, when a reaction does occur, a metal oxide can be expected to form a *base* when added to water. A metal oxide that will react with water to form a base is called a **basic anhydride**. Literally, this term means "a base without water."

11:7 Reactions with Nonmetal Oxides Carbon dioxide (CO_2) bubbled through water gives carbonic acid (H_2CO_3).

$$CO_2 + H_2O \longrightarrow H_2CO_3$$

Carbonic acid is a very weak acid and an unstable compound, which will readily break down into CO_2 and H_2O.

Stronger acids are formed with sulfur dioxide (SO_2) and phosphorous pentoxide (P_2O_5).

$$SO_2 + H_2O \longrightarrow H_2SO_3$$
$$P_2O_5 + 3H_2O \longrightarrow 2H_3PO_4$$

The first acid formed is called sulfurous acid (H_2SO_3); the second is phosphoric acid (H_3PO_4). Both are considered to be moderately strong acids.

A nonmetal oxide that will react with water to form an acid is called an **acid anhydride**, meaning "acid without water."

11:8 Water of Crystallization You will notice on the labels of some chemicals that water is included in the formula. For example, copper sulfate is usually sold in the form

$$CuSO_4 \cdot 5H_2O$$

The dot signifies that 5 parts of water are chemically united with 1 part of copper sulfate. This water comes from the original solution in which the crystals were formed. Water of crystallization (**water of hydration**) is not strongly attached to the compound and can be driven off by heat (see Figure 11:12). This dehydration process can be summarized by the equation

$$CuSO_4 \cdot 5H_2O \xrightarrow{\Delta} CuSO_4 + 5H_2O\uparrow$$

Some of the water vapor condenses near the mouth of the test tube. The copper sulfate loses its blue color when all the water has been driven off.

Experiment 11:2 *Water of Hydration*

1. Weigh a clean, large test tube to the nearest tenth of a gram.
2. Place about 5 grams of sodium sulfate or copper sulfate crystals in the test tube and weigh again.
3. Support the tube by a clamp and ringstand as shown in Figure 11:13. Heat gently for 5 minutes, then strongly for 10 minutes.
4. Cool, then wipe the outside of the test tube, and weigh. Calculate the percent of water in the original compound.

Data:

1. Weight of test tube .... ____g
2. Weight of test tube + hydrate .... ____g
3. Weight after heating .... ____g
4. Weight of hydrate (2. – 1.) .... ____g
5. Weight of water lost (2. – 3.) .... ____g
6. Percent of water ($\frac{5.}{4.} \times 100$) .... ____%

11:12 *Water of hydration being driven off from crystals of copper sulfate.*

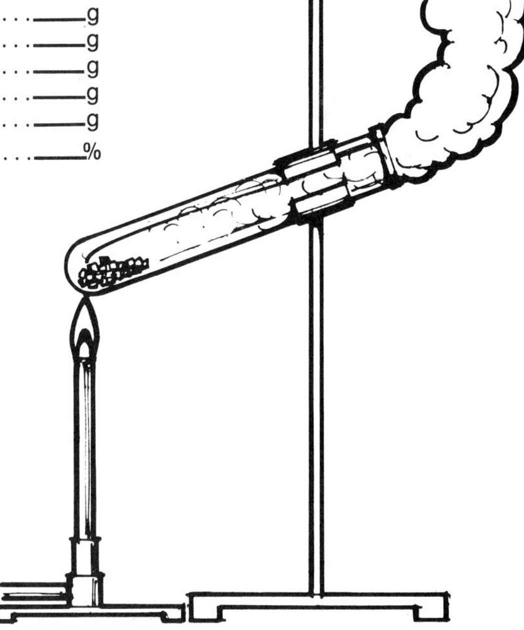

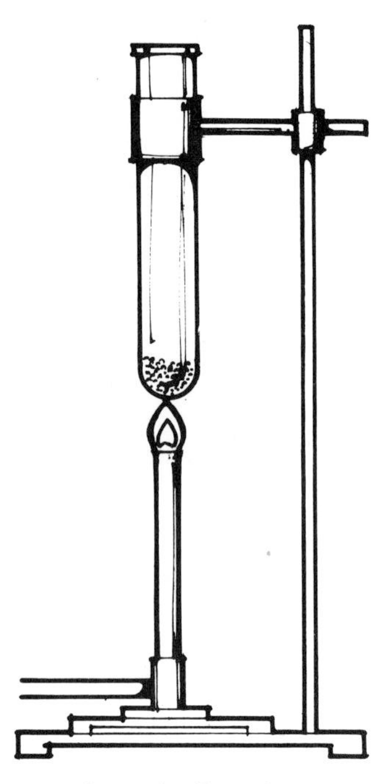

11:13 *Setup for Experiment 11:2.*

11:14 *Electrolysis of water. Water is decomposed into its elements by a low-voltage, direct current. Twice as much hydrogen (left) as oxygen (right) is produced, giving experimental evidence that the formula for water is H_2O.*

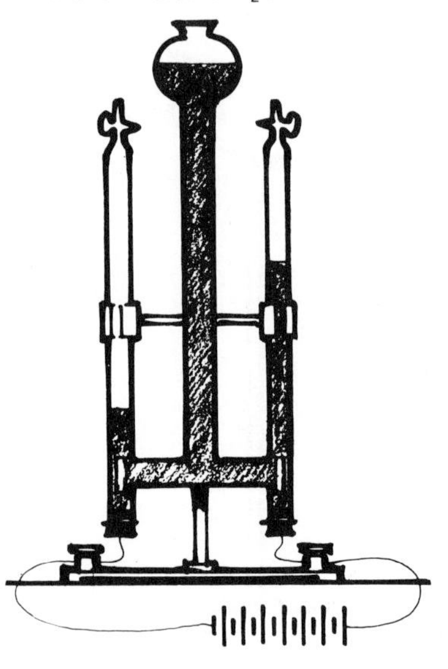

11:9 Decomposition of Water Water can be decomposed (separated into its elements) by means of an electric current. This process is called **electrolysis** (see Figure 11:14). A small quantity of sulfuric acid is added to the water so that the solution will conduct an electrical current. When a low voltage direct current source is connected to the terminals, bubbles can be seen to rise on either side of the apparatus. Oxygen forms at the positive electrode (called the **anode**); hydrogen collects at the negative electrode (called the **cathode**). Two parts of hydrogen are produced for every one part of oxygen.

The gases formed can be readily identified. Hydrogen either burns or produces a slight explosion, depending on its purity. Oxygen causes a glowing wood splint to burst into flame. Water is formed by the reaction and can be seen condensing as droplets on the inner surface of the container.

11:10 Synthesis of Water Water can be formed from its elements by burning hydrogen in air or in pure oxygen.

$$2H_2 + O_2 \longrightarrow 2H_2O\uparrow$$

11:11 Water as a Catalyst Water furnishes a scene of action where ions or molecules can come together and react. When dry sodium carbonate and dry potassium hydrogen tartrate are mixed together on a watch glass, there is no reaction. But when a few drops of water are added, a neutralization reaction takes place and carbon dioxide is given off. A noticeable **effervescence** (bubbling) occurs in the water.

List of Terms

acid anhydride
anode
basic anhydride
cathode
distillation
effervescence
electrolysis
hydrogen bond
hygrometer
ionic character
polar
solvent
stable
synthesis
vacuum distillation
water of hydration
water vapor

Questions

1. Name five physical properties of water.
2. Under what conditions does one gram of water have a volume of exactly one cm^3?
3. How does the boiling point of water depend upon atmospheric pressure?
4. Which has the greater mass—a cm^3 of ice or a cm^3 of water?
5. What is a hydrogen bond?

6. How could pure water be obtained from a mixture of sugar and water?
7. Name a chemical reaction that is promoted by the presence of water.
8. Which common metals react vigorously with water at room temperature?
9. Write chemical equations for the reactions in Question 8.
10. Is water a chemically stable compound?
11. Write the balanced equation for the burning of hydrogen in air.
12. What is the meaning of the dot in $CuSO_4 \cdot 5H_2O$?
13. Name four Biblical miracles that were connected with water.
14. What volume of steam would be formed by 10 cm^3 of water boiled at normal atmospheric pressure?
15. What valences do hydrogen and oxygen exhibit in water?
16. Which occupies more space—a gram of hot water or a gram of cold water? Why?
17. How much does water expand when it freezes? How do we explain the expansion on the basis of the crystalline structure of ice?
18. When water is decomposed to its elements (electrolyzed), which gas is produced in greater volume?
19. What percent of the earth's surface is covered with water? Where else does water occur in nature?
20. How is water used symbolically in Scripture?
21. What unique property of water is evidence of God's design? What would happen if water behaved like other compounds of comparable molecular weight?
22. How are the atoms thought to be arranged in the water molecule?
23. Write the complete balanced equation for the reaction between hot iron and steam.
24. When chlorine is bubbled into water, what change is observed in the pH of the water?
25. Give names and formulas for (a) three basic anhydrides (b) three acid anhydrides.
26. Write complete balanced equations to show how the anhydrides of Question 25 react with water.
27. What are the tests for the presence of oxygen and hydrogen?

Student Activities

1. Make a study of the Canopy Theory and report on it in class.
2. Find out from your local waterworks what substances are added to the water in your community. What is the purpose of each substance added?
3. Test the following substances to see which dissolve in water: sodium nitrate, calcium carbonate, calcium nitrate, benzene, denatured ethyl alcohol, isopropyl alcohol, iron filings, sucrose, glucose, carbon tetrachloride, lead shot, iodine, glacial acetic acid, glycerol, and any other substances your teacher suggests. In each case, use enough of the substance to fill the round part of a small test tube, add water until the test tube is half full, and shake the test tube vigorously, holding your thumb over its mouth. If the substance is water soluble, its quantity in the bottom of the test tube will become noticeably smaller. In general, like dissolves like: those substances which are similar enough to water in their makeup will be dissolved by water.

Photo next page
Oxygen is converted to ozone by lightning.

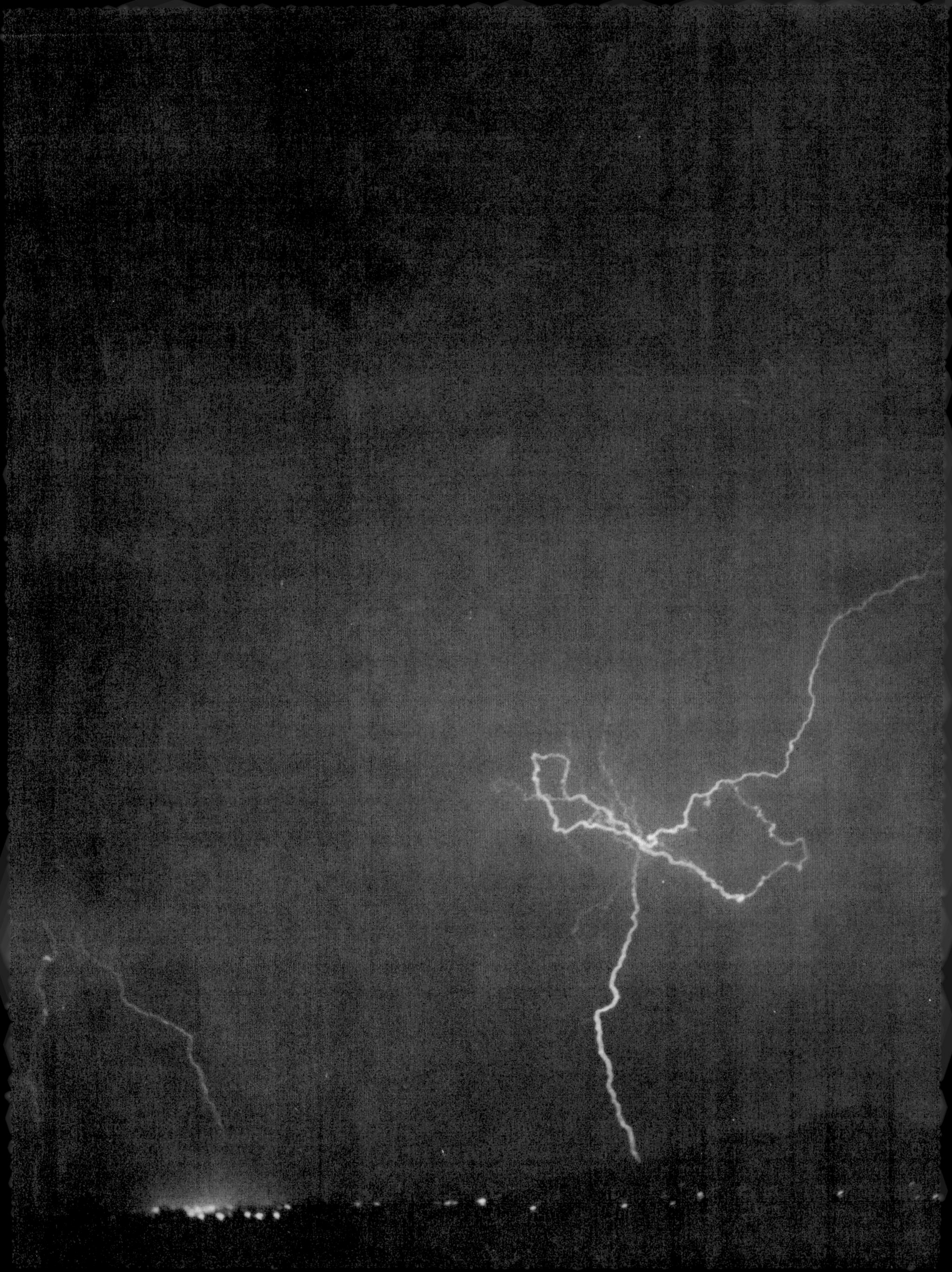

UNIT IV WATER AND ITS ELEMENTS

Chapter 12 Oxygen and Hydrogen

12:1 Occurrence of Oxygen Oxygen is an extremely abundant element; half the weight of the earth's crust is oxygen. Among the 105 known elements, oxygen ranks first in abundance in the earth and third in the universe, being surpassed only by hydrogen and helium. Oxygen is the active ingredient in the earth's atmosphere; it constitutes about 21 percent of the atmosphere's volume. Also, oxygen combines with other elements to form substances important to life, such as proteins, fats, and carbohydrates.

Evolutionary theories cannot tell us where, when, or how our original atmospheric oxygen was formed. The most popular theory holds that oxygen was generated by simple plants in the "primitive oceans." Unfortunately (for the theory), however, oxygen had to be present *before* the plants could exist. Without it there would not have been an **ozone layer** in the stratosphere to protect the plants from the harmful ultraviolet rays of the sun. Photosynthesis is a good means of maintaining the oxygen in the atmosphere, but it cannot account for the original formation of oxygen.

12:1 *Joseph Priestley, discoverer of oxygen.*

12:2 Discovery of Oxygen Oxygen was discovered by an English clergyman, Joseph Priestley (1733-1804), in 1774. Using a newly acquired 12-inch lens, Priestley focused the sun's rays on various substances to see which substances would decompose (break down into their elements). When he heated mercury(II) oxide (HgO) in a closed container, the oxide decomposed, giving mercury and oxygen. The equation for this reaction is

$$2HgO \longrightarrow 2Hg + O_2\uparrow$$

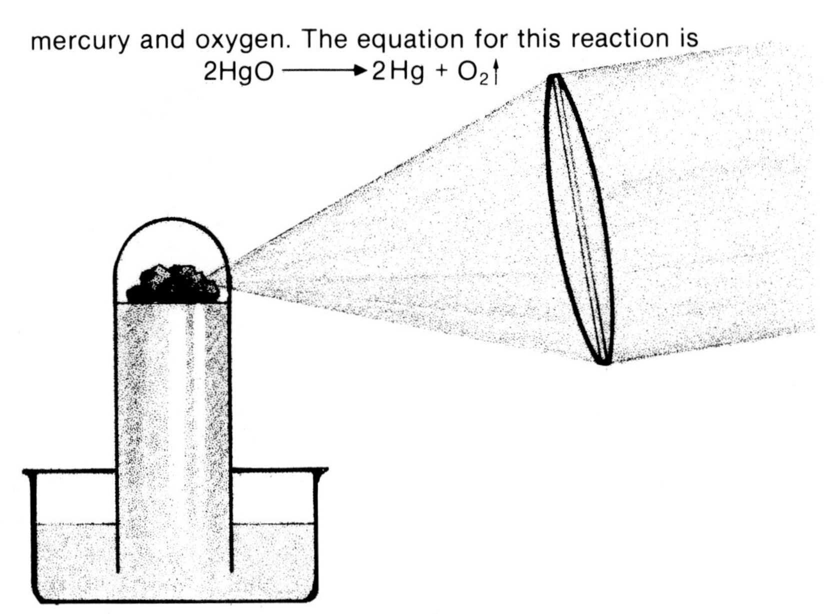

12:2 *Apparatus similar to that used by Priestley. Rays from the sun are focused on mercuric oxide in an enclosed container. Oxygen is formed as the oxide decomposes.*

Priestley found that combustible substances burned much more brilliantly in "perfect air," the term he used for oxygen. Mice placed in containers of oxygen became livelier, and Priestley himself felt invigorated when he breathed it. The Englishman did not recognize oxygen as a new element but considered it to be merely a very pure type of air. Like other scientists of his time, Priestley believed in the **phlogiston theory** (see Section 2:3), which explained combustion using the imaginary substance phlogiston. The real importance of oxygen was not appreciated until the work of Antoine Lavoisier became generally known at the end of the eighteenth century.

The liberal-minded Priestley attacked many of the doctrines of the Christian faith from the pulpit. Persecuted in England for his sympathy with the American Revolution, he emigrated to America in 1794, settling in Northumberland, Pennsylvania. His home and scientific apparatus have been preserved by Penn State University.

You can try the Priestley method of oxygen preparation by heating a small amount of mercury (II) oxide in a test tube. Mercury will collect on the sides of the test tube, and the oxygen can be detected escaping from the mouth of the tube. The simplest way to test for oxygen is to light a wood splint, blow it out, then hold the splint in an unknown gas while the splint is still glowing. If the splint bursts into flame again, the gas is assumed to be oxygen.

12:3 Laboratory Preparation of Oxygen Although oxygen can be prepared in many different ways in the laboratory, the most commonly used method is the **catalytic decomposition** of potassium chlorate ($KClO_3$). The apparatus often used is shown in Figure 12:3; the gas is collected over water. This method of collection, called the **water displacement method**, can be used for any gas that does not dissolve very much in water. Ammonia and hydrogen chloride are examples of gases that dissolve readily in water; they must be collected in some other way. Oxygen, however, is only slightly **soluble** (able to be dissolved) in water, and the water displacement method is quite useful.

Potassium chlorate, when heated to a high temperature, will decompose to produce oxygen, but you can get the same result at a lower temperature if you add manganese dioxide. Manganese dioxide serves as a **catalyst** or catalytic agent but does not actually enter into the reaction. *A catalyst is a substance that changes the speed of a chemical reaction.* In this case, manganese dioxide speeds up the reaction. In chemical equations, the catalyst is sometimes placed above the arrow to indicate that it is not one of the reactants or products:

$$2KClO_3 \xrightarrow{MnO_2} 2KCl + 3O_2\uparrow$$

Experiment 12:1 *Preparation of Oxygen*

1. Support a 1-inch diameter test tube at a 45-degree angle, as shown in Figure 12:3.
2. Mix thoroughly 6 grams of potassium chlorate ($KClO_3$) and 3 grams of manganese dioxide (MnO_2) on a piece of paper and transfer this mixture to the test tube.
3. Place a short length of glass tubing in a 1-hole #4 rubber stopper; then insert the stopper tightly into the test tube.
4. Connect a 3-foot rubber tube to the glass tube.
5. Fill a **pneumatic trough** with water and put the crosspiece in place. Fill a bottle with water and invert it on top of the crosspiece, taking care to keep the water from running out of the bottle.
6. Place the end of the rubber tubing through the hole in the crosspiece and about an inch into the bottle.
7. Fill 3 other bottles with water and keep them in the trough ready for use.
8. Light a Bunsen burner and heat the test tube, first gently, then more strongly. When the first bottle is full of oxygen, place a glass plate over the mouth of the bottle, remove it from the trough, and put the second bottle in its place. Place the bottle of oxygen right side up on the table, keeping the glass plate over it. Repeat the procedure until all 4 bottles are full of oxygen.
9. Remove the Bunsen burner from the test tube and remove the rubber tubing from the pneumatic trough. (Why should you do this?)
10. Suspend a small birthday-type candle from a wire wrapped around its base and light the candle; then lower it into the first bottle. Observe the increased rate of burning.
11. Holding a small piece of charcoal with a pair of forceps, ignite the charcoal with the burner and lower it into the second bottle of oxygen. Again note the increased rate of reaction.

12. Put a small piece of sulfur about the size of a pea into a deflagrating spoon, ignite in the burner flame, and lower into the third bottle of oxygen. You should be able to smell the pungent odor of sulfur dioxide. (The equation for the burning of sulfur is $S + O_2 \longrightarrow SO_2\uparrow$).
13. Add some water to the bottle, replace the cover, and shake it. Test the water with both red and blue litmus paper. Sulfurous acid (H_2SO_3) is formed by the reaction

$$SO_2 + H_2O \longrightarrow H_2SO_3$$

Sulfurous acid is a bleaching agent as well as an acid.
14. Wrap one end of a piece of magnesium ribbon 3 inches long around the end of the deflagrating spoon. Hold the other end of the ribbon in the flame until it starts to burn, and thrust it quickly into the fourth bottle of oxygen. Do not look directly at the burning magnesium ribbon; the light is much brighter than when it burns in air.
15. Clean out the test tube using detergent and a large test tube brush. If you do not clean it soon after the experiment, the test tube may be ruined.

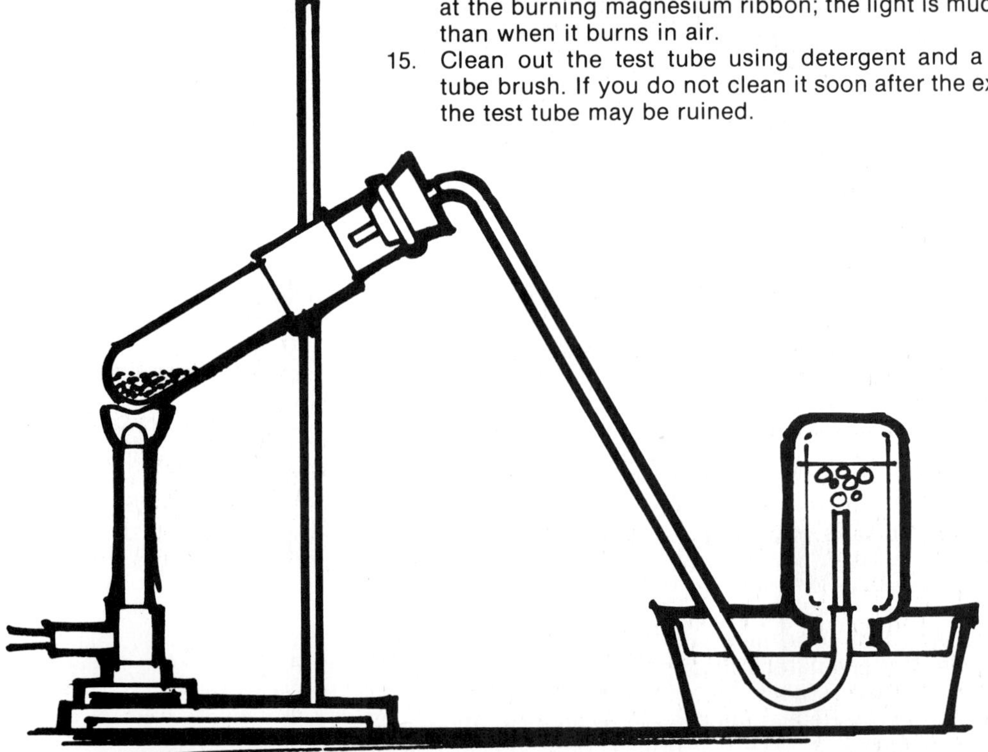

12:3 *Setup for Experiment 12:1 (laboratory preparation of oxygen).*

12:4 Commercial Preparation of Oxygen For large-scale production of the various elements, laboratory methods are almost always too costly. Less expensive methods can usually be developed using cheaper raw materials.

Oxygen can be prepared from the abundant substances, water and air. A very pure grade of oxygen is obtained by the electrolysis of water (see Chapter 11). Oxygen almost as pure (95 percent pure) is produced by the **fractional distillation** of liquid air. Air liquefies under pressure at about -200° C. If the liquid air is then warmed slightly, liquid nitrogen will boil off at about -196° C, leaving mostly liquid oxygen. The boiling point of liquid oxygen is −183° C.

12:4 *Air-separation complex used for the commercial preparation of oxygen.*

12:5 Physical Properties of Oxygen Oxygen is a colorless, odorless, tasteless gas, and is slightly heavier than air. As we have already mentioned, oxygen is only very sparingly soluble in water. Enough oxygen dissolves in water, however, so that fish and other aquatic creatures are able to obtain a sufficient supply from the rivers, lakes, and oceans they inhabit. Between −183° C and −218° C, at atmospheric pressure, oxygen exists as a pale blue liquid; below −218° C it is a pale blue crystalline solid (see Figure 12:5).

12:5 *Oxygen in the solid state exhibits magnetic properties. A chunk of frozen oxygen suspended on a string is attracted by a strong magnet.*

12:6 *Although divers must furnish their own oxygen, there is enough dissolved oxygen in water to support aquatic life.*

12:6 Chemical Properties of Oxygen Oxygen is one of the most active elements; it will form compounds with all the other elements except the rare gases. Oxygen's binary compounds (those containing only oxygen and one other element) are called **oxides**. Oxygen is the second most **electronegative** element, being surpassed only by fluorine. By electronegativity we mean the vigor with which an element accepts electrons from other elements as compounds are formed.

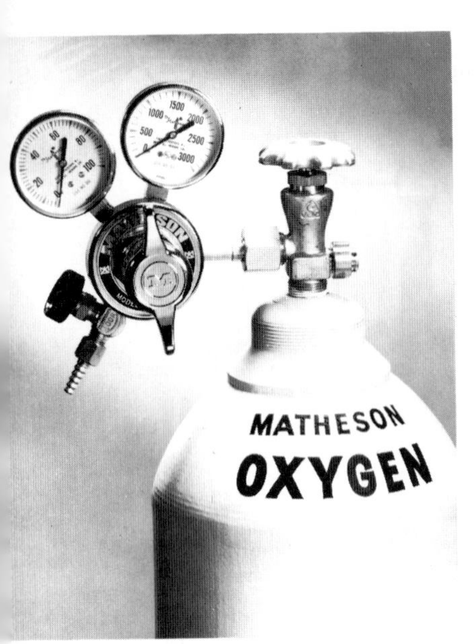

12:7 *Tanks of compressed oxygen are commonly used in chemical laboratories.*

Active metals, such as calcium, often form oxides at room temperature.

$$2Ca + O_2 \longrightarrow 2CaO$$

Less active metals oxidize less vigorously. Iron reacts slowly at room temperature (if a slight amount of moisture is present) to form iron (III) oxide (Fe_2O_3), commonly called iron rust.

$$4Fe + 3O_2 \longrightarrow 2Fe_2O_3$$

This reaction is usually undesirable, and various methods are used to combat rust formation. Rust **inhibitor**, a negative catalyst, is used in automobile radiators to retard the formation of iron rust.

Nonmetals usually require high temperatures before they react. Sulfur, phosphorus, carbon, and hydrogen burn in oxygen to give these corresponding oxides:

$$S + O_2 \longrightarrow SO_2\uparrow \text{ (sulfur dioxide)}$$
$$4P + 5O_2 \longrightarrow P_4O_{10}\uparrow \text{ (phosphorus pentoxide)}$$
$$C + O_2 \longrightarrow CO_2\uparrow \text{ (carbon dioxide)}$$
$$2H_2 + O_2 \longrightarrow 2H_2O\uparrow \text{ (water vapor)}$$

The French chemist, Antoine Lavoisier (1743-1794), unraveled the riddle of combustion in experiments conducted between 1772 and 1777. He found that the combustion (burning) of a substance did not involve a *loss* of material, as the old phlogiston theory had held, but rather the *gain* of something. That something was none other than Priestley's "perfect air." In 1777 Lavoisier gave this "perfect air" the name "oxygine." This name, meaning "acid former," was based on the erroneous idea that oxygen was the vital ingredient that gave acids their acidity. We retain the name "oxygen" today, however, with a slightly altered spelling.

12:8 *Combustion involves the chemical combination of a fuel with oxygen. One method of fighting a fire is to deprive it of oxygen by smothering it with a cloud of carbon dioxide.*

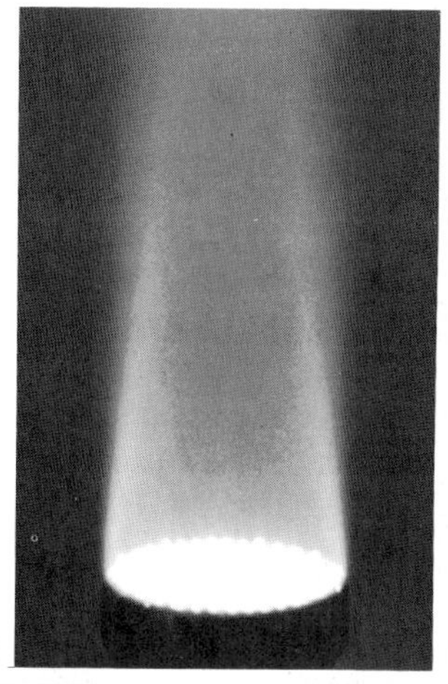

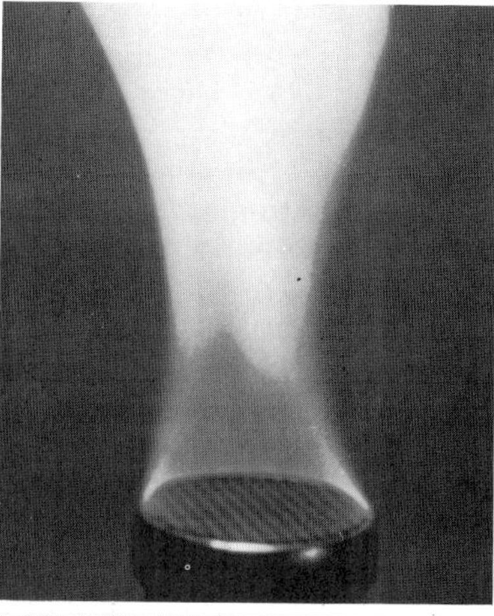

(a) (b)

12:9 *(a) and (b) Proper and improper adjustment of laboratory burner flame. Flame at right has insufficient air. Yellow color is caused by incandescent unburned carbon.*

12:10 *Two different gases are required for welding—one, a fuel such as acetylene, the other, oxygen.*

Although oxygen is usually involved when something burns, it is not the only gas that supports combustion. Hydrogen burns in chlorine, for example, to form hydrogen chloride (HCl). Magnesium burns in nitrogen to yield magnesium nitride (Mg_3N_2). And certain *compounds* of oxygen, such as laughing gas (N_2O), support combustion by decomposing into a mixture of oxygen (O_2) and nitrogen (N_2) as the reaction proceeds.

Oxygen usually exhibits a valence of -2 in its compounds. The oxygen atom has six electrons in its outer shell but needs eight for a stable octet. After it has acquired these two needed electrons, it will have two extra negative charges. Even though the oxygen atom may only be *sharing* these electrons, rather than "owning" them outright, for the sake of simplicity, we still call the

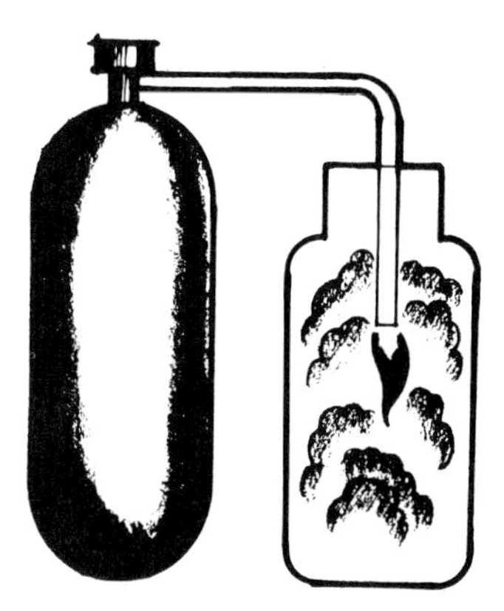

12:11 *Hydrogen burns in chlorine to form hydrogen chloride.*

valence or oxidation number "–2."

Oxygen atoms pair up with their own kind to form oxygen molecules when more reactive elements are not available. This helps to satisfy their "craving" for electrons. All common oxygen gas, whether obtained from chemicals or from the air, is in the form of O_2. **Atomic oxygen** (written [O]), on the other hand, is so active that it cannot be stored. It must be produced "on the spot" whenever it is needed. Oxygen is one of seven common elements that form **diatomic** (two-atom) molecules at room temperature. You should memorize this list and keep it in mind when writing equations:

O_2	oxygen
H_2	hydrogen
N_2	nitrogen
F_2	fluorine
Cl_2	chlorine
Br_2	bromine
I_2	iodine

12:7 Ozone Ozone, another form of the element oxygen, consists of **triatomic** (three-atom) molecules. Ozone is an allotropic form of oxygen. (By **allotropes** or *allotropic* forms we mean two or more varieties of the same element differing in the number of atoms contained in each molecule.) Ozone can be made from regular oxygen by an electric spark; three molecules of O_2 (regular oxygen) yield two molecules of O_3 (ozone).

12:12 *Oxygen is converted to ozone by lightning.* *(See chapter title page)*

$$3O_2 \rightleftharpoons 2O_3$$

Ozone is unstable and tends to change back to ordinary oxygen, as indicated by the reverse arrow in the equation.

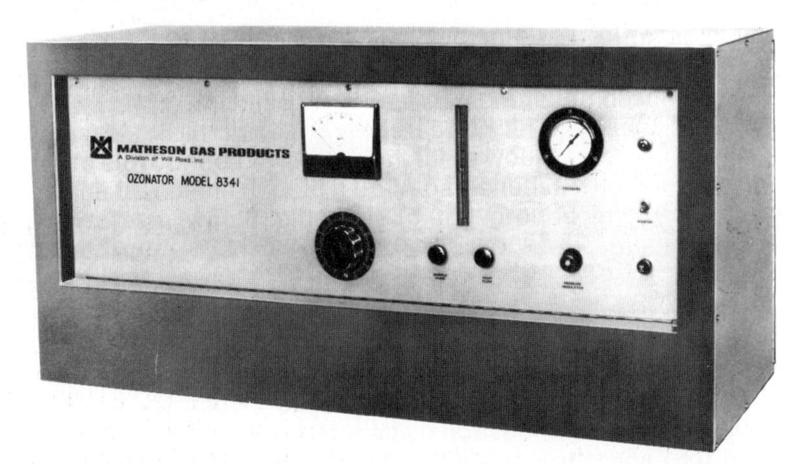

12:13 *Machine used to generate ozone for laboratory use.*

Ozone has a pungent odor. This odor can sometimes be detected in the air after a thunderstorm or in the vicinity of a static electricity generator. Ozone is 50 percent more dense than regular oxygen because of the additional atom in each molecule. Because of its chemical instability, ozone is a more powerful oxidizing agent than ordinary oxygen. It is used for sterilizing foods, deodorizing air, and bleaching fabrics.

It is the ozone layer in the stratosphere that protects us from the ultraviolet rays of the sun. Again, this is clear evidence of God's wise design of the earth and its near surroundings. Although the total quantity of ozone in this layer is small, it effectively absorbs just those wavelengths which would be most damaging to living organisms. The absorbed energy warms the air from approximately 6 miles above the earth to 50 miles up; the greatest heating occurs at about 30 miles above the earth. According to information from our space probes, no other planet has such an ozone layer.

12:8 Occurrence of Hydrogen Hydrogen is the most abundant element in the universe. Although there is no way to make accurate measurements at every point in space, it has been estimated that more than 90 percent of all the atoms in existence are hydrogen atoms. But in the earth's crust, hydrogen ranks only tenth in order of abundance, and it constitutes less than one part per million of the earth's atmosphere by volume.

In many high schools and colleges, students are being taught that plants, animals, and people "evolved"—through many stages, over billions of years—from original clouds of hydrogen gas! Usually no attempt is made to explain where the hydrogen clouds came from. People who make such evolutionary claims are considered by some to be scholars. Yet it does not take great knowledge, either spiritual *or* scientific, to realize the errors in such speculations. Had our world developed in this way, the process would have had to violate the second law of thermodynamics repeatedly. (This will be discussed in Chapter 29.) Keep firmly in mind the fact that there can be no science without *observations*, and there was no scientist present to make observations of *anything* when the world began.

Rather than evolutionary improvement, what we actually observe are *degenerative* processes wherever we look in the universe. This is exactly opposite from the teaching that hydrogen evolved into plants, animals, and people!

12:9 Discovery of Hydrogen Hydrogen was discovered by the English scientist Henry Cavendish (1731-1810) in 1766. He prepared the gas by reacting a metal with an acid. Cavendish found that several different metals, such as zinc, iron, or tin, could be used, and that dilute sulfuric acid (H_2SO_4) and hydrochloric acid (HCl) worked equally well. Typical reactions are as follows:

$$Zn + H_2SO_4 \longrightarrow ZnSO_4 + H_2\uparrow$$
$$Zn + 2HCl \longrightarrow ZnCl_2 + H_2\uparrow$$

In each case hydrogen is replaced by the metal and released in the form of diatomic molecules.

12:14 *Henry Cavendish, discoverer of hydrogen.*

Although other experimenters had undoubtedly prepared hydrogen before Cavendish, the Englishman was the first to collect the gas and study its properties. He found hydrogen to be considerably lighter than air and highly combustible. He gave it the name "inflammable air." A few years later Lavoisier renamed the gas "hydrogen." Because this gas yields water when it burns, the Frenchman chose a name which means "water former."

12:10 Laboratory Preparation of Hydrogen Hydrogen is still prepared in laboratories by Cavendish's method, although the apparatus for preparing it has been changed somewhat. Zinc (Zn) and dilute hydrochloric acid (HCl) or sulfuric acid (H_2SO_4) react in a bottle fitted with a 2-hole rubber stopper (see Figure 12:15). One hole is used for a *thistle tube*, through which additional acid is added. The thistle tube also acts as a safety valve, in case pressure builds up inside the bottle. The other hole in the stopper accommodates a tube which conducts hydrogen gas to a pneumatic trough. Like oxygen, hydrogen is only slightly soluble in water and is collected by the water displacement method. The reaction takes place at room temperature. Caution: Never use heat and be very careful to keep flames away from the apparatus. Also, never light the gas which comes out of a hydrogen generator; even a small air impurity in the hydrogen can cause a violent explosion.

Experiment 12:2 *Preparation of Hydrogen*

CAUTION: At no time should the hydrogen coming from the generator be ignited directly. Ignite only the bottles as directed after they have been removed from the pneumatic trough.

1. Set up the apparatus as shown in Figure 12:15. Fill a bottle with water and invert it on the crosspiece of a pneumatic trough, taking care to keep the water from running out. Fill two other bottles with water and keep them in the trough ready to use. Place the end of a rubber tube (2-3 feet in length) through the hole in the crosspiece and about an inch into the bottle.
2. Place 5 grams of mossy zinc in the reaction bottle and replace the stopper tightly.
3. Pour dilute sulfuric acid (H_2SO_4) through the thistle tube until the level of the liquid in the bottle is above the end of the thistle tube. Hydrogen should now be forming in the generator and collecting in the first bottle.
4. When the first bottle is full, replace it with a second one.
5. Keeping the first bottle upside down, ignite it with a burning wood splint. The first bottle is a mixture of air and hydrogen; hence, there will be a slight explosion.
6. Remove the second bottle, keeping it upside down, and replace with the third bottle.
7. Ignite the second bottle. It should burn more quietly because it is purer.
8. Remove the third bottle and hold it upside down for five minutes, allowing the mouth of the bottle to remain open. Now attempt to ignite it. Due to the rapidity with which hydrogen diffuses, there will be little (if any) of the gas left in the bottle.
9. Bubble the hydrogen coming from the rubber tubing into about 10 ml of soap solution in an evaporating dish. If the concentration of the solution is correct, lighter-than-air bubbles will be produced and will rise toward the ceiling.

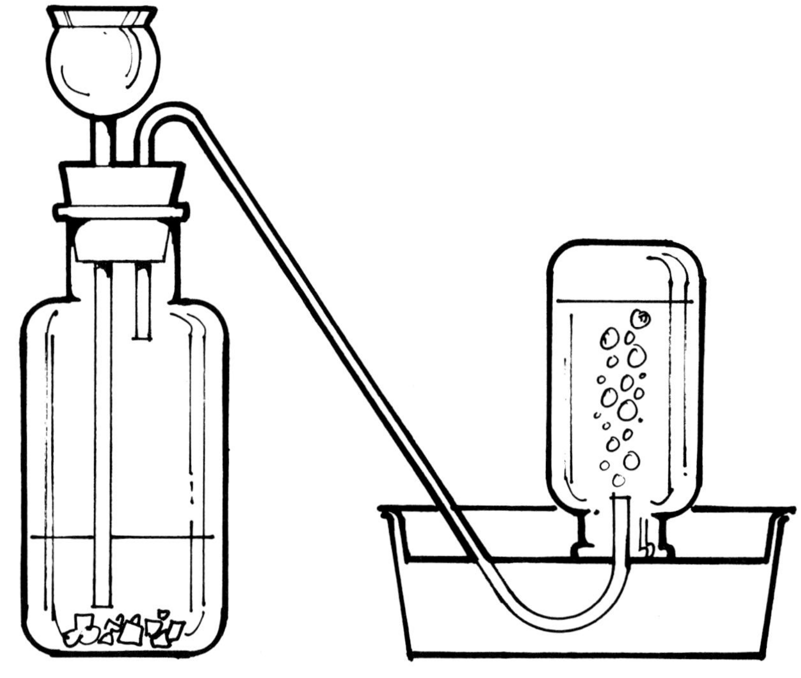

12:15 *Setup for Experiment 12:2 (laboratory preparation of hydrogen).*

Hydrogen may be prepared in smaller quantities by reacting an active metal with water. Calcium gives a reasonably rapid reaction and can be used to produce test tube quantities of hydrogen gas (see Figure 12:16). This is a *replacement* reaction as is the first method. In this case calcium replaces hydrogen in water.

$$Ca + 2H_2O \longrightarrow Ca(OH)_2 + H_2\uparrow$$

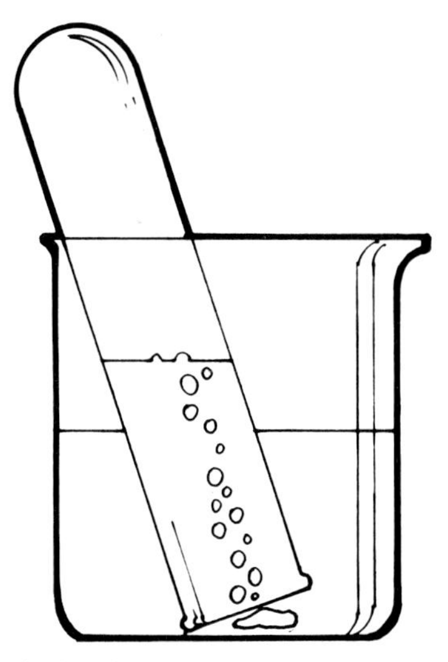

12:16 *Hydrogen can easily be made by reacting calcium with water and collecting the product in a test tube by water displacement.*

12:11 Commercial Preparation of Hydrogen Where cheap electricity is available, hydrogen may be produced by the electrolysis of water. The hydrogen forms at the *cathode* (negative electrode) as a pure product.

$$2H_2O \longrightarrow 2H_2\uparrow + O_2\uparrow$$

In another method, steam is passed over hot coke, forming carbon monoxide and hydrogen gases. (Coke, an impure form of carbon, is made by the destructive distillation of soft coal.)

$$C(coke) + H_2O \longrightarrow CO\uparrow + H_2\uparrow$$

When the mixture of carbon monoxide and hydrogen is compressed and cooled, the carbon monoxide (CO) liquefies, leaving hydrogen (H_2) of acceptable purity.

12:12 Physical Properties of Hydrogen Hydrogen is a colorless, odorless, tasteless gas whose density is only 7 percent that of air. It is the lightest of all gases. Helium, the second lightest gas, is approximately twice as dense as hydrogen.

12:17 *Although helium is used for blimps and dirigibles, the less expensive hydrogen gas is used in weather balloons.*

At atmospheric pressure, a temperature of −253°C is required to liquefy hydrogen. Six degrees lower, at −259°C, liquid hydrogen freezes to form transparent solid hydrogen, the least dense of all solids.

Because of their low mass, hydrogen molecules travel at exceedingly high speeds. The speed of a typical hydrogen molecule is 3.8 times as great as that of a typical nitrogen molecule in the air. As a general rule, *the lighter the molecule of a gas the higher its speed*. Hydrogen, therefore, diffuses more readily than any other gas.

12:13 Chemical Properties of Hydrogen Hydrogen is the simplest element. A hydrogen atom consists of a single electron orbiting around a single proton, according to the Bohr model. Chemical reactions involving hydrogen must be understood on the basis of this electron and the outer electrons of the element or elements with which hydrogen is uniting.

1. *Reactions with nonmetals:* As we have seen before, hydrogen burns in oxygen to form water vapor, and in chlorine to give hydrogen chloride (HCl) (see Figure 12:11).

$$2H_2 + O_2 \longrightarrow 2H_2O\uparrow \quad \text{(water)}$$

$$H_2 + Cl_2 \longrightarrow 2HCl\uparrow \quad \text{(hydrogen chloride)}$$

Hydrogen chloride is extremely soluble in water, combining with it to form hydrochloric acid. (Hydrochloric acid has the same formula as hydrogen chloride, HCl.)

Hydrogen forms binary (two-element) compounds with many other nonmetals. The accompanying table gives a sampling of these compounds:

HF	hydrogen fluoride
HBr	hydrogen bromide
HI	hydrogen iodide
H_2S	hydrogen sulfide
NH_3	ammonia
CH_4	methane
SiH_4	silane
PH_3	phosphine

The first three compounds listed are similar to hydrogen chloride. They are highly soluble gases that combine with water to form acids. Hydrogen sulfide is a combustible gas having the odor of rotten eggs. Ammonia is a lighter-than-air gas that combines with water to form the important base ammonium hydroxide (NH_4OH). Methane is one of the hundreds of binary compounds formed between carbon and hydrogen (see Section 15:1). Compounds of hydrogen and carbon are called **hydrocarbons**.

2. *Reactions with metals:* In the compounds just discussed, hydrogen donates its only electron to a nonmetal. The donation is either an outright gift or some kind of covalent bonding. There is another possibility, though. The hydrogen atom can *gain* an electron from an active metal, giving hydrogen *two* electrons. Since only two electrons are needed for a complete K shell, hydrogen can become chemically stable in this manner. Compounds formed between a metal and hydrogen are called ionic **hydrides**, a few of which are listed here:

LiH	lithium hydride
NaH	sodium hydride
KH	potassium hydride
MgH_2	magnesium hydride
CaH_2	calcium hydride
BaH_2	barium hydride

Hydrogen behaves like a metal when reacting with nonmetals, and like a nonmetal when reacting with metals.

12:18 *Lithium hydride, a typical ionic hydride, is handled by means of a special hooded machine. It must be kept in a dry nitrogen atmosphere because of its high reactivity.*

3. *Reducing action:* **Reduction** means lowering the valence of an atom or group of atoms by the *addition* of electrons. When hydrogen is passed over hot copper (II) oxide, a replacement reaction occurs in which the copper is replaced by hydrogen. Hydrogen is *above* copper in the activity series and is therefore capable of replacing it. The reaction is

$$CuO + H_2 \xrightarrow{\Delta} Cu + H_2O$$

The valence of the copper is lowered in this reaction from +2 in the compound to zero.

When something is reduced, something else must be oxidized. By **oxidation** we mean raising the valence by the *removal* of electrons. In the above reaction hydrogen has been oxidized from a valence of zero (in the free element) to +1 (in the water). Oxidation and reduction *always* occur together; one substance loses electrons while another substance acquires them. *The substance that is oxidized is called the* ***reducing agent****; the substance that is reduced is the* ***oxidizing agent****.*

4. *Acidity and the hydrogen ion:* We have defined an acid as a proton donor (Chapter 7). An acid, therefore, must contain hydrogen. An acid also must be able to make the hydrogen available as a hydrogen ion (H^+). In water solution, this positively charged hydrogen ion attaches itself to the negatively charged part of the water molecule to form a hydronium ion (H_3O^+).

$$H^+ + H_2O \longrightarrow H_3O^+$$

The greater the hydronium ion concentration in a solution, the lower the pH of that solution.

5. *The diatomic hydrogen molecule:* In the absence of something more reactive, hydrogen atoms pair together to form the hydrogen molecule, H_2. Each atom contributes one electron to the shared pair, a covalent bond. The molecule is sometimes represented as H:H or H-H. Because both nuclei are the same, the electrons are shared equally and the bond has zero percent ionic character. The negative electron cloud between the two protons serves to bind them together by the attraction of unlike charges.

12:14 Isotopes of Hydrogen **Isotopes** are different forms of an element; these forms differ only in the number of neutrons in the nucleus. Regular hydrogen has no neutrons, but it has two other isotopes—**deuterium** and **tritium**—which have one neutron and two neutrons, respectively. We have represented the three forms of hydrogen in a very crude way in Figure 12:19.

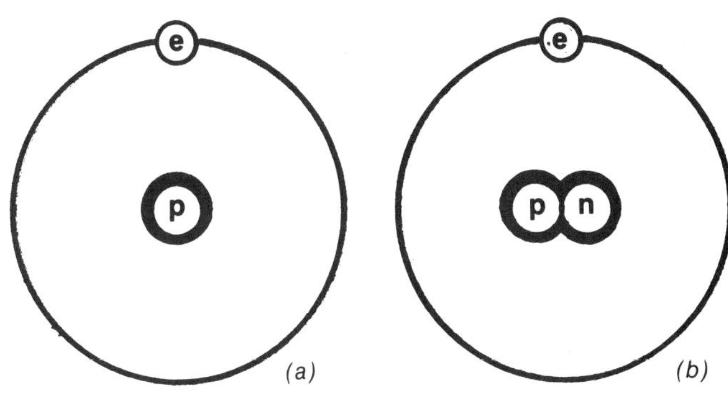

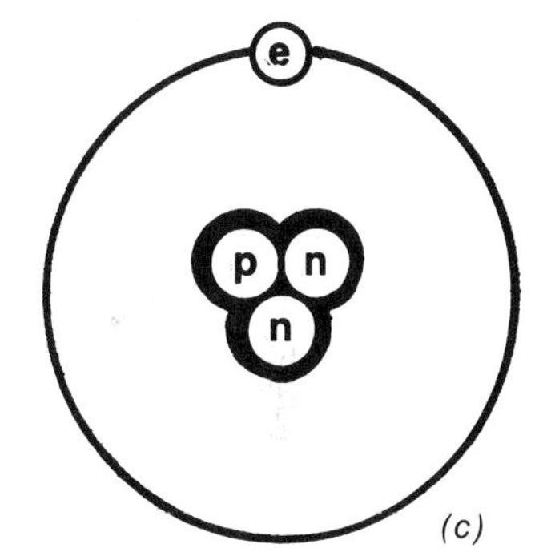

12:19 *The three isotopes of hydrogen: (a) regular hydrogen (b) deuterium (c) tritium. Protons, neutrons, and electrons are designated by "p," "n," and "e," respectively.*

Deuterium is a naturally occurring, stable isotope. It is present in nature to the extent of about one atom in every 6000 hydrogen atoms. Water which contains deuterium in place of hydrogen is called **heavy water**, represented by the formula D_2O. Heavy water is used as a moderator in nuclear reactors.

Tritium is much less abundant in nature than deuterium, but it can be produced artificially in nuclear reactors. Tritium has been used in the manufacture of luminous paints and in the hydrogen bomb. This isotope is unstable, having a half-life of about 12.5 years. By **half-life** we mean the length of time required for one half of an original number of atoms to decay into something else. After two half-lives, one quarter of the original number of atoms remain. After three half-lives one eighth of the original number of atoms remain. In this way the number of atoms of the unstable isotope is gradually reduced.

List of Terms

allotrope
atomic oxygen
catalyst
catalytic decomposition
deuterium
diatomic
electronegativity
fractional distillation
half-life
heavy water
hydride
hydrocarbon
inhibitor
isotope
oxidation
oxides
oxidizing agent
ozone
ozone layer
phlogiston
pneumatic trough
reducing agent
reduction
soluble
triatomic
tritium
water displacement method

Questions

1. Compare the relative abundance of oxygen and hydrogen in the earth's crust and in the universe.
2. (a) Name the discoverer of oxygen. (b) Name the discoverer of hydrogen.
3. Show by balanced equations how oxygen combines with (a) K, (b) H_2, (c) S, (d) P, (e) C, (f) Mg, (g) Ca.
4. Explain the difference between allotropes and isotopes.
5. Who was the first scientist to understand that burning involves a chemical union with oxygen? What theory was believed prior to that time?
6. What is a catalyst? What two kinds of catalysts exist?
7. What is the valence of chlorine in potassium chlorate?
8. Why are laboratory preparations usually different from commercial preparations?
9. What kind of gases can be collected by the water displacement method?
10. Is oxygen necessary for combustion?
11. Why does oxygen have a valence of -2 in most compounds?
12. Name the seven common diatomic molecules.
13. How can ozone be produced from ordinary oxygen?
14. Of what importance is the ozone layer in the stratosphere?
15. Give a scientific argument against the "hydrogen-to-man" theory of evolution.
16. (a) What metals will replace hydrogen from water? (b) Write an equation for the reaction of potassium (K) with water (H_2O). (c) Write an equation for the reaction of calcium (Ca) with water (H_2O).
17. What do all acids have in common?
18. When magnesium (Mg) burns in nitrogen (N_2), the compound magnesium nitride (Mg_3N_2) is formed. Give the complete balanced equation for this reaction.
19. How is the hydronium ion represented? How is it formed?
20. Give the equation for the reaction between ammonia gas (NH_3) and water (H_2O).
21. What type of compound is formed when oxygen combines with another element?
22. What type of compound is formed when hydrogen reacts with a metal?
23. When a fuel burns in air, what is the oxidizing agent?

24. In the reaction $Zn + S \longrightarrow ZnS$
    (a) Which element is oxidized?
    (b) Which element is reduced?
    (c) What is the oxidizing agent?
    (d) What is the reducing agent?
    (e) What is the name of the product?
    (f) What are the valences of Zn and S in ZnS?
25. Give three chemical properties of oxygen.
26. Give three physical properties of hydrogen.
27. What is the best gas for lighter-than-air craft? Why?
28. Give a word description of the equation $2H_2 + O_2 \longrightarrow 2H_2O\uparrow$.
29. What solid has the lowest density?
30. What type of compound is formed when hydrogen reacts with carbon?
31. (a) Under what circumstances does a bottle of hydrogen gas explode upon ignition? (b) When does a bottle of hydrogen gas burn quietly?
32. What salt is formed when sulfuric acid reacts with (a) zinc, (b) calcium, (c) magnesium, (d) iron, (e) nickel?
33. Can hydrogen support combustion? Explain.
34. Name the three isotopes of hydrogen. Which of the three isotopes are stable?
35. Could hydrogen be formed from sulfuric acid and copper? Explain. (See Appendix III.)
36. Could hydrogen be formed from hydrochloric acid and zinc?
37. Give the complete balanced equation for the decomposition of mercury (II) oxide (HgO).
38. Explain why atmospheric oxygen could not have developed by natural processes.

Student Activities

1. Prepare a report on the Hindenburg dirigible.
2. Show that oxygen can be prepared from hydrogen peroxide (H_2O_2) by adding a small quantity of manganese dioxide (MnO_2) to a test tube half full of hydrogen peroxide. Test the gas coming from the test tube with a glowing wood splint. The reaction is a catalytic decomposition
$$2H_2O_2 \xrightarrow{MnO_2} 2H_2O + O_2\uparrow$$
3. Write a report on Antoine Lavoisier's "twelve-day experiment."
4. Prepare a biographical sketch of Joseph Priestley or Henry Cavendish.
5. Priestley's preparation of oxygen may be duplicated by heating a small amount of mercury (II) oxide (HgO) in an open test tube. Use just enough of the oxide to fill the *round* part of the test tube. Test the gas with a glowing wood splint.
6. Hydrogen may be easily prepared by placing a piece of calcium (Ca) in a beaker full of water. The gas may be collected in a test tube as in Figure 12:16. Fill a test tube with water and invert it while holding your thumb over the mouth of the tube. Push the test tube full of water down into the water in the beaker until the tube is directly over the calcium. Allow the test tube to fill with hydrogen. Test the gas collected with a burning wood splint.

Further Reading

Faraday, Michael. *The Chemical History of a Candle.* New York: Collier Books, 1962. This is a collection of the famous Faraday Christmas lectures for young people.

Photo next page
The Princess Column stalactite in the Luray Caverns.

UNIT V DESCRIPTIVE CHEMISTRY
Groups IA-IVA

Chapter 13
The Active Metals

13:1 Electron Arrangement Looking at a periodic table, you will see that from left to right the first two vertical groups are metals. This first family is called the **alkali metals**. Listed below are the alkali metals and their symbols.

lithium	Li
sodium	Na
potassium	K
rubidium	Rb
cesium	Cs
francium	Fr

The second group is called the **alkaline-earth metals**. Listed below are the alkaline-earth metals and their symbols.

beryllium	Be
magnesium	Mg
calcium	Ca
strontium	Sr
barium	Ba
radium	Ra

Both groups are called the **active metals** because they are the most chemically reactive metals known. The alkali metals are the *more* reactive of the two groups.

According to current atomic models, the arrangement of the electrons (particularly the valence electrons) determines the chemical properties of an element. Elements with the same number of valence electrons and similarly arranged valence electrons should, therefore, react in the same manner. Table 13-1 is a chart of the electron arrangement of the atoms of the elements that make up the active metals.

Table 13-1 *Electron Arrangement in the Atoms of the Active Metals (Group IA and Group IIA Elements)*

Element	**Quantum Levels and Electron Orbitals**															
	1	2		3			4				5			6		7
	s	s	p	s	p	d	s	p	d	f	s	p	d	s	p	s
Alkali metals																
Li	2	1														
Na	2	2	6	1												
K	2	2	6	2	6		1									
Rb	2	2	6	2	6	10	2	6			1					
Cs	2	2	6	2	6	10	2	6	10		2	6		1		
Fr	2	2	6	2	6	10	2	6	10	14	2	6	10	2	6	1
Alkaline-earth metals																
Be	2	2														
Mg	2	2	6	2												
Ca	2	2	6	2	6		2									
Sr	2	2	6	2	6	10	2	6			2					
Ba	2	2	6	2	6	10	2	6	10		2	6		2		
Ra	2	2	6	2	6	10	2	6	10	14	2	6	10	2	6	2

13:2 Alkali Metals You will notice in Table 13-1 that all of the alkali metals have a single electron in the valence shell. These are all *s* orbital electrons. If any alkali metal can give away this electron in a chemical reaction, it will then have a filled outer shell of electrons, which is a very stable state. Each metal in this group can ionize by releasing one electron; in this way alkali metals are chemically similar.

$$Na \longrightarrow Na^+ + e^-$$
$$K \longrightarrow K^+ + e^-$$
$$Cs \longrightarrow Cs^+ + e^-$$

We have been using the term **metal**. What do we mean by this in the chemical sense? The characteristic chemical property of a metal is *its ability to lose or give away electrons*. The alkali metals are very good at giving away an electron in a chemical reaction. In fact, they are so good at giving away an electron that the free (uncombined) metals are never found in nature.

As the atomic weight of the elements in the group increases, the chemical activity also increases. The exceptions are lithium and francium (see Section 8:7). Lithium is the most reactive of the group. Generally, as the number of electrons in the atomic orbitals is increased, the *screening effect* causes the outermost electrons in the larger atoms to be very loosely bound. The activity of the group is as follows:

$$Li > Cs > Rb > K > Na$$

This means that lithium is more active than cesium, which is more active than rubidium, which is more active than potassium, which is more active than sodium.

Alkali metals are so reactive that they must be stored in a liquid such as kerosene, away from moisture and air.

Ions of alkali metals are found in sea water, and all of the compounds of these metals are very soluble. Many of the salts of alkali metals are found in crystalline form in many areas of the world. The metal salts were obviously deposited there by water; the Great Salt Lake is an example of this. These waters were probably trapped inland as the Flood waters began to recede from the face of the earth. Movements of the upper crust of the earth immediately after the Flood could have blocked the recession of the waters into the oceans and underground. The water evaporated, or is now evaporating, leaving large deposits of alkali metal salts. Also, compounds of sodium, potassium, and lithium are found in a great many minerals in the earth's crust.

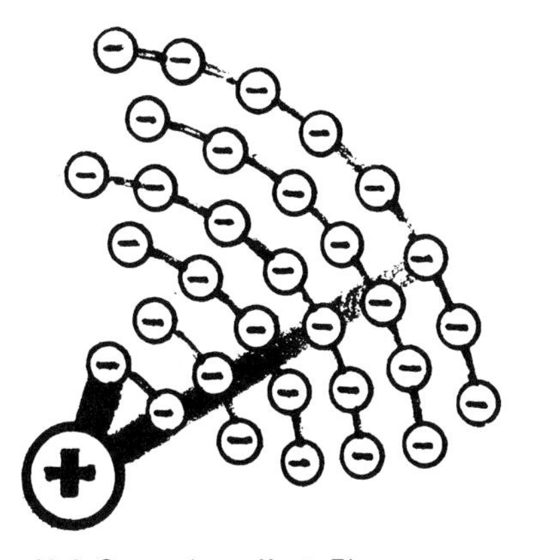

13:1 *Screening effect. Electrons closer to the nucleus are more tightly held. Electrons in outer orbits are screened from the nucleus by electrons.*

13:3 Physical Properties of Alkali Metals Alkali metals are very light (low density) and very soft. You can crush most of them, with the exception of lithium, with your fingers. They have low melting points and a bright metallic luster when freshly cut. Alkali metals are excellent conductors of electricity.

13:4 Reactions of Alkali Metals These metals readily react with oxygen in the air; alkali metals burn violently. When alkali metals react with oxygen, however, the normal oxides are not formed; **peroxides** are formed. A peroxide is a compound that contains more than the normal amount of oxygen. (The exception is lithium, which forms Li_2O.) The tendency to form a peroxide increases from sodium to cesium.

$$2Na + O_2 \longrightarrow Na_2O_2 \text{ (sodium peroxide)}$$

Alkali metals also react explosively with water to form bases and give off hydrogen gas. Normally, any metal will react with water to form a base.

$$\text{Metal} + \text{Water} \longrightarrow \text{Base} + \text{Hydrogen}$$
$$2Na + 2H_2O \longrightarrow 2NaOH + H_2\uparrow$$
$$2K + 2H_2O \longrightarrow 2KOH + H_2\uparrow$$

13:5 Flame Tests for Alkali Metals If salts of the alkali metals are placed on the end of a metal wire and held in a Bunsen burner flame, they will yield a characteristic color.

Salt Containing	*Flame Color*
sodium	yellow
potassium	violet
lithium	crimson
rubidium	dark red
cesium	sky blue

13:6 Compounds of Sodium The most commonly known sodium compound is table salt, sodium chloride (NaCl). Salt, necessary in the diet of humans and animals to help maintain blood pressure, is used as a food preservative and adds flavor to many foods. Salt is mentioned many times in the Bible. Christians are pictured as the spiritual "salt of the earth" (Matthew 5:13); as salt adds flavor to food, so Christians should add the proper spiritual flavor (savor) to a world that lies in the arms of the wicked one. We are God's salt in this world.

Sodium chloride can be obtained from deposits like the Great Salt Lake, evaporating sea water, or underground mines. Some natural salt crystals are very beautiful and have the characteristic sodium chloride cubic structure (see Figure 13:5).

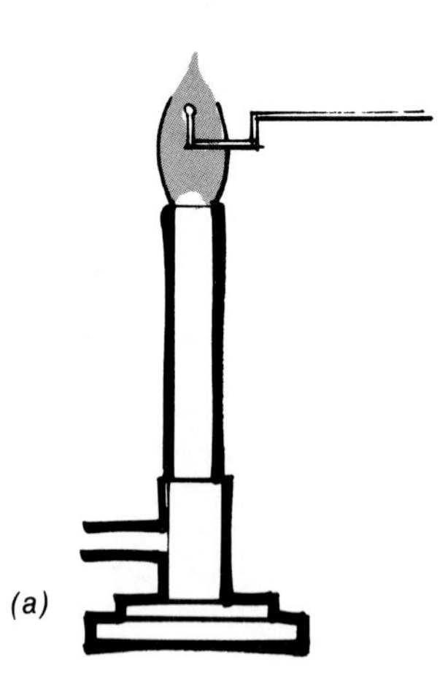

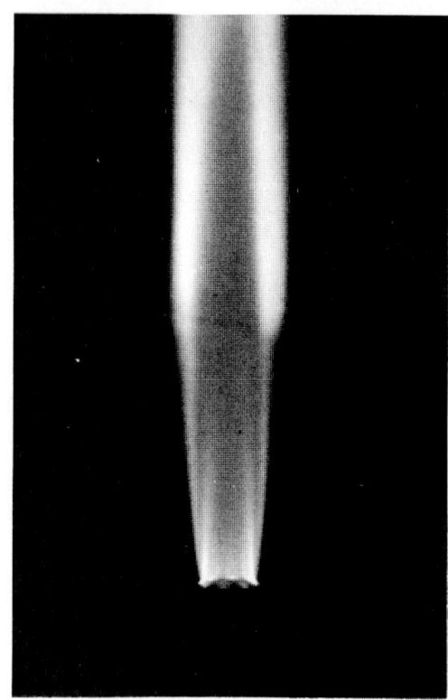

13:2 *(a) Sodium salt on a wire is placed in a Bunsen burner flame for flame test. (b) Sodium flame.*

13:3 *The Great Salt Lake as seen from Gunnison Island.*

13:4 *Underground salt mine, Grand Saline, Texas.*

13:5 *Magnified cubic crystals of evaporated table salt.*

Many sodium compounds are used in cleaning solutions. Sodium carbonate (Na_2CO_3) is used both in cleaning preparations and in the manufacture of glass. Sodium hydroxide (NaOH, also known as lye) is the major ingredient of commercial products used to unstop sink and bathtub drains, and it also plays an important part in the manufacture of soap. Borax ($Na_2B_4O_7 \cdot 10H_2O$) is another well-known sodium compound used for cleaning.

Sodium bicarbonate ($NaHCO_3$) is known as baking soda. In the presence of an acid, the bicarbonate ion (HCO_3^-) is converted to carbonic acid (H_2CO_3),

$$NaHCO_3 + H^+ \longrightarrow Na^+ + H_2CO_3$$

which immediately decomposes, giving off carbon dioxide (CO_2).

$$H_2CO_3 \longrightarrow CO_2\uparrow + H_2O$$

The formation of this carbon dioxide gas causes dough to rise (leavening) and makes bread and other baked goods porous and light. Indeed a little bit of leaven leavens the whole lump of dough (I Corinthians 5:6)! Sour milk, vinegar, or lemon juice are all acidic enough to react with sodium bicarbonate. Commercial baking powders are usually a mixture of sodium bicarbonate and a solid acid salt. The addition of water to the mixture will start the chemical action. When sodium bicarbonate is heated, it decomposes into sodium carbonate, water, and carbon dioxide.

$$2NaHCO_3 \longrightarrow Na_2CO_3 + H_2O + CO_2\uparrow$$

13:6 *Cleaning is one of the many applications of baking soda.*

Sodium nitrate ($NaNO_3$) is used in commercial fertilizers as a source of nitrogen. Nitrogen is necessary for plant growth. Large deposits of sodium nitrate are found in Chile.

13:7 Use of Sodium Metal The ability of liquid sodium to conduct heat is of particular value in nuclear reactors. It will, however, completely remove the carbon from some steels; therefore, it must be contained in vessels made of a special type of stainless steel.

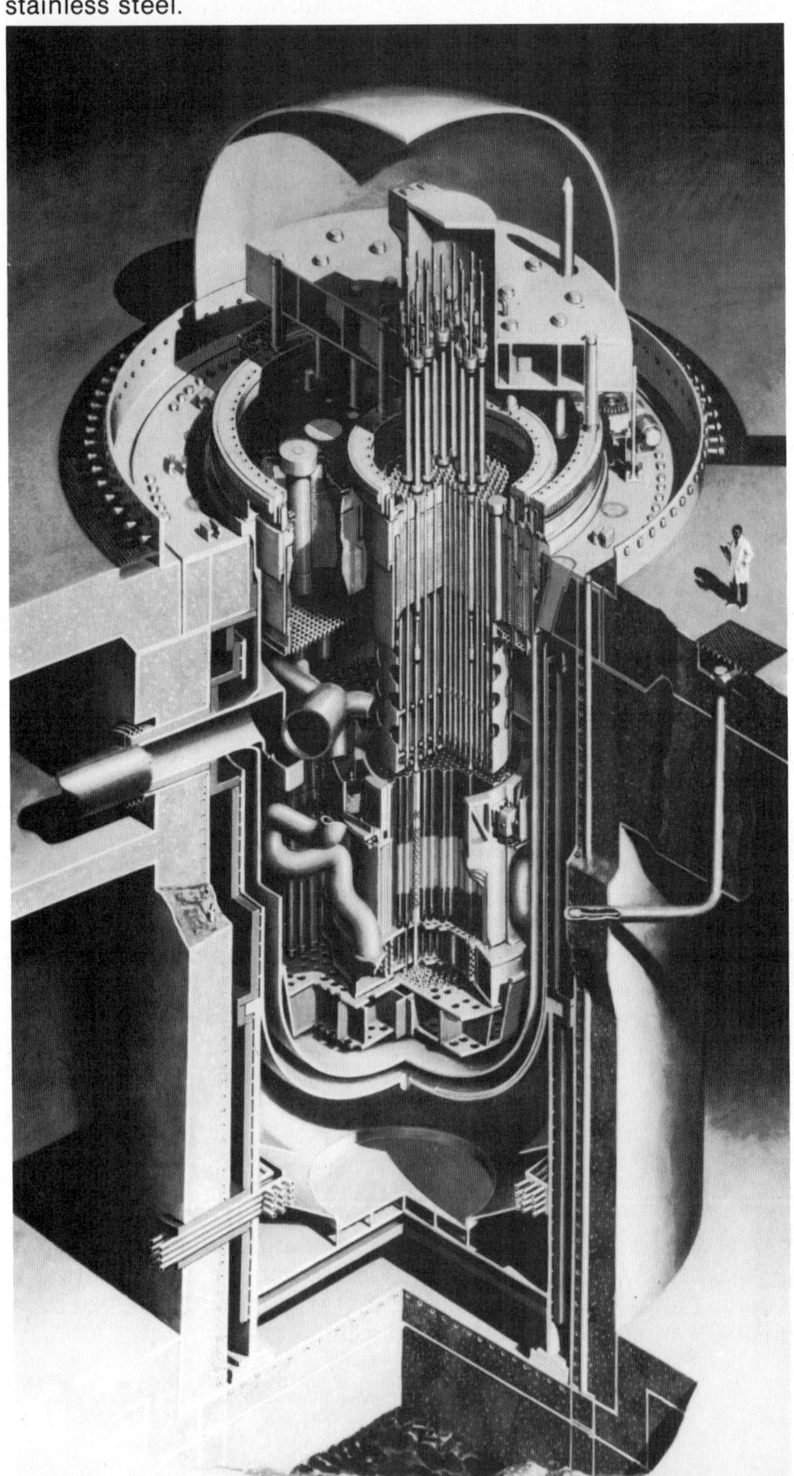

13:7 *Model of a liquid-metal, fast-breeder reactor using liquid sodium metal as a coolant.*

13:8 Potassium Compounds Potassium nitrate (KNO_3) is also useful as a fertilizer because potassium is another of the elements essential to plant life. At least 1/10 of one percent potassium is necessary in the soil to insure healthy plant growth. Potassium nitrate is used in the making of black powder and is used today as an aid to combustion.

Valuable potassium compounds can be obtained from evaporation of the waters of the Dead Sea in Israel. This large inland lake is highly prized for the wealth of its mineral content.

13:8 *The Dead Sea, showing mineral deposits.*

13:9 Lithium Compounds Lithium salts are used in the glass industry and in the manufacture of ceramic glazes. Many organic compounds of lithium are added to greases to improve their stability at high temperatures. Some lithium compounds are used in nuclear devices.

13:10 The Alkaline-Earth Metals Looking back at Table 13-1, you will see that the alkaline-earth metals have two electrons to donate in a chemical reaction. If all of the alkaline-earth metals (except beryllium) lose two *s* orbital electrons, they will have a stable, filled outer shell of electrons. Each one (except beryllium) ionizes, releasing two electrons.

$$Mg \longrightarrow Mg^{+2} + 2e^-$$
$$Ca \longrightarrow Ca^{+2} + 2e^-$$
$$Ba \longrightarrow Ba^{+2} + 2e^-$$

Beryllium forms covalent bonds when it reacts chemically.

The chemical activity of the alkaline-earth metals increases as the metals increase in atomic size. This can be "explained" again by the *screening effect.*

$$Ba > Sr > Ca > Mg > Be$$

Barium is the most chemically active, and beryllium is the least active. Alkaline-earth metals are not as reactive as the alkali metals because it is more difficult to donate two electrons at a time than it is to donate one at a time.

13:9 *The lithium hydride shield for the SNAP-10A space reactor. The shield, shown in its stainless steel container, protects the reactor's instrument package from the system's radiation.*

Unlike the alkali metal salts, many compounds of the alkaline-earth elements are insoluble (cannot be dissolved) in water. Also, the alkaline-earth metals differ more from each other chemically than alkali metals do. Examples of this varied activity can be shown in the way they react with water.

$$Be + H_2O \xrightarrow{\Delta} \text{No Reaction}$$
$$Mg + 2H_2O \xrightarrow{\Delta} Mg(OH)_2 + H_2\uparrow$$
$$Ca + 2H_2O \longrightarrow Ca(OH)_2 + H_2\uparrow$$

Beryllium will not react with boiling water, but magnesium will. Calcium will react with cold water.

All of the metals except beryllium will tarnish readily in air.

13:11 Physical Properties of Alkaline-Earth Metals The alkaline-earth metals have a white luster when freshly cut. They are much harder than the alkali metals. Beryllium can scratch glass, although barium is about as soft as lead. All of the alkaline-earth metals are malleable (can be hammered into various shapes without breaking). The densities and melting points of these elements are higher than those of the alkali metals.

13:12 Occurrence of the Alkaline-Earth Metals None of the alkaline-earth metals are found free in nature. The primary sources of the metals are listed in Table 13-2.

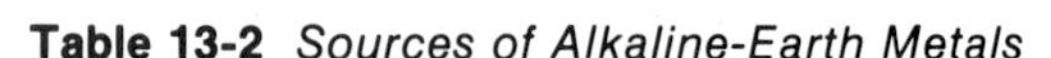

Table 13-2 *Sources of Alkaline-Earth Metals*

Metal	Major Sources
calcium	limestone ($CaCO_3$), gypsum ($CaSO_4 \cdot 2H_2O$)
magnesium	ocean, dolomite ($MgCO_3 \cdot CaCO_3$), magnesite ($MgCO_3$), many other minerals.
beryllium	beryl ($Be_3Al_2(SiO_3)_6$)
barium	barite ($BaSO_4$)
strontium	celestite ($SrSO_4$)
radium	uranium ores (product of radioactive decay)

13:10 *Burruss Hall on the Virginia Tech campus was constructed from limestone blocks.*

13:13 Calcium and Its Compounds The biggest source of calcium compounds is **limestone** (calcium carbonate—$CaCO_3$). Limestone is a water-laid sedimentary material; many marine fossils have been found buried in it. It is believed that limestone was formed during and after the Flood.

Limestone can be used as a construction material.

Mortar is made by mixing one part slaked lime ($Ca(OH)_2$) and three parts sand with enough water to make a paste. The hydroxide dries, bonding the sand particles together. As the mortar ages, carbon dioxide in the air reacts with the slaked lime to form $CaCO_3$, limestone.

$$Ca(OH)_2 + CO_2 \longrightarrow CaCO_3 + H_2O$$

Thus, the mortar becomes stronger with time.

Calcium oxide (CaO), or lime, is usually produced by the decomposition of limestone.

$$CaCO_3 \xrightarrow{\Delta} \underset{\text{lime}}{CaO} + CO_2\uparrow$$

The carbon dioxide must be removed from the reaction site, or the gas will react with the freshly formed lime and revert back to calcium carbonate.

Besides being used in mortar, lime is also used in plaster, concrete, glass, stucco, sand-lime brick, and cold water paint. Lime and limestone are also used in farming to neutralize soils that are too acidic.

Also, calcium will react with oxygen to form calcium oxide.

$$2Ca + O_2 \longrightarrow 2CaO$$

Calcium oxide reacts with water to form calcium hydroxide ($Ca(OH)_2$).

$$CaO + H_2O \longrightarrow Ca(OH)_2 + \text{heat}$$

This reaction is **exothermic**, which means it gives off heat. Pure calcium reacted with water also forms slaked lime.

$$Ca + 2H_2O \longrightarrow Ca(OH)_2 + H_2\uparrow$$

When **gypsum** is heated, it forms **plaster of Paris.**

$$\underset{\text{gypsum}}{2CaSO_4 \cdot 2H_2O} \xrightarrow{\Delta} \underset{\text{plaster of Paris}}{(CaSO_4)_2 \cdot H_2O} + 3H_2O$$

When the plaster is added to water, the reverse reaction takes place—the crystals of gypsum that form bond together to produce a rigid substance.

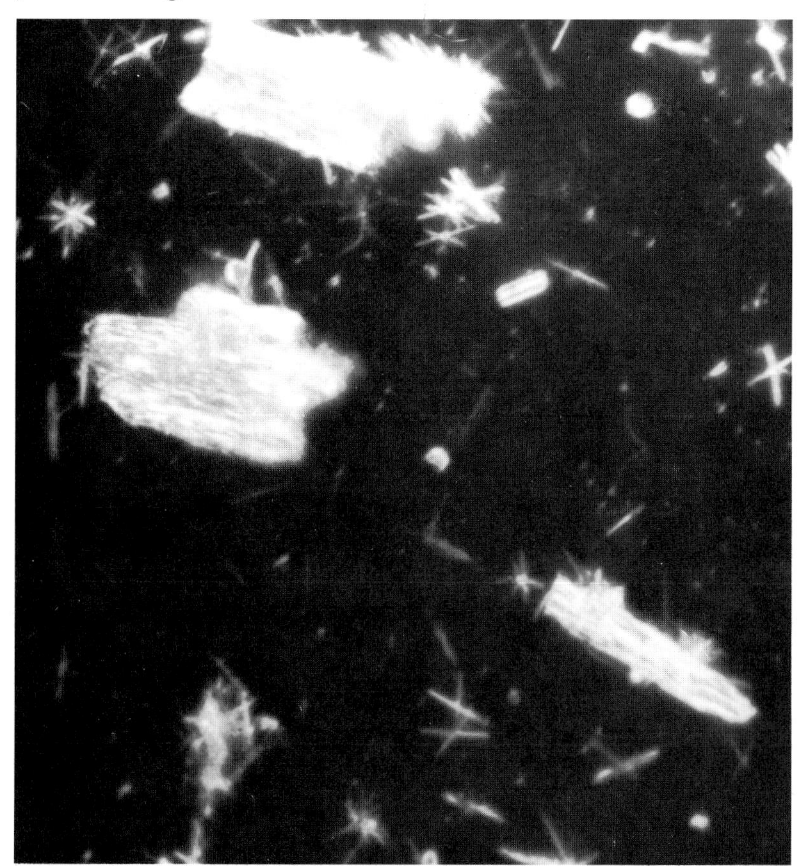

13:11 *Gypsum crystals are beginning to harden out of solution to form a cement.*

Portland cement is formed by heating limestone and certain clays together in large rotary kilns (ovens). This cement is mixed with sand and water to form a plastic (gooey) mass that gradually hardens. It is obvious from these examples that calcium compounds are very important in the construction industry.

13:12 *A rotary kiln used in preparation of cements.*

13:13 *Imagine how much cement was used to construct the Shasta Dam.*

Calcium chloride ($CaCl_2$) is a very interesting compound. Since it will pick up water from the atmosphere and dissolve in it, it is referred to as being **deliquescent**. Calcium chloride is used as a drying agent for some gases and is useful in melting ice on roads in the winter. The salt that you throw on the sidewalk in winter to melt the ice is probably calcium chloride.

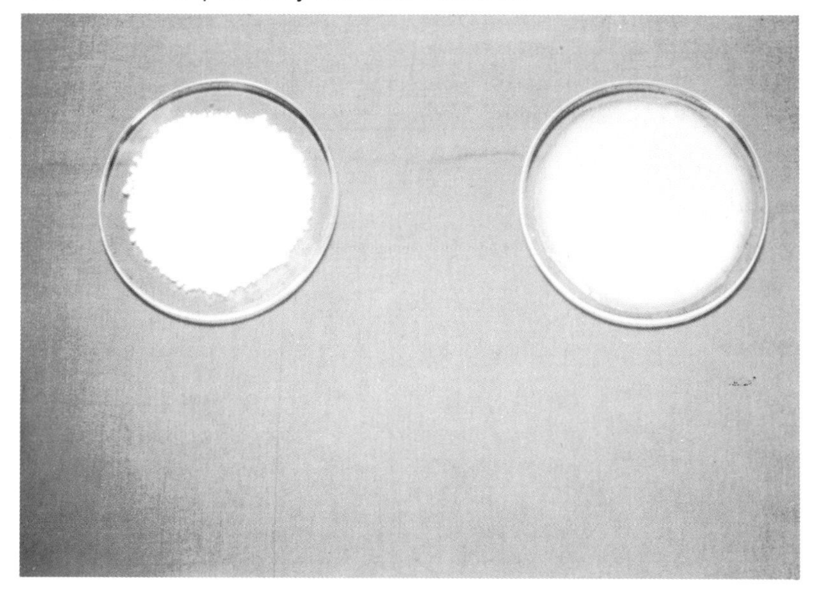

13:14 *Deliquescence. The watch glass on the left contains fresh $CaCl_2$. The watch glass on the right contains $CaCl_2$ that has been exposed to a moist environment.*

13:14 Stalactites and Stalagmites In areas where there are heavy limestone deposits, many beautiful caverns have been formed by the chemical action of water containing dissolved carbon dioxide with limestone.

$$CO_2 + H_2O \longrightarrow H_2CO_3$$

Water laden with CO_2 seeps through limestone, dissolving some of the calcium carbonate.

$$CaCO_3 + H_2CO_3 \longrightarrow Ca(HCO_3)_2$$

Further along in the cave the dissolved calcium bicarbonate comes out of solution as calcium carbonate as CO_2 escapes from the water (the reverse of the first two reactions).

$$Ca^{+2} + 2HCO_3^- \longrightarrow CaCO_3 + H_2CO_3$$

$$H_2CO_3 \longrightarrow CO_2\uparrow + H_2O$$

The results of this action are beautiful **stalactites**, limestone formations which hang from the roof of the cavern, and **stalagmites**, formations which rise up from the floor of the cavern.

The formation of stalactites and stalagmites is not as simple as explained here, but this will give you a satisfactory idea of what happens. Many evolutionists say that the formation of stalactites and stalagmites is evidence that the earth is very old because in many places the rate of growth of these structures is very slow. However, there are also caves where stalactites and stalagmites

13:15 *The Princess Column stalactite in the Luray Caverns.* *(See chapter title page)*

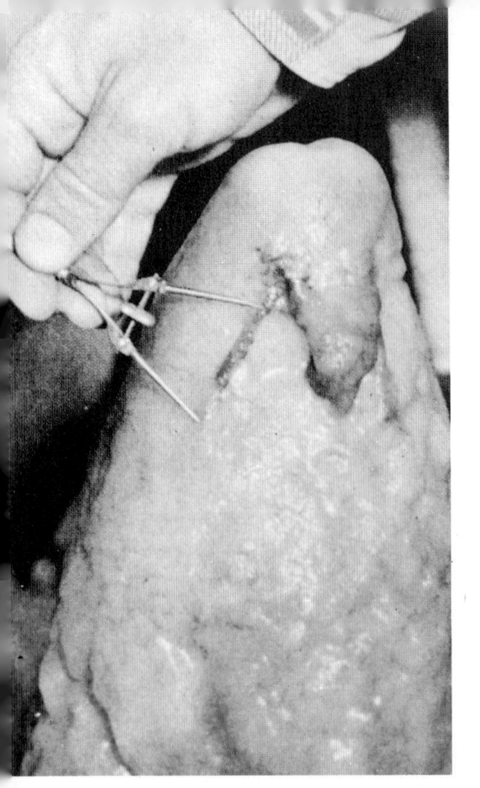

13:16 *A bat encased in a stalagmite. The stalagmite must have formed rapidly around the bat to have preserved it so well.*

grow very rapidly. It has been shown in experiments run in the laboratory that the rate of growth of stalactites can be quite rapid.

After the Flood there was a great deal of water remaining in many places in the earth. As this water seeped through recently deposited limestone, rapid formation and growth of stalactites and stalagmites would have been possible because of the large quantities of water flowing back into the earth.

13:17 *Notice the curd that forms after soap has been added to hard water.*

13:15 Hard Water Water containing a high concentration of calcium and magnesium ions is called **hard water**. Water with these dissolved ions is frequently found in areas with large limestone deposits. It is extremely difficult to wash dishes or clothing in hard water. Soaps are made up of sodium or potassium salts of certain organic acids; an example is sodium oleate ($NaC_{18}H_{33}O_2$). The magnesium and calcium ions in hard water react with the anion in the soap to form an insoluble scum (see Figure 13:17). (The arrow downward in the chemical reaction represents a **precipitate** coming out of solution.) A precipitate is a substance that falls out of a solution or a suspension by chemical or physical action.

$$Ca^{+2} + 2C_{18}H_{33}O_2^- \longrightarrow Ca(C_{18}H_{33}O_2)_2\downarrow$$

This ruins the action of the soap—no lather (soap suds) can be produced. More soap, therefore, must be added to hard water to get a sufficient lather. Also, if hard water is used in boilers, it tends to plug up the piping with calcium salts.

Hard water can be "softened" by several methods. If the calcium is in the form of a bicarbonate, such as $Ca(HCO_3)_2$, heating the water will cause the bicarbonate to decompose into the carbonate ion (CO_3^{-2}), which in turn reacts with the dissolved calcium to form insoluble calcium carbonate ($CaCO_3$). Thus, the Ca^{+2} ions are removed from the water.

$$Ca(HCO_3)_2 \longrightarrow CaCO_3\downarrow + H_2O + CO_2\uparrow$$

13:18 *Boiler scale in a pipe.*

Sodium carbonate, or soda (Na_2CO_3), can be added to hard water to remove the bivalent ions (ions with a valence of +2). The resulting reaction is

$$Na_2CO_3 + Ca^{+2} \longrightarrow CaCO_3\downarrow + 2Na^+$$

Another effective way to "soften" water is by the use of **ion-exchange** techniques. Hard water can be passed through a column of some substance that has sodium or hydrogen ions in it. The calcium ions in the water will replace the hydrogen or sodium ions in the column material. The exit water is then "soft." The column material can be made of various materials with the univalent positive ions (+1 ions) held in the structure by negative ions. Calcium (Ca^{+2}), which is more strongly positive, is more easily held by the negative ions and therefore replaces the sodium or hydrogen ions.

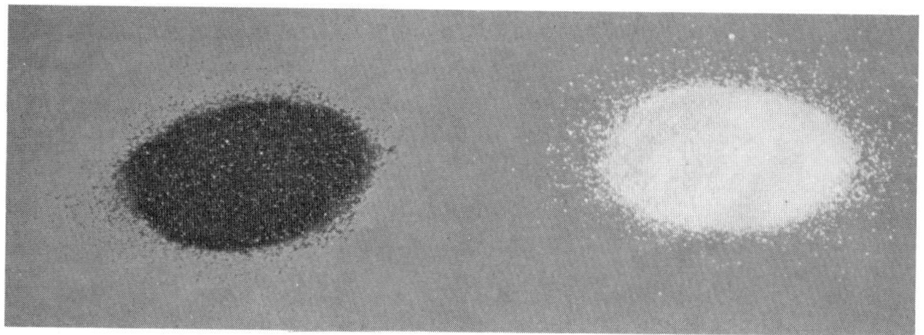

(b)

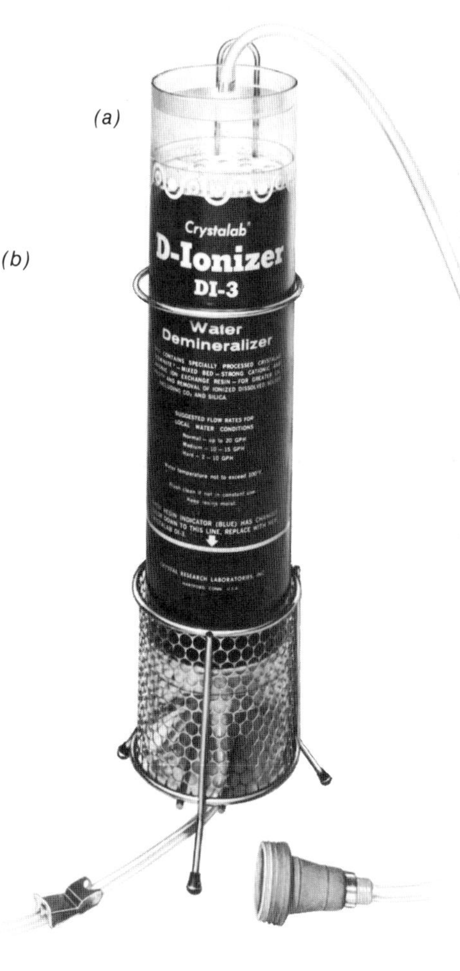

(a)

13:19 *(a) An ion-exchange column used to remove bivalent ions from water. (b) Ion-exchange resins.*

Experiment 13:1 *Temporary Hard Water*

If you do not have temporary hard water in your area, it can be prepared in the laboratory. Your teacher will furnish you with a sample of temporary hard water.

1. Fill a small aluminum pan half full of temporary hard water.
2. Boil the water. Continue heating until most of the water has evaporated.
3. Notice the deposit left on the aluminum pan. What is it? Write the chemical equation for the reaction.

Experiment 13:2 *Permanent Hard Water*

Permanent hard water is made by adding soluble calcium and magnesium salts (other than the bicarbonates) to water. Calcium sulfate ($CaSO_4$), calcium chloride ($CaCl_2$), magnesium sulfate ($MgSO_4$), or magnesium chloride ($MgCl_2$) are satisfactory to use. 1 gram of any of these compounds added to 100 ml of water will make a sufficiently hard water for this experiment.

1. Add a soap solution, drop by drop, into 5 ml of hard water in a test tube. What is the nasty-looking precipitate? Write the chemical equation for the reaction.
2. After adding a few drops of soap solution, shake the test tube. Does a lather form?
3. Continue adding soap solution until no more precipitate forms; then add a few more drops.
4. Shake the test tube. Does a lather form now? Make some comments on the waste of soap in hard water.

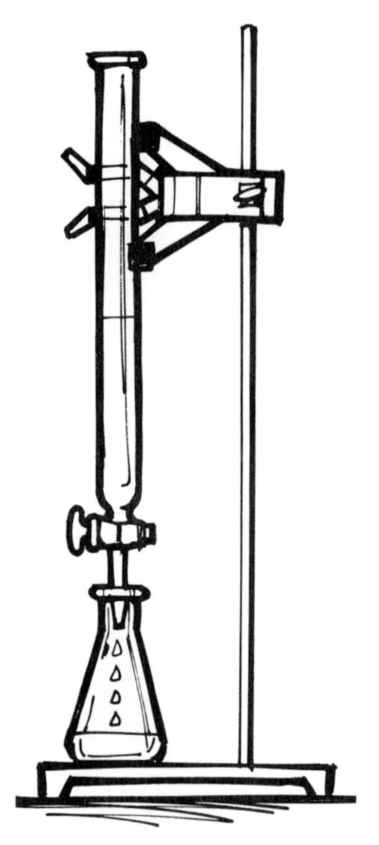

13:20 *Buret containing soap solution being titrated into Erlenmeyer flask.*

Experiment 13:3 *Hard Water and Cooking*

1. Heat 25 ml of distilled water in a beaker until the water boils.
2. Stop heating and add a few tea leaves; then stir the water.
3. Let stand until a good color has developed in the water; then strain the tea to remove the leaves.
4. Place the tea solution in a bottle or flask that you can close or stopper.
5. Repeat steps 1-4 using 25 ml of hard water.
6. Leave both samples in your laboratory desk or in a refrigerator until the next day. When you return, examine the color and clearness of the 2 solutions. Record your results. Which would you choose to make tea, hard or soft water?

Experiment 13:4 *Water Softeners*

1. Place some soap solution in a buret.
2. Place 20 ml of hard water in an Erlenmeyer flask.
3. Slowly add the soap solution to the hard water, shaking the flask at all times. Stop adding soap solution when a lather about 1 cm high has formed. Record how much soap solution was used.
4. Repeat steps 1-3 using 20 ml of boiled temporary hard water.
5. Repeat steps 1-3 using 20 ml of distilled water. (Be sure to record how much soap solution you used.)
6. To 20 ml of hard water add 0.1g Na_2CO_3; shake the flask until the Na_2CO_3 dissolves.
7. Add soap solution until a lather 1 cm high has formed.
8. Repeat steps 6 and 7 using 20 ml of hard water plus 0.1g Na_3PO_4.
9. Repeat steps 6 and 7 using 20 ml of hard water plus 0.1g borax ($Na_2B_4O_7 \cdot 10H_2O$).

Make a data sheet so that you can record your experimental results on it. Which water sample required the least amount of soap solution? Of the three water softeners used, which one required the least amount of soap? Which of these water softeners produce precipitates with hard water? What is the disadvantage of a water softener which forms a precipitate?

13:16 Magnesium and Its Compounds Because of its lightness, magnesium can be **alloyed**, or mixed, with other metals. It is used structurally, especially in the construction of aircraft, where weight is of primary importance. However, the low density and lack of strength of magnesium alloys prevent their use where heavy loads are involved. The brightness with which magnesium burns makes it valuable for use in flares and flashbulbs.

13:21 *Magnesium alloy bats.*

Experiment 13:5 *Magnesium Flame*

1. Burn some Mg ribbon.
2. Observe the brightness of the flame.
3. Observe the formation of MgO (the powder product).

Magnesium carbonate ($MgCO_3$) can be decomposed into magnesium oxide (MgO).

$$MgCO_3 \xrightarrow{\Delta} MgO + CO_2\uparrow$$

Magnesium oxide has excellent heat-resistant properties and is used to make refractory (heat resistant) bricks for the lining in furnaces.

Magnesium oxide (MgO) reacts with water to form magnesium hydroxide ($Mg(OH)_2$).

$$MgO + H_2O \longrightarrow Mg(OH)_2$$

A suspension of magnesium hydroxide is known as **milk of magnesia**. Perhaps your mother has given you a dose of milk of magnesia when you complained of an upset stomach.

Another magnesium compound, **epsom salt** ($MgSO_4 \cdot 7H_2O$), is also used for medical purposes.

Magnesium oxide (MgO) can be mixed with an aqueous solution of magnesium chloride ($MgCl_2$). As the paste hardens, it forms a cement of magnesium oxychloride.

Experiment 13:6 *Magnesium Oxychloride Cement*

1. Mix magnesium oxide powder (MgO) and magnesium chloride ($MgCl_2$) crystals with enough water to make a thick paste. Use 80 percent MgO and 20 percent $MgCl_2$ by weight.
2. Place the mixture in a mold and allow it to harden. The resulting hardened mass is a magnesium oxychloride cement.

13:22 *Elaborate vacuum systems suck up beryllium dust and chips to protect equipment operators during the fabrication of beryllium parts.*

13:17 Beryllium and Its Compounds It is very difficult to extract beryllium from its ore and costly to fabricate the metal because each machine must have an efficient vacuum system to clean up beryllium particles so that the operator will not inhale them. Swallowing any quantity of beryllium is very dangerous, and breathing in beryllium dust can cause a respiratory infection.

A chief use of beryllium is its combination with copper to produce beryllium-copper alloy springs. Beryllium has also been used in the nuclear field as a moderator in reactors.

Beryllium forms covalent bonds when it reacts. Some binary beryllium compounds exhibit a chain-type structure.

13:18 Barium - Strontium - Radium Barium compounds are used for various useful products. A barium sulfate-sulfide mixture is used as white paint pigment. Barium carbonate ($BaCO_3$) is a rat poison. Barium nitrate ($Ba(NO_3)_2$) and barium chlorate ($Ba(ClO_3)_2$) give a green color in fireworks. Strontium compounds are also used in fireworks. Strontium nitrate ($Sr(NO_3)_2$), strontium carbonate ($SrCO_3$), and strontium chlorate ($Sr(ClO_3)_2$) give red colors in fireworks.

Radium chloride ($RaCl_2 \cdot 2H_2O$) is used with a fluorescent material to make luminous paint for watch and instrument dials. Also, radium chloride is used in the treatment of cancer. The chemical properties of radium are seldom utilized; the interest in the metal is mostly because of its radioactivity.

13:19 Flame Tests for Alkaline-Earth Elements Beryllium and magnesium do not give a positive flame test. However, calcium, strontium, and barium give orange-red, crimson, and light green flame colors, respectively, in a Bunsen burner.

Metal	*Flame Color*
beryllium (Be)	none
magnesium (Mg)	none
calcium (Ca)	orange-red
strontium (Sr)	crimson
barium (Ba)	light green

List of Terms

active metals	limestone
alkali metals	metal
alkaline-earth metals	milk of magnesia
alloy	mortar
deliquescent	peroxide
epsom salt	plaster of Paris
exothermic	Portland cement
gypsum	precipitate
hard water	stalactite
ion-exchange	stalagmite

Questions

1. Write the names and chemical symbols for all of the alkali metals.
2. Write the names and chemical symbols for all of the alkaline-earth metals.
3. How many valence electrons does each alkali metal have?
4. How many valence electrons does each alkaline-earth metal have?
5. (a) When sodium reacts with a nonmetal, how many electrons does it lose?
   (b) When barium reacts with a nonmetal, how many electrons does it lose?
6. Define metal (chemically).
7. Explain how many of the salts of the alkali metals could possibly have been left in such large deposits.
8. (a) Are the alkali metals usually hard or soft?
   (b) Do alkali metals have a bright metallic luster?

★9. Since the alkali metals are excellent conductors of electricity, why are they not used for this application?

10. True or False:
    (a) Alkali metals react readily with oxygen.
    (b) Alkali metals form acids when they react with water.
    (c) Oxygen is given off when alkali metals react with water.
11. What color is rubidium when it is placed in a Bunsen burner flame?
12. What is the most common compound of sodium found in nature? (Write its common name, chemical name, and chemical formula.)
13. List three Bible verses which mention salt.
14. List three sodium compounds and a common usage of each.
15. Is sodium metal a good conductor of heat?
16. What are two uses of potassium nitrate (KNO_3)?
17. Where can rich deposits of potassium compounds be found?
18. Name two applications for lithium compounds.
19. Why does the chemical activity of the alkali or alkaline-earth metals increase with increasing size of the atom of the metals?
20. (a) Which one of the alkaline-earth metals does not act like the others? Why?
    (b) Is this metal more or less reactive than the rest of the group?
21. What is the major source of radium?.
22. True or False:
    (a) Alkali metals are more dense than alkaline-earth metals.
    (b) The melting points of the alkali metals are higher than those of the alkaline-earth metals.
23. When did limestone probably form?
24. Name some uses of lime, slaked lime, and limestone.
25. Name all of the calcium compounds used in the building industry.
26. How is Portland cement manufactured and hardened?
27. Define deliquescent.
28. Chemically, how do stalactites and stalagmites form?
29. Give some evidences that stalactite and stalagmite growth can be very rapid.
30. (a) What is hard water?
    (b) What is one problem with hard water?
    (c) How can this problem be corrected? (Suggest two methods.)
31. What is ion exchange?
32. Name two uses of magnesium alloys.
33. Name two uses of magnesium compounds.

34. What is the hazard in manufacturing beryllium?
35. Beryllium is alloyed with what metal to make springs?
36. Define alloy.

Student Activities

1. Write a history of one of the alkali metals or one of the alkaline-earth metals. Trace its various uses up to the present.
2. Write a history of the Portland cement industry and the uses of Portland cement. Make up various mixtures of cement, adding just enough water, too much water, too little water, too much sand, too little sand, and just enough sand. Display these using posterboard, and explain the various specimens. A cubic mold in which to make specimens can be made of wood or metal (see Figure 13:23). Oil the inside of the mold so that the specimen can be removed after it hardens. Try to devise a way to test the strength of your specimens.

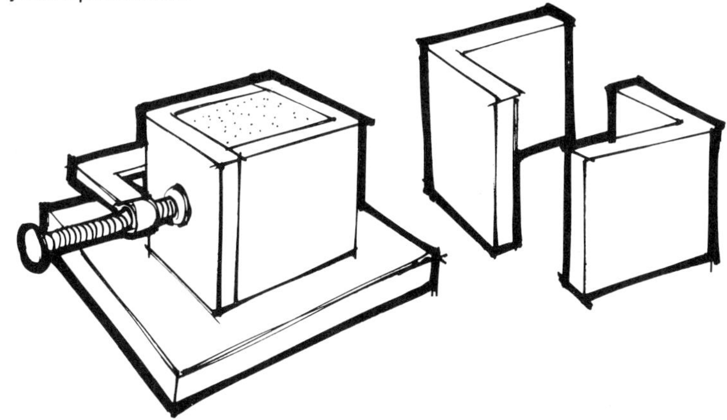

13:23 *Cement mold.*

Place your illustrated history in a notebook in front of the display. If you have any experimental data, place this in the back part of the same notebook.

For help with this activity, write to the Portland Cement Association, Old Orchard Road, Skokie, Illinois 60076. Two books that might be in your school or community library are *The Chemistry of Portland Cement* by R. H. Bogue, 2nd edition (1955), Reinhold Publishing Corporation, New York, and *The Chemistry of Cement and Concrete* by F. M. Lea and C. H. Desch, revised edition (1956), Edward Arnold, Limited, London. These have histories and much helpful information in them.

3. Write a paper from a creationist standpoint on the formation of stalactites and stalagmites. Show how these could form in a very short time. If you are interested in doing some experiments on the formation of flowstone, write for experimental details to the Bob Jones University Science Department in Greenville, South Carolina 29614. The experiments are easy to do if you have a supply of limestone and carbon dioxide gas.
4. Write a report on the ion-exchange method. Make an ion-exchange column using a plastic or glass pipe filled with an ion-exchange medium. Experiment using an ion-exchange column in the same way as the hard water experiments (Experiments 13:2, 3, 4) were conducted.

Photo next page
Many beautiful containers can be made from glass.

UNIT V DESCRIPTIVE CHEMISTRY Groups IA-IVA

Chapter 14 Chemistry of Carbon and Silicon

14:1 Metalloids On the left-hand side of the periodic table are metals (electron donors), and on the right-hand side are nonmetals (electron acceptors). This is a very general classification. In the middle of the periodic table are elements that could be metal, nonmetal, or intermediate (between metals and nonmetals). Those elements that have properties intermediate between metals and nonmetals are called **metalloids**, which means "like metals."

The categories metal, nonmetal, and intermediate are a reflection of the kinds of people we find in the world. On the two extremes are saved and unsaved people. A saved person should lead a separated life; others should be able to see easily that he is a Christian. An unsaved person leads a worldly life; and, although he may seem like a good, moral person, he is lost if he has not accepted Jesus Christ as his Saviour.

Unfortunately, there is an intermediate group, the metalloid Christian—the worldly or carnal Christian. This type of person says he is saved but acts like an unsaved person at times. You should not be a metalloid Christian; a child of God should act properly so that others can definitely see that he is what he claims to be.

A chart of the electron orbitals of metalloid elements is shown in Table 14-1. The atoms of each element have two *s* electrons and two *p* electrons in the valence shell. The metallic characteristics increase as you move down the group. Carbon is considered to be a nonmetal, whereas lead is a metal. We will discuss only carbon and silicon now. Some chemists call silicon a nonmetal while others refer to it as a metalloid.

Table 14-1 *Electron Distribution in the Atoms of Group IVA Elements*

Element	Quantum Level and Electron Orbitals														
	1	2		3			4				5			6	
	s	s	p	s	p	d	s	p	d	f	s	p	d	s	p
C	2	2	2												
Si	2	2	6	2	2										
Ge	2	2	6	2	6	10	2	2							
Sn	2	2	6	2	6	10	2	6	10		2	2			
Pb	2	2	6	2	6	10	2	6	10	14	2	6	10	2	2

14:2 Bonding Habits To get a stable octet of electrons, both carbon and silicon share electrons with other elements. Neither of these elements forms simple ions; each is always found in complex ions or molecules. Carbon has the unusual ability to form countless compounds in which many carbon atoms are bonded to each other covalently to form chains, branched chains, closed chains, and rings. The most important bond of silicon is the one it makes with oxygen in forming silicon dioxide (SiO_2) and many **silicates**. Crystals of silicon dioxide are actually giant molecules in which each silicon atom is bonded to four oxygen atoms.

14:3 Occurrence of Carbon About three hundredths (0.03) of the earth's crust is made up of carbon. Although this is a small amount, carbon is present in many substances, such as **petroleum**, **natural gas**, and **limestone**. A great deal of carbon was deposited in the earth's crust during the Flood. This carbon was contained in living organisms which were buried during the Flood. Uncombined carbon is found in forms such as **coal**, **graphite**, and **diamond**.

All living organisms have carbon in them. Carbon is one of the most important elements listed in the periodic table. An entire field of study, organic chemistry, is based on the chemistry of carbon-hydrogen compounds and compounds in which the hydrogen has been replaced by other elements.

14:4 Forms of Carbon When the word "diamond" is mentioned, every woman's heart skips a beat. Diamonds are quite beautiful and are used in jewelry. Few things seem lovelier to a woman than the diamond in her engagement ring. More mundane uses of the diamond capitalize on its hardness. Diamond needles for record players, grinding powder for polishing, and cutting tools for steels are just a few of the uses of this valuable substance.

Diamonds are mentioned four times in the Bible, all in the Old Testament.

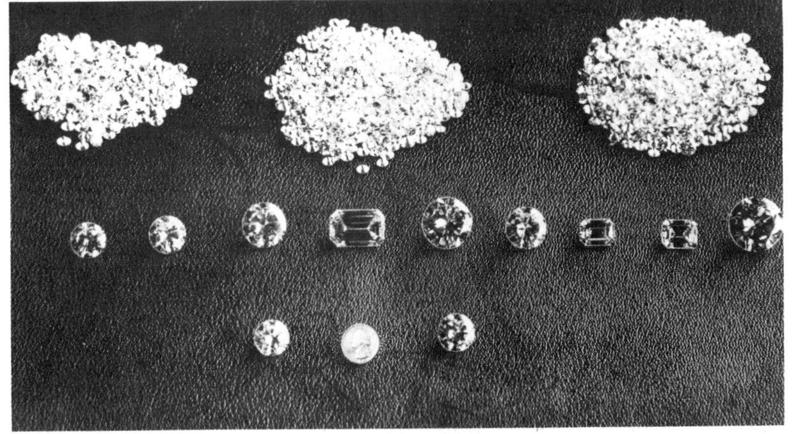

(a)

(b)

(c)

14:1 *(a,b) Various sizes of uncut and faceted diamonds. (c) Diamond cutting drill.*

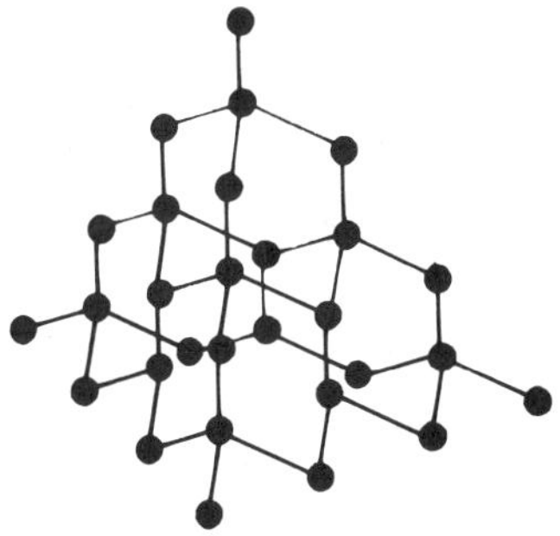

It seems strange that carbon can form into something as hard as a diamond and also something as soft as graphite. The difference between these substances is the way the carbon atoms are arranged within the solids. In diamond, the carbon atoms are bonded covalently to one another in the arrangement of a **tetrahedron** (see Figure 14:2). In graphite, the atoms are arranged in hexagonal layers. The bonds between carbon atoms within these hexagonal layers are very strong covalent linkages; however, the bonds that hold the layers themselves together are very weak. The layers slide across one another very easily; thus, graphite is quite soft.

14:2 *Crystal structure of diamond.*

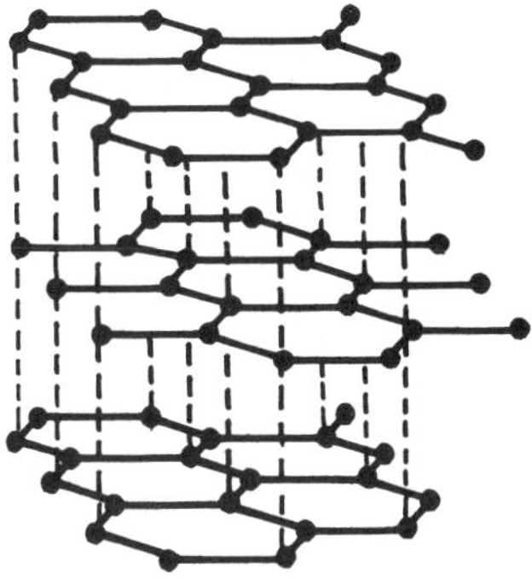

14:3 *Crystal structure of graphite.*

14:4 *Graphite is used in batteries, electrodes, and pencils.*

This sliding property makes graphite useful as pencil lead (a clay and graphite mixture), for it will smear easily. Graphite is also an excellent lubricant because of this property.

Graphite has a high enough electrical conductivity to be useful in several electrical applications, too (see Section 32:11).

Coal is a particular form of impure carbon. Coal possibly formed after the Flood when a great deal of vegetation had been buried and subjected to high pressures by the sedimentary layers on top of the vegetation. The various forms of coal are **lignite**, **bituminous**, and **anthracite**.

Lignite, sometimes referred to as brown coal, has a fibrous appearance. Lignite is about 50 percent moisture. Bituminous, or soft coal, has less moisture in it than lignite. It contains substances which form flammable gases when heated. Anthracite, or hard coal, contains a higher percentage of carbon and a lower percentage of **volatile** (easily vaporized) matter than bituminous coal.

Some man-made carbon products and their uses are shown in Table 14-2.

Table 14-2 *Man-made Carbon Products*

Form of Carbon	Preparation	Use
charcoal	heating wood or bones	fuel, absorbent
coke	high temperature distillation of coal	fuel, absorbent
carbon black	incomplete distillation of natural gas	pigment, compounding of rubber for tires
activated charcoal	charcoal heated in steam	absorbent

Experiment 14:1 *Ash Content in Coal*

1. Crush a sample of coal.
2. Weigh the sample, then heat it in a porcelain crucible until it is completely burned.
3. Weigh the ash remaining in the crucible.
4. Determine the percentage of ash in your sample of coal. The greater the amount of ash in a coal, the less energy you get from the coal per unit of weight and the greater the air pollution you will have if the ash escapes up a smokestack.

Experiment 14:2 *Destructive Distillation of Coal*

1. Obtain some bituminous coal and hammer it into small pieces.
2. Place the pieces of coal into a test tube until the tube is about two-thirds full.
3. Put the test tube into a clamp, then connect the tube to a gas bottle as shown in Figure 14:5.
4. Heat the tube strongly. Note the escaping vapor. Some of the vapor condenses in the jar to form a tar. The gas that is escaping from the bottle can be ignited and will burn. The black residue left behind in the test tube is coke.

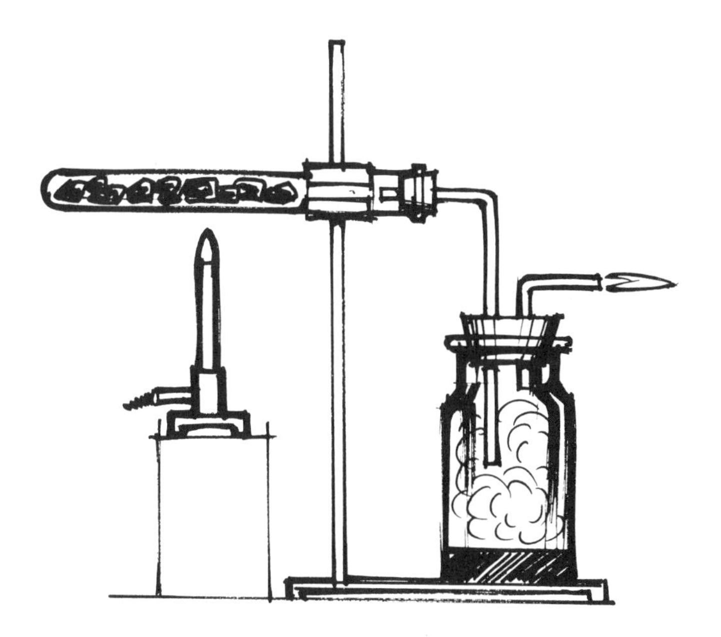

14:5 *Destructive distillation of coal.*

14:5 Fossil Fuel Reserves The world's reserves of the so-called *fossil fuels*—coal, oil, and gas—are steadily being depleted. These fuels have played an important role in the development of our present technology and continue to be essential for maintaining our standard of living. While coal is being replaced by oil and gas for home use, it is still needed for making coke, which is necessary in the steel industry. Therefore, all three—coal, oil, and gas—are still very vital resources.

Some alarmists have stated that the fuel reserves will last only a few more years at the most. Actually, it is impossible to take an exact inventory of all the world's fossil fuels. Recognized geologists seem to think that the fossil fuel reserves will last at least another 150 years at the present rate of consumption. Meanwhile they are being supplemented more and more with nuclear power. Nuclear energy is extremely efficient; a few tons of uranium fuel will yield enough electricity to supply all the homes in a city the size of Boston for an entire year.

14:6 *The Chisholm Mine at Phelps, Kentucky.*

14:7 *Drilling for oil.*

14:8 *Offshore drilling for oil.*

14:9 *A view of the Arizona wing of the Hoover Dam power plant. The total capacity of Hoover Dam's 17 generators is 1,344,800 kilowatts.*

What are the prospects that *new* sources of power might be developed in the future? Solar energy (utilizing the sun's rays for direct heating or for generating electricity with solar cells) has been demonstrated to be workable on a limited basis. However, it does not appear that solar energy will achieve widespread use.

Scientists have been tempted to try to develop a thermonuclear reactor, which would use fusion reactions similar to those of the sun and the hydrogen bomb. The efficiency of a thermonuclear reactor promises to be excellent, but the problems of building such a device are formidable. For example, the enormous temperatures generated by thermonuclear reactions might destroy the reactor and everything around it the first time the reactor is turned on. We will discuss thermonuclear reactions further in Chapter 39.

Another idea for supplying power involves the use of antiparticles. Every *particle*, such as a proton or electron, has an *antiparticle*—an "opposite" with which it can combine. The combination of the particle and antiparticle would theoretically convert the mass of both particles into energy. However, there is also a very serious problem with this idea. It would be extremely difficult to store the antiparticles prior to their use. If their container were made of ordinary matter, there would be an immediate explosion; the antiparticles and container would react and destroy each other. If the container were made of antiparticles, it would destroy any ordinary matter it touched, including the table on which it was placed and the surrounding

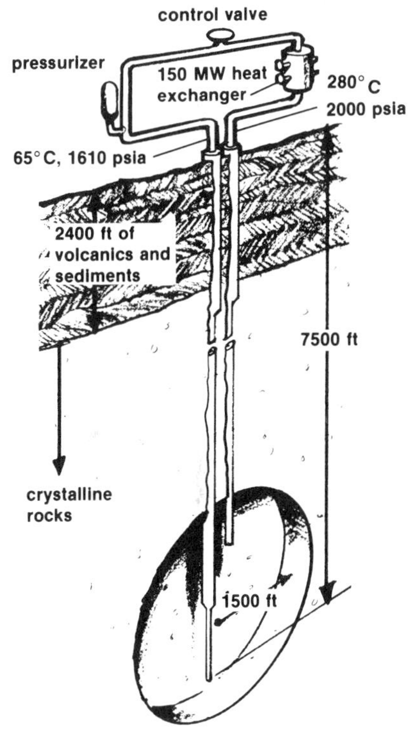

14:10 *The geothermal concept. Extracting heat from dry, hot rocks deep in the earth is currently under investigation. Water would be pumped down one pipe to take the heat from the rocks and return to the surface to run electrical generators. Such a well could last for 30 years or more.*

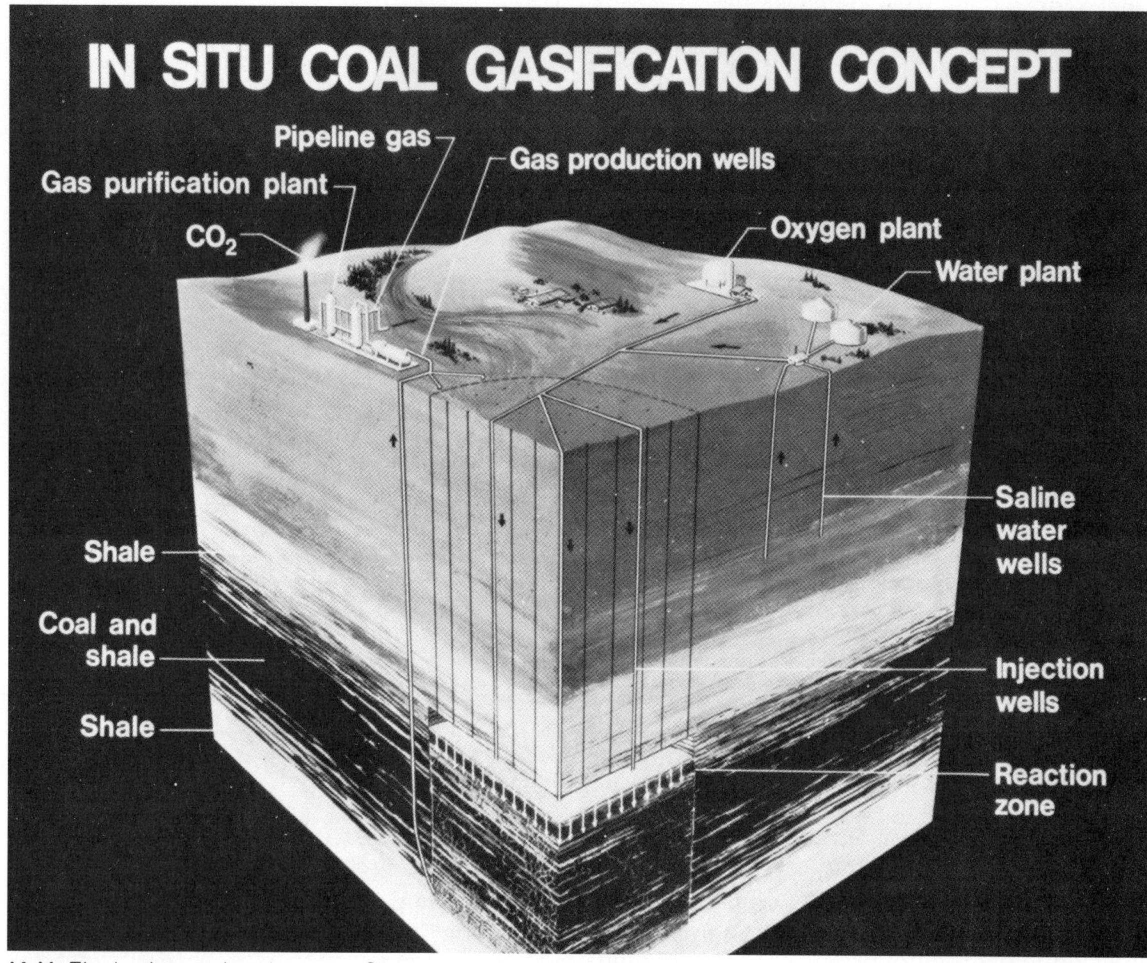

14:11 *The in situ coal gasification concept. Gasification of coal in its natural state would be accomplished by shattering the coal with conventional explosives, retorting it from the top of the bed down by starting a methane flame and sustaining combustion by water and oxygen spray, and piping the resulting methane gas from the bottom. Sulfur and other pollutants would remain underground.*

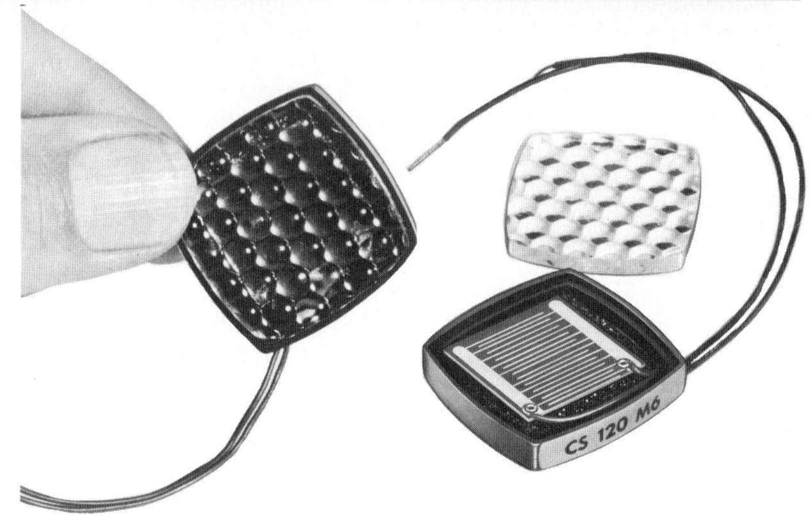

14:12 *Solar cells made from cadmium sulfide.*

air. One suggestion that has been made is to keep the antiparticles "trapped" in a magnetic field, so that they would be unable to come into contact with anything. At present, the matter-antimatter principle is nowhere near being "harnessed."

14:6 Carbon Dioxide and Carbon Monoxide Carbon dioxide (CO_2) is found in the exhaled respiratory gases of men and animals. Oxygen is carried by the blood from the lungs to the cells. Sugars in the cells are then oxidized, producing carbon dioxide, water, and energy. Here is a sample reaction of sugar with oxygen:

$$\underset{\text{sugar}}{C_6H_{12}O_6} + 6O_2 \longrightarrow 6CO_2 + 6H_2O + \text{energy}$$

Thus energy is made available so that men and animals can function. The carbon dioxide (CO_2) is carried by the blood to the lungs, where it is exhaled.

Green plants absorb CO_2 from the air and, with the aid of sunlight, form sugar. This is the reverse of the previous reaction.

$$6CO_2 + 6H_2O \xrightarrow{\text{sunlight}} C_6H_{12}O_6 + 6O_2\uparrow$$

This interesting balance between plants and animals is necessary. Plants cannot exist without animals, and animals cannot exist without plants. This relationship shows purposeful design in nature, not an accidental evolutionary blunder. Only an omnipotent Creator could have planned this; the idea of natural processes blindly stumbling on such a balance is a fanciful dream.

Carbon dioxide is a major product of **combustion**.

$$C + O_2 \longrightarrow CO_2\uparrow$$

Because of the increased use of fossil fuels, the concentration of CO_2 gas in the atmosphere is steadily increasing. There is no immediate danger from this buildup, but it may cause atmospheric temperature to increase, resulting in warmer weather throughout the world.

Certain types of fire extinguishers contain carbon dioxide. When CO_2 is blown over a fire, the gas literally smothers the flames. Solid carbon dioxide is known as **dry ice**. Dry ice does not melt; it *sublimes*, leaving no liquid mess. Soft drinks contain an excess of carbon dioxide; the gas is responsible for their tart taste.

Carbon monoxide is formed when incomplete combustion of some carbon-containing substance occurs. Two typical reactions are

$$2C + O_2 \longrightarrow 2CO\uparrow$$

$$2CH_4 + 3O_2 \longrightarrow 2CO\uparrow + 4H_2O\uparrow$$

Carbon monoxide gas is poisonous. Deadly CO gas is often present in the exhaust fumes of an automobile. No one should be in a closed garage while an auto engine is running.

Further combustion of carbon monoxide is possible. Carbon monoxide will react with oxygen to form carbon dioxide; this reaction gives off heat:

$$2CO + O_2 \longrightarrow 2CO_2\uparrow + \text{heat}$$

Therefore, carbon monoxide can be used as a commercial fuel.

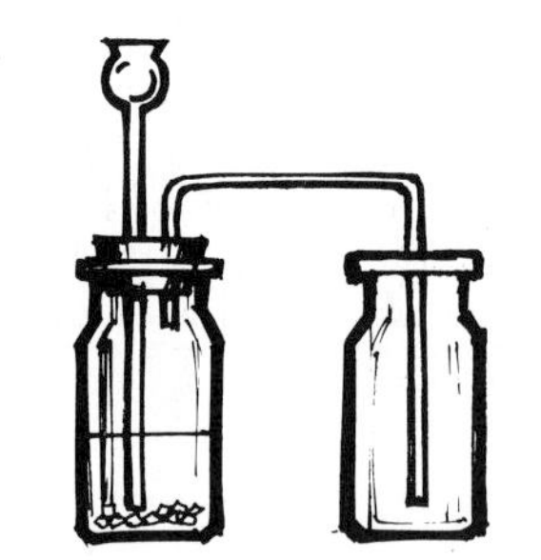

14:13 *Preparation of carbon dioxide.*

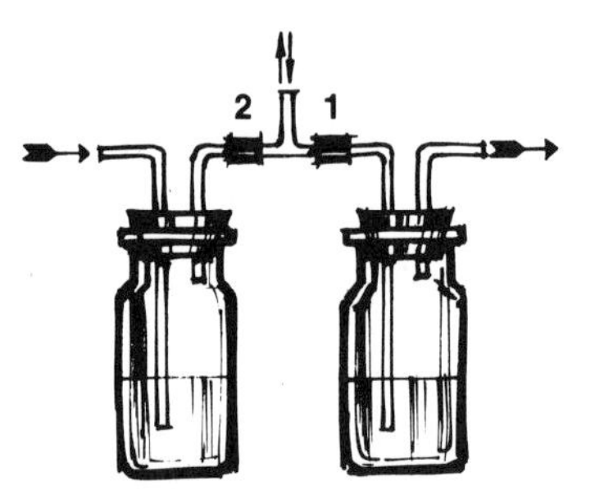

14:14 *Inhalation apparatus.*

Experiment 14:3 *Preparation of Carbon Dioxide*

1. Fill the bottom of a bottle with marble chips.
2. Insert a thistle tube and glass tubing into a two-hole stopper, setting up the apparatus as shown in Figure 14:13.
3. Pour some dilute hydrochloric acid (HCl) down the thistle tube. (Make sure that the tube is down far enough into the bottle so that the end of the tube is well beneath the liquid level.) A vigorous reaction occurs when the acid is added.
4. When you feel the other gas bottle is filled with carbon dioxide (CO_2), thrust a burning splinter into the bottle and observe what happens. Why do you suppose this happened?
5. Remove the splinter and fill the bottle again with CO_2.
6. Pour some limewater (water plus calcium hydroxide) into the bottle; then shake the bottle.
7. Record your results. Write the chemical equation for the reaction.
8. Place a piece of blue litmus paper in a test tube containing water.
9. Bubble some CO_2 into the water. What color does the litmus paper turn? Why?

Experiment 14:4 *Carbon Dioxide Content*

1. Set up an apparatus of 2 bottles, glass tubing, and rubber tubing connectors as shown in Figure 14:14.
2. Pour limewater into each bottle.
3. Inhale and exhale air from the T tube between the 2 bottles. When inhaling, pinch the rubber joint 1; when exhaling, pinch the rubber joint 2. This way you will inhale through the bottle on the left and exhale through the bottle on the right.
4. Inhale and exhale about 10 times. Notice the degree of "milkiness" in each bottle. Does inhaled or exhaled air have more CO_2 in it? Why do you suppose this is true?

Experiment 14:5 *Preparation of Carbon Monoxide*

1. Mix 10 ml of formic acid (CH_2O_2) and 10 ml of concentrated sulfuric acid in a large test tube.
2. Into the mouth of the test tube, place a no. 4 rubber stopper with a 4-inch piece of glass tubing in the hole.
3. Support the test tube on a ringstand and heat gently. The reaction that occurs is

$$CH_2O_2 \xrightarrow{H_2SO_4} CO\uparrow + H_2O$$

4. Ignite the CO as it comes out of the glass tubing.
5. Write the equation for the combustion of carbon monoxide.

14:7 Occurrence of Silicon Free silicon does not occur in nature, but it is the second most abundant element in the earth's crust (approximately 28 percent). Silicon and oxygen are the principal elements in soil and rocks. Silicon is as important to the mineral world as carbon is to the animal and vegetable world.

14:8 Silicon Dioxide As carbon atoms form a tetrahedral structure in diamond, so silicon forms a tetrahedral structure with oxygen (see Figure 14:15). This is known as **silica** (SiO_2), or sand. Actually, sand is chains of $(SiO_2)n$, where n is a large number. Silica can occur in three distinct crystalline forms—**quartz**, **cristobalite**, and **tridymite**. Some forms of silica found in nature are given in Table 14-3.

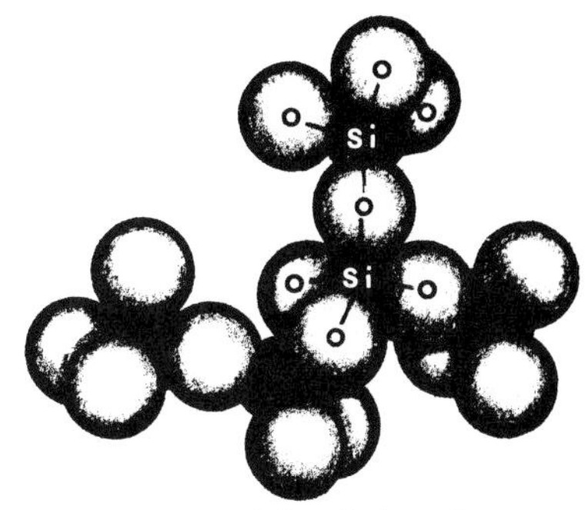

14:15 *Model of the chains of silicon dioxide.*

Table 14-3 *Natural Forms of Silica*

quartz	pure, transparent SiO_2
flint, agate	impure, opaque SiO_2
jasper, onyx	SiO_2 colored by Fe_2O_3
amethyst	SiO_2 colored by Mn impurities
opal	partly hydrated SiO_2

Quartz is the crystalline form of silica that is stable at room temperature. Quartz can be used to make lenses, and because of its high temperature strength it is used to make furnace tubes and temperature-resistant glassware.

The only acid that will chemically attack silica is hydrofluoric acid (HF). Silica is resistant to most bases.

14:9 Silicates The minerals in the earth's crust are mainly rock-forming silicates (materials that are built around the $(SiO_4)^{-4}$ anion). Some mineral silicates are listed in Table 14-4.

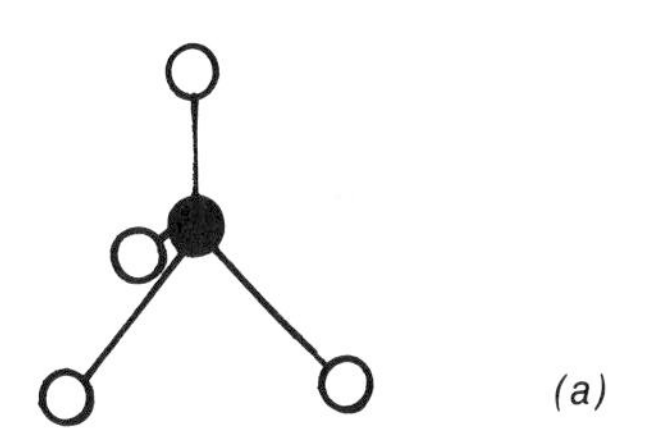

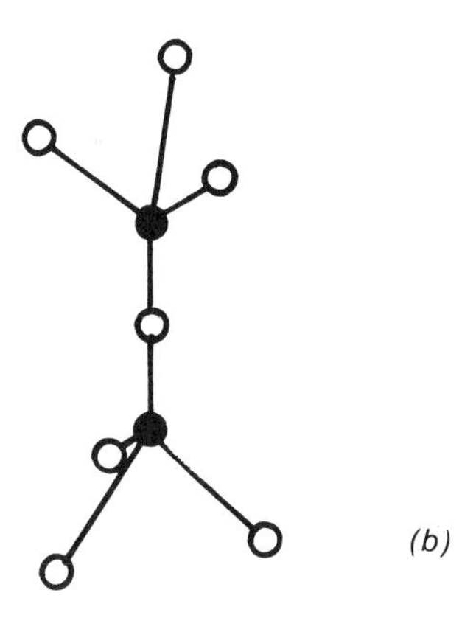

14:16 *(a) Model of a $(SiO_4)^{-4}$ anion. (b) Model of a $(Si_2O_7)^{-6}$ anion.*

Table 14-4 *Mineral Silicates*

willemite	Zn_2SiO_4
beryl	$Be_3Al_2Si_6O_{18}$
benitoite	$BaTiSi_3O_9$
diopside	$CaMgSi_2O_6$
tremolite (asbestos)	$Ca_2Mg_5Si_8O_{22}(OH)_2$
chrysotile (asbestos)	$Mg_6Si_4O_{11}(OH)_6 \cdot H_2O$
muscovite (mica)	$KAl_3Si_3O_{10}(OH)_2$
talc	$Mg_3Si_4O_{10}(OH)_2$
orthoclase	$KAlSi_3O_8$
chabozite	$CaAl_2Si_4O_{12} \cdot 6H_2O$
sodalite	$Na_4Al_3Si_3O_{12}Cl$

The tetrahedral arrangement of four oxygen atoms around a silicon atom is found in many silicates. Some silicate anions, extended silicate chains, and silicate sheets are shown in Figures 14:16 and 14:17.

14:17 *Model of a silicate anion chain.*

Tetrahedral symmetry seems to be a common feature in God's creation. Remember that when design is evident in nature, the Creator simply used the same form over and over again as it pleased Him. This is why skeletons in the animal world are similar; each animal was created by the same Person.

Silicate chains (negatively charged) are held together by positive metal ions. In the mineral **asbestos**, the silicate chains are parallel to each other, forming a fibrous structure.

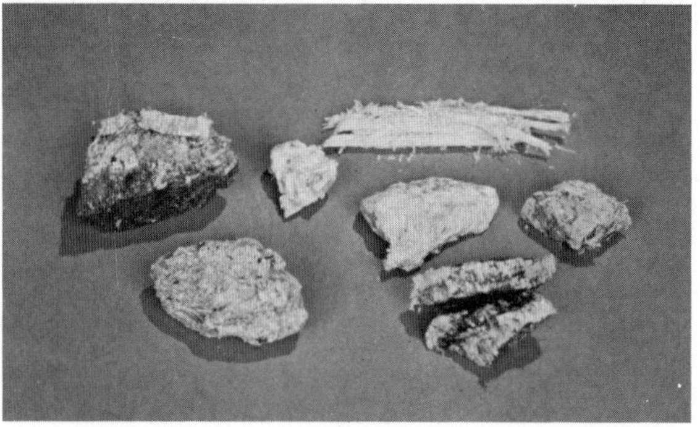

14:18 *Some asbestos minerals.*

Sheets of silicates are found in minerals such as **mica** and **talc**. Other cations may be present in the sheet structure, but the basic atom arrangement is the same.

Clays are made up of very small particles of silicates containing, among other things, aluminum oxide, iron oxide, and organic matter. Because clay particles are small, they have a great deal of surface area and will absorb many different substances. Rocks are also basically silicates; **granite** is 70 percent silica.

Experiment 14:6a *The Settling Rate of Clays*

1. Fill a pan with clay; then dry it in the oven.
2. After it is dried, crush the pieces of clay using a hammer.
3. Suspend about 2 grams of clay in 100 ml of water in a beaker. Mix by shaking well.
4. Turn off the lights in the room; then shine a flashlight into the beaker. What do you notice about the size of the suspended clay particles? Record your observation.

★ **Experiment 14:6b**

1. Mix 20 grams of clay into 1000 ml of cold water (the water should be as close to 0° C as possible). Mix 20 grams of clay into 1000 ml of warm water (the water should be as close to 50° C as possible). 1000 ml graduated cylinders are excellent for this purpose.
2. Observe the rate of settling of the clay particles in each of the containers for about 15 minutes. The particles settle faster in which container? Why? Record your observations and explanations.

A water-soluble silicate, known as **water glass**, can be formed by reacting sodium carbonate and silica:

$$Na_2CO_3 + SiO_2 \xrightarrow{\text{water}} \underset{\text{water glass}}{Na_2SiO_3} + CO_2\uparrow$$

This syrupy liquid, sodium silicate (Na_2SiO_3), has been used as an adhesive, particularly for holding cardboard together. Water glass can be used also in cleaning agents, waterproofing, and fireproofing.

14:10 Glass When most liquids freeze, they form a definite arrangement of atoms or molecules called **crystals**. If a liquid hardens, but not in a definite crystalline pattern, it is said to be a **glass**. A glass is a supercooled liquid, not a true crystalline solid. Glass hardens as it cools; it does not have a definite freezing point.

When most people speak of glass, they usually mean a transparent silicate used in bottles and window panes. The composition of typical glass is given below:

Compound	*Percent in glass*
SiO_2	80.5
B_2O_3	12.8
Na_2O	4.3
Al_2O_3	2.0
K_2O	0.4

The basic "structure" of glass is considered to be a *random* arrangement of tetrahedral silicates.

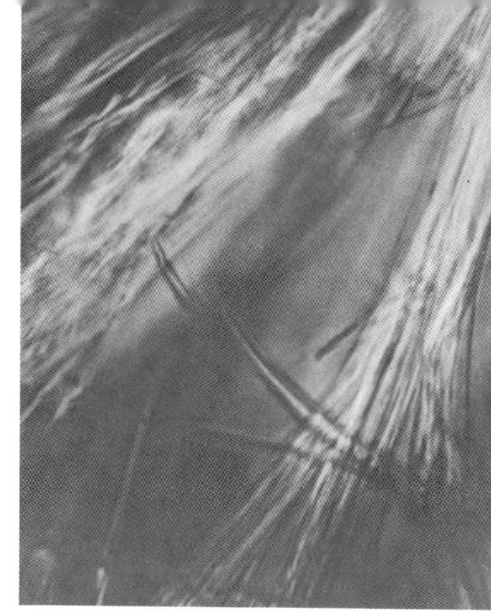

14:19 *A crystalline formation in glass.*

14:20 *This is not a giant pizza oven but is part of the world's largest gas hearth tempering system at PPG Industries' Mount Vernon, Ohio, plant. A worker in a special suit watches as glass is conveyed into the "oven," where it is softened. Then the glass is cooled at a controlled rate to produce tempered safety glass.*

14:21 *Many beautiful containers can be made from glass. (See chapter title page)*

There are many practical and ornamental uses of glass. Imagine a house without any windows! Glasses of many compositions, sizes, shapes, and colors have been made today. Anything from a small glass bead to the 200-inch Palomar telescope mirror can be made of glass. High temperature furnaces are required to melt and soften glass. Did you ever watch a glassblower? Using an open flame, he melts and shapes the glass into beautiful and intricate objects. Glassblowing is an art requiring much skill.

14:22 *Pyroceran glass-ceramic nose cones.*

14:23 *Molten brass is poured into a Vycor glass crucible on a block of ice. This glass has excellent resistance to thermal shock.*

14:11 Ceramics The ceramics industry includes the manufacture of cement, brick, tile, stoneware, china, pottery, terra-cotta, porcelain, and glass. The basic materials used in the ceramic industry are metal oxides and silicates, particularly clays. The red color of brick, terra-cotta, and many other clay products comes from the iron oxide impurity in the clay.

Much high-temperature equipment is needed to dry and harden ceramic items. Many ceramic items must be heated to 1200-1500°C in order to be "fired." Ceramic furnaces are electrically powered or gas-fired (the term "firing" comes from this type of furnace).

Porcelain glazes are applied to many ceramic items for protection and beauty. Porcelain is also glazed over metal to make bathtubs, sinks, stoves, refrigerators, and washing machines. The **glaze**, a mixture of oxides and silicates, fuses during firing to form a glass-like coating over the ceramic or metal surface.

14:24 *Bricks being removed from a kiln.*

14:25 *Spray and bake. Plastic truck box covers, reinforced with fiberglas, a ceramic product, move on an assembly line into spray paint booths and baking ovens.*

14:12 Silicones Chains of silicon, oxygen, carbon, and hydrogen are called **silicones.**

Typical silicone structure

$$-O-\underset{CH_3}{\overset{CH_3}{\underset{|}{\overset{|}{Si}}}}-O-\underset{CH_3}{\overset{CH_3}{\underset{|}{\overset{|}{Si}}}}-O-$$

Silicone products have excellent temperature and chemical stability. Silicones are used as gasket lubricants, electrical insulation, "super balls," golf-ball centers, waxes, and calking compounds. When astronaut Neil Armstrong made the first footprints on the moon, he was wearing silicone rubber boots. This was the only material that could withstand the extremes of temperatures on the lunar surface.

14:26 *Coil insulation (mica and glass tapes) for nuclear power plant pump motors is impregnated with a silicone resin. This insulation enables the system to be stable in heat, moisture, high voltage, and radiation.*

14:13 Semiconductors Both silicon and germanium are used in **semiconductor** devices, such as **transistors**, which have revolutionized the electronics industry. A great deal of semiconductor technology has been developed for the space and computer industry.

Silicon and germanium are "doped" (treated with small amounts of impurities) to make these electronic devices. One type of semiconductor is electron-rich, the other is electron-poor. An amazing variety of electronic devices can be made from semiconductors at great savings of space and weight.

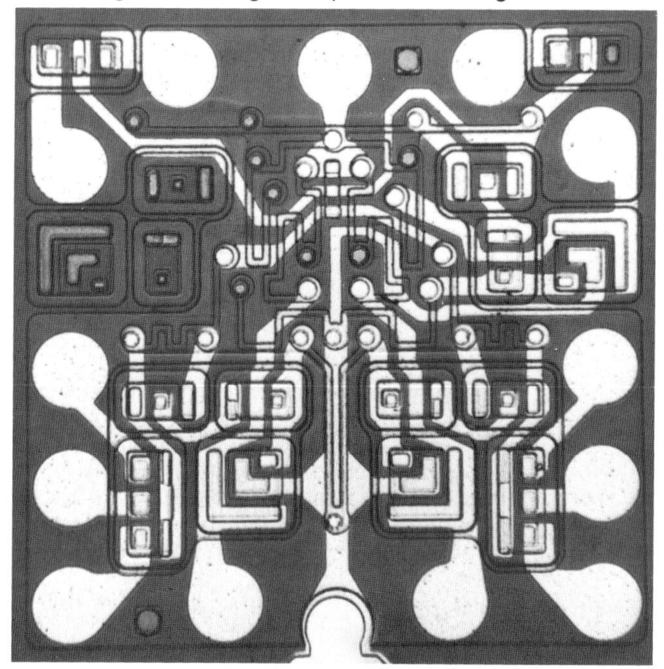

14:27 *A simple integrated circuit containing 14 transistors and 10 resistors.*

14:28 *A single silicon chip inside an integrated circuit package—contrasted with a silicon wafer with over 100 chips screened prior to scribing with diamond or laser techniques.*

14:29 *Various sizes and shapes of semiconductor devices.*

List of Terms

activated charcoal
anthracite
asbestos
bituminous
ceramics
charcoal
clay
coal
combustion
cristobalite
crystal
diamond
dry ice
glass
glaze
granite
graphite
lignite
limestone
metalloid
mica
natural gas
petroleum
quartz
semiconductor
silica
silicate
silicone
talc
tetrahedron
transistor
tridymite
volatile
water glass

Questions

1. Define metalloid.
2. What is the similarity of electron structure in the elements of the IVA group?
3. (a) Which element is the most metallic of the IVA group?
   (b) Which element acts more like a nonmetal than the rest?
4. What type of bonds do carbon and silicon normally form?
5. Name three sources of uncombined carbon.
6. List four Bible verses in which diamonds are mentioned.
7. What is the explanation given for why diamond is so much harder than graphite, even though both consist of carbon atoms?
8. Name three uses of graphite.
9. How did fossil fuel form?

10. (a) What gas is found in the exhaled gases of animals?
    (b) How does this gas form in the animal?
    (c) What type of organisms use this gas to produce food for themselves?
11. Carbon dioxide gas aids combustion. Is this statement true or false?
12. What is dry ice?
13. (a) Suggest one way carbon monoxide can be formed.
    (b) Is carbon monoxide poisonous?
    (c) Carbon monoxide can be burned in oxygen. Is this statement true or false?
14. Name three types of coal found naturally.
15. What naturally occurring form of carbon is the most expensive?
16. What are the two principal elements in soil and rock?
17. What is the geometric shape of the silica (SiO_2) molecule?
18. What are two uses of the quartz form of SiO_2?
19. What is the only acid that will attack silica (SiO_2)? Is silica resistant to most bases? Since glass consists of a great deal of silica, what would you suspect about the attack of acids and bases on glass?
20. What anion or radical is present in most minerals?
21. What is the geometric shape of the silicate structure?
22. What is meant by silicate sheets?
23. What is a positive Christian answer as to why there are so many tetrahedrally-arranged atoms in molecules in nature?
24. How do you account for the fibrous structure of asbestos?
25. Mica is a sheet-like mineral. How do you account for this sheet structure?
★26. Why do clays have such a large amount of surface area for a given volume?
27. Name five items around your house which are made of glass.
28. Glass is a supercooled liquid. Is this statement true or false?
29. Why is high-temperature equipment needed in the ceramics industry?
30. What gives red brick its color?
31. What are the raw materials for porcelain?
★32. Why is the arrangement of the molecules in many ceramic structures chain-like?
33. What are silicones?
34. Name two uses of silicones.
35. Silicon is used in what type of electronic equipment?

Student Activities

1. Compare the percent of ash in hard coal, soft coal, coke, and charcoal. Pulverize about five grams of each material using a mortar and pestle. Place each sample in a porcelain crucible and heat until it is completely burned. Calculate the percent of ash for each sample.
2. Write a paper on the structure and properties of clays and their uses. Some production techniques should be included.
3. Write a paper on the properties of various glasses and their uses. Some production techniques should be included.
4. Write a paper on the production of silicon for use in transistors. Discuss what is meant by the term semiconductor. Obtain some pictures of semiconductor devices to use in the report.

Photo next page
Giant roll of cord fabric is shown in the weaving operation at a textile plant.

170
FRONT

UNIT V DESCRIPTIVE CHEMISTRY Groups IA-IVA

Chapter 15 ★ Organic Chemistry

15:1 Alkanes *Hydrocarbons are compounds made up of carbon and hydrogen only.* The simplest hydrocarbons known are called **alkanes.** The carbon-hydrogen and carbon-carbon *single* covalent bonds are the only ones present in alkanes. The carbon-carbon electron sharing is represented this way:

C — C

The "stick" between the two carbon symbols represents the sharing of a pair of electrons between the two atoms; an electron is contributed by each atom. The carbon-hydrogen sharing is similarly shown:

C — H

Again the "stick" means a shared electron pair.

The simplest alkane is **methane** (CH_4). The **molecular orbital** representation of CH_4 is shown in Figure 15:1. A simpler way of illustrating the structure is

```
      H
      |
H  —  C  —  H
      |
      H
```

Remember what the "sticks" mean. This model on a flat page should not lead you to believe that methane is a "flat" molecule. The angles between the carbon-hydrogen bonds are about 109 degrees. The molecule is shaped like a tetrahedron.

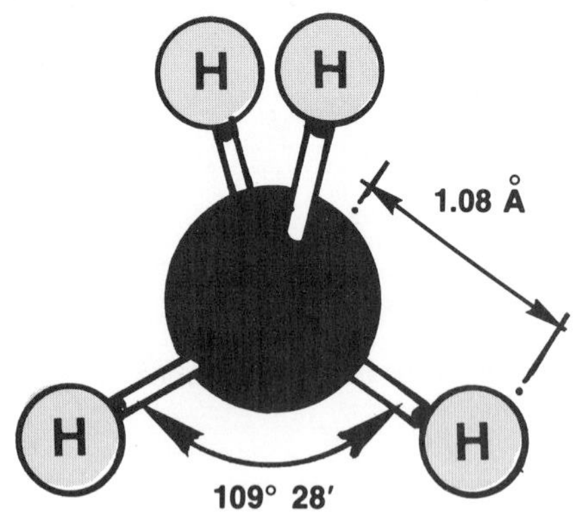

15:1 *Model of a methane molecule.*

Experiment 15:1 *Preparation of Methane*

1. Mix 1 gram of anhydrous sodium acetate ($NaC_2H_3O_2$) with 6 grams of solid soda lime ($CaO + NaOH$) in a clean, dry test tube.
2. Clamp the tube to a ringstand; then insert a one-hole rubber stopper and glass tubing into it. Connect it to a water-filled gas bottle as shown in Figure 15:2.
3. Heat the tube strongly at the *top* of the solid mixture. As gas is given off, move the flame gradually down the mixture.
4. When you remove the bottle from the trough, cover the mouth of the bottle with a glass plate *while it is still under the water.* The chemical reaction for the production of methane is

$$NaC_2H_3O_2 + NaOH \xrightarrow{CaO} CH_4\uparrow + Na_2CO_3$$

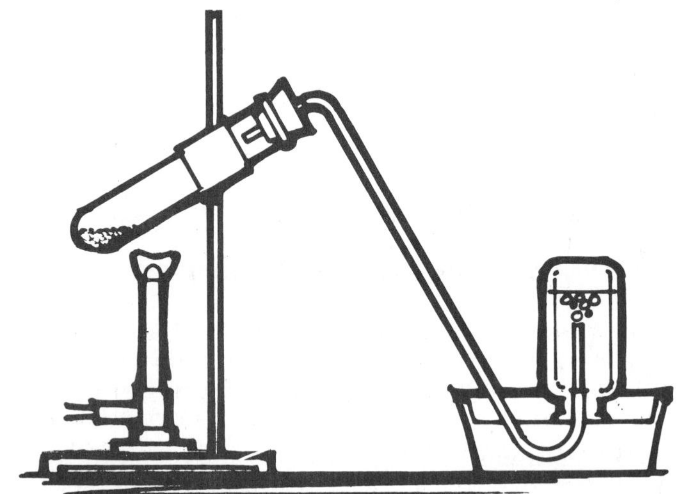

15:2 *Laboratory rig for the preparation of methane.*

5. Record the color, odor, and solubility in water of methane.
6. Slide the glass cover off the gas bottle and hold a burning splint over the mouth of the jar as shown in Figure 15:3. This will show you how combustible the gas is. *Do not hold your hands over the bottle.*

15:3 *Testing for the combustibility of methane.*

The second compound in the alkane series is the gas, ethane. **Ethane** has two carbon atoms and six hydrogens. The model for ethane is

```
    H   H
    |   |
H — C — C — H
    |   |
    H   H
```

Next in the alkane series is propane (three carbons), then butane (four carbons). These compounds are used as bottled gas.

For the fourth compound in the alkane series, butane, two structures are possible. The first is the model for normal butane (n-butane):

$$\begin{array}{ccccccccccc} & & H & & H & & H & & H & & \\ & & | & & | & & | & & | & & \\ H & - & C & - & C & - & C & - & C & - & H \\ & & | & & | & & | & & | & & \\ & & H & & H & & H & & H & & \end{array} \qquad (C_4H_{10})$$

$$\text{or} \quad CH_3CH_2CH_2CH_3$$

The second possibility, isobutane, is

$$\begin{array}{ccccccccccccc} & & & & & & H & & & & & & \\ & & & & & & | & & & & & & \\ & & H & & H & - & C & - & H & & H & & \\ & & | & & & & | & & & & | & & \\ H & - & C & — & — & — & C & — & — & — & C & - & H \\ & & | & & & & | & & & & | & & \\ & & H & & & & H & & & & H & & \end{array} \qquad (C_4H_{10})$$

$$\text{or} \quad \begin{array}{c} \quad CH_3 \\ \quad | \\ CH_3CHCH_3 \end{array}$$

Both compounds have the same molecular formula (C_4H_{10}) but different carbon and hydrogen arrangements. Compounds like these, which have the same molecular formula, are referred to as **isomers**.

Farther along in the alkane series are the straight chain compounds: pentane (five carbons), hexane (six carbons), heptane (seven carbons), and octane (eight carbons). You may recognize the name octane; isooctane, an isomer of octane, is used as fuel in piston-driven machines, such as airplanes (but not jets or prop jets) and automobiles. The straight chain compounds are always named *normal* (abbreviated "n-"), such as n-hexane (CH_3-CH_2-CH_2-CH_2-CH_2-CH_3).

The alkane compounds with only a few carbon atoms are gases at room temperature (methane, for example). As the number of carbons per molecule increases, the compounds are liquids at room temperature (n-pentane, for example). When the number of carbons per molecule increases to 18, the alkane is solid at room temperature (octadecane, $C_{18}H_{38}$, for example). The solids are waxy in nature; as a matter of fact, the alkanes are referred to as paraffin hydrocarbons.

All of the alkane compounds are insoluble in water. They are soluble, however, in certain organic compounds which we will discuss later.

15:2 Petroleum Most of the hydrocarbons come from coal, petroleum, and natural gas. Petroleum in particular must be refined to separate the various hydrocarbons.

15:4 *Fluid catalytic cracking unit at an oil refinery.*

Table 15-1 *Petroleum Distillates**

Specific Distillates	Boiling Temperature (°C)	Carbon Chain Length	Uses
gas	below 20°	C_1 to C_4	fuel
petroleum ether	20-60°	C_5 to C_6	solvent, dry cleaning, and refrigerant
ligroin	60-100	C_6 to C_7	solvent, dry cleaning, and refrigerant
gasoline	40-205°	C_6 to C_{12}	motor fuel
kerosene	175-235°	C_{12} to C_{20}	diesel fuel
gas oil	above 275°		furnace oils
lubricating oil	nonvolatile liquids		lubrication
paraffin	melts 51-55°	C_{20} up	candles, waterproofing
asphalt or petroleum coke	residue		fuel, road construction material

* *The boiling temperature and carbon chain length can vary depending on whether the petroleum is asphalt- or paraffin-based.*

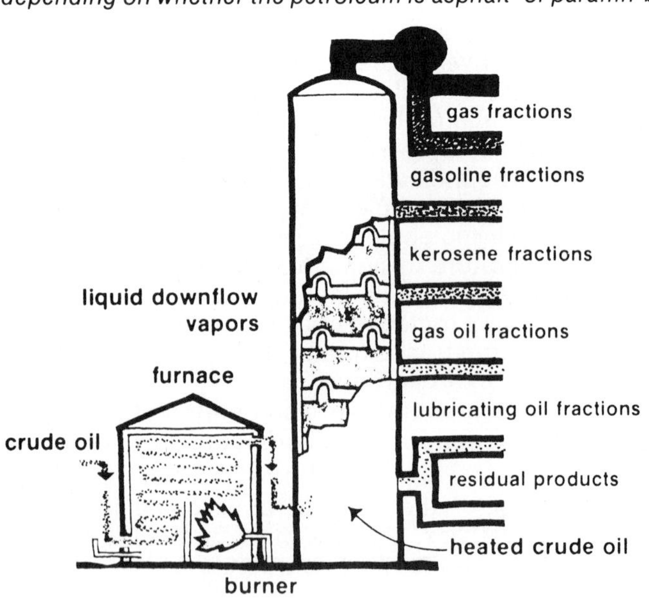

15:5 *Fractioning tower used for petroleum distillation.*

15:6 *An oil refinery at night.*

Experiment 15:2 *Distillation of Petroleum*

This entire experiment should be done by the teacher with the class observing.

1. Set up the apparatus according to Figure 15:7 using a dry distilling flask and a dry air condenser. Use 3 clean, dry beakers to collect the distillate.
2. Pour 200 ml of petroleum through a funnel into the distilling flask.
3. Heat with a low flame so that the distillate comes out at a rate of about 1 drop per second.
4. Collect the first fraction coming out from 40 to 150° C, the second from 150 to 200° C, the third from 200 to 300° C. What remains in the flask is the fourth fraction.
5. Check Table 15-1 to identify the fractions.

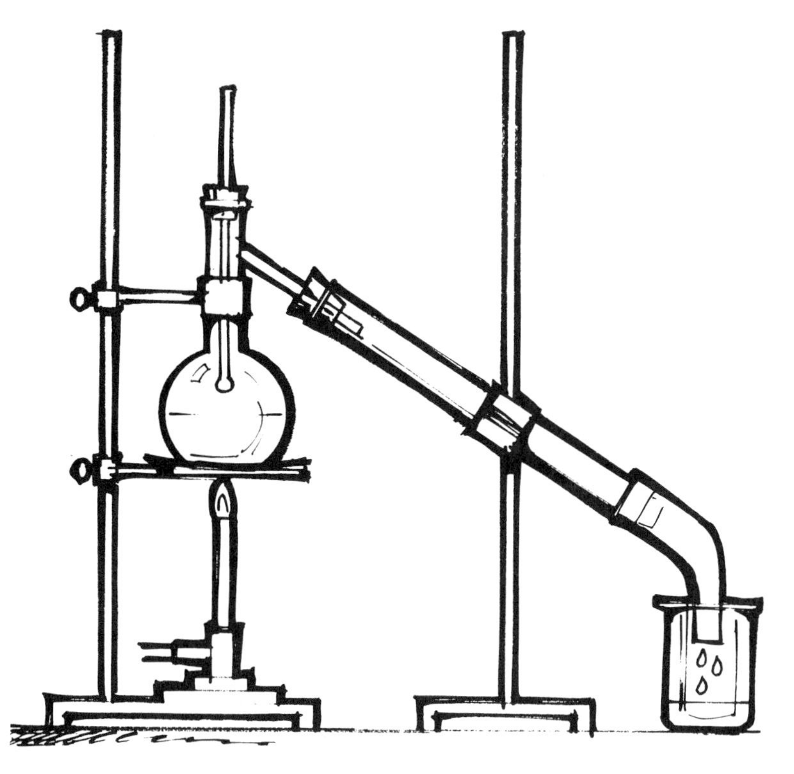

15:7 *Apparatus for the distillation of petroleum.*

15:3 Reactions of Alkanes Most of the alkanes are used for fuel. Therefore, *combustion* is the primary reaction of interest with any of the alkane compounds. Examples of complete combustion are

$$\underset{\text{methane}}{CH_4} + 2O_2 \longrightarrow CO_2\uparrow + 2H_2O\uparrow + \text{heat}$$

$$\underset{\text{n-octane}}{2C_8H_{18}} + 25O_2 \longrightarrow 16CO_2\uparrow + 18H_2O\uparrow + \text{heat}$$

Incomplete combustion (caused by a limited supply of oxygen) would result in the following reactions:

$$2CH_4 + 3O_2 \longrightarrow 2CO\uparrow + 4H_2O\uparrow + \text{heat}$$

$$2C_8H_{18} + 17O_2 \longrightarrow 16CO\uparrow + 18H_2O\uparrow + \text{heat}$$

or

$$CH_4 + O_2 \longrightarrow C + 2H_2O\uparrow + \text{heat}$$

$$2C_8H_{18} + 9O_2 \longrightarrow 16C + 18H_2O\uparrow + \text{heat}$$

The last two reactions produce carbon, which can be seen in the blackened exhaust fumes from engines. When someone "guns" his car, it is a gross waste of fuel, as evidenced by the exhaust fumes that are emitted. If complete combustion occurs in efficient engines, the exhaust gas is invisible. The less you see in exhaust smoke, the more efficient the engine is.

All of the hydrogens in methane can be replaced by chlorines to form the compound carbon tetrachloride (CCl_4).

```
      Cl
      |
Cl —  C  — Cl
      |
      Cl
```

Chlorine substitutes for hydrogen in the compound. Carbon tetrachloride is a useful cleaning fluid. The fluid evaporates rapidly after the cleaning operation.

15:4 Alkenes We have discussed only carbon-carbon **single bonds**. In some organic compounds there are **double bonds** between carbon atoms. They share four electrons together in the same bond; each atom contributes two electrons. This is referred to as a double bond; any hydrocarbon with this type of bond is called an **alkene**.

Ethylene is the simplest of the alkenes

```
H         H
 \       /
  C === C
 /       \
H         H
```

or

$$CH_2{=}CH_2 \quad (C_2H_4)$$

ethylene

The molecular orbital model of the ethylene compound is shown in Figure 15:8.

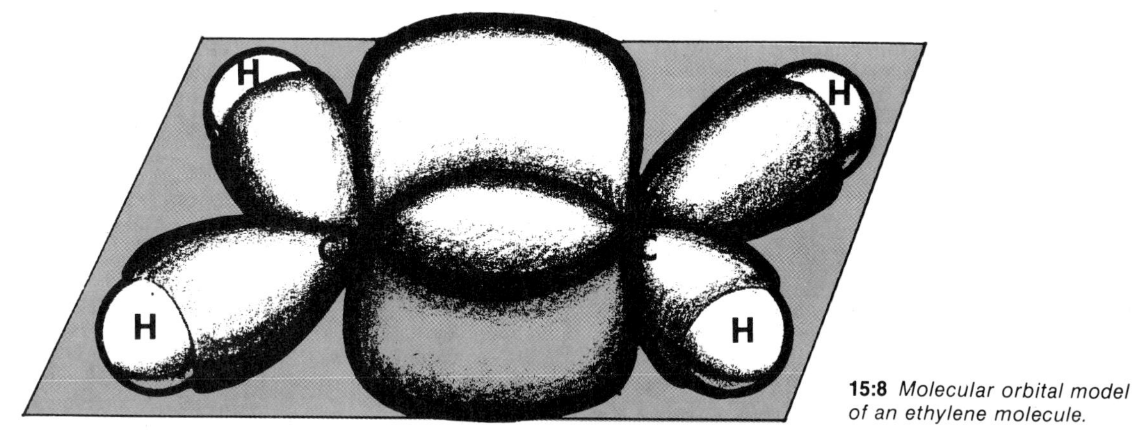

15:8 *Molecular orbital model of an ethylene molecule.*

Carbon atoms have four electrons to share. Ethane carbons can have three hydrogens per carbon, the ethylene only two. If two electrons per atom are already shared with another carbon atom, they have only two electrons each to share with hydrogen atoms.

$$\mathrm{H-\overset{\displaystyle H}{\underset{\displaystyle H}{C}}-\overset{\displaystyle H}{\underset{\displaystyle H}{C}}-H}$$

ethane

$$\mathrm{H-\overset{\displaystyle H}{C}{=}\overset{\displaystyle H}{C}-H}$$

ethylene

$$\mathrm{H-\overset{\displaystyle H}{\underset{\displaystyle H}{C}}-H} \qquad \mathrm{C-\overset{\displaystyle H}{\underset{\displaystyle H}{C}}-H} \qquad \mathrm{C-\overset{\displaystyle H}{\underset{\displaystyle H}{C}}-C} \qquad \mathrm{{}^{H\diagdown}_{H\diagup}C\overset{1}{\underset{2}{=}}C}$$

(a) (b) (c) (d)

15:9 *Some possibilities of four bonds per carbon.*

15:5 Alkynes The group of hydrocarbons which exhibits a carbon-carbon **triple bond** is called **alkynes**. The simplest alkyne compound is **acetylene** (C_2H_2), the gas used in welding:

$$\mathrm{H-C{\equiv}C-H}$$

Since the carbons share six electrons, they can only accommodate one hydrogen each. A molecular orbital model of acetylene is shown in Figure 15:10. Any time there are carbons with double or triple bonds, they are referred to as **unsaturated hydrocarbons**.

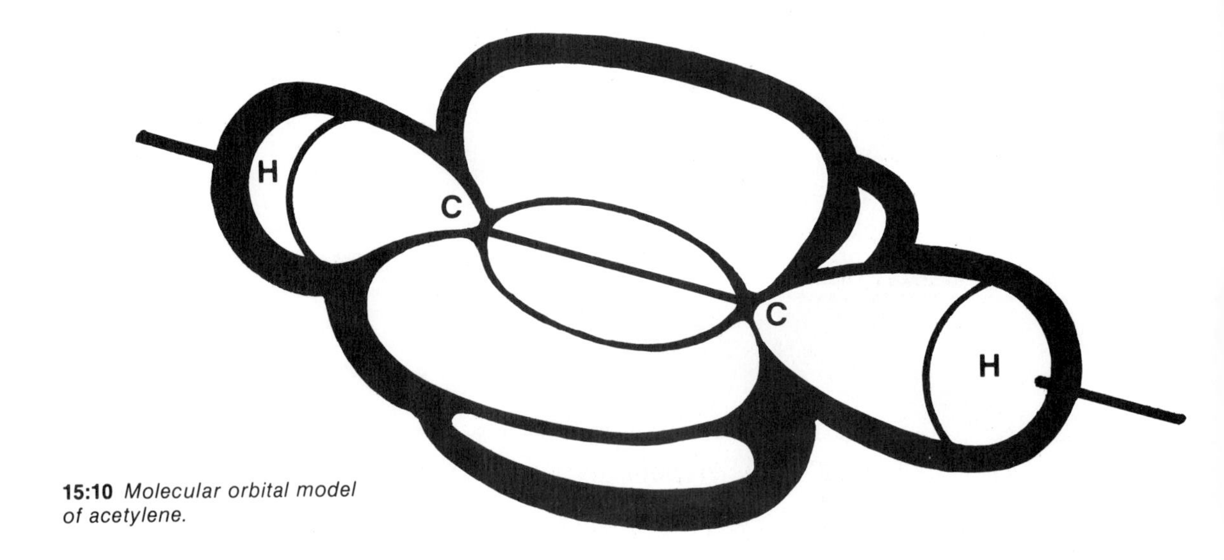

15:10 *Molecular orbital model of acetylene.*

Experiment 15:3 *Preparation of Acetylene*

1. Place 1 gram of calcium carbide (CaC_2) in a test tube and set up the experiment the same way as in the preparation of methane shown in Figure 15:2. However, fill only a test tube, not a bottle, with gas.
2. Add 10 ml of water to the calcium carbide, stoppering the tube immediately afterward.
3. Let the air bubble out of the delivery tube for 1 or 2 seconds before placing it in the test tube full of water. This will prevent the formation of an explosive mixture.
4. Write the reaction for the formation of acetylene (C_2H_2).
5. Remove the test tube full of gas, slide the glass plate off it, and hold a burning splint over the top of the tube. This will illustrate how combustible the gas is. *Do not point the tube toward you or anyone else nearby.* Why is there so much soot in the tube?

In our discussion of hydrocarbons, we have introduced simple compounds only. Many interesting ring compounds have also been discovered. Some of their shapes are illustrated in Figure 15:11.

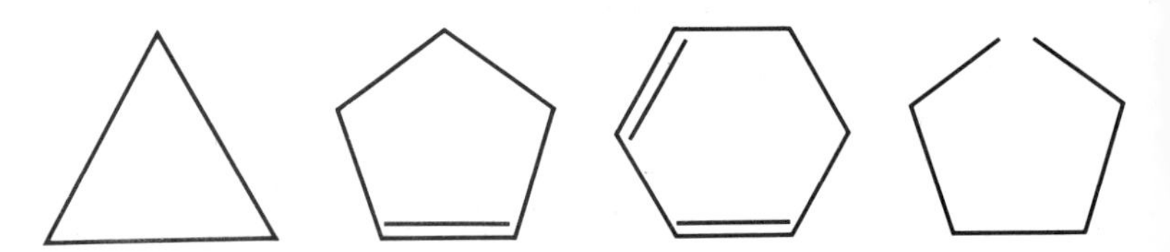

15:11 *Some possible shapes of the molecules of different hydrocarbons.*

15:6 Aromatic Hydrocarbons The group of ring compounds known as **aromatic hydrocarbons** is given that name because of their characteristic fragrance. The simplest compound of this group is **benzene** (C_6H_6), which is a highly volatile liquid at room temperature. The benzene structure suggested by Friedrich August Kekulé (1829-1896), a German chemist, is

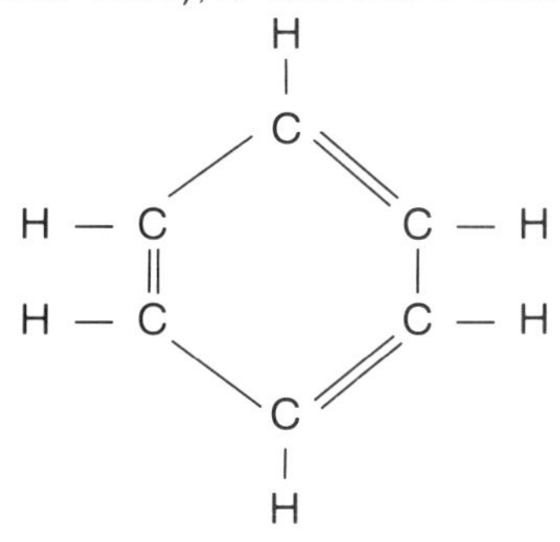

The modern molecular orbital model for benzene, Figure 15:12, yields a hexagonal shape similar to the **Kekulé model**. The outer electron orbitals (top and bottom doughnuts) are circular with the carbon-connecting orbitals (between the doughnuts) maintaining a hexagonal relationship.

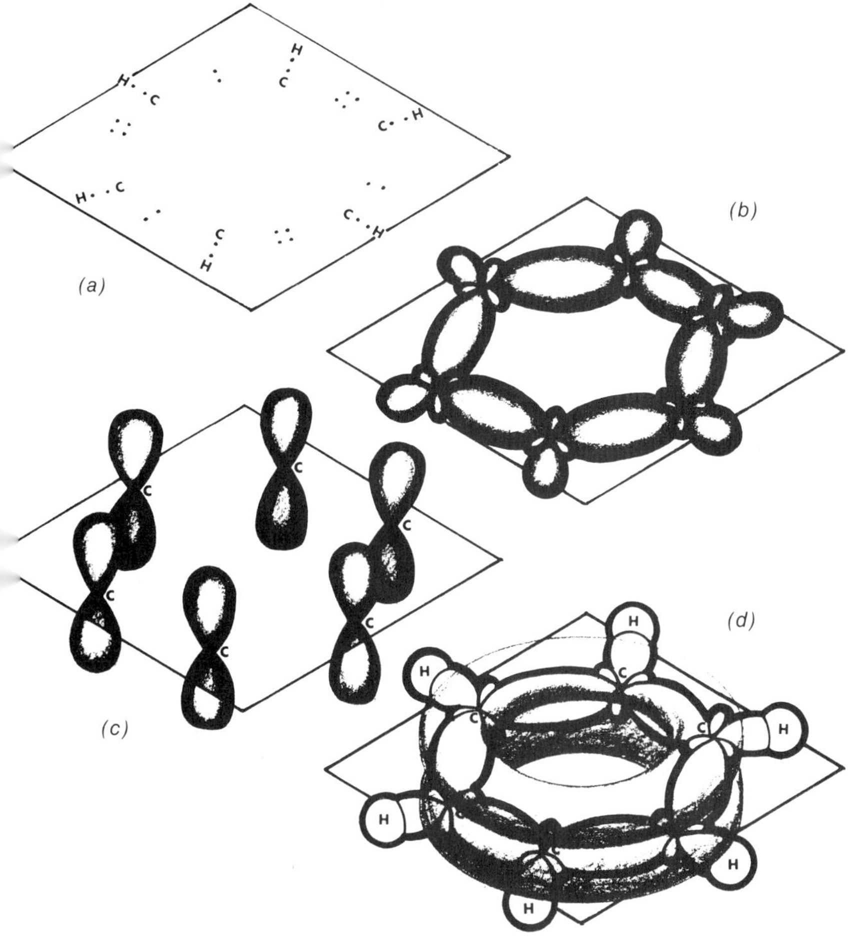

15:12 *(a) Planar arrangement of carbon and hydrogen atoms with bonding electrons in a model of a benzene molecule. (b) Planar molecular orbitals in the model of a benzene molecule. (c) Electron orbitals that make up the outer molecular orbitals (doughnuts) in the model of the benzene molecule. (d) Complete molecular orbital model of benzene molecule showing outer electron "doughnuts."*

Benzene is an excellent solvent for many organic compounds. It is also a good cleaning fluid; but do not use it on plastics or painted surfaces!

The benzene ring is very stable, and some of the reactions involving benzene are simply *additions* to the carbon. However, most of the reactions involve *substitution* for a hydrogen.

Addition

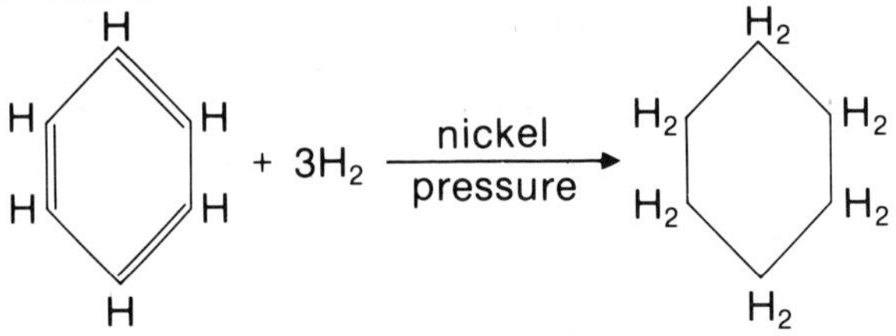

(benzene ring with H on each carbon) $+ 3H_2 \xrightarrow[\text{pressure}]{\text{nickel}}$ (ring with H_2 on each carbon)

cyclohexane

Substitution

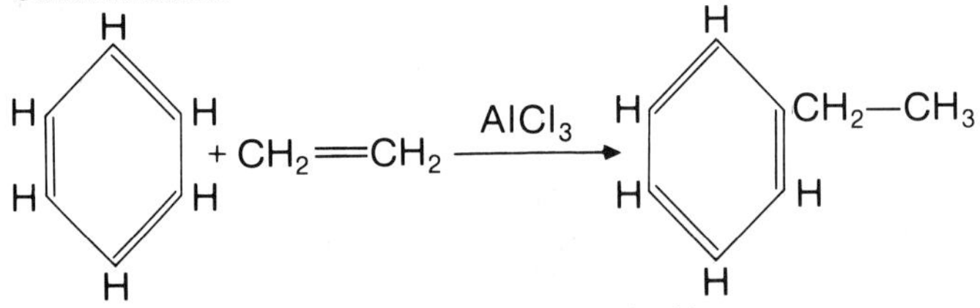

(benzene ring with H on each carbon) $+ CH_2{=}CH_2 \xrightarrow{AlCl_3}$ (benzene ring with H on five carbons and $CH_2{-}CH_3$ on one)

ethylbenzene

In the above reactions the carbons are understood to be in the benzene ring.

15:7 Alcohols The alcohols are thought of as carbon compounds formed from water. If one of the hydrogens in water is replaced by a hydrocarbon group, an alcohol is formed:

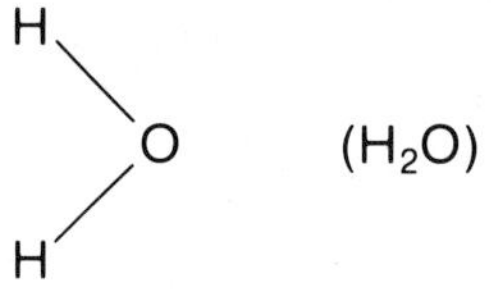

H–O–H (H_2O)

water

CH_3–O–H (CH_3OH)

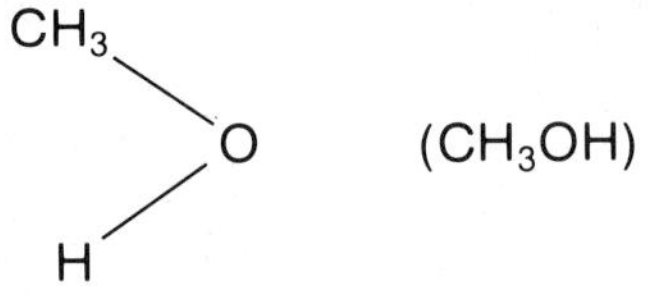

methyl alcohol

Small chain alcohols are soluble in water and in a few hydrocarbons. Many drugs are dissolved in alcohols because of their excellent solvent properties. Also, alcohols are used as solvents in many of the pressurized canned sprays. After the solution is sprayed on an object, the alcohol will evaporate. The simplest alcohol is methyl alcohol. It is commonly known as wood alcohol and is very poisonous.

Ethyl alcohol (C_2H_5OH) can be produced by fermentation. Sugar is converted into alcohol and carbon dioxide by biological catalysts called enzymes. But first, starch must be transformed into sugar by an enzyme found in malt (freshly sprouted barley).

$$\text{starch} \xrightarrow{\text{malt, water}} \text{sugar}$$

$$\text{sugar} \xrightarrow[\text{of yeast}]{\text{enzymes}} C_2H_5OH + CO_2\uparrow$$

The alcohol in many beers, whiskeys, and liquors is ethyl alcohol. Scripture strongly condemns the misuse of this chemical as a beverage (Proverbs 20:1; Ephesians 5:18). Noah made a fool of himself after the Flood by getting drunk (Genesis 9:21). A great many crimes are committed by people who have been drinking alcoholic beverages. The world would be a better place to live if it could rid itself of alcoholic drinks.

The consumption of too much alcohol will eventually destroy a person's body. Alcohol acts as a depressant and robs a person of his quick reflexes. It tends to dehydrate the body, and it produces more sugar than the body can use.

Clever advertising convinces many people of the "joys" to be found by drinking liquor. But the truth is that the liquor business is responsible for the ruin of thousands of people. The true picture of the effects of liquor is given by such verses as Proverbs 20:1 and Proverbs 23:19-21, 29-35. These verses paint a sordid portrait of anyone who overloads his system with alcohol by drinking liquor. Thirty percent of all hospital beds are occupied by people with alcohol-related illnesses.

15:13 *Litter thrown on a highway by beer drinkers.*

Another familiar alcohol is isopropyl, or rubbing alcohol.

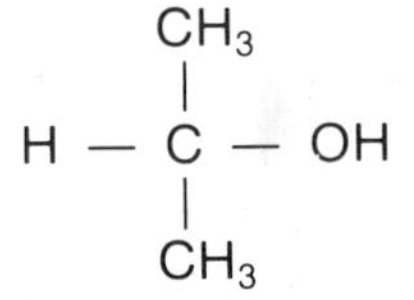

Alcohols can be used as **antiseptics** to kill bacteria.

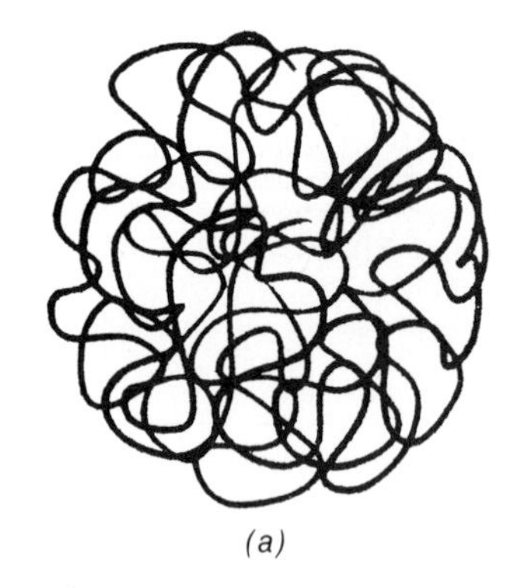

(a)

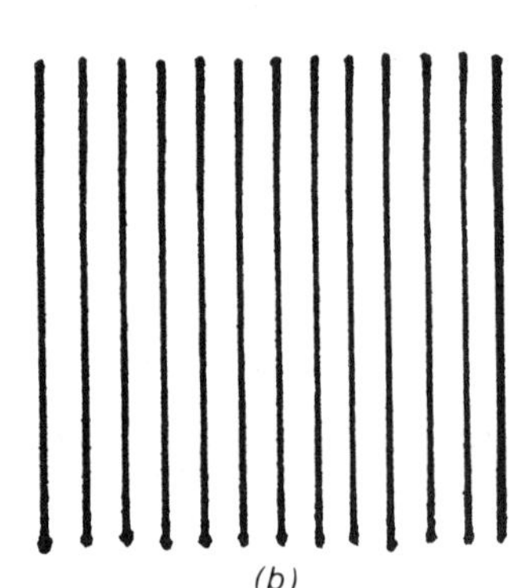

(b)

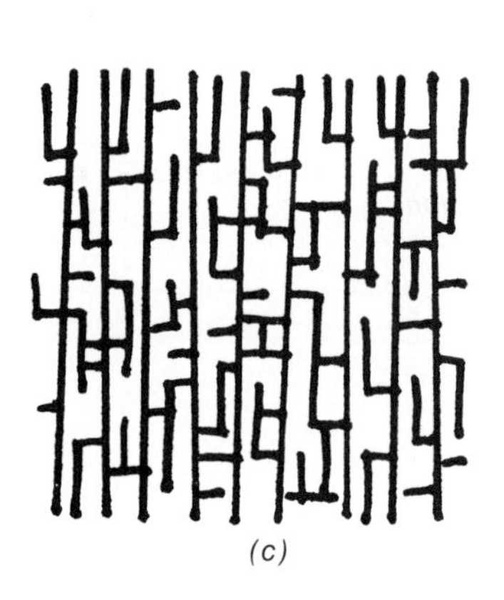

(c)

15:14 *Possible shapes of polymer chains.*

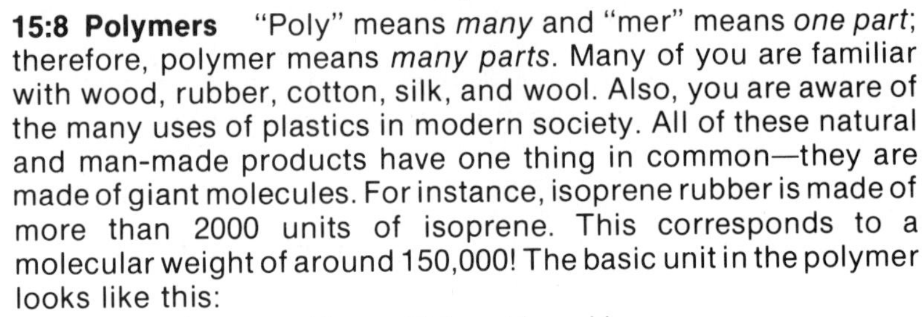

15:8 Polymers "Poly" means *many* and "mer" means *one part*; therefore, polymer means *many parts*. Many of you are familiar with wood, rubber, cotton, silk, and wool. Also, you are aware of the many uses of plastics in modern society. All of these natural and man-made products have one thing in common—they are made of giant molecules. For instance, isoprene rubber is made of more than 2000 units of isoprene. This corresponds to a molecular weight of around 150,000! The basic unit in the polymer looks like this:

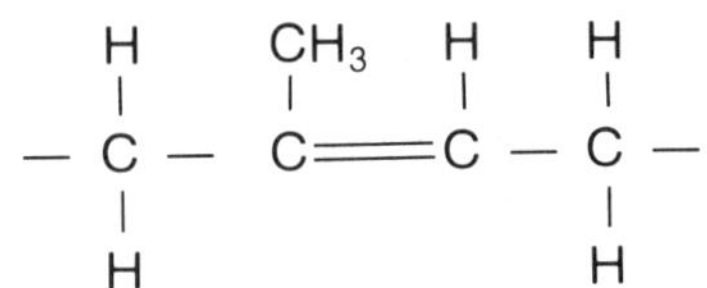

Hydrocarbon units can be added together to make giant molecules. Normally, the process begins with an unsaturated hydrocarbon, such as ethylene.

$$CH_2{=}CH_2 + CH_2{=}CH_2 \xrightarrow{\text{catalyst}} CH_3CH_2CH{=}CH_2$$

This reaction destroys a double bond and results in a dimer (one molecule made up of 2 smaller molecules). Continuing the process:

$$CH_3CH_2CH{=}CH_2 + CH_2{=}CH_2 \xrightarrow{\text{catalyst}} CH_3CH_2CH_2CH_2CH{=}CH_2$$

Another double bond is eliminated and a trimer is formed. If this process were continued, the final product would be polyethylene, a well-known, tough, flexible plastic. The "part" or "mer" which is repeated over and over is

$$-\ CH_2CH_2\ -$$

The individual polymers may take various shapes after they are formed into an object. The actual polymer chain may be wound within itself like spaghetti or a ball of yarn (see Figure 15:14a). It could also resemble straight chains (see Figure 15:14b), or branched chains (see Figure 15:14c). Some well-known polymers are nylon, celluloid, neoprene, bakelite, and styrene. Polymer chemistry is a special, fascinating branch of technology. Think of all the things you use today which are made of polymers!

15:15 *Natural and man-made fibers are polymers. Notice how many textile products are shown in this illustration.*

15:16 *The fibers represented in these textile end-products have been dyed. Much organic chemistry is involved in the production of textile dyes and the dyeing of fibers.*

15:17 *Giant roll of cord fabric is shown in the weaving operation at a textile plant.* *(See chapter title page)*

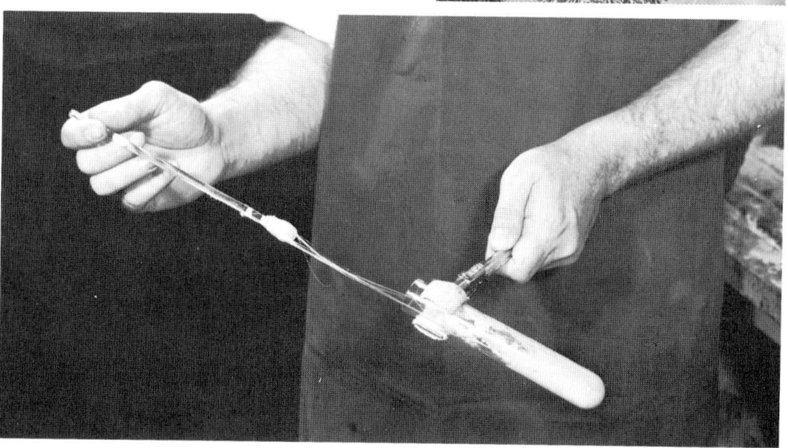

15:18 *It was an experiment such as this that suggested the fiber-forming possibilities of nylon. Here, in its most elementary form, is the birth of a man-made fiber (nylon, a polymer) in a research chemist's test tube.*

15:19 *The extrusion of polyvinyl chloride (PVC) polymer pipe.*

List of Terms

acetylene
alcohol
alkane
alkene
alkyne
antiseptic
aromatic hydrocarbons
benzene
double bond
ethane
isomer
Kekulé model
methane
molecular orbital
petroleum
polymer
single bond
triple bond
unsaturated hydrocarbons

Questions

1. What is the group of simplest hydrocarbons called?
2. Draw a model of the propane molecule using "sticks" as bonds.
3. Draw models for all the possible isomers of pentane.
4. What is the geometric shape of the methane molecule?
5. Name four alkanes which are gases at room temperature.
6. Give the total number of carbon atoms in each of the following alkanes: propane, methane, butane, ethane, and heptane.
7. Which two alkanes are used as bottled gas?
8. What is a common use of isooctane?
9. Write the combustion equation for hexane (C_6H_{14}). Assume complete combustion.
10. Chemically, what is wrong with fast starts in an automobile?
11. What is a use of carbon tetrachloride (CCl_4)?
12. What are the structural differences between alkanes and alkenes?
13. How is a carbon-carbon double bond represented on paper?
14. When electron sharing is mentioned, what type of bond is the chemist talking about?
15. What types of carbon-carbon bonds are characteristic of alkynes? How are they represented on paper?
16. What is the primary use of acetylene?
17. List two uses of the compound benzene.
18. Are both of these models for the benzene ring equally correct? Explain.

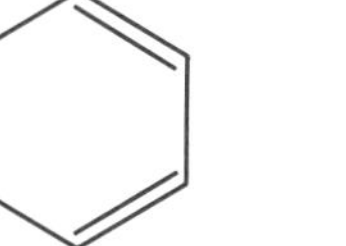
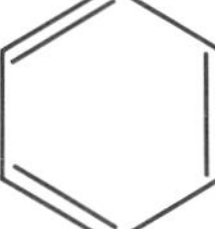

19. What is the advantage of the molecular orbital model of benzene over the Kekulé model?
20. List two uses of alcohols.
21. What is the common name of methyl alcohol? Is it poisonous?
22. Write out three Bible verses which condemn the use of alcohol as a beverage.
23. List five harmful physiologic effects of drinking too much alcohol.
24. What percentage of hospital bed occupancy can be attributed to the treatment of ailments and accidents caused by liquor consumption?
25. Define antiseptic.
26. Define "mer" and "poly."
27. Name five useful substances that consist of polymers.
28. Illustrate straight-chained and branched polymers.

Student Activities

1. Make models of several organic compounds discussed in this chapter. You may use styrofoam, or paper and sticks. Discuss briefly the production of the compounds and their uses.
2. Write a paper on the plastics industry. Discuss the chemistry and formation of several different polymers. Include diagrams or photographs of various products. Trace out the industrial processes necessary to get the raw material into the usable polymer and into the fabricated part.
3. Obtain medical and scientific information and make a report on the harmful effects of alcohol on the human body.

Photo next page
Cellulose makes attractive cardboard containers.

UNIT V DESCRIPTIVE CHEMISTRY
Groups IA-IVA

Chapter 16 ★ Biochemistry

16:1 Chemistry of Living Organisms The branch of chemistry relating to the chemical processes that occur in living organisms is called biochemistry. The fields of biology and chemistry overlap in biochemistry. Because of the complexity of vital processes and the complexity of substances found in living organisms, we can only introduce a few general subject areas and give you some simple examples.

The organic material in living cells is made up mainly of **carbohydrates, lipids** (fats and oils), and **proteins**. Certain similarities of structure and chemical behavior are common to each group.

16:2 Carbohydrates Carbohydrates are compounds of carbon, hydrogen, and oxygen. This name is given because these compounds contain hydrogen and oxygen in the same proportions as water. Table sugar, with the formula $C_{12}H_{22}O_{11}$, can be written as $C_{12}(H_2O)_{11}$. However, carbohydrates are not true hydrates, because if the water is driven off by heating, it cannot be restored to the molecule. Heating to remove the water will cause sugar to decompose. (The removal of water by means of concentrated sulfuric acid is shown in the photographs of Figure 19:10.)

Carbohydrates are the main products of life processes in plants. The supporting structure of plants is made of carbohydrates. Also, carbohydrates provide the main source of energy for animals.

16:1 *Production biochemistry. These huge fermentation tanks, filled with a sterile nutrient medium, are inoculated with the high-yielding culture grown in seed tanks.*

Carbohydrates can be roughly classified into sugars and starches. This is not absolutely correct, but for our discussion it will be satisfactory.

16:3 Sugars (Monosaccharides and Disaccharides) The **sugars** are colorless, crystalline, water-soluble substances with a sweet taste. The most widely distributed sugar is **glucose** (also called dextrose, grape sugar, or corn sugar). Glucose is found in many plants in the leaves, sap, flowers, honey, and fruits. Ripe grapes may contain as much as 30 percent glucose. Glucose is the source of energy for most living organisms. Blood contains 0.08 to 1.0 percent glucose.

The chemical structure of glucose ($C_6H_{12}O_6$) is

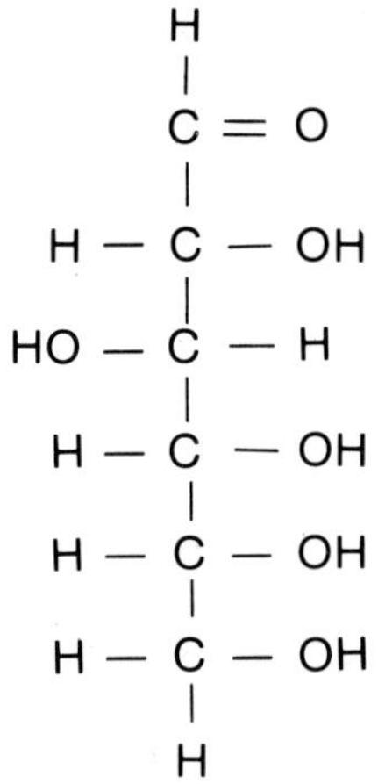

```
         H
         |
         C = O
         |
    H  — C — OH
         |
    HO — C — H
         |
    H  — C — OH
         |
    H  — C — OH
         |
    H  — C — OH
         |
         H
```

D-glucose

An interesting reaction, called **mutarotation**, takes place internally in glucose if the sugar is placed in water. The following ring structures form:

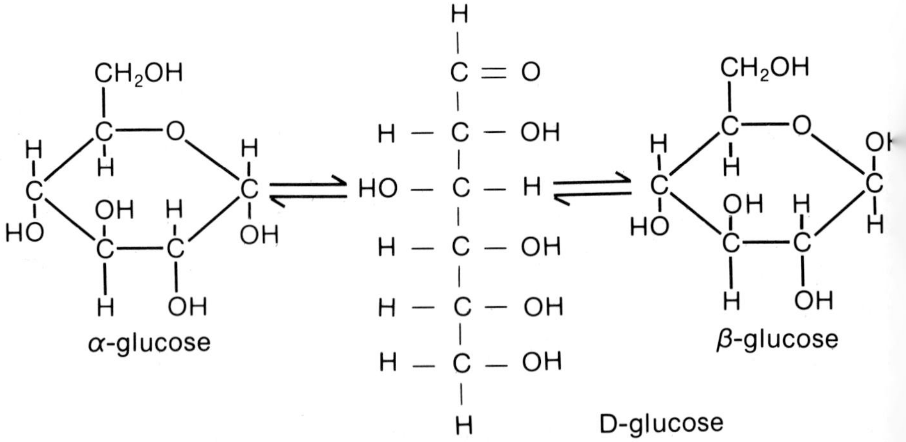

```
      CH2OH                    H                      CH2OH
      |                        |                      |
      C———O                    C = O                  C———O
  H  / H   \   H               |                  H  / H   \   OH
  | /       \  |          H  — C — OH             | /       \  |
  C          C  ⇌    HO — C — H   ⇌                C          C
  | \ OH  H /  |          H  — C — OH             | \ OH  H /  |
 HO  C———C    OH          H  — C — OH            HO  C———C    H
     |   |                H  — C — OH                |   |
     H   OH                    |                     H   OH
                               H
 α-glucose                 D-glucose              β-glucose
```

The second most important sugar is fructose ($C_6H_{12}O_6$), also called levulose, or fruit sugar. Fructose is found in fruit juices and honey; it is the compound that makes honey so sweet.

fructose

Fructose reacts internally, like glucose, to form the ring structure shown above.

A sugar that is a combination of glucose and fructose is sucrose ($C_{12}H_{22}O_{11}$), also called cane sugar or beet sugar. Sucrose is found in sugar cane, sugar beets, sugar maple, and in the sap of many plants. The structure of sucrose is

sucrose

Upon eating sucrose, the body will *hydrolyze* it into glucose and fructose.

$$\underset{\text{sucrose}}{C_{12}H_{22}O_{11}} + H_2O \longrightarrow \underset{\text{glucose}}{C_6H_{12}O_6} + \underset{\text{fructose}}{C_6H_{12}O_6}$$

Glucose and fructose are considered **monosaccharides** because they will not react with water. Sucrose is called a **disaccharide** because it reacts with water to form glucose and fructose. **Polysaccharides** react with water to form a large number of monosaccharides.

16:2 *Production of table sugar. Raw sugar is mixed with a dose of warm syrup in a mingler in order to loosen the film of molasses on the crystal. This mixture is poured into centrifugal machines (shown below). The liquid is forced through a fine screen while the crystals remain behind. As the spin cycle nears its end, a stream of water is sprayed onto the crystals to complete the washing.*

16;3 *A Laue transmission X-ray pattern of sucrose. Each spot represents a particular set of molecular planes.*

Experiment 16:1 *Dehydration of Sugar*

1. Heat some table sugar in an evaporating dish with a Bunsen burner flame. What happens? What is the residue?
2. Very carefully pour some concentrated sulfuric acid onto a small amount of sugar in a beaker. What are the black particles in the foam? (See Figure 19:10.)

Experiment 16:2 *Caramelization of Sugar*

When sugar is heated slowly a series of reactions occurs and "caramel" is formed.

1. Place 1 tablespoonful of sugar in a clean evaporating dish.
2. Heat the dish slowly, stirring the sugar with a stirring rod.
3. When the sugar melts and turns light brown, stop heating the dish.
4. Cool the dish and add 10 ml of water.
5. Heat again and stir to dissolve the "caramel."
6. Test the solution with pink and blue litmus paper. Record the result. Taste the solution. Is it bitter or sweet?
7. Add a lump of $NaHCO_3$ to the solution. Is CO_2 given off?

If caramelization is done properly, flavorful compounds are formed. If the sugar is overheated, acids form and give the mixture a bitter taste.

16:4 Starch (Polysaccharide) A polymer formed from the union of many sugar molecules is called **starch**. The general structural formula of starch is $(C_6H_{10}O_5)_n$, where n is very large. Starch is formed in plants and is stored as a reserve food supply, usually in the seeds and tubers. Seedlings use this starch for nutrition until they are able to care for themselves. Starch is obtained from corn, rice, wheat, and potatoes.

Starch will form a fine suspension of particles in hot water. Further hydrolysis will cause starch to decompose into **dextrins**, which are sticky substances used as glue for envelopes and postage stamps. If a fine suspension of starch is sprayed on clothing, the heat from an iron causes the hydrolysis products to form. This leaves the clothes stiff and shiny.

16:4 *Some starch products displayed in a supermarket.*

Experiment 16:3 *Hydrolysis of Starch*

1. Mix 0.2 grams of starch with 3 ml of water.
2. Add this mixture to 20 ml of boiling water.
3. Let this stand until it cools. Does the "thickness" of the solution change? Why?

16:5 Cellulose (Polysaccharide) The most abundant carbohydrate is **cellulose**. Cellulose makes up the skeletal material in all plants.

Table 16-1 *Cellulose Content of Certain Materials*

Substance	Percent Cellulose
cotton	95
coniferous wood	60
cereal straws	35
absorbent cotton	100
filter paper	100

Cellulose is composed of 1500 to 3000 glucose units; it can be hydrolyzed to glucose by boiling it in dilute acids. Cellulose is the raw material for many products, such as cotton textile, lumber, paper, rayon, cellophane, gunpowder, and dynamite. **Paper** consists of a mat of short cellulose fibers.

16:5 *Continuous digester. Wood chips are cooked in an alkaline solution of sodium sulfide and sodium hydroxide as they travel down through a series of screens inside the digester to be converted into pulp. 250,000 gallons of chemical solution are needed to fill the digester. The used liquor is recovered, reconstituted, and recycled through the digester.*

16:6 *Fourdrinier paper machine. A slurry of pulp and water is deposited on the forming fabric (foreground), where most of the water is drained during sheet formation. The press section (center) removes more water by squeezing the sheet between large rolls. Hot air completes the drying of the continuous sheet of paper in the dryer section (right rear).*

16:7 Cellulose makes attractive cardboard containers. (See chapter title page)

16:6 Fats (Lipids) The term **fats**, or lipids, refers to a certain group of structurally similar substances found in both plants and animals. The term **oil** is used to designate a liquid fat. Most fats are also called **glycerides**, and the acids derived from them by hydrolysis are called fatty acids. The structure of a typical fat, glyceryl tristearate, is

$$
\begin{array}{ccccccccc}
 & & H & & & & & & \\
 & & | & & & & & & \\
H & - & C & - & O & - & C & - & (CH_2)_{16} - CH_3 \\
 & & | & & & & \| & & \\
 & & | & & & & O & & \\
 & & | & & & & & & \\
H & - & C & - & O & - & C & - & (CH_2)_{16} - CH_3 \\
 & & | & & & & \| & & \\
 & & | & & & & O & & \\
 & & | & & & & & & \\
H & - & C & - & O & - & C & - & (CH_2)_{16} - CH_3 \\
 & & | & & & & \| & & \\
 & & H & & & & O & &
\end{array}
$$

glyceryl tristearate

Table 16-2 *Natural Sources of Fats and Oils*

Substances	**Examples**
seeds and nuts	olives, peanuts, cotton, corn, soybeans
meat	suet, lard, beef tallow, whale blubber
dairy products	cream, butter
vegetables	oleomargarine, vegetable shortenings

16:8 *How many different types of carbohydrates do you see?*

Fats and oils feel greasy to the touch and are insoluble in water; fats and oils are lighter than water. They dissolve in benzene, gasoline, and other organic solvents. When pure, fats and oils are almost without color, odor, or taste.

When oils are shaken together with water, they form **emulsions** (refer to Table 17-2); the oil breaks into minute droplets throughout the water. If left alone, the water and oil will separate into two distinct layers. The emulsion can be stabilized by adding soap, gum, or proteins. These substances produce a film about the oil drops and prevent them from coalescing. Typical emulsions are listed in Table 16-3.

16:9 *A scene in a typical dairy—the processing of milk.*

Table 16-3 *Common Emulsions*

milk	butter fat in water stabilized by casein
mayonnaise	oil in vinegar stabilized by egg yolk
photographic film	silver bromide in gelatin

16:7 Soap Fats and oils are insoluble in water, as we have seen. However, when they are hydrolyzed in sodium hydroxide, glycerol and sodium salts of long chain fatty acids (**soaps**) are formed. A typical soap formula is $NaC_{16}H_{31}O_2$ (sodium palmitate).

If we call the sodium (Na) end of this sodium palmitate molecule the head and the hydrocarbon end the tail, we can describe the action of soap. The ionic head is soluble in water, whereas the hydrocarbon tail makes grease soluble in water. The soap acts as a link between the water and organic grease. As the soap in water contacts grease on an object, the tail **adsorbs** the grease (the head is dissolved in the water). The washing action causes the grease attached to the soap to go into the water. The grease is emulsified by the soap and is held in suspension in the water. **Detergents** are nonsoap cleaning agents.

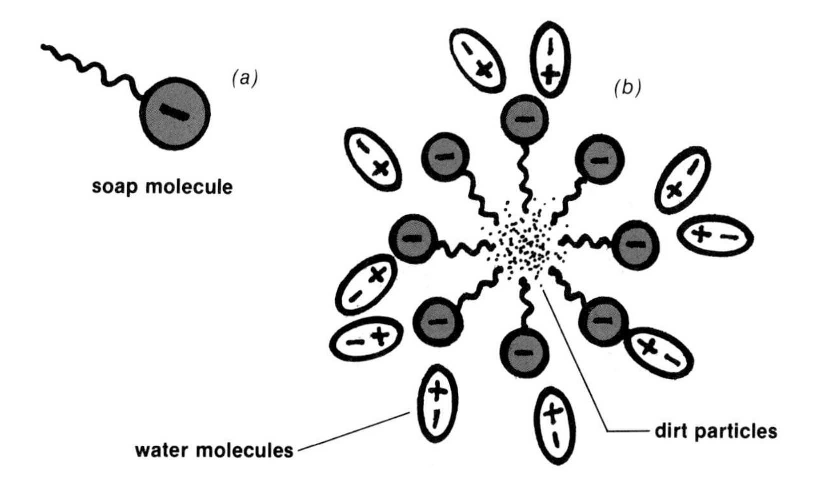

16:10 *Cleaning action of soap. (a) Model of soap molecule—ionic "head" with an organic "tail." (b) Ionic head of soap molecule is attracted to polar water molecule. Dirt is held by the organic tail.*

Experiment 16:4 *Soap Making*

1. Put 2 tablespoons of fat or oil in an evaporating dish. Add 20 ml of saturated NaOH solution.
2. Boil the mixture gently until it becomes as thick as honey. Do *not* allow the mixture to spatter.
3. Add 50 ml of hot distilled water to the mixture; then stir until it is well dispersed.
4. Add 15 ml of saturated NaCl solution. Stir to mix well.
5. The soap should float to the top of the solution. Skim it off; then let the soap stand until it hardens.

Experiment 16:5 *Water Content of Soap*

1. Bring 2 bars of unused soap of different brands to class with you. Weigh each bar and record the weight and cost.
2. Cut the bars of soap in half.
3. Take a slice of soap off a freshly cut surface and get a sample of at least 3 grams. Do not slice off the outside of the bar because it contains less water than the inside.
4. Place each sample from the 2 bars into evaporating dishes and weigh.
5. Dry the samples in an oven overnight. Be sure not to mix up the samples and dishes.
6. Cool the samples and dishes and weigh again. The difference in weight is the water content. Calculate your results with other members of the class. Which brand gives you the most soap for your money?

16:8 Fats in the Body As a body fuel, fats supply more energy than carbohydrates or proteins. Fat is deposited in various parts of the body to serve as a reserve supply of energy and as a protective padding for nerve endings and delicate organs.

When fats are eaten, they pass through the stomach into the intestine unchanged chemically. In the intestine they are hydrolyzed by the action of enzymes into fatty acids and glycerol. These fatty acids and glycerol pass through the intestine wall, recombine to form fats, and then are carried through the body by the blood.

16:9 Proteins Proteins are very complex nitrogen-containing organic compounds which are found in every living organism. They serve as structural material in animals. Proteins are formed in plants from carbon dioxide, water, nitrates, sulfates, phosphates, and other substances obtained from the air or soil.

Animal bodies manufacture their own protein but must start with proteins from plants or other animals. Proteins are necessary in the animal and human diet for growth and for repair of tissue. About one fifth of the human body is protein. The parts of the animal body which are essentially protein include muscles, feathers, nerve tissue, horns, skin, hoofs, hair, solid parts of the blood, and wool.

16:11 *Some foods with high protein content.*

Seeds in plants contain the highest concentration of protein. Foods with a high protein content include lean meat, milk, fish, cheese, eggs, and cereals. Different types of proteins serve various functions in living organisms. Table 16-4 shows some of the functions of proteins.

Table 16-4 *Functions of Protein*

Type	Job
enzymes	catalysts to speed up a variety of chemical reactions in living cells
hormones	control processes in living organisms
hemoglobin	oxygen carrier in blood
antibodies	produced by protein in response to foreign proteins introduced into blood

16:10 Amino Acids and Structure of Proteins Proteins are hydrolyzed by strong acids and become **amino acids**. "Amino" means *ammonia containing*. The simplest amino acid is glycine.

$$\begin{array}{c} \text{H} \\ | \\ \text{H} - \text{C} - \text{CO}_2\text{H} \\ | \\ \text{NH}_2 \end{array}$$

glycine

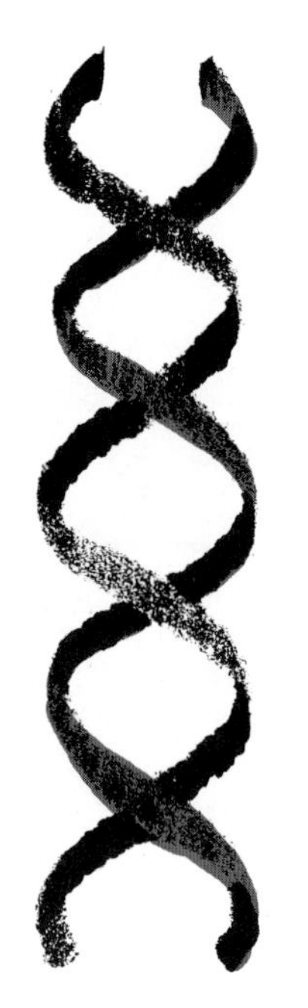

16:12 *Many protein molecules have a double helix shape.*

Chemists consider proteins to be polymers of amino acids. Through the use of an experimental technique called X-ray diffraction analysis (see Section 35:5), scientists have found that some crystalline protein molecules may have a spiral or *helical* shape like a coiled spring.

Nucleoproteins, an important type of protein, are found in chromosomes, cell nuclei, viruses, and the heads of sperm cells. These transmit hereditary characteristics in living organisms. You have probably heard of **DNA** and **RNA**; both are nucleoproteins. The action of DNA in the genes is the mechanism God has ordained to insure that living organisms will reproduce "after their kind."

16:11 Industrial Uses of Proteins Proteins form the raw material for many industries. Examples include meat packing, glue, cheese making, some plastics, wool textiles, some synthetic textile fibers, and leather objects.

16:12 Vitamins Naturally occurring organic substances, called **vitamins**, must be included in the diet if normal life processes are to continue properly. In Table 16-5 are listed various vitamins, the sources of each, and the resulting ailment from a lack of each one.

Table 16-5 *Vitamin Sources and Ailments*

Vitamin	Major Sources	Ailment Caused by Lack of Vitamin
A	fish, liver, fish liver oils, butter, cheese, eggs, carrots, squash, sweet potatoes, spinach, broccoli	night blindness
D	irradiated milk	bone deformation (rickets) and bone degeneration
E	wheat germ oil, leafy vegetables, vegetable oils	unknown in humans
K	green leaves	blood will not clot properly
C	citrus fruits tomatoes, green peppers	scurvy, used in prevention of colds
B_1	grains, liver, lean meat	beriberi
B_2	milk, liver, lean meat, eggs, leafy vegetables, cheese	skin disorders, sores on mouth and tongue
niacin	liver, lean meat, whole cereal grains, eggs	pellagra.
B_6	fish, eggs, whole cereal grains, meat	convulsions in infants
B_{12}	liver, eggs, seafood, meat	anemia

16:13 *Model of a DNA helix.*

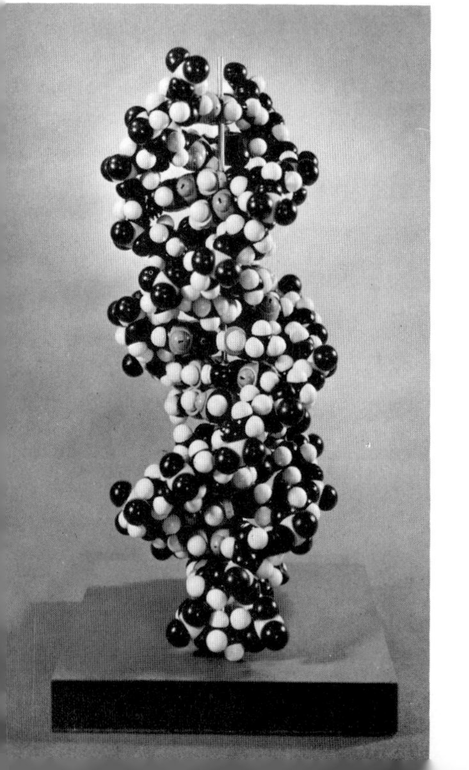

16:14 *How many different vitamins do you think are contained in this delicious-looking meal?*

16:13 Chemical Evolution Many people think that evolution involves only the field of biology. That is not true; evolution is an all-encompassing world-view that dogmatically rules all phases of scientific study. According to the theory, before biological evolution could occur, chemical evolution must first have occurred.

As we have noted, the compounds found in living organisms are extremely complex. For evolution to be true, these complex compounds would have had to "evolve" from simpler chemical compounds. Is this chemical evolution possible? No!

A great deal of time and money have been spent in trying to develop a workable scheme of chemical evolution. Some scientists realized that the *oxidizing* conditions which exist on the earth today would hinder the formation of amino acids and more complex chemical substances. So the evolutionists decided that the "prebiotic" earth had a "reducing" atmosphere. The reason this conclusion is necessary is that the only way that evolutionists can form amino acids in their laboratories is with a reducing atmosphere. Of course, this is not science; it is a make-believe game. And the evolutionists make up new rules as they play the game.

Continuing their imaginative guesswork, these evolutionists found that the major source of energy for the earth, the sun, was also detrimental to the formation of complex compounds. The ultraviolet rays of the sun cause complex amino acids to decompose. Thus, to aid the evolutionary process, it is necessary to imagine that the energy from lightning in a prebiotic atmospheric "soup" of methane and ammonia (previously evolved, of course) caused the formation of amino acids. These acids supposedly drifted downward to the ancient oceans to collect so that they could form proteins when life spontaneously generated (something which has been proved to be scientifically unworkable).

Even though the evolutionist sets up his own rules to play his game, his scheme still fails. There must be high concentrations of very complex amino acids in the oceans for the formation of any kind of proteins. The reactions of formation are *reversible* reactions. As complex compounds form, the reverse reactions cause them to decompose. As a matter of fact, the *decomposition reaction is more likely to occur than the synthesis.* Coupled with this are the destructive effects of the ultraviolet rays of the sun decomposing the compounds in the prebiotic atmosphere and oceans. Not enough complex amino acids, carbohydrates, or anything could form for life to generate spontaneously in them. Chemical evolution is unreasonable.

How could such complex chemical compounds come into being? Only by being created by an omnipotent God. Genesis chapters 1 and 2 contain the only reasonable account of the origin of complex chemical compounds. The chemistry is not divulged; however, the Bible is clear that God created everything by direct acts, as opposed to any lengthy natural processes.

chemical evolution

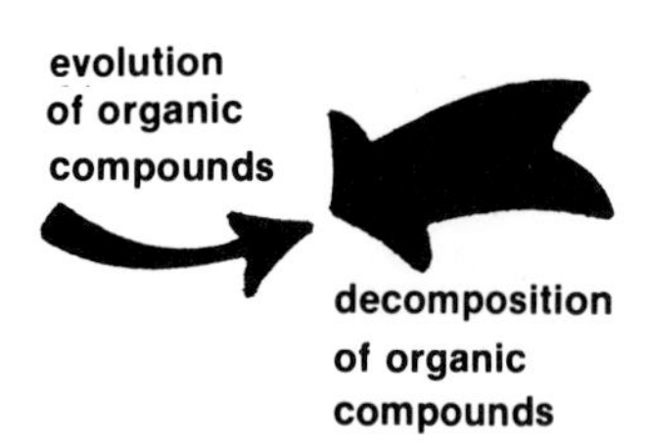

Which one will win?

16:15 *The chemical reactions causing decomposition of organic compounds are more likely than the reactions that form organic compounds.*

List of Terms

adsorbs
amino acid
carbohydrates
cellulose
chemical evolution
detergent
dextrins
disaccharide
DNA
emulsion
fats
glucose
glycerides
lipid
monosaccharide
mutarotation
nucleoprotein
oil
paper
polysaccharide
proteins
RNA
soaps
starch
sugar
vitamin

Questions

1. Define biochemistry.
2. What two fields of science overlap in biochemistry?
3. What makes the organic material in living cells?
4. What elements are found in carbohydrates?
5. Why are carbohydrates given this name?
6. What is a well-known carbohydrate? What is its chemical formula?
7. If you remove the elements which compose water from sugar, can you put them back into the sugar in your laboratory? Why? Write the chemical reaction for the dehydration of sugar.
8. Carbohydrates can be divided into what two general categories?
9. (a) Name three sugars mentioned in this chapter.
   (b) Where do they come from?
   (c) Which one is a combination of the other two?
10. Define mutarotation.
11. Why are glucose and fructose called monosaccharides?
12. What is the general structural formula for starch?
13. From what plants is starch obtained?
14. What part of plants does cellulose form?
15. Name three products made from cellulose.

16. What does the term *oil* mean when referring to a particular kind of liquid?
17. Name some sources of fats and oils.
18. Are fats soluble in water?
19. How is soap produced?
20. What are soaps chemically?
21. (a) What are proteins chemically?
    (b) Where are they found?
22. (a) Name five parts of living organisms that are essentially protein.
    (b) What part of a plant has the highest concentration of protein?
23. What is the function of an enzyme?
24. How are amino acids formed?
25. (a) What kind of proteins are DNA and RNA?
    (b) Where are they found?
    (c) What is their function?
26. (a) Are vitamins necessary for life processes?
    (b) Name a source of vitamin A.
    (c) Name a source of vitamin E.
    (d) Name a source of vitamin B_2.
27. (a) Briefly, what is chemical evolution?
    (b) What is wrong with the present earth's atmosphere as far as chemical evolution is concerned?
    (c) Why is it not scientific to say that the prebiotic (before life) earth had a reducing atmosphere?
28. Why is lightning instead of sunlight supposed to be what was used in the process of evolution as an energy source for chemical evolution?
29. What great scientist cast serious doubt on the spontaneous generation of life in his experiments?
30. How does the evolutionist fail in chemical evolution, even though he makes up the rules?
31. (a) How did the first complex compounds form?
    (b) How did God create?

Student Activities

1. The water content of the soap in Experiment 16:5 can be extended to check all available brands of soap on the market. Write a report on your findings.
2. Write a history of soap production. Be sure to cover thoroughly modern-day soap production.
3. To investigate the lathering power of soap, cut slices of soap from bars of various different brands the same way as you were instructed in Experiment 16:5. A sample of two grams will be sufficient. Weigh your samples as accurately as possible. Record the weights of each sample. Be careful so that you do not mix up the samples!

   Add enough hot distilled water to completely dissolve the soap. Place the soap solution in a buret. Add 25 ml of water to a large flask or beaker. Add the soap solution, drop by drop, into the flask until you can produce (by shaking or stirring) a lather that is stable for one minute. Record how much soap solution had to be used to attain this condition.

   The less soap solution you have to add, the greater the lathering power of the soap. Arrange the soap in order of lathering power. The lather formed in water by soap is the place where dirt particles can accumulate and be carried away from the washed item.

Photo next page
Smoke and steam from a volcanic eruption off the coast of Iceland. Smoke is a common example of a colloid.

UNIT V DESCRIPTIVE CHEMISTRY
Groups VA-VIIIA

Chapter 17 Colloids

17:1 Colloid Mixtures We have discussed *solutions*, in which the solute dissolves in the solvent. The particle size of the solute is so small that there is no magnifying device powerful enough to enable us to see the particles.

A **suspension** contains very large particles of one substance dispersed in (distributed through) another phase (substance). The *dispersed phase* will settle quickly unless the mixture is shaken. Medicines that must be shaken well before use are examples of suspensions. Muddy water is a suspension of clay in water.

Between the extremes of solutions and suspensions are **colloidal dispersions** (colloids). The dispersed particles do not go into solution, nor are they big enough to settle out rapidly. For example, when a beam of light is shined into a darkened room, you do not actually see the beam; you see light reflected from the dust particles suspended in the air. This effect is characteristic of colloidal suspensions. This is referred to as the **Tyndall effect**, named after the British scientist John Tyndall (1820-1893). Beautiful sunsets are actually caused by dust particles in our atmosphere which reflect light rays from the sun. Table 17-1 will help you distinguish among the various types of mixtures.

17:2 Types of Colloids There are eight types of possible colloidal dispersions.

Colloidal particles have a great deal of surface area for the amount of material present because of their tiny size. Thus, they are capable of **adsorbing** large amounts of other substances. Activated charcoal is an example of a colloid which is used to remove odors from the air.

Table 17-1 *Solutions, Suspensions, and Colloidal Dispersions*

	Solution	Colloid	Suspension
particle size of solute or dispersed phase	$<1 \times 10^{-7}$ cm*	1×10^{-7} to 1×10^{-5} cm	$>1 \times 10^{-5}$ cm
visibility	invisible to any microscope	visible in ultramicroscope	visible to naked eye
settling rate	particles never settle	particles settle very slowly	particles settle rapidly
ease of filtration	particles pass through filter paper	special filters needed to trap particles	particles easily filtered

* *Refer to Appendix II for an explanation of exponential numbers.*

Table 17-2 *Classification of Colloids*

Dispersed Phase	Dispersion Medium	Examples
gas (bubbles)	in a gas	impossible
	in a liquid	whipped cream, many foams
	in a solid	meerschaum
liquid (drops)	in a gas	fogs, aerosols
	in a liquid[1]	homogenized milk
	in a solid[2]	mayonnaise, cheese, jelly
solid (particles)	in a gas	smoke
	in a liquid[3]	glue, paints
	in a solid	many metallic alloys, colored gems

1. called an **emulsion** (colloidal suspension of a liquid in a liquid).
2. called a **gel** (colloidal suspension of a liquid in a solid).
3. called a **sol** (colloidal suspension of a solid in a liquid).

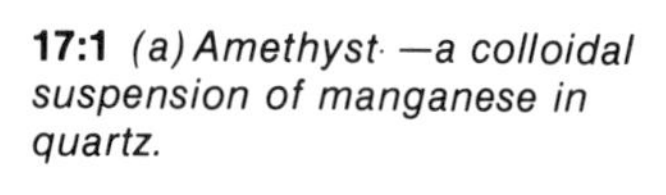

17:1 *(a) Amethyst —a colloidal suspension of manganese in quartz.*

17:1 *(b) Typical colloidal dispersions.*

It is necessary to use a special instrument to observe colloidal particles. The **ultramicroscope** for viewing colloidal suspensions is simply a regular microscope with a powerful light beam passing through the mixture being observed.

Experiment 17:1 *Preparation of an Emulsion*

1. Obtain some oil and put it in a flask or beaker with water.
2. Shake well. What happens after you stop shaking the mixture?
3. Make a soap solution by adding 1 gram of soap to 100 ml of warm water.
4. Pour some oil into the soap solution and shake the mixture well. What happens after you stop shaking? Why?

Experiment 17:2 *Preparation of Ferric Hydroxide Gel*

1. Dissolve 6 grams of ferric chloride ($FeCl_3$) into 100 ml of distilled water.
2. Add this solution slowly to a beaker containing 100 ml of boiling distilled water until a dark red colloid results. The reaction for this process is
   $$FeCl_3 + 3H_2O \longrightarrow Fe(OH)_3 + HCl$$
   Ferric hydroxide ($Fe(OH)_3$) is the colloid formed. The acid in solution stablizes the colloidal dispersion.
3. You can precipitate the colloid by adding NaOH to the dispersion.

17:2 *This lady is using a colloid, egg white, in her cooking. Many of the metal substances seen in this photograph contain colloidal particles.*

17:3 *Clay being moved by this road grader is made of many colloidal particles.*

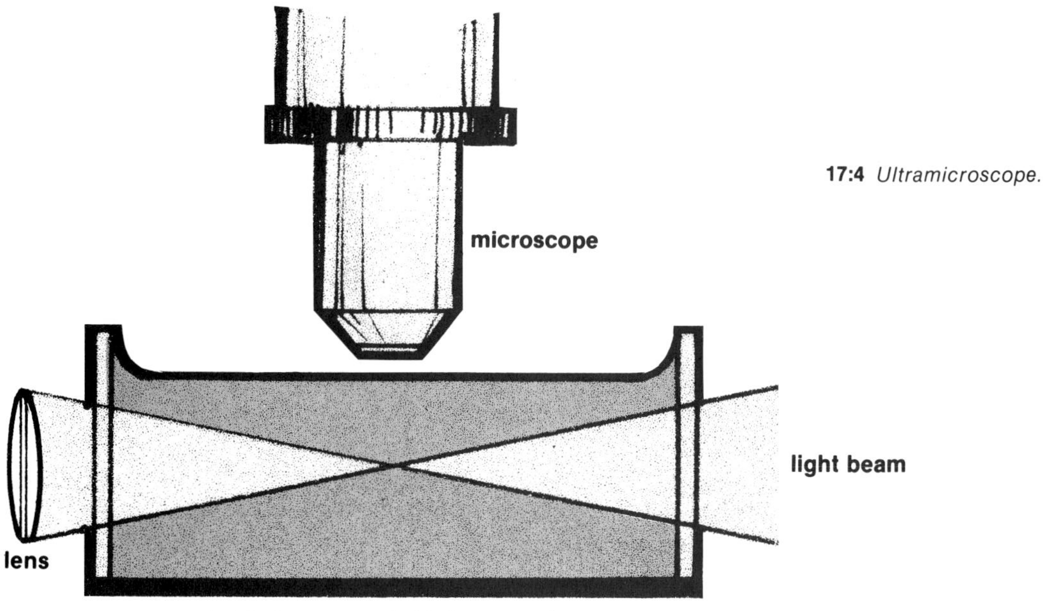

17:4 *Ultramicroscope.*

Experiment 17:3 *Preparation of a fruit gel (jelly)*

1. Add 1 gram of dry pectin, 0.5 gram of tartaric acid, and 62 grams of sugar to 100 ml of water.
2. Heat the mixture slowly until it boils.
3. Allow the mixture to cool, and jelly will form. Sometimes it is necessary to add acids such as lemon juice to certain jellies. Why? (Refer to Experiment 17:2.)

List of Terms

adsorb	sol
colloidal dispersion	suspension
emulsion	Tyndall effect
gel	ultramicroscope

Questions

1. Can the solute particles in a solution be seen with a powerful microscope?
2. What is a suspension?
3. Are suspensions stable?
4. What is a colloidal dispersion?
5. What is the size range of colloidal particles?
6. What is an emulsion?
7. What is a gel?
8. What is a sol?
9. How can you use activated charcoal?

Student Activity

Many foods are colloidal suspensions. Prepare a report on how at least two of these foods are processed.

Photo next page
Pellet-forming tower in fertilizer plant.

UNIT VI DESCRIPTIVE CHEMISTRY Groups VA-VIIIA

Chapter 18 Nitrogen and Phosphorus

18:1 Electronic Structure The electronic structure of the elements in group VA are shown in Table 18-1. Each element has five electrons in the valence shell. All of these elements prefer covalent bonding to fill this outer shell. And again, the metallic properties of the elements increase as you move down the group. One of the interesting features of the elements in group VA is their ability to assume different valences when forming compounds. The valences can vary between +5 and −3.

Table 18 -1 *Electron Distribution in the Atoms of Group VA*

Element	**Quantum Levels and Electron Orbitals**													
	1	2		3			4				5			6	
	s	s	p	s	p	d	s	p	d	f	s	p	d	s	p
N	2	2	3												
P	2	2	6	2	3										
As	2	2	6	2	6	10	2	3							
Sb	2	2	6	2	6	10	2	6	10		2	3			
Bi	2	2	6	2	6	10	2	6	10	14	2	6	10	2	3

Nitrogen (N) and phosphorous (P) are nonmetals, arsenic (As) and antimony (Sb) are metalloids, and bismuth (Bi) is a metal. This is definitely a mixed group!

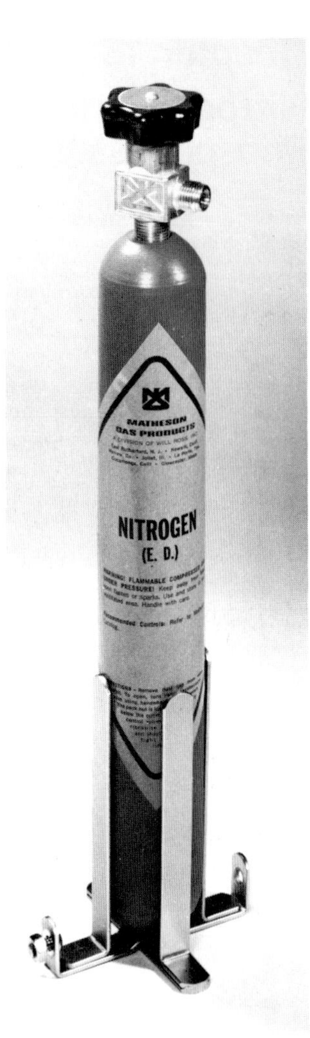

18:1 *A cylinder of nitrogen gas.*

18:2 Nitrogen The atmosphere of the earth is about 78 percent nitrogen by volume. Nitrogen is a relatively inactive gas; we can tell this by its presence as an uncombined element in air. It is the nitrogen in the air that dilutes the oxygen to the proper concentration for use by living organisms. If nitrogen were more reactive, it would combine with the oxygen. Life as we know it would be impossible on the earth if our atmosphere were made up of some oxide of nitrogen rather than elemental oxygen and nitrogen. The fact that nitrogen is uncombined in air is not an evolutionary accident; it is evidence of God's design.

Nitrogen is present in **nitrates** (salts of nitric acid) in many parts of the earth. The largest single deposit is the sodium nitrate (saltpeter) found in Chile. Nitrogen is found in living organisms in amino acids and proteins.

18:3 Nitrogen Cycle Nitrogen is necessary in all living organisms. Plants must have it to grow; animals must have it to build body proteins. Through their roots plants absorb nitrogen from the soil in the form of nitrates and ammonium salts. Animals then eat the plants and get the protein-building nitrogen that they need.

If nitrogen were not placed back into the soil by some means, the supply would soon be depleted. Decayed plant and animal tissue and animal waste are natural sources for nitrogen. These substances are decomposed by bacteria. Eventually, nitrates from further decomposition are absorbed by plants. This complicated process is known as the **nitrogen cycle** and is illustrated in Figure 18:2.

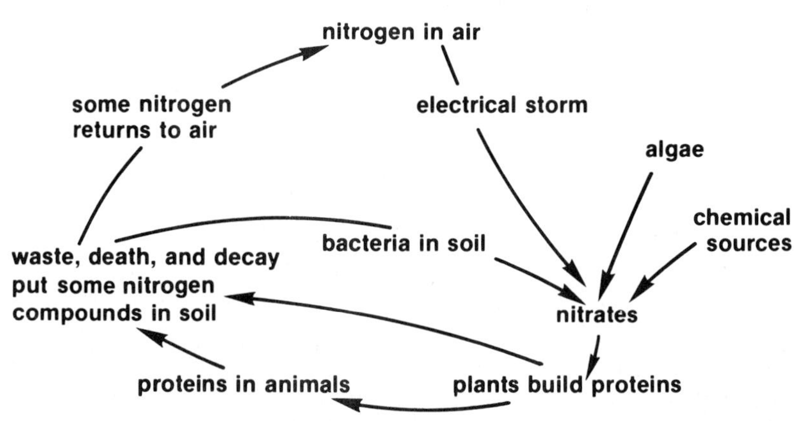

18:2 *Nitrogen cycle.*

The nitrogen cycle so necessary for good soil is another evidence that all things work together in the physical world.

Man must work the soil "by the sweat of his brow." The plants grown in soil provide man with food. However, as man works, he wears out his body. Eventually he dies. The decay and waste products from his dead body then return necessary organic materials to the soil and help sustain vegetation. Thus, the soil can continue to grow plants which provide other men with food.

In an area where the land is farmed, not enough time is available between plantings for the nitrogen in the soil to be replaced naturally. Chemical sources of nitrogen **(fertilizers)** are then used to replenish the soil with this necessary element. Soil that loses all of its decayed organic matter **(humus)** and is not fertilized will not successfully support abundant vegetation. Soil without nitrogen is useless for farming; the land is considered "worn out." This happened in the early development of America when farmers used very shallow-digging plows. Much of the top soil (full of decayed matter) was washed away by rain. Few, if any, organic or inorganic nitrogen compounds were placed back into the soil, and large tracts of land eventually became "worn out."

18:3 *A farmer spraying a nitrogen-containing fertilizer on his field.*

God commanded the Jews to allow their lands to "rest" every seven years. This is very sensible; soil can easily be depleted of the food that vegetation needs. "Resting" the soil enables it to regain some of the necessary nutrients for plant growth.

(a)

(b)

18:4 *Compare eroded farm land in (a) with (b) the rich, deep loam soil capable of producing excellent crops such as the alfalfa shown. The land in (a) is depleted of nitrogen as well as other soil nutrients.*

Nitrogen can be put into the soil by other methods. Nitrogen-fixing bacteria grow on the roots of *leguminous* plants, such as clover, peas, and alfalfa. These bacteria absorb nitrogen from the air and combine it chemically for their own use. They store up more nitrogen than they need and, therefore, build up the nitrogen content of the soil. Figure 18:5 is an illustration of a plant with nodules of bacteria in its root system.

18:5 *Black-eyed peas with well-developed nitrogen nodules on their roots.*

Also, algae can take nitrogen from the air and thereby increase the fertility of the soil. Finally, lightning causes nitrogen and oxygen to combine. The resulting oxides are dissolved in rain and absorbed by the soil. These three processes are referred to as **nitrogen-fixing** processes. "Fixing" simply means to cause the nitrogen to react to form compounds with other substances.

18:4 Chemistry of Nitrogen In its elemental state, nitrogen forms a molecule, N_2. The two nitrogen atoms are covalently bonded together. The molecular model can be illustrated with "sticks" (electron pairs) or molecular orbitals, as in Figure 18:6.

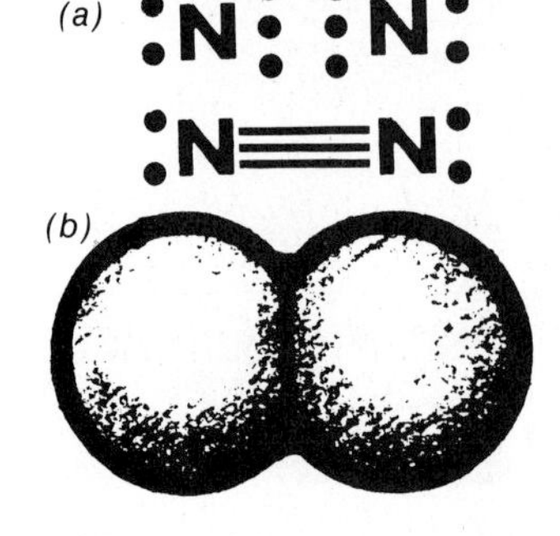

18:6 *Various models of the nitrogen molecule: (a) shows electron pairs and (b) illustrates overlapping spheres. (c) Molecular orbital model of nitrogen molecule. (Dots represent the nitrogen nuclei.)*

Nitrogen exhibits more valence states than any other element in Group VA. It has nine possible valence states; these are listed in Table 18-2.

Table 18-2 *Valence States of Nitrogen*

Valence	Example Compounds
+5	nitrogen pentoxide (N_2O_5), nitrates
+4	nitrogen dioxide (NO_2)
+3	nitrogen trioxide (N_2O_3)
+2	nitric oxide (NO)
+1	nitrous oxide (N_2O)
0	elementary nitrogen (N_2)
-1	hydroxylamine (NH_2OH)
-2	hydrazine (NH_2NH_2)
-3	ammonia (NH_3)

As mentioned earlier, nitrogen can react with oxygen if air is passed through an electric arc.

$$N_2 + O_2 \longrightarrow 2NO\uparrow$$

Nitrogen can also combine with certain metals to form **nitrides**.

$$3Mg + N_2 \longrightarrow Mg_3N_2$$

Magnesium nitride (Mg_3N_2) is actually an ionically bonded compound; the nitrogen has a valence of -3.

18:5 Ammonia Nitrogen and hydrogen are odorless gases in their uncombined states, but when they form the gas **ammonia** (NH_3) they give off a powerful, repugnant odor. Ammonia gas can be detected even in very slight amounts because of this odor.

Ammonia can be produced by the **Haber process**, named for Fritz Haber (1868-1934), a German chemist, in which elemental nitrogen and hydrogen are used.

$$N_2 + 3H_2 \xrightarrow[\text{pressure}]{\text{catalyst, } \Delta} 2NH_3\uparrow$$

18:7 *A farmer loading a fertilizer spreader with an ammonia-rich fertilizer.*

Ammonia and ammonium salts are used as a source of nitrogen in chemical fertilizers. Ammonia will dissolve in water to form a base, ammonium hydroxide (NH_4OH), which is used in many cleaning solutions.

$$NH_3 + H_2O \longrightarrow NH_4OH$$

18:8 *An ammonia plant with the control house in the foreground and the gas synthesis and purification towers in the background.*

Ammonia has a low boiling point and can be liquefied very easily under pressure at room temperature. This makes the gas useful as a refrigerant. In a refrigerator cycle, ammonia gas is compressed to a liquid, then circulated through coils where it is allowed to expand back into the gaseous state. In changing from a liquid back to a gas, ammonia absorbs heat (heat of vaporization). This heat is removed from the contents of the refrigerator and is released to the atmosphere through a series of heat exchanger coils. The ammonia gas is then cycled back to the condenser to be reliquefied. In this way, ammonia keeps the items in the refrigerator continually cool.

The Gas Refrigerator

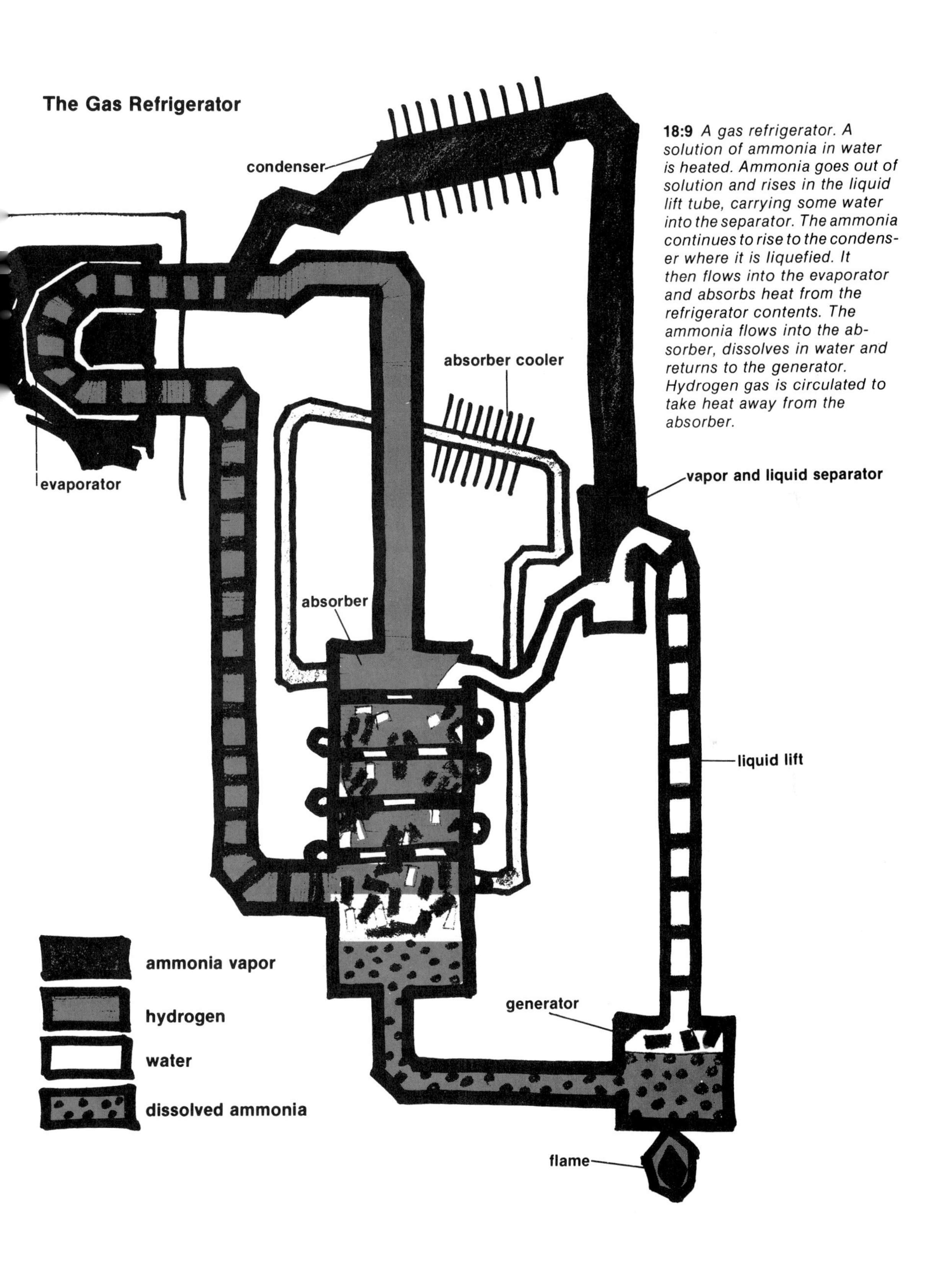

18:9 *A gas refrigerator. A solution of ammonia in water is heated. Ammonia goes out of solution and rises in the liquid lift tube, carrying some water into the separator. The ammonia continues to rise to the condenser where it is liquefied. It then flows into the evaporator and absorbs heat from the refrigerator contents. The ammonia flows into the absorber, dissolves in water and returns to the generator. Hydrogen gas is circulated to take heat away from the absorber.*

Ammonia forms complex ions with many metals. The aqueous solutions of these ions are usually deep blue in color. The reaction of ammonia with compounds containing the copper (II) ion is as follows:

$$Cu^{+2} + 4NH_3 \longrightarrow \underset{\text{deep blue}}{Cu(NH_3)_4^{+2}}$$

Experiment 18:1 *Properties of Ammonia (NH_3)*

1. Obtain a few drops of the dilute ammonium hydroxide (NH_4OH) solution prepared for you by the teacher. Very carefully sniff the solution. You will notice the characteristic smell of ammonia.
2. Put 1 ml of concentrated ammonium hydroxide in a test tube.
3. Dip a clean glass stirring rod into a bottle of dilute hydrochloric acid. Hold the wet end of the rod over the top of the tube containing the ammonium hydroxide. Record your observation. Write the equation for the chemical reaction.
4. Add 1 gram of nickelous nitrate ($Ni(NO_3)_2$) to 20 ml of water in a beaker. Stir to dissolve the salt. Does the water heat up or cool down as you dissolve the $Ni(NO_3)_2$? Hold your hand on the bottom of the beaker as you stir the mixture.
5. What is the color of the solution? Add concentrated ammonium hydroxide slowly to the solution until it changes color. Record the color change. This is the color of the nickel-ammonia complex ion (similar to the copper-ammonia complex ion). Write the equation for the formation of the nickel-ammonia complex ion.

18:6 Oxides of Nitrogen Nitrous oxide (N_2O) is the anesthetic commonly known as **laughing gas**. It can be prepared by the careful heating of ammonium nitrate

$$NH_4NO_3 \xrightarrow{\Delta} N_2O\uparrow + 2H_2O$$

Laughing gas supports combustion very well. A glowing splint will burst into flames in this gas.

Nitrogen dioxide (NO_2) is a red-brown poisonous gas with a very sharp odor. It is the gas given off by fuming nitric acid. Nitrogen dioxide also forms when nitric acid decomposes. The two gases, NO and NO_2, are involved in air pollution, which will be discussed in Chapter 23.

Nitrogen trioxide (N_2O_3) and nitrogen pentoxide (N_2O_5) are very unstable, and we will not discuss them further.

Experiment 18:2 *Properties of Nitrous Oxide (Laughing Gas)*

1. Place 5 grams of ammonium nitrate (NH_4NO_3) into a Florence flask and connect to a pneumatic trough containing hot water as shown in Figure 18:10.
2. Heat the flask gently. The following reaction occurs:
$$NH_4NO_3 \longrightarrow N_2O\uparrow + 2H_2O$$
3. Fill 2 gas bottles with N_2O. Plunge a glowing wood splint into 1 bottle. Record your observations.

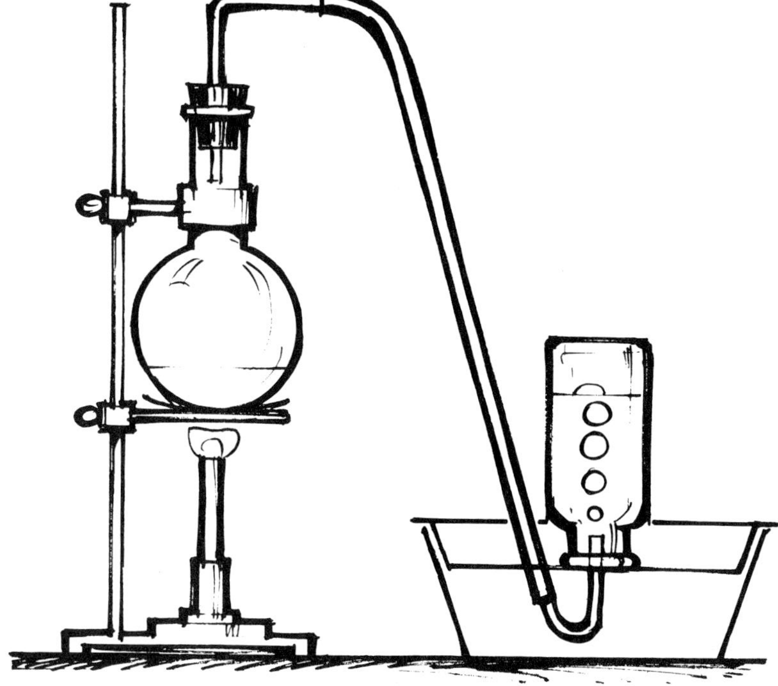

18:10 *Preparation of nitrous oxide.*

4. Using forceps, dip a small piece of steel wool into powdered sulfur.
5. Ignite the steel wool and sulfur in a Bunsen burner and plunge it into the second bottle. Record your observations. Based on the observations from this experiment, does N_2O support combustion?

18:7 Nitric Acid No chemistry laboratory is complete without a bottle of strong **nitric acid** (HNO_3). The concentrated commercial acid is about 68 percent HNO_3. If any of the nitric acid turns brown in color, it is due to slight decomposition with the formation of NO_2. Nitric acid itself is colorless.

Nitric acid can be commercially produced from ammonia by the **Ostwald process**, named for Wilhelm Ostwald (1853-1932), a German chemist. Extremely hot ammonia and air are passed over a platinum catalyst to form nitric oxide.

$$4NH_3 + 5O_2 \text{ (in air)} \xrightarrow{\Delta,\ \text{catalyst}} 4NO\uparrow + 6H_2O + \text{heat}$$

This oxide then reacts with more oxygen to form nitrogen dioxide.

$$2NO + O_2 \text{ (in air)} \longrightarrow 2NO_2\uparrow$$

The dioxide is dissolved in water to form HNO_3.

$$3NO_2 + H_2O \longrightarrow 2HNO_3 + NO\uparrow$$

Nitric oxide is recycled in the process.

The hydrolysis of HNO_3 produces the hydronium and **nitrate ion**.

$$HNO_3 + H_2O \longrightarrow H_3O^+ + NO_3^-$$

A mixture of the two strong acids, nitric and hydrochloric, is called **aqua regia** and will dissolve gold. Silver and copper will dissolve in HNO_3, but not in sulfuric and hydrochloric acids.

Nitric acid is used in many commercial processes and in the manufacture of fertilizers, dyes, and explosives.

18:11 *Many nitrogen-containing compounds are used in explosives.*

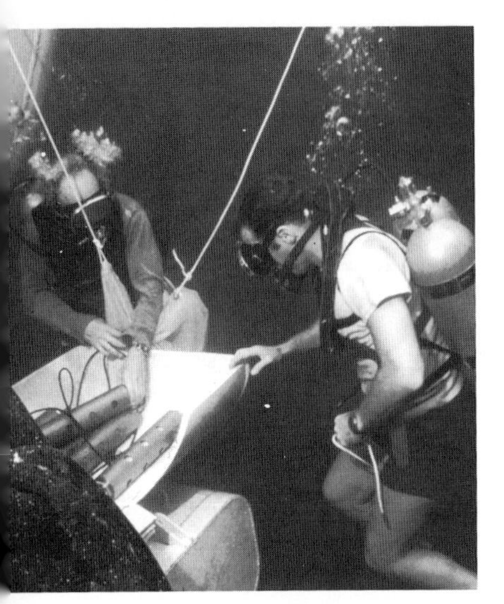

18:12 *Naval salvage operation utilizing explosives.*

Experiment 18:3 *Properties of Nitric Acid (HNO_3)*

1. Fill a glass-stoppered acid bottle half-full of concentrated nitric acid. Set it on a shelf in the laboratory for several days. What color does it turn? Why?
2. Your teacher will supply you with 3 bottles of unidentified acids and pieces of copper. Take 10 ml of each acid and pour them into 3 different test tubes. Identify which acid is nitric.

18:8 Phosphorus Phosphorus is much more reactive in the elemental state than nitrogen and is never found uncombined. Pure phosphorus exists in several forms, including red, white, and black phosphorus. Each form reacts differently from the others, but white phosphorus is the most active. White phosphorus is a deadly poison. Red phosphorus is the form commonly used in the chemical laboratory.

Elemental white phosphorus consists of P_4 molecules. Each phosphorus atom can obtain a stable octet of electrons if it shares electrons with three other phosphorus atoms. A model of this structure is shown in Figure 18:13. Note that white phosphorus has a tetrahedral structure.

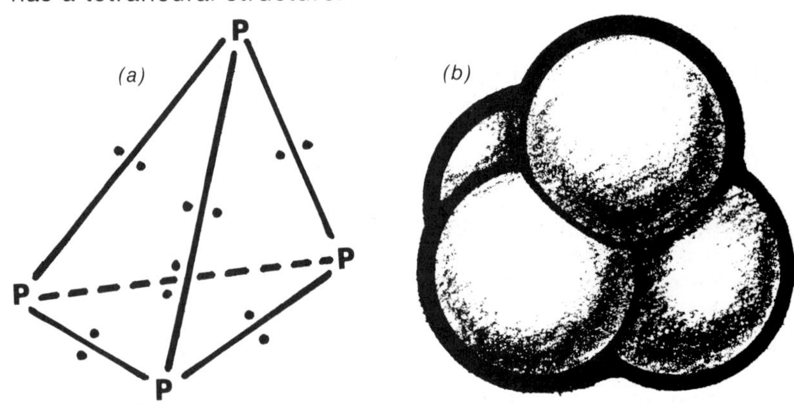

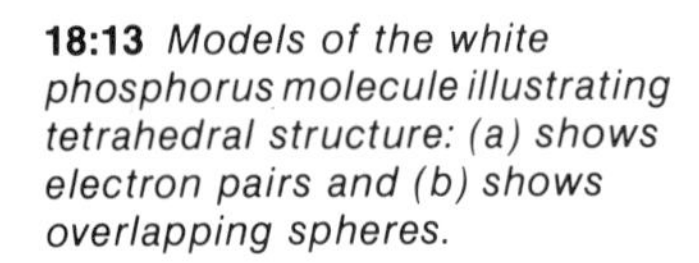

18:13 *Models of the white phosphorus molecule illustrating tetrahedral structure: (a) shows electron pairs and (b) shows overlapping spheres.*

18:9 Chemistry of Phosphorus Phosphorus, like nitrogen, exhibits a multiple valence, although it does not have quite as many different valences as nitrogen. The valence states of phosphorus are listed in Table 18-3.

Table 18-3 *Valence States of Phosphorus*

Valence	Example Compounds
+5	phosphorus pentoxide (P_4O_{10})
+3	phosphorus trioxide (P_4O_6)
0	elemental phosphorus (P_4)
-2	phosphorus dihydride (P_2H_4)
-3	phosphine (PH_3)

White phosphorus readily combines with the oxygen in air to form oxides. P_4O_{10} forms in excess air, whereas P_4O_6 forms when there is a deficiency of air.

$$4P + 5O_2 \longrightarrow P_4O_{10}$$

$$4P + 3O_2 \longrightarrow P_4O_6$$

Phosphorus burns brightly, giving off a dense white oxide smoke which has been used in shells to produce smoke screens, incendiary bombs, and tracer bullets.

18:14 *Explosion of phosphorus incendiary bomb.*

Experiment 18:4 *Burning of White Phosphorus*

1. Your teacher will supply you with a solution of white phosphorus dissolved in carbon disulfide (CS_2).
2. Place a few drops of solution on a piece of filter paper and quickly place the paper on top of a graduated cylinder or beaker. Describe what happens. (Wash your hands if you handled the filter paper or any of the solution.)

Phosphorus reacts with the halogens to form trihalides or pentahalides. Phosphorus and chlorine react to form either phosphorus trichloride or phosphorus pentachloride.

$$2P + 3Cl_2 \longrightarrow 2PCl_3$$

$$2P + 5Cl_2 \longrightarrow 2PCl_5$$

Phosphorus reacts with sulfur to form various sulfides, such as P_2S_3, P_4S_7, and P_4S_3.

18:10 Compounds of Phosphorus Phosphorus pentoxide has a high affinity for water and is used as a drying agent. When this oxide combines with water, it forms phosphoric acid (H_3PO_4).

$$P_4O_{10} + 6H_2O \longrightarrow 4H_3PO_4$$

Phosphoric acid is considered a weak acid, and the concentrated acid is somewhat syrupy in consistency. Phosphoric acid is used in many metal cleaning solutions. Most metals can be cleaned to a brilliant metallic luster in phosphoric acid electrocleaning baths. As a matter of fact, many phosphate compounds are used in cleaning applications—trisodium phosphate (Na_3PO_4) is used as a cleaning powder and a water softener, and polyphosphates such as sodium tripolyphosphate ($Na_5P_3O_{10}$) are used in detergents.

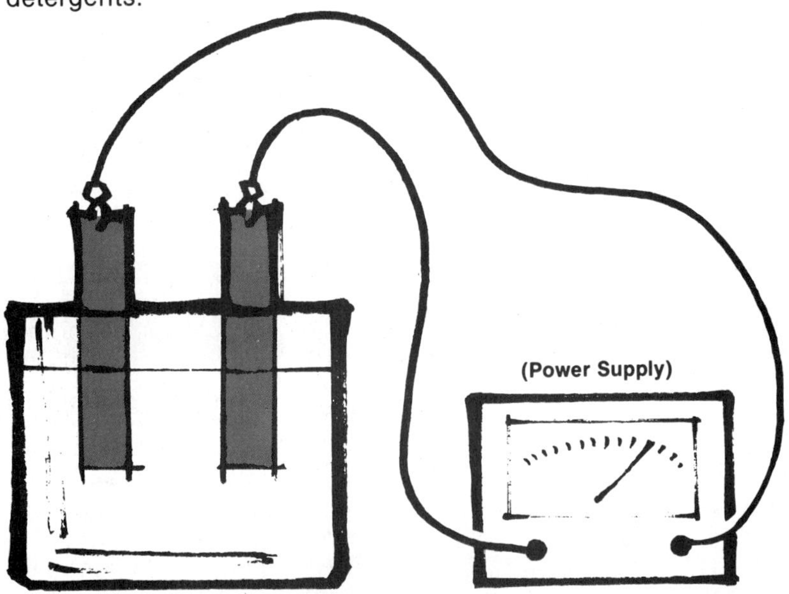

18:15 *Phosphoric acid electrocleaning bath.*

Experiment 18:5 *Electrocleaning Ability of Phosphoric Acid*

1. Obtain 2 pieces of dirty copper. Connect them to a direct current power supply by using electrical lines with alligator clips, as shown in Figure 18:15.
2. Submerge the copper pieces into a beaker of concentrated phosphoric acid.
3. Turn the power supply up slowly. Watch the copper pieces in the acid. What happens?

Phosphine (PH_3) is a colorless gas with an unpleasant odor. It is highly poisonous and was used in chemical warfare during World War I. This phosphorus compound (PH_3) corresponds to the nitrogen compound, ammonia (NH_3).

The name "phosphorus" means *light bearer*. Phosphorus is

utilized in matches because it ignites easily. The first successful matches contained white phosphorus. Because of the danger involved in using the element, **safety matches** were developed. **Kitchen matches** have a tip which can be ignited by friction. The heat of friction causes the tetraphosphorus trisulfide (P_4S_3) to react with lead dioxide (PbO_2). This reaction ignites the burnable material in the match head, which in turn ignites the stick.

18:16 *(a) Kitchen matches. (b) Safety matches.*

(a)

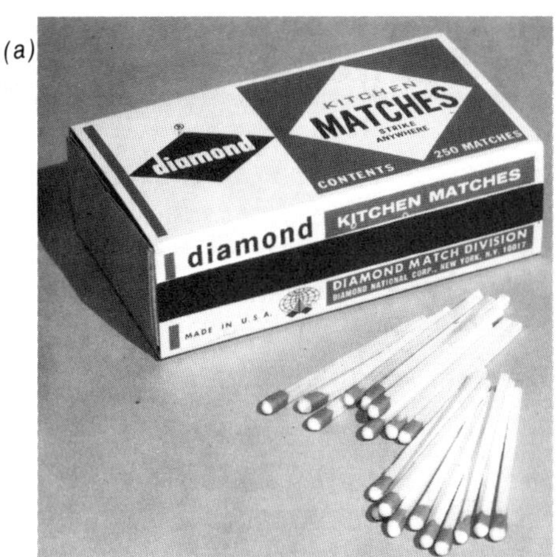

(b)

No phosphorus is contained in safety matches, but red phosphorus is put in the striking surface. The heat of friction causes the red phosphorus on the striking surface to burn. The burning phosphorus then causes the antimony trisulfide (Sb_2S_3) to react with the potassium chlorate ($KClO_3$). This reaction ignites the head of the match.

18:11 Phosphate Fertilizers Phosphorus in various forms is found in the blood, muscles, brains, and nerves of animals. Calcium phosphate is a principal component of bones and teeth. Animals obtain phosphorus from plants; plants get phosphorus from the soil. Phosphorus is necessary for plant growth. When the soil is depleted of this element, phosphorus must be added in fertilizers. A quick-acting fertilizer, **superphosphate**, is made by reacting sulfuric acid with phosphate rock.

$$\underset{\text{rock}}{Ca_3(PO_4)_2} + 2H_2SO_4 \longrightarrow \underset{\text{superphosphate}}{Ca(H_2PO_4)_2} + 2CaSO_4$$

The rock is not soluble, but the superphosphate (dihydrogen phosphate) *is* soluble. If phosphate rock is reacted with phosphoric acid, it forms an even higher phosphorus content fertilizer, triple superphosphate.

$$\underset{\text{rock}}{Ca_3(PO_4)_2} + 4H_3PO_4 \longrightarrow \underset{\text{triple superphosphate}}{3Ca(H_2PO_4)_2}$$

18:12 Arsenic, Antimony, and Bismuth Arsenic (As) and antimony (Sb) are used as alloying additions to harden lead (Pb).

Arsenic compounds are highly poisonous and are used as rat poisons, wood preservatives, paint pigments, and insecticides.

18:17 *Differences between (a) the kitchen match and (b) the safety match are shown.*

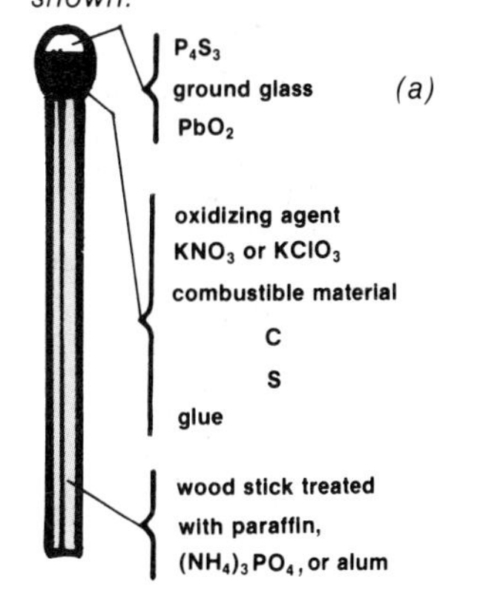

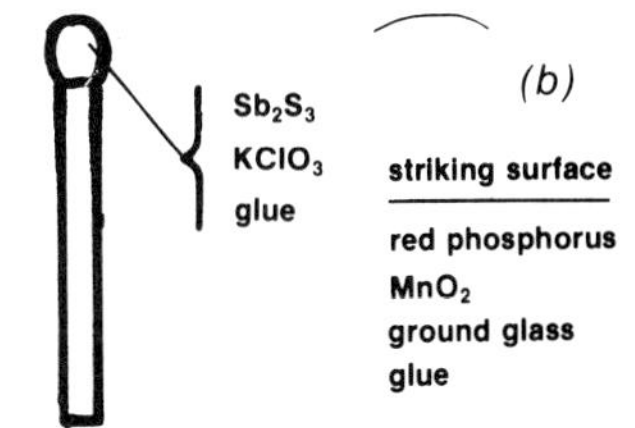

18:18 *Pellet-forming tower in fertilizer plant.*
(See chapter title page)

Antimony is used in type metal because it expands when it solidifies and takes the shape of a mold better.

Bismuth (Bi) is used as an alloying addition in low-melting-point alloys for safety devices, such as electrical fuse strips and release parts for automatic fire extinguishers. Some bismuth compounds are used in medicines to calm stomach disorders.

18:19 *A sprinkler head utilizing a low-melting-point alloy for a fire extinguisher system.*

18:13 Occurrence of Group VA Elements The natural sources for elements we have discussed in this chapter are shown in Table 18-4.

Table 18-4 *Occurrence of Group VA Elements*

Element	Important Sources
nitrogen	air
phosphorus	calcium phosphate ($Ca_3(PO_4)_2$) apatite ($CaF_2 \cdot 3Ca_3(PO_4)_2$)
arsenic	realgar (As_2S_2) orpiment (As_2S_3)
antimony	stibnite (Sb_2S_3)
bismuth	uncombined bismite ($Bi_2O_3 \cdot H_2O$)

List of Terms

ammonia
aqua regia
fertilizer
Haber process
humus
kitchen match
laughing gas
nitrate
nitrate ion
nitric acid
nitride
nitrogen cycle
nitrogen-fixing
Ostwald process
safety match
superphosphate

Questions

1. What similarity is there in the electronic structure of the group VA elements?
2. Why is group VA called a mixed group?
3. Is the valence of the group VA elements fixed?
4. Name two sources of phosphorus.
5. How does the chemical activity of nitrogen in the atmosphere show evidence of God's design?
6. Where is the largest deposit of nitrates found?
7. What elements do plants need for growth?
8. Considering the nitrogen cycle, is death in the animal and plant world needless?

★9. Are chemical fertilizers the complete answer for continual soil enrichment? Why?

10. Why is letting the land "rest" every few years a good idea?
11. What is the advantage of planting leguminous plants?
12. Name three nitrogen-fixing processes for soil use.
13. How many electron pairs are bonded together in a nitrogen molecule?

14. (a) How many valence states does nitrogen have?
(b) Name an oxide in which nitrogen has a valence of +1. What is its chemical formula?
15. What is a nitride?
16. (a) What gas that is made of nitrogen and hydrogen has a very powerful odor?
(b) What is the chemical formula of this gas?
(c) Does this gas, added to water, form an acid or base?
(d) What can the liquid form of this gas be used for?
(e) In the refrigerator cycle, when does this gas absorb heat? Why?
17. (a) What nitrogen oxide is called laughing gas? Write its chemical formula.
(b) What gas is given off when fuming nitric acid decomposes? Write its chemical formula.
18. (a) What is the commercial process for the production of nitric acid (HNO_3) called?
(b) Write the equation for the hydrolysis of nitric acid.
(c) Suppose you had three unidentified acids on the shelf. You knew they were either sulfuric acid (H_2SO_4), hydrochloric acid (HCl), or nitric acid (HNO_3). How could you identify the nitric acid, assuming all three are colorless?
19. (a) Why is phosphorus never found in the elemental state?
(b) Name three kinds of phosphorus.
(c) Why does white phosphorus form P_4 molecules?
(d) What is the geometric shape of the P_4 molecules?
(e) Name two other compounds with the same structure as the P_4 molecule.
20. (a) How many valence states does phosphorus exhibit?
(b) Why is phosphorus used in tracer bullets?
(c) What acid does P_4O_{10} produce when it combines with water? Write the chemical formula of this acid.
21. What is a use of trisodium phosphate (Na_3PO_4)?
22. What is the difference between kitchen matches and safety matches?
23. (a) What is the principal compound of bones and teeth?
(b) Is phosphorus necessary for plant growth?
24. (a) Why are arsenic (As) and antimony (Sb) alloyed with lead (Pb)?
(b) Why is antimony used in type metal?
(c) What is the use of bismuth (Bi) in fire alarm systems?

Student Activities

1. Write a paper on the need of nitrogen in the soil for the healthy growth of plants. Cover all methods of fertilization from chemical to organic.
2. Write a paper on the need of phosphorus in the soil for the healthy growth of plants. Cover all methods of fertilization from chemical to organic.
3. Write a paper on the industrial uses of nitrogen and its compounds. Trace the various processes necessary to produce at least two different nitrogen products.
4. Write a paper on the chemistry of ammonia, its compounds, and its complexes. Be sure to include chemical reactions and as many illustrations of industrial chemistry as you can find.

Photo next page
Paricutin volcano in Mexico, photographed at night. More sulfur-containing pollutants are added to the atmosphere by volcanos than by all of man's industries combined.

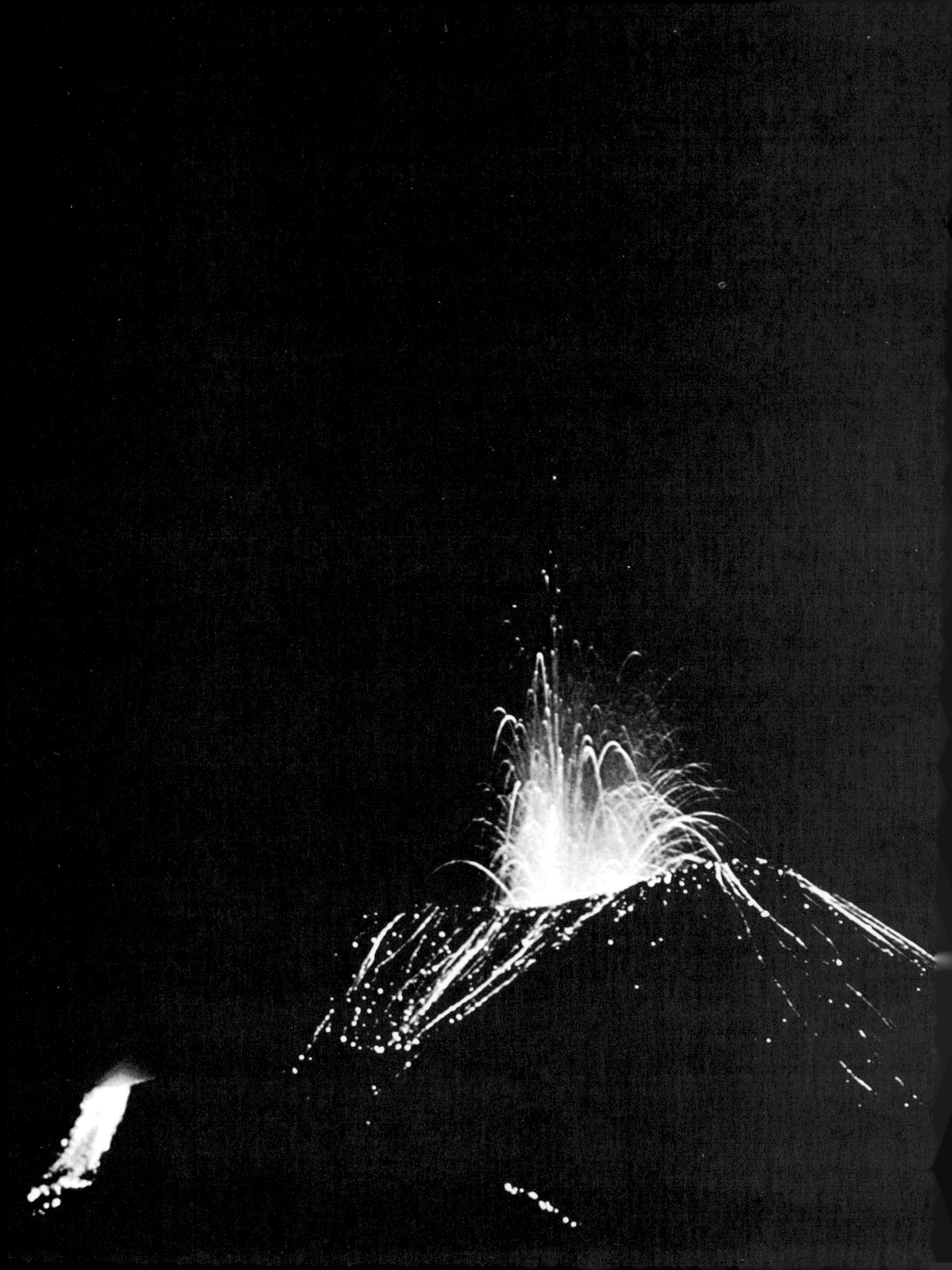

UNIT VI DESCRIPTIVE CHEMISTRY
Groups VA-VIIIA

Chapter 19
Sulfur

19:1 Brimstone Sulfur is referred to in the Bible as **brimstone**—the stone that burns. Brimstone is mentioned 15 times, seven times in the book of Revelation. Often it is used in connection with Hell, the lake of fire and brimstone (see Revelation 19:20). Sulfur burns with an interesting blue flame; its odor is strongly repulsive. Smelling burning sulfur may give you some idea of what Hell is like.

Sulfur and its compounds were used in ancient times for various purposes, such as in medicines, yellow paint pigment, fumigants, and bleach.

19:2 Occurrence of Sulfur Elemental (native) sulfur is found in mineral springs, volcanic deposits, and gypsum deposits. Combined sulfur is found in the following minerals: gypsum ($CaSO_4 \cdot 2H_2O$), barite ($BaSO_4$), iron pyrites (FeS_2), galena (PbS), and cinnabar (HgS). Sulfur occurs as an impurity in coal and petroleum. It is also found in some other organic substances.

Large deposits of native sulfur are found in salt domes in Louisiana and Texas at depths of 500 to 3000 feet. An unusual method of extracting the element from these domes was developed in the 1890's by Herman Frasch (1851-1914), a German who immigrated to the United States. It is know as the **Frasch process**. First, a well is dug to the depth of the sulfur deposit. Then superheated water is allowed to flow into the well; this melts the sulfur. After it is melted, the molten sulfur is pumped to the surface by means of compressed air.

19:1 *Pouring molten sulfur into a vat.*

19.2 *Frasch process for mining sulfur.*

19:3 Forms of Sulfur Sulfur can exist in two different crystalline states, the rhombic state and the monoclinic state. These are referred to as **allotropic** forms. Anytime a substance can assume different crystal structures, the properties of the substance will be different. Normally, the crystal structures are stable at different temperatures. For instance, rhombic sulfur is stable *below* about 96° C, whereas the monoclinic form is stable *above* this temperature.

Both allotropic forms of sulfur are made up of S_8 molecules. The eight atoms form a ring resembling a badly shaped hammock (see Figure 19:4). Each "stick" represents an electron pair.

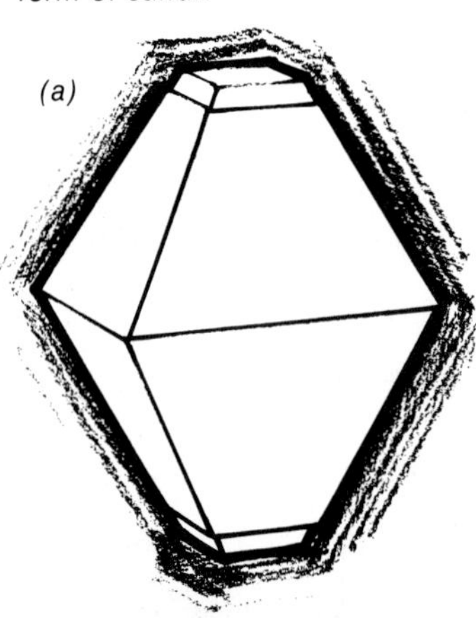

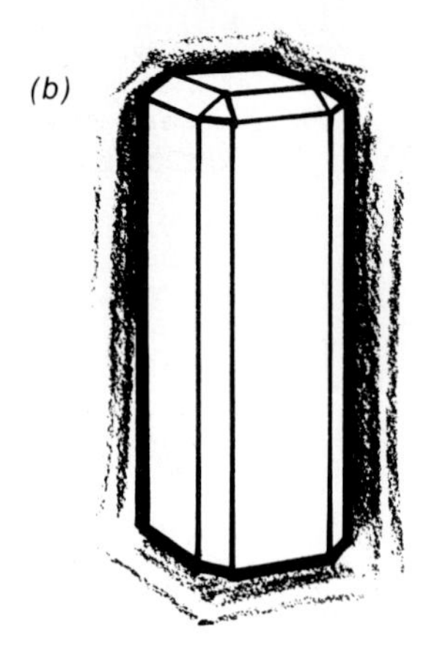

19:3 *(a) The rhombic form of sulfur. (b) The monoclinic form of sulfur.*

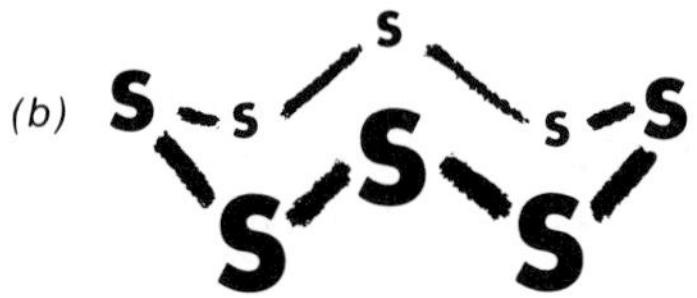

19:4 *(a,b) Models of the S_8 molecule.*

When sulfur is heated above 200° C, then poured into cold water, a form of sulfur results that has no crystal structure. It is **amorphous** sulfur. "Amorphous" means *without structure.*

The cooling operation is referred to as **quenching**. If the sulfur were *slowly* cooled from above 200° C, the atoms would have time to assume the S_8 ring structure. However, the *sudden* cooling does two things. One, it shortens the time the atoms have to diffuse into a ring. Two, it removes the energy (heat) necessary for the atoms to diffuse. As a result, the high temperature form (amorphous) remains at the low temperature.

As the amorphous sulfur sets at room temperature, the atoms gradually diffuse into an S_8 molecule, and the sulfur becomes rhombic in form.

Gaseous sulfur can exist as S_2, S_6, or S_8 molecules. It seems as if sulfur can "twist" itself into more shapes than a worm!

19:5 *The formation of amorphous sulfur.*

19:4 Chemistry of Sulfur Sulfur is a nonmetal; its electronic structure is shown below:

Filled atomic orbitals of sulfur

1s	2s	2p	3s	3p
2	2	6	2	4

Sulfur needs two electrons to have a stable octet in the valence shell. It can gain two electrons and become a negatively charged ion, or it can share some or all of its valence electrons in covalent bonding.

Sulfur exhibits four valence states: −2, 0, +4, and +6. Typical compounds of these valence states are shown in Table 19-1.

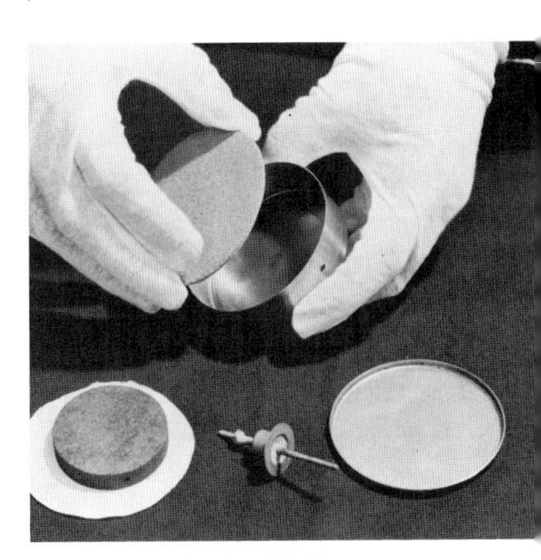

19:6 *A use of sulfur. Lithium-sulfur storage cell. The lithium electrode is installed in the case. The sulfur electrode (left) is on top of a thin ceramic cloth which separates the two electrodes. A current feed-through (center) connects to the sulfur electrode to conduct electrical current through the case when sealed with the cap (right).*

Table 19-1 *Valence States of Sulfur*

Valence	Typical Compounds
−2	hydrogen sulfide (H_2S) calcium sulfide (CaS)
0	sulfur (S_8)
+4	sulfur dioxide (SO_2) sulfur tetrafluoride (SF_4)
+6	sulfur trioxide (SO_3) sulfuric acid (H_2SO_4)

The −2 valence state in a compound represents ion formation. The +4 and +6 states represent covalent bonding. In **sulfuric acid**, the positive hydrogen ions are covalently bonded to the negative **sulfate ion**; the sulfur shares six electrons with four oxygen atoms in the radical. The oxygens are tetrahedrally arranged about sulfur in the SO_4^{-2} radical.

19:7 *A model of a sulfate tetrahedron.*

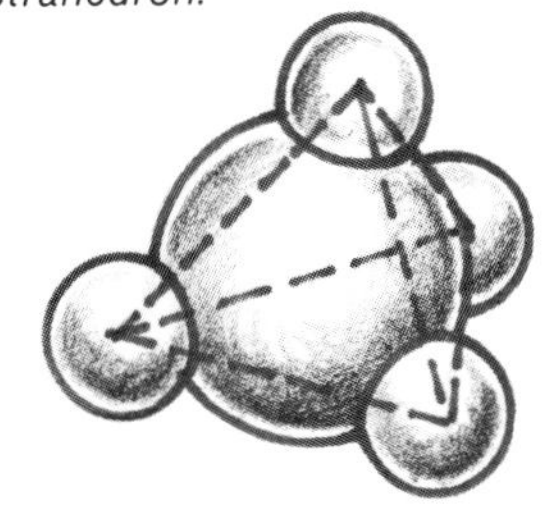

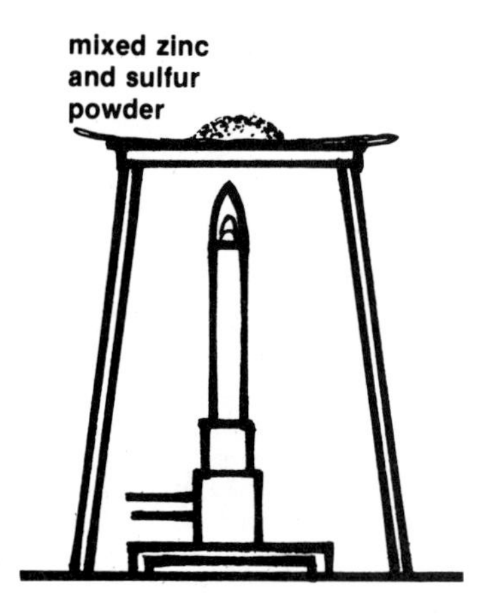

19:8 *How to react zinc and sulfur—stand back!*

Sulfur readily reacts with metals to form **sulfides**. If a mixture of zinc dust and sulfur is ignited with the flame of a Bunsen burner (see Figure 19:8), a violent reaction takes place.

$$Zn + S \xrightarrow{\Delta} ZnS$$

Sulfur reacts with oxygen to form sulfur dioxide (SO_2). This is the foul-smelling gas that is given off when sulfur burns.

$$S + O_2 \longrightarrow SO_2\uparrow$$

Oxygen and sulfur form many similar compounds, since they are in the same periodic table group (VIA). Some similar compounds are

Sulfur:	H_2S	CS_2	ZnS
Oxygen:	H_2O	CO_2	ZnO

19:5 Compounds of Sulfur and Their Uses The gas that smells like rotten eggs, hydrogen sulfide (H_2S), is dangerously poisonous. When any sulfur-containing organic substance decays, such as egg albumen, hydrogen sulfide is formed. Hydrogen can be reacted directly with heated sulfur to form the gas

$$H_2 + S \xrightarrow{\Delta} H_2S\uparrow$$

Since most metallic sulfides are not soluble in water, an interesting application has been found for hydrogen sulfide. If a chemist is analyzing an unknown substance, he may wish to precipitate it out of a test solution. He can do this using H_2S. For instance, zinc in solution can be precipitated as zinc sulfide.

$$Zn^{+2} + H_2S \longrightarrow ZnS\downarrow + 2H^+$$

Hydrogen sulfide has found great use in analytical laboratories in the identification of elements.

The most important use of the colorless sulfur dioxide (SO_2) is in the manufacture of sulfuric acid. Sulfur dioxide can be used as a **bleach** for organic materials; articles that would be injured by chlorine can be bleached using moist SO_2. However, the bleaching is not permanent, as it would be if chlorine were used. Sulfur dioxide is used as a disinfectant, too, and can prevent the growth of molds. SO_2 is used in the paper industry (with calcium hydroxide) to free the cellulose fibers from the bulk wood.

19:9 *The bleaching power of sulfur dioxide. Put a fresh carnation in a bottle containing SO_2 gas. After a time the carnation will be bleached.*

Experiment 19:1 *Preparation of Sulfur Dioxide*

1. Burn some sulfur in a deflagrating spoon. What is the color of the flame? Cautiously smell the vapor. Record your observations.
2. Dissolve 2 grams of sodium sulfite (Na_2SO_3) in 15 ml of water in a large test tube.
3. Add a few drops of dilute HCl and stir. Notice the odor. What gas is it?
4. Place a one-hole stopper into the tube and bubble the gas into a bottle where a moist, colored carnation has been placed in water.
5. Bubble a large amount of the gas into the water and seal the bottle. What happens to the carnation after a short time?

The most important sulfur-containing compound is definitely sulfuric acid (H_2SO_4), a thick, syrupy, colorless liquid when concentrated. Most of the sulfur produced is used to manufacture this acid.

Sulfuric acid has countless industrial uses. It is used in the production of nitric acid (HNO_3) and hydrochloric acid (HCl) because of its high boiling point. Sulfuric acid is an excellent dehydrating agent. When any of it comes into contact with organic material, sulfuric acid removes the elements that make up water and leaves a black carbon residue. An excellent demonstration of this is the dehydration of sugar by sulfuric acid.

$$C_{12}H_{22}O_{11} \xrightarrow{H_2SO_4} 12C + 11H_2O$$

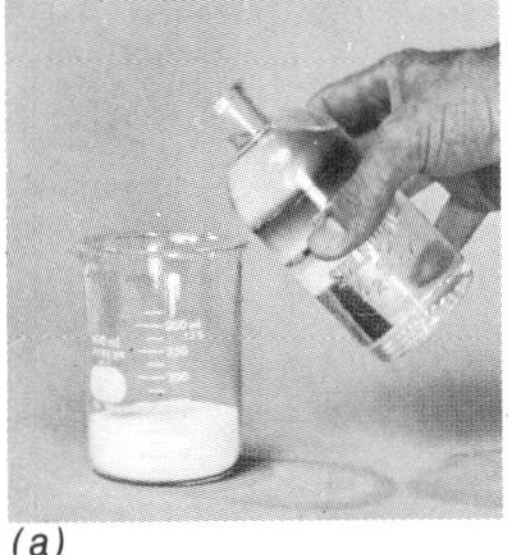

(a)

(b)

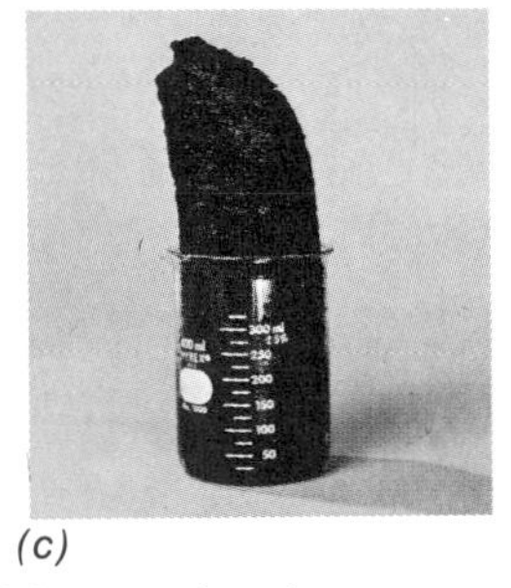

(c)

19:10 *The reaction of sugar with sulfuric acid. (a) Pouring the acid into a beaker containing sugar. (b) The reaction begins—some carbon is forming. (c) The reaction is complete—a foamy column of carbon results.*

Sulfur can form polymeric sulfides with metals; these are referred to as **polysulfides**. The polysulfide ions are simply chains of sulfur atoms bonded together to form a negatively charged ion.

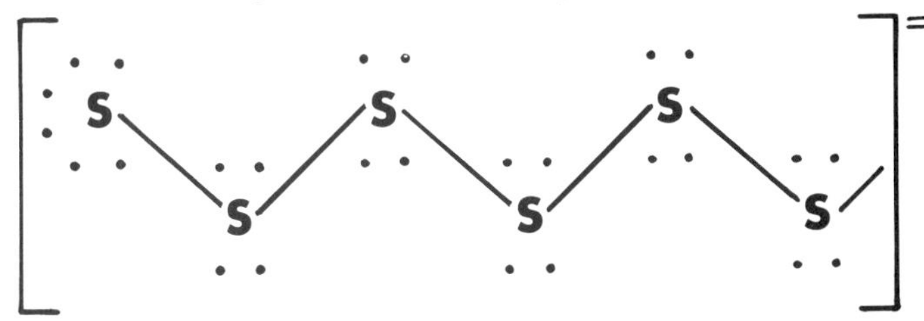

19:11 *Model of part of polysulfide ion.*

Craftsmen use certain polysulfides on copper engravings to react with the metal slowly and form a controlled, dark sulfide stain in certain areas. This enhances the appearance of the art work.

19:12 *Copper panels darkened by polysulfides.*

19:6 Vulcanization of Rubber Rubber tends to become sticky when it is hot, and hard and brittle when it is cold. If rubber is vulcanized with sulfur, the rubber becomes less susceptible to temperature changes. It is then more durable and serviceable. The chemical change that takes place between the rubber and sulfur is called **vulcanization**. Charles Goodyear (1800-1860), the American inventor, discovered this important process in 1839.

19:13 *Vulcanization of passenger car tires. A green tire is placed in a mold. When the mold closes, internal pressure forces a curing bladder into the tire, shaping it. Live steam causes the mold to heat, softens the rubber, and forces it into the non-skid pattern of the tire mold.*

List of Terms

- allotropic
- amorphous
- bleach
- brimstone
- Frasch process
- polysulfides
- quenching
- sulfate ion
- sulfide
- sulfuric acid
- vulcanization

Questions

1. (a) What is the name for sulfur given in the Bible?
   (b) What is the color of a sulfur flame?
2. Name three sources of sulfur.
3. Describe the Frasch method of removing native sulfur from salt domes.
4. Name three forms of solid sulfur.
5. Define amorphous.
6. How many sulfur atoms make up a molecule of each of the allotropic forms?
7. Why does amorphous sulfur form?
8. (a) What is the arrangement of electrons in the outermost electron shell of sulfur?
   (b) How does sulfur act in chemical bonding?
9. (a) How many different valences can sulfur assume?
   (b) Name a compound where sulfur exhibits a valence of -2.
   (c) Name a compound where sulfur exhibits a valence of +4.
10. What sulfur radical has a tetrahedral structure?
11. (a) What gas smells like rotten eggs? Give its name and chemical formula.
    (b) Where has this compound been found to be of great use?
    (c) Why is it so useful?

★12. Which of the following reactions are possible?
   (a) $Ni^{+2} + H_2S \rightarrow NiS\downarrow + 2H^+$
   (b) $Sn^{+2} + H_2S \rightarrow SnS\downarrow + 2H^+$
   (c) $Hg^{+2} + H_2S \rightarrow HgS\downarrow + 2H^+$

13. (a) What sulfur compound can be used as a bleach? Give its name and chemical formula.
    (b) How is it used in the paper industry?
14. (a) Write the chemical formula for sulfuric acid.
    (b) Name three of its possible uses.
15. What are polysulfides?
16. (a) What is vulcanization?
    (b) Why is it done?

★17. What does a valence state of zero mean?

★18. Complete the following reaction.

$$SO_3 + H_2O \longrightarrow$$

19. What is the name of SO_3?

Student Activities

1. Write a paper on the various structural forms of sulfur. Draw models of the various crystalline forms. Produce all of the forms you can by experiment and mount them on a display board.
2. Write a paper on the chemistry of sulfuric acid. Perform various experiments with the acid under your teacher's direction. Carefully record all of your experimental results.

Photo next page
A vault containing silver to be used in the production of photographic film.

UNIT VI DESCRIPTIVE CHEMISTRY Groups VA-VIIIA

Chapter 20 The Halogens

20:1 Salt Formers The word **halogen** comes from two Greek words and means *salt former*. All of the members of the halogen group, except Astatine (At), are definitely nonmetals. Therefore, there is no trouble in classifying the halogens chemically. This is a pleasant change after having considered so many mixed groups.

20:2 Electron Distribution A chart of the electronic distribution in the halogen elements is shown in Table 20-1. Each Group VIIA element has seven electrons in its outermost electron shell. Therefore, each element can accept one electron to have a stable octet.

Table 20-1 *Electron Arrangement in the Halogen Atoms*

Element	Quantum Level and Electron Orbitals														
	1	2		3			4				5			6	
	s	s	p	s	p	d	s	p	d	f	s	p	d	s	p
F	2	2	5												
Cl	2	2	6	2	5										
Br	2	2	6	2	6	10	2	5							
I	2	2	6	2	6	10	2	6	10		2	5			
At	2	2	6	2	6	10	2	6	10	14	2	6	10	2	5

20:1 *Halogen child "begging" for one electron.*

All I want is one electron!

20:3 Chemical and Physical Properties Fluorine is the most non-metallic member of the halogens. Fluorine is also the most chemically active of the group.

non-metallic character ←
most active
F > Cl > Br > I
→ least active
metallic character

This diagram indicates that fluorine is more active than chlorine, which is more active than bromine, which is more active than iodine. The chemical activity decreases as you move down the periodic table within Group VIIA. Fluorine can replace chlorine and other members of the group in their compounds; chlorine can replace bromine and iodine from their compounds; and bromine can replace iodine from its compounds.

In a chemical reaction, when the halogens accept an electron (the characteristic property of a nonmetal), they become negative ions and are bonded to metals (electron donors) by positive-negative electrical attraction. This ionic bonding is between a positive metal ion and a negative halogen ion. Table 20-2 shows the ions of the halogens and the ion names.

Table 20-2 *Halogen Ions*

Element	Ion	Ion Name
fluorine	F^-	fluoride
chlorine	Cl^-	chloride
bromine	Br^-	bromide
iodine	I^-	iodide

20:2 *The sublimation of iodine.*

Iodine is a substance that will *sublime*. If solid iodine is heated, the solid will form a vapor, which will recrystallize on the bottom side of a cool watch glass (see Figure 20:2).

Fluorine and chlorine are gases at room temperature, whereas bromine is a liquid and iodine is a solid. Table 20-3 shows the colors and states of these elements at room temperature. The gaseous form of each halogen element is **diatomic** and is represented thus: fluorine (F_2), chlorine (Cl_2), bromine (Br_2), and iodine (I_2).

Table 20-3 *Color of Halogens at Room Temperature*

Element	Color	State
fluorine	light yellow	gas
chlorine	greenish yellow	gas
bromine	dark red	liquid
iodine	gray	solid

20:4 Van der Waals' Bonding You may be interested in how gaseous atoms "stick together" to form a molecule. We do not know for certain how this happens, but we can present a bonding model that will give you a picture to "explain" the forces. We will imagine that the electrons are in a "cloud" around the nucleus of an atom; we do not try to pinpoint the position of each electron. This is a physical impossibility.

The arrangement of this electron cloud about the nucleus does not always need to be symmetrical, as shown in Figure 20:3a. As the electrons move about the nucleus within the cloud, the cloud may become "skewed," as shown in Figure 20:3b; that is, the majority of the electrons are on the "right-hand side" (R) of the molecule, and the positive nucleus is no longer in the middle of the cloud. This means that the charge distribution within the atom is not balanced, as it is in Figure 20:3a. Since the positive nucleus is now close to the "left-hand" side (L) of the atom, L has a *net positive charge*. R has a *net negative charge* because of the large number of electrons there. Thus, the atom has positive and negative poles.

An atom with positive and negative poles is called an electrical **dipole**; the "di" means *two*. The positive pole of one atom then attracts the negative pole of another atom to form a weak electrical bond, as illustrated in Figure 20:3c. These very weak forces are called **van der Waals' forces**, named for the Dutch scientist J.D. van der Waals (1837-1923). They are thought to exist only momentarily, since the electrons are continually moving and changing the charge distribution of the atom. However, two halogen atoms can be associated with each other at any time in this van der Waals' bonding.

(a)

20:3 *(a) Symmetrical arrangement of electron clouds about the nucleus. (b) Skewed arrangement of electron clouds about the nucleus. (c) Model of van der Waals' bonding.*

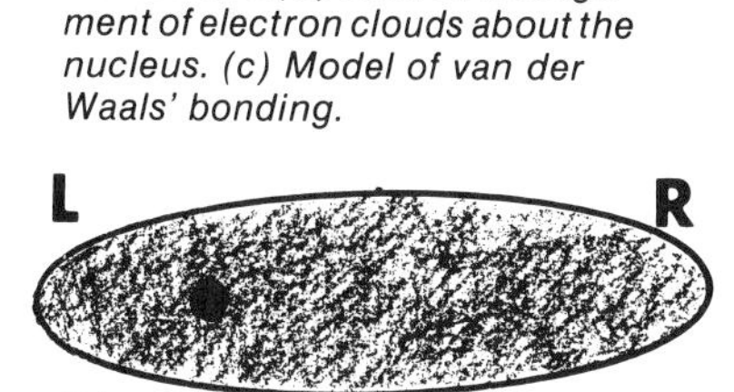

(b)

(c)

20:5 Sources of Halogens Since most of the salts of the halogens are soluble, the ocean is a good source of these elements. Table 20-4 lists some of the principal sources of the halogen elements. Also, many deposits of the halogen salts may have resulted from the Flood. As the Flood waters evaporated, large concentrations of halogen compounds—particularly sodium chloride—probably remained.

Table 20-4 *Principal Sources of the Halogens*

Element	Form and Source
chlorine	as dissolved or solid sodium chloride in underground salt beds, brine wells, and sea water
fluorine	fluorspar (CaF_2) cryolite (Na_3AlF_6)
bromine	dissolved in sea water
iodine	as sodium iodate ($NaIO_3$), an impurity in sodium nitrate deposits in Chile; also, dissolved sodium iodide in seaweed and brine wells
astatine	formed by transmutation of francium-87

20:6 Commercial Preparation of Fluorine and Chlorine Pure halogens are difficult to obtain because their salts are so easily soluble in water.

Fluoride compounds are so stable that there is no substance which can be reacted with them to release pure fluorine. To obtain fluorine, an electrical current must be passed through molten fluorides. Fluorine gas is then released at the anode during this electrolysis.

Chlorine is prepared by the electrolysis of an aqueous solution of sodium chloride. Pure chlorine is released at the anode. These electrolysis operations are similar to the electrolysis of water, which was discussed in Chapter 11.

20:4 *Paul Bunyan's Waffle Iron? This photo actually shows less than half the interior of the cell room at a single-circuit mercury cell chlorine plant. The unit is wider than a regulation football field and nearly as long. It consumes enough electrical power to serve a community of 175,000 homes, while producing more than 600 tons of chlorine and 650 tons of caustic soda each day.*

20:7 Reactions of Halogens The halogens react readily (and often violently) with metals to form halide salts. Basically, the reaction is

$$\text{Metal} + \text{Nonmetal} \longrightarrow \text{Salt.}$$

Specifically, the reactions of halogen elements with sodium are

$$2Na + F_2 \longrightarrow 2NaF \quad \text{sodium fluoride}$$
$$2Na + Cl_2 \longrightarrow 2NaCl \quad \text{sodium chloride}$$
$$2Na + Br_2 \longrightarrow 2NaBr \quad \text{sodium bromide}$$
$$2Na + I_2 \longrightarrow 2NaI \quad \text{sodium iodide}$$

The halogens also react with hydrogen to form acids. The more familiar acids are hydrochloric acid and hydrofluoric acid:

$$H_2 + Cl_2 \longrightarrow 2HCl \quad \text{hydrochloric acid}$$
$$H_2 + F_2 \longrightarrow 2HF \quad \text{hydrofluoric acid}$$

Experiment 20:1 *Laboratory Preparation and Chemical Properties of Chlorine*

1. Place 15 grams of manganese dioxide (MnO_2) in a round-bottom flask.
2. Using a two-hole stopper, insert a thistle tube and glass tubing in the stopper as shown in Figure 20:5.
3. Add 30 ml of dilute hydrochloric acid (HCl) through the thistle tube. (Be sure that the bottom end of the tube is below the liquid level.)
4. Warm the flask gently. The chlorine gas may be collected by displacing air, since chlorine gas is heavier than normal air. Be careful; *chlorine gas is poisonous.* The reaction can be represented as

$$MnO_2 + 4HCl \longrightarrow MnCl_2 + Cl_2\uparrow + 2H_2O$$

5. Collect 3 bottles of chlorine gas.
6. Sprinkle some antimony (Sb) powder into a gas bottle containing chlorine. Write down your observations and the chemical equation for the reaction that occurred.
7. Add some water and copper to a gas bottle containing chlorine. Write down your observations and the chemical equations for the reactions that occurred.
8. Place a moist, colored cloth into a gas bottle containing chlorine. What happens? Write down your observation.

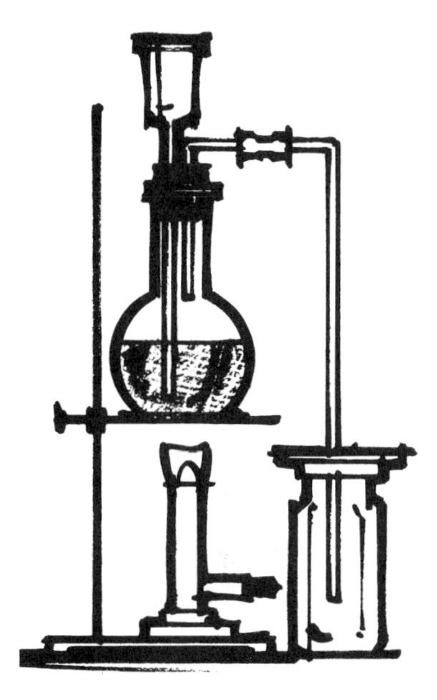

20:5 *Laboratory rig for the preparation of chlorine.*

Hydrochloric acid (HCl) is a common laboratory acid. It has a very piercing odor from the hydrogen chloride gas coming out of solution. Hydrochloric acid is one of the most industrially useful of all of the acids. Some of its many uses are in the preparation of metal chlorides and the manufacture of many organic chemicals. Hydrochloric acid is also pumped down oil wells to "attack" rocks where the oil is trapped in pores or cracks of the rock. HCl is an excellent cleaner; it will remove scale (rust) from steel. This cleaning operation is known as **pickling**.

20:6 *Cold-strip-acid-pickling line in a steel mill. Hydrochloric acid is used to clean steel sheet.*

The acid in your stomach, which is used to digest food, is hydrochloric acid. If you overeat or eat foods which are hard to digest, your stomach may secrete too much acid. A mildly basic substance or antacid must be taken to neutralize this excess acid.

Hydrofluoric acid (HF) is used to etch glass. It is one of the few substances that will react with silicon dioxide or silicates. The reaction can be represented as

$$SiO_2\text{(in glass)} + 4HF \longrightarrow SiF_4\uparrow + 2H_2O$$

or

$$Na_2SiO_3 + 6HF \longrightarrow 2NaF + SiF_4\uparrow + 3H_2O$$

You may wonder why glass should be chemically etched, rather than scratched, if you wanted to write on it. A scratch on glass causes many invisible internal stresses in the object; any slight load or tap will then break the glass. When glass breaks on the slightest touch, it has probably been previously dropped, hit, or scratched. Chemical etching does not leave any weakening internal stresses in the glass.

Experiment 20:2 *Properties of Hydrofluoric Acid (HF) in Etching Glass*

1. Pour 10 ml of concentrated sulfuric acid (H_2SO_4) into a clean test tube. Allow it to stand about 5 minutes. Does it etch the glass?
2. Add 1 gram of calcium fluoride (CaF_2) to the acid. The following reaction takes place:

$$CaF_2 + H_2SO_4 \longrightarrow CaSO_4 + 2HF\uparrow$$

Allow it to stand about 5 minutes. Is the glass etched now? Why?

20:7 *Teflon-coated cookware.*

Table 20-5 *Common Uses of the Halogens and Their Compounds*

Element	Use
fluorine	glass etchant; Teflon plastic; UF_6 gas for separation of U-235 and U-238; refrigerants—freons; propellent in spray cans—freons; fluoridation of water, toothpastes
chlorine	manufacture of organic chemicals; matches; textile, paper-pulp bleaching; purification of water by killing bacteria; manufacture of bromine and iodine; chlorinated solvents such as carbon tetrachloride (CCl_4) and chloroform ($CHCl_3$); production of plastics—vinyl chloride; insecticide—DDT; weed killer—2,4-D; lamprey eel killer—3-4-6; laundry bleach
bromine	"antiknock" gasoline; headache powders; photographic film; disinfectants; fumigants
iodine	antiseptic solutions such as tincture of iodine (iodine in alcohol); iodized salt; emergency disinfectant for drinking water; necessary in diet

20:8 Uses of Halogens and Their Compounds Uses of the halide compounds and the halogens are numerous. Some of these uses are cataloged in Table 20-5. For example, a plastic material known as Teflon contains fluorine. **Teflon** is a highly stable material that resists chemical attack from acids and bases. It is used as a coating for kitchenware to provide a "non-stick" surface. Foods and greases will not stick to Teflon as they would to metals. Teflon is made by polymerizing a fluorocarbon, tetrafluorethylene.

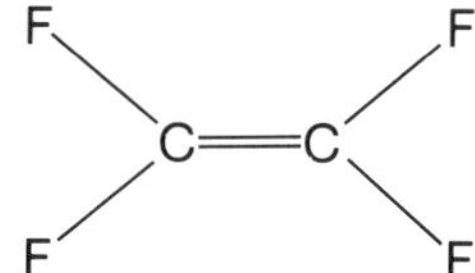

Tetrafluorethylene

The double bonds between the carbons are broken and a large molecule (polymer) of carbons with attached fluorines is formed. It is similar to polyethylene, except that fluorine takes the place of hydrogen.

```
     F   F   F   F
     |   |   |   |
   — C — C — C — C —
     |   |   |   |
     F   F   F   F
```

or

$—CF_2CF_2CF_2CF_2—$

Model of part of a Teflon Molecule

Sodium hypochlorite (NaClO) is used as a bleach. It is sold under several names and is normally used in the washing of white fabrics.

The halides of silver—silver chloride (AgCl), silver bromide (AgBr), and silver iodide (AgI), are sensitive to light. When "particles" of light collide with any of the above compounds, the compounds slowly decompose into silver and the halogen.

20:8 *Positive.*

20:9 *Negative.*

20:10 *A vault containing silver to be used in the production of photographic film. (See chapter title page)*

Photographic film is an emulsion of colloidal silver bromide particles dispersed in gelatin and spread over a glass plate or sheet of plastic. The film is kept in darkness until it is exposed. Areas exposed to light are more easily decomposed in the **developer** solution than unexposed areas of the film. The silver bromide areas exposed to the most light decompose faster than areas exposed to less light. The more exposed areas will appear darker on the film than the less exposed areas. If the film is left in the developer too long, all of the silver bromide will decompose and leave the entire film completely darkened.

Once the film is developed, it is placed in an intermediate solution (acetic acid) to "stop" the developer action. It is then placed in a **hypo** bath, a solution of sodium thiosulfate ($Na_2S_2O_3$), to dissolve the remaining silver bromide. This hypo bath is called the "fixer."

Up to this point, the film can also be called the "negative." After the negative is washed and dried, a "positive" is made from it. The positive, or print, has a silver halide emulsion on its surface. The dark areas which appear on the print correspond to the light areas which appear on the negative. The print is a true representation of the object photographed.

20:11 *The examination of developed photographic film.*

Experiment 20:3 *The Action of Light on Silver Chloride (AgCl)*

1. Add dilute HCl to a solution of silver nitrate ($AgNO_3$) in a beaker. A precipitate of AgCl forms according to the following reaction:

$$AgNO_3 + HCl \longrightarrow AgCl\downarrow + HNO_3$$

2. Place the beaker in a strong light (bulb or sunlight). What happens to the AgCl precipitate? Why? Record your observations.

List of Terms

developer
diatomic
dipole
halogen
hydrochloric acid
hydrofluoric acid
hypo
pickling
sodium hypochlorite
Teflon
van der Waals' forces

Questions

1. What does the word "halogen" mean?
2. What is the similarity in the electronic structure of the halogens?
3. (a) What is the total number of electrons in the outermost electron shell of each element?
   (b) What does this mean chemically?
4. (a) Which element of the halogens acts the most like a nonmetal?
   (b) Which element of the halogens is most chemically active?
5. Are the following reactions possible? Why?
   (a) $2NaI + F_2 \longrightarrow 2NaF + I_2$
   (b) $2NaF + Br_2 \longrightarrow 2NaBr + F_2$
6. (a) What type of bond do the halogens form when combining with metals?
   (b) Do they accept or donate electrons?
7. Briefly explain van der Waals' bonding.
8. (a) Suggest two natural sources of chlorine.
   (b) Suggest two natural sources of iodine.
9. Complete the following reactions.
   (a) $2Na + Br_2 \longrightarrow$ ____________
   (b) $Mg + Cl_2 \longrightarrow$ ____________
   (c) $2Al + 3F_2 \longrightarrow$ ____________
10. To what does "pickling" refer?
11. (a) What acid digests the food in your stomach?
    (b) What is the name of the acid used to etch glass? Write the chemical formula for it.
12. (a) Fluorine is contained in a well-known polymer material. What is the trade name for this material?
    (b) Write the chemical formula for this material.
    (c) Write the chemical formula for polyethylene.
13. Write the chemical equation for the decomposition of silver chloride by light.
14. (a) What is photographic film?
    (b) Explain the light and dark areas on exposed, developed photographic film.
15. What is the purpose of the "hypo" bath or "fixer" in photographic developing?
16. Name four uses of chlorine compounds.
17. Write the ionization reactions of HCl and HF in water.
18. Define sublimation.

Student Activities

1. Write a paper on the various industrial uses of chlorine and its compounds. Be sure to include the chemical reactions of interest.
2. Write a paper on the chemistry of the photographic process, including negative and print development.

Photo next page
A high-sensitivity rare-gas mass spectrometer, the modern counterpart of Thomson's spectrograph.

UNIT VI DESCRIPTIVE CHEMISTRY Groups VA-VIIIA

Chapter 21 The Rare Gases

21:1 A Unique Family In this chapter we shall concentrate on the extreme right-hand column of the periodic table. These elements, called the **rare gases**, are different from the other elements discussed so far. The rare gases have full outer shells of electrons. Helium has a full *K* shell of two electrons, and the other members of the family have octets in their outermost shells. Since the rare gases are already chemically satisfied, they participate in very few chemical reactions. In some older texts they are sometimes called "inert" gases, but in modern usage the term **noble gases** is more often applied to this family of elements.

The rare gas atoms do not even combine with their own kind; they exist as *monatomic* molecules. The atoms and the molecules of these elements are, therefore, one and the same thing. All of the rare gases have low melting points and boiling points for the weight of their atoms. There is very little difference *between* the boiling and melting points. This indicates that there is little interaction of any kind between the atoms. Helium has the lowest melting point of any known element and is widely used in **cryogenic** (low-temperature) research (see Figures 21:4, 21:5).

He
Ne
Ar
Kr
Xe
Rn

21:1 *The Rare Gases.*

21:2 Helium The second lightest element, helium, was first discovered in an analysis of sunlight in 1868 through the combined efforts of French astronomer Pierre Janssen (1824-1907), English astronomer Sir Norman Lockyer (1836-1920), and English chemist Edward Frankland (1825-1899). The name "helium" is derived from the Greek word "helios," which means *sun*. It was not until 1895 that Sir William Ramsay (1852-1916) discovered helium on the earth by heating the uranium mineral clevite.

Although helium is the second most abundant element *in the universe*, there is relatively little of it *on the earth*. Evolutionists say that the radioactive decay processes that produce helium (alpha decay reactions) have been going on in the earth's crust for *billions* of years. Yet helium is present in the atmosphere only to the extent of 1 part in 200,000. Where, then, has all the helium gone? Calculations show that helium cannot escape into space the way hydrogen does. Something must be wrong with the evolutionists' statement. Realistic calculations based on available figures show that the span of time involved for natural alpha decay processes to have produced the observed helium is much more likely to be measured in *thousands* of years, rather than billions of years. This is excellent evidence for a young earth.

Helium cannot be obtained from compounds because it has no compounds! It is produced commercially by the fractional distillation of natural gas obtained from wells in Texas, Kansas, and Oklahoma.

21:2 *Helium is used to fill lighter-than-air craft such as blimps and dirigibles.*

Helium is used to fill balloons and lighter-than-air craft such as blimps. This gas combines the desirable properties of good lifting power with non-combustibility. Another use of helium is in deep-sea diving air tanks: a mixture of 80 percent helium and 20 percent oxygen is used instead of the usual nitrogen-oxygen mixture of ordinary air. Because helium is less soluble in the blood than nitrogen, it creates less trouble when a diver attempts to surface. Ordinary air can produce a condition called "the bends" when a diver surfaces. "The bends" occur when the nitrogen which has dissolved in the blood comes out of solution and forms bubbles. Since these bubbles have nowhere to go, they cause the diver great pain. By contrast, very little helium dissolves in the blood in the first place.

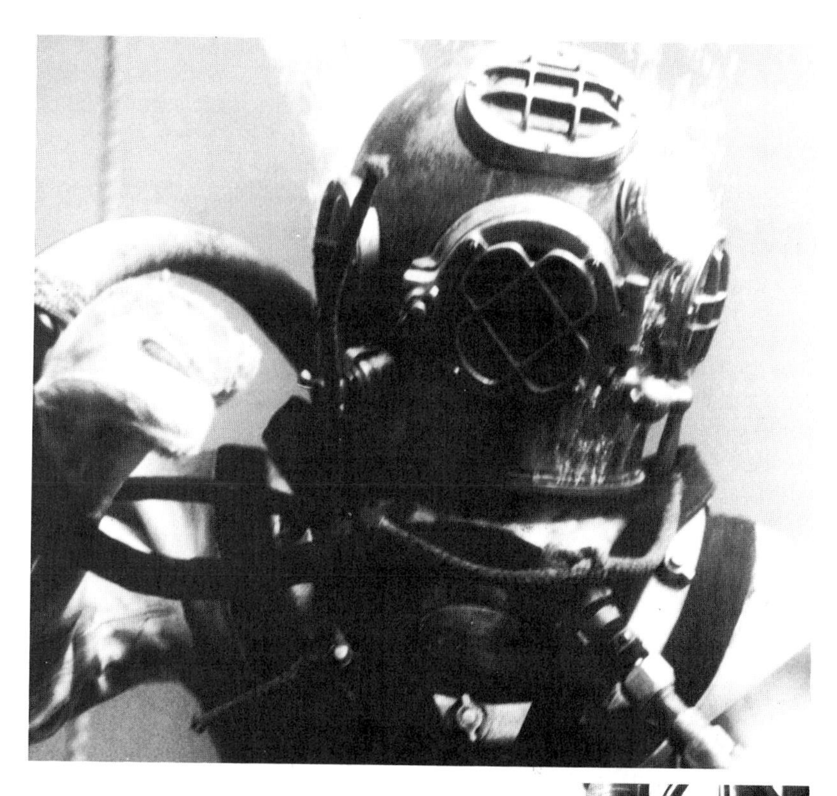

21:3 *Deep-sea divers breathe a mixture of oxygen and helium to avoid the condition called "the bends." Here a hard-hat diver searches for wreckage.*

21:4 *Liquid helium is used to cool a magnetism experiment. The temperature of the helium is approximately 4° K.*

21:5 *A liquid helium fountain—one of the natural phenomena of supercooled liquid helium. The energy from a light source excites the liquid helium to such an extent that it climbs the small tube in the center, spurting out at the top.*

21:6 *Sir William Ramsay, discoverer of several of the rare gases.*

21:3 Neon Sir William Ramsay discovered neon in 1898. Several of the rare gases were discovered by Ramsay and his co-workers when they removed oxygen and nitrogen from air and subjected the remaining gases to careful analysis. Neon is present in the atmosphere to the extent of only 1 part in 65,000. It *is* commercially feasible to remove neon from the air by **fractional distillation**, but it is an expensive process.

The primary use of neon is for neon advertising signs. Of all the known gases, neon gives the most light for a given voltage and current. A neon light has a characteristic orange-red color. Neon, at a reduced pressure, is placed in a nearly empty glass tube that has been bent to the desired shape. Metal electrodes are sealed into the ends of the tubing. When the electrodes are connected to a suitable high-voltage power supply, outer electrons in the neon atoms become "excited" and move to higher levels (more energetic orbitals). As these atoms fall back to their normal levels, their excess energy is given off in the form of light.

Light bulbs can be made using neon gas instead of the conventional filament. Neon bulbs have various applications, such as low wattage night lights, high voltage indicators, pilot lights, lightning arrestors, and tuning indicators for transmitters.

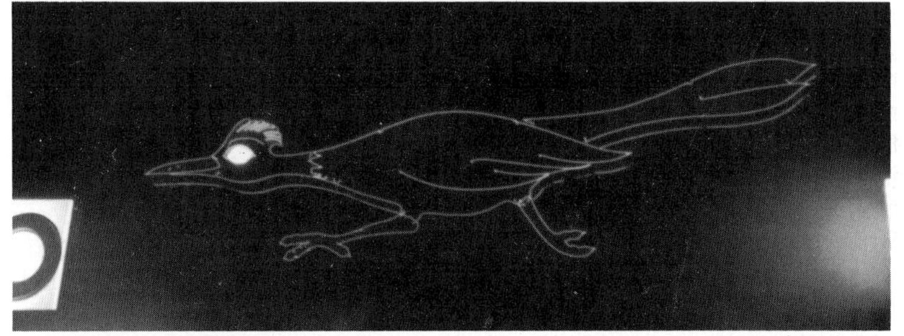

21:7 *A neon sign.*

21:8 *J.J. Thomson, who discovered isotopes while investigating neon.*

An interesting historical note is the fact that **isotopes** were first observed in the element neon. An English physicist, Joseph John Thomson (1856-1940), while using a device of his own invention called the **mass spectrograph**, observed two different patterns for neon on a fluorescent screen—one corresponding to a mass of 20, the other corresponding to a mass of 22. Since the known mass of naturally occurring neon was 20.2, Thomson reasoned that the neon found in nature consists of 90 percent neon-20 and 10 percent neon-22. His deduction was later shown to be correct. Neon-22 is identical with neon-20 in every respect, except that it has two extra neutrons.

21:4 Argon In 1894, Ramsay and English physicist Lord Rayleigh (1842-1919) discovered argon. The name "argon" means *lazy one*; it is related to the word "lethargic." Argon is the most plentiful of the rare gases; it makes up 0.94 percent of the earth's atmosphere by volume and 1.3 percent by weight. Like neon, argon is prepared commercially from liquid air. Like neon, argon is a colorless, odorless, tasteless gas. But unlike neon, argon is heavier than air. (A few years ago, the symbol for argon was

21:9 *A high-sensitivity rare-gas mass spectrometer, the modern counterpart of Thomson's mass spectrograph.*
(See chapter title page)

changed from "A" to "Ar." Some of the reference books you use may list the old symbol.)

The most important use for argon is in the manufacture of incandescent light bulbs. Mixed with nitrogen, argon forms an inert atmosphere that keeps the hot filament from oxidizing or vaporizing. Argon also finds application as an inert gas shield for arc welding and cutting.

21:5 The Potassium-Argon Dating Method You have perhaps heard of the **potassium-argon dating method**. This is one of several methods used in an attempt to justify the five- or six-billion-year age of the earth which is required by the theory of evolution. It has been proven that the potassium-argon method gives totally wrong answers on rock samples of known age. Yet the potassium-argon method is still believed to be valid by evolutionists. Why? Because the potassium-argon method and other methods like it are needed to make evolution seem reasonable. This, of course, is not science. Evolution, the myth to which the potassium-argon dating method has become such an important handmaid, is not science either.

First of all, how is the potassium-argon method *supposed* to work? The nucleus of the potassium-40 atom is slightly unstable; it decays with an exceedingly long half-life to form argon-40. (There is also a side reaction producing calcium-40, but this does not concern us here.) Now the idea is to find a sample of igneous rock which contains *both* K-40 and Ar-40. The quantities of these two isotopes present in the rock sample are measured, and an age is calculated for the rock. The age is calculated *assuming that there was no argon present in the rock when the rock was formed.* But who was there when the rock was formed to verify this assumption? If observations are lacking, there can be no science! As a matter of fact, God could have put *any* mixture of His choosing into the original rock. In assuming that there was no argon present in the original rock, the evolutionists have automatically (in their own thinking) pushed the date of creation back several billion years. Yet this assumption is the very thing they claim to be proving! Man must be careful that he does not attempt to limit the Creator to a certain course of action simply because that course of action appeals to him intellectually.

Let's put the potassium-argon dating method to a test. This method is supposed to tell us how much time has elapsed since a given rock sample solidified. It just so happens that there are igneous rocks in Hawaii for which the age is known. These rocks were actually observed in the process of formation as volcanic lava flowed and hardened. In the *Journal of Geophysical Research* for July 15, 1968, it was reported that rocks formed in the years 1800 and 1801 gave potassium-argon ages of formation ranging from 160 million to 3 billion years. But the rocks were known to be less than 170 years old! And this is not an isolated report. In *Science* for October 11, 1968, it was stated that volcanic rocks known to be less than 200 years old gave potassium-argon ages ranging from 12 to 22 million years! This proves beyond a

21:10 *Incandescent light bulbs use an inert atmosphere of argon and nitrogen to protect the filament.*

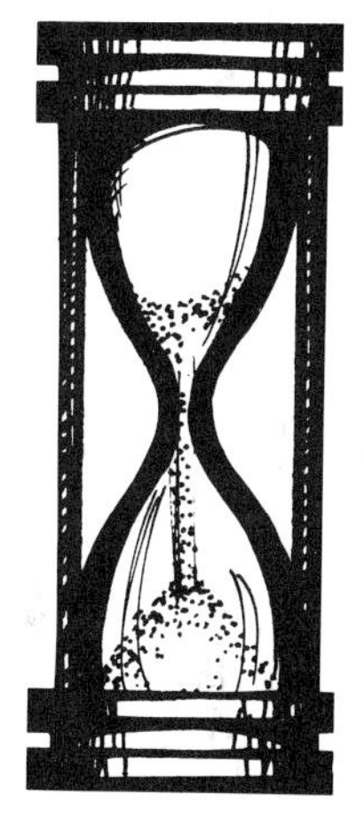

21:11 *The potassium-argon fallacy. The operation of the potassium-argon dating method is said to be similar to that of an hourglass. Potassium-40 (represented by the sand in the upper part of the glass) changes by radioactive decay to argon-40 (represented by the sand in the lower part). Age is estimated by comparing the amount of argon to the amount of potassium in a given sample. However, there is no guarantee that the sample was argon-free when the rock was formed. Recent studies on lava in Hawaii indicate that a considerable quantity of argon was already present when the rocks solidified. (See text.)*

21:12 *Laboratory-type xenon stroboscope (flashing light) used to time moving objects.*

doubt that the potassium-argon dating method is quite wrong. Yet the method is still used and trusted by evolutionists, and the results are called "science." This is one of the most ironic paradoxes of twentieth-century man: claiming to be completely scientific, in the final analysis he trusts only those methods that appeal to him emotionally.

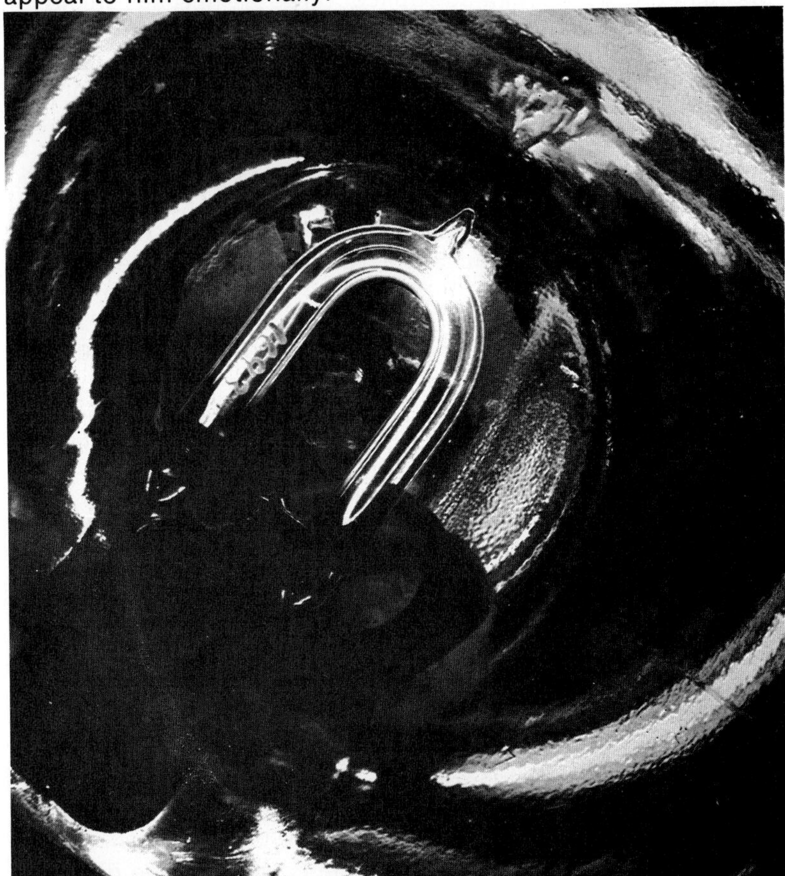

21:13 *A xenon strobe used for photography. A daylight-quality burst of light is produced when it is connected to a source of high-voltage direct current.*

21:6 Krypton Discovered by Ramsay and his co-workers in 1898, krypton is one of the rarer rare gases; it is present in the atmosphere to the extent of only 1 part per million. One significant failing of the various evolutionary theories of the solar system is their inability to explain why the heavier noble gases are so scarce in the earth's atmosphere. According to current theories, the noble gases should be considerably more abundant.

You will recall that the orange-red light of krypton-86 is used to define the standard meter (see Chapter 9). Krypton is also used in certain special application lamps and electronic tubes.

21:7 Xenon Like most of the other members of the family, xenon was discovered by analyzing the residue left when oxygen and nitrogen are removed from liquid air. Xenon is extremely rare; it is found in the atmosphere to the extent of only about one molecule in 20 million. Xenon is used in special application vacuum tubes, bactericidal lamps, stroboscope bulbs, and lamps to excite ruby lasers. It is also used in atomic-energy bubble chambers and probes. Xenon is a very heavy gas, having a density about 4.5 times the density of air.

For many years chemistry textbooks erroneously stated that the rare gases do not form compounds with other elements. But in 1962, Neil Bartlett at Argonne National Laboratory near Chicago reported the formation of the first rare gas compound, xenon hexafluoroplatinate ($XePtF_6$). Since that time several other rare gas compounds have been produced (see Table 21-1).

Xenon tetrafluoride (XeF_4) is made by heating a mixture of xenon and excess fluorine to 400° C at 60 atmospheres pressure in a nickel vessel. The product is a white solid with a melting point of just under 100° C. As yet there is no acceptable theory to explain the bonding in such compounds. The compounds listed are relatively stable, with the exception of xenon trioxide, which explodes violently when it is rubbed, heated, or pressed gently.

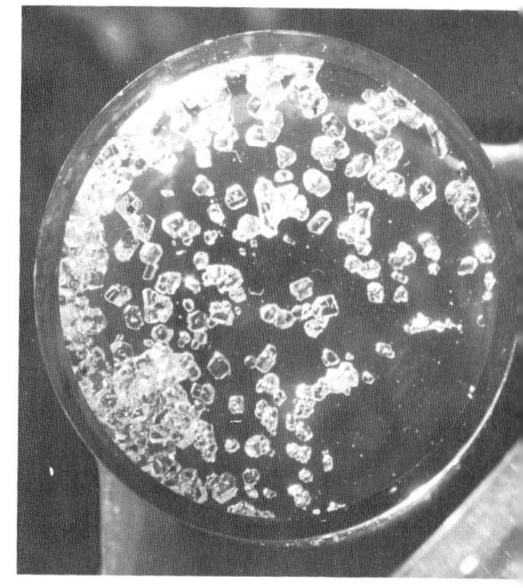

21:14 *Crystals of xenon tetrafluoride, XeF_4.*

21:15 *Apparatus used for preparation of xenon tetrafluoride.*

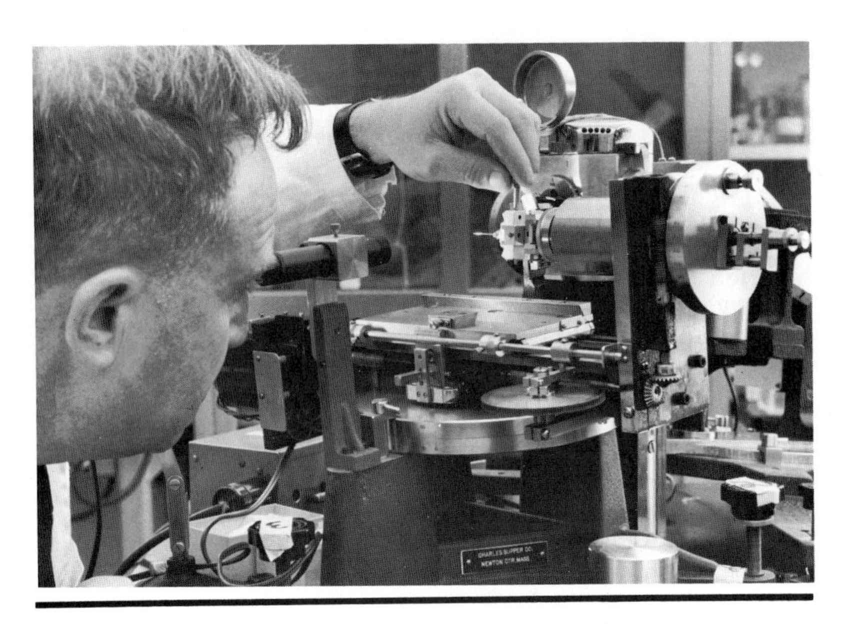

21:16 *Chemist positioning capillary containing XeF_4 in an X-ray camera.*

Table 21-1 *Rare Gas Compounds*

Name	Formula
xenon hexafluoroplatinate	$XePtF_6$
xenon hexafluororhodinate	$XeRhF_6$
xenon difluoride	XeF_2
xenon tetrafluoride	XeF_4
xenon hexafluoride	XeF_6
xenon trioxide	XeO_3
krypton tetrafluoride	KrF_4
radon fluoride	RnF_x

21:8 Radon The heaviest known gas is radon. Radon is approximately 7.7 times as dense as air! Radon was discovered in 1900 by Friedrich Dorn (1848-1916) and later isolated and studied by Ramsay and his co-workers in 1908. There is only a very tiny amount of radon in the atmosphere—about one part in a sextillion!!

Radon is continually being formed in nature by the radioactive decay of radium. Radium decays spontaneously into radon and helium with a half-life of 1620 years. (The half-life, as you will recall, is the length of time required for one half of an original number of atoms to decay into something else.)

The chemistry of radon has been difficult to study because radon *itself* is radioactive. Radon-222, the isotope formed from radium, has a half-life of only 3.8 days, decaying into polonium by releasing an **alpha particle** (helium nucleus). In spite of this complicating factor, radon has been successfully reacted with fluorine to form radon fluoride. There is some question as to the exact composition of radon fluoride; so it is represented as RnF_x.

Because it emits alpha particles at a fairly rapid rate, radon has been found to be effective for cancer therapy.

List of Terms

alpha particle
cryogenic
fractional distillation
isotopes
mass spectrograph
noble gases
potassium–argon dating method
rare gases

Questions

1. Name the six rare gases.
2. What is "different" about the rare gas elements?
3. What was unusual about the discovery of helium?
4. Which of the rare gases is radioactive?
5. Which compound of a rare gas is explosive?
6. What is the heaviest known gas? How does its density compare with the density of air?
7. What other name is applied to the rare gases?
8. What observation concerning helium contradicts the idea of a very old earth?
9. Which of the rare gases does not have an octet in its outer shell?
10. Which is the most abundant of the rare gases?
11. Which rare gas has the lowest melting point of any element?
12. What is the function of the argon in an incandescent light bulb?
13. How is helium obtained commercially?
14. Why is it better for deep-sea divers to breathe a helium-oxygen mixture than a nitrogen-oxygen mixture?
15. Which is the least abundant of the rare gases?
16. How do neon-20 and neon-22 differ from each other?
17. What observations show decisively that ages determined by the potassium-argon dating method are not trustworthy?
18. Is there any relationship between the arrangement of the rare gases in the periodic table and their densities?
19. What property of fluorine enables it to react with xenon?
20. What very questionable assumption underlies the potassium-argon dating method?
21. Give the symbols for the rare gases.
22. What is the chief use of neon?
23. Explain how a rare gas atom can produce light when electrically excited.
24. What is clevite?
25. What property of radon makes it effective for cancer therapy?

Student Activities

1. Make a report on the derivations of the names of the six rare gases. List some words that are related to each..
2. Prepare a report on the life and work of Sir William Ramsay.

Further Reading

Chernick, Cedric L. *The Chemistry of the Noble Gases*. United States Atomic Energy Commission, Division of Technical Information (obtainable from USAEC, Post Office Box 62, Oak Ridge, Tennessee 37830).

Photo next page
An electric steelmaking furnace used primarily in producing alloy, tool, stainless and other specialty steels.

UNIT VII CHEMICAL TECHNOLOGY

Chapter 22 Metallurgy and Metals

22:1 The Science of Metallurgy In this chapter we shall discuss some commercially important **metals** and **alloys**. The science of metals, called **metallurgy**, involves both physics and chemistry.

Metallurgy is usually divided into two general categories, chemical metallurgy and physical metallurgy. **Chemical metallurgy** *is the science that deals with the extraction of metals from their ores.* **Physical metallurgy** *is the science that deals with the adaption of metals for commercial use.* Physical metallurgy is by far the larger field. For every person employed as a chemical metallurgist, there are about nine people employed as physical metallurgists.

Metallurgy is a fascinating subject. If your teacher showed you a microscope and asked you, "In what branch of science is such an instrument used?" you probably would answer, "biology." However, a special type of microscope is a necessary tool for the physical metallurgist, too. An experienced metallurgist can look at the microstructure of a metal and then tell you a great deal about the history of the metal.

The use of metals can be traced throughout the history of man. In Genesis 4:22, Tubal-cain is mentioned as an artificer (craftsman) in brass and iron. You can see from this Scriptural example that it did not take man very long to realize the value of metals. Early man was quite skilled; he was not an ape-like creature at all.

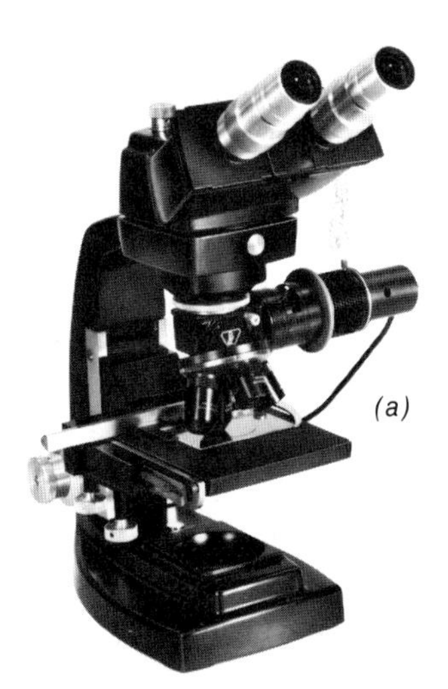

22:1 *(a) A* **metallurgical microscope** *that depends upon reflected light from the surface of a metallic specimen. It differs from a biological microscope in that the latter uses light transmitted through the specimen.*

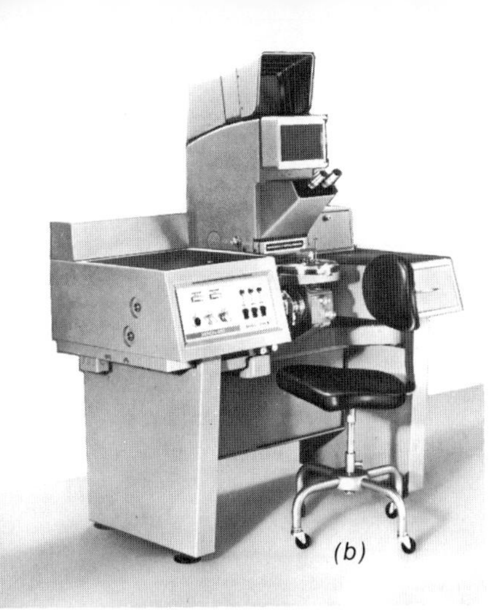

22:1 *(b) A metallograph is used to take photographs of the microstructure of a metal sample.*

(a)

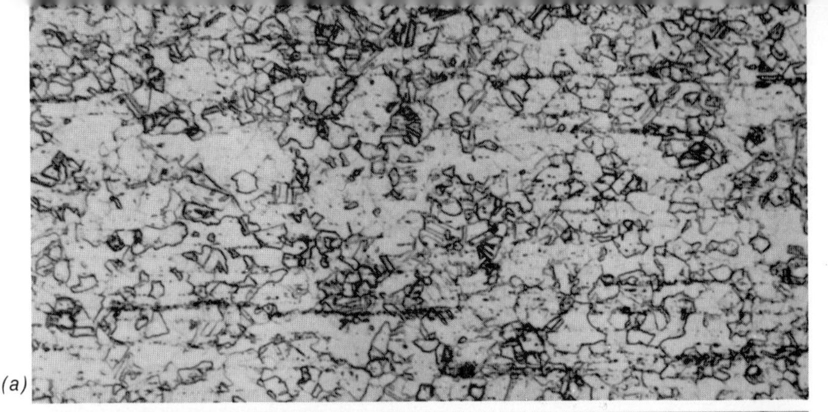

(b)

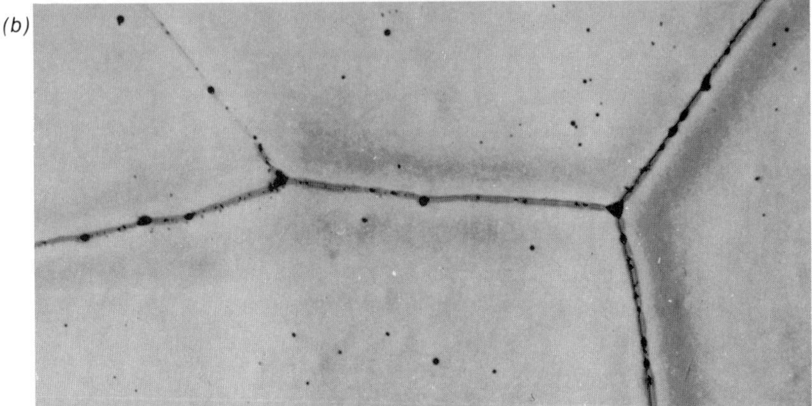

22:2 *You are looking at two different microstructures of tough-pitch copper. Both structures have been magnified 150 times. (a) Hot-rolled tough-pitch copper. The direction of rolling is parallel to the stringers of copper oxide. (b) As-cast tough-pitch copper exhibits a very large grain size.*

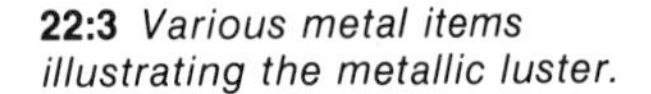

22:3 *Various metal items illustrating the metallic luster.*

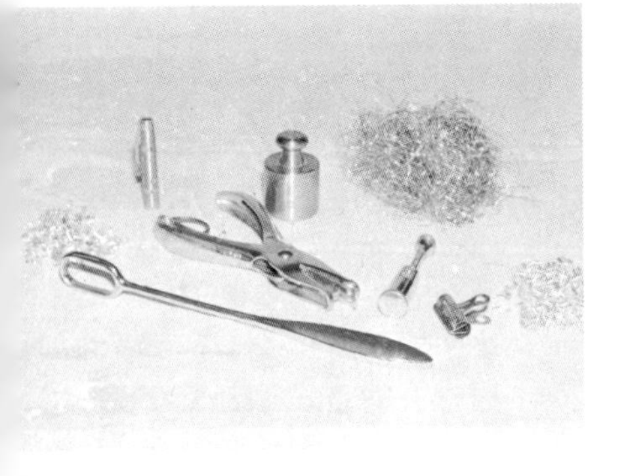

Our modern economy is based on metals and their uses; industry could not survive without metals.

22:2 Physical Properties of Metals In Chapter 7 we discussed the *metallic bond.* Now we will concentrate on some **metallic properties**. The first thing you should notice about a freshly polished metal is its shiny luster. Metals reflect light, and a freshly polished metal surface is mirror-like. This is called the **metallic luster**.

Metals are hard, but they are also **ductile**, or **malleable**; that is, they will change their shape drastically under a sufficient load. Take a rubber band, stretch it, then let it go. What happens? It stretches, but it returns rather rapidly to its original shape. You have **elastically deformed** the rubber band. However, you can plastically deform solid metals by working with the metal until it changes its shape permanently. This is called **plastic**, or permanent, **deformation**. If you try to deform a rubber band plastically, you will break it before it changes shape permanently.

This combination of properties, *hardness* and *ductility*, is one reason metals are so useful. No other substance has this combination of properties. Most substances that are able to be deformed plastically are so soft that they are useless as structural materials.

Metals are also good conductors of electricity and heat. Electricity and heat will flow through metals more easily than through other substances. When you combine the properties of luster, hardness, ductility, and good thermal (heat) and electrical conductivity, you obtain the unique capabilities of metals. We could discuss several other metallic properties, but these few will give you an understanding of just what metals are.

22:3 Natural Sources of Metals When a mineral deposit is found to contain a measurable quantity of a metal, it is referred to as the **ore** of the metal. Some typical metal ores are listed in Table 22-1.

22:4 *Metals have excellent thermal conductivity.*

Table 22-1 *Typical Metal Ores*

Type of Ore	Examples
carbonates	$CaCO_3$, $MgCO_3$, $FeCO_3$
halides	halides of Na, K, Mg, Ag
native metals	Ag, Au, Hg, Cu, Bi, As, Sb, Pt
oxides	oxides of Fe, Mn, Cr, Al, Cu, Sn
silicates	silicates of Ni and Zn
sulfates	$CaSO_4 \cdot 2H_2O$, $SrSO_4$, $BaSO_4$, $PbSO_4$
sulfides	sulfides of Hg, Cu, Zn, Pb, Ni, Co, Sb, Bi, Ag

22:4 Extraction of Metals from Their Ores Once an ore of a metal has been discovered and it has been determined that the metal can be extracted from the ore, several processing steps are usually involved to obtain a commercial product.

First, the metal must be mined. When a metal is mined, there are normally undesirable substances, such as rock, which must be separated from the ore. The ore must be **concentrated** by removing these impurities. Usually the mined material is ground into small particles. The ore then can be separated from other substances in any of several ways. Sometimes the densities of the ore and rock are different enough that one will float in a particular liquid while the other sinks. This process is called **flotation**.

22:5 *Diamond drilling in a mine.*

22:6 *Zinc and lead flotation cells.*

Many times, a chemical processing is necessary to separate the ore from the rock. Silicate rocks are reacted with limestone in concentrating iron ores. The reacted impurity is then removed. This undesirable material is called **slag** or **dross**.

22:7 *Electrolytic refining tanks used for the production of 99.8 percent pure copper.*

Once the ore is concentrated, it must be **reduced** (lowered in valence) to metal. The reduction operation usually involves *both* chemical and thermal (heat) processes. Sometimes reduction can be done electrolytically.

Experiment 22:1 *Reduction of Copper Oxide to Copper*

1. Weigh out 5g of powdered cupric oxide and mix it with an equal bulk of powdered wood charcoal.
2. Heat the mixture strongly for 4 or 5 minutes in a Pyrex test tube. You should observe a reaction occurring in the test tube during this heating.
3. After the test tube cools, pour its contents into a beaker.
4. Hold the beaker under the faucet at the sink, and let a very slow stream of water wash away the excess charcoal so the product remaining can be identified. How can you tell that free copper has been formed?
5. In order to help prove that the product is copper, let your instructor add concentrated nitric acid (HNO_3) to a few pieces of the product in a test tube and observe the gas given off. What is its color? If the brown nitrogen dioxide is formed, it has reacted properly.

(a)

(b)

22:8 *(a) Overall view of a blast furnace plant showing two furnaces and the stoves that operate in conjunction with them. Iron ore is piled in the foreground area. (b) Identification of the principal dimensions and components of a blast furnace.*

After reduction comes **refining**. The reduced metal often is very impure; it must be refined to remove the impurities in the form of slag or dross. Spiritually, the Holy Spirit continually encourages and chastens believers to refine their faith and testimony by removing the "dross" from their lives. The term "refining" is used in the Bible to describe this spiritual operation. Does your life need refining? The wicked people of the earth are considered as dross; they are to be thrown away. But believers are as refined precious metals to the Lord. Usually, the refining of metal requires heat as well as chemical action. The Lord Jesus Christ can refine a person by fire as well; this is usually a painful testing experience (see I Peter 1:7).

After a metal has been refined, it can be shaped into a final product, or it can be alloyed and then shaped. This shaping is referred to as "heat and beat" in the metallurgical profession.

22:5 Iron and Steel

Iron is the most commercially important of all metals. When carbon is added to iron, the resulting alloy is called **steel**.

The reduction of iron ore, hematite (Fe_2O_3), is accomplished in a **blast furnace** (see Figure 22:8). The gas used to reduce the ore is carbon monoxide.

$$Fe_2O_3 + 3CO \longrightarrow 2Fe + 3CO_2\uparrow$$

The removal of silica out of iron ore is done by adding limestone, calcium carbonate ($CaCO_3$), as a **flux**. A slag forms and can be poured off.

$$\underset{\text{flux}}{CaCO_3} + SiO_2 \longrightarrow \underset{\text{slag}}{CaSiO_3} + CO_2\uparrow$$

The blast furnace is designed to operate continuously, with the metal and slag being removed at regular intervals. The molten metal, or **pig iron**, from the blast furnace is very high in impurities (silicon, sulfur, manganese, and carbon) and is too brittle for commercial use. It is as undesirable as the feet of Nebuchadnezzar's image (Daniel 2).

The pig iron is refined and properly alloyed in basic oxygen or open hearth furnaces. The highest quality steels are refined in electric furnaces or vacuum furnaces.

22:6 Basic Open Hearth Furnace The huge oven referred to as the **basic open hearth furnace** is so named because of the material used for lining the furnace. The refractory, or heat-resistant, brick used as lining contains two chemicals—calcium oxide and magnesum oxide—both basic materials. There are also acid open hearth furnaces in which acidic materials are used in the refractory brick lining; these acidic furnaces are used mainly in Europe.

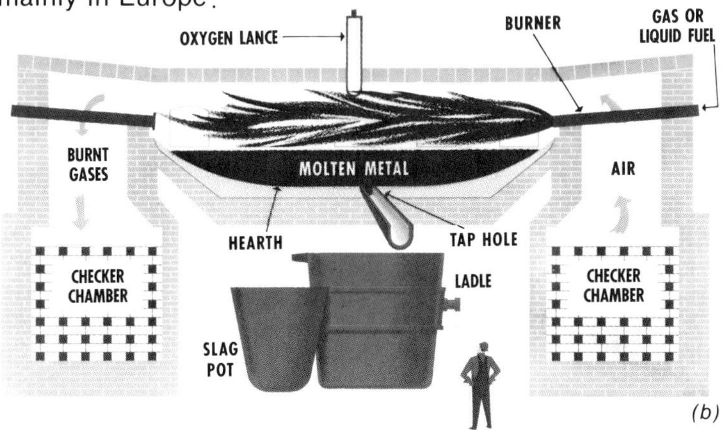

(b)

22:8 *(c) Slag being withdrawn from a blast furnace.*

22:9 *(a) Open hearth furnaces. A helper checks a "heat" of steel in one of the furnaces. Materials are charged into the furnace through the seven doors marked by the small circles of light. (b) The name "open hearth" comes from the fact that the pool of molten metal covered with slag lies on the hearth of the furnace, exposed to the sweep of flames.* (a)

Pig iron, scrap steel, iron ore, and limestone are loaded into the open hearth furnace. A flame is maintained across the surface of the load; a slag layer protects the steel being refined from the flame. Some reactions that occur in the open hearth furnace can be represented as follows:

Removal of carbon

($\underline{C}$ represents carbon in solution in the metal.)

$$3\underline{C} + 2Fe_2O_3 \longrightarrow 3CO_2\uparrow + 4Fe$$

iron ore

Removal of silicon

$$3\underline{Si} + 2Fe_2O_3 \longrightarrow 4Fe + 3SiO_2$$

Removal of silica

$$CaCO_3 + SiO_2 \longrightarrow CaSiO_3 + CO_2\uparrow$$

These reactions are similar to the blast furnace reactions. The additions of other metals for making alloys can be done during refining, preferably toward the end of the refining process. Many different steel compositions can be made in an open hearth furnace.

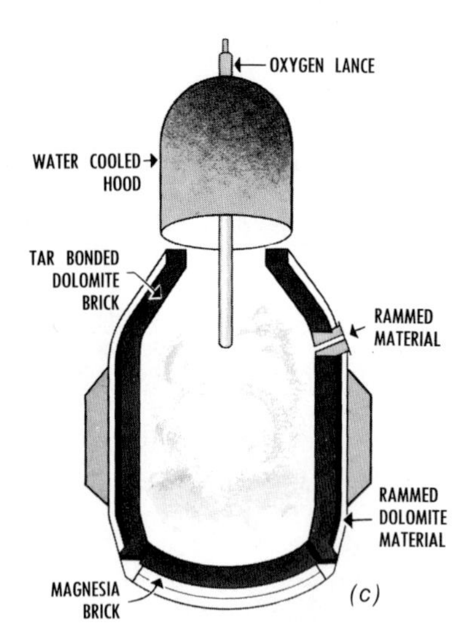

22:7 Basic Oxygen Furnace Shortly after World War II ended, a fast, efficient method for refining steel was developed. This method uses oxygen gas. An oxygen lance (see Figure 22:10c) is lowered into the furnace and the oxygen gas jet blown over the surface of the molten metal charge, or load. The refining is accomplished by the reaction of the oxygen with the impurities.

Steels tend to pick up unwanted sulfur from the fuels used in open hearth refining, but this is not a problem in the basic oxygen process. In the open hearth operation the oxygen needed to refine the steel must be transmitted through the slag. This means that time must be taken to build up a good slag. This build-up time is unnecessary in the basic oxygen process because oxygen is furnished directly. Thus the basic oxygen refining is faster. The quality of steel produced is about the same as that produced in the open hearth furnace.

Molten pig iron, solid pig iron, scrap, and iron oxide are used to charge (load) the oxygen furnace; molten pig iron is the most important source. After charging, the oxygen lance is lowered and the gas turned on. A visible reaction flame begins to leave the furnace mouth almost immediately. Lime (CaO) and fluorspar (CaF_2) are then added to the furnace by means of an overhead chute; these substances aid the refining process.

22:10 *(a) Molten pig iron, one of the basic ingredients in steelmaking, is poured into this basic oxygen furnace. (b) Steel scrap slides from an uptilted charging box into the top of a basic oxygen furnace in the first step of the refining process. (c) Basic oxygen furnace. A charge of iron ore, steel scrap, and molten iron is refined into steel by blowing oxygen down from the top through a vertical lance extending to within five feet or so from the bath.*

(a)

(b)

Some of the chemical reactions which occur in the refining of steel by the basic oxygen process are

$$\underline{Si} + O_2 \longrightarrow SiO_2$$
$$\underline{Mn} + O_2 \longrightarrow MnO_2$$
$$\underline{C} + O_2 \longrightarrow CO_2\uparrow$$
$$\underline{S} + O_2 \longrightarrow SO_2\uparrow$$
$$\underset{\text{lime}}{CaO} + SiO_2 \longrightarrow \underset{\text{slag}}{CaSiO_3}$$

Fluorspar acts as a flux and aids the formation of a molten slag. The complete operation from charging to pouring the refined steel requires about one hour. Alloying additions are made toward the end of the refining process. The tonnage of steel production from different types of refining furnaces is shown in Table 22-2. We have not discussed electric furnace refining, but it is used to produce very high quality steels. Many times steels from the open hearth and basic oxygen furnaces are refined further in electric furnaces. Because of the large amount of power used in refining, electric furnace steels are usually more expensive than other types of steel.

22:11 *An electric steelmaking furnace used primarily in producing alloy, tool, stainless and other specialty steels.* *(See chapter title page)*

Table 22-2 *Raw Steel Production by Type of Refining Furnace (thousands of tons)*

	Open Hearth	Basic Oxygen	Electric
1966	85,025	33,928	14,870
1971	35,559	63,943	20,941

22:8 Heat Treatment of Steel The refined steel is poured into large crucibles (pots) or ladles. Aluminum is then added to the steel to react with the remaining oxygen in solution so that no further reduction occurs.

$$2Al + 3\underline{O} \longrightarrow Al_2O_3$$
or
$$2Al + Fe_2O_3 \longrightarrow Al_2O_3 + 2Fe$$

The molten steel is poured from the ladle into the ingot molds, where it is allowed to solidify. Still red hot, the steel is placed in ovens, called soaking pits. The purpose of this operation is to allow the alloying additions to diffuse throughout the steel ingot so that the ingot is homogeneous (uniform throughout). This is the first step in the "heat and beat" operations.

22:12 *Glowing ingot, uniformly heated to a temperature of about 2200° F, is removed from the soaking pit furnace. Next it will travel to the rolling mills where it will be shaped into a bloom, billet, or slab.*

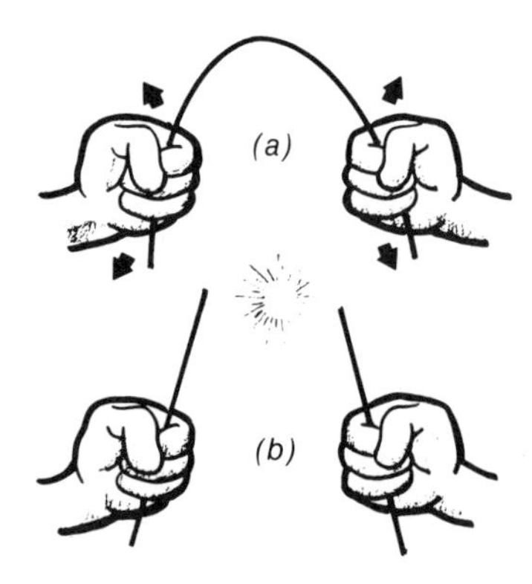

22:13 *(a) When you first bend a copper wire, it is very ductile. (b) However, after much bending it becomes brittle and breaks.*

22:9 Mechanical Deformation of Steel Steel can be **hot-worked** or **cold-worked** to get it into a desired shape. Hot-working is exactly what the term implies, the mechanical deforming of a metal while it is very hot. Cold-working, then, is mechanically deforming a metal at temperatures close to room temperature.

Cold-working eventually makes a metal harder and harder until the metal becomes so brittle it will crack. To prevent this a metal is *annealed* after a sequence of cold work to soften it again. The metal is heated so that the metal atoms can diffuse and relieve the stresses built up during cold-working. Annealing is unnecessary when you are hot-working, because the metal is hot enough to self-anneal.

Steels can be heat-treated in many ways. One interesting property of steel is its ability to dissolve more carbon at high temperatures than at low temperatures. As steel cools, carbon precipitates out of the solid metal as iron carbide (Fe_3C). Also, the iron changes crystal structure at a certain temperature upon cooling (or heating). If you rapidly cool the steel, the carbide precipitation and crystal structure change cannot occur; the steel becomes very hard as a result of this quenching.

22:14 *(a) Glowing slab entering a roughing stand. Roughing stands squeeze slabs to one fifth their original thickness and stretch them about five times their original length. (b) Blooming mill has rolled an ingot down to bloom size. Blooms, which are rectangular or square in cross-section, are further processed into structural steel and railroad rails. (c) Five-stand tandem mill for rolling out steel sheet. (d) Machining of steel parts.*

(a)

(b)

(c)

(d)

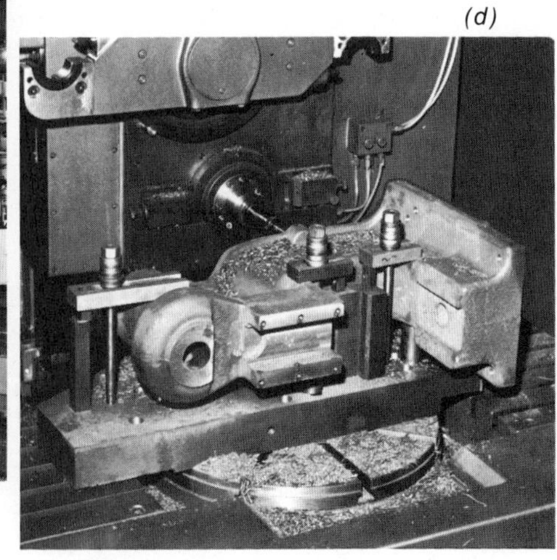

Steels can be formed and shaped by rolling, forging, extrusion, drawing, stretching, machining, and so forth. There are many available forming operations to shape metals. We have mentioned only a few "heat and beat" operations that are possible to use on metals. The furnaces, ladles, and rolling mills used in the steel industry are much larger than you probably realize—some rolling mill systems are as long as a football field, and a single blast furnace may be two to three times as big as your house.

22:15 *One of the many uses of steel. "Spinning wheel" carries wires to waiting bridgemen during cable-spinning operations on second Delaware Memorial Bridge.*

22:10 Steel Products The products of steel are almost too numerous to mention. Everything from nails to ocean liners, from hand tools to the structural materials used in the tallest buildings, from razor blades to automobile bodies, can be made of steel. An infinite variety of steels is available, also—hard steels to resist wear, durable steels to resist corrosion, strong steels to hold heavy loads, and steels to withstand high temperatures. Many different substances are used as alloying additions to steels, too. Table 22-3 lists some of these alloying additions and their purposes.

Table 22-3 *Alloying Additions to Steel*

Element	Purpose
carbon	to increase strength
chromium	to increase corrosion resistance
nickel	to increase high temperature resistance
molybdenum	to increase high temperature resistance
vanadium	to increase wear resistance
manganese	to increase hardness
tungsten	to increase toughness and heat resistance
silicon	to increase hardness and corrosion resistance
lead	to increase machinability
cobalt	to increase hardness and wear resistance

(a)

22:16 *(a) Reactor well. This tank, 38 feet in diameter and 28 feet tall, is made from 1/4-inch thick stainless steel plates. The tank will be assembled directly above a nuclear reactor vessel to provide a vertical water-filled column. (b) The increasing number of knee injuries resulting from sports and other accidents prompted the design of this new brace fabricated from stainless steel.*

(b)

22:11 Chemistry of Iron Corrosion Steels corrode or rust, particularly in the presence of moisture. All of the complex chemistry cannot be covered here, but basically a brown oxide (Fe_2O_3) forms.

$$4Fe + 3O_2 \xrightarrow{H_2O} 2Fe_2O_3$$

This would not be so bad if the oxide would adhere to the metal surface, but the oxide flakes off, exposing more metal to **corrosion.** This is one reason chromium is added to steels. When a chromium steel oxidizes, a corrosion product of chromium oxide forms on the surface of steel. This product adheres to the metal as a thin film and prevents further corrosion. High chromium-nickel alloy steels, called **stainless steels**, are very corrosion resistant. You may have a set of stainless steel tableware at home.

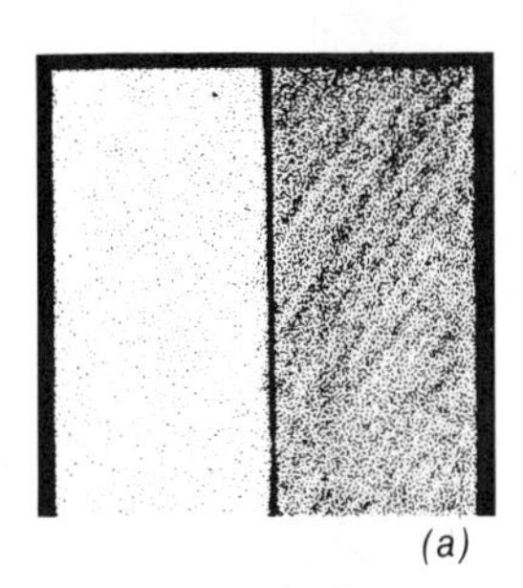

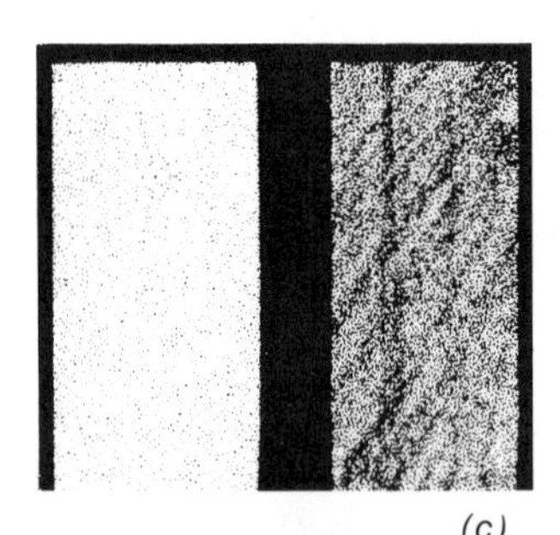

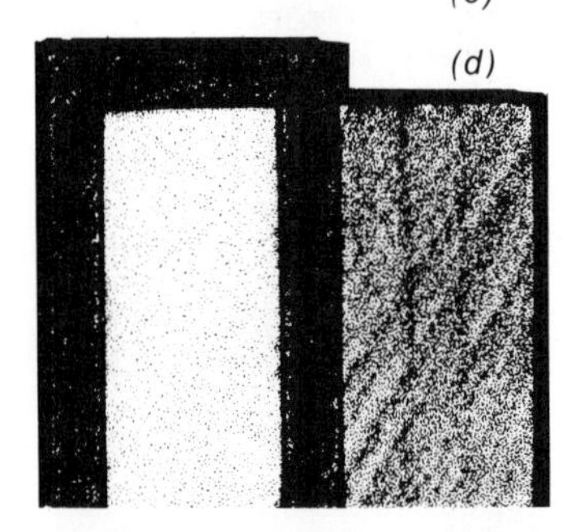

22:17 *(a) Two dissimilar metals in contact. (b) The less noble metal corrodes in the presence of water. Galvanic corrosion is prevented by placing a non-conductive spacer (c) between the metals or coating one of the metals as in (d).*

Another way to prevent corrosion of steels is to coat them with another metal. Chromium plating makes a steel highly resistant to corrosion, as well as a very attractive item. "Chrome" automobile bumpers and trim are an example of this. Steel can be coated with molten zinc, as well. This is called **galvanizing** and usually is done on steel sheet. The zinc is put on as a sacrificial coating. The zinc will corrode, rather than the iron. Even if some steel is exposed, the zinc is attacked rather than the base metal.

$$2Zn + O_2 \xrightarrow{H_2O} 2ZnO$$

The appearance of white areas on old galvanized steel is the zinc oxide (ZnO) that forms as a result of corrosion.

Any two different metals placed in contact in the presence of moisture will create a situation called **galvanic corrosion**. In galvanic corrosion, the less noble metal will be attacked—look at the activity table in Appendix III again. If a more active (less noble) metal is touching a less active (more noble) metal in water, the more active metal will corrode. Each drop of water sets up an electrolytic cell; the more active metal acts as the anode and goes into solution (corrodes). Corrosion in water is, therefore, an **electrochemical** process. To prevent this corrosion, the two metals must be separated by something that is not a conductor, or else one of the metals must be coated with some kind of resistant coating.

Experiment 22:2 *Corrosion of Steel in Water*

1. Obtain a piece of plain carbon steel. Clean it well with soap and water, rinsing it well in distilled water. Dry it thoroughly.
2. Place a drop of tap water on the steel and leave it overnight. When you return the next day, observe the spot where the water was. Record your observation.

Experiment 22:3 *Galvanic Corrosion*

1. Obtain specimens of two different metals. Lay them in a beaker full of tap water. Make sure the metals touch.
2. After several days remove the specimens and check them, especially in the area of contact. Which specimen appears to be the most affected? Why?

22:12 Concentration and Reduction of Aluminum The primary mineral of aluminum is **bauxite**, which is hydrated aluminum oxide ($Al_2O_3 \cdot xH_2O$), where $1 \leq x \leq 3$). Bauxite is crushed and ground to a fine powder; then it is reacted with sodium hydroxide to get the aluminum into solution as an aluminate.

$$Al_2O_3 \cdot 3H_2O + 2NaOH \xrightarrow[\text{pressure}]{H_2O} 2Na^+ + 2Al(OH)_4^-$$

This solution is filtered from the undissolved impurities in the bauxite. Carbon dioxide is bubbled through the solution to convert the aluminate into a pure aluminum hydroxide.

$$Al(OH)_4^- + CO_2 \longrightarrow Al(OH)_3\downarrow + HCO_3^-$$

The ore is now sufficiently concentrated to be reduced. Reduction occurs after the water is removed from the hydroxide.

$$2Al(OH)_3 \xrightarrow{\Delta} Al_2O_3 + 3H_2O$$

The dehydrated powder is then dissolved into molten cryolite (Na_3AlF_6) at 1000° C and an electrical current is passed through it. The oxide is decomposed electrically. Molten aluminum flows out of the bottom of the cell. The reactions may be represented as

$$Al^{+3} + 3e^- \longrightarrow Al$$
$$2O^{-2} - 4e^- \longrightarrow O_2$$

22:18 *Cells for reducing bauxite to aluminum.*

Since the electrodes are made of carbon, the oxygen reacts with them, forming carbon monoxide.

$$2C + O_2 \longrightarrow 2CO\uparrow$$

This complete operation is called the **Hall process**, named for Charles Martin Hall (1863-1914), the American chemist, and produces commercially pure aluminum (99 percent). If higher purity aluminum is desired, further refining is necessary.

22:13 Properties of Aluminum A freshly exposed aluminum surface will be attacked by oxygen in the air immediately. However, unlike the film on steels, the oxide film on aluminum is very thin and holds tightly to the metal so that no further oxidation occurs.

$$4Al + 3O_2(\text{air}) \longrightarrow 2Al_2O_3 \text{ (thin layer on surface of metal)}$$

This oxide film enables the metal to resist corrosive attack. Although aluminum is a fairly reactive metal, the film of oxide renders it inactive, and it has good corrosion resistance because of this.

Aluminum and its alloys have been used for years in the aircraft industry because of their light weight. Pure aluminum is not very strong; it must be alloyed with other metals to increase its strength. Once alloyed and properly heat-treated, aluminum alloys are moderately strong.

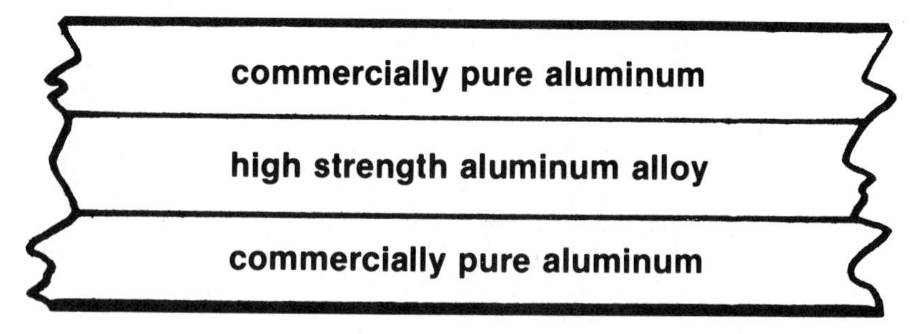

22:19 *Al-clad "sandwich."*

However, when aluminum is alloyed to strengthen it, its corrosion resistance is lowered. Therefore, a product of commercially pure aluminum and an aluminum alloy "sandwiched" together combines strength with resistance to chemical attack (see Figure 22:19). Because aluminum conducts heat and resists corrosion so well, it is used in cookware. Its malleability allows it to be rolled into very thin sheets, making aluminum foil, another useful kitchen product. Aluminum is also used in cables to conduct electricity; aluminum wires are wound around a steel core (the steel is for strength) to form a cable. And aluminum siding, coated with a specially formed oxide, is valuable to the construction industry.

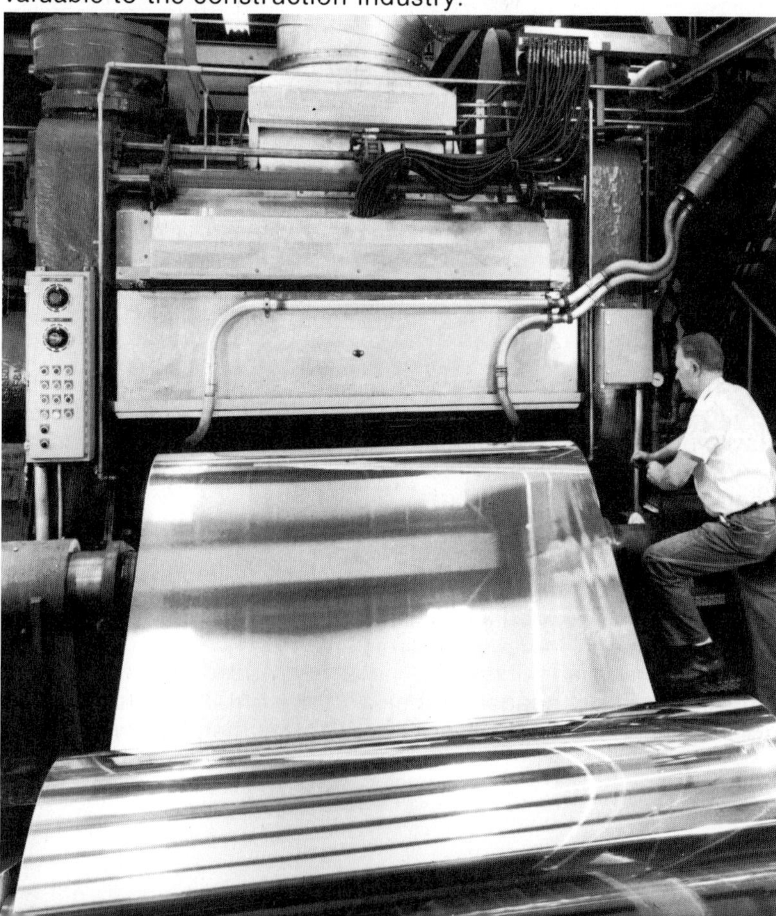

22:20 *Aluminum foil for the nation's kitchens, packaging, and other uses rolls off an 84-inch rolling mill.*

(a) (b)

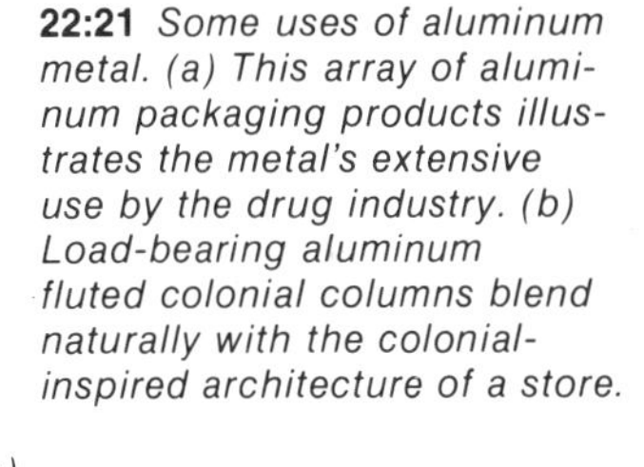

22:21 *Some uses of aluminum metal. (a) This array of aluminum packaging products illustrates the metal's extensive use by the drug industry. (b) Load-bearing aluminum fluted colonial columns blend naturally with the colonial-inspired architecture of a store.*

22:14 Copper Copper sulfide ores are concentrated by a flotation process. Some of the sulfide is converted to an oxide by heating the ore in air.

$$2Cu_2S + 3O_2 \longrightarrow 2Cu_2O + 2SO_2\uparrow$$

More air is then blown into the Cu_2S-Cu_2O mixture to reduce the compounds to metal.

$$Cu_2S + 2Cu_2O \longrightarrow 6Cu + SO_2\uparrow$$

The metal product is 99.5 percent copper. It is then electrolytically refined in a copper sulfate solution.

Copper conducts electricity and heat very well and has good resistance to corrosion. It is used in electrical cable and wiring, cookware, piping, and coins. **Brass** is a copper-zinc alloy; **bronze** is a copper-tin alloy.

22:15 Other Metals Lead has a high resistance to corrosion. It is quite useful in sinks and piping for chemical laboratories, where many strongly acidic and basic substances are poured down the drain.

Tin is used as a coating on food cans because of its resistance to chemical attack. When tin and lead are mixed, they

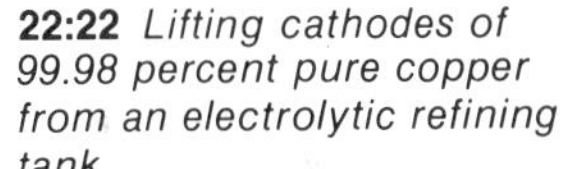

22:22 *Lifting cathodes of 99.98 percent pure copper from an electrolytic refining tank.*

form many low-melting point alloys which are used as solders (to join metallic surfaces).

Silver and gold, being **precious metals**, are used in jewelry and ornamental ware. Precious metals are the most inert of metals; they do not corrode or tarnish easily. This property makes them very valuable. Also, the appearance of precious metals makes them very desirable. Silver and gold, by themselves, are very weak and must be alloyed to increase their strength. For example, sterling silver contains 7.5 percent copper. Coin silver and coin gold contain 10 percent copper. Gold is still considered a primary monetary standard.

List of Terms

alloy
basic open hearth furnace
basic oxygen furnace
bauxite
blast furnace
brass
bronze
chemical metallurgy
cold-working
concentration (ore)
corrosion
deformation
dross
ductile
elastic
elastic deformation
electrochemical
flotation
flux
galvanic corrosion
galvanizing
Hall process
hot-working
malleable
metal
metallic luster
metallic properties
metallurgical microscope
metallurgy
ore
ore reduction
physical metallurgy
pig iron
plastic
plastic deformation
precious metals
refine
slag
stainless steel
steel
sterling silver

Questions

1. (a) Define chemical metallurgy.
   (b) Define physical metallurgy.
2. What might looking at the microstructure of a metal reveal to you?
3. Where can we find the first mention of metalworking in the history of man? Who was the man who worked with metals?
4. What is the importance of metals to the modern economy?
5. Compare metals and rubber.
6. (a) Define plastic deformation.
   (b) Define elastic deformation.
7. What two physical properties are combined in metals that make metals particularly useful?
8. Are metals generally good conductors of electricity? Why?
9. What is meant by the ore of a metal?

10. Write down the steps (in proper sequence) for extracting a metal from its ore.
11. An ore is mined with an impurity physically mixed with it. The ore has a density of $7.0 g/cm^3$; the impurity has a density of $4.0 g/cm^3$. How would you separate the ore from the impurity?
12. What is refining?
13. What is steel?
14. (a) What is a blast furnace?
    (b) Write a chemical reaction for the reduction process.
15. What is the purpose of slag in a metallurgical operation?
16. Is pig iron normally a usable product? Why?
17. (a) Why are the open hearth and oxygen furnaces referred to as basic?
    (b) Refining in a basic oxygen furnace is faster than in an open hearth furnace. Is this statement true or false?
18. What impurities are removed in the refining of pig iron?
19. How is refining accomplished in a basic oxygen furnace?
20. Why are lime and fluorspar added to a basic oxygen furnace charge?
21. Why is refined steel placed in soaking pits (furnaces)?
22. Can steel be cold-worked indefinitely? Why?
23. (a) Why is molybdenum alloyed with steel?
    (b) Why is chromium alloyed with steel?
24. (a) What is rust on steels?
    (b) Name two methods that are used to increase the corrosion resistance of steel.
25. What is meant by a sacrificial coating on steel?
26. Why is corrosion considered an electrochemical process?
27. What is the difference in the films that form on most steels and aluminum?
28. Why are aluminum and its alloys used in the aircraft industry?
29. Name three uses of copper. Why is it used in the applications you suggested?
30. (a) Brass is copper alloyed with ______.
    (b) Bronze is copper alloyed with ______.
31. Why are metals like gold and silver called precious metals?

Student Activities

1. Obtain as many different metal and alloy specimens as possible. Clean the specimens, dry them, then dip them in some degreasing fluid, such as acetone. Weigh the specimens before placing them in the corrosive medium. Expose each specimen to the same corrosive medium, such as a dilute acid or base. Allow them to remain in the fluid for several days. Remove the specimens; then clean and dry them. Weigh them to determine how much metal was lost in the corrosive attack. Collect as much of the corrosive product as possible to display with your specimens. Arrange the specimens and corrosion products on a display board in the order of increasing resistance to corrosion.

   In a written report, discuss your results. Why do you think certain specimens corroded faster than others? Consult the chemical activity table.
2. Choose a metal or alloy and trace its history and development. Obtain as many illustrations of industrial processes as possible. Also, discuss the modern usage of the metal or alloy.

Photo next page
A fishing scene in Montana.

UNIT VII CHEMICAL TECHNOLOGY

Chapter 23 Pollution and Chemistry

23:1 Divinely Ordained Pollution *"Ecology now!"* has been the cry of a great many young people lately. We are being warned that if something is not done soon, man will pollute himself right off the earth. But before you become too frightened about being devoured by this strange monster called **pollution**, check your Bible to see if Scripture predicts such a fate for mankind.

As you read the book of Revelation, you will find that *man* will not destroy the earth and himself with pollution, but that *God* will eventually destroy sinful man. Man has polluted this world with sin, and God will pollute the environment because of man's sin. This judgment will come as a punishment for past sins and as a warning to turn to God. Pollution has been used to some degree during past periods of divine judgment. When Moses was dealing with Pharaoh to let the children of Israel leave Egypt, many of the curses God brought on Egypt were forms of pollution (Exodus 7-10). The following table lists some of the God-ordained pollutions of the earth which will occur in the future.

Table 23-1 *Some Judgments That Will Cause Pollution*

"Pollution" Judgment	Results	Reference
Hail and fire, mixed with blood, will fall upon the earth	One third of the trees and all the grass will be destroyed	Revelation 8:7
A large burning mountain will fall into the sea	One third of sea life and one third of all ships will be destroyed	Revelation 8:8-9
A star (Wormwood) will fall into certain rivers and waters	Many people who drink the polluted water will die	Revelation 8:10-11
Smoke will permeate the atmosphere and obscure the sun	Result not mentioned in Scripture	Revelation 9:2
Pollution of the sea	All animal life in the sea will be destroyed	Revelation 16:3
Pollution of all rivers and waters	Result not mentioned in Scripture	Revelation 16:4

23:2 Man-Made Pollution Where there are people, there will be pollution. The only way to completely rid the earth of pollution would be to rid the earth of people. This is hardly a suitable solution.

23:1 *(a) Air pollution near a heavily-populated urban area. (b) Solid waste. Unsightly roadside area used for disposing of solid waste. (c) Water pollution and resulting fish-kill.*

(a)

(b)

(c)

Pollution is not a problem of recent origin. And certain types of pollution may have been greater in past times than they are now. In 1728, for instance, the colonial governments of Virginia and North Carolina sent out a survey team to resolve the disputed boundary line between the two colonies. One of the Virginians appointed to the team was William Byrd of Westover. He kept official and personal diaries of the two expeditions. As the survey team proceeded toward the mountains, they moved into territory controlled by Indians.

When the Indians hunted, they would burn a large portion of the woods in a circle and then kill the animals trapped by the fire. Also, as they moved from place to place, they would leave their campfires burning. Byrd recorded in his diary for October 20, 1728, "The atmosphere was so smoaky all round us that the Mountains were again growing invisible. This happen'd not from the Hazyness of the Sky, but from the firing of the Woods by the Indians . . . . On their way the Fires they make in their camps are left burning, which, catching the dry leaves they ly near, soon put the adjacent Woods into a flame." This was air pollution in America in 1728! All that was necessary to eliminate this pollution was a hard, steady rain, and the same thing will take care of most present air pollution.

But pollution *can* become a serious problem. Air pollution can cause the death of physically weak people, particularly people with lung and heart diseases. And polluted water can cause the death of many water-dwelling creatures, as well as humans.

The area of Copperhill, Tennessee, is a glaring example of what pollution can do. Since 1847 the mining and smelting of copper ore has been the principal industry there. You will remember from Chapter 19 that sulfur dioxide is a by-product of the smelting operation. Until 1917 this sulfur dioxide gas was released into the atmosphere, and vegetation was killed for miles around. With no vegetation to hold the soil together, erosion set in; and the area became as barren as a desert. Most of the area has now been revegetated, but some of the barren landscape remains.

(a)

(b)

23:2 *(a) Erosion, the result of killing a forest by smelter fumes, near Cherokee National Forest, Tennessee. (b) General view of erosion and denudation due to smelter action in Shasta County, California*

23:3 *Tye River looking north from United States Highway 29 Photo taken in November 1972.*

The Tye River in Virginia affords another example of pollution. One of the authors was fishing in the river without having much luck when a farmer drove by and told him there had been no fish in that river for a long time. An industrial plant had been dumping its waste into the river for years, killing all the fish below the plant location. So you see, pollution can change the ecology of an entire area.

The deaths of many people have been blamed on air pollution. In October 1948 in Donora, Pennsylvania, 20 people and 800 domestic animals died, and 5910 people became ill as a result of five days of extreme air pollution.

Reasonable and well-planned pollution control is certainly a worthy goal. However, if more deaths and diseases occur because of pollution-control measures than from pollution itself, then nothing has been accomplished; a step backwards has been taken by society.

The clamor for an immediate end to all air and water pollution can be carried too far too fast. For instance, the use of **DDT** as an insecticide has been condemned. This chemical accumulates in animal fats, particularly in fish. Because birds of prey, such as eagles, ospreys, and pelicans, feed mainly on fish, high DDT concentrations have been found in these birds. The DDT has been found to interfere with their egg formation. This fact has set off a cry among many people that the use of DDT means wholesale destruction of wild life, and these ecology-minded people pressured various government agencies to pass legal bans on the use of DDT. But DDT applications kill the mosquito and other insects which carry malaria and viral encephalitis. Without DDT there will undoubtedly be an increase in these diseases. The sudden stoppage of the use of this insecticide, or replacement by another insecticide whose side effects *are not known*, may be even more dangerous. The sensible approach, then, would be to *phase out* DDT as *suitable alternatives* are developed.

Yes, the ecological balance can be upset if many birds of prey are killed. However, other species of animals have become extinct in the past; and with time nature tends to rebalance itself.

What about people, though? Are they more important than eagles? What should be done about the diseases which are inflicted upon the human race by germ-carrying insects? A single insect, the tsetse fly, transmits diseases to men and cattle and has retarded the development of more than four million square miles of Africa. Many harmful insects will increase drastically if a means of control is not developed and used soon.

There are many misguided environmentalists who would like to ban the use of all herbicides, insecticides, fungicides, and chemical fertilizers and "return to nature." They want only organic gardening with no synthetic chemicals added to the soil. But there are problems with this attitude, too. First, not enough animal waste is available to fertilize all the ground that would have to be cultivated to feed the population of the earth. Second, the price of organically grown food would be so high that many poor people could not afford to eat.

In a study conducted in Missouri a farmer grew corn in two separate gardens. He sprayed insecticides, fungicides, and herbicides on one garden, but he did not spray anything on the other. Here are the results of the study.

Treated Garden		Untreated Garden
$547	Production costs per acre	$327
$1300	Yield per acre	$520
$.42	Production costs (dozen ears)	$.63
$.79	Retail cost (dozen ears) to consumer to provide same reasonable income to grower under both conditions, plus retailer markup to handle marketing costs	$1.31

Corn grown in the untreated garden, therefore, required a retail price 66 percent higher than the retail price of corn grown in the treated garden.

The same type of study was done on a peach orchard, except that the sprays recommended for peach trees were used. The results were

Treated Orchard		Untreated Orchard
$981	Production costs per acre	$281
500 bu.	Yield per acre (50 lb. bu.)	52 bu.
$1.96 (4¢ per lb.)	Production costs per bu.	$4.19 (8¢ per lb.)
19¢ per lb.	Retail cost to consumer to provide same reasonable income to grower under both conditions, plus retailer markup to handle marketing costs	52¢ per lb.

23:4 Gypsy moth damage in a New England wooded area.

Peaches grown in the untreated orchard, therefore, required a retail price 63 percent higher than the retail price of peaches grown in the treated garden to produce the same income for the grower.

Dr. Earl L. Butz, vice-president of the Purdue Research Foundation, has said, "We can go back to organic agriculture in this country if we must—we once farmed that way 75 years ago. . . .However, before we move in that direction, someone must decide which 50 million of our people will starve. We simply cannot feed, even at subsistence levels, our 205 million Americans without a large production input of chemicals and antibiotics."

Dr. Robert White-Stevens, chairman of the Bureau of Conservation and Environmental Sciences at Rutgers University, has predicted that " if pesticides were eliminated entirely, crop yields would drop more than 50 percent, and the cost of food would increase four or five fold for food that would be totally unacceptable to the modern American housewife." He added that over the past years extensive tests of "the food supply, the water sources," and "the bodies of people" have revealed no buildup of agricultural chemicals, fertilizers, or pesticides in them.

23:3 The Wrong Solutions to Pollution The larva of the **gypsy moth** is a brown caterpillar with blue and red spots. A single caterpillar can consume one square foot of tree leaves every day. Because the gypsy moth is an imported insect, it has no natural enemies. For several years the larvae were kept under control by aerial spraying of DDT. When this spraying was stopped, the gypsy moth population literally exploded. Gypsy-moth larvae completely stripped hardwood trees of their leaves, destroying many hardwood forests in New England. Once restricted mainly to New England, this dangerous moth is now spreading throughout the Eastern United States. Continued spraying of DDT could have prevented this.

Many people are opposed to the use of DDT because the chemical is stable and its effects are long-lasting. However, Gordon Edwards, professor of entomology at San Jose State College, California, insists that the use of DDT has "been of great value to birds, mammals, fish, and honeybees" by increasing the yield not only of cultivated crops but of wild plants as well, and by saving millions of acres of forests from destruction by insects. Dr. Roy Kottman, dean of the College of Agriculture, Ohio State University, compares the use of DDT to the use of modern drugs. He notes that allergic reactlon to penicillin kills a great many more people each year than pesticides do, yet no one wants to ban penicillin from use.

Several pesticides which have a much shorter residue time have been recommended to replace DDT. Environmentalists are opposed to many of these "short-lived" substances, however, because they kill many helpful insects and certain young fish. One of the insecticides that has been welcomed by the environmentalists is **parathion**, an organic phosphate used as a nerve gas.

$$(C_2H_5O)_2P(=S) - O - C_6H_4 - NO_2$$

parathion

But parathion is lethal to humans! The use of this chemical will certainly cause many human deaths. On the other hand, DDT has never been known to cause death or any harmful effects to humans. Again we ask, which is most important, humans, fish, or eagle eggs?

23:5 *A fishing scene in Montana. (See chapter title page)*

Right now the choice is either to continue the use of synthetic chemicals in agriculture and "pollute" the environment or to choose which people in the world must starve because of the problems of producing organic foods. The best solution to the problem of pollution must be a reasonable one. Man's involvement with his natural environment can no longer be regarded as all good or all bad. "Pollution-free" solutions are impossible. Pollution should be held to a minimum but at a reasonable cost.

23:4 A Humorous Pollution Problem Recently methods have been developed whereby we are able to detect when small quantities of mercury are present in organic matter. Through chemical analysis a swordfish was found to contain a level of mercury which some considered to be dangerous for human consumption. Immediately, swordfish were taken off the food market. Fish that had been frozen for 40 years were analyzed chemically and found to contain the same level of mercury as the recently caught fish. That means that during 40 years of man's pollution of the water, the mercury content in fish has not increased at all. This was a case of jumping to the wrong conclusions with a minimum of scientific data. There was no justifiable reason for removing swordfish from the food market.

23:5 What is Pollution? How do we decide what pollution is? What is the level of acceptable pollution content? If we completely rid the atmosphere and water of certain chemicals, what will be the effect of this on life? Will the effect be beneficial, or will it be worse than if some of the chemicals were present? These questions must be considered before we leap too quickly into unsatisfactory solutions. Much research needs to be done in this area.

Table 23-2 *Composition of Clean Dry Air Near Sea Level (parts per million)*

Component	Content	Component	Content
nitrogen	780,900	nitrous oxide	0.25
oxygen	209,400	carbon monoxide	0.10
argon	9,300	xenon	0.08
carbon dioxide	318	ozone	0.02
neon	18	ammonia	0.01
helium	5.2	acetone	0.001
methane	1.5	nitrogen dioxide	0.001
krypton	1.0	sulfur dioxide	0.0002
hydrogen	0.5	lead	0.00000013

From *Chemistry, Man, and Society* by Jones, Netterville, Johnson and Wood, W. B. Saunders, Philadelphia, Pennsylvania, 1972. Used by permission.

23:6 *Notice the silt washing down the street in a heavy rain.*

Table 23-2 supposedly shows the composition of clean, dry air near sea level. If any of these substances becomes excessive, it is then considered a pollutant. This is a rather arbitrary judgment. By what standard can the composition of this table be called *clean* air? Is this composition actually realistic? Does air need to be this pure? Are some of these so-called impurities actually helpful? These are questions for which we do not have answers now.

23:6 Sources of Water Pollution

Silt

A primary source of water pollution has been the great amounts of **silt** (soil and rock particles) coming into the waters during rains. This happens where the land has been stripped of its vegetation and is eroding badly. In most places where the land is farmed and the soil will not absorb the rain, the run-off carries much silt into streams, completely clouding them. In many places the streams never clear up; there is much silt in them at all times. Some of this silt buildup is natural and cannot be prevented. However, excessive silt formation may reduce or destroy certain forms of life in water.

Sewage

If raw **sewage** is dumped into a stream or lake, the sewage will take oxygen from the water as it decomposes. This can be harmful to aquatic animals that need the oxygen to breathe. Sewage disposal plants eliminate a great deal of the problem. Some human waste which is allowed to go into water is biodegradable; that is, natural enzymes in the water will decompose the organic wastes. However, many industrial wastes and household products are not biodegradable and thus offer serious pollution problems.

Human waste could be put to good use. It has definite value

as an organic fertilizer. The waste can be treated in such a manner that it could be used to fertilize food crops.

Thermal Pollution

Water used as a cooling fluid for many industrial operations is often pumped back into a river or lake while it is still very hot. The higher the temperature of the water, the less oxygen it will dissolve. This **thermal pollution**, like sewage pollution, lowers the amount of oxygen available to the marine life in the water.

Industries can recycle water, rather than dump it back into natural water. "Cooling ponds" can be built to allow waters to lose some of their heat. The problem *can* be solved; however, it will be expensive to the industrial plant involved.

Phosphates

Fertilizers and detergents are the major sources of phosphates that enter lakes and streams. Phosphates are not actually a pollutant in the sense that there is any danger to people. Although there are conflicting opinions on the subject, they appear to encourage the growth of large amounts of algae in water, which can choke out desirable life forms. The phosphates in detergents are not biodegradable.

The ban on the use of phosphates in detergents led to the use of **NTA** (Nitrolotriacetic acid), a biodegradable replacement.

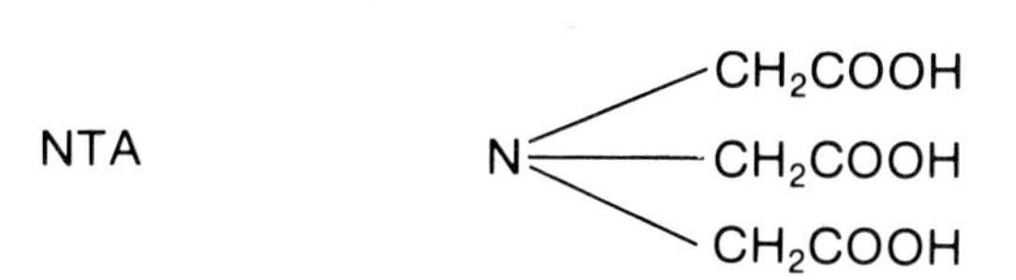

However, NTA solubilizes heavy metal ions and produces toxic effects in many living organisms. In this case, the cure for pollution was worse than the ailment. NTA has now been banned from use and is being subjected to further testing.

A subcommittee of the National Industrial Pollution Control Council reported that if the detergent industry decreases the phosphate level in detergents, housewives add more detergent to their wash to get the clothes cleaner. This, of course, means that the same amount of phosphate goes into the water, but it costs the housewife more to do the family wash.

Two alternatives have been advanced in the phosphate controversy. The W. R. Grace Company has announced that they have developed a process to remove phosphates from sewage waters at the treatment plant. The product of the removal operation is magnesium ammonium phosphate ($MgNH_4PO_4$).

Newer fertilizers have been developed that retain their plant nutritional value longer. This enables the vegetation to utilize it over a long period of time rather than have it washed away rapidly in surface water run-off.

Nitrates

The major source of nitrates is fertilizers. Nitrates appear to encourage plant growth in water, also. Nitrates in drinking water

23:7 *Scum and plant growth on Island Grover Pond, Abington, Massachusetts.*

can act as a poison by destroying the ability of the human hemoglobin to carry oxygen. Furthermore, normal purifying procedures for water do not remove the nitrates. Slow-acting fertilizers are necessary to control the influx of nitrogen into rivers and lakes from surface water run-off.

Pesticides

We have discussed the problems involved with pesticides. The long-lived or persistent pesticides, such as DDT, appear to be harmful to certain birds through the food chain mentioned earlier (DDT accumulates in fish, birds eat the fish, the chemicals may interfere with egg formation). The short-lived pesticides will decompose; but some, like parathion, are highly poisonous to humans.

23:8 *City dump contaminates the Solomon River near Delphos, Kansas. Farmers along the river are rightfully annoyed when the high waters spread debris over their land.*

Much research is necessary to develop pesticides that will control pests and then be quickly biodegraded. The continued use of some long-lived pesticides, such as DDT, may be necessary until then; or we may face ecological problems of horrible magnitudes, such as the destruction of complete forests or food crops by insects.

Industrial Wastes

The problem of industrial wastes is probably the most serious water pollution problem. Many industries dump waste products into available waters. Many factories may have been built near water for that very reason. Their waste products consist of such things as acids, bases, solvents, and greases.

Industrial wastes can be cleaned up. However, the cost in many cases is unreasonable. Many industrial operations would have to be closed if they tried to clean up their wastes; their operation would become too expensive to maintain. It has been estimated that the cost of achieving clean water by 1985 will be 2.3 *trillion* dollars.

If the environment becomes overly important to the American people, the economy of the whole nation may be affected. Factories, unable to meet the costly standards for pollution control, may have to close, putting many workers out of jobs and products off the market. If a sane balance is maintained between environmentalism and industrial development, the problems of pollution can be overcome with time.

Experiment 23-1 *Water Pollution*

1. Collect several samples of water from streams, ponds, and lakes.
2. Place a drop of each on a microscope slide and view each sample under the microscope.
3. Compare each with a drop of tap water. Draw pictures representing what you see in each sample.
4. Evaporate 10 ml of each sample, including tap water, and check the residue under the microscope also. Draw pictures of the residues.

23:7 Air Pollution Air pollution can be classified into two broad categories—gases and particulate matter. The sources also are twofold. *Man-made* pollutants come from automobiles, industry, heaters, and the combustion of refuse. *Natural* pollutants come from volcanic action, forest fires, and dust storms.

Particulate Pollutants

Particulates found in the atmosphere range from one to ten microns in diameter. (We will not discuss aerosols, which have a particle size from 0.001 to one micron in diameter.) Particles which appear as dust are abrasive and can damage the finish on any painted object when blown by the wind. Particles which settle into electrical gear can cause electrical shorting, especially around contact points. Air conditioning is one solution to **particulate pollution**. Well-filtered air does not contain much dust.

Small fluoride particles, which can be emitted from aluminum-ore reduction facilities and fertilizer plants, have been known to cause weakening of bones and loss of mobility in animals after the animals have eaten plants covered with the particles. Most scientists do not seem to think this is a problem for humans, because they do not normally take in such large amounts of fluoride. Many dentists recommend treatments of fluoride compounds in prescribed amounts as a means of preventing tooth decay in children. Some cities have added controlled amounts of fluorides to the drinking water to help children build up more resistance to tooth decay.

Any person suffering from respiratory ailments will be bothered by natural particulates, such as airborne pollen and man-made dust. Dust particles can be removed from industrial smoke by filtering the exit gases through bags, by centrifuging the dust out, and by electrostatic precipitation. Although the number of industrial plants has increased since 1930, dust pollution has actually decreased. The decreased use of coal as a fuel is responsible for part of this reduction. More efficient ways of burning fuels have been developed, and newly built plants have better equipment for controlling pollution problems.

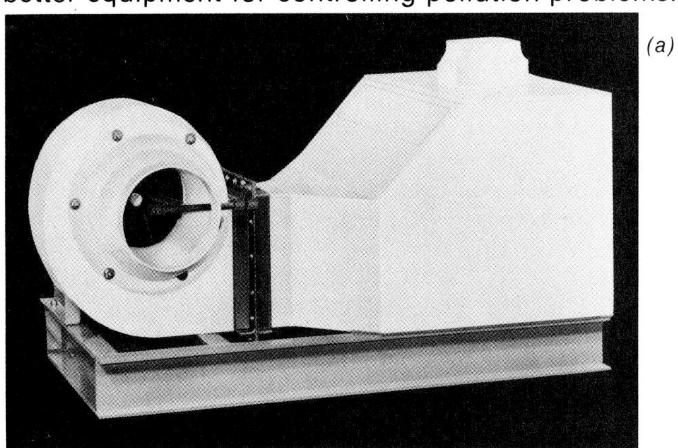
(a)

23:9 *(a) Laboratory exhaust system. (b) Shows the path of laboratory exhaust air through the water spray and filter system.*

(b)

Table 23-3 *National Particulate Pollution*

Year	
1930	— 519 micrograms/meter^3 of air
1957	— 120 micrograms/meter^3 of air
1968	— 96 micrograms/meter^3 of air
1969	— 92 micrograms/meter^3 of air

23:10 *(a) Smoke from plant before electrostatic precipitators were added. (b) After the addition of the precipitators, the air pollution has been minimized.*

(a)

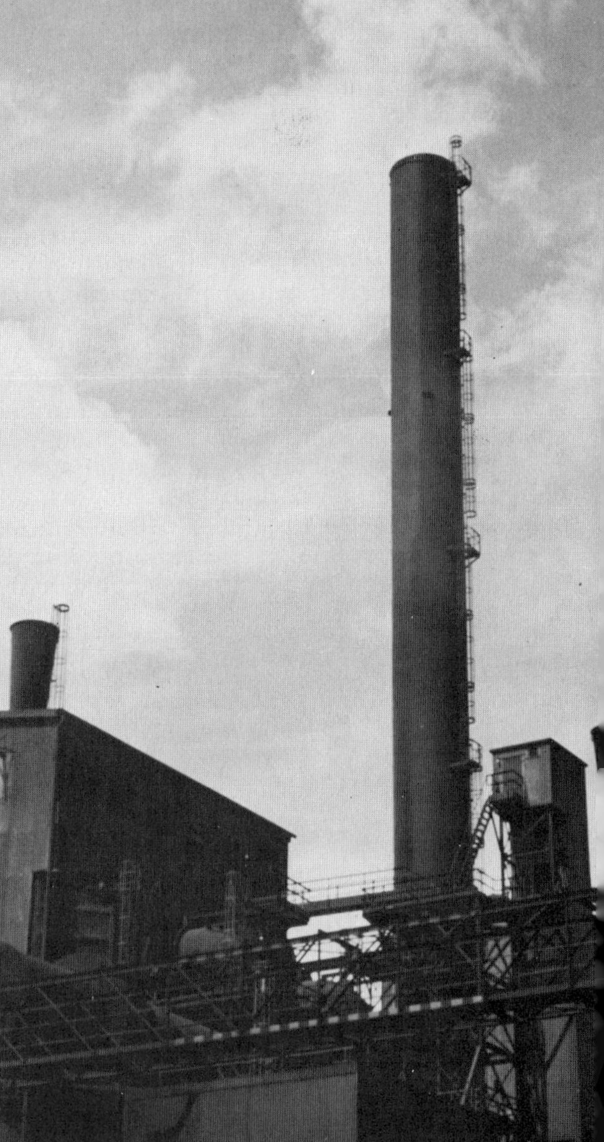

(b)

Experiment 23-2 *Air Pollution Detection*

1. Place a piece of filter paper over the end of a vacuum cleaner hose.
2. Turn on the cleaner and hold the hose in mid-air. Do this in a room and out-of-doors.
3. Let the motor run for 5 minutes; then turn off the cleaner, remove the filter paper, and inspect it under a bright light. What pollutants are you picking up?

Experiment 23-3 *Air Pollution Detection*

1. Place some white petroleum jelly on a microscope slide.
2. Put the slide outside and leave it there during dry weather for 3 days.
3. Examine the jelly with a magnifying glass. What pollutants are you picking up?

Experiment 23-4 *Air Pollution Detection*

1. Fill a jar one-third full of distilled water.
2. Put an insect screen over it and place the jar out-of-doors (see Figure 23:11). Leave it for 2 to 3 weeks, adding water when necessary to keep the same level of liquid.
3. Weigh a beaker. Pour the collecting water into it.
4. Rinse the bottle well with distilled water. Add the rinse water to the beaker.
5. Evaporate the water by slow boiling.
6. Weigh the beaker again with a sensitive balance after the water has evaporated. What type of pollutant is this?

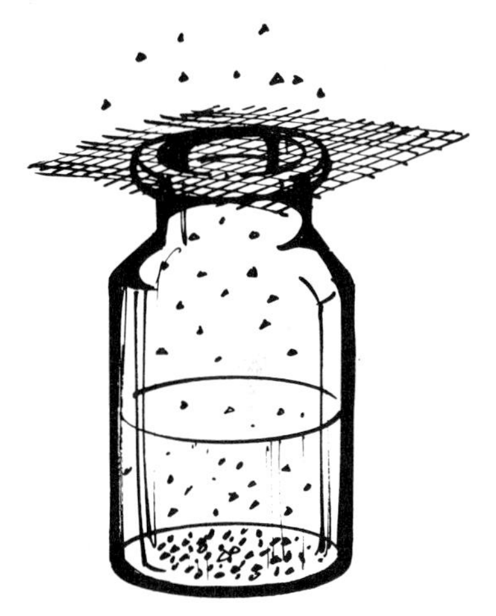

23:11 *Bottle for collecting dust and other airborne particles.*

Gas Pollution

Sulfur dioxide

There are varying opinions on the major cause of sulfur dioxide pollution. Volcanic activity contributes a major amount of SO_2 contamination in the atmosphere. The burning of sulfur-containing fuels by man is another major source of SO_2 in the air. Too much sulfur dioxide emission in any area can cause considerable damage, as was noted in the smelting of copper ore at Copperhill, Tennessee.

Low sulfur fuels can be burned, but these are much more expensive than other fuels. Chemical removal of sulfur dioxide is also expensive.

Sulfur dioxide does not last very long in the atmosphere. It is converted to sulfurous acid in the presence of water.

$$SO_2 + H_2O \longrightarrow H_2SO_3$$

This acid can cause considerable damage if it is highly concentrated in one area.

Oxides of nitrogen

Nitrogen dioxide is considered a major cause of haze in

urban and industrial atmospheres. Man-made oxides of nitrogen make up a very small part of the total amount in the atmosphere. Natural means, such as lightning, produce far more nitric oxide and nitrogen dioxide (the nitrogen cycle is discussed in Chapter 18).

Ozone

Ozone gas is not considered a primary pollutant. However, if the concentration of ozone becomes very high in an urban smog, the effect could be dangerous.

Carbon monoxide

Carbon monoxide, a poisonous gas, is produced by the incomplete combustion of carbon-containing fuels, as we saw in Chapter 14. However, the concentration of this gas in the atmosphere has not been increasing.

Hydrocarbons

The principal source of hydrocarbon pollutants is from automobile exhausts. Hydrocarbons appear to contribute to the smog problem in certain urban areas.

Photochemical smog

A temperature inversion at high altitudes can cause smoke and exhaust fumes to "hover" over a city. Normally, hot waste gases continue to rise into the cool atmosphere and scatter. But if there is a warmer layer of air above the gases in an area blocked in by hills or mountains, the waste gases cannot rise and are trapped, forming a smog. The smog over Los Angeles is typical of this.

Smog is referred to as being *photochemical* because light from the sun starts many of the reactions necessary for the chemistry of the smog. The smog process is still not well understood, and very little has been accomplished to prevent it.

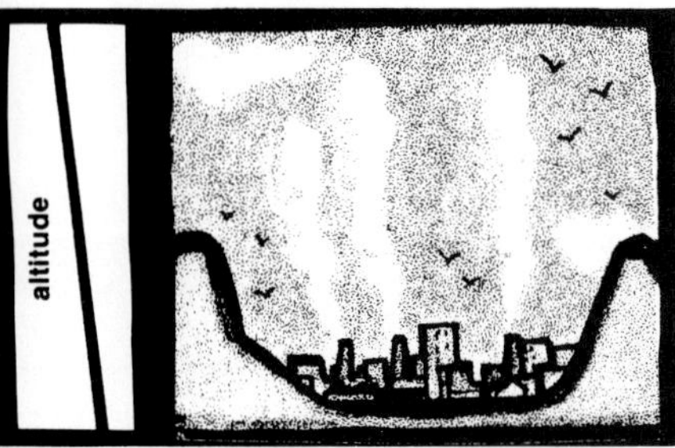

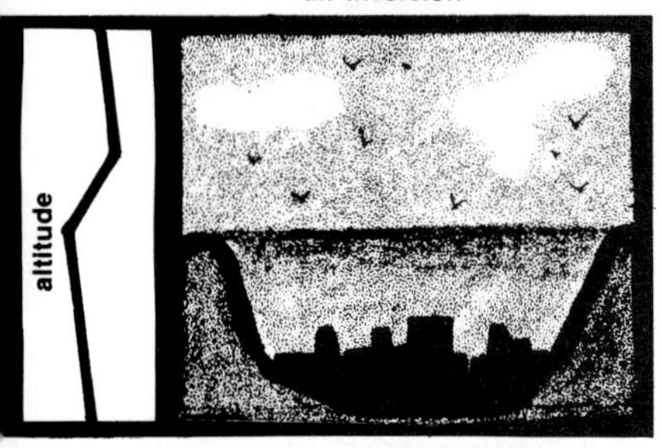

23:12 *How photochemical smog can develop. Warm air over cooler air can cause the retention of smoke and auto exhausts.*

23:8 Tobacco Smoke Tobacco smoke is another type of air pollution. It contains a variety of poisons—acetone, acrolein, ammonia, carbon monoxide, hydrocyanic acid, nicotine, nitric oxide, and nitrogen dioxide, to mention just a few. The carbon monoxide is especially harmful; its concentration in cigarette smoke is 640 times the level considered safe for industrial plants. Because carbon monoxide interferes with the red blood cells and their ability to transport oxygen to the cells of the body, smokers tend to become "winded" more quickly than nonsmokers; that is, they run out of breath in less time. A cigarette smoker at sea level is getting only as much oxygen as a nonsmoker living at an altitude of 8000 feet!

A considerable portion of cigarette smoke takes the form of tiny particles. This can be demonstrated by the "smoking machine" shown in Figure 23:13.

The results of this experiment with the "smoking machine" are startling. A single cigarette produces a dark deposit of tar on the filter paper. Even a filter-tip cigarette produces an easily observed deposit. What looks to the human eye like pure white smoke is in reality dirty brown. This material, called tar, contains about 30 **carcinogens**—cancer-producing chemicals. Much of this tar is trapped in the smoker's mouth and throat and then swallowed. The rest of the tar passes on to the lungs.

Several kinds of cancer can be caused by smoking—cancer of the lung, mouth, lip, tongue, larynx, esophagus, pancreas, and bladder. Chronic bronchitis and emphysema are two other serious diseases that have been traced to cigarette smoking. Because smoke paralyzes the lining of the nose and throat and interferes with the body's normal defenses, smokers find that they have colds and other respiratory diseases more often than nonsmokers. Moreover, it has been estimated that cigarette smoking accounts for about one fifth of all deaths due to heart attack in the United States. It has even been demonstrated that an expectant mother who smokes can cause damage to her unborn child. Smoking mothers have a higher percentage than nonsmokers for miscarriages, stillbirths, and infants who die within the first month.

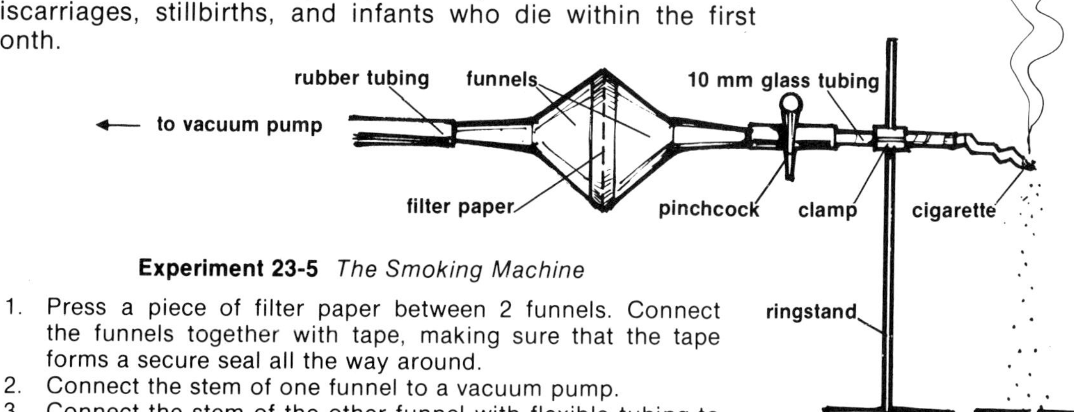

23:13 *The smoking machine.*

Experiment 23-5 *The Smoking Machine*

1. Press a piece of filter paper between 2 funnels. Connect the funnels together with tape, making sure that the tape forms a secure seal all the way around.
2. Connect the stem of one funnel to a vacuum pump.
3. Connect the stem of the other funnel with flexible tubing to a short length of 10-millimeter glass tubing.
4. Place a cigarette in the end of the glass tubing, and light it as the pump is turned on. A pinchcock can be used as shown in the drawing to adjust the rate of burning.
5. Shut off the pump and examine the filter paper when the cigarette has burned to within an inch of the end.

The Christian has a sacred responsibility to keep his body pure. His body is the temple of the Holy Spirit. We are told that we are not our own; we have been bought with a price: "What? Know ye not that your body is the temple of the Holy Ghost which is in you, which ye have of God, and ye are not your own? For ye are bought with a price: therefore glorify God in your body and in your spirit, which are God's" (I Corinthians 6:19-20).

It is ironic that many of the environmentalists who make dogmatic statements against air pollution do so in smoke-filled rooms. This kind of inconsistency may lead you to wonder whether these people are really interested in pure air. It makes little sense to get overly concerned about trace impurities in the air when many of the same pollutants are continually being introduced into the air in far greater quantities by cigarette smoking.

23:14 *(a) A littered campsite. (b) A clean, well-kept campsite. Which one should a Christian try to maintain?*

23:9 What God Has Given Us Several of the astronauts, on returning to earth from the moon, have expressed their thankful appreciation for the earth and its provisions for our comforts. In their travels these space explorers have never seen another place like the planet earth. Should we not try to show our respect for what God has given us and do our utmost to preserve and maintain it?

The Christian should be exemplary in his life, abstaining from anything that would cause unnecessary deterioration of the earth or its near surroundings. It is a poor testimony for a Christian, for example, to be a **litterbug**. The Christian should be different from the unregenerate individual, who has no particular guiding principle in life, or reason to do things decently and in order. In many places there are specific laws against littering. Therefore, it would be both a *sin* and a *crime* to deface the area with trash.

If a Christian citizen knows of industrial abuses, such as the dumping of harmful wastes into a river, he should do something. He should write letters to company officials, to newspapers, and to government officials—whatever is within reason—in an attempt to make others aware of the situation and to correct the problem.

Society should be concerned about pollution. It is necessary that we try to eliminate as much pollution as possible. However, we should allow time for the proper methods to be developed so that pollution control can be economical and effective. Hurried schemes instituted with very little thought may result in greater pollution than is now present.

As Christians we should be concerned about our community and its needs. One of the greatest needs is to clean up personal pollution. We have mentioned the fact that many people pollute not only the air but also their bodies with tobacco smoke. Other personal pollutants are alcoholic beverages (discussed in Chapter 15) and drugs, or dope. The use of these pollutes both body and mind. The Christian should keep his "temple" free from these pollutants and should be concerned with telling people about the dangers of using tobacco, alcohol, and drugs. At the same time, however, we must not lose sight of the far more important spiritual needs of the community. The pollution problem is only temporal; spiritual matters have eternal consequences.

List of Terms

carcinogens
DDT
gypsy moth
litterbug
NTA
parathion
particulate pollution (air)
pollution
sewage
silt
smog (photochemical)
thermal pollution (water)
tobacco smoke

Questions

1. (a) Will man pollute himself off the earth?
   (b) Name the pollution judgments in the book of Revelation.
   (c) What is the primary pollution in this world?
2. How did the Indians pollute the environment?
3. What can eliminate most air pollution?
4. Name some specific instances of pollution that affect the ecology of an area.
5. (a) What is the harmful effect of DDT?
   (b) Is DDT harmful to man?
6. (a) What moth has literally destroyed many forests in New England?
   (b) What is the problem with the use of parathion as an insecticide?
7. What is wrong with exclusively organic gardening?
8. How is the reasoning for banning DDT similar to the reasoning that one would use in banning the use of penicillin?
9. What would be an absolutely clean environment? Why?
10. (a) What is silt? What is the pollution problem it causes in water?
    (b) What is the pollution danger of raw human waste put into water?
    (c) What is a possible use of human waste?
11. What is thermal pollution? What is the danger of this?
12. (a) How can excessive phosphates affect water?
    (b) Suggest a solution to this problem.
    (c) Where do most of the nitrates that enter natural bodies of water come from?
13. What are some industrial wastes?
14. (a) How is air pollution classified?
    (b) What are some possible harms of particulate air pollution?
    (c) Is particulate air pollution decreasing or increasing? Why?
15. (a) Name three specific compounds that are considered gas pollutants.
    (b) What particularly can nitrogen dioxide in air cause?
    (c) How is carbon monoxide in the air usually formed?
16. Name the dangerous compounds in tobacco smoke.
17. Why do smokers tend to get "winded"?
18. (a) What are carcinogens?
    (b) How many carcinogens does cigarette tar usually contain?
19. (a) Name some diseases caused by cigarette smoking.
    (b) Why is cigarette smoking wrong for a Christian?
20. (a) Why should a Christian be concerned about pollution?
    (b) Suggest some ways you can fight pollution.
21. What is photochemical smog?
22. What are the major pollution problems?

Student Activity

Pick a specific form of pollution, such as air pollution by sulfur dioxide or water pollution by nitrates, and write a paper on it. Suggest solutions to the problem.

Photo next page

Pulleys form an important part of this surface mining shovel.

UNIT VIII MECHANICS

Chapter 24 Introduction to Physics; Simple Machines

24:1 Physics The study of energy is called **physics**. The **physicist** attempts to understand the fundamental laws and processes of nature. Typical general physics topics include *mechanical energy, heat, light, sound, electricity, magnetism,* and *nuclear energy*.

In recent decades, physics as a vocation has increased in popularity. Nuclear physics and space travel have contributed to this upswing. We do not know all there is to know about physics; there are many challenging questions remaining to be answered by enterprising researchers. Gravity and magnetism are only poorly understood. The structure of matter is still a great puzzle (for instance, what are protons and electrons made of?). Thermonuclear reactions and particle-antiparticle interactions have not been harnessed as sources of energy. And even when one question in the study of physics is answered, several new questions quickly take its place. If nothing else, modern physics has shown us how woefully little we really know about the world we live in.

The physicist studies and attempts to understand the *orderliness of nature*. Many people take for granted the order in the universe, as though order were a necessary property of matter. The cosmic evolutionist (a person who would argue that the universe "evolved") is guilty of *assuming* that the laws of nature were present at the beginning; he never actually *accounts* for these laws. To attempt an explanation by saying that the laws

24:1 *Physics as a vocation can be fascinating and challenging. Here a physicist is performing experiments which hopefully will lead to a new use for the laser beam.*

themselves evolved from something simpler would create more problems than the attempted explanation would solve. Instead, evolutionists accept the order which God has created into the universe and use it freely in their theories; but they refuse to acknowledge the Power that put the order into operation. They fail to realize, it seems, that where there is a law, there must be a lawgiver.

The orderly natural events which scientists observe and describe form a background against which God occasionally performs the mighty wonders that we call miracles. A miracle, when performed against a background of chaos, might go unnoticed. For instance, if it were normal for objects to fall upward on occasion instead of downward, there would have been no particular wonderment caused by the axe head that floated to the surface of the Jordan River (see II Kings 6:1-7). Because all things are subservient to God's will and purpose, He may choose to overrule any of His laws with higher laws.

Those people who deny the reality of miracles misunderstand the interrelationship between God and nature. They have exalted nature to an unreasonably high position (sometimes they even spell nature with a capital "N"), and they have tried to reduce God to a position of unimportance. These people apparently feel (if they believe in God at all) that He lacks the power to alter the normal routine of the world. The atheistic scientist is in a particularly double-minded position. On the one hand, he denies the existence of any supernatural ordering force in the world; on the other hand, he relies on the created orderliness in nature every time he performs an experiment. "A double minded man is unstable in all his ways" (James 1:8).

24:2 The Physics of the Ancients The ancients made use of several simple inventions to make life easier for themselves. For example, they used the lever, pulley, wheel and axle, inclined plane, wedge, and screw to move or lift heavy objects. These devices are called **simple machines**. Because they are still very important today, we shall study them in some detail in the next few sections.

The history of physics begins with Archimedes of Syracuse (287-212 B.C.). What did Archimedes do that was different from those before him? He applied mathematics to nature in a manner which not only *described* the events that he observed, but permitted him to *predict* what would happen under certain conditions. In addition to his extensive work in pure mathematics, Archimedes studied floating bodies, submerged bodies, levers, pulleys, and many other machines, both simple and complex.

From 215 to 212 B.C., the Romans attempted to capture the city of Syracuse, which is on the southeast coast of Sicily. It is said that Archimedes' clever inventions kept the Roman navy at bay for three years. As the Roman invaders entered the harbor at Syracuse, they were greeted by rocks launched from catapults. Also, ropes with hooks were swung out on booms (long poles) from the cliffs overlooking the harbor. When a hook had attached itself to a Roman ship, the Syracuse defenders reeled in the rope, causing the ship to capsize. Some historians state that Archimedes was able to set fire to some of the Roman ships by focusing the sun's rays on the ships through the use of large mirrors. Finally the Roman general Marcellus succeeded in taking the city, and Archimedes was immediately put to death.

24:2 *The ancient Greek physicist Archimedes used his scientific ingenuity to defend the harbor of Syracuse, Sicily, against the attacking Roman navy.*

Fortunately, some of the writings of Archimedes have been preserved. (For an account of the part Archimedes played in the defense of Syracuse, see the section on Marcellus in *Plutarch's Lives*, Loeb Classical Library, Volume 5, pages 469-479.)

24:3 The Lever A **lever** is *a rigid bar capable of turning about a fixed point*, called the **fulcrum**. The ancients, even many years before Archimedes, had used levers. In Old Testament times, money was weighed on an equal-arm lever balance (see, for example, Jeremiah 32:9-10). But until Archimedes made a quantitative study of the lever, the principle of unequal arms was not understood.

24:3 *In Old Testament times money was weighed out on an equal-arm balance.*

Imagine a see-saw on which you are trying to balance a heavy boy and a light boy. You know by intuition that the heavy boy should sit closer to the fulcrum than the light boy. But how much closer should the heavy boy sit? The best way to solve this problem is to perform an experiment, gather data, and then analyze the data. First weigh both boys. Suppose you find that their weights are 100 pounds and 50 pounds, respectively. Next, place the boys on a see-saw (making sure that the board balances by itself *before* they get on). Move them around until they

24:4 *The 100-pound boy must sit closer to the fulcrum than the 50-pound boy to balance.*

balance. You will find that there will be a number of "right" combinations. For example, when the 100-pound boy is 1 foot from the fulcrum, the 50-pound boy will balance him by being 2 feet from the fulcrum. When the 100-pound boy is 2 feet from the fulcrum, the 50-pound boy will balance him by being 4 feet from the fulcrum. Also, with the heavy boy at 3 feet from the fulcrum, the light boy has to move 6 feet from the fulcrum. The data are summarized in the following table:

Feet from fulcrum in order to balance

100-pound boy	*50-pound boy*
1	2
2	4
3	6

The physicist usually looks for some *unifying* principle or law in a table to tie his data together. With enough "head scratching," you should be able to see the general principle involved in your own data: In each case the first boy's weight multiplied by his distance from the fulcrum is equal to the second boy's weight multiplied by *his* distance from the fulcrum:

$$(100)(1) = (50)(2)$$
$$(100)(2) = (50)(4)$$
$$(100)(3) = (50)(6)$$

This relationship certainly seems to "work," at least for these two boys. We would like to test further, however, to make sure that this is a general principle that works for objects of any weight. Try the experiment again, using different people. This time, use a 40-pound boy and a 120-pound boy. You should obtain the following table of data:

Feet from fulcrum in order to balance

40-pound boy	*120-pound boy*
3	1
$4\frac{1}{2}$	$1\frac{1}{2}$
6	2

Again, the products of weight and distance are equal in each case:

$$(40)(3) = (120)(1)$$
$$(40)(4\tfrac{1}{2}) = (120)(1\tfrac{1}{2})$$
$$(40)(6) = (120)(2)$$

There might be a question in your mind as to whether this principle is also true for inanimate (non-living) objects. To settle this question, you could experiment with such things as a sack of flour, a keg of nails, or any other heavy objects you might have. After enough testing, you will be able to confidently "generalize" for all kinds of objects. What objects you use does not really matter. The *weights* and their *distances* from the fulcrum are the essential items of information.

We can state the principle we have just discussed in the form of an algebraic equation.

$$w_1d_1 = w_2d_2$$

In words, this means that the first weight (w_1) times the first distance (d_1) equals the second weight (w_2) times the second distance (d_2). This is called the **law of moments**. If any three of these quantities are known, the fourth quantity can be found. This principle, or general law, applies to all three classes of levers, so long as you are careful to measure distances from the fulcrum to the weight.

EXPERIMENT 24:1 *The Law of Moments*

1. Support a meterstick at its center by means of a rectangular ink eraser turned sideways. The eraser will serve as a fulcrum.
2. Place a 100-gram weight and a 200-gram weight on opposite sides of the meterstick and move them to any position in which they balance. (The system may be considered to be in balance when the meterstick is level.)
3. Measure the distance from the center of the eraser to the center of each weight and record it.
4. Verify the law of moments ($w_1d_1 = w_2d_2$) by computing w_1d_1 and comparing it to w_2d_2.
5. Repeat the procedure using a 200-gram weight and a 500-gram weight.
6. Finally, attempt to find out whether three weights follow the relationship $w_1d_1 = w_2d_2 + w_3d_3$. Place the 500-gram weight on the left-hand side of the stick and the 200- and 100-gram weights on the right-hand side. Adjust their positions until the system is in balance.
7. Measure the distances and calculate the **moment** (wd product) for each weight.
8. Determine whether the moment of the 500-gram weight is equal to the sum of the moments of the two smaller weights.

24:4 First-Class Levers Suppose you try to lift a large stone. One way to do this would be to use a crowbar and another stone to form a fulcrum (see Figure 24:5). You would push down at the upper end of the lever with a **force** known as **effort (E)**. The object being lifted, in this case the rock, is called the **resistance (R)** because it resists your efforts to lift it. *If the fulcrum (f) of a lever is* *located between the effort and the resistance, it is called a* **first-class lever**. The see-saw and equal-arm balance are examples of first-class levers.

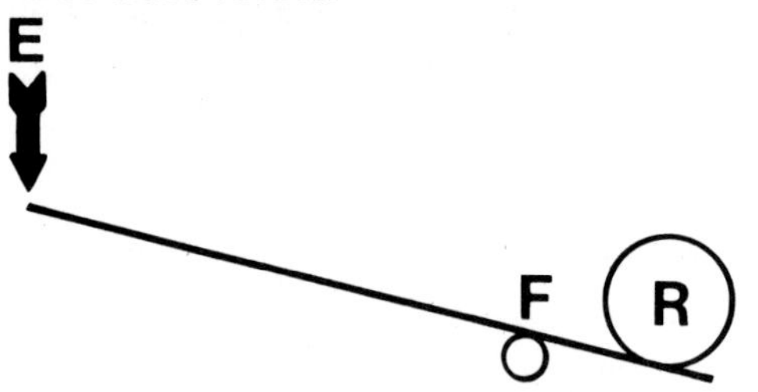

24:5 *Diagram of a crowbar lifting a stone.*

The first-class lever is probably the most common type. Other examples can be found readily around the home, in the classroom, in the family car, and even in the human body. Brooms, shovels, knives, forks, spoons, pins, and needles are used as first-class levers at least some of the time.

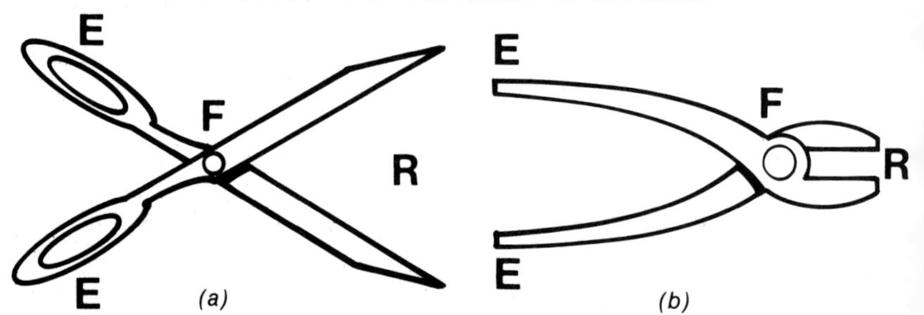

24:6 *(a) A pair of scissors is a combination of two first-class levers sharing a common fulcrum. (b) A pair of pliers is also made up of two first-class levers. Because of the placement of the fulcrum, the pliers have a higher mechanical advantage than the scissors.*

A pair of scissors is a combination of two first-class levers. The effort is applied by your hand pressing on the handle. The fulcrum is the screw holding the blades together. The resistance is the object being cut. A pair of pliers is a similar combination of two first-class levers.

Now let us assume that the crowbar is 5 feet long and that the fulcrum is 1 foot from the right-hand end (4 feet from the left-hand end). How heavy a rock could be lifted using an effort of 50 pounds? To find the answer, simply substitute the appropriate numbers in the law of moments:

$$w_1 d_1 = w_2 d_2$$
$$(50)(4) = (w_2)(1)$$
$$200 = w_2$$

Solving for w_2, you would get *200 pounds* for the weight of the stone that can be lifted.

Your effort, in this particular problem, has been quadrupled (multiplied by four). You can now begin to appreciate the advantage that the lever gives you. If you had to lift the stone by yourself, you would have to have an effort equaling the full 200 pounds. Furthermore, you would have to lift *upward* rather than push *downward*, as you can do with the lever. The amount that the force (effort) is multiplied in a simple machine, such as a lever, is called the **mechanical advantage (M.A.)**. In this case the M.A. is 4.

24:5 Mechanical Advantage The mechanical advantage of a lever can be calculated two ways. First, the M.A. can be determined by the *ratio* of the effort arm to the resistance arm. The **effort arm** is *the distance from the fulcrum to the effort*; the **resistance arm** is *the distance from the fulcrum to the resistance.*

$$\text{M.A.} = \frac{\text{effort arm}}{\text{resistance arm}}$$

If you were to insert the numbers from the problem in Section 24:4, it would look like this:

$$\text{M.A.} = \frac{4\text{ ft}}{1\text{ ft}} = 4$$

Note that the units cancel to give a pure number for the answer. Mechanical advantage, being a ratio of two numbers of like units, has no units of its own.

The other method of calculating the mechanical advantage is more general. Simply determine the ratio of the weight that is being lifted (the resistance) to the force that is being applied (the effort).

$$\text{M.A.} = \frac{\text{resistance}}{\text{effort}} = \frac{R}{E}$$

If you were to insert the numbers from the problem in Section 24:4, it would look like this:

$$\text{M.A.} = \frac{200\text{ lb}}{50\text{ lb}} = 4$$

Both methods should, of course, give the same answer.

24:6 A Word About Friction Throughout this chapter, we will try to make simple machines and their operation easier to understand by ignoring **friction**—the rubbing together of parts of a machine which wastes some of the energy that is put into the machine. Friction is a more serious problem in some simple machines than in others. You can lubricate the parts which rub together, but there will always be *some* friction wherever moving parts are involved. Realistically, then, we must be prepared to put in somewhat more effort than the effort indicated by the simple mechanical advantage formula for each device.

24:7 Second-Class Levers *A lever in which the resistance is located between the fulcrum and the effort is a* **second-class lever**. Most doors are second-class levers. The door hinges form the fulcrum, the effort is applied to the handle, and the weight of

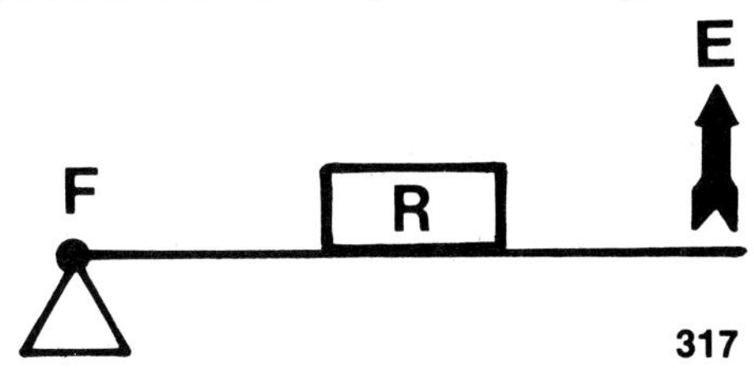

24:7 *A second-class lever. The resistance is located between the fulcrum and the effort.*

the door is the resistance. The wheelbarrow is another good example of a second-class lever. A nutcracker is *two* second-class levers joined by a short connecting bar.

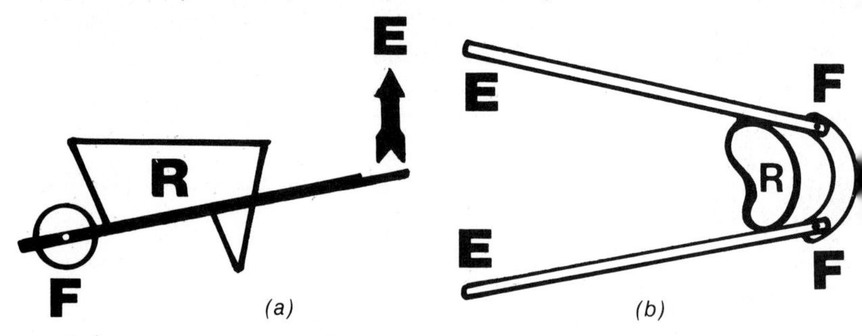

24:8 *(a) The wheelbarrow, an example of a second-class lever. (b) A nutcracker consists of two second-class levers joined by a short connecting bar.*

24:8 Third-Class Levers *A lever in which the effort is located between the fulcrum and resistance is a* **third-class lever**. This type of arrangement is effective for producing speed, rather than mechanical advantage, especially if the effort is placed very near the fulcrum. A flyswatter is used most successfully as a third-class lever. Your forearm is another example of a third-class lever.

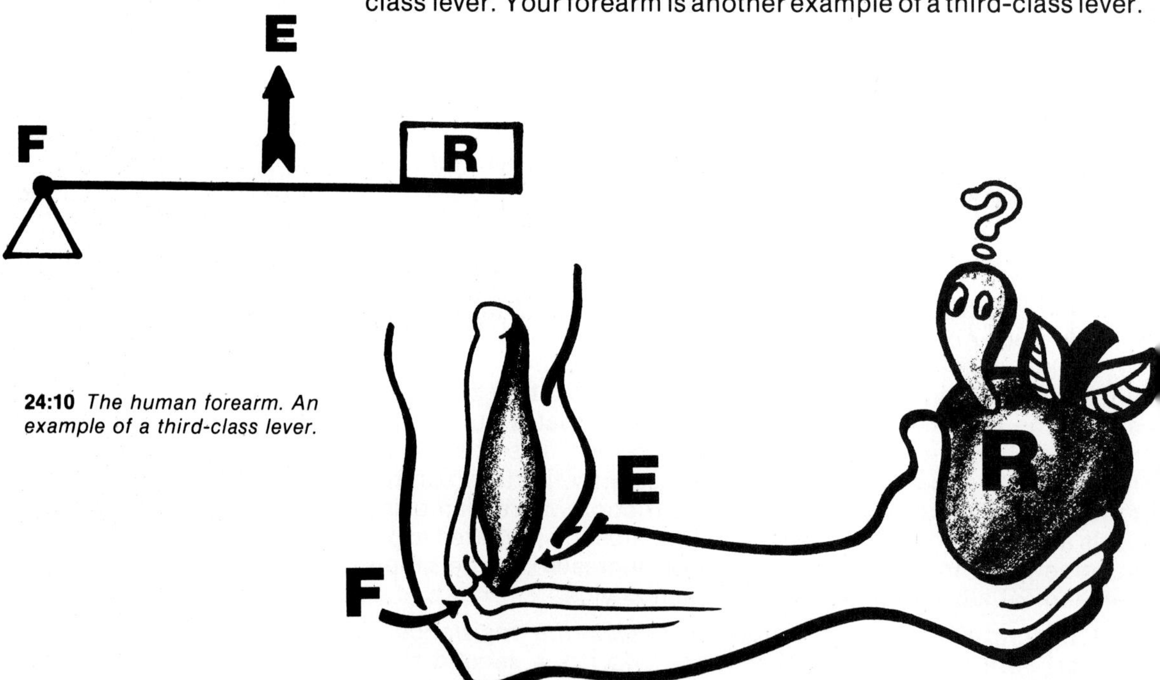

24:9 *A third-class lever. The effort is located between the fulcrum and the resistance.*

24:10 *The human forearm. An example of a third-class lever.*

The point where the biceps muscle attaches to the bone of the forearm is close enough to the elbow joint (fulcrum) to produce considerable speed when the biceps contracts. Pens and pencils are usually used as third-class levers. A pair of tweezers is a combination of *two* third-class levers. Some typical uses of levers are shown in Figure 24:12.

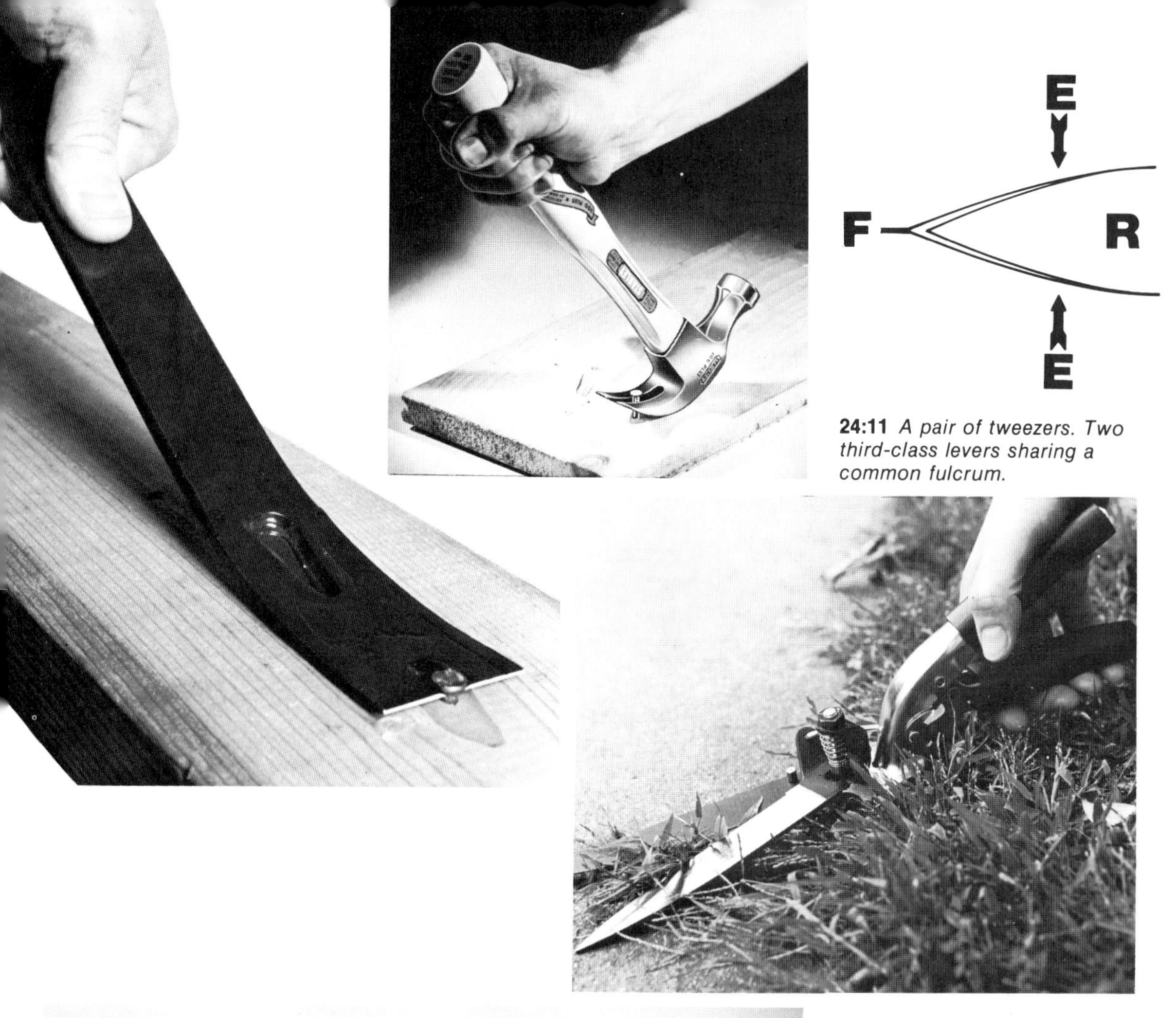

24:11 *A pair of tweezers. Two third-class levers sharing a common fulcrum.*

24:12 *Some uses of levers.*

24:9 The Distance Principle You must not get the idea that simple machines, such as the lever, give us "something for nothing"; they do not. All machines are less than 100 percent efficient and really give *less* than you put into them. In the examples given, however, we are ignoring the losses to make the subject easier to understand.

Let us return to the example of the crowbar having a mechanical advantage of 4. We saw that a force of 50 pounds lifted a stone weighing 200 pounds. Do not be misled into thinking that the lever in some way "supplies" the other 150 pounds of force. What actually happens is that the *effort* must travel *4 times* as much distance as the resistance. When the effort moves 2 feet, for example, the stone only moves 6 inches. This is known as the **distance principle**; that is, if force is gained, distance must be sacrificed. Conversely, if speed or distance is gained, force must be sacrificed. **Work** is defined as *force times distance*. Therefore, you can readily see that the work done by the effort is equal to the work done by the resistance:

$$W = Fd$$
$$W = (50\text{ lb})\ (2\text{ ft})$$
$$W = 100\text{ ft-lb}$$

Effort: $(50\text{ lb})\ (2\text{ ft}) = 100$ ft-lb of work
Resistance: $(200\text{ lb})\ (\tfrac{1}{2}\text{ ft}) = 100$ ft-lb of work

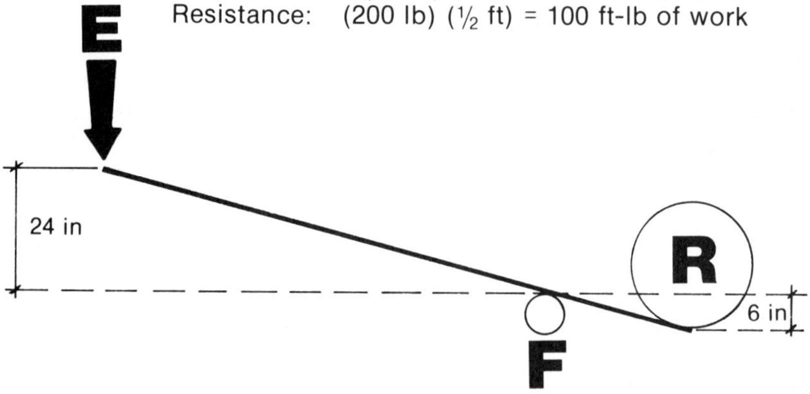

E
24 in
R
6 in
F

24:13 *The distance principle. When force is gained, distance must be sacrificed. Although the force is being multiplied by a factor of 4 in this illustration, the resistance moves only one fourth as far as the effort.*

Notice from this result that work is expressed in the English system in **foot-pounds**.

The distance principle applies to all simple machines—the lever, the wheel and axle, the pulley, the wedge, the screw, and the inclined plane. If force is gained, distance must be sacrificed. If speed is gained, force must be sacrificed. In no case can the work gotten out of the machine exceed the work put into it.

24:10 The Pulley The simple machine known as a **pulley** is a wheel with a grooved rim. The groove is shaped so that a rope or cable will fit into it. Sometimes you may be interested in changing only the direction of motion of the cable. With a **single fixed pulley** (shown in Figure 24:15a), the direction of the motion is changed from downward to upward. There is no multiplication of force, but it is usually easier to pull down than up.

24:14 *Pulleys form an important part of this surface mining shovel.*
(See chapter title page)

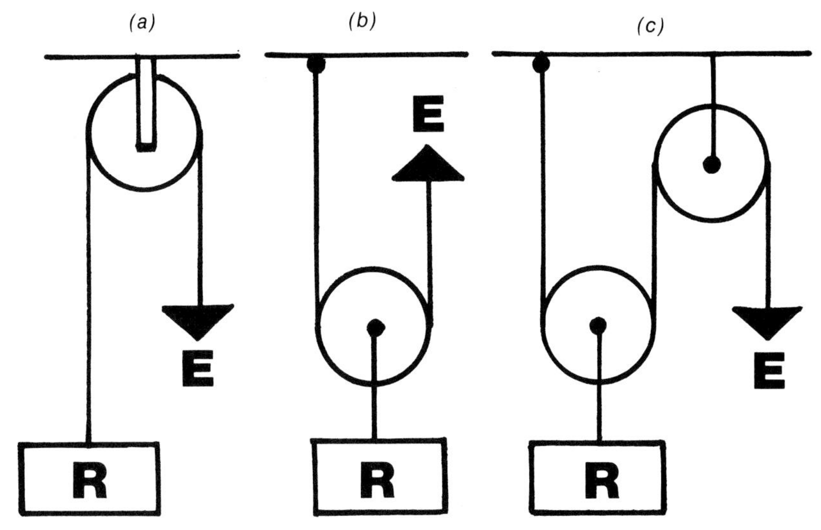

24:15 *(a,b,c) Some simple pulley arrangements.*

24:16 *Block and tackle arrangements having mechanical advantages of (a) 4 and (b) 5.*

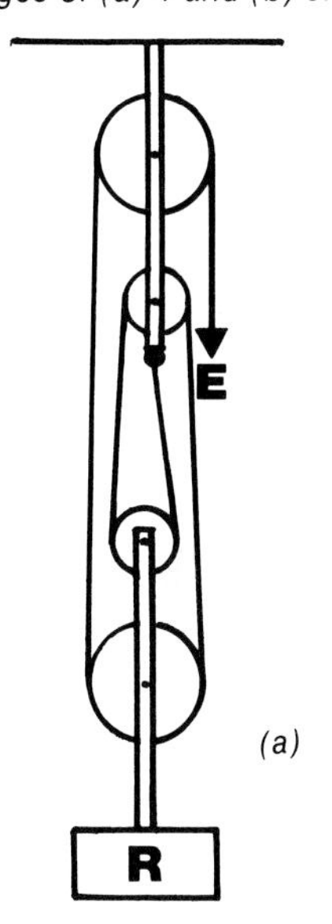

Figure 24:15b shows a **single movable pulley**. As you pull up on the rope, the pulley and resistance move upward, but at a slower rate than the rope. Unlike the single fixed pulley, the single movable pulley is a device that *does* multiply force. The single movable pulley has a mechanical advantage of 2, whereas the M.A. of the fixed pulley is only 1.

It is easy to find the mechanical advantage of a pulley system: *Simply count the number of strands directly supporting the resistance.* In the single fixed pulley, there is only one strand directly supporting the load; the M.A. is therefore 1. In the case of the single movable pulley, there are *two* strands directly supporting the resistance, one on either side of the pulley; the M.A. is therefore 2.

Figure 24:15c shows a combination of the devices shown in Figure 24:15a and Figure 24:15b. Can you tell what the M.A. of the combination should be? Do not be fooled by the fact that there are three strands. Only two of the strands actually support the resistance directly; so the M.A. is 2. This pulley arrangement combines the good points of the other two. The force is doubled *and* we are able to pull down rather than up.

A more complicated arrangement is shown in Figure 24:16a. A device consisting of a number of fixed and movable pulleys is called a **block and tackle**. The block and tackle in the illustration has 4 strands directly supporting the resistance; its mechanical advantage is therefore 4. A 500-pound weight could be lifted with this block and tackle using only 125 pounds of force.

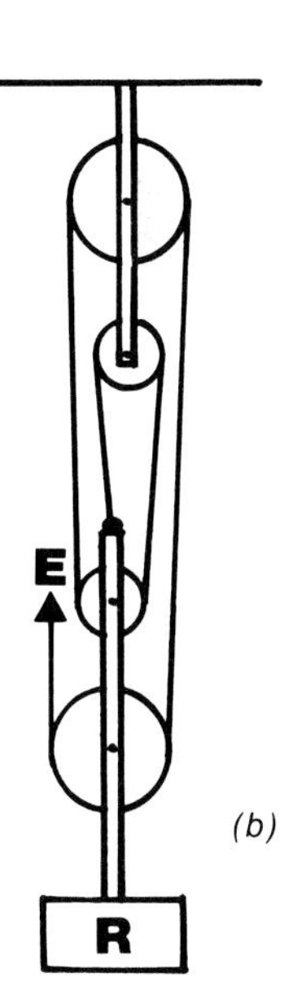

Figure 24:16b shows a block and tackle with a mechanical advantage of 5. More complex arrangements than this are used for lifting extremely heavy objects. An automobile mechanic, for example, can move a large engine from one part of his shop to another through use of a suitable block and tackle on an overhead rail.

The distance principle applies to the pulley just as it does to the lever. When force is multiplied, distance must be sacrificed. In

24:17 *The pedals of a bicycle and the shaft connecting them furnish an example of a wheel and axle of the windlass type.*

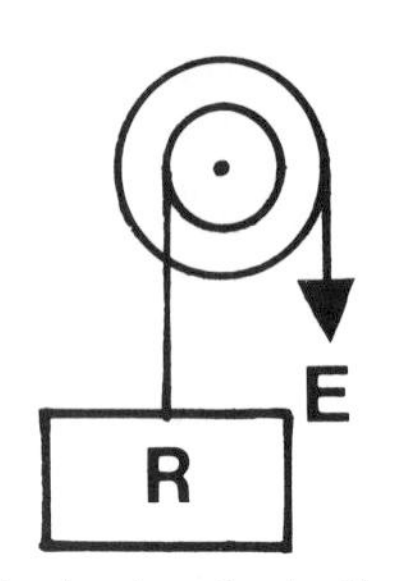

24:18 *A wheel and axle. Generally the effort is applied to the wheel and the resistance is connected to the axle.*

24:19 *Old water wells used a windlass to draw up the bucket.*

a system having a mechanical advantage of 5, the effort must move 5 times as far as the resistance. If you have ever operated a pulley system, you may have been surprised at the length of rope that had to pass through your hands to get the job done.

24:11 The Wheel and Axle Another method of combining pulleys is to fasten two pulleys of different sizes to the same shaft. This forms the simple machine known as the **wheel and axle**. The larger pulley is termed the **wheel**; the smaller pulley is known as the **axle**. Generally, *the effort is applied to the wheel*, and the *resistance is connected to the axle.* The mechanical advantage of a wheel and axle can be figured by dividing the radius of the wheel by the radius of the axle.

$$\text{M.A.} = \frac{\text{radius of wheel}}{\text{radius of axle}}$$

A special kind of wheel and axle having an unusually high mechanical advantage is the **windlass**. You may know it better by the name "crank." Old-fashioned water wells had a windlass to pull up the bucket. A well crank may not look like a wheel and axle, but as you turn it, your hand passes around in a large circle. This circle is the *wheel*. The rope winds onto the shaft, or *axle*. If you had a 15-inch crank and a shaft having a radius of ½ *inch, the mechanical advantage would be 30:*

$$\text{M.A} = \frac{\text{radius of wheel}}{\text{radius of axle}} = \frac{15\text{ in}}{\frac{1}{2}\text{ in}} = 30$$

Some common examples of the wheel and axle are doorknobs, steering wheels, screwdrivers, controls on a radio or television, and the cranks on can openers, ice cream freezers, pencil sharpeners, and egg beaters.

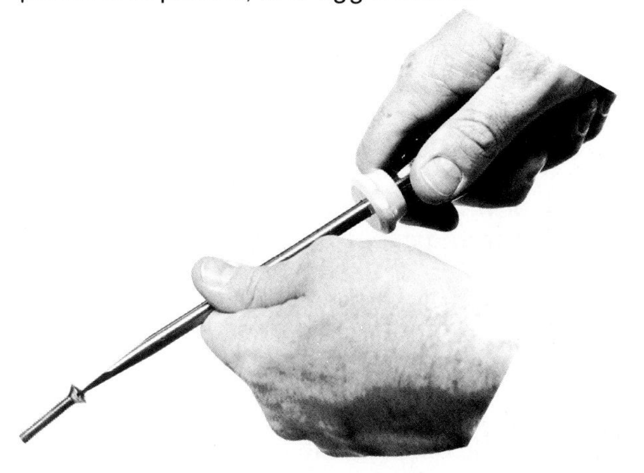

24:20 *The screwdriver is a common example of a wheel and axle.*

24:12 The Inclined Plane A *slanted surface that enables heavy objects to be raised more easily than they could be lifted* is called an **inclined plane**. The inclined plane allows the work to be spread out over a much greater distance than would be used in lifting. Hence, the force required is considerably less.

24:21 *Although this individual is experiencing some difficulty in pushing the object up the ramp, he would be completely unable to accomplish his task without the ramp.*

Suppose you wish to move a 350-pound oil drum onto a loading dock 4 feet high by using a ramp 20 feet long. First, figure how much work you would have to do if you were to lift the drum. Since you are exerting a force of 350 pounds over a distance of 4 feet, and work is the product of force and distance, the total work done is 1400 foot-pounds.

$$W = Fd$$
$$W = (350 \text{ lb})(4 \text{ ft})$$
$$W = 1400 \text{ ft-lb}$$

By rolling the drum up the 20-foot ramp, you are still doing 1400 foot-pounds of work, but the increased distance cuts down the force needed. By substituting 1400 foot-pounds in the formula for work and 20 feet for the distance, we can find out the force needed:

$$W = Fd$$
$$1400 \text{ ft-lb} = (F)\ (20 \text{ ft})$$
$$F = \frac{1400 \text{ ft-lb}}{20 \text{ ft}}$$
$$F = 70 \text{ lb}$$

The force needed to *roll* the drum up the ramp is 70 pounds, only one-fifth of the 350 pounds required to *lift* the drum. The mechanical advantage of this inclined plane can easily be calculated.

$$\text{M.A.} = \frac{\text{resistance}}{\text{effort}} = \frac{350 \text{ lb}}{70 \text{ lb}} = 5$$

As you might guess, different inclined planes give different mechanical advantages. A short, steep ramp will have a lower mechanical advantage than a long ramp with a more gradual slope. Another method for figuring the mechanical advantage of an inclined plane is to divide the slant-length of the ramp by its height.

$$\text{M.A.} = \frac{\text{slant-length}}{\text{height}}$$

In the problem above

$$\text{M.A.} = \frac{20 \text{ ft}}{4 \text{ ft}} = 5$$

24:22 *The mechanical advantage of an inclined plane is found by dividing its slant length (L) by its height (H).*

This is the same answer as that obtained by figuring the mechanical advantage as the resistance divided by the effort.

Inclined planes are fairly common. A stairway is a modified inclined plane. You might find it difficult to shinny up a pole or rope to get to the next floor, but walking up a stairway is relatively easy. A road winding up the side of a 100-foot hill provides a convenient means of raising a car to that height; lifting the car 100 feet with a crane or an elevator would be considerably more difficult. Yet the work accomplished in each case would be the same.

24:23 *The bridge crossing this superhighway is an example of an inclined plane.*

EXPERIMENT 24:2 *Mechanical Advantage of an Inclined Plane*

1. Place one end of a 4-foot board on a stack of books to form an inclined plane.
2. Measure the height and slant-length of the inclined plane and determine its mechanical advantage by the relationship
$$\text{M.A.} = \frac{\text{Slant-length}}{\text{height}}$$
3. Determine the combined weight of a cart and a 1-kilogram weight. Record this value as the *resistance.*
4. Place the weight in the cart, and measure the force needed to pull it up the inclined plane at constant speed using a spring balance. Record this value as the *effort.*
5. Calculate the mechanical advantage by the relationship
$$\text{M.A.} = \frac{\text{resistance}}{\text{effort}}$$
6. Compare with the M.A. calculated by the other method. Ideally, the two results should be equal; but there may be some difference between them because of inaccuracies in the spring balance and friction in the wheels of the cart.

24:13 The Wedge Two inclined planes placed "back to back" form a **wedge**. The wedge helps a woodsman to split a log more easily (see Figure 24:24). While a mechanical advantage could be calculated for a wedge in a manner similar to that of the inclined plane, we will not attempt to do so because the large amount of friction caused by the use of wedges makes the result somewhat meaningless. Some common examples of wedges are axes, hatchets, chisels, and knife blades.

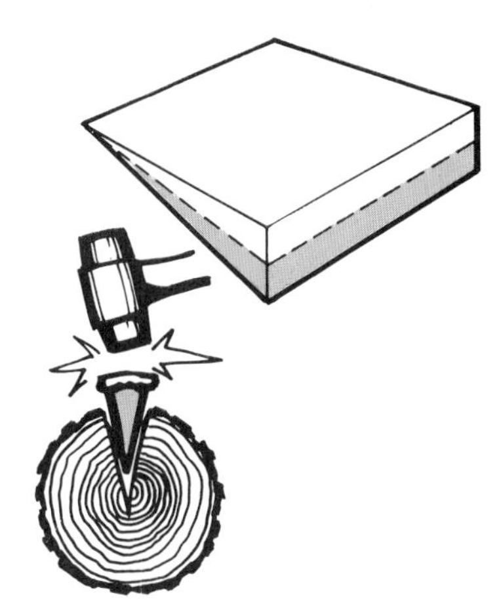

24:24 *The wedge, a combination of two inclined planes, is commonly used for splitting logs.*

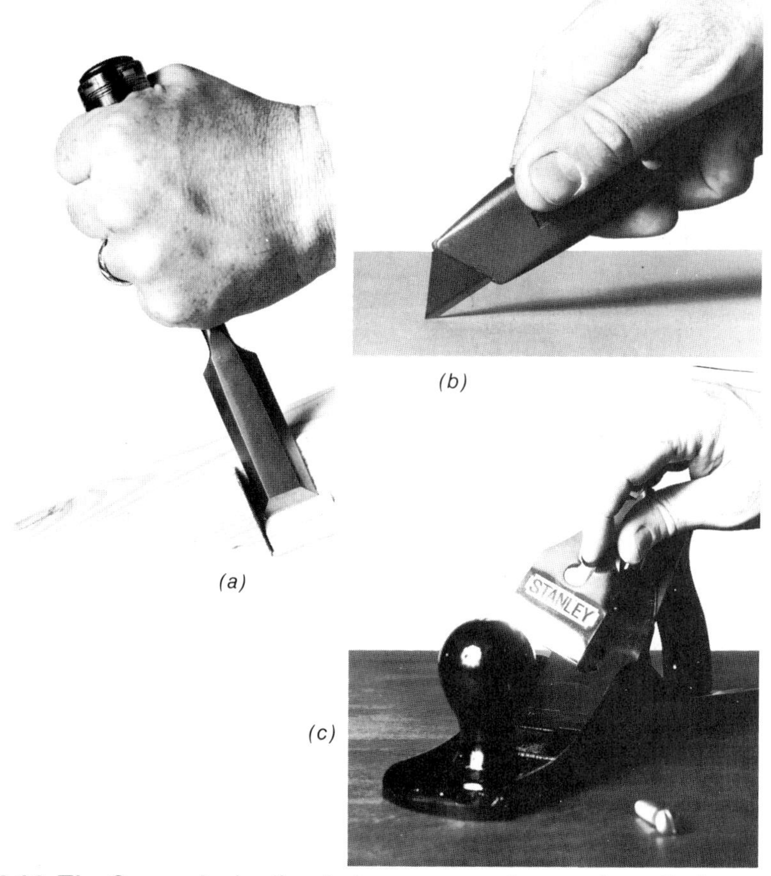

24:25 *Some useful examples of wedges: (a) a chisel (b) a knife blade (c) the cutting edge of a plane.*

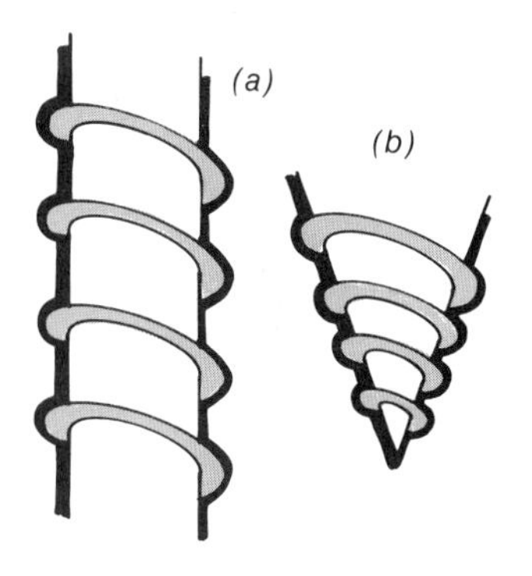

24:26 *A screw is an inclined plane wrapped around a cylinder (a) or a cone (b).*

24:14 The Screw An inclined plane wrapped around a cylinder or cone is a **screw**. The *cylindrical* type of screw called a bolt (see Figure 24:26a) is often designed to accept a nut and is threaded in such a way as to fit tightly into the threads of the nut. The *conical* type of screw (see Figure 24:26b), more often than not, is designed to force its way into wood or other compressible materials.

The distance between two threads which are next to each other in a screw is called its **pitch**. The smaller the pitch, the higher the mechanical advantage. You can notice a great deal of difference in pitch from one type of screw or bolt to the next (see Figure 24:27).

24:27 *Assorted screws.*

Vices and clamps used for holding objects while they are being tooled or glued are another application of the basic screw. A spiral staircase is a modified screw, as is a road that spirals its way to the top of a mountain.

The **jackscrew** is a *combination of the screw and the wheel and axle.* Because it is two simple machines in one, the jackscrew is capable of giving tremendously high mechanical advantages. In fact, jackscrews are used to lift houses. Four jackscrews, one at each corner of the house, are usually sufficient to accomplish this task.

24:28 *Spiral staircase, a modified screw.*

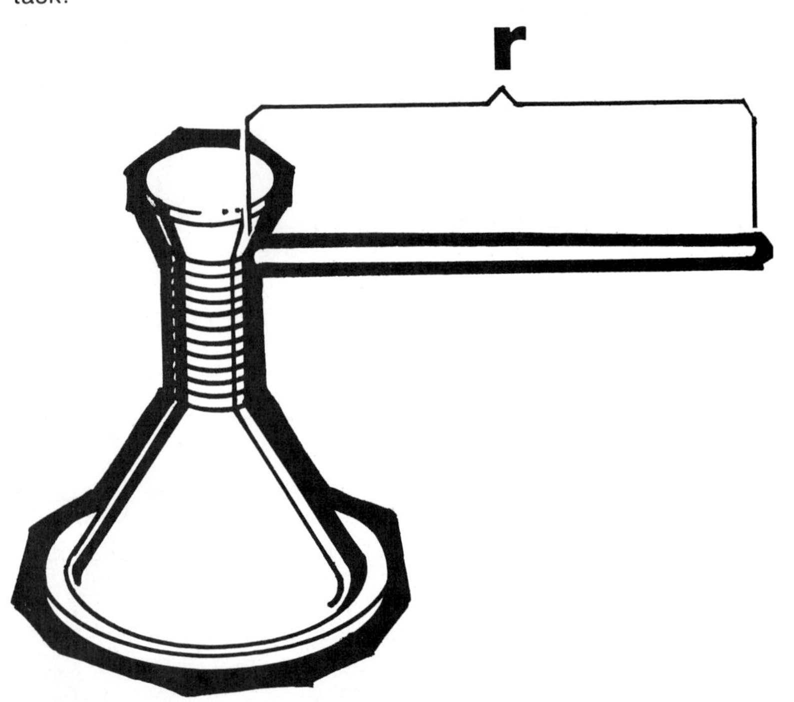

24:29 *The jackscrew is a combination of the screw and the wheel and axle.*

★ The mechanical advantage of a jackscrew is calculated by multiplying 2 times pi times the radius, then dividing by the pitch.

$$\text{M.A.} = \frac{2\pi r}{\text{pitch}}$$

A typical jackscrew might have a radius of 12 inches and a pitch of 1/4 of an inch. Putting these numbers into the formula, we obtain a mechanical advantage of 301:

$$\text{M.A.} = \frac{(2)\ (3.14)\ (12)}{1/4}$$

$$\text{M.A.} = \frac{75.36}{1/4}$$

$$\text{M.A.} = 301$$

Therefore, a man using a jackscrew and applying 70 pounds of force can lift more than 21,000 pounds!

24:30 *The power being developed by this grader could be determined by measuring what weight of earth it transports, how far it moves it, and how long it takes to move it.*

24:15 Power We have defined work as force times distance, but so far we have said nothing about *how fast* the work gets done. *The rate of doing work is called* **power.** Or, we could say that power is the amount of work in a period of time. Written as an equation, *power equals work divided by time:*

$$P = \frac{W}{t}$$

In the English system, work is expressed in foot-pounds, and time is expressed in seconds; so you might expect that the unit of power would be foot-pounds per second. However, physicists have found it more convenient to use a larger unit called **horsepower**, which is equal to 550 foot-pounds per second.

In the metric system, the unit of power is the **watt**, defined as *the amount of power developed when one joule of work* (about 0.74 foot-pounds) *is done in one second.* One horsepower is equal to 746 watts.

In order to do work at the rate of one horsepower, you would have to lift 550 pounds one foot in one second, or 275 pounds *two* feet in one second, or 137½ pounds *four* feet in one second. There will be one combination at which your rate of doing work will be a maximum.

★ Suppose you were interested in measuring your power in running up a stairway. You first need to determine how high you are climbing, how many seconds it takes you to do it, and your weight. Then substitute the data in this formula:

$$hp = \frac{wh}{550t}$$

w is your weight in pounds, h is how high you are climbing in feet, and t is the time in seconds. For example, a man who weighs 180 pounds runs up an 8-foot stairway in 2½ seconds. This figures out to 1.05 horsepower:

$$hp = \frac{(180)\ (8)}{(550)\ (2.5)}$$

$$hp = \frac{1440}{1375}$$

$$hp = 1.05\ hp$$

How fast would *you* have to run up an 8-foot stairway to develop 1.05 horsepower? If your weight is less than 180 pounds, you will have to do it in less than 2½ seconds to produce the same amount of power.

List of Terms

block and tackle
distance principle
effort
effort arm
first-class lever
foot-pounds
force
friction
fulcrum
horsepower
inclined plane
jackscrew
law of moments
lever
mechanical advantage
moment
physics
physicist
pitch
power
pulley
resistance
resistance arm
screw
second-class lever
simple machines
single fixed pulley
single movable pulley
third-class lever
watt
wedge
wheel and axle
windlass
work

Questions

1. A pair of tweezers is a combination of two ______________ class levers.
2. In using a simple machine to lift an object, a smaller force can be used than is required to lift the object without the machine, but the force must be applied over a greater ______________.
3. The highest mechanical advantage that can be realized using a single pulley to lift an object is ______________.
4. A doorknob is an example of the simple machine called the ______________.
5. A wheelbarrow is an example of a ______________-class lever.
6. Which machine would be most effective for lifting a house?
7. Multiplying force by distance gives a quantity called ______________.
8. The jackscrew is a combination of the ______________ and the ______________.
9. The wedge is a combination of two ______________.
10. The mechanical advantage of the human forearm is ______________ (less than, equal to, greater than) one.
11. Dividing work by time gives a quantity called ______________.
12. The unit that is defined as 550 foot-pounds of work per second is the ______________.

Problems

(Ignore Friction in All Problems)

1. A force of 100 pounds is needed to roll a barrel up an inclined plane which has a mechanical advantage of 4. What is the weight of the object?
2. A man pushes a 200-pound object with a force of 40 pounds, sliding it a distance of 15 feet. How much work is accomplished?
3. A windlass has a handle 4 feet long and an axle with a radius of 4 inches. How much force is needed to lift a load of 1000 pounds using this windlass?

4. A ramp on a moving van has a length of 18 feet and a height of 5 feet. What is the mechanical advantage of the ramp?
5. A force of 10 pounds is needed to roll an object up an inclined plane which has a mechanical advantage of 4. The weight of the object must be ______________.
6. A wheel and axle arrangement has a wheel radius of 20 cm and an axle radius of 8 cm. Its mechanical advantage is therefore ______________.
7. A 40-pound boy is sitting 8 feet from the fulcrum of a see-saw. How far from the fulcrum should a 64-pound boy sit to exactly balance the see-saw?
8. A crowbar 6 feet long is used to lift a rock weighing 600 pounds. The fulcrum is located 6 inches from the rock and the force is applied at the other end. Calculate the force which needs to be applied to lift the rock.
9. If you can lift 1000 pounds 3 feet with an effort of 200 pounds with the aid of a machine, you will have to work through a distance of ______________.
10. What is the mechanical advantage of the pulley system shown in Figure 24:31?
11. A 200-pound barrel is rolled up an inclined plane to a height of 9 feet, using a force of 40 pounds. How long is the inclined plane?
12. Assume that the biceps attaches to the forearm 1 inch from the elbow joint, and that a 5-pound weight is held in the hand at a distance of 18 inches from the elbow joint. In order to lift the weight, the biceps must pull on the forearm with a force of ______________. (Ignore the weight of the arm itself.)
13. A tack remover is used to pull a tack out of a board (see Figure 24:32). Using the dimensions shown, find how much force is applied to the tack when the effort is 10 pounds.
14. How much work is done by a person pushing against a brick wall with a force of 35 pounds for 2 minutes? How much power is developed?
15. How much power is developed by an elevator that lifts three 200-pound men to a height of 50 feet in 8 seconds?

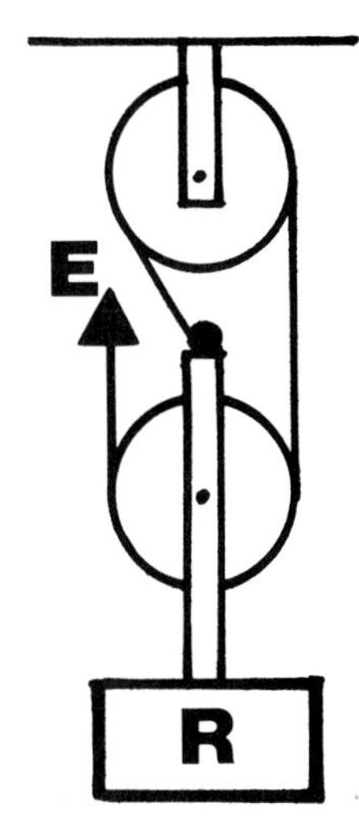

24:31 *Pulley arrangement referred to in Problem 10.*

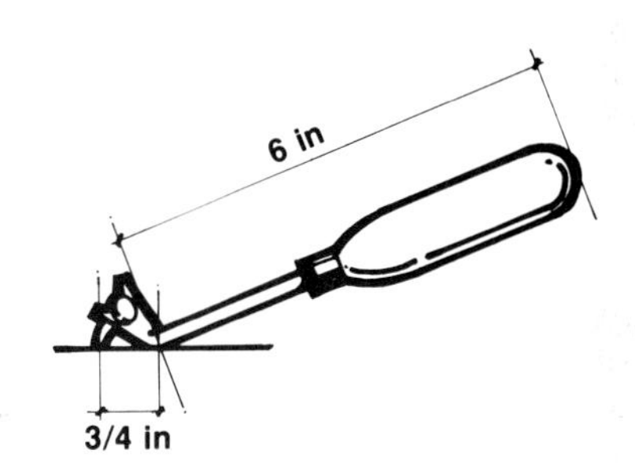

24:32 *Tack remover referred to in Problem 13.*

Student Activities

1. Make a list of all the simple machines you can identify in an automobile.
2. Prepare a report on the life and work of Archimedes.
3. Make a study of the various job opportunities in physics. Your local librarian should be able to help you find plenty of material.
4. Measure the horsepower developed by a wind-up toy car or tractor as it travels up an inclined plane. Use the following procedure:
   (a) Determine the weight of the car (w) in pounds.
   (b) Measure the height of the inclined plane (h) in feet.
   (c) Determine the time in seconds (t) for the car to travel the length of the inclined plane.
   (d) Calculate the horsepower of the car by the formula

$$hp = \frac{wh}{550t}$$

Photo next page
In accordance with Newton's second law, all excess mass has been eliminated from this hypersonic research aircraft to achieve the maximum possible acceleration.

U S AIR FORCE
USAF

UNIT VIII MECHANICS

Chapter 25

Newton's Laws of Motion; Mechanical Energy

25:1 A Basic Misconception While some of the ancients were quite capable of dealing with certain kinds of physical problems, they were stumped when it came to the problem of motion. One major error in their thinking, an error which persisted through the Middle Ages, was the assumption that it was natural for objects in motion to slow down and stop if left alone. Galileo Galilei (1564-1642), an Italian physicist, challenged this assumption. On the basis of his experiments, he instead chose to believe that it was natural for objects in motion *to keep on moving indefinitely* until they are slowed or stopped by friction. Galileo's radical idea opened up a whole new area of physics; several decades later Sir Isaac Newton (1642-1727) incorporated this idea into his first law of motion.

25:2 Newton's First Law of Motion *Objects at rest tend to remain at rest unless acted upon by an unbalanced force. Objects in motion tend to remain in motion in the same direction and at the same speed unless acted upon by an unbalanced force.* This is Newton's First Law of Motion.

A **force**, you will recall, is a push or a pull. By an **unbalanced force,** we mean *one that does not have another force of equal strength acting against it.* Suppose two people are pulling on you; one person is pulling on your right arm, while the other person is pulling on your left arm. You are caught in the middle of a tug-of-war. As long as both people pull with equal force, you will not move—the forces are balanced. But suppose one person lets go;

25:1 *Sir Isaac Newton, probably the greatest genius in the history of science. His three laws of motion are covered in this chapter, his law of gravitation in the next chapter.*

you are then subjected to an unbalanced force, and you quickly move toward the person who is still pulling.

(a)

25:2 *Balanced forces (a) and an unbalanced force (b).*

(b)

If you have ever tried to push a car using your own muscles to supply the power, you have experienced a good demonstration of Newton's first law of motion. It is much more difficult to get the car moving than it is to *keep* it in motion—*objects at rest tend to remain at rest.* Once you have the car moving, however, you find that it is difficult to stop the car—*objects in motion tend to remain in motion.* Notice that Newton's first law of motion does not "explain" what is happening; the law only *describes* what is happening.

25:3 *An object at rest tends to remain at rest.*

Sometimes the first law of motion is referred to as the *law of inertia*. The word **inertia** means *laziness*, *sluggishness*, or *unwillingness to change*. An object at rest is said to exhibit inertia of rest; an object in motion is said to have inertia of motion.

You have undoubtedly noticed that when you are in a car that starts rapidly, you are thrown back into your seat (your inertia of rest). When the car stops suddenly, you are thrown forward (your inertia of motion). When the car rounds a bend in the road, you are thrown toward the outside of the curve (your inertia of motion tends to keep you moving in the same direction).

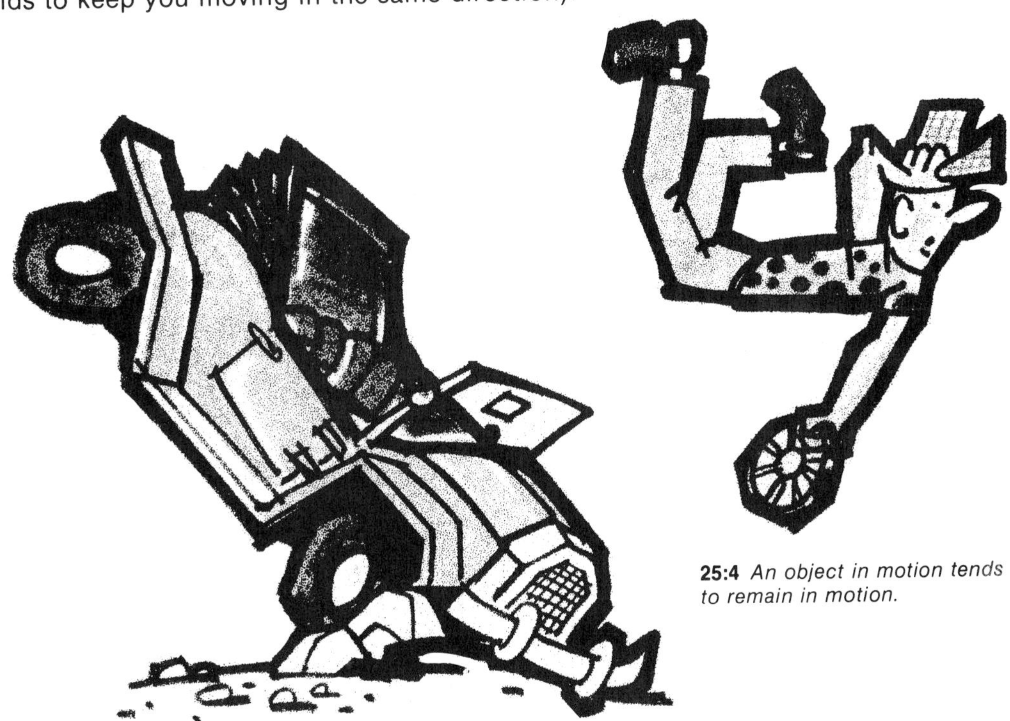

25:4 *An object in motion tends to remain in motion.*

25:5 *A popular fad of bygone days—grabbing pennies from the elbow. Inertia of rest keeps the pennies in approximately the same place for a brief instant before they fall.*

At one time it was a popular fad to see how many pennies a person could grab from his elbow. The arm is bent, as shown in Figure 25:5, and pennies are stacked on the elbow. With a sudden motion, the hand grabs for the pennies. Inertia of rest keeps the pennies in approximately the same place for a brief instant before they fall.

The little demonstration shown in Figure 25:6 also illustrates inertia of rest. A postcard is placed on top of an inverted glass; then a nickel is placed on top of the card. The card can easily be snapped out from under the coin with your index finger, leaving the coin in place.

You may know someone who has tried to pull the tablecloth out from under the dishes of an elaborately set table. The idea behind this stunt may be "scientific," but it rarely works. Friction between the tablecloth and the dishes will (more often than not) carry at least some of the dishes off the table and onto the floor.

Before the sixteenth century, most people believed that the earth was stationary. If the earth moved, they reasoned, we should be able to observe some definite effects of that motion. For instance, if a person jumped a foot off the ground, he should come down in a different place, because while he was in the air, the earth would have shifted its position under him. You know from experience, however, that you come down in the same place. How can we understand this?

Newton's first law of motion makes this easy to understand. The earth is moving, but so is the person; and the person is moving in the same direction as the earth. The person is also moving at the same speed as the earth. Therefore, his inertia of motion will *appear* to be keeping him right in the same place.

The cart-and-ball demonstration shown in Figure 25:7 illustrates this principle. As the cart moves forward at constant speed, a spring release suddenly projects the ball upward. You might expect that the ball would land *behind* the cart. But in actuality, the ball lands right back in the cart because it has the same forward motion as the cart.

25:6 *This card can be snapped out from under the coin with the index finger. Inertia of rest keeps the coin in place.*

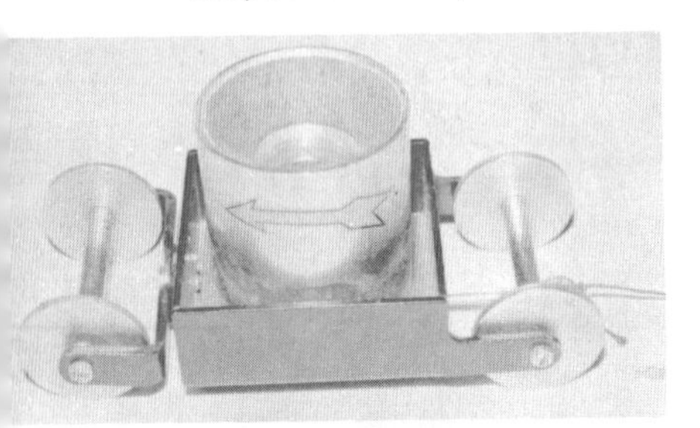

25:7 *Cart for demonstrating Newton's first law. While the cart is in motion a spring projects a steel ball straight up into the air. Because it retains the cart's forward motion while in the air, the ball lands back in the cart.*

25:8 *If the boy throws the ball straight up, it will land back in his hands—assuming no wind resistance. The explanation for this is the same as in Figure 25:7.*

25:3 Newton's Second Law of Motion An important quantity connected with motion is **velocity**. "Velocity" means *how fast something is moving*, or, stated in more technical terms, *the change in location in a given period of time*. Velocity means almost the same thing as **speed**, except that velocity also implies the idea of *direction*. A speed might be given as "35 miles per hour," whereas a velocity would have a more specific designation: "35 miles per hour due west."

The physicist calls quantities that have both size (magnitude) and direction associated with them vector quantities, or **vectors**. Those quantities with no direction, such as mass and density, are called scalar quantities, or **scalars**.

Velocity (v) is computed by dividing distance by time:

$$v = \frac{d}{t}$$

The small *d* represents distance and *t* represents time. If *d* is in miles and *t* is in hours, the velocity will be in miles per hour. If *d* is in meters and *t* is in seconds, the velocity will be in meters per second. Units that are commonly used to designate velocity are miles per hour, miles per second, feet per second, meters per second, centimeters per second, and kilometers per second.

Acceleration is another important quantity connected with motion. Acceleration means *the change in velocity in a given period of time*. Acceleration is a vector quantity; its direction can be forward (when speeding up), backward (when slowing down), or sideways (when turning).

For an object which starts at rest and then speeds up, we can calculate the acceleration (a) in any given length of time by dividing the velocity (v) it is able to attain by the length of time (t) it takes to attain it:

25:9 *In accordance with Newton's second law, all excess mass has been eliminated from this hypersonic research aircraft to achieve the maximum possible acceleration.*
(See chapter title page)

$$a = \frac{v}{t}$$

Suppose that a soapbox derby racer starting from rest is able to reach a velocity of 20 feet per second in 5 seconds. His acceleration is

$$a = \frac{v}{t} = \frac{20 \text{ ft/sec}}{5 \text{ sec}} = 4 \text{ ft/sec}^2$$

The average acceleration for the 5-second interval is 4 feet per second *each second*. This can be expressed as *4 feet per second per second*, or, more simply, *4 feet per second squared*, as shown in the answer.

We are now ready to state Newton's second law of motion: *Force equals mass times acceleration*, or, written as an equation,

$$F = ma$$

F is the unbalanced force applied to an object, *m* is its mass, and *a* is its acceleration. Mass, you may recall, is a measure of the amount of material in an object.

From this equation we can derive two very logical predictions:

1. The harder an object is pushed by an unbalanced force, the more it will accelerate.
2. The more massive a given object is, the less it will accelerate with a given applied force.

Model rocket enthusiasts are aware that there are two main ways that the acceleration of a rocket can be improved—by *increasing the force* or by *decreasing the mass.* Increasing the force can be accomplished by using a larger engine; decreasing the mass can be accomplished by eliminating any weight in the rocket that is not absolutely necessary.

Try this numerical example using Newton's second law of motion:

Problem: How much force is needed to give a football, which has a mass of 0.40 kilograms, an acceleration of 10 meters per second squared?

Solution: When the mass is in kilograms and the acceleration is in meters per second squared, the force will come out in **newtons**, a unit named in honor of Sir Isaac Newton, the man who originated the laws we are discussing. A newton is equal to 1 kg-m/sec^2. Substituting in the equation:

$$F = ma$$
$$F = (0.40 \text{ kg})\ (10 \text{ m/sec}^2)$$
$$F = 4.0 \text{ newtons or } 4.0 \text{ N}$$

A force of 4.0 newtons must be applied to the football to give it an acceleration of 10 m/sec^2. A newton is equal to about a quarter of a pound of force; therefore, four newtons is roughly equivalent to one pound.

25:10 *A scientific law is demonstrated when a basketball is thrown. According to Newton's second law, the acceleration of the ball will be equal to the total unbalanced force applied to it divided by its mass.*

Experiment 25:1 *Newton's Second Law of Motion*

In this experiment we shall establish the workability of the relationship $F = ma$. Solving the equation for acceleration (a), we obtain $a = F/m$. This predicts that the acceleration should increase if the force is increased and decrease if the mass is increased.

1. Place a 1-kilogram weight on a roller skate.
2. Pull the roller skate across a level surface using a spring balance. Try to pull it in such a way that the force remains constant.
3. Repeat this procedure using a larger force, again keeping the force constant as you pull.
4. Repeat the procedure again using a still greater force. Note that the acceleration increases as the force is increased.
5. Try different weights on the roller skate, keeping the force indicated on the spring balance the same for each weight. Observe how little acceleration is realized when larger weights are used.

25:4 Newton's Third Law of Motion *For every action there is an equal and opposite reaction;* this is Newton's third law of motion. If you have ever tried jumping from a rowboat, you may have noticed that the boat moved back somewhat as you moved forward. Your forward motion could be considered the "action"; the backward motion of the boat would then be considered the "reaction."

25:11 *An illustration of Newton's third law (action and reaction). Jumping from a rowboat causes the boat to move backward.*

A toy balloon that is blown up and released to fly around the room is another demonstration of the third law of motion. The air rushing out of the balloon is the "action"; the motion of the balloon itself is the "reaction." The jet airplane and the rocket illustrate the same principle, but on a much larger scale.

25:12 *If a balloon is blown up and released, its behavior furnishes a demonstration of Newton's third law.*

25:13 *X-15 hypersonic research aircraft shown immediately after release from B-52 mother plane. Both planes operate on the principle of action and reaction (Newton's third law).*

The rotary lawn sprinkler is another common example of action and reaction. Water leaving the sprinkler is the "action"; the backward turning of the sprinkler is the "reaction." Still another example is the recoil of a rifle or cannon. Note again that the devices we have mentioned do not work "because of" Newton's third law. This law, like every scientific law, is only a *description* of what happens.

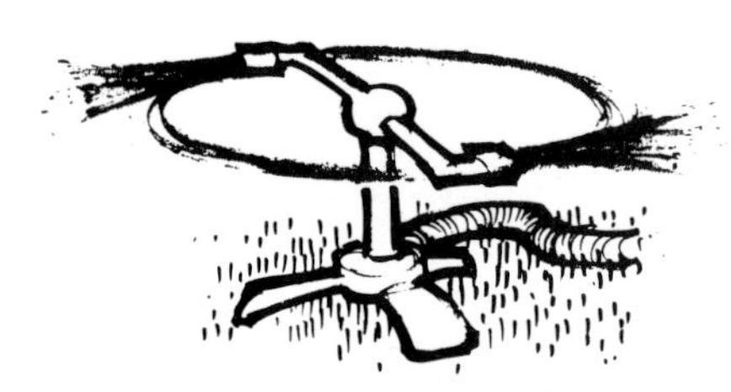

25:14 *The rotary lawn sprinkler offers still another application of the action-reaction principle.*

25:15 *Laboratory carts used to demonstrate Newton's third law. A spring-operated plunger causes the carts to push apart.*

There are many examples of the third law in which nothing moves. When you sit in a chair, your body presses down against the chair with a certain number of pounds of force. However, the chair pushes up against *you* with exactly the *same* force. If the chair were unable to push back with the same amount of force, it would collapse. This is just as much an action-reaction pair as the examples given in which something moves.

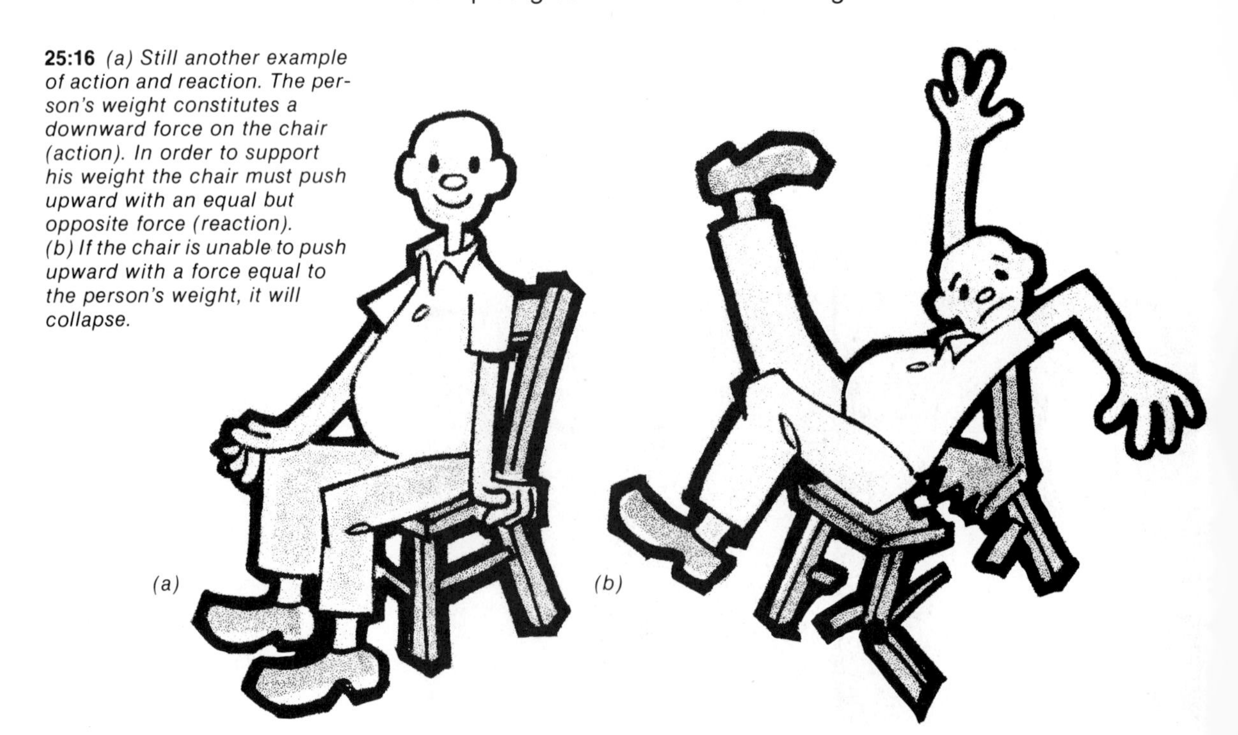

25:16 *(a) Still another example of action and reaction. The person's weight constitutes a downward force on the chair (action). In order to support his weight the chair must push upward with an equal but opposite force (reaction). (b) If the chair is unable to push upward with a force equal to the person's weight, it will collapse.*

(a) *(b)*

25:5 Momentum An important concept connected with motion is momentum. "Momentum" is defined as *mass times velocity*. Expressed as an equation

$$p = mv$$

The p is momentum, m is mass, and v is velocity. Like velocity, momentum is a vector quantity—a quantity having a direction associated with it.

25:17 *A slowly moving truck can easily possess more momentum than a speeding motorcycle.*

A slowly moving truck can easily have more momentum than a speeding motorcycle. For example, an 8-ton truck moving at 5 miles per hour has *twice as much momentum* as a 1/4 ton motorcycle traveling at 80 miles per hour! For the motorcycle's momentum to equal the truck's momentum, the motorcycle would have to be moving at 160 miles per hour!

Mass, then, is just as important a factor in momentum as velocity. To illustrate this more clearly, let us compare the action of a bowling ball knocking down pins to the action of a volleyball traveling at an equal speed and aimed at the same pins. The volleyball will bounce away after knocking down only one or two pins; it is simply too light. And because of its low mass, the volleyball has little momentum.

25:18 *The concept of momentum can be appreciated by comparing the effectiveness of a bowling ball with that of a volleyball for knocking down pins.*

25:6 Conservation of Momentum Newton's third law of motion predicts that a rifle should recoil as a bullet is fired from it. The "action" and "reaction" here are, in reality, the momentum of the bullet and the momentum of the rifle. The *mv* product for the bullet must equal the *mv* product for the rifle; this is known as the **law of conservation of momentum**:

$$m_1v_1 = m_2v_2$$

The mass of the bullet is m_1; v_1 is the velocity of the bullet; m_2 is the mass of the rifle; and v_2 is the velocity of the rifle.

In normal use, some of the recoil of the rifle is absorbed by the person holding it. To see what its full recoil speed would be, the rifle should be suspended from the ceiling by cords as shown in Figure 25:19.

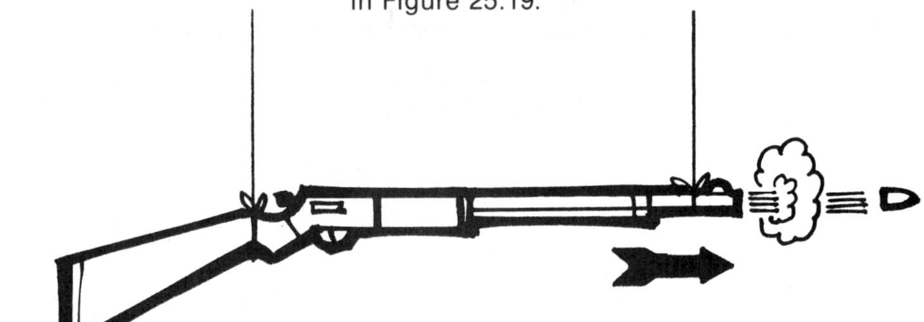

25:19 *The recoil of a rifle illustrates conservation of momentum.*

Problem: Assume that a bullet having a mass of 0.002 kg is fired from a rifle having a mass of 2.5 kg. If the bullet leaves the gun at 350 m/sec, how fast does the gun recoil?

Solution: Solve the equation for v_2 and substitute the values given:

$$v_2 = \frac{m_1v_1}{m_2} = \frac{(0.002\text{ kg})\ (350\text{ m/sec})}{2.5\text{ kg}}$$

$$v_2 = \frac{.7}{2.5} = 0.28\text{ m/sec}$$

The recoil velocity is 0.28 m/sec (or 28 cm/sec) directed exactly opposite to the direction of the bullet.

Not only does the law of conservation of momentum apply to objects moving away from each other, such as the rifle and bullet, but it also applies to *colliding* objects. The total momentum of all objects taking part in the collision must be the same before and after the impact.

The following demonstration is best performed with large steel balls on a metal track. Place six balls in a row in the track, making sure that they are all in contact. Roll a single ball from the left toward the group of six. As this ball hits the first ball of the group of six, it stops dead. Simultaneously, a single ball leaves from the right-hand end of the group. The other five balls do not appear to move at all. Remarkably, they are able to *transmit* the motion to the ball that *does* move. When two balls are rolled in from the left, two balls leave from the right. Three balls entering from the left will release three balls from the right, four will release four, five will release five, and six will release six. But what happens when seven moving balls hit the six stationary ones? In this case, one of the seven moving balls keeps moving. Added to the six balls that start from the rest, this makes seven in motion after the impact, giving the same momentum as before the impact.

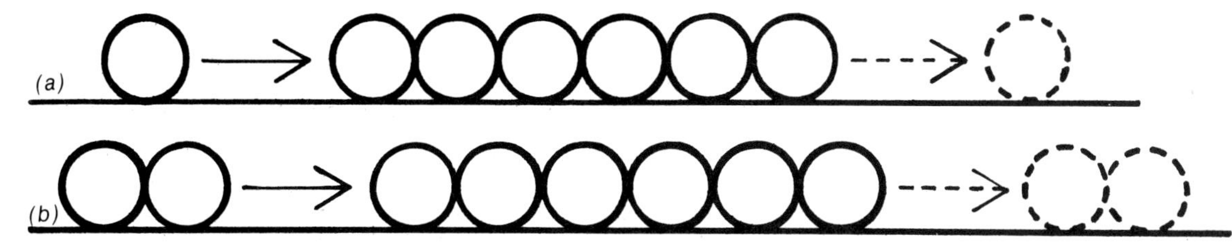

25:20 *(a) Conservation of momentum. When one steel ball is rolled into a group of six balls, all in contact with each other, one ball leaves from the far end. (b) When two balls are rolled into the group, two leave from the far end.*

The above demonstration uses objects which all have the same mass. What happens with *unequal* masses? In general, when two objects of unequal mass collide, the lighter object ends up moving faster than the heavier object. Suppose a "100-pound weakling" collides with a 250-pound football player who is running in the opposite direction. Assuming that they were moving at the same speed before the collision, the weakling's speed after the collision will be considerably greater than that of the football player.

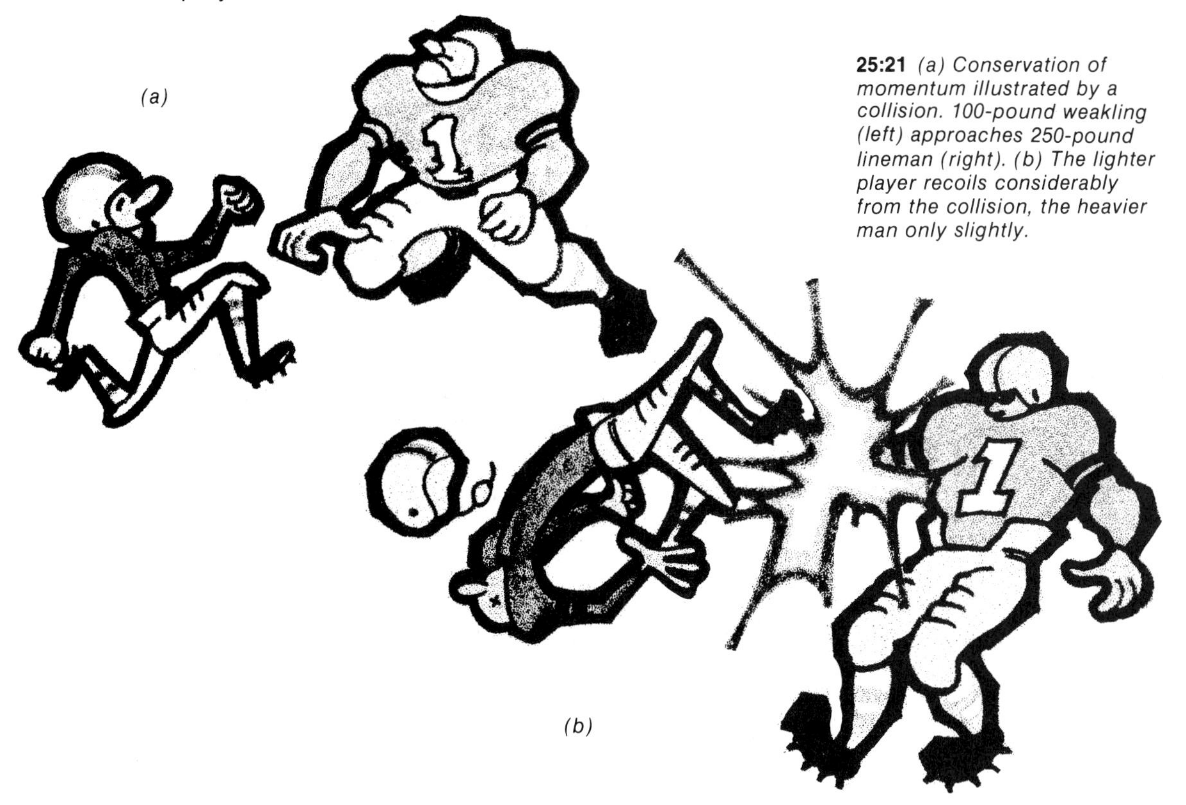

25:21 *(a) Conservation of momentum illustrated by a collision. 100-pound weakling (left) approaches 250-pound lineman (right). (b) The lighter player recoils considerably from the collision, the heavier man only slightly.*

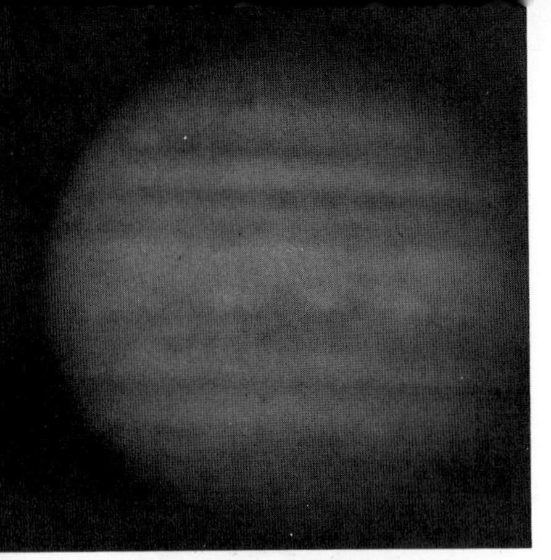

25:22 *The massive planet Jupiter makes one rotation every ten hours. The angular momentum due to its rotation is therefore very great.*

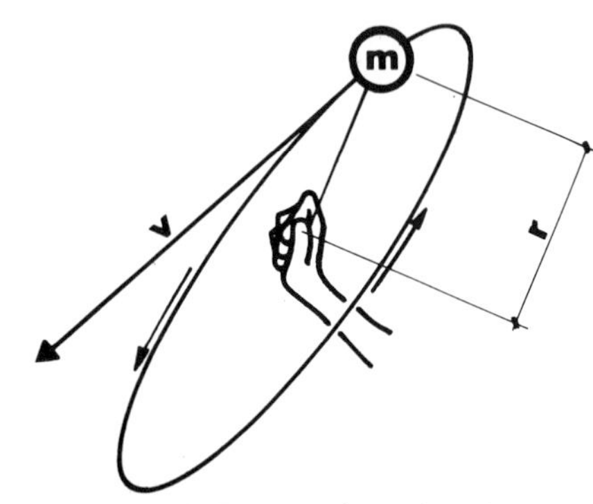

25:23 *The angular momentum of a point with mass m describing a circle of radius r with a velocity v is computed by the product mvr.*

25:7 Conservation of Angular Momentum Objects that are undergoing circular motion possess a type of momentum called **angular momentum**. Turning wheels, satellites in orbit about the earth, and spinning tops all possess angular momentum in varying amounts.

Different formulas are used to compute the angular momentum for different-shaped objects and their different ways of turning. The simplest situation is the circular motion of a **point mass.** Picture a stone on a string being swung around in a circle. The angular momentum in this case is *mvr*—mass times velocity times the radius *(r)* of the circle.

Angular momentum is conserved in isolated systems, that is, any system that is not exchanging matter or energy with its surroundings. Suppose you get the stone swinging around fairly rapidly; then with your other hand you take hold of the loose end of the string and pull it, shortening the radius to half its original length. What happens to the velocity of the stone? Since the *mvr* product must remain the same as before and the mass cannot change, the velocity must double. In this way, angular momentum is conserved. Interestingly, you have speeded up the stone without supplying any additional energy to it.

Conservation of angular momentum is also demonstrated by a person standing on a rotating platform holding weights with outstretched arms (see Figure 25:24). Suppose someone starts him turning slowly. Whatever his rate of rotation at first, the rate of rotation will greatly increase when he suddenly pulls the weights in toward his body. Figure skaters often use this principle to add effect to their performance.

25:24 *Conservation of angular momentum. The person's speed of rotation increases as he pulls the weights toward his body.*

The angular momentum principle has been an important one in the history of astronomy. Many evolutionary theories of how planets and stars were formed have had to be discarded because they failed to meet the demands of this law. The main problem is the evolutionists' insistence that large clouds of interstellar material (dust and gas) condensed and contracted into stars and planets. This has never been observed. There are many other difficulties connected with this thought, both scientific and Scriptural. But if the contraction *could* ever take place, the object being formed would end up spinning so rapidly that it would be unlike any of the known stars and planets that we observe. The law of conservation of angular momentum, therefore, is strongly opposed to evolutionary theories which say that the stars and planets were formed by unguided forces. Creation by direct acts of God is still the most logical answer.

25:25 *The principle of conservation of angular momentum has proved disastrous for many theories of the origin of the solar system. The sun is rotating far too slowly to have contracted from a cloud of dust and gas as is often claimed.*

25:8 Kinetic Energy **Energy** in general is defined as *the ability to do work.* Energy of motion is called **kinetic energy**. Kinetic energy is calculated by multiplying one-half times the mass of the object in motion times its velocity squared:

$$K.E. = \frac{1}{2}mv^2$$

A car traveling at 60 miles per hour has *four times* as much kinetic energy as a car moving at 30 miles per hour.

Now you can begin to understand why the distance required to stop a car moving at 60 miles per hour is so much greater than the distance needed to stop a car moving at 30 miles per hour. The additional energy has to "go somewhere." It is normally converted to heat in the brakes. If a car comes to a skidding halt, some of the heat is developed in the tires and on the road.

Kinetic energy is a measure of the work that can be accomplished by a moving body as it comes to rest. The kinetic energy of a moving hammer is a measure of how much work the hammer can do in driving a nail into a board. The kinetic energy of a speeding bullet is a measure of how much "work" the bullet can accomplish in damaging its target.

25:9 Potential Energy Energy that an object has because of its position or condition is called **potential energy**. Potential energy is stored energy. A skier at the top of a hill, because of his position, has potential energy. This potential energy can be converted into kinetic energy as he skis down the hill. This is an example of potential energy because of position.

Winding a wristwatch gives potential energy to its mainspring, simply because of its wound condition. This potential energy is converted back into kinetic energy as the mainspring drives the gears, the hairspring, and the hands of the watch. In this case, we have an example of potential energy because of condition—the state of strain on the mainspring. Flexible materials, such as spring steel and rubber, can be deformed in such a way that they store energy.

Kinetic energy and potential energy can be readily converted back and forth. A swinging pendulum is constantly exchanging kinetic and potential energy. At the top of its swing, the pendulum's energy is all potential. At the lowest point of its swing, the pendulum's energy is all kinetic. For points in between, some of the energy is kinetic, some potential.

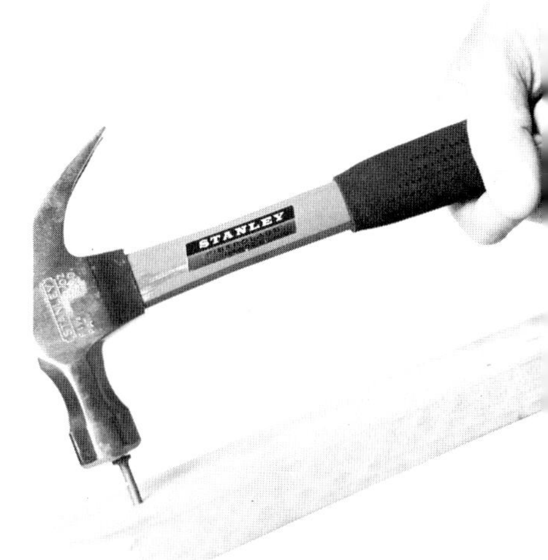

25:26 *The kinetic energy of a moving hammer is a measure of how much work it can accomplish in driving a nail into a board.*

25:27 *Shasta Dam and Reservoir in California. The water in the reservoir possesses potential energy—energy stored by virtue of its position.*

Potential energy because of an object's position is calculated in the English system by multiplying the weight of the object by the height to which it is carried:

$$P.E. = wh$$

The weight in pounds is w, and h is the height in feet. Multiplying pounds by feet gives us foot-pounds.

You have already encountered the unit known as the foot-pound in Chapter 24 when we calculated the work done by a simple machine. It is not by accident that these units are the same. Since energy is the ability to do work, both *work* and *energy* must have the same units. In the English system the *foot-pound* is the most commonly used unit; in the metric system the **erg** and the **joule** are used. The erg is a very tiny unit; it is *the amount of work done when a force of one* **dyne** (1/100,000 of a newton) *acts through a distance of one centimeter.* The joule is *the work done when a force of one newton acts through a distance of one meter.* The joule is by far the larger of the two units, being equal to 10 million ergs.

Suppose you want to calculate the potential energy of a 10-pound bag of sugar located on a table 3 feet high. Substituting in the potential energy equation:

$$P.E. = wh \quad = \quad (10 \text{ lb})\,(3 \text{ ft}) \quad = \quad 30 \text{ ft-lb}$$

The bag of sugar has 30 foot-pounds of potential energy *with respect to the floor.* You must always specify what you are using as your base level, or reference level. The answer would be entirely different if you specified potential energy *with respect to street level* or *with respect to sea level*.

It is interesting to note how much work it would take to lift the bag of sugar from the floor to the table. Work is computed as force times distance. It takes 10 pounds of force to lift a 10-pound object. You are lifting it 3 feet; therefore

$$W = Fd = (10 \text{ lb})(3 \text{ ft}) = 30 \text{ ft-lb}$$

You get the same answer for the work done as you did for potential energy. It is only logical that the work required in lifting an object should be the same as the energy stored in the object.

Potential and kinetic energy are both forms of **mechanical energy**. Later we shall discuss heat, electricity, magnetism, light, sound, and nuclear energy, as well as some of the ways in which energy is converted from one kind to another.

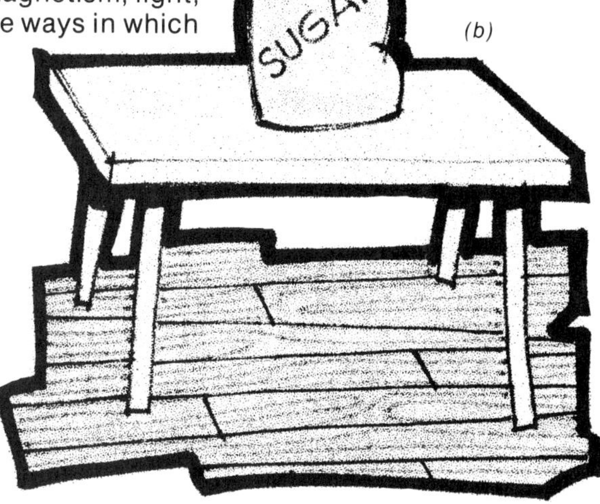

25:28 *It takes 30 foot-pounds of work to lift a 10-pound bag of sugar from the floor (a) to a table 3 feet high (b). When on the table the sugar possesses 30 foot-pounds more potential energy than when on the floor.*

Experiment 25:2 *Kinetic and Potential Energy*

1. Lift a 1-kilogram weight (equipped with a hook) from the floor to a table with a spring balance. Record the force (in pounds) used in lifting the object.
2. Measure the distance from the floor to the table top (in feet). Record.
3. Calculate the potential energy imparted to the weight by the relationship $P.E. = wh$.
4. Place a piece of cardboard on the floor below the weight. Push the weight off the edge of the table and estimate the speed with which it strikes the cardboard.

Just before the weight hits the floor, all of its potential energy will have been converted to kinetic energy (an illustration of the principle of conservation of energy): $P.E. = K.E.$ Since we know the potential energy, we also know the kinetic energy. We can, therefore, compute the speed of impact by the equation $K.E. = \frac{1}{2} mv^2$. To do this, we must first know the mass of the weight in slugs. There are 32 pounds in one slug. Use the value thus obtained for m in the equation and solve for v.

List of Terms

acceleration
angular momentum
dyne
energy
erg
force
inertia
joule
kinetic energy
law of conservation of momentum
mechanical energy
momentum
newton
point mass
potential energy
scalar
speed
unbalanced force
vector
velocity

Questions

1. Multiplying force by distance gives a quantity called ________.
2. Newton's first law of motion is sometimes called the law of ________.
3. Change in location in a given period of time is called ________.
4. ________ is defined as a push or a pull.
5. Change in velocity in a given period of time is called ________.
6. Energy of motion is ________ energy.
7. In calculating with Newton's second law of motion, when the mass is expressed in kilograms and the force in newtons, the acceleration must be in ________.
8. Stored energy is called ________ energy.
9. Momentum is defined as mass times ________.
10. ________ is defined as the ability to do work.
11. Units of force in the metric system are the newton and the ________.
12. Units of energy in the metric system are the erg and the ________.
13. Angular momentum for a point mass is computed by multiplying the mass times the velocity times the ________.
14. Which of Newton's laws of motion is involved in each instance?
    (a) A newspaper article stated that a car traveling at 125 miles per hour struck a tree and stopped, but the car's battery kept going another 200 feet.
    (b) Coal is shoveled into a furnace by thrusting the shovel into the furnace, then suddenly stopping the shovel.
    (c) A chandelier exerts a downward force on the chain from which it hangs; the chain exerts an upward force on the chandelier.
    (d) A bird is unable to fly away from a small branch without moving the branch.
    (e) A man, lying on his back with an extremely large stone on top of him, is unharmed when the stone is hit with a sledge hammer.
    (f) A car and a truck have identical engines. It is found that the car can accelerate more rapidly than the truck.
    (g) Seat belts have been found to reduce the seriousness of injuries in automobile accidents.
    (h) The world's record for the running broad jump is greater than the record for the standing broad jump.
    (i) Two rockets have different engines but identical overall masses. The one with the larger engine exhibits the greater acceleration.
    (j) It is difficult for a person on roller skates to remain stationary while throwing a heavy object.

(k) All unnecessary hardware is thrown overboard during a prolonged space flight.

(l) Bombs are released *before* a bomber is directly over its target area.

Problems

1. Calculate the force necessary to give a 4-kilogram rock an acceleration of 18 meters/sec^2.
2. How much potential energy is gained by a 150-pound person who takes an elevator to the roof of a 200-foot building?
3. Find the acceleration of a bicycle that starts from rest and reaches a speed of 22 ft/sec in 8 seconds.
4. Calculate the acceleration imparted to a 750-gram mass by a 5-newton force.
5. How much work is done when a 2-pound book is raised from the floor to a shelf 5 feet above the floor? How much potential energy is imparted to it?
6. A 6-kilogram doll carriage rolls down a sidewalk at 2.3 meters per second. Find the momentum of the doll carriage.
7. A 100-pound boy on roller skates throws a 15-pound weight straight ahead at 12 ft/sec. With what speed does he recoil? (Ignore friction. Also, note that weight can be used instead of mass in this case because weight appears on both sides of the equation.)
8. A 1-kg stone at the end of a string 3 meters long is swinging in a circle at a velocity of 4 meters/sec. Find the new velocity when the string is increased to 1.25 times its original length.
9. Two identical airplanes travel at speeds of 640 miles per hour and 520 miles per hour, respectively.
   (a) How many times greater is the momentum of the faster airplane than the momentum of the slower airplane?
   (b) How many times greater is the kinetic energy of the faster airplane than the kinetic energy of the slower airplane?
10. A bicycle starts from rest and accelerates at the rate of 2 ft/sec^2. How fast will it be traveling at the end of 8 seconds?
11. A man on horseback travels 6 miles due south in 1½ hours.
    (a) What is his speed?
    (b) What is his velocity?
12. A stone on a 3.50-meter string is swung in a circle at a speed of 2.00 meters per second. If the string is suddenly shortened to 2.30 meters, what will be the velocity of the stone?

Student Activities

1. Demonstrate conservation of momentum by experimenting with marbles or steel balls on a track as described in Section 25:6.
2. Prepare a report on Galileo Galilei or Sir Isaac Newton.
3. Demonstrate conservation of angular momentum by swinging a weight on a string, then suddenly shortening the string with your other hand. Observe the increased speed of the weight.

   If a rotating platform is available, try standing on the platform with outstretched arms, holding a 1-kilogram weight in each hand. Have someone start you turning slowly. After you have made two or three complete turns, pull the weights in to your body. Be prepared for a sudden increase in rotational velocity. Extend your arms again to slow down.

Photo next page
Earth photographed from the moon. The force of attraction between the earth and moon is proportional to the product of their masses, and inversely proportional to the distance between them squared.

UNIT VIII MECHANICS

Chapter 26 Gravitation

26:1 The Mystery of Gravitation As we noted in Chapter 5, science cannot tell us *why* an object falls to the ground. It is really quite mysterious how there can be an attraction between two objects that are not physically connected. How can a *pull* be transmitted through space? No immediate solution to this puzzle seems likely. Although we cannot tell just *what* the force or pull is, we can calculate how strong the force is and predict how a falling object should behave.

Galileo, the sixteenth-century Italian physicist mentioned in Chapter 25, was a pioneer in the field of **gravitation**. He experimented with spheres and cylinders rolling down an inclined plane, which "diluted" the effect of gravity so that its force could be more easily measured. Galileo also experimented with falling objects. Tradition says that he dropped various articles from the Leaning Tower of Pisa to investigate their rate of fall. As with many early traditions, we have no way to prove or disprove whether Galileo actually performed such an experiment. Some historians believe that a student of Galileo's, Vicenzo Viviani, performed the actual experiment. Regardless of who dropped the objects from the Leaning Tower, Galileo was able to arrive at a surprising number of right answers. His conclusions are especially noteworthy because he had inherited so much misinformation from the ancients. For example, the Greek philosopher Aristotle (384-322 B.C.) had proclaimed that heavier objects fall faster than lighter objects. This idea went unchallenged for many centuries. Galileo was able to disprove Aristotle's assertion in two ways—by a "thought experiment," and by an actual physical experiment.

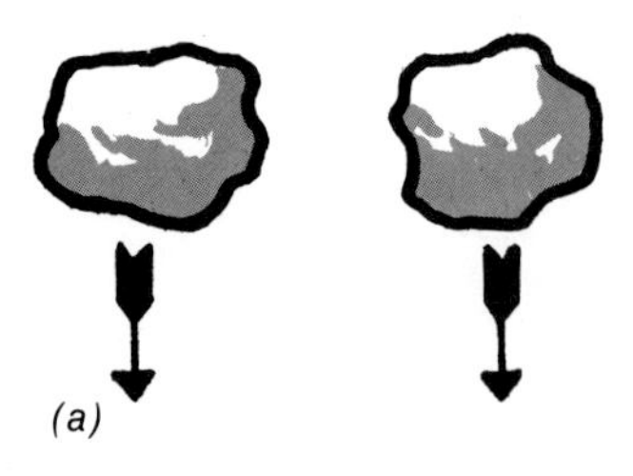

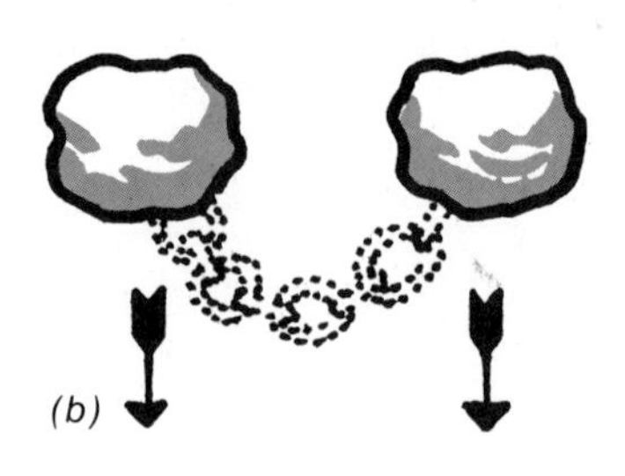

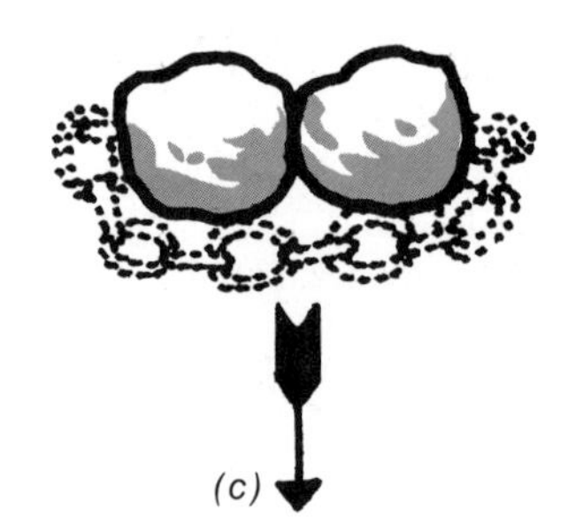

26:1 *Galileo's thought experiment to show that a 10-pound weight falls no faster than a 5-pound weight. He first envisioned two 5-pound weights falling side by side (a). He then imagined a weightless chain loosely binding the weights together as they fall (b). Finally, the chain is tightened (c) to form a 10-pound weight which still falls at the same rate as either of the 5-pound weights.*

Galileo's "thought experiment" is a notable example of clear thinking and originality. He imagined two 5-pound weights falling side by side. If the weights are released from the same height at the same time, they will strike the ground at the same instant (see Figure 26:1). Next, the experiment is repeated with the two weights connected by a loose chain. As the weights are repeatedly dropped, the chain is made shorter and tighter. Eventually, the two 5-pound weights are closely bound together, making, in effect, a 10-pound weight that still accelerates at the same rate as each of the separate 5-pound weights. Aristotle said that a 10-pound weight should fall twice as fast as a 5-pound weight. But Galileo's scientific insight told him that there should be no difference between them.

A simple experiment would have settled the question of whether a heavier object falls faster than a light object. But the early Greek philosophers avoided actual experimentation; they considered it beneath their dignity to work with their hands. (There were exceptions, of course, such as Archimedes.)

Galileo later tried dropping objects of various sizes, shapes, and weights. He found that there was virtually no difference in the way they fell, provided their weight was great enough compared to their surface area. You can prove this for yourself. Try dropping a heavy book and a pencil from the same height. They will fall side by side and land together. We can understand this by noting that the earth attracts the book more, and hence pulls harder on it. The book, however, has more inertia than the pencil and therefore has more resistance to acceleration. These two factors exactly offset each other so that the behavior of the pencil matches that of the book.

Now try dropping a pencil and a piece of notebook paper together. The pencil hits the floor first. In this case, something else enters into the picture—**air resistance**. The paper is greatly slowed down by the air because of the paper's small weight and large surface area. Now wad the paper into a ball and drop the paper and pencil together again. What happens now that the air resistance against the paper is decreased?

Although Galileo had no way to test his theory, he predicted that in such a case as the paper and the pencil, if the air could be removed, the two objects would fall at the same rate. Several years after his death, a vacuum pump was invented that enabled two such objects to be dropped in an evacuated tube (a container from which most of the air has been removed). Galileo was right; in the absence of air, the two objects behave identically. The astronauts tried this experiment on the surface of the moon and obtained the same result. A classic apparatus used to demonstrate this principle consists of a feather and a coin in a long, almost empty tube (see Figure 26:2). Many people are surprised when they see the feather keeping pace with the coin.

★ **26:2 Free Fall Relationships** Suppose you release a 1-inch diameter steel ball from the top of a tall building. Suppose also that a number of observers standing at the windows past which the steel ball will fall have the equipment with which to make accurate velocity measurements. At the end of 1 second, it is found that the steel ball is traveling at 32 feet per second. At the end of 2 seconds, its velocity is 64 feet per second, and at 3 seconds, 96 feet per second. Figure 26:3 is a graph of these data. You can see that the velocity is directly proportional to the time. That is, velocity increases as time increases. When the time is doubled, the velocity is doubled; when the time is tripled, the velocity is tripled, and so on. We can write this using a proportionality sign:

$$v \propto t$$

26:2 *A classic experiment. In an evacuated tube a feather and a coin fall at the same rate.*

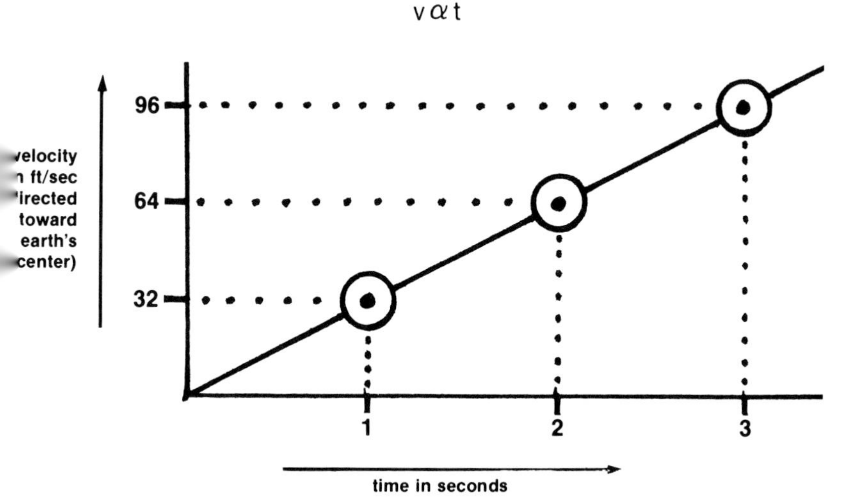

26:3 *Increase of velocity with time for a freely falling object.*

Or, we can write it as an equation replacing the proportionality sign with an equal sign and a **constant**:

$$v = gt$$

A constant, in general, is *a quantity that remains the same throughout a given calculation or experiment.* This is the opposite of a **variable**, *a quantity whose value can change during the course of a given calculation or experiment.*

In this case, we have a very special constant—the **acceleration due to gravity** at the surface of the earth, designated g. Its value is 32 ft/sec^2 in the English system, meaning that a body in **free fall** increases its velocity by 32 feet per second every second. (An object is said to be in "free fall" when we make the simplifying assumption that it experiences no air resistance.) In the metric system, g is 9.8 m/sec^2 or 980 cm/sec^2. These are standard values measured at sea level at a latitude of 45 degrees. There is a slight variation in g with latitude, as can be seen from the figures in Table 26-1. The higher the latitude, the greater the acceleration due to gravity. This is explained by the fact that the earth *bulges* at the equator and is *flattened* at the poles. The poles are, therefore, closer to the center of the earth than the equator, resulting in a greater gravitational pull at the poles (see Figure 26:4).

26:4 *The earth is not a perfect sphere. Objects at the poles are closer to the earth's center, so they experience a greater gravitational pull. A person who weighs 200 pounds at the North Pole would weigh 199 pounds at the equator.*

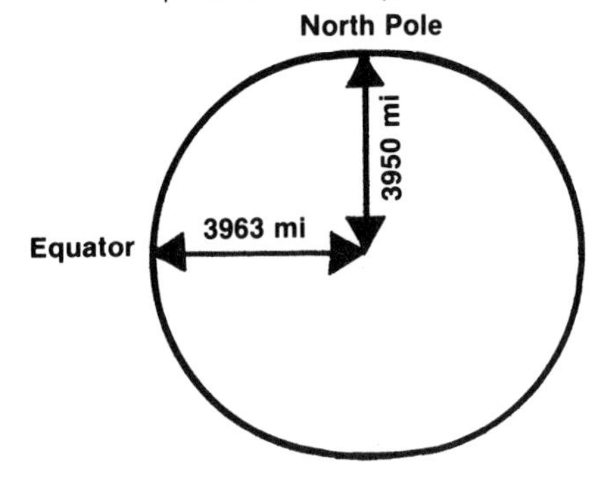

Table 26-1 *Acceleration Due to Gravity*

Location	g in m/sec^2
Equator, sea level	9.780
45° latitude, sea level	9.807
North Pole, sea level	9.832

There is also an altitude effect. The higher the altitude, the smaller the acceleration due to gravity. On top of Pike's Peak, which is 14,108 feet above sea level, the value of g is only 9.789 m/sec^2. Again, this is due to increased distance from the center of the earth.

Let us work out a relationship to describe how far an object falls in a given length of time. For simple experiments, time measurements are best made with a *metronome*—a device that "ticks" at a predetermined time interval. It is most convenient to set the metronome at 120 ticks per minute, or one tick every half-second. We will use a steel ball because of its high density and low surface area.

It is our desire to drop the steel ball in such a way that its free fall starts on one tick and ends on another. After some trial and error, we find that if we release the ball from a height of 4 feet on one tick, it strikes the ground on the next tick. Thus, an object starting at rest travels 4 feet the first half-second. With additional experimentation, we can determine that the distance is 16 feet for 1 second, 36 feet for 1½ seconds, and 64 feet for 2 seconds. We can display this information graphically, as shown in Figure 26:5. The curve is part of a **parabola**. However, the left-hand part of the parabola is non-existent because time cannot be negative.

26:5 *Total distance traveled by a body in free fall at various times. The curve is part of a parabola.*

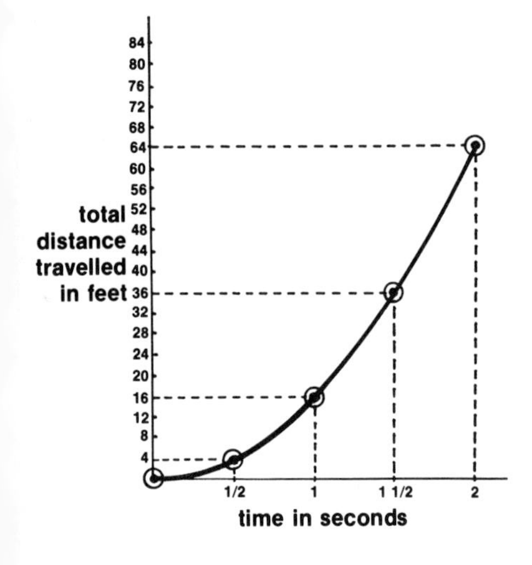

The parabola is always the *graph of a square relationship*. In this case, the distance is proportional to the time squared. We can express this with a proportionality sign:

$$d \propto t^2$$

Or, we can express it as an equation with a constant:

$$d = kt^2$$

The *k* is a constant which, with the equal sign, takes the place of the proportionality sign. The actual value of *k* can be found by substituting the experimental data from any point on the graph into the equation. For example, when *d* is 16 and *t* is 1

$$16 = (k)(1)^2$$

Therefore, *k* must be 16. The equation of the parabola, then, is

$$d = 16t^2$$

We can check it at the other points to see if the equation "works." When *t* is 1/2 second,

$$d = (16)(1/2)^2 = (16)(1/4) = 4 \text{ ft}$$

This does check with our experimental findings. When *t* is 1½ seconds,

$$d = (16)(3/2)^2 = (16)(9/4) = 36 \text{ ft}$$

Again, the result agrees. When *t* is 2 seconds

$$d = (16)(2)^2 = (16)(4) = 64 \text{ ft}$$

The equation, then, describes the relationship between distance and time quite satisfactorily. However, note carefully that it can be used only when the distance is in feet and the time is in seconds.

A more general form of the relationship is

$$d = \tfrac{1}{2}gt^2$$

Here, *g* is again the acceleration due to gravity. Since *g* is 32 ft/sec² in the English system, ½g will be 16 ft/sec², the same value we have been using. If we desire to obtain the distance traveled in meters, 9.8 m/sec² should be used for *g*. Time is expressed in seconds in both systems. This equation can be used only if the object starts from rest.

Problem: How far will an object fall in 6 seconds? How fast will it be traveling at the end of that time? (Assume free fall.)
Solution: To find the distance covered, we can use $d = 16t^2$:

$$d = (16)(6)^2 = (16)(36) = 576 \text{ ft}$$

The velocity at the end of 6 seconds is found by $v = gt$:

$$v = (32)(6) = 192 \text{ ft/sec}$$

You should remember that the above relationships do not take air resistance into consideration. As an object continues to fall over a longer period of time, air resistance increases because of the increasing velocity. The retarding force experienced by the falling body keeps increasing until it finally equals the weight of the object. The downward force is then equal to the upward force, and the object falls at a constant velocity called the **terminal velocity**. Long before the terminal velocity is reached, however, the simple equations given above will have ceased to describe accurately what is happening.

26:3 The Pendulum It is said that Galileo first became interested in studying the pendulum as he watched a chandelier swinging back and forth in a cathedral in Italy. He noted that the **period of the pendulum** (how long it takes to swing back and forth to make one complete cycle) was the same regardless of whether the chandelier was making large swings or small swings. Later, Galileo made experimental pendulums of different weights; these may have been no more than weights hanging on pieces of string. Interestingly, he found that the weight of the pendulum does not affect its period. The only variable that seemed to matter was the *length* of the pendulum. This is true as long as all experiments are performed at the same location. There is another factor, however, that can change the period of a pendulum. If its location is changed, the acceleration may change because of gravity. At a location where g is greater, the pendulum will swing faster. Where g is less, the pendulum will swing more slowly.

At the surface of a massive planet such as Jupiter, a pendulum would swing considerably faster than it swings on the earth; its period would be much shorter. At the surface of a very small astronomical body having a low mass, such as the asteroid Ceres, a pendulum would swing very slowly; it would have an extremely long period. We calculate that Ceres (diameter, less than 500 miles) has such a weak gravitational pull that if a person walking on its surface should trip and fall, it would take him 17 seconds to hit the ground! A person attempting to play golf on Ceres would lose the golf balls into the sky when he struck them because their speed would exceed the **escape velocity** (the velocity at which an object breaks free of the gravitational pull).

Experiment 26:1 *The Pendulum*

1. Make pendulums by hanging various weights of different materials (lead, wood, or iron) from a 3-foot-long string.
2. Using a watch with a second hand, count the number of swings for each pendulum in 30 seconds. Compute the average period for a 3-foot pendulum.
3. Try different lengths of string using the same weight. Use lengths of 9", 16", 25", 36", and 49" counting the number of swings for each in 30 seconds.
4. Find the period for each. Make a graph of the period versus the length.
5. Make a second graph of the period versus the *square root* of the length.

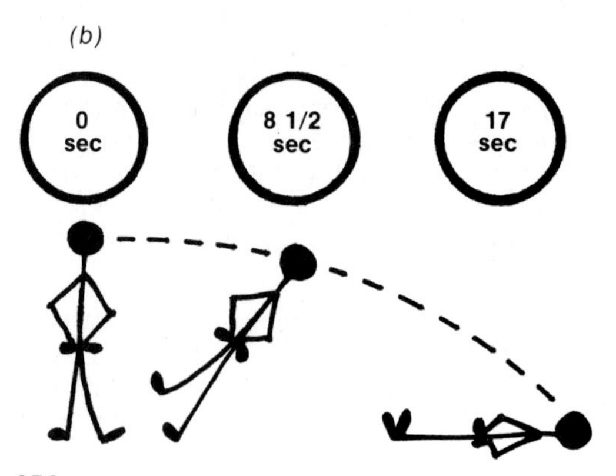

26:6 *(a) A person attempting to play golf on the asteroid Ceres would lose the golf balls into the sky. (b) A person who tripped and fell on Ceres would take 17 seconds to hit the ground.*

Table 26-2 shows the wide variation of g at the surface of various astronomical bodies in the solar system.

Table 26-2 *g at Different Places in Solar System*

Body	**g at surface in m / sec²**
Sun	274.4
Mercury	3.92
Venus	8.82
Earth	9.80
Moon	1.67
Mars	3.92
Jupiter	26.46
Saturn	11.76
Uranus	9.80
Neptune	9.80

If you knew the mathematical equation relating the period, the length of the pendulum, and g, you could determine these values at the surface of each planet with a simple pendulum.

26:7 *The pull of gravity on the moon's surface is only one sixth that of earth gravity.*

26:8 *Earth photographed from the moon. The force of attraction between the earth and moon is proportional to the product of their masses, and inversely proportional to the distance between them squared. (See chapter title page)*

★ **26:4 The Law of the Pendulum** The equation used to describe the motion of a pendulum exactly would be extremely complicated. In fact, no equation would give a perfect description of the actual phenomenon. However, we can use certain simplifying assumptions that make the problem more manageable. Let us assume that the weight of the pendulum is all concentrated at the end of a weightless string. Furthermore, let us agree to use only short swings. With these restrictions, the pendulum is satisfactorily described by the equation:

$$T = 2\pi\sqrt{\frac{\ell}{g}}$$

T is the period of the pendulum in seconds, ℓ is its length, and *g* is the acceleration due to gravity. For use on the planet earth, if ℓ is in feet, then 32 ft/sec^2 should be used for *g*; if ℓ is in meters, 9.8 m/sec^2 is the value used for *g*.

Problem: What is the period of a pendulum whose length is 3.2 feet, if it is measured at sea level and a latitude of 45 degrees?
Solution: For the location described, *g* would be 32 ft/sec^2. Substituting in the formula

$$T = (2)\ (3.14)\sqrt{\frac{3.2}{32}}$$

The fraction inside the square root sign reduces to 1/10. The square root of 1/10 is 1/3.16. We have then

$$T = (2)\ (3.14)\ \frac{1}{3.16} = 2.0 \text{ seconds}$$

The period of a pendulum 3.2 feet long, then, is approximately 2.0 seconds.

★ **26:5 Newton's Law of Gravitation** Upon graduation from Trinity College of Cambridge University in 1665, Isaac Newton was appointed a fellow of the college. However, because of a severe plague in England that year and the next, the schools were closed and Newton returned home to his family's country estate at Woolsthorpe.

It was during this extended "vacation" of about a year and a half that Newton did some of his most important scientific work. For instance, while sitting in the garden under an apple tree one day, he received the inspiration for his law of gravitation. As he sat thinking about the moon and other celestial bodies, Newton saw an apple fall to the ground. He began to wonder whether the force that made the apple fall might not be the same force that keeps the moon in its orbit. His preliminary calculations seemed to bear this idea out; Newton was encouraged. But it was not until a much later time that he had overcome all the obstacles in his way. Finally, Newton stated the law that is so familiar to students of physics everywhere: *Any two bodies attract each other with a force proportional to the product of their masses and inversely proportional to the square of the distance between them.*

Expressed as a proportionality, Newton's law of gravitation is

$$F \alpha \frac{m_1 m_2}{d^2}$$

F is the force of attraction, m_1 and m_2 are the masses of the two objects, and *d* is the distance between their centers. To express the relationship as an equation, it is necessary to use a constant. This constant is represented by *G* and is called the **universal gravitational constant**. The equation for Newton's law of gravitation, then, is

$$F = \frac{G\, m_1 m_2}{d^2}$$

Let us think through some imaginary situations to become more familiar with the nature of this inverse square law. How would the gravitational force between the earth and the moon be affected:

1. If the moon were twice as far away from the earth?
   Answer: The force would be only $1/2^2$ as great, or 1/4 of its present value.
2. If the moon were three times as far away from the earth?
   Answer: $1/3^2$ or 1/9 of its present value.
3. If the moon were four times as far away from the earth?
   Answer: $1/4^2$ or 1/16 of its present value.

The above examples are shown on the graph of Figure 26:9. The shape of the curve is characteristic of **inverse square laws**—the force falls off as the square of the distance.

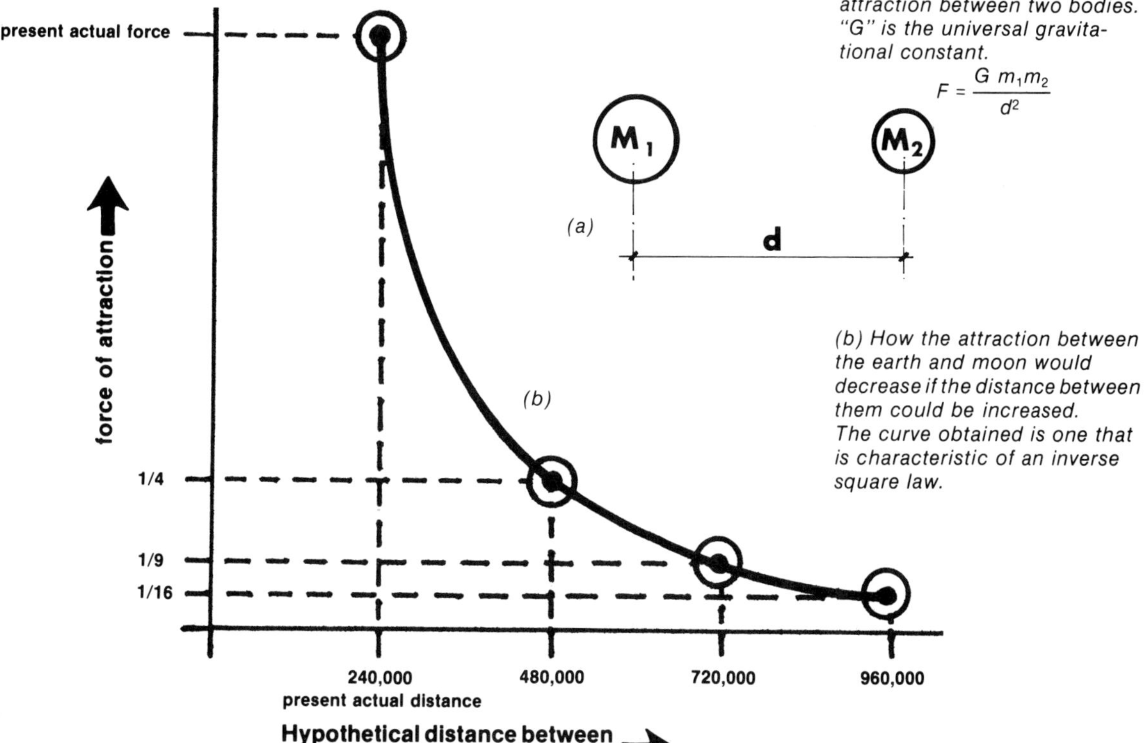

26:9 *(a) The three quantities that must be known to calculate the force of gravitational attraction between two bodies. "G" is the universal gravitational constant.*

$$F = \frac{G\, m_1 m_2}{d^2}$$

(b) How the attraction between the earth and moon would decrease if the distance between them could be increased. The curve obtained is one that is characteristic of an inverse square law.

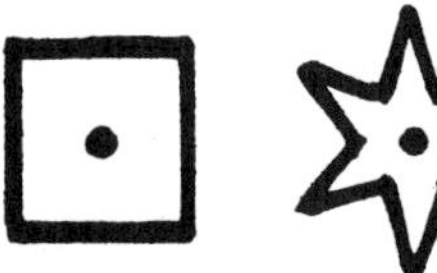
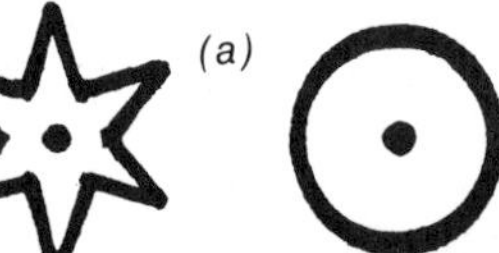

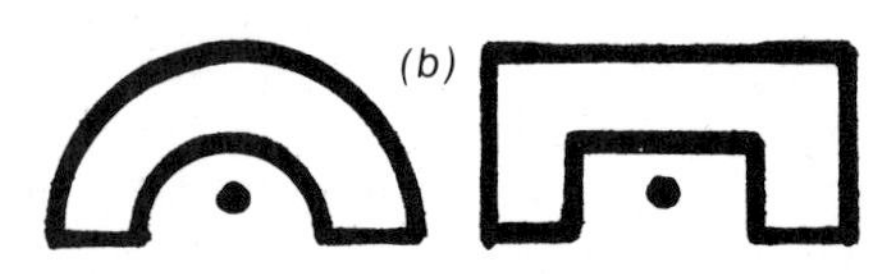

26:10 *(a) Figures whose center of gravity lies at the geometric center of the object. (b) Figures whose center of gravity lies outside the object.*

26:6 Center of Gravity While the law of gravitation was still forming in Newton's mind, one thing especially bothered him. He realized that every particle in the universe attracted every other particle. Therefore, in order to calculate the force of attraction between, say, a 2-ton boulder on the earth's surface and the earth itself, he would presumably have to compute the force of attraction between every particle in the boulder and every particle in the earth individually, then add them all together. This, of course, is too tedious a task to undertake.

Newton finally struck upon an easy way to solve this problem. Scientists still use it today. We pretend that all of the mass in an object is concentrated at some central point, called its **center of gravity**. For example, we consider the earth as 6.6 sextillion tons, all located at its center. The boulder is treated as 2 tons, all concentrated at *its* center. *One* calculation of the force between these two point masses, then, will give us the force of attraction between the boulder and the earth.

The center of gravity is easy to locate in some of the simpler geometric figures. Often it is at the center. Occasionally it lies outside the figure. Two forks stuck together at an obtuse angle can be supported on the tip of a pencil. The reason for this is that the center of gravity lies outside the forks themselves and is low enough to stablize the system. The actual location of the center of gravity of the forks (or for that matter, for the center of gravity of anything, regularly or irregularly shaped) can be ascertained by the following method: 1. The object is suspended from one end and a plumb line is drawn from the point of suspension through the object (see Figure 26:12a). 2. The process is repeated, using any desired second point of suspension, and a plumb line is again drawn. The center of gravity will be located at the point of intersection of the two plumb lines.

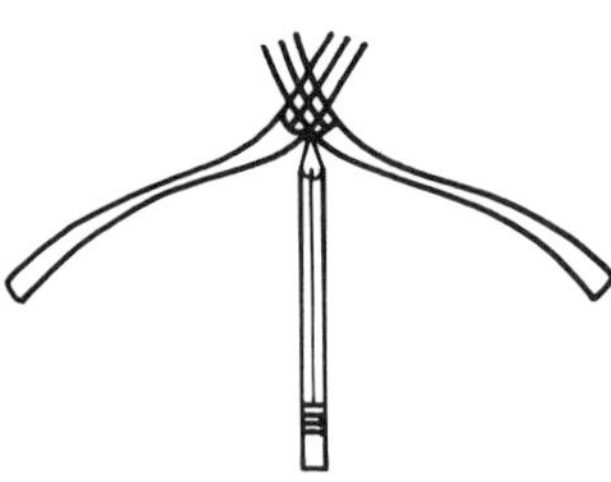

26:11 *Two forks stuck together can be easily balanced on the tip of a pencil.*

26:12 *One method of finding the center of gravity of an object. (a)The object is suspended from one end and a plumb line is drawn from the point of suspension through the object. (b) The process is repeated using any desired second point of suspension, and a second plumb line is drawn. The center of gravity is located at the point of intersection of the two plumb lines.*

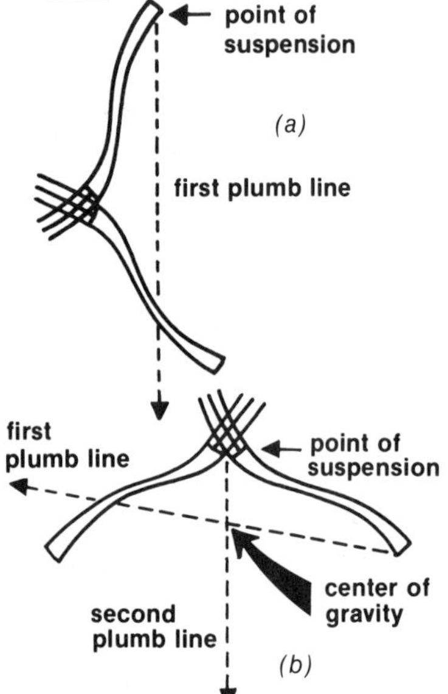

26:13 *Plants tend to grow in such a way that their center of gravity lies directly above the point at which the plant comes out of the ground. Note the mass distribution in the cactus at the right.*

Plants grow so that their center of gravity is just above the point where the stem comes out of the ground. Similarly, a person stands best when his center of gravity is directly above his feet. If your center of gravity is too far out of line, you will fall over. Try this amusing demonstration: Stand with your back against a wall, making sure that your heels are touching the wall. Now, try to touch your toes *without bending your knees*. As you lean forward, you find that your center of gravity is forced forward to such an extent that you fall, regardless of how determined you are not to.

26:14 *An impossible feat. Stand with your heels against a wall. Now try to touch your toes without bending your knees. You will fall forward regardless of how determined you might be to prevent it.*

26:7 "Weighing the Earth" Newton did not have any way of determining the actual numerical value of the universal gravitational constant (G). In order to calculate it, he would have to have known all the other terms in the equation for any two objects. What two bodies could be used—the earth and the moon? No, their masses were not known even roughly at the time. The earth and a falling body? No, only one of the two masses would be known. Both had to be known.

Henry Cavendish (1731-1810), an English chemist and physicist, believed the only way to solve the problem was to measure the force of attraction between two laboratory weights. This is much more easily said than done. Gravitation is such a weak force that you must have a very sensitive measuring device, even if large lead weights are used. Cavendish made a remarkably accurate torsion balance that recorded the very small force between the two weights. He substituted the force he obtained, the masses of the two weights, and the distance between their centers in Newton's equation $F = \frac{G\,m_1 m_2}{d^2}$ and obtained a reasonably good value for G. It then became possible to calculate the mass of the earth, and finally the weight of the earth. For this reason, Cavendish's experiment is sometimes called "weighing the earth." It was carried out in the years 1797 and 1798, much after Newton's time. The weight of the earth turns out to be about 6.6 sextillion tons!

We become more appreciative of the place the Creator has given us when we realize that, not only must the earth be made of the right materials, but the total quantity of matter composing it is crucial. If the mass of the earth were too great, our bodies would experience difficulty in trying to overcome the effects of gravity. Our hearts would have trouble pumping blood against its pull, and it would be harder to move around. Just standing up would even be harder! If the earth had too little mass, it would be unable to retain a proper atmosphere and unable to keep the moon in its orbit.

26:15 *An analytic balance of the single pan type shown here compares the mass of the object to be weighed with the masses of built-in weights.*

★ **26:8 Relationship Between Mass and Weight** In Chapter 25 we studied Newton's second law of motion:

$$F = ma$$

F is the unbalanced force applied to an object of mass m, and a is the acceleration it undergoes as a result of the force. For a falling body, the acceleration is g, the acceleration due to gravity. We may therefore write

$$F = mg$$

But the downward force on a falling object is nothing more or less than its weight (w). (**Weight** is the force with which the earth pulls an object downward.) Making this substitution, we have

$$w = mg$$

This gives us a means of finding the weight of an object if its mass is known, and of finding the mass of an object if its weight is known.

In the English system *mass* is expressed in **slugs** and *weight* is expressed in **pounds**. The value of *g* in the English system is 32 ft/sec^2.

In the metric system there is a choice of units. The mks (meter, kilogram, second) or large units are more convenient for most common applications of the relationship than the cgs (centimeter, gram, second) or small units. In the mks units, *mass* is expressed in **kilograms** and *weight* in **newtons**. The value of *g* is 9.8 m/sec^2.

Problem: What is the weight of a bar of gold which has a mass of 1 kilogram?
Solution: $w = mg$
$w = (1\text{kg})\ (9.8\ \text{m sec}^2) = 9.8$ newtons

26:16 *The condition called weightlessness is experienced in or near a satellite in orbit around the earth.*

You can see from the relationship $w = mg$ that mass and weight are directly proportional. An object that is twice as massive as another will also be twice as heavy, so long as they are both weighed at the same location. On the moon both objects would be reduced to about 1/6 their earth weight because the value of *g* is correspondingly less on the moon. But they would retain the same 2 to 1 ratio.

Of the two quantities, mass and weight, mass is the more constant. Mass is a measure of the amount of material in a body. This remains the same whether it is on the earth, the moon, or the planet Jupiter. Weight, however, depends on an object's location and the gravitational pull at that location. A bag of sugar that weighs 10 pounds on the earth would weigh 1.7 pounds on the moon and 27 pounds on the planet Jupiter. It could even become completely **weightless** if placed in a satellite in orbit around the earth. But its mass would remain constant at 0.31 slugs under all these conditions.

Problem: What is the mass of a 3200-pound car?
Solution: $w = mg$
$m = w/g = 3200\ \text{lb}/32\ \text{ft sec}^2 = 100$ slugs

26:9 Archimedes' Principles If you have ever tried to lift a heavy stone under water, you probably observed that it seemed lighter than it would out of the water. Archimedes, the early Greek physicist, made a study of the fact that objects submerged in water lose weight. After a thorough investigation of many different objects of various sizes and weights, he arrived at the following generalization:

Archimedes' Principle for Submerged Bodies—*The weight lost by a submerged object is equal to the weight of the fluid it displaces.*

According to this principle, a small object, because it displaces very little fluid, will lose very little weight; a large object, because of its greater displacement, will lose correspondingly more weight. Of course, the density of the object is also a factor. If its density is low enough, the object will float.

Archimedes' principle applies to objects which are submerged in a gas, as well as those submerged in a liquid. In reality, we live at the bottom of an "ocean" of air. As a result, we are buoyed (lifted) up by a force equal to the weight of the air we displace. This is not a very great upward force, but we would, in fact, weigh slightly more in a vacuum. In very accurate weighing operations performed on an analytical balance, an allowance is often made for the buoyant effect of the air.

26:17 *Deep submergence rescue vehicle—an example of a submerged body. It is buoyed upward by a force equal to the weight of the displaced water.*

26:18 *A blimp is buoyed up with a force equal to the weight of the displaced air.*

We shall illustrate Archimedes' principle for submerged bodies with two numerical examples, one using the English system, the other the metric system.

Problem: A stone weighs 110 pounds in air and has a volume of 0.5 cubic feet. Find the weight of the stone when it is submerged in water. (The density of water is 62.4 lb/ft^3.)

Solution: The stone displaces 0.5 cu ft of water. Using the relationship that *density* = weight/volume, you can find the weight of the displaced water by multiplying density times volume. This is 62.4 lb/ft^3 times 0.5 cu ft, or 31.2 lb. This value represents the weight lost by the stone. Therefore, its weight under water is 110 minus 31.2, or approximately 79 pounds.

26:19 *Archimedes formulated separate laws for floating bodies and submerged bodies.*

Problems of this nature are generally easier in the metric system because the density of water is a handy round number: 1 g/cm^3 as opposed to 62.4 lb/ft^3 in the English system.

Problem: A 200-gram bar of metal has a density of 8.0 g/cm^3. Find its "weight" in grams under water. (Note: It is customary in this type of problem to call the mass in grams the "weight" of the object.)
Solution: The volume of the bar can be found by $D = m/V$ Rearranging this relationship to $V = m/D$ and substituting the numbers, we have $V = 200g/8.0\ g/cm^3 = 25\ cm^3$. The volume of the bar is 25 cm^3; therefore, it will displace 25 cm^3 of water. Because the density of water is 1 g/cm^3, the "weight" of the displaced water is 25g. This is the "weight" lost by the bar; so its "weight" under water is 200g minus 25g, or 175g.

26:20 *A 10,000-ton ship displaces 10,000 tons of water.*

Archimedes made a separate study of objects which rise to the surface. In the course of this investigation, he became aware of another principle which he formulated as follows:

Archimedes' Principle for Floating Bodies—*When a solid object floats in a liquid, the weight of the displaced liquid equals the weight of the object.*

Thus, a ship which weighs 10,000 tons will float in such a way that it displaces 10,000 tons of water. As a result, the downward pull of gravity is exactly offset by the buoyant effect of the water, and the ship neither rises nor sinks. Similarly, an iceberg displaces a quantity of water having a weight equal to its own weight.

26:21 *The weight of the water displaced by an iceberg is equal to its own weight.*

It is remarkable that Archimedes' principles have withstood some 22 centuries of testing and have never had to be modified or corrected in any way.

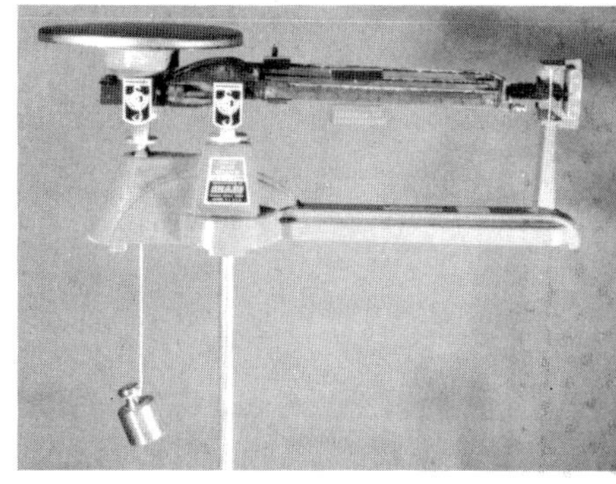

26:22 *Method of attaching weight to pan hook of a triple beam balance.*

Experiment 26:2 *Archimedes' Principle for Submerged Bodies*

1. Using a piece of heavy thread, suspend a 100-gram brass weight from the pan hook of a triple-beam balance (see Figure 26:22). Make sure that its "weight" in air indicates 100 grams on the balance.
2. Support a 250-ml beaker in such a way that it surrounds the weight.
3. Add water to the beaker until the weight is completely submerged.
4. Determine its "weight" under water and record this value. What "weight" of water was displaced? What volume of water was displaced? (Remember that each gram of water has a volume of 1 cubic centimeter.) What is the volume of the brass weight? Calculate its density by the relationship

$$D = m/V$$

JOHANNES KEPLER *(1571-1630)*

German astronomer

"Knowest thou the ordinances of heaven? Canst thou set the dominion thereof in the earth?" (Job 38:33)

Sir Isaac Newton once commented, "If I have seen further, it is by standing on the shoulders of giants." Johannes Kepler was one of those giants. Kepler was the first man to demonstrate that the motions of the planets are precise and predictable, obeying definite rules. The German astronomer's three famous laws of planetary motion are used extensively today in calculating the orbits of satellites and in mapping routes for space travel. Kepler is also called "the father of modern optics" because of his masterful mathematical analysis of lenses and mirrors.

Kepler was a devout Lutheran, having trusted Christ as his Saviour at an early age. Although he was persecuted for his Protestant beliefs, he remained true to the Lord throughout his entire life. Kepler saw himself as an instrument in God's hands for revealing more of the details of His handiwork to men. "Since we astronomers are priests of the highest God in regard to the book of nature," he once wrote, "it befits us to be thoughtful not of the glory of our minds, but rather, above all else, of the glory of God." A devoted family man who sought to give his children a thoroughly Christian upbringing, Kepler wrote Bible study guides to aid their understanding. One of these guides, "The Body and Blood of Jesus Christ Our Saviour," is still preserved in the University of Tubingen library.

At the age of seven Johannes entered the Latin school at

Leonberg, an institution established by the dukes of Wurttemberg. He showed promise of a brilliant future, but he was forced to leave school for several months at a time to support his family by working on a farm. As a result it took him five years to complete the three-year course of study. Nevertheless, because of his outstanding work and Christian testimony, Kepler's teachers encouraged him to go on to a seminary to study for the ministry.

Kepler entered the Lutheran lower seminary, Adelberg, when he was 13. There he was subjected to a rigorous schedule of study that included Bible, Latin, Greek, rhetoric, grammar, logic, and music. Kepler completed the two-year curriculum, then moved on to the higher seminary, Maulbronn, where he received his bachelor's degree in 1588.

Michael Maestlin, whom Kepler met at the University of Tübingen, was a major source of inspiration for his life's work. An instructor in mathematics and astronomy, while still teaching the old geocentric (earth-centered) theory of the solar system, Maestlin exposed his students to the newer ideas of Copernicus in a way that kindled the imagination. Kepler was compelled by a consuming curiosity to study the subject on his own. Many after-class discussions with Maestlin provided Kepler with new sparks of thought. Maestlin, a fellow Christian, became a lifelong friend and later helped Kepler with the publication of his first book.

When Kepler was graduated from the Faculty of Arts at Tübingen in 1591 (the equivalent of today's master's degree), he felt certain that he was destined for the Lutheran ministry. Accordingly, he enrolled at the Theological Faculty and continued his studies. In 1594, however, before he was able to complete his examinations, an unexpected opportunity arose. The death of the mathematics and astronomy teacher at the Protestant seminary in Graz (the capital of the Austrian province of Styria) created a vacancy that was most difficult to fill. The senate of the University of Tübingen was asked to recommend a candidate. Their choice: Kepler. His extraordinary mathematical ability had not gone unnoticed.

Needless to say, this offer came as a surprise to the young divinity student. He had been most successful in his theological studies and felt assured of a good position in the Lutheran Church. Now came the prospect of an appointment of lower esteem and considerably less pay in faraway Austria. It was a difficult decision that required much thought and prayer. Kepler finally became convinced that it was the Lord's will for him to accept the position at Graz, and he did so with the understanding that he could return to Tübingen to finish his theological studies at some later date. As it turned out, he never did return. It was an important turning point, not only in his own life, but also in the history of astronomy.

Kepler's years at Graz were both challenging and rewarding. It was here that he produced his first serious scientific treatise,

The Mystery of the Universe, an ambitious geometric description of the solar system. The work closed with a magnificent hymn of praise to the Creator. The treatise was well received, and, as a result of his shrewdness in placing it in the hands of several leading astronomers (including Galileo and Tycho Brahe), Kepler's name became known in several of the scientific centers of Europe. That same year Kepler married a young widow of 23, Barbara Müller, the oldest daughter of a wealthy mill owner.

As the century drew to a close, the political and religious conditions in Graz grew increasingly more turbulent. As a result of severe religious persecution Kepler and his family were forced to flee from Graz in 1600. Earlier the same year he had accepted an invitation to visit Tycho Brahe, imperial mathematician to Rudolph II, Emperor of Bohemia, in Prague. Recognized as the world's leading astronomer, Tycho (as he is called) had been making remarkably accurate observations of the positions of the planets for 20 years. It now remained for someone to "make sense" out of the massive columns of data he had assembled. Kepler desired to have access to these observations, and Tycho was eager to add a theoretician of Kepler's caliber to his staff of assistants. It was indeed providential that circumstances brought together these two giants whose abilities complemented each other so perfectly. Kepler was not unaware of the unique opportunity that had been afforded him nor unthankful for the strange turn of events that had brought it to pass, for he later said, "I see how God let me be bound with Tycho through an unalterable fate and did not let me be separated from him by the most oppressive hardships."

When Kepler arrived in Prague, Tycho assigned him the task of interpreting the observations on the planet Mars. These calculations occupied the German astronomer for many years. Kepler explored numerous blind alleys and was often forced to start over again. However, his perseverance eventually bore fruit. One by one, he was able to formulate the three laws that are now so familiar to every student of physics and astronomy:

1. *Planets move in ellipses with the sun at one focus.*
2. *An imaginary line from the center of the sun to the center of a planet always sweeps over an equal area in equal time.*
3. *The squares of the periods of revolution are to each other as the cubes of the mean distances from the sun.*

The third law formed a foundation for Newton's law of gravitation just half a century later. Kepler was well ahead of his time; he had no rivals who claimed to have anticipated him. In fact, to other astronomers of his day, the area and ellipse laws were new, unorthodox, and quite difficult to understand.

Kepler's later life was one of many hardships—sickness and death in his family, religious persecution, war, a fire, and a trial in which his mother was falsely accused of witchcraft. But his faith in Christ brought him triumphantly through these tribulations, and he was able to give God the glory for His sustaining grace.

Kepler's name has been immortalized by his three laws of planetary motion. A prominent crater on the moon has been named in his honor. And his native Germany has paid him homage by erecting elaborate monuments to him in Regensburg and Weil der Stadt. His birthplace in Weil der Stadt has been converted into an attractive museum. Any fame he achieved, however, was simply a by-product of his lifelong endeavor to glorify the name of the heavenly Father. "Let also my name perish," Kepler stated, "if only the name of God the Father . . . is thereby elevated."

List of Terms

- acceleration due to gravity
- Archimedes' principles
- air resistance
- center of gravity
- constant
- escape velocity
- free fall
- gravitation
- ★ inverse square law
- ★ kilogram
- mass
- ★ newton
- parabola
- pendulum
- period of a pendulum
- ★ pound
- ★ slug
- terminal velocity
- ★ universal gravitational constant
- variable
- ★ weight
- ★ weightless

Questions

1. As long as all measurements are made at the same location, the time required for a pendulum to swing back and forth is dependent only on its ________________.
2. On a large massive planet such as Jupiter, the period of a pendulum will be ________________ (greater, less) than it is on the earth.
3. When the relationship $S = \frac{1}{2}gt^2$ is plotted on a graph, the shape of the curve obtained is ________________.
4. Where would an object weigh more—in a deep well or on a high mountain? How would its mass compare at these two locations?
5. A stream of oil flowing from a can becomes thinner as it falls farther. How can this be explained?
6. It was reported during World War II that an American soldier in Europe survived a fall from a plane when his parachute failed to open. This is partly explained by the fact that he landed in a tree. What other factor must enter into the explanation?
7. The retarding force on a falling body due to air resistance is directly proportional to ________________.
8. What names are given to the two constants *g* and *G*?
9. What important simplification did Newton use for calculating the force of attraction between two bodies?
10. Is it possible for the center of gravity to lie outside a body? Why?
11. How can we explain the fact that a heavy object and a light object accelerate at the same rate as they fall to the earth?

★ 12. Pretend that you have landed on an unknown planet. Describe how you could use a pendulum to determine the value of g on its surface.
13. What factors determine the escape velocity at the surface of a planet?
14. Where on the earth's surface would you find the lowest value of g?
15. Specifically, what is it about gravity that we still do not understand?

Problems

1. A parachute jumper waits five seconds before opening his chute. How far has he fallen in this time?
2. A stone dropped from the roof of a department store strikes the sidewalk in 1.75 seconds.
   (a) How high is the roof?
   (b) How fast is the stone traveling at the instant it strikes the sidewalk?
3. If the moon were four times as close to the earth as it is now, how many times greater would the force between the two bodies be than it is now?
4. Find to the nearest 1/10 of a second the time required for an acorn to fall from the top of a 77-foot oak tree.
5. What would be the weight of a 200-pound person 4000 miles above the earth's surface? 8000 miles above the earth's surface? (The earth's radius is about 4000 miles.)
6. How long would it take for an object to fall from the top of the Empire State Building to the ground—a distance of 1248 feet?
★ 7. Find the mass of an 8-ton truck. (1 ton = 2000 pounds.)
★ 8. Find the weight of a 5-kilogram brass weight.
★ 9. Find the weight of a boulder which has a mass of 190 slugs.
10. How fast is a freely falling object traveling at the end of half a second?
11. An object thrown straight up takes 8 seconds to strike the ground. How high did it rise?
12. A 500-pound oil drum has a volume of 7 cubic feet. What does it weigh under water? (The density of water is 62.4 lb/ft^3.)
13. A block of aluminum has a "weight" of 170 grams and a density of 2.7 g/cm^3. Find its "weight" under water.
14. A sphere made of an unknown material "weighs" 36 grams in air and 12 grams in water. Find its density.

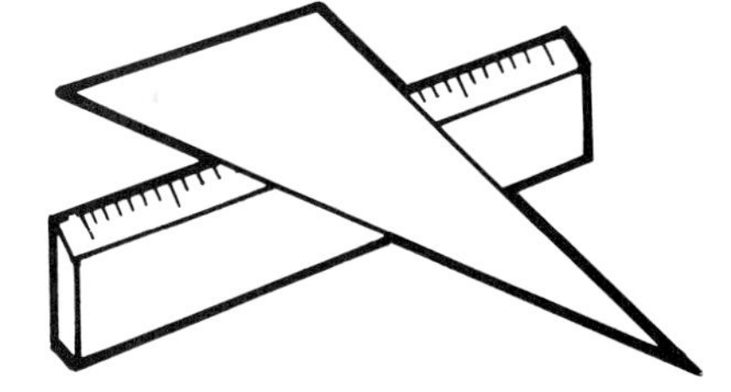

26:23 *Finding the center of gravity of an isosceles triangle. (Activity #1).*

Student Activities

1. Cut several cardboard isosceles triangles having various base angles. Locate the center of gravity of each by balancing it on a ruler. What generalization can you make concerning the location of the center of gravity in isosceles triangles? Hint: measure the fraction of the distance between the base and the vertex (angle opposite the base) in each case.
2. Which should fall faster—a table tennis ball or a solid ball of the same size? Why? Try it and see.
3. Prepare a report on the latest research being conducted on "gravitational waves." Your teacher may be able to suggest a magazine article to consult.

Photo next page
Fossil fuel steam generating plant.

UNIT IX HEAT

Chapter 27 Heat as Energy

27:1 *A housewife using heat to cook food.*

27:1 Introduction You can describe the painful sensation of putting your hand on a hot piece of metal; you can tell about the discomfort of a hot, humid summer day; you are able to remember the shock of a swim in cold water; and you can relate the numb feeling in your hands after you made snowballs for awhile. But do you really know what **heat** is?

We have a general idea of *hot* and *cold*. We also speak of something being *warm* and something else being *cool*. Some people like a hot shower, others a warm one, others a cool one, and still others a cold shower.

Similar expressions of the relative heat, or lack of it, are used frequently in the Bible. When we suffer because of our faith, we are "tried by fire," meaning that it is "hot" for us spiritually. A person who is zealous in his Bible study and his desire to serve the Lord is said to be "on fire for the Lord." A person who is backslidden, who neglects spiritual things, is said to be "cold." In the book of Revelation, John describes the church at Laodicea as being "lukewarm" (Revelation 3:14-15). There is nothing more insipid than a cup of hot chocolate or a glass of ice water that has become lukewarm. Being "neither hot nor cold" is also a sad spiritual condition.

27:2 Practical Uses We use heat to cook and to keep our houses warm during cold weather. Modern industry uses heat to melt metals and plastics and for many thousands of other operations. Heat is used to generate electricity, which runs

27:2 *Heat from a fireplace has warmed many families over the years.*

countless millions of machines—from the largest metal presses to the smallest mixer in your kitchen.

27:3 *This industrial machine requires electricity for its operation.*

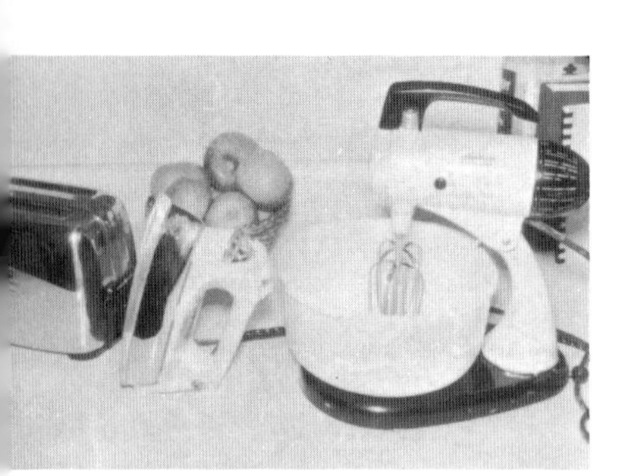

27:4 *Electrical appliances used in the home.*

27:3 How Hot or How Cold? We cannot be very scientific using the terms hot, warm, cool, and cold. They are too dependent on personal feelings. If someone asks you, "How much does John weigh?" you might say, "He is very heavy." "Heavy" could mean different things to different people. But if you say, "He weighs 201 pounds," you have given an exact amount and a more complete description of John's weight.

Likewise, if someone asks you how cold a cup of water is, and you say "extremely cold," this means different things to different people. But if you tell the person that the temperature of the water is 40° F, you have given him an exact description of the temperature of the water in the cup.

Temperature is *a measure of the heat content of a body*. Two popular temperature scales are used today, the **Fahrenheit (F)** and Centigrade or **Celsius (C)** scales. The sign ° F means degrees Fahrenheit; the sign °C means degrees Celsius.

°C	°F
100	212
80	176
60	140
40	104
20	68
0	32
-20	-4
-40	-40

27:5 *Comparison of Celsius and Fahrenheit temperature scales.*

The two temperature scales are based on the boiling point and freezing point of water. The boiling point of water is 100°C or

212° F. By using a very simple formula, you can calculate °F from °C or °C from °F:

$$9C = 5(F - 32)$$

Problem: If the temperature is 72°F, what is the reading in °C?

Solution: Substituting in the formula 9C = 5(F - 32):

$$9C = 5(72-32)$$

$$9C = 5(40)$$

$$\frac{9C}{9} = \frac{200}{9}$$

$$C = 22.2°$$

Problem: If the temperature is 75°C, what is the reading in °F?

Solution: Substituting in the formula 9C = 5(F - 32):

$$9(75) = 5(F - 32)$$

$$\frac{675}{5} = \frac{5(F-32)}{5}$$

$$135 = F - 32$$

$$+32 \quad\quad +32$$

$$167° = F$$

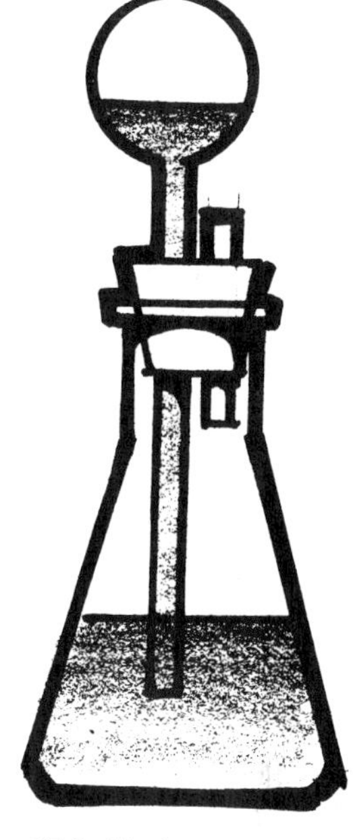

27:6 *Air thermometer.*

The Celsius scale is used universally in the scientific world; therefore, we use it in this book. Remember, using numbers in science gives you a *quantitative* understanding.

Experiment 27:1 *Making a Thermometer to Measure Temperature in Air*

1. The thermometer to be made is shown in Figure 27:6. Glass tubing with a bulb in the end of it can be obtained from several scientific supply companies. Otherwise, a piece of glass tubing can be heated on one end until it melts together and seals.
2. Heat this end until the glass becomes quite soft; then very carefully blow from the cool end and make a bubble in the soft glass.
3. Insert the warm tube into a rubber stopper. Be sure that the cool end of the tube extends into the colored water. As the glass cools, the liquid will rise into the tube.
4. Get a good mercury thermometer to check your water thermometer. Make a mark (or use a rubber band) to record the place where room temperature is on your thermometer. This level of water is the temperature that is shown on the mercury thermometer.
5. Place both thermometers into a refrigerator and leave them there for an hour. Then mark the water level and record this as the temperature shown on the mercury thermometer (see Figure 27:7). The thermometer is now calibrated. The temperature range you have marked off may be divided into as many divisions as you wish.

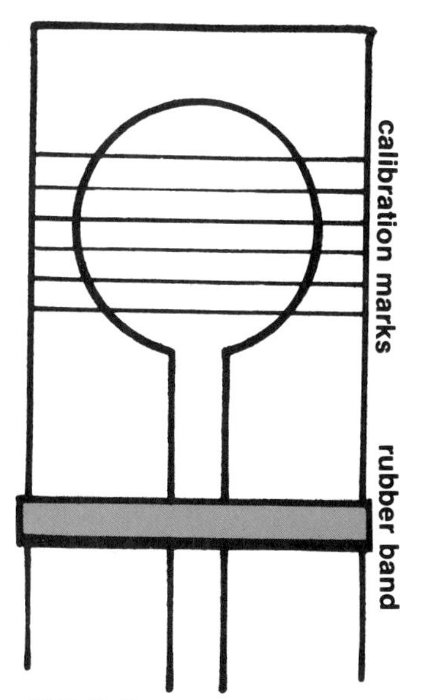

27:7 *Calibration board for air thermometer.*

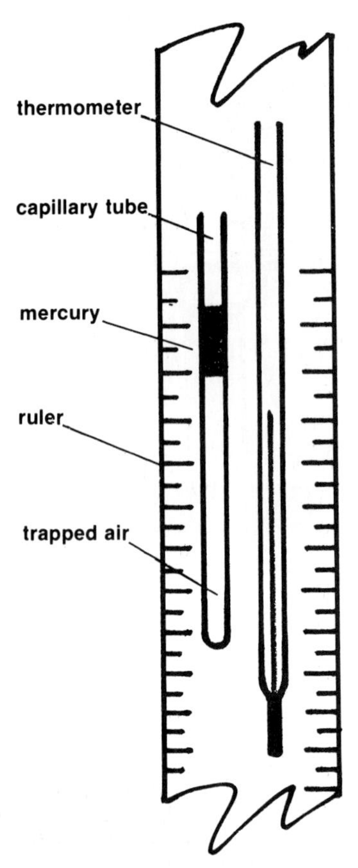

27:8 *Capillary air thermometer.*

Experiment 27:2 *Making a Capillary Air Thermometer*

1. Obtain a 12-inch piece of capillary tube and heat one end to seal, as was done in Experiment 27:1.
2. As this end cools, place the other end under a pool of mercury. Allow about an inch of mercury to go into the tube.
3. Calibrate this thermometer according to steps 4 and 5 in Experiment 27:1.
4. You can use this air thermometer in the same manner as the water thermometer, or the air thermometer can be used as an immersion thermometer. It can be placed in various fluids; and, if the height of the air column underneath the mercury can be measured, you can know the temperature. Make sure the entire column of air under the mercury is submerged when you measure temperature in this way.

27:4 Measuring Temperature It is convenient to have a temperature scale so that you can tell how hot or cold something is, but how do you measure temperature? Generally, the measuring of temperature is based on a selected property of matter that changes when heat is added or taken away from the object. But what property of matter changes as something gets hotter or colder? Think back to Chapter 10, where we discussed gases. Do you remember **Charles' law**? Charles' law says that as the temperature of an ideal gas increases, the volume of the gas increases, or as the temperature decreases, the volume decreases.

$$V \propto T$$

To measure temperature according to the expansion and contraction of a gas, we could design a gas thermometer. A **thermometer** is *a device used to measure temperature.* An actual gas thermometer is shown in Figure 27:9. Constant pressure gas thermometers are very accurate; however, they are complicated and bulky. Also, they do not respond very rapidly to changes in temperature. Because they are so accurate, though, constant pressure gas thermometers can be used to calibrate other types of thermometers.

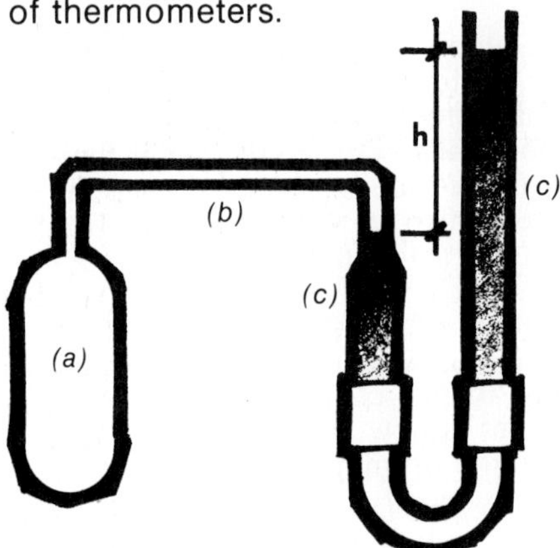

27:9 *Gas thermometer. Glass vessel (a) containing a gas is placed in a heated system. The heated gas expands along (b) forcing the mercury column (c) up the U-tube. The difference in the mercury levels is adjusted to h by moving the flexible tube up and down. The volume of the gas recorded and the temperature thus can be determined.*

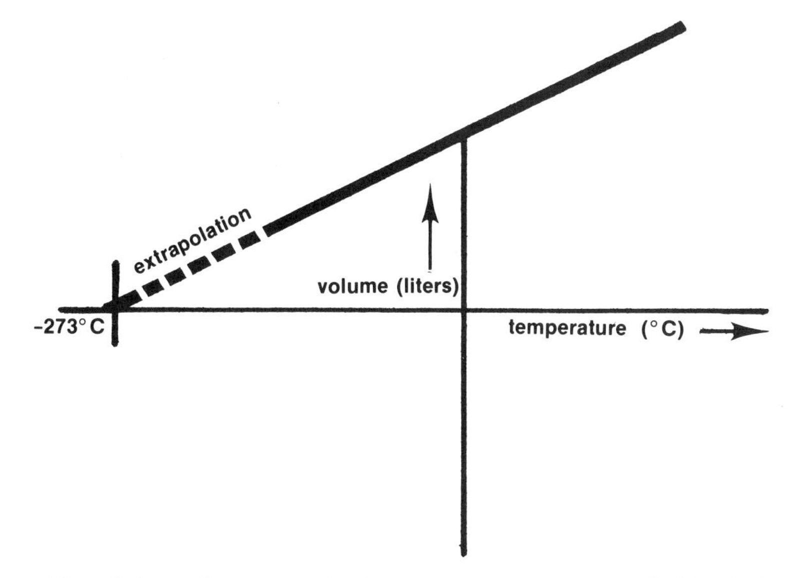

27:10 *The variation of volume with temperature of an ideal gas.*

One interesting aspect of gas thermometers and Charles' law is that of an **absolute temperature scale**. Look at the volume versus temperature graph of an ideal gas at a constant pressure in Figure 27:10. According to Charles' law, as the gas cools down farther and farther, it approaches a zero volume. If we extend the curve downward to the position of zero volume, it meets the temperature line at -273° C. This temperature is called **absolute zero**. We can now define a new temperature scale based on this information. The absolute temperature scale is named for Lord Kelvin, and the temperature readings are referred to as degrees Kelvin (°K). To calculate °K from a reading in °C, simply add 273:

$$K = C + 273$$

Problem: If a gas is at 75°C, what is its temperature on the absolute scale?
Solution: Substituting in the formula $K = C + 273$:

$$K = 75 + 273$$
$$K = 348°$$

Problem: If a gas is at 140°K, what is the temperature on the Celsius scale?
Solution: Substituting in the formula $K = C + 273$:

$$140 = C + 273$$
$$140 - 273 = C$$
$$-133° = C$$

Of course, no gas volume ever decreases to zero. Gases liquefy before they reach -273° C or 0 °K. Also, no gases are ideal; this is just a theoretical model that works quite well.

★ **Experiment 27:3** *Calculation of Absolute Zero Using Air Capillary Thermometer*

If you know the size of the hole in the capillary tube, you can calculate the volume of air in the tube at any temperature using the formula

$$V = \frac{\pi d^2 h}{4}$$

V is the volume of air; *d* is the inside diameter of the tubing; and *h* is the height of the air column under the mercury.

1. Measure the temperature of several mixtures of ice and water. Record the volume and temperature in each case. (If you can obtain some dry ice, mix it with alcohol for a colder bath. You may need to use a lower range thermometer to record this low temperature.) Get as many low temperature readings as possible.
2. Plot a graph similar to Figure 27:10. The volume can be expressed in ^{3} or cm^3.
3. Extend the volume-temperature line back until it crosses the abscissa (horizontal axis). The point where it crosses should be absolute zero (—273° C).
4. You can use the equation for a straight line and determine the T-coordinate at V = 0.

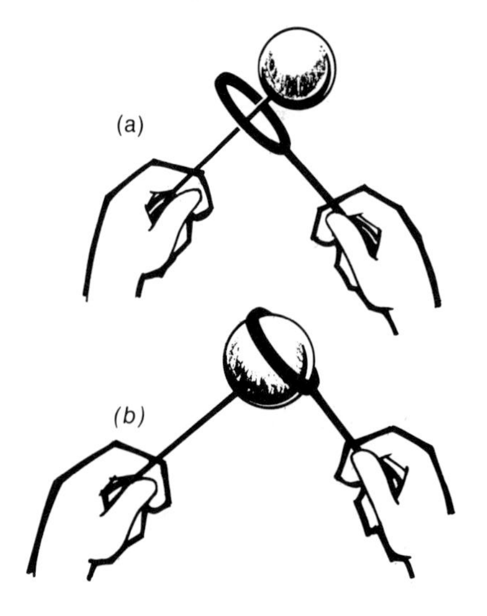

27:11 *(a) Cold ball goes through cold ring. (b) When ball is heated, it will not fit through the ring.*

27:5 Solid Expansion and Thermostats Since gas thermometers are not easy to use, other types of thermometers are employed for general laboratory purposes. As a general rule, liquids and solids expand on heating and contract on cooling and are much easier to handle than gases.

First, let us look at solids. It is easy to demonstrate the **thermal expansion** of solids by using a ball and ring apparatus (see Figure 27:11). When the ball and ring are at room temperature, the ball will go through the ring (even though it is a snug fit). If you heat the ball so that it expands, the ball will not go through the ring. If you quench, or cool, the ball in very cold water, it will go through the ring easily.

Solids do not expand at the same rate. Some solids expand a great deal with an increase in temperature, whereas others do not expand so much. One of the measurable properties of a solid is the **coefficient of linear expansion**. Materials with a high coefficient expand more than materials with a low coefficient as the temperature goes up. For instance, the coefficient of linear expansion of brass is 19×10^{-6}. (See Appendix I for explanation of exponential numbers.) This means that a brass rod one meter long will expand 19×10^{-6} meters every time it is raised one ° C. It will increase 38×10^{-6} meters in length if it is heated up two ° C.

A glass rod one meter long will expand only 3.3×10^{-6} meters when it is heated up one ° C. The coefficient of linear expansion is always a very small number. The larger it is, the more the solid will expand when heated. If you know the coefficient of linear expansion, you can calculate how much a solid will expand along a linear (straight-line) dimension. We will use the symbol $\Delta \ell$ to mean the change in length of the solid with temperature; ℓ_0 is the original length of the solid before heating; t_0 is the temperature in °C before heating; and t_f is the final temperature in °C to which the solid is heated. α is the coefficient of linear expansion. The formula is

$$\Delta \ell = \ell_0 \alpha (t_f - t_0)$$

Table 27-1 *Coefficients of Expansion of Certain Substances*

Substance	Coefficient of linear expansion α, $(C°)^{-1}$	Coefficient of volume expansion β, $(C°)^{-1}$
Solids		
aluminum	26×10^{-6}	77×10^{-6}
brass	19×10^{-6}	56×10^{-6}
concrete	12×10^{-6}	36×10^{-6}
iron or steel	11×10^{-6}	33×10^{-6}
lead	29×10^{-6}	87×10^{-6}
ordinary glass	8.5×10^{-6}	26×10^{-6}
platinum	9.0×10^{-6}	27×10^{-6}
Pyrex glass	3.3×10^{-6}	10×10^{-6}
Liquids		
carbon tetrachloride		581×10^{-6}
gasoline		960×10^{-6}
mercury		182×10^{-6}
methyl alcohol		1134×10^{-6}
turpentine		900×10^{-6}
Gases		
air		3670×10^{-6}
carbon dioxide		3740×10^{-6}
hydrogen		3660×10^{-6}

★**Problem:** How much does a 20-meter brass rod expand when it is heated from 20°C to 300°C?

Solution: Checking Table 27-1, we find that $\alpha_{brass} = 0.000019/°C$

$$\Delta \ell = \ell_0 \alpha (t_f - t_0)$$
$$= 20m(.000019/°C)(300-20)°C$$

(Notice that the units °C/°C cancel out.)

$$= .00038(280)m$$
$$= .1064m$$

★**Problem:** A brass rod of length 50 meters at 200°C is cooled to 10°C. What is the length of the rod at 10°C?

Solution: $\Delta \ell = \ell_0 \alpha (t_f - t_0)$

$$\Delta \ell = 50 \text{ meters } (.000019/°C)(10-200)°C$$
$$= .00095(-190m).$$
$$= -.1805m$$
$$\ell_{10°} = (50.0000 - 0.1805)m$$
$$= 49.8195 \text{ meters}$$

★**Experiment 27:4** *Measuring the Coefficient of Thermal Expansion of a Metal*

1. Obtain a metal rod from your teacher. Measure the length (ℓ_o) of the rod with a meter stick and place it into the expansion rig shown in Figure 27:12.
2. Place a thermometer into the rig and read the temperature (t_o). Record where the pointer arm is at this temperature.
3. Connect a steam generator to the expansion rig with a rubber hose.
4. Boil the water in the steam generator. Allow the rod to get as hot as possible. Record this temperature (t_f) and the new reading of the pointer.

The difference in the 2 scale readings, divided by 10, is $\Delta\ell$. Using $\Delta\ell = \ell_o\alpha(t_f - t_o)$, calculate α for the metal or alloy.

27:12 *Coefficient of thermal expansion apparatus.*

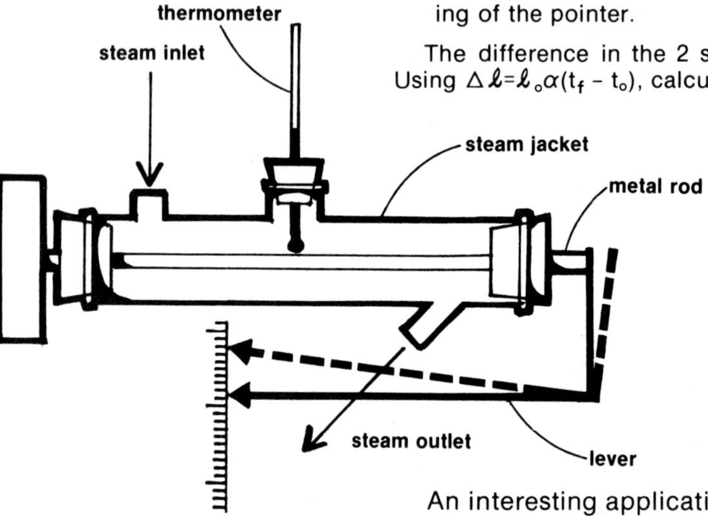

An interesting application of the expansion of metals is the use of bimetal strips in **thermostats** to control the temperature of an oven or a house. Steel and brass expand at different rates when heated. If strips of the materials are placed back-to-back and bonded together, they will bend when heated because one strip expands more than the other.

27:13 *A bimetallic strip. (a) Brass contracts more than steel when the strip is cooled. (b) Brass expands more than steel when the strip is heated.*

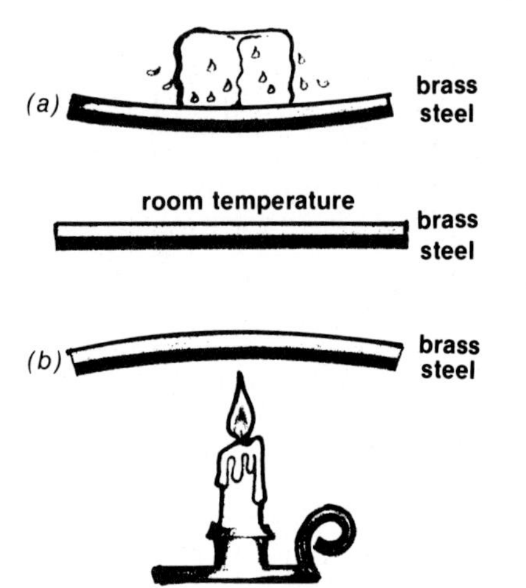

27:14 *A thermostat. As the temperature falls, the coil unwinds, touching the right contact (R) which closes a switch and starts electrical current flowing into the heater coils. When the unit is hot enough the coil winds up, touching the left contact (L) which closes a switch, turning off the electrical current.*

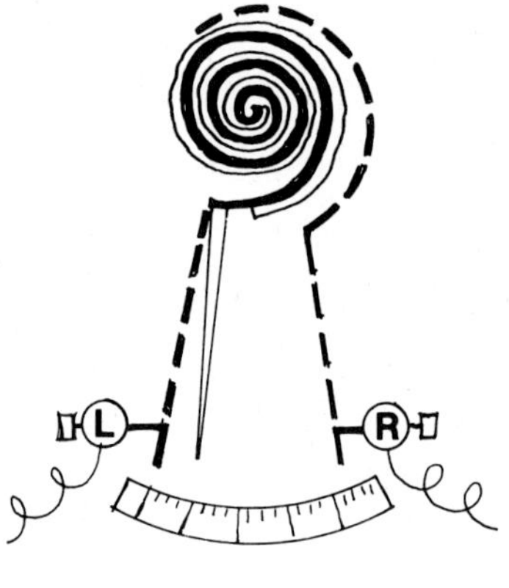

The purpose of the thermostat is to maintain a constant temperature. The bimetallic strip is coiled in a spring-like arrangement, as shown in Figure 27:14. As the temperature of the oven (or house) drops, the coil unwinds, bending the movable arm outward until it hits a contact and turns on the electricity to the coils. As the oven heats, the coil winds up. When the oven gets to the proper temperature, the movable arm moves away from the contact. The thermostat is a very simple way to control the temperature.

The expansion of solids caused by heat can become a problem and must be taken into consideration when designing certain structures. A civil engineer must design a bridge or highway to allow for contraction during the winter and expansion during the summer. Spaces are left between sections in a concrete highway to allow for expansion. The next time you drive across a steel bridge, look closely at the bridge to see how sections are fitted to allow for expansion and contraction.

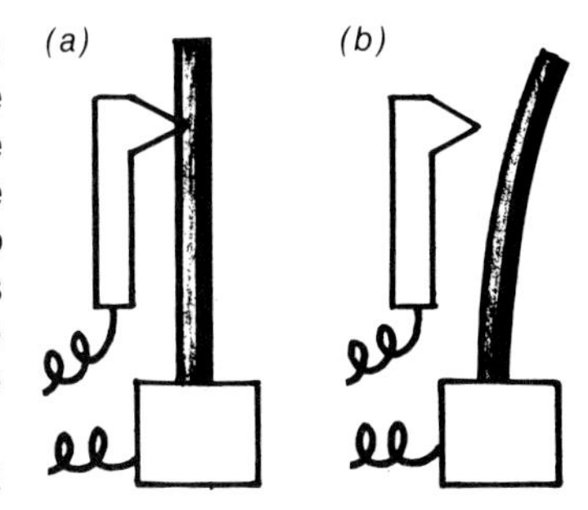

27:15 *A thermostatic relay. (a) The relay is in the closed position and current flows through the unit. (b) The relay is open and no current can flow through the unit.*

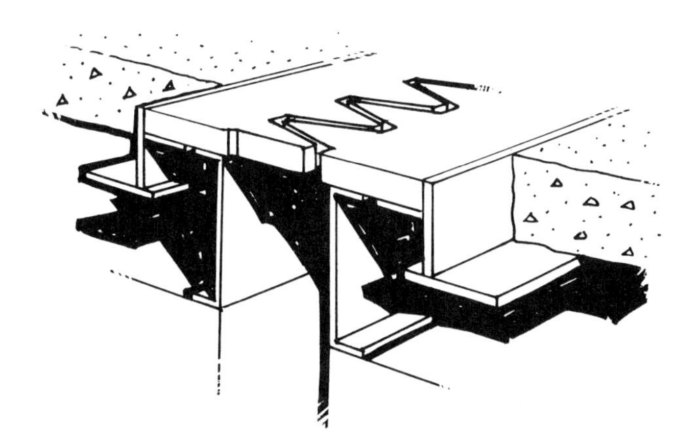

27:16 *Finger plate expansion joints used in highway and bridge construction.*

(a)

(b)

27:17 *(a,b) Think of all the expansion problems that hinder the design of piping systems in an oil refinery!*

★ **27:6 Volume Expansion** Actually, a solid or a liquid expands in all three dimensions when heated. For a solid, you generally can multiply the coefficient of linear expansion by three (for the 3 dimensions) to obtain the **coefficient of volume expansion**.

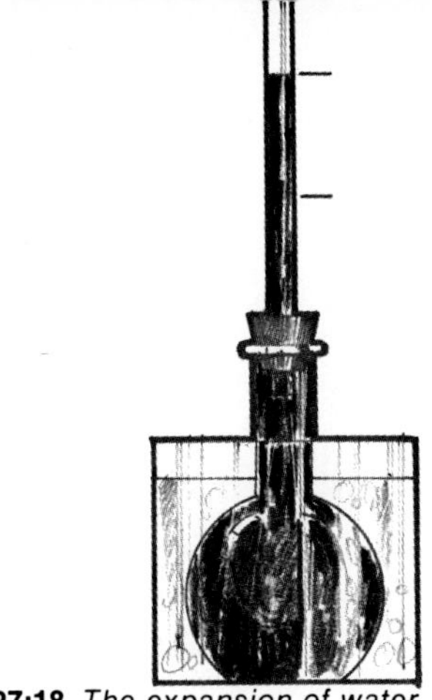

27:18 *The expansion of water.*

Experiment 27:5 *Volume Expansion of Water*

1. Add some red ink or food coloring to water and fill a small flask with the colored water.
2. Insert a rubber stopper and glass tube as shown in Figure 27:18.
3. Immerse the flask in a beaker of hot water. What happens? Why do you suppose this happens?

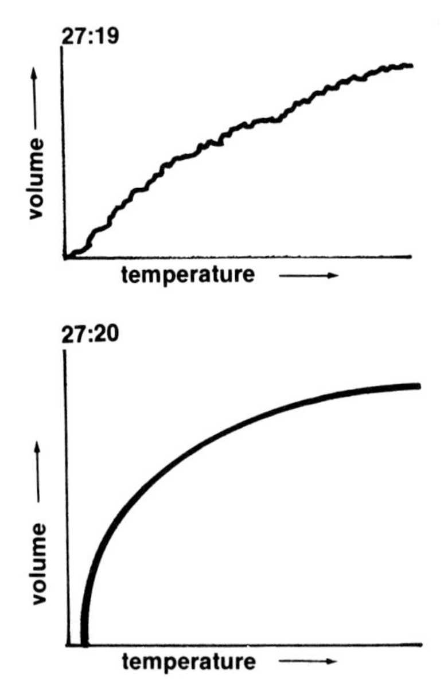

27:7 Liquids and Thermometers Liquids make the best thermometer materials for indicating temperature. You can confine a liquid in a small capillary tube so that it can expand or contract only up and down the tube. But what liquid is best? If a liquid has an expansion curve as shown in Figure 27:19, it would be almost impossible to get a sensible scale on a capillary tube. Also, if the liquid has expansion characteristics as shown in the scale in Figure 27:20, the capillary tube would probably look like the one in Figure 27:21. It would be impossible to read it at the higher temperatures.

27:21

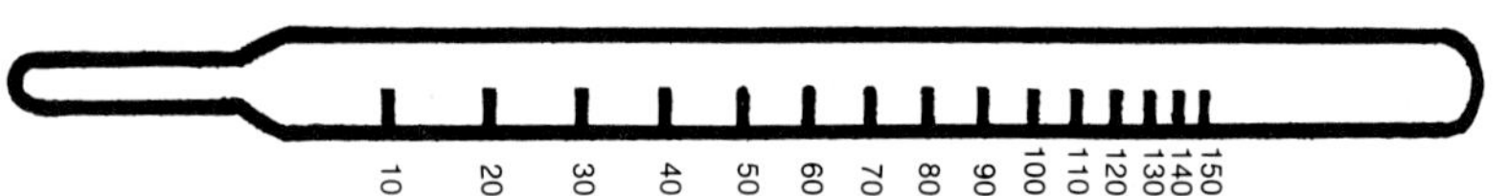

The liquid used in a thermometer needs to expand *uniformly*. Mercury does expand linearly and is, therefore, an ideal thermometer fluid. Mercury can be used up to about 350° C; it boils at 360° C. Above 350° C, other temperature-measuring devices must be used instead of mercury thermometers.

27:22 *Uniform expansion of mercury.*

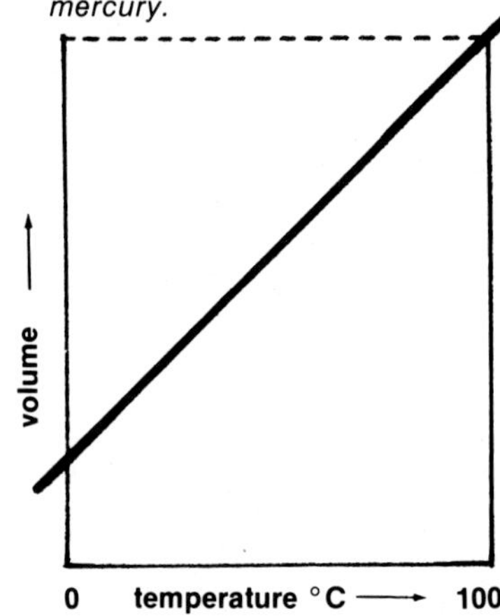

27:8 Water, the Exception Water expands as it freezes; thus, solid water (ice) is less dense than liquid water. As ice thaws, the water continues to contract as it heats up to 4° C. Above 4° C, the water begins to expand (see Figure 27:23). Water is most dense at 4° C. As you can see, water would not make a good thermometer fluid from 0° C to 10° C because of this behavior.

In the winter as the weather gets colder and colder, the water on the surface of a fresh-water lake reaches 4° C, becomes denser, and sinks. Warmer water is continually forced to the surface until it, too, reaches 4° C and sinks. Eventually, in colder climates the top layer of water will freeze, but the water deep in the lakes remains at 4° C. This is shown in Figure 27:24. Thus, water freezes *from the top down* in a lake.

The fact that water freezes from the top down in a lake is vital to the survival of living organisms in the water. If the water froze from the bottom up, life on the bottom of the lake would die; and creatures such as fish would be forced closer and closer to the top, where they would die because of extremely low temperatures. When a lake freezes on top, the ice helps prevent a mixing of warm and cold water, which would cool the deep water below 4°C. Normally, a lake will never freeze completely; thus, living organisms can survive.

Life as we know it could not exist in a lake or pond if it were not for this property of water. This is not a fortunate evolutionary accident; it is evidence of planning and design by a Creator.

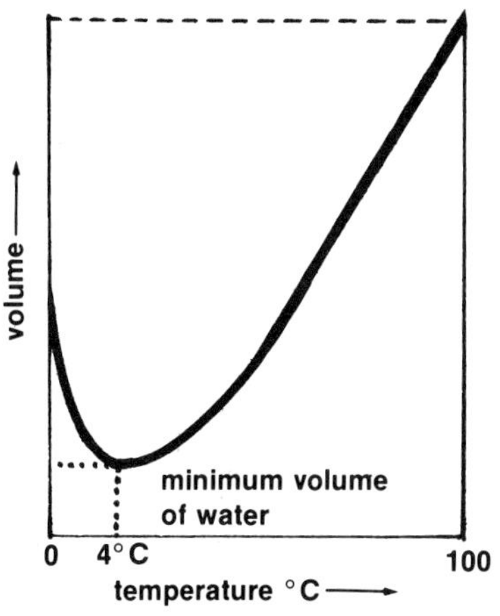

27:23 *Expansion of water.*

Experiment 27:6 *Expansion of Water from 4° C to 0° C*

1. Connect a glass capillary tube to a glass bulb as shown in Figure 27:25. Fill the apparatus with water to about half-way up the capillary tube.
2. Submerge the bulb in a large beaker partially filled with water. Keep a thermometer in the large beaker also, so that you can observe the temperature.
3. Add crushed ice slowly to cool the water. The water will continually contract, as shown by its retreat down the capillary tube until 4° C. From 4° C to 0° C, it should advance back up the capillary tube.

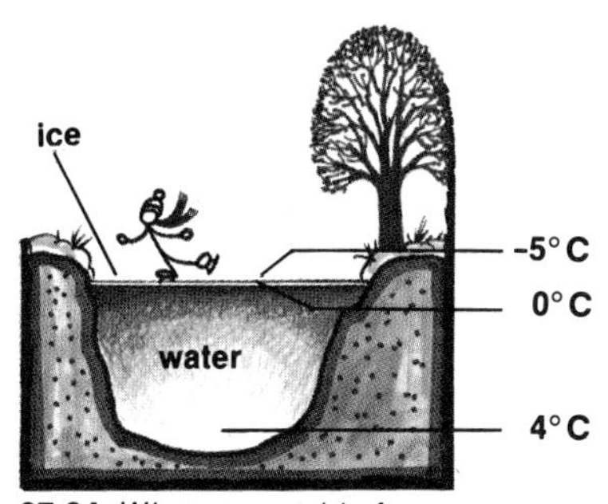

27:24 *When a pond is frozen over, maximum density (warmest) water is at the bottom of the pond.*

27:9 Heat as Energy Now that we can measure the *degree of heat* a substance has, what is heat itself? To help explain this, we will do three experiments.

capillary tube

27:25 *As the temperature drops from 4° C to 0° C, what happens to the water level in the capillary tube?*

Experiment 27:7 *Heat from Mechanical Work*

1. Fill a beaker half full of room-temperature water.
2. Insert a thermometer into the water.
3. Stir the water vigorously for 5 minutes.
4. Note that the temperature of the water rises.
5. You have generated heat by doing mechanical work.

Experiment 27:8 *Heat from Friction*

1. Rub two pieces of wood together rapidly.
2. Feel the surfaces that have been in contact. Are they warm? Again you have generated heat by doing mechanical work.

27:26 *(a) Temperature of cool water in a beaker. (b) Temperature increases after considerable stirring.*

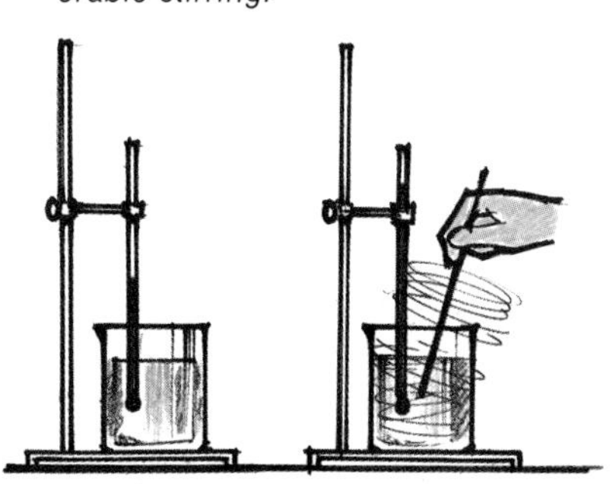

Experiment 27:9 *Mechanical Work From Heat*

1. Fill a metal can half full of water.
2. Insert a rubber stopper with a glass tube through it into the top of the can.
3. Boil the water.
4. Place a plastic pinwheel fan directly over the jet of steam. The fan will be turned by the exiting steam. The burning of the gas generates heat. The heat causes the water to boil; the boiling water does work by turning the fan blades. Heat can be changed into work.

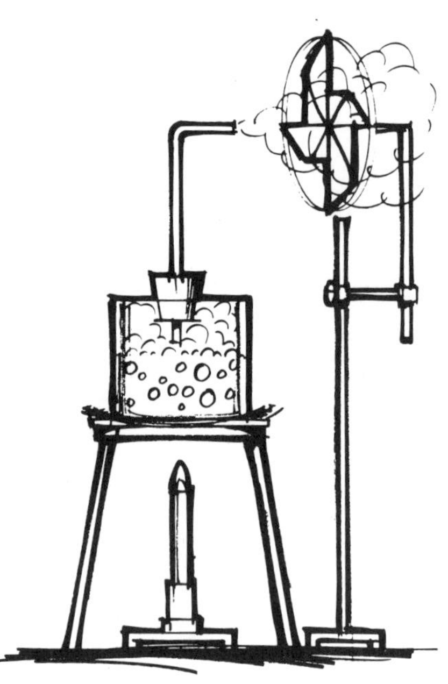

27:27 *Boiling water gives off steam, which can do work (turn a fan blade).*

27:28 *Fossil fuel steam generating plant.*
(See chapter title page)

27:29 *Steam produced by boiling water will turn these turbine blades to generate electricity.*

Heat and work are both forms of **energy** and are interchangeable. Energy is *the ability to do work*. Heat can do work, and for that reason it is **thermal energy**. That is all we know about heat; we can use it and change it into other forms of energy, but besides giving a few definitions, we know very little about heat. But do we really need to know everything about heat? No, as long as we know enough to be able to use heat for our benefit, that is all that counts.

One of the discarded ideas of science was that heat was a fluid called **caloric**. Today, we consider both heat and matter as energy, although they are different forms of energy.

Using the kinetic theory of gases (Chapter 5) that matter is thought to be made up of tiny particles called atoms and molecules, scientists have tried to discover how an object could accept heat on the molecular level. As you heat a gas, it expands; but what is actually happening on the molecular level? We cannot see anything that small, but we can use our imaginations to construct a mental model.

How does the heat go into the gas? We imagine that each molecule goes faster and faster as we continue to heat the gas (raise the temperature of the gas). The heat is converted into **kinetic energy** of the molecules. Heat causes the particles of matter to move faster. Remember this is just a workable scheme; we do not know what actually happens.

27:10 Heat and Electricity Heat is used to generate electricity. Heat energy is transformed into electrical energy. As the steam turns the fan (pinwheel) in Experiment 27:9, so steam is used to turn a blade in a turbine to generate electricity. Our modern way of life depends upon electricity; electricity is used to do work.

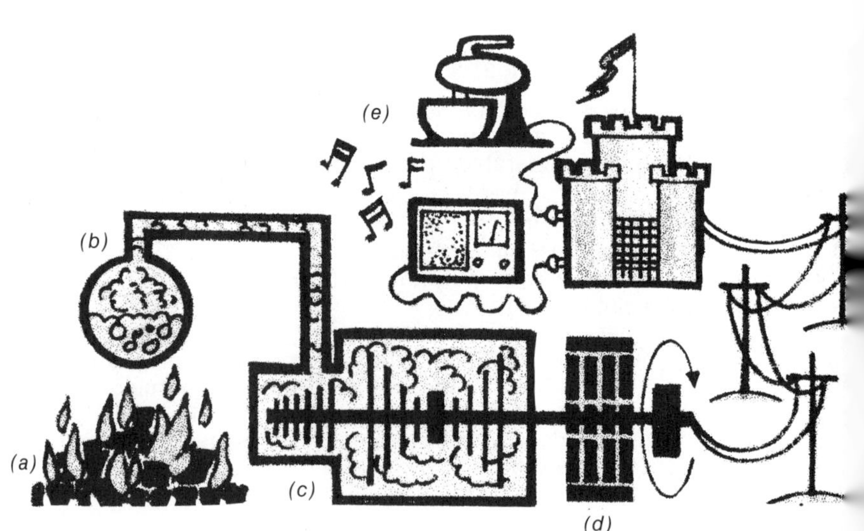

27:30 *How you may get electricity to operate your castle (home): (a) Coal or oil is burned (b) to generate steam (c) to turn turbine blades (d) which generate electricity (e) which mixes a cake or plays a radio.*

List of Terms

absolute temperature scale	Fahrenheit temperature scale
absolute zero	heat
caloric	kinetic energy
Celsius temperature scale	temperature
Charles' law	thermal energy
coefficient of linear expansion	thermal expansion
★ coefficient of volume expansion	thermometer
energy	thermostat

Questions

1. Name three ways heat affects your life.
2. Why are the terms hot, cold, warm, and cool not very scientific?
3. Define temperature.
4. What is Charles' law?
5. What does "calibrate" mean?
6. According to the theory of ideal gases, what happens to a gas at absolute zero?
7. Calculate the following.
   (a) 14° C = ________° F
   (b) 387° F = ________° C
8. Calculate the following.
   (a) 72° C = ________° K
   (b) 840° K = ________° C
9. Calculate the following.
   (a) -20° F = ________° K
   (b) 280° K = ________° F
10. (a) Do liquids normally expand or contract (shrink) when heated?
    (b) Do solids normally expand or contract (shrink) when cooled?
11. ★ Calculate how much linear expansion can be expected when a 10-meter pyrex glass rod is heated from 25° C to 432° C.
12. ★ A brass rod three meters long is cooled from 300° F to -30° F. How much did the rod contract along its length?
13. ★ A 200 cm lead rod is heated from 0° C to 100° C. What is the length of the rod at 100° C?
14. What material in a thermostat is used to control temperature?
15. Why do expansion joints have to be placed in bridges?
16. (a) What is unusual about the behavior of water around 4° C? Explain.
    (b) Why is there design behind this?
17. (a) The volume expansion of a substance is about ________ times that of its linear expansion.
    (b) Why is mercury such a good thermometer fluid?
18. Is heat a form of energy? How can you prove this? Can heat do work?
19. ★ Why does filling a balloon with hot air cause it to rise?
20. Name five items in your home that run on electricity.

Student Activities

1. Write a paper on thermostats. Get as many pictures of different kinds of thermostats as you can to include in the report.
2. Write a paper on gas thermometers. Discuss how and when they are used. Write to the National Bureau of Standards in Washington, D.C. 20234 for information.

Photo next page
Attractive fountains on the Bob Jones University campus form part of an efficient air conditioning system. Water is cooled by spraying it into the air.

UNIT IX HEAT

Chapter 28 Heat Flow

28:1 Units of Thermal Energy Since heat is a form of energy, we must develop some way to measure how much energy is used in certain operations. Just as we use the pound as a unit of weight and the gram as a unit of mass, so we can specify units of energy. The unit of heat energy is the **calorie**. The calorie is *that quantity of energy that will raise the temperature of one gram of water 1° C.*

Different substances require differing amounts of heat for the same increase in temperature. For instance, one gram of iron requires only about 1/10 of a calorie to raise its temperature 1° C. This difference in heat absorption is designated by a quantity called the **specific heat**. The specific heat is *the heat absorbed by 1 gram of a substance when its temperature is raised 1°C.* Specific heat (c) is normally expressed in calories per gram per degree Celsius, or cal/g° C. Table 28-1 lists the specific heat of some common substances.

When you calculate work in *joules*, there is a way to convert to calories:

$$1 \text{ calorie} = 4.18 \text{ joules}$$

or

$$0.239 \text{ calorie} = 1 \text{ joule}$$

In the English system a unit called the British Thermal Unit (Btu) is used. One Btu is the amount of heat needed to raise the temperature of a pound of water one degree Fahrenheit. It is related to the calorie as follows:

$$1 \text{ Btu} = 252 \text{ calories}$$

or

$$1 \text{ calorie} = 0.00397 \text{ Btu}$$

Table 28-1 *Specific Heats of Some Substances*

Substance	Temperature or temperature range. °C	Specific heat, cal/g°C
Solids		
aluminum	15-100	0.22
asbestos	20- 98	0.195
copper	15-100	0.093
glass	20-100	0.20
ice	-3	0.50
iron or steel	20	0.11
lead	20	0.0306
silver	20	0.056
Liquids		
benzene	20	0.41
mercury	20	0.033
methyl alcohol	20	0.60
water	15	1.00
Gases		
air	50	0.25
steam	110	0.48

Problem: How much heat energy is required to raise 20 grams of copper from 15°C to 75°C?
Solution: The formula to be used is

$$Q = mc(t_f - t_i)$$

Q is the amount of heat absorbed by a substance in calories; m is the mass of the substance; c is the specific heat of the substance; t_f is the final temperature; and t_i is the initial temperature.

For this problem, m = 20 grams; c = 0.093 cal/gm°C; t_f = 75°C; t_i = 15°C.

Therefore,

$$Q = 20(.093)\ (75-15)$$
$$Q = 1.86\ (60)$$
$$Q = 111.6 \text{ calories}$$

Problem: 50 grams of water absorbed 200 calories of heat. If the initial temperature is 1°C, find the final temperature.
Solution: Q = 200 cal; c = 1 cal/gm°C; t_i = 1°C; t_f = ?

Therefore,

$$200 = 50\ (1)\ (t_f-1)$$
$$200 = 50t_f - 50$$
$$50t_f = 250$$
$$t_f = 5°C$$

Experiment 28:1 *The Specific Heat of Lead*

1. Weigh out about 200 grams of lead shot.
2. Place the shot in a metal cup; then insert the cup into a water boiler as shown in Figure 28:1.
3. Insert a thermometer into the shot.
4. Heat the water to boiling.
5. Place the lead shot over the boiling water for at least 10 minutes. Record the temperature of the lead.
6. Weigh the aluminum inner cup and stirring rod of the calorimeter shown in Figure 28:2.
7. Fill the inner cup about two-thirds full of water and weigh the cup, stirring rod, and water together.
8. Cool this water to at least 5°C below room temperature.
9. Stir the water in the calorimeter (not vigorously). Record the temperature of this water to the nearest 1/10 of a degree, if possible.
10. Pour the lead shot into the calorimeter.
11. Stir slowly by moving the stirrer up and down. Record the highest temperature of the calorimeter water after the lead shot has been poured into it. Be sure the thermometer does not touch the shot or the sides of the cup. Use $Q = mc(t_f - t_i)$ to find the specific heat of lead.

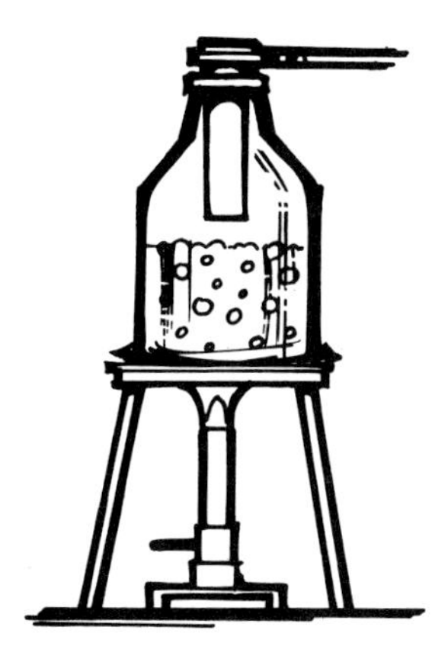

28:1 *Lead shot is heated in a water boiler.*

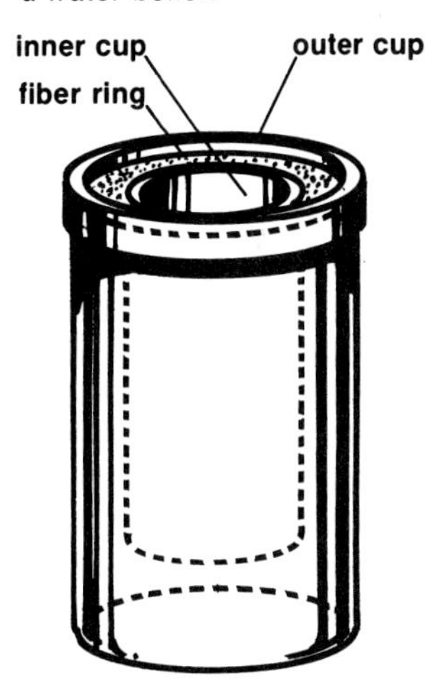

28:2 *A calorimeter. A fiber ring supports the inner cup and insulates it.*

★ **Experiment 28:2** *The Temperature of a Body*

1. Weigh a small metal object with a known specific heat. Set up the calorimeter as explained in Experiment 28:1.
2. Using a rod as shown in Figure 28:3, hold the metal piece in a Bunsen burner flame for about 30 seconds.
3. Drop the metal piece carefully into the calorimeter water and measure the temperature increase.
4. Stir as in Experiment 28:1. Using $Q = mc(t_f - t_i)$, find the initial temperature of the metal object.

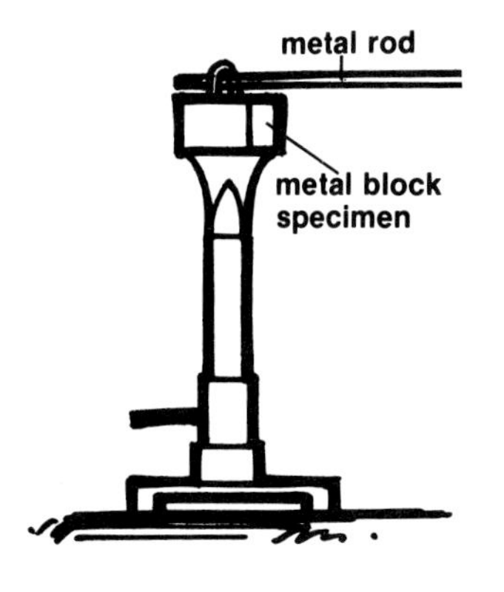

28:3 *Heating a block of metal.*

28:2 Heat Movement by Conduction If you hold one end of a copper rod in a Bunsen burner flame, the end of the rod that you are holding warms up rapidly. This happens because heat is transferred through the rod. This type of heat flow is called **conduction**.

There are two ways to visualize thermal conduction (heat flowing) in the rod. One way is to imagine that, as heat is applied to the end of the rod, the atoms on that end start vibrating rapidly; they bump or collide with the atoms nearest to them, transferring some of their (kinetic) energy to them. This bumping continues down the length of the rod; and the heat, therefore, moves down the rod in the form of kinetic energy of the atoms.

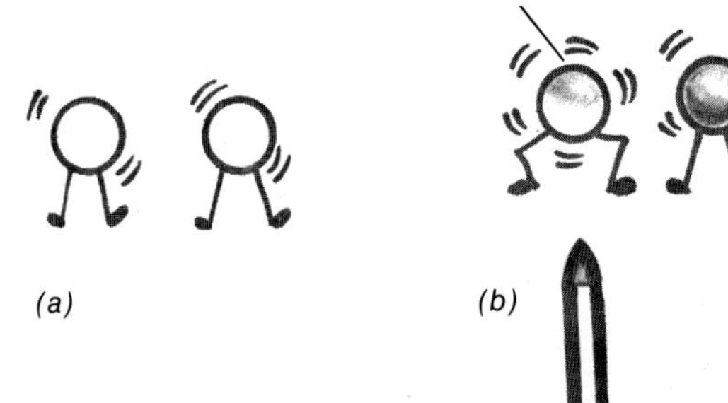

28:4 *Model of conduction through a solid by atomic collisions. (a) unheated end of rod atoms rocking slightly (b) heat applied (c) heat transfer (d) heat transfer*

Another way of looking at this is to imagine heat as energy waves being transmitted through the rod in the same way that a wave moves along a string. The atoms connected together by imaginary springs *oscillate* (rock back and forth) in all three dimensions. As heat is applied, the atoms on the end begin to oscillate violently. These oscillations are picked up by the atoms nearest to them through the springs; and the heat, therefore, travels down the rod in wave-like oscillations.

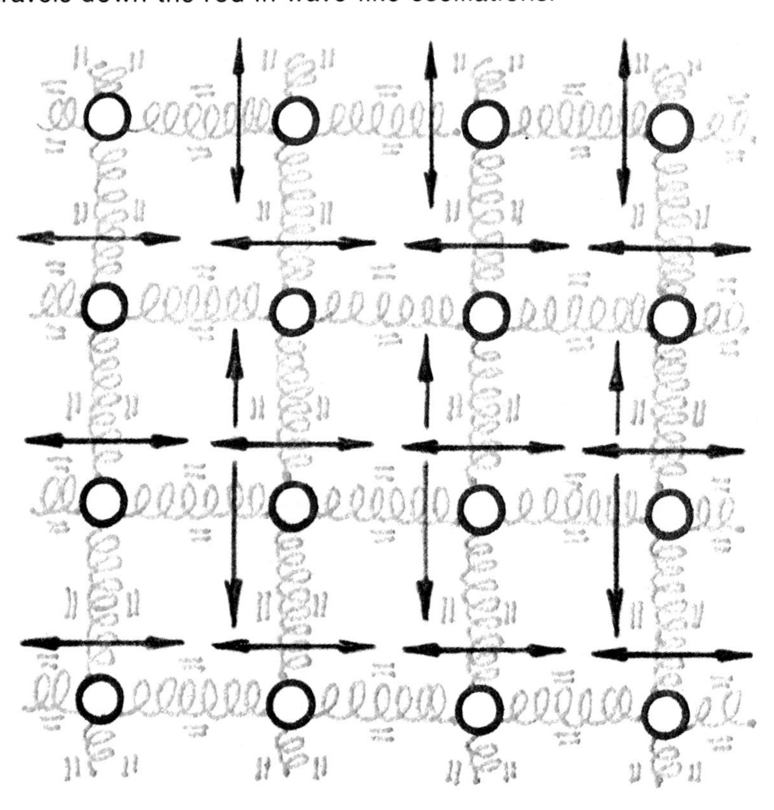

28:5 *Model of heat conduction through a solid by wave motion. Heat is applied to a solid, and the atoms begin to vibrate. These vibrations are carried through the solid much the same as balls (atoms) vibrate when hooked together by springs (interatomic forces).*

The springs can be considered the bonding forces between the atoms. We can illustrate this action by tying equal weights to a stretched wire at equal intervals. As you shake or vibrate the end weight, the vibration is carried down the entire wire.

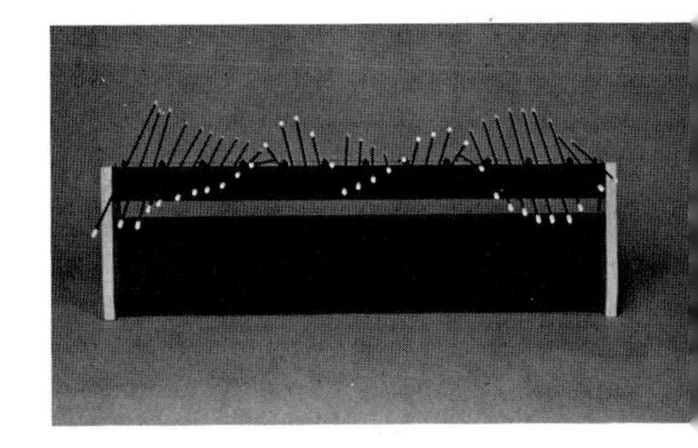

28:6 *The wave demonstrator was set into motion by shaking one of the end rods. The vibration is carried down the entire wire.*

Experiment 28:3 *Difference in Thermal Conductivity of Materials (1)*

1. Obtain a glass rod and a metal rod.
2. Hold each rod the same distance from one end and place the end into a Bunsen burner flame as shown in Figure 28:7. Which rod gets hot first? Why do you suppose this happens?

Heat flows through different substances at different speeds. Glass is not a good conductor of heat; this is why glass may break when a hot liquid touches it. One side of the glass expands from the heat; however, the other side of the glass is still fairly cool and does not expand because the heat has not been conducted fast enough. This sets up large thermal stresses within the glass causing it to shatter.

Pyrex and Vycor glass have been developed to overcome much of this problem. They do not expand as much when heated as regular glass does.

Substances can be classified according to their ability to conduct heat. Most metals, such as iron, aluminum, and copper, are good conductors of heat. It is thought that the free electrons aid in the conduction of heat in metals by carrying added kinetic energy away from the source of heat, thus transferring the energy through the metal structure.

Insulators, such as asbestos, cork, and gases, are poor conductors of heat. Since gases are poor conductors, any material that has dead air spaces (places where air is unable to flow) in it will be a good insulator. Rock wool, asbestos, and fiberglass house insulation are deliberately manufactured so that there will be many air spaces in the material.

28:7 *Heating a glass and metal rod.*

Experiment 28:4 *Difference in Thermal Conductivity of Materials (2)*

1. Wrap one layer of paper tightly around a copper rod and a wood rod butted together as shown in Figure 28:8.
2. Move the paper-covered section back and forth in a Bunsen burner flame. Keep the flame as close to the wood-copper joint as possible.
3. Remove from the flame and inspect. Is the paper over the wood rod scorched? Is it scorched over the copper rod? Explain.

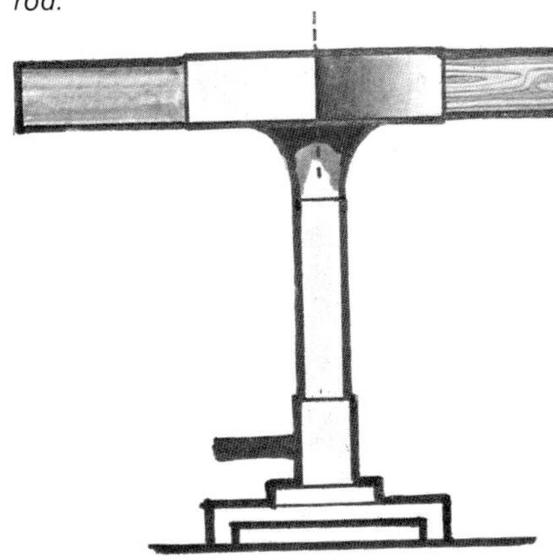

28:8 *Heating a copper and wood rod wrapped in paper.*

28:9 *A workman putting heat-insulation material in the walls of a house.*

28:3 Heat Movement by Convection Heat transfer by convection occurs mainly in gases and liquids. **Convection** means that when a certain amount of gas or liquid is heated, the entire amount (matter and heat) moves—the heat is in the form of kinetic energy of the molecules.

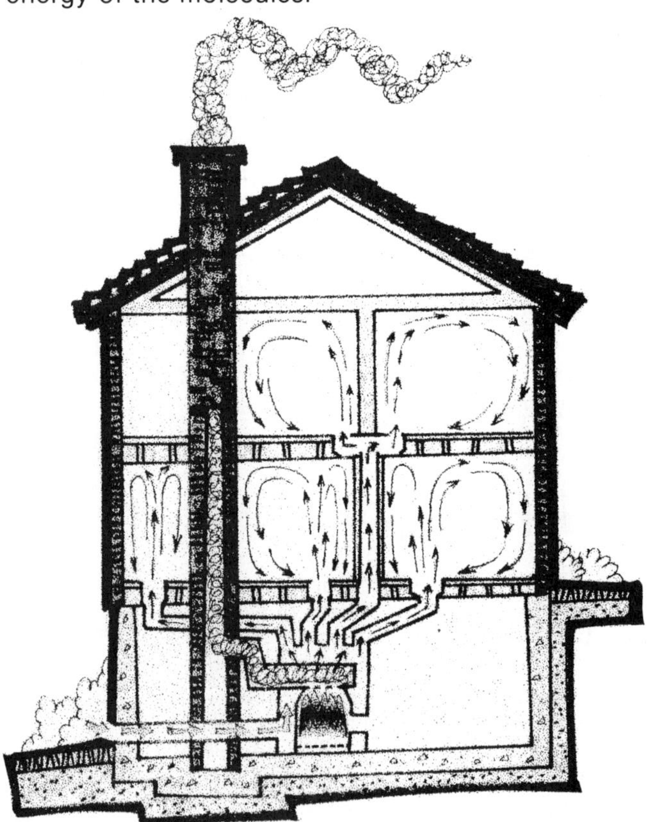

28:10 *Convection currents in a house caused by heating with a hot-air unit.*

In conduction, heat is transferred from molecule to molecule; in convection, the heat remains *with* the molecule. Some conduction occurs during convection; however, the amount is so small that it is not very important in most cases. If you heat water in a beaker with a Bunsen burner, you should be able to see the **convection currents** circulating within the water. To see them, your eyes must be on the same level as the beaker.

One of the best illustrations of convection is a forced-air heating system for a house. The heated air is less dense than the cold air and tends to rise, displacing the cold air above it. The hot air strikes the ceiling and spreads throughout the room. More warm air coming into the room pushes cooler air downward to the floor. This cooler air returns to the furnace through a return vent, usually located in the floor of the house.

Winds are simply convection currents in the earth's atmosphere. The heating effect of the sun causes these currents. The land receives heat from the sun's rays; then the air which comes in contact with the land heats up. This heated air rises because it is less dense than the cooler air. The cooler, heavier air is forced down toward the earth; gravity encourages the movement of the heavier air downward. These convection currents, combined with the rotation of the earth, account for nearly all of the winds. **Trade winds** and **sea breezes** are examples of this kind of wind.

A sea breeze is a clue to the fact that land heats up much faster than water. This heated air over the land rises and is replaced by cooler air which blows in from the water to equalize the pressure.

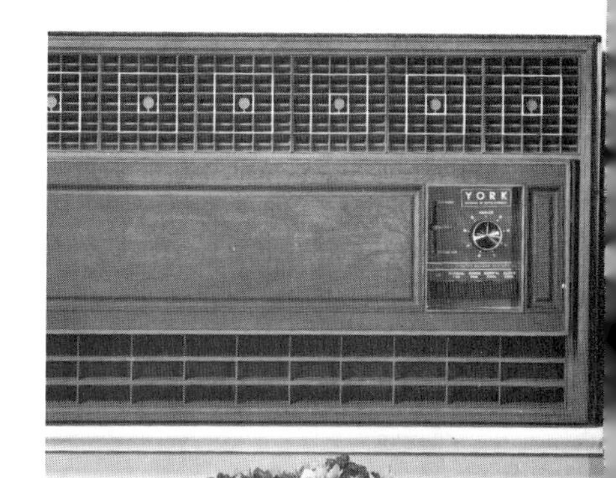

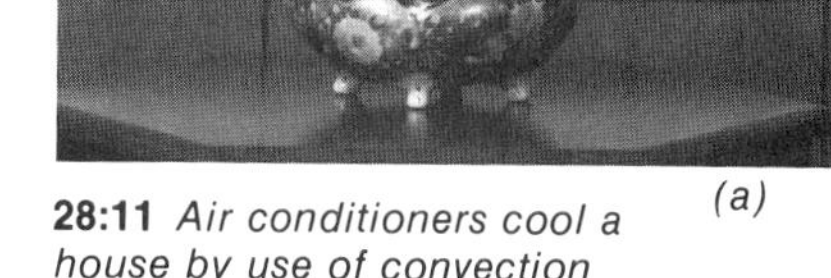

28:11 *Air conditioners cool a house by use of convection currents also. (a) outside of unit (b) inside of unit*

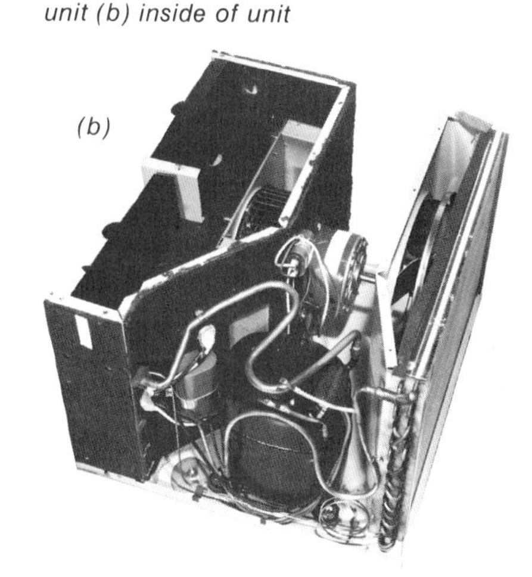

28:12 *Prevailing surface winds in North and South America.*

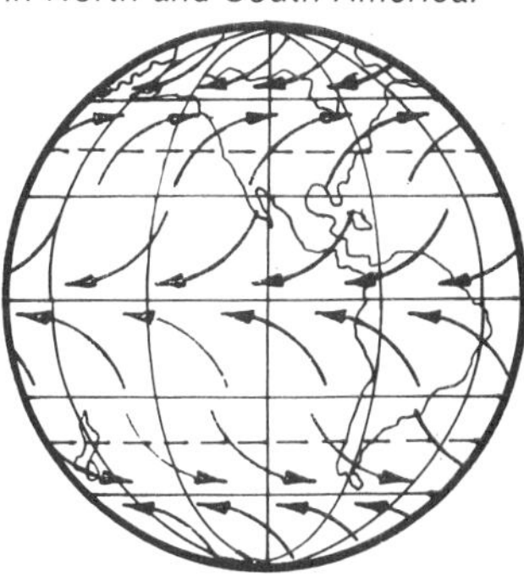

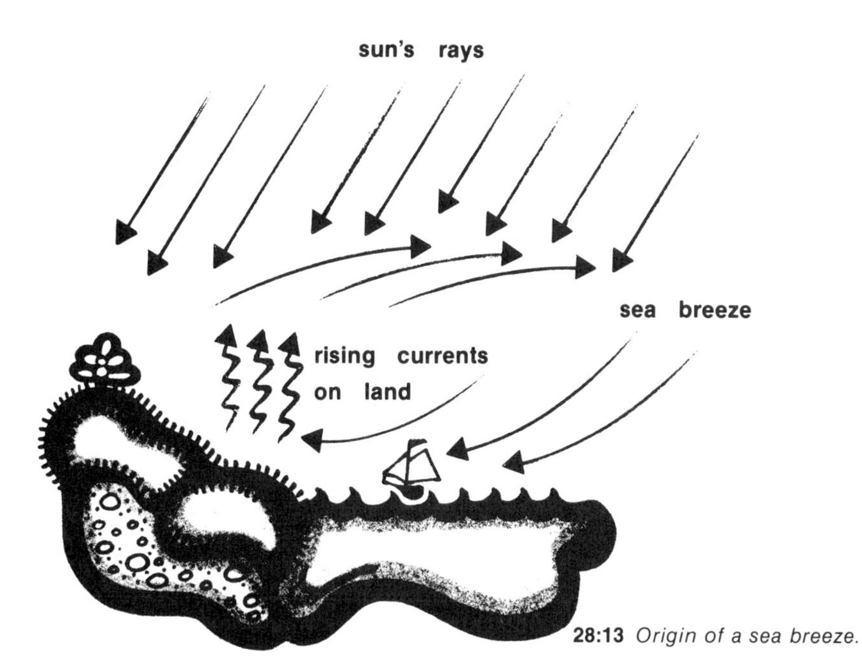

28:13 *Origin of a sea breeze.*

28:14 *This old farmer depends mainly on the process of radiation to warm himself by the stove.*

28:4 Heat Movement by Radiation When matter is not involved in the transfer of heat, the energy transportation process is called **radiation**. Picture the heat from the sun hitting the surface of the earth. The heat must travel through "outer space," which is almost a vacuum. The farmer in Figure 28:14 backing up to a pot-bellied stove is heated by radiation. Light is radiant energy. Other types of radiant energy, also called **electromagnetic radiation**, are shown in Table 28-2.

Table 28-2 *Electromagnetic Radiation*

Type	Usual Source
radio waves	electric circuits
infrared radiation (heat)	hot objects
visible light	hot objects, electric arcs
ultraviolet radiation	electric arcs
X-rays	rapid deceleration of electrons
gamma rays	radioactive nuclei

If you leave a cup of water uncovered on a clear, cool night when the temperature is above freezing, many times a small amount of ice will form on the surface of the water. The water has actually been radiating its heat into "outer space," which has a temperature of about 3° K. The water actually became cooler than the atmospheric temperature through the process of radiation.

Dark objects absorb more radiant energy than light objects. Experiments 28:5, 28:6, and 28:7 should convince you of this.

Experiment 28:5 *Effect of Color on Heat Transfer*

1. Obtain 2 flasks, 2 thermometers, and 2 one-hole stoppers.
2. Paint one of the flasks black and the other white.
3. Fill each flask to overflowing with warm water; then insert the stoppers and thermometers as shown in Figure 28:15.
4. Place both flasks in the shade. Record the temperature initially and once every 5 minutes for 20 minutes. Which flask cooled faster? Why?

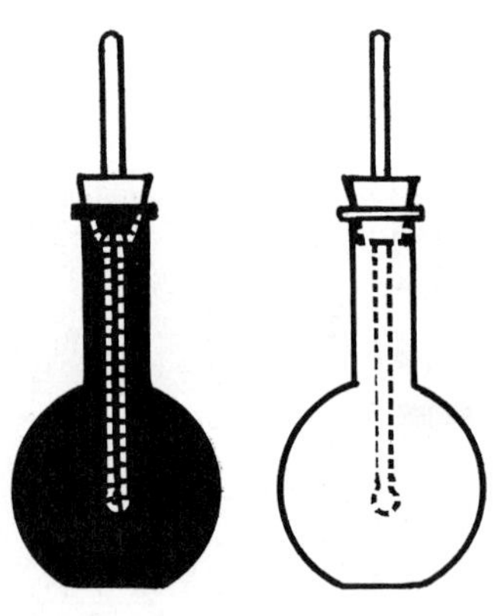

28:15 *Which thermometer records the higher temperature?*

Experiment 28:6 *Effect of Color on Radiant Energy Absorption (1)*

1. Obtain 2 thermometers, a strip of black paper, and a strip of white paper.
2. Place the strips of paper side by side underneath a heat lamp.
3. Place a thermometer under each strip.
4. Turn the lamp on for 10 minutes. Record the temperatures every minute. Explain why one thermometer records a higher temperature than the other.

Experiment 28:7 *Effect of Color on Radiant Energy Absorption (2)*

1. Obtain 2 thermometers, and wrap black paper tightly around the bulb of one of the thermometers.
2. Place them in a window in the sun as shown in Figure 28:16. Why does the black bulb thermometer read higher?
3. Place both thermometers in the shade. Why do they both read the same temperatures?

28:16 *Why is the thermometer surrounded by the black bulb hotter in the sun?*

Radiant energy can be detected with the use of a **radiometer**. A radiometer is constructed by mounting four vanes on an almost frictionless rotor. One side of the vane is black, while the other side is white. The black side absorbs the radiant energy; the white side gives off radiation to the air molecules inside the bulb. These energetic molecules then strike the black side of the adjacent vane with a force strong enough to cause the vanes to rotate (see Figure 28:17).

28:5 The Thermos Bottle The **thermos** or **vacuum bottle** is an interesting example of how the three mechanisms of heat transfer—conduction, convection, and radiation—are slowed down. The vacuum bottle is made of double-walled glass with a vacuum between the walls. Conduction through the vacuum is impossible. (Heat can be conducted up the glass walls and through the plastic or cork stopper, but this is extremely slow, because glass, plastic, and cork are poor conductors.) The vacuum also prevents any heat loss by convection, because convection requires matter. The silvered surfaces on the vacuum bottle reflect heat inward or outward, depending upon the direction from which the heat is coming. This prevents heat loss or gain by radiation.

28:17 *A radiometer.*

28:6 Fossil Fuels Much of our heat today is still generated by the so-called **fossil fuels**—coal, oil, and gas. The important elements in these fuels are *carbon* (in coal) and *carbon and hydrogen* (in gas and oil). Large quantities of heat are released as carbon is oxidized to form carbon dioxide, and hydrogen is oxidized to form water vapor.

How did this energy come to be stored in the fossil fuels? Presumably, plants and animals that lived a long time ago became buried in such a way that their decay was, to a great extent, prevented; their energy was thus locked in and preserved. Scientists disagree on how and when this burial process took place. Evolutionists, who think that slow, gradual processes were responsible for the formation of the earth, say that the burial process occurred over vast periods of time. Even using an imaginary time scale of millions of years, however, they are unable to explain how the burial of the plant and animal remains could occur quickly enough to prevent their decay and deeply enough to change them chemically into their present form. Scientists who believe the Biblical framework of earth history, on

28:18 *Thermos bottles.*

the other hand, say that the formation of fossil fuels is due to the *sudden, deep* burial of plants and animals at the time of the Flood (see Genesis 6-8). The Flood model (explanation) appears to be superior in at least three significant respects: its ability to supply the necessary raw materials, the rapidity of their burial, and the depth of their burial.

One proof that the evolutionists are wrong when they talk in terms of millions of years comes from the radiocarbon dating method. When the three fossil fuels—coal, oil, and gas—are tested for age using this method, they are found to be only thousands of years old, not millions. You probably have not heard about these findings. Because they are so contrary to the conventional thinking of today, these test results are generally ignored. One of the more unfortunate aspects of "modern science" is that too often facts are ignored because they do not fit the widely believed theories.

Coal

28:19 *A crane that can handle large amounts of coal.*

You have no doubt seen pictures of "prehistoric peat bogs." These are nothing more than *drawings*, often very imaginatively done. Evolutionists hire artists and tell them what they wish to be drawn. The art work is then displayed in museums and printed in books as though it were absolute fact. Remember, no human was there to observe these "prehistoric peat bogs." They are purely conjectural (guesswork) and, therefore, do not come under the heading of true science. Even if the existence of these bogs could be positively established, there is no mechanism known by which they could change to coal.

A number of items found embedded in coal tend to cast doubt on the evolutionary peat bog theory. The boulders, cobbles, and petrified tree trunks so frequently found in coal seams, which were obviously washed into place from other localities, indicate rapid currents which occur under flood conditions, rather than the tranquil, undisturbed growth demanded by the peat bog theory. Such rapid currents are also evidenced by the intermingling of the layers of coal with other sedimentary rocks. Finally, artifacts (objects made by human hands) that have been found in coal clearly show that the evolutionary time scale is an outrageous lie. For example, an iron pot and a gold chain have been found completely embedded in coal (see the *Creation Research Society Quarterly*, March 1971, page 201, and the *Bible-Science Newsletter*, July 1970, page 2).

Evolutionary theories cannot explain discoveries such as these. The coal is supposed to have formed millions of years before man "evolved," yet the evidence is clear that civilized man was on the scene *before* the coal was formed. These facts make evolutionary theories look like nonsense, but they fit quite well into a Biblical time scale. In Genesis 4, metalworking was already highly developed. Tubal-cain was an instructor of every artificer in brass and iron (Genesis 4:22). In Genesis 7 and 8, the Flood buried the antediluvian (pre-Flood) civilizations, encasing many of their artifacts in coal and other sedimentary rocks.

Recent research has shown that *heat* and *pressure*, rather than *time*, are the important factors in the formation of coal. It has long been known that the grade of coal (that is, whether it is anthracite, bituminous, or of some lower rank) is proportional to the depth of burial (*Hilt's law*). The greater the pressure due to the overburden, the higher the grade of the coal. Two samples of coal having the same evolutionary "age," but located at different depths, can be quite different in their grade.

A very significant fact is that a high grade of coal has been produced from wood in the laboratory by the application of heat and pressure; the time required for this transformation is not millions of years, but only a few hours (see *Chemical Technology*, May 1972, page 296; or *Acts and Facts*, Volume I, Number 4). While this in itself does not *prove* the correctness of the Flood model, it does demonstrate that there is nothing magic about "time" for bringing about the conversion of wood to coal.

Oil

Walter Sullivan, science editor of the *New York Times*, says in his book, *We Are Not Alone*, that six major universities worked on the problem of the origin of petroleum for 10 years with no success. Those who conducted the research became so desperate for an explanation that they even considered the theory that vapor from pine trees had condensed and fallen to the ground in the form of rain, where it accumulated over long periods of time to form oil. This theory, like all the others, has numerous defects and has not been taken seriously by the scientific world.

28:20 *Offshore oil rig.*

Other individuals with even stranger ideas have speculated that petroleum was in some way dumped onto the earth from outer space. This idea simply removes the problem of petroleum's origin from the earth. However, no petroleum has been observed in or on other astronomical bodies. Therefore, no one has any right to speak about its existence in space just because he cannot determine its earthly source.

Still another theory, one that has gained popularity with some Russian scientists, is that oil can form from a *non-biological* source. However, a majority of geologists feel that oil and gas came largely from marine organisms, such as fish. But within the framework of conventional geology there is no means known whereby enough of these creatures could have been trapped and encased in sedimentary rock to produce what is present in the oilfields today.

Many of the things that used to be "known" are now being challenged and restudied. Evolutionary speculations have proved to be unfounded. Christian men of science realize that the evidence demands a huge catastrophe, such as the Flood.

Dr. Melvin Cook, a physical chemist who has studied the fossil fuels from a creationist standpoint, points out that the pressure under which oil is found could not have continued for millions of years. Such pressures would be reduced after only a

28:21 *A natural gas pipeline is being fitted together.*

few thousand years, he believes, by leakage into the surrounding rocks. This agrees with the radiocarbon dates of *thousands* rather than millions.

With oil as with coal, heat and pressure are the key factors rather than time. United States Bureau of Mines scientists have performed experiments in which organic materials, such as garbage and manure, have been converted into oil by the application of heat and pressure (see *Chemical and Engineering News*, May 29, 1972, page 14; also, Reginald Daly, *Earth's Most Challenging Mysteries*, page 231).

Gas

Many scientists are of the opinion that natural gas forms as a by-product of the oil formation process. But again, there is no general agreement among them. Presumably, oil and gas are what remains when organic materials decay until only hydrocarbons (compounds of hydrogen and carbon) are left. Oil and gas are usually found together in thick marine (ocean-derived) sediments.

Recently, a new theory has been advanced that gas can also form *inorganically* by the reaction of molten igneous rock with underground limestone (see *Oilweek*, April 5, 1971; the *Wall Street Journal*, March 31, 1971; and the *Bible-Science Newsletter*, June 1971, page 5). Heat from the igneous rock is said to break down the limestone to form carbon dioxide. This in turn reacts with the hydrogen in the igneous material to produce hydrocarbons.

Research still continues on many of these interesting questions. In view of all the evidence against it, it is shameful that evolution is taught in many places as the true answer and the only answer. The very things that are still in question are often used as "proof" *for* evolution, thus completing an illogical circle of reasoning. Future research, we believe, should demonstrate even more convincingly how superior the Flood model is as a means of understanding the formation of the fossil fuels.

28:22 *Tanks used for storing butane or propane gas.*

List of Terms

calorie
coal
conduction
convection
convection currents
electromagnetic radiation
fossil fuels
gas
oil
radiation
radiometer
sea breezes
specific heat
thermos or vacuum bottle
trade winds

Questions

1. What is a calorie?
2. Define specific heat.
3. How much heat energy is needed to raise the temperature of 80 grams of aluminum from 20° C to 95° C?
★ 4. It was determined that 200 grams of benzene absorbed 800 calories of heat. If the final temperature of the benzene is 37° C, what was its initial temperature?
5. Define heat conduction.
6. Why do dead air spaces make for good insulation?
7. (a) Define heat convection.
   (b) Convection occurs usually in what two states of matter?
8. Explain how a forced air heating system in a house works.
9. Define radiation.
10. (a) What is light?
    (b) Do dark or light colors absorb more radiant energy?
11. How does the thermos bottle prevent heat loss?
12. What is supposed to be the original material from which coal is formed?
13. What is the evolutionist's explanation for how coal formed?
14. What is the creationist's explanation for how coal formed?
15. Why is the Flood model superior to the evolutionary model?
16. (a) Has any man ever observed a prehistoric peat bog forming coal?
    (b) List several items found in coal that would tend to disprove the peat bog theory.
17. (a) Of the factors heat, time, and pressure, which are the important factors in the formation of coal?
    (b) Give an example to show that one of the above factors is not important.
18. What does radiocarbon dating of oil indicate?

Student Activities

1. Obtain as many different kinds of metal shot as you can. Try to determine the specific heat of each sample, using the procedure from Experiment 28:1, *The Specific Heat of Lead*. Keep good records of your experimental results.
2. Write a paper on conduction, convection, or radiation. Discuss various devices that employ the particular mechanism of heat transfer.

Photo next page
An oil refinery. Think of all the energy used to produce fuel, which in turn will provide more energy for automobiles, machines, and homes.

UNIT IX HEAT

Chapter 29 Thermodynamics

29:2 *Matter, heat, and work are just different forms of energy.*

29:1 Energy Transformations By now you realize the great importance of heat and its ability to do work. You also know that it is possible to transform (change) one form of energy into another. For instance, heat energy can be transformed into mechanical energy or electrical energy. Matter is even considered to be a form of energy. Figure 29:1 shows the relationship of some forms of energy.

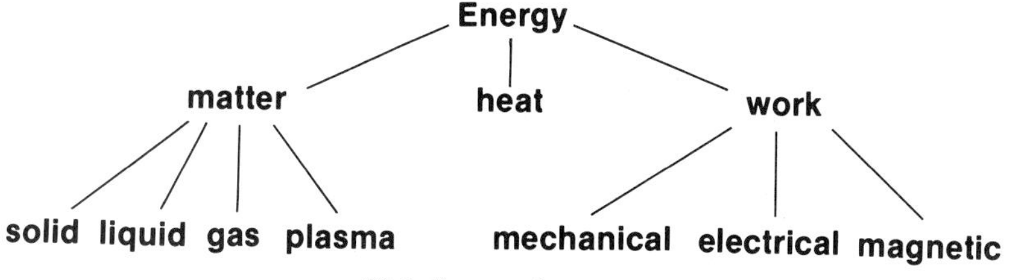

29:1 *Forms of energy.*

All of these transformations are included in the "queen" of sciences, **thermodynamics**. This big word comes from two Greek words—"therme," meaning *heat*, and "dynamis," meaning *power*.

This field of study has taken its place in the center of science; out of it have come two basic principles of modern science, the first and second laws of thermodynamics. These laws deal with energy and its transformations. Every process in the physical world uses energy; if there were no available energy, there would be no processes or events occurring in the universe. Everything

"I'm Mr. Heat Power."

29:3 *Heat is capable of doing much work.*

we do and everything that happens in nature requires energy. Scientists have tested the laws of thermodynamics under many different conditions in many types of experiments and have found them to be scientifically reliable. They are the crowning achievement of scientific endeavor.

29:2 Energy Conservation In the last century, scientists and engineers realized that there has to be a source of energy. Energy can be taken from something and used somewhere else, and energy can be changed in form, but first it has to come from something.

For example, if 100 calories of heat were released by burning wood, they would raise the temperature of a pot of water or the air surrounding the pot. Now, the heat energy did not magically appear to warm the water; it had to come from something. Likewise, the 100 calories would not disappear; they would heat something, if nothing more than a space of air. It is impossible to destroy the energy—it came from somewhere and it went somewhere.

Thus, we can state the **first law of thermodynamics:** *In any process, energy is conserved; it can neither be created nor destroyed.* Julius Robert Mayer (1814-1878), a German physician, was the first scientist to suggest this general principle of energy conservation. This idea has tremendous scientific and philosophical consequences.

If you ignite and burn one gram of pure carbon in enough pure oxygen to form pure carbon dioxide, you will always release about 2138 calories of energy. You can expect this result every time (assuming the experiment behaves properly). Likewise, when you drive your car, you expect a certain amount of energy out of every gallon of gasoline. You may say, "I get 12 miles per gallon of gasoline," or "I get 15 miles per gallon of gasoline with my car."

If the first law of thermodynamics were not operating, the following might happen to you. Suppose you filled your tank with gasoline and were able to drive 10,000 miles before you had to fill the tank again. You filled the tank the second time with exactly the same amount of gasoline as the first time. You drove 10 miles and ran out of gas. What happened? One time you got a very high energy release out of the gasoline; the next time you got a very low energy release. Nothing was wrong with the car. However, energy seemed either to be *created* in one case or *destroyed* in the other. Actually, this could not have happened because gasoline will release the same energy time after time; you cannot get any extra energy, nor will you destroy any.

If you go into a laboratory and do an experiment today, you can go into the laboratory and duplicate that experiment tomorrow or one year from today or 100 years from today. Time has no effect on the experiment, so long as the conditions of the experiment are the same. A reaction such as

$$2H_2 + O_2 \xrightarrow{\text{spark}} 2H_2O$$

will occur the same way each time. This ability to duplicate a result has to do with the **symmetry of time**; which means, simply, that time has no effect on the result of an experiment. Symmetry of time involves the principles of conservation, also. The same thing happens under the same conditions because it takes the same amount of energy to do the same thing over and over again. Time does not affect this.

We would have no science if it were not for this conservation principle. We could never tell how much work would be necessary to raise a 10 kg weight. We might design a machine to do this work but find out that on certain days the machine could not lift the load. On other days the machine might be capable of lifting 1,000,000 kg. You would never know what to expect! You could never predict with any confidence anything in science.

Life as well as science depends upon the same thing happening under the same conditions. If conservation principles were not operating, the following might happen. Suppose you ate a big breakfast one morning and could work for three straight days before you ate again. Yet another day you ate the same breakfast and went to do the same type of work, but fainted from hunger within three hours. This could not happen, because the same amount of food will provide you with the same amount of energy (assuming you are in the identical physical and mental condition as before). This conservation principle takes its scientific form as the first law of thermodynamics.

29:3 Conservation and Quantification When calculating how much work is done by a machine or how much energy is released by a chemical reaction, scientists automatically assume conservation of energy. There would be no reason to make calculations that would be valid only one time. Suppose you could never be certain how much energy you could get out of a tank of gasoline. It would be somewhat of a gamble to take a long trip; you might run out of gas between filling stations. Mathematics would be meaningless in science if we could not depend on conservation processes. Actually, science itself would be meaningless. Science depends on the ability to reproduce similar conditions.

Since we *must* assume conservation in nature, and since this assumption works, we can make some simple calculations. If energy is conserved, we simply need to figure out where the energy goes.

Problem: Heat a gas (giving it 1000 calories of heat) and allow the gas to expand. The gas does work in expanding so that it loses energy, 400 calories in this case. What is the energy change of the gas?

Solution: Energy Change = Heat Change + Work Change
= 1000 calories − 400 calories
= 600 calories

"You can't destroy Mr. Energy."

"But you can make me weak."

29:4 *Energy cannot be destroyed, but it can be broken down into smaller units of energy.*

Problem: Suppose you heat a gas (giving it 500 calories of thermal energy), and then compress it. When you compress the gas, you perform work on it, increasing its energy. Suppose it took 200 calories of work to compress the gas. What was the energy increase of the gas?

Solution: Energy Change = Heat Change + Work Change
= 500 calories + 200 calories
= 700 calories

In the process of heating and compressing the gas, it gained 700 calories of energy. The gas is capable of doing more work.

Problem: Suppose you cool the gas (extract 300 calories of heat) and compress the gas as before. What is the energy change of the gas?

Solution: Energy Change = Heat Change + Work Change
= –300 calories + 200 calories
= –100 calories

From the initial state to the final state the gas lost energy. The energy has not been destroyed; it has simply gone somewhere else. It is not in the gas.

The gas in two of the problems gained energy. This energy was not created; it was obtained from another source. If a substance gains energy, a plus (+) sign is used. If it loses energy, a minus (–) sign is used. We can keep up with energy transformations this way. This enables us to design machines and engines capable of doing certain jobs because we know the energy requirements of the jobs.

Note that we can never know or measure the exact energy content of anything. However, this is of no interest to us anyway. We simply want to know how much energy we can get out of a machine or an engine.

29:4 Scripture and Conservation You might ask, "Why are conservation principles universal and repeatedly successful in science?" Our answer would be "because they are Scripturally based." Nehemiah 9:6 says, *Thou, even thou, art Lord alone; thou hast made heaven, the heaven of heavens, with all their host, the earth, and all things that are therein, the seas, and all that is therein, and thou preservest them all.* And II Peter 3:7 declares, *But the heavens and the earth, which are now, by the same word are kept in store.* These two passages, one from the Old Testament and the other from the New Testament, indicate that preservation processes are operating—the physical world is "kept in store" (conserved).

Hebrews 1:3 says that the Lord Jesus Christ is *upholding all things by the word of his power.* The Lord is *"upholding"* or *"conserving"* all things, including the physical world. Thus, conservation processes operate because they are God-ordained and God-controlled.

That energy can be neither created nor destroyed, but only used or changed in form, is a powerful argument for the fact that creation is finished. Only conservation processes are possible: no further creation can occur. We cannot create energy; we can use only what God has given us.

Thus the heavens and the earth were finished, and all the host of them. And on the seventh day God ended his work which he had made (Genesis 2:1-2);

For in six days the Lord made heaven and earth, the sea, and all that in them is (Exodus 20:11);

By the word of the Lord were the heavens made; and all the host of them by the breath of his mouth . . . . For he spake, and it was done; he commanded, and it stood fast (Psalm 33:6, 9).

Creation was finished in six days. No further creation has taken place since that time. Thus, no new energy is coming into being at the present time. And energy cannot be destroyed, because God is keeping everything "in store." Other verses which indicate a finished creation are Exodus 31:17 and Hebrews 4:3, 10.

29:5 The First Law and Evolution The Bible and results of scientific investigation indicate that conservation processes presently operate in nature. The Bible states that creation is finished. However, the evolutionist says that things in nature are still evolving, not being conserved. He says that new *kinds* are coming into being. Creation is not finished and has been going on for 4½ billion years, according to the evolutionist; man and nature are still evolving. Evolutionary processes are not conservation processes. Evolutionary processes cannot operate under conservation conditions. Whatever the evolutionist imagines as his starting condition for the universe (a gas cloud, for example), it will remain the same and not evolve into anything else. It is only possible to conserve what is present at the beginning. The theory of **evolution** violates scientific law and is completely contrary to Scripture.

29:5 *Like produces like. Dogs have puppies that grow up to be dogs. Apple trees yield apples, and when the seeds are planted, they grow into apple trees. People have babies which grow up to be adults.*

29:6 Energy Waste—The Second Law of Thermodynamics The second law of thermodynamics came about when scientists observed that natural processes go in a particular direction. One statement of the second law is that *heat spontaneously (naturally) goes from a hot object to a cold object.* If we heat a steel ball, then place it next to a cold steel ball, the heat will flow from the hot ball to the cold ball. However, the reverse will not occur; heat will not flow from the cold ball to the hot ball. This would result in the cold ball becoming colder and the hot ball becoming hotter! Heat never flows this way. All natural processes are "one way." This "one-wayness" of natural processes is expressed in all statements of the second law.

In developing heat engines and machines, scientists and engineers found that when energy in some form flows from one place to another, some energy is wasted. Energy is not destroyed, but it is lost for human use and goes into the atmosphere.

Let us look at a very simplified diagram of a power plant operation that uses heat flow (see Figure 29:6). Water is boiled to produce steam. The steam turns **turbine** blades to generate electricity. Then the electricity is used in an industrial plant to stamp a metal sheet into a desired shape.

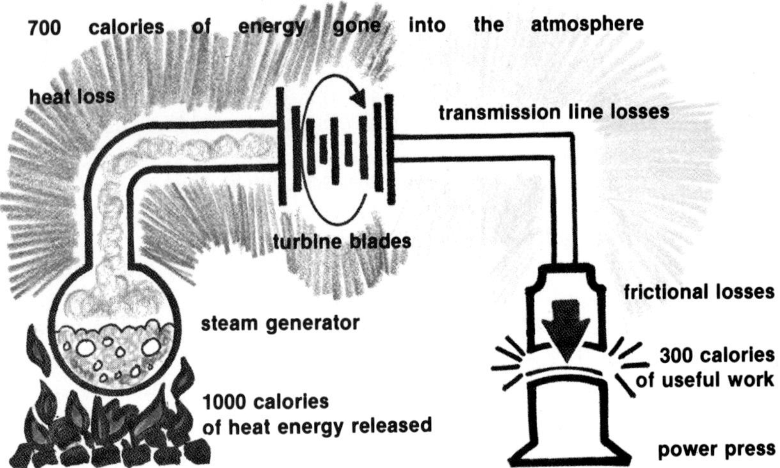

29:6 *Energy transfer process with unavoidable degeneration.*

The water which was boiled to produce steam was heated with coal. Suppose 1000 calories of heat energy are released from the coal. If you approach the steam generator, the connecting pipe, and the turbine, you will notice that they are hot. They are losing heat to the atmosphere. This is a waste of energy. After the turbine transforms the remaining heat energy into electrical energy, some of this electrical energy is lost as it is carried through the electrical lines. The power press changes the remaining electrical energy into mechanical work; some of this work is lost as frictional heating as parts of the machine rub together. This heat energy dissipates (scatters) into the surroundings.

All of this wasted energy is not destroyed. It becomes subdivided into small units of energy that are of no use. Most of the original 1000 calories are wasted. Only about 300 calories worth of useful work is accomplished; 700 calories of energy has been degenerated into a nonusable form.

Also, if you strike a match and hold your hand just over the match, you can feel the heat given off. Next, hold your hand a foot above the match; then hold it three feet above the match. At three feet away from the flame, you probably can no longer feel any heat from the match. Why? The large concentration of heat at the match now has been "diluted" into small "bundles of heat" and your hand cannot feel any warming effect. In this case, the warm, energetic molecules have lost much of their energy to cooler molecules. The energy is so subdivided into smaller and smaller quantities that it has lost its "power." It is a degenerated form of energy.

The same thing happens in natural processes. Every time anything occurs in nature, there is a degeneration (breaking down) and waste of energy. The nonusable energy in the universe is increasing, whereas the amount of usable energy is decreasing. Natural processes tend to change nature from conditions of order to conditions of disorder.

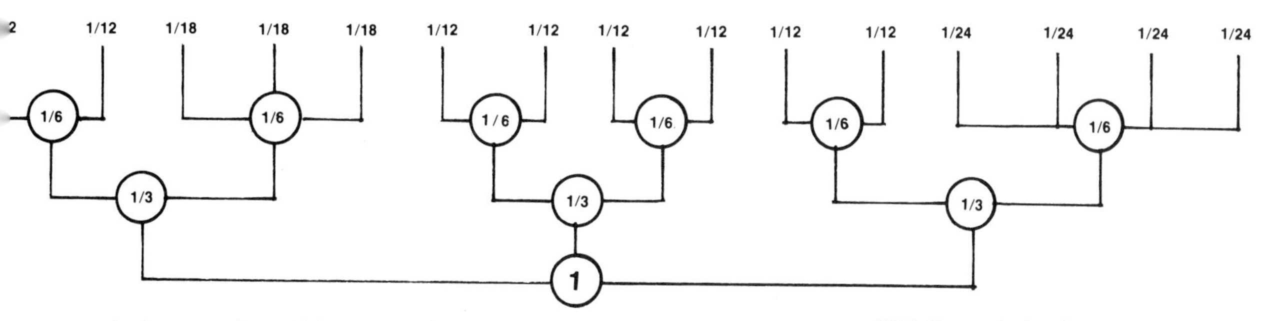

29:7 *One calorie of energy may be dissipated into smaller and smaller quantities in the atmosphere.*

Let us explore this more closely by using a natural process such as dropping bobby pins on the floor. Try it yourself. How do the bobby pins land on the floor? Do they land in a neat, ordered arrangement as shown in Figure 29:8a? No, they will land in a disordered arrangement, such as Figure 29:8b. This happens every time. You can drop them as many times as you wish; they will always end in a disordered array. Natural processes go in a direction toward wastage and decay! Degeneration in nature (things wear out, die, and decay) is stated scientifically by the second law of thermodynamics.

29:7 Scripture and the Second Law Why is the second law of thermodynamics a universal principle in science? It is because this law is Scripturally based. Psalm 102:25-26 says, *Of old hast thou laid the foundation of the earth: and the heavens are the work of thy hands. They shall perish, but thou shalt endure; yea, all of them shall wax old like a garment; as a vesture shalt thou change them, and they shall be changed.* And Ecclesiastes 3:20 declares, *All go unto one place; all are of the dust, and all turn to dust again* (a state of disorder).

These verses give us a vivid picture of the wearing-out and degeneration of the physical world. Living organisms eventually return to dust, a state of disorder. Other verses that indicate the degeneration of the world are Isaiah 51:6; Romans 8:20-22; and I Peter 1:24.

29:8 The Second Law and Evolution Molecules-to-man evolution, as an imagined process, produces higher and higher order, more kinds, and more varieties. The physical world, particularly the organic world, is getting better and better (according to evolution). This is ever-onward, ever-upward evolution! This would be like the dropped bobby pins falling on the floor in an ordered arrangement (see Figures 29:8a and 29:8b). However, the best scientific evidence we have available indicates that nature moves in an ever-downward

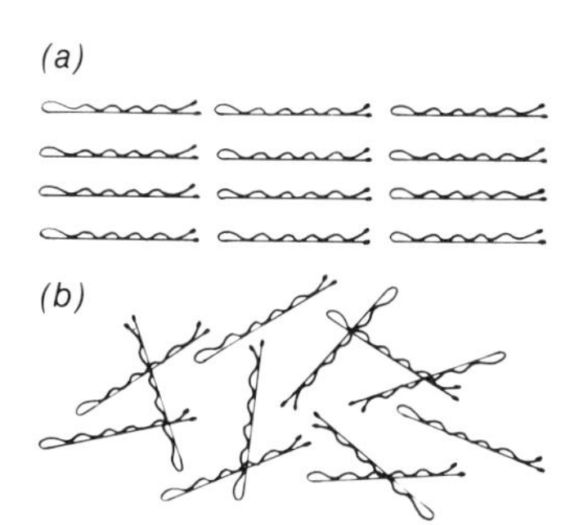

29:8 *If you dropped a handful of bobby pins, would they land in a neat, ordered arrangement as in (a) or in a disordered arrangement as in (b)?*

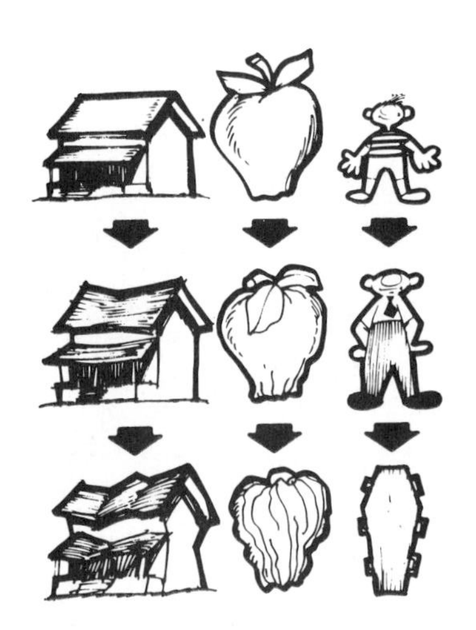

29:9 *Things wear out and decay. Houses degenerate and fall apart. Apples decay; humans die.*

manner. Living organisms wear out (grow old), die, and decay. Living organisms can produce other living organisms (conservation), but each organism wears out and dies. Also, heavenly bodies such as stars are "dying" as they lose thermal energy. The whole universe is wearing out.

Consider an automobile made by the best engineering skill in the world. Will the automobile last forever? Will it later evolve into an airplane? No, the car is new only once, and then it wears out. Eventually, every part will have to be replaced to keep the car running.

Suppose you buy a new dress or a new suit. The dress or suit will last forever, right? Wrong, the garment wears out! People age and die. Everything in nature degenerates. You cannot observe upward evolution, but you can observe downward degeneration. Things change, but still they degenerate. The second law of thermodynamics is opposed to the theory of evolution. Could evolution have occurred? From everything we know and understand in science today, evolution is scientifically impossible.

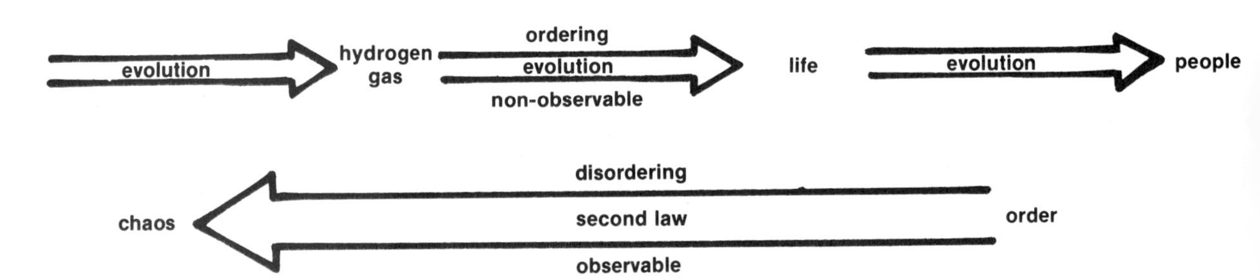

29:10 *The second law of thermodynamics versus evolution.*

29:9 Conclusion The most reliable laws of science we have, the first and second laws of thermodynamics, indicate that conservative and degenerative processes operate at the same time in nature. Evolution is a creative, ordering, bettering process. The first and second laws of thermodynamics are demonstrable, reproducible, scientific principles. Evolution is non-demonstrable and non-reproducible; it is neither conservative nor degenerative. The first and second laws of thermodynamics have a basis in Scripture, whereas evolution is unscriptural. The evolutionary world-view is unreasonable and is not based on any observable facts. The Christian world-view is reasonable and better fits *true* science than evolution.

For further study, ask your teacher for a list of articles and books on this subject.

List of Terms

conservation
evolution
first law of thermodynamics
second law of thermodynamics
symmetry of time
thermodynamics
turbine

Questions

1. Divide the word thermodynamics into parts and define the word.
2. With what do the laws of thermodynamics deal?
3. Define transformation.
4. Why is the study of energy and its transformations so important?
5. What is the best science we have today?
6. Can energy be created or destroyed?
7. What is the first law of thermodynamics?
8. Suppose you do an experiment today; then you do the same experiment tomorrow and get the same result. What property of time insures this?
9. How much energy must a system have if it has to do 2000 calories worth of work?
10. Why are we not concerned about the absolute energy content of a body, only the changes in energy that occur?
11. A gas is heated and absorbs 30 calories of energy. Work is done on the gas and it absorbs 100 calories. What is the energy change in the gas?
12. A gas is heated 75° C and absorbs 3000 calories of energy. However, the gas does work and loses 790 calories in the same operation. What is the energy change in the gas?
13. A gas drops in temperature and loses 900 calories of energy. It also does 1200 calories of work. What is the energy change in the gas?
14. ★ 30 grams of a gas with a specific heat of .039 cal/g/° C is heated from 100° K to 600° K. 100 cal of work is done on the gas also. What is the energy change in the gas?
15. Name one verse in the Old Testament and one in the New Testament that indicate conservation processes operate in nature.
16. Who is the Conserver? On what verse of Scripture is this based?
17. If conservation processes operate, is further creation possible? Why?
18. Name two Scripture references indicating that creation is finished.
19. How is the idea of conservation processes in nature opposed to the idea of molecules-to-man evolution?
20. (a) Do natural processes usually take a particular direction?
    (b) Give an example to demonstrate this.
    (c) In what law of science is the direction of natural processes taken into account?
21. (a) When heat goes from a hot object to a cold object, is any energy wasted?
    (b) Is energy destroyed?
    (c) Where does the energy go?
22. Do natural processes normally generate more order or more disorder?
23. List two Scripture verses that indicate that the world is degenerating.
24. Is death a result of the second law of thermodynamics? Why?
25. Are degeneration processes in nature opposed to molecules-to-man evolution? Why?

Student Activity

Using creationist materials supplied by your teacher, write a paper on the first and second laws of thermodynamics and their relationship to the theory of evolution.

Photo next page
One form of man-made lightning.

UNIT X ELECTRICITY AND MAGNETISM

Chapter 30 Static Electricity

30:1 The Electrical Properties of Matter Electricity may be divided into two broad categories—**static electricity** and **current electricity**. Static electricity consists of *charges at rest*; current electricity is produced by *charges in motion*. In this chapter we will deal with charges at rest, static electricity.

You have no doubt noticed that you can acquire a static charge by walking across a thick rug. If you then touch an object that is a good conductor of electricity, there is a slight spark at the point of contact. Also, you may experience a shock. How can we understand this?

You will recall that scientists picture an atom as consisting of negatively charged particles (electrons) in motion around a positively charged nucleus. When two different materials are rubbed together, electrons from one material can be transferred to the other material. The material from which the electrons are removed is then left with a positive charge, while the material to which they are transferred becomes negative. What we are using, then, to visualize what is happening, is a *model* involving *movable negative charges* and *fixed positive charges*.

As you walk across a rug, electrons are believed to be transferred from the rug to your shoes. Since the electrons all possess the same electrical charge, they repel one another and spread over your body (which is a fairly good conductor), giving your body a negative charge. When you touch a good conductor of electricity, such as a metal door knob, these extra electrons suddenly leave your body through the point of contact. If enough electrons are present, there will be a noticeable spark.

(a)

(b)

30:1 *(a) You can acquire a static charge by walking across a thick rug. (b) You get a shock as the charge leaves your body.*

(a)

(b)

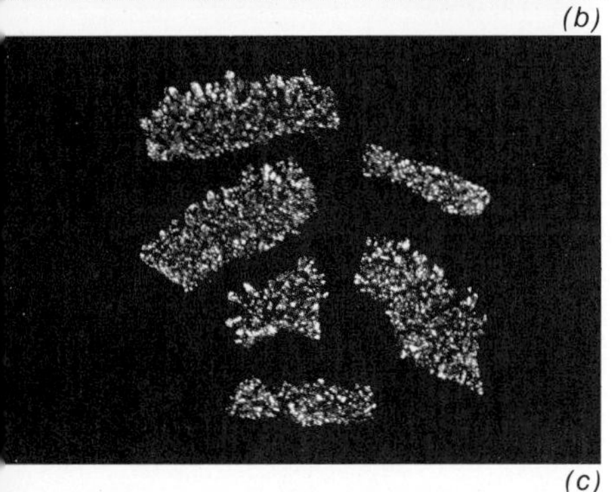

(c)

30:2 *(a) Two examples of conductors—copper wire and iron wire. (b) Some insulators used in radio work. (c) Crystals of silicon. Silicon is a semiconductor.*

From the standpoint of electrical behavior, substances are classified into three kinds: **conductors, insulators,** and **semiconductors.** As a general rule, the elements on the left-hand side of the periodic table (metals) are conductors, those on the right-hand side (nonmetals) are insulators, while a few near the middle of the main-group elements (metalloids) are semiconductors.

For uncombined elements, the electrical properties appear to be determined by how loosely or how tightly the electrons are bound to the nuclei of their atoms. If the electrons are able to move about freely—that is, if they are able to roam from one end of the material to the other end—the material will be a good conductor. If the electrons are tightly bound—that is, if they are tied to their own atoms—it will be an insulator. If only a few of the electrons are free to move about within the material, it will be a semiconductor.

For compounds, the electrical properties depend upon the type of bonding between atoms. In general, metallic bonding contributes to conductivity, while other types of bonding tend to hinder conductivity.

Some typical materials are given in Table 30-1. Most of those listed are elements; a few are compounds and mixtures.

Table 30-1 *Conductivity*

Conductors	Insulators	Semiconductors
aluminum	bakelite	germanium
brass	celluloid	silicon
copper	dry air	
gold	glass	
graphite *	lucite	
iron	mica	
mercury	paper	
nickel	polyvinyl chloride	
silver	porcelain	
tin	rubber	
zinc	textiles	
	wood	

*Graphite is an exception to the rule that nonmetals are insulators. Although it is a nonmetal (carbon), it is a good conductor of electricity.

30:2 Generating Charges Walking across a rug is an example of the general method of producing a static charge by friction. This method works best when two different insulators are rubbed together. Through trial-and-error experimentation, using a wide range of materials, the authors have found the following combinations to work most efficiently:

Positive Charge—produced on a lucite or celluloid rod by rubbing it with wool.

Negative Charge—produced on an ebonite or polyvinyl chloride rod by rubbing it with wool.

30:3 Attraction and Repulsion Using a relatively simple experimental setup, you can easily observe how positive (+) and negative (−) charges attract and repel each other. Suspend a 12-inch *ebonite* rod so that it is free to swing about a vertical axis. Charge one end of the rod negatively (−) by rubbing it with wool. When you hold a second negatively charged (−) *ebonite* rod near the charged end, the suspended rod is repelled (see Figure 30:3).

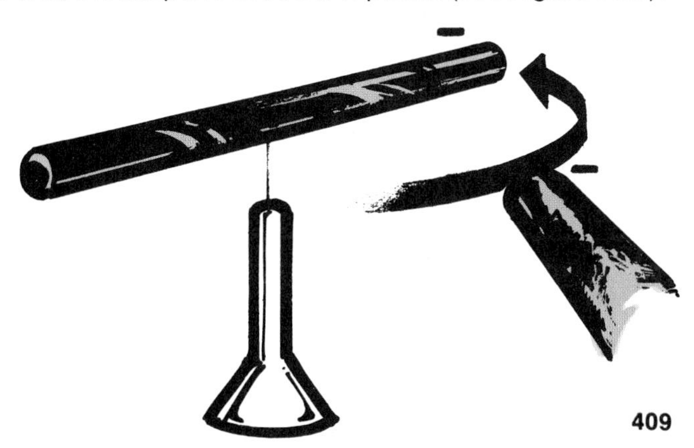

30:3 *Like charges repel. A negatively-charged hard-rubber rod suspended on a pivot will be repelled by another negatively-charged rod.*

Attraction may be demonstrated by holding a positively charged (+) *lucite* rod near the negatively charged (−) end of the suspended *ebonite* rod. We have, therefore, proven experimentally that *like charges repel and unlike charges attract.*

Experiment 30:1 *Static Charges*

1. Charge a rubber rod negatively by rubbing it with wool.
2. Brush the rod against the plate of a Braun electroscope. Observe how the charge remains on the instrument when the rod is removed.
3. Ground the electroscope by touching the plate with your hand.
4. Repeat steps 1, 2, and 3 using a positive charge. Generate the charge by rubbing a lucite or celluloid rod with wool.
5. Place a positive charge on the electroscope.
6. Using a negatively charged rod, attempt to add enough negative charge to bring the overall charge on the electroscope exactly to neutral. It may be necessary to alternate positive and negative charges until an exact balance is achieved.
7. Use a charged rod to pick up small bits of paper. The paper should be no larger than ½" in any dimension.
8. Use a charged rod to pick up a small quantity of cork dust. The particles are first attracted to the rod, then repelled as they acquire the same charge as the rod.
9. Hold a positively charged rod as near to the plate of the electroscope as you can without actually touching it. While doing this, touch the plate with your other hand; then remove the rod and your hand. If you have done this correctly, a negative charge will remain on the electroscope because of the electrons attracted onto the electroscope from your hand. This process is called charging by induction.
10. Repeat step 9, using a negatively charged rod. A positive charge should remain on the electroscope. Explain what has happened.

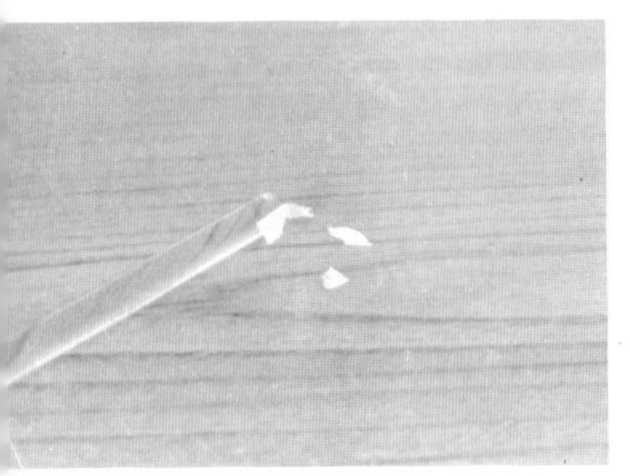

30:4 *A charged rod will pick up small bits of paper.*

It may come as somewhat of a surprise to many people, however, to learn that there can be attraction between a *charged* object and a *neutral* object. A small piece of paper which has no electrical charge can be picked up using *either* a negatively or a positively charged rod (see Figure 30:4). How can we account for this?

Remember, paper is an insulator, and as such its electrons are *not* free to move about. However, within each paper molecule it is thought that the electrons can be displaced slightly, either toward or away from the charged object producing the disturbance. If the rod is negative, the electrons in the paper will be pushed farther from the rod than their normal positions, while their nuclei remain fixed. On the average, then, the positive nuclei will be closer to the rod than their electrons. The attractive force between the rod and the nuclei of the paper will then be greater than the repulsive force between the rod and the electrons of the paper. The attractive force predominates, and the paper is attracted to the rod.

When a positively charged rod is held near the paper, electrons are displaced to the *near side* of their respective molecules. The average positions of the electrons in the paper are then closer to the rod than to their nuclei. And their attractive force for the rod more than offsets the repulsive force between the rod and the nuclei of the paper. Thus the paper is attracted by a charged rod, regardless of whether the rod is positive or negative.

The attraction of an uncharged substance can also be demonstrated by holding a charged rod near a thin stream of water (see Figure 30:5). This has a spectacular effect. If a thin enough stream of water is used with a strong enough charge on the rod, the water can be attracted upward, making it travel horizontally for a short distance.

Still another example is the attraction between a charged rod and a live plant leaf or flower (see Figure 30:6). Some types of leaves are especially susceptible to electrostatic attraction. Various types can be tried out-of-doors (blades of grass, tree leaves, and the like) or a small potted plant can be used indoors.

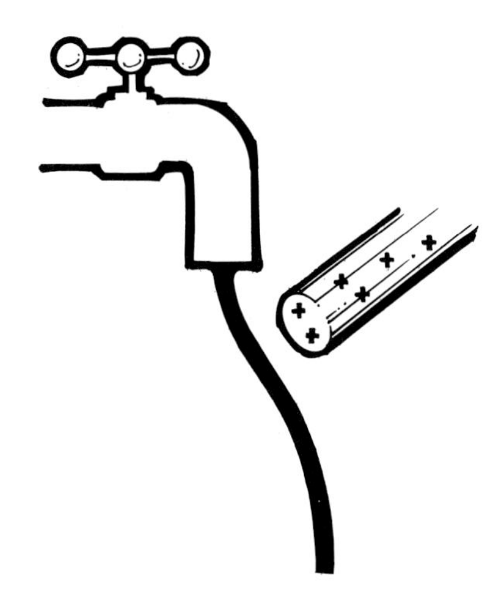

30:5 *A thin stream of water is deflected by a nearby charge.*

30:6 *Some types of plant leaves and flowers are attracted to a charged rod.*

30:4 Coulomb's Law We have not yet discussed *how much* attraction or repulsion there is between two materials. French physicist Charles Augustin de Coulomb (1736-1806) worked out a mathematical relationship of the force of attraction or repulsion, the charges on the objects, and the distance between them (see Figure 30:7). His formula is called **Coulomb's law**.

The equation for Coulomb's law looks surprisingly similar to the equation for Newton's law of gravitation. In this case, however, we have charge instead of mass. Also, the constant is much larger, reflecting the fact that the electrostatic force between any two charged objects will be many times greater than the gravitational force. Another important difference is that gravitation is always an attractive force, whereas the electrostatic force can be either attractive *or* repulsive. The relationship is

$$F = \frac{kq_1q_2}{d^2}$$

F is the force of attraction or repulsion between the two bodies; k is a constant; q_1 and q_2 are the two charges; and d is the distance between the centers of the bodies. If the charges are specified in units called *coulombs* and the distance is in *meters*, the constant is 9×10^9. The force comes out in *newtons*. The **coulomb** is *a quantity of electrical charge equal to* 6.25×10^{18} *times the charge on one electron.*

This type of relationship occurs frequently in nature. Like the law of gravitation, Coulomb's law is an **inverse square law**; that is, *the force falls off as the square of the distance*. It predicts that the force will be reduced to 1/4 when the distance is doubled, 1/9 when the distance is tripled, and 1/16 when it is quadrupled. If you are like most people, you will find that the force falls off far more rapidly than you would expect.

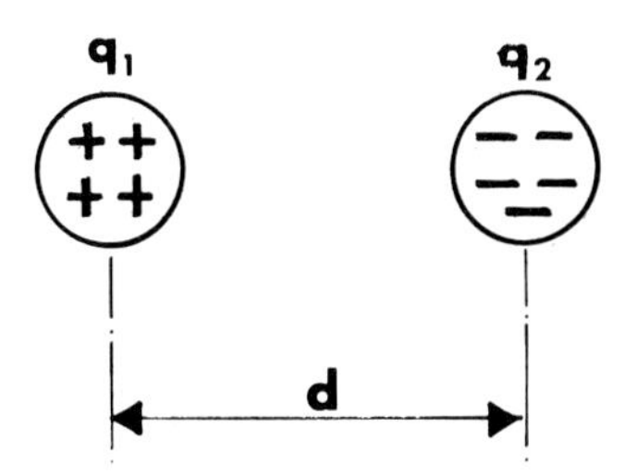

30:7 *Coulomb's law states that the force of attraction or repulsion between two charged bodies will be proportional to the product of their charges* ($q_1 \times q_2$) *and inversely proportional to the distance between them squared.*

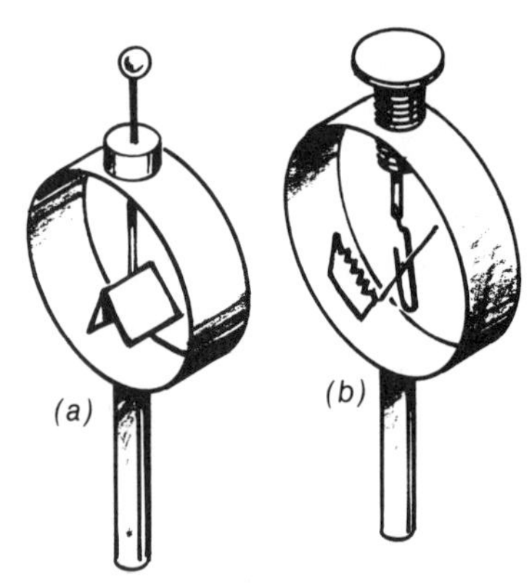

30:8 *(a) Gold-leaf electroscope and (b) Braun electroscope.*

30:9 *The Braun electroscope, with only one moving part, is easier to maintain than a gold-leaf electroscope.*

30:10 *Transistorized electroscope designed and built at Bob Jones University. This instrument gives an upward deflection of the meter for positive charges, a downward deflection for negative charges.*

30:5 The Electroscope The **electroscope**, an instrument for detecting electrical charges, has been made in a great many varieties. One of the older forms uses two gold leaves suspended side by side in an enclosure such as that shown in Figure 30:8a. The metal rod supporting the leaves extends upward through a rubber stopper to the metal sphere outside the case. When a charge is placed on the sphere, the leaves swing apart.

Because gold leaves disintegrate with use, the arrangement known as the **Braun electroscope** (see Figures 30:8b and 30:9) is more satisfactory. In this case there is only one movable "leaf," a small wire pointer. The other "leaf," called the *stem*, is fixed in its position. A round metal plate takes the place of the sphere of the older model. When a charged rod is placed on or near the plate, the pointer swings away from the stem. If the rod actually touches the plate, the pointer will retain its new position even after the rod is taken away. A negatively charged rod placed in contact with the plate transfers electrons to the plate. The electrons are quickly conducted to the stem and pointer. The repulsion of like charges accounts for the behavior of the pointer. When a positively charged rod is placed in contact with the plate, electrons are transferred from the plate to the rod. Other electrons are immediately attracted up from the stem and pointer, leaving a positive charge on both. Again, the pointer swings out because of the repulsion of like charges.

If a burning match is held near the plate of a charged electroscope, the electroscope will quickly discharge. Combustion produces ions which neutralize electrical charges in their vicinity. Radioactive materials also discharge a charged electroscope, but much more slowly than a flame. The strength of a radioactive source can be measured by observing how rapidly the electroscope is discharged. A strong source generates more charged particles, which neutralize the electroscope more rapidly.

Other more elaborate types of electroscopes use electronic devices such as vacuum tubes and transistors to increase sensitivity. From several feet away, a vacuum tube electroscope can respond to the charges created by a person combing his hair. Another advantage of these instruments is their ability to distinguish directly between positive and negative charges. Whereas the simpler units give the same indication for either charge, electronic electroscopes can be arranged to give an upward meter reading for a plus charge, and a downward reading for a minus charge (see Figure 30:10).

An **electrometer** is a calibrated electroscope; that is, it has a numbered scale for reading units of charge. The electrometer was the detection device used by Marie Curie in analyzing the uranium ore called pitchblende. The ore showed more radioactivity than she could account for with uranium alone. Her feeling that something else must be present in the ore eventually led to the discovery of the element radium.

Workers in plants or laboratories where radioactive materials are used wear a very tiny electrometer called a pocket **dosimeter** (see Figure 30:11). This instrument is only slightly larger than a fountain pen and can be worn in a shirt pocket. It is charged at the beginning of the day with a battery or power supply. The amount it has discharged by the end of the day is a measure of the radiation to which the person has been exposed.

30:6 Radio Static As you may have guessed, there is a connection between static electricity and the noise called "**static**" that is heard on a radio. You may verify this for yourself by charging a plastic rod near an AM radio ("AM" means *amplitude modulation*). Set the dial *between* stations and turn the volume to maximum. A crackling noise will be heard each time the rod is stroked. You will be unable to observe this effect with an FM (*frequency modulation*) radio. One of the advantages of FM is that static is eliminated.

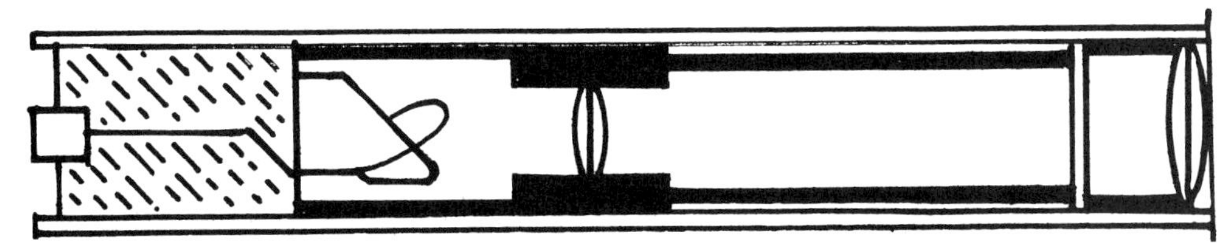

30:11 *A pocket dosimeter used to detect radioactivity is actually a small calibrated electroscope.*

30:7 The Faraday Ice-Pail Experiment In the early 1800's, it became understood that the charge of an electrified conductor is located only on its outer surface. This was demonstrated by Michael Faraday's famous "ice-pail experiment" in 1810 (see Figure 30:13).

A small metal pail resting on a good insulator is connected by a wire to an electroscope, as shown in the diagram. At the beginning of the experiment there is no charge on either the pail or the electroscope. But when a charged metal ball is lowered into the pail (but not touching the pail), the leaves of the electroscope go apart. There is no change in the amount of charge indicated as the ball is moved around to different positions inside the pail. Surprisingly, there is still no change, even when the ball is touched to the inside of the pail. When the ball is completely withdrawn from the pail, the charge on the electroscope remains, but the inside of the pail and the ball are both found to be neutral! At the end of the experiment the charges are located on the outside of the pail only.

Let us assume that the charge on the ball was negative, due to a certain number of excess electrons. When the ball was placed inside the pail, an equal number of electrons were repelled to the outside of the pail, leaving the same number of positive charges on the inside. When the ball was touched to the pail, the electrons from the ball exactly neutralized the positive charges. But the electrons that were repelled to the outside of the pail still remain on the outside when the ball is removed.

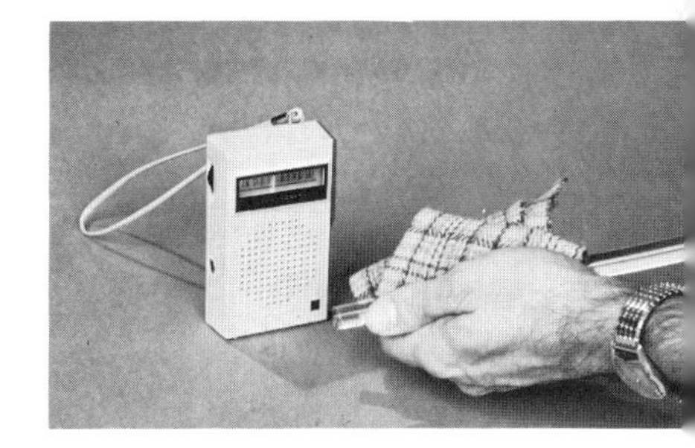

30:12 *Radio static is a manifestation of static electricity.*

30:13 *Faraday's ice-pail experiment.*

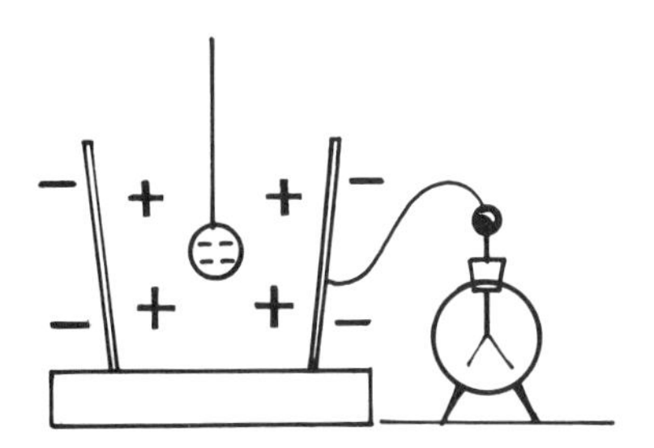

The experiment works equally well using a positively charged metal ball. Can you reason out what is happening in this case?

30:8 The Wimshurst Generator When you electrify a plastic rod with wool, you are limited in how much charge you can give to the rod. For one thing, you are not able to achieve a very high velocity to your motion. Another limitation is your inability to produce a continuous motion in the same direction. Suppose, however, that the rod could be coiled up into a perfectly round hoop and rotated very rapidly against the wool. With this arrangement a much higher voltage could be developed.

30:14 *A Wimshurst generator.*

The principle of rotary motion is used in the **Wimshurst generator**, shown in Figure 30:14. As the crank is turned, two large plastic discs rotate in opposite directions. Charges are picked off by small metal brushes and transferred at the same time to Leyden jars (Section 30:9) on either side for storage, and to the two steel spheres at the front of the machine. If the spheres are properly spaced, sparks will jump between them. You can get some idea of the voltage being generated from the fact that about 30,000 volts are needed for each centimeter of spark produced. Fortunately for the experimenter, the Wimshurst generator is not hazardous. Although the voltage is quite large, the very tiny amount of current available at its terminals is far below the level which could kill a person.

The sparks from a Wimshurst machine will punch microscopic holes through a piece of paper or thin cardboard. When a glass plate is held between the spheres, the spark is unable to go through the glass and must detour around its edge. The voltage developed by the generator will light a gas discharge tube. When a candle flame is placed between the charged spheres, positive ions in the flame cause it to be attracted to the negative sphere. This is an easy way to determine which sphere is which.

You may be wondering how, if dry air is an insulator as indicated in the table at the beginning of the chapter, it is capable of *conducting* a spark from one terminal of the Wimshurst machine to the other. This is an interesting question. From many experiments with a great variety of materials, scientists have arrived at this general principle: *anything will conduct if the voltage is high enough.* Metals and graphite conduct readily; semiconductors conduct with difficulty. Some of the better insulators often require millions of volts before they "break down" and conduct.

30:9 The Leyden Jar We noted that two **Leyden jars** were used in the Wimshurst machine to store electrical charges. The Leyden jar—named for the University of Leyden in the Netherlands where it was invented in 1746—is simply a bottle that has been coated both inside and out with tin or lead foil.

Contact is made with the inner conductor by means of a chain which connects to the sphere at the top (see Figure 30:15). A charge is placed on the jar by putting excess electrons onto the one conductor and removing electrons from the other. This general arrangement of two conductors separated by a thin insulator is called a **condenser** or **capacitor**. (The term "capacitor" is preferred today.) Capacitors are extremely important components of electronic devices such as radios, televisions, and tape recorders. The capacitor also forms a vital part of the ignition system of a car. Some of the different types of capacitors are shown in Figures 30:16a and 30:16b.

30:15 *(a) A Leyden jar. (b) Construction of the Leyden jar. A glass jar is coated both inside and outside with a thin layer of metal. Contact is made with the inner conductor by means of a chain.*

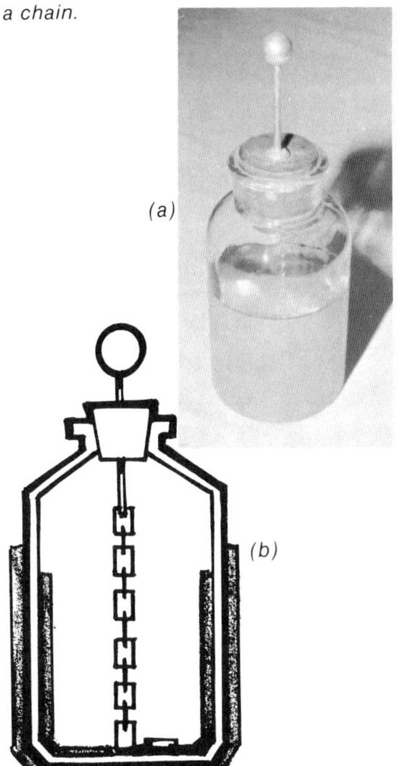

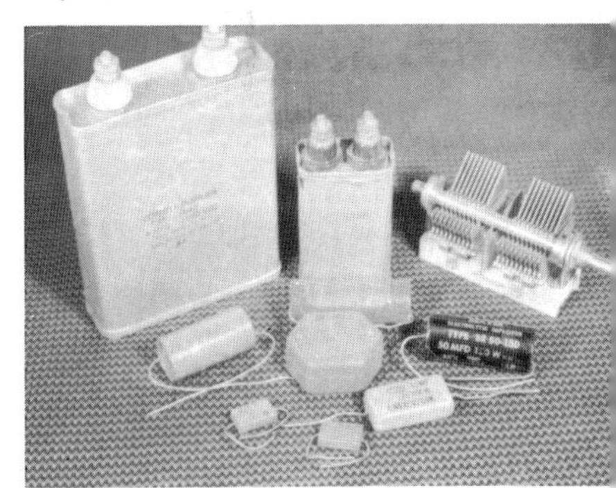

30:16 *(a) Some types of capacitors (condensers) in common use today. (b) Showing the use of capacitors in an electronic circuit. The larger cylindrical components are capacitors; the smaller parts are resistors.*

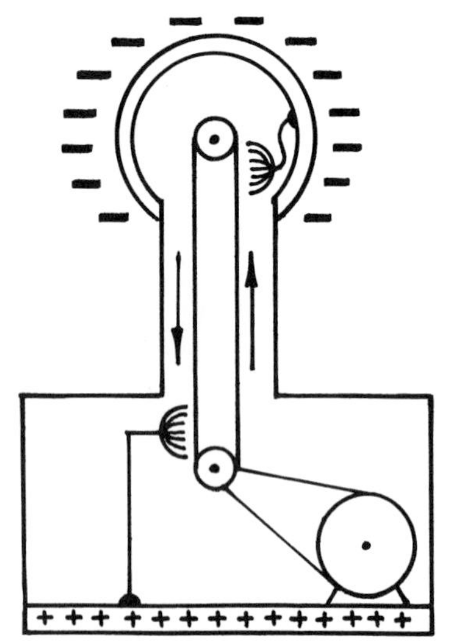

30:17 *The essential parts of a simple Van de Graaff generator. Charges are carried to the dome by a moving belt.*

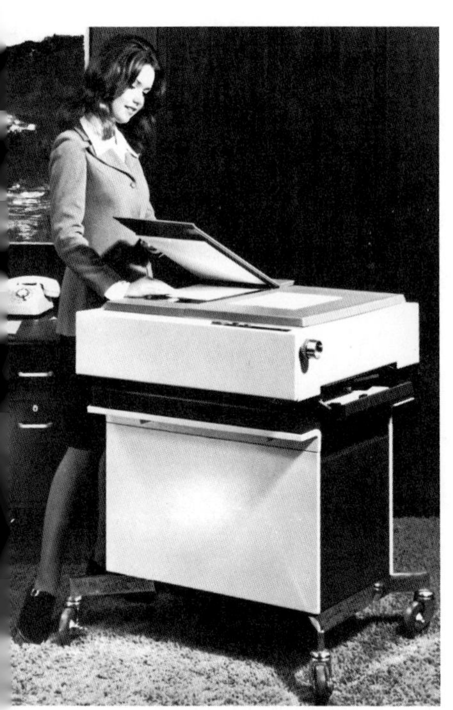

30:18 *An important application of static electricity—electrostatic copying machines.*

30:10 The Van de Graaff Generator Even higher voltages than those produced by the Wimshurst generator can be attained with a machine called the **Van de Graaff generator**. It bears the name of Robert J. Van de Graaff (1901-1967), a Princeton University scientist who built the first unit of its kind in 1931. Van de Graaff generators in use today range in size from the small classroom model to huge research machines 25 feet or more in height. The research machines can produce up to several million volts for use in accelerating charged particles.

The essential parts of the simple Van de Graaff generator used in schools are shown in Figure 30:17. It is still another variation on the age-old method of producing an electrostatic charge by friction. A moving belt driven by a motor passes over two pulleys made of different insulating materials. The belt carries one kind of charge up to the dome and the other kind of charge down to the base. The charges are transferred to or from the belt at both ends by metal "fingers," located close to but not touching the belt. The charges at the upper end are conducted to the dome where they accumulate on its outer surface; the opposite charges collect on the metal base. If electrons are being accumulated on the dome, for example, they are being removed from the base at the same time.

While the unit described here produces many tens of thousands of volts and may cause annoying shocks, it is still, like the Wimshurst generator, perfectly safe to use because of its extremely small current-producing capability.

30:11 Some Uses of Static Electricity Static electricity has been put to work in many different ways. The process by which photocopying machines use electrostatic charges is most ingenious, for example. The image from the original to be reproduced is first transferred by means of light to a positively charged drum, which in turn picks up negatively charged powdered ink. Finally, the ink is deposited on a positively charged piece of paper and bonded to the paper by heat.

Static electricity has also been used to combat air pollution. Many of the particles in industrial smoke can be removed by adding an electrostatic charge to the particles, then collecting them on oppositely charged metal plates even before the smoke enters the smokestack. Often the particles contain valuable by-products which can be recovered and reused in the plant. Several different types of **electrostatic precipitators**, as these devices are called, have been proven to be successful. Probably the best known is the Cottrell precipitator, invented by American chemist Frederick G. Cottrell (1877-1948) (see Section 23:7).

For several years stereo music enthusiasts have been able to obtain electrostatic loudspeakers capable of reproducing with great clarity sounds that are far above the frequency range of the human ear. These units are built like a large capacitor—one electrode is non-movable; the other, which is free to vibrate, causes the air to move, which , in turn, produces the sounds.

30:12 Lightning In 1752 Benjamin Franklin (1706-1790) demonstrated by means of his famous kite experiment that lightning is a form of electricity. During a thunderstorm Franklin flew a kite using a long length of wet twine (a moderately good conductor). One end of the twine was connected to an ordinary key. Stationing himself under a dry shed, Franklin held the key by means of a dry silk string (a good insulator). He observed that sparks flew from the key when he held it near a *grounded object* (an object electrically connected to the ground). He then attempted to charge a Leyden jar by touching the key to one terminal of the jar and grounding the other. As Franklin had suspected, the charge produced in the jar was identical with that developed by a friction-type static generator. His conclusion: lightning *is* static electricity. Don't go out during the next thunderstorm and try to duplicate Franklin's experiment with a kite and key, however. It is not as easy to do as it sounds. In fact, several persons have been killed attempting to duplicate Franklin's success. Lightning is unpredictable. Don't invite trouble by deliberately trying to attract it.

30:19 *The famous kite experiment.*

Lightning is mentioned several times in the Bible. Exodus 19:16 speaks of "thunders and lightnings and a thick cloud" on Mount Sinai prior to the giving of the Ten Commandments. The display was so fearful and awesome that the Israelites "removed and stood afar off" from the mountain (Exodus 20:18). There are references to lightning in the book of Job (28:26; 37:3; 38:25, 35) and in several of the Psalms (18:14; 77:18; 97:4; 135:7; 144:6). Lightning is frequently mentioned in connection with prophetic visions.

The source of the electrical energy in clouds is the sun. Air rises when heated by the sun. When droplets of water in clouds fall through a rising column of air, they are broken into smaller droplets of unequal size. As they divide, the smaller portions that split off acquire a negative charge; the larger portions that remain, therefore, assume a positive charge. (The same phenomenon has been observed in the spray droplets in waterfalls.) The heavier, positive droplets continue to fall, but the lighter, negative droplets are carried upward by the rising air. Thus, a separation of charge occurs. The process can continue until extremely high voltages exist. A flash of lightning occurs when the potential is high enough to ionize the air and force it to conduct. As it does so, a large number of the charges in the air are neutralized.

30:20 *Static charges occur in the spray droplets of a waterfall. The mechanism is similar to that found in clouds.*

Lightning can strike from a cloud to the ground, from the ground to a cloud, from one cloud to another, or between parts of the same cloud. **Thunder** is the sound that accompanies lightning. When a stroke of lightning occurs, the air through which it passes is heated. The sudden increase in pressure due to the expansion of this channel of heated air generates a clap of thunder. After the first lightning stroke, there is usually a return stroke along a slightly different path. It is possible to have as many as 40 strokes and return strokes, all taking place in less than a second.

30:21 *Lightning over Phoenix, Arizona.*

Most authorities consider that there are at least three types of lightning—**forked lightning**, **sheet lightning**, and **ball lightning**. Forked lightning (also called zigzag, or chain lightning) is the type most commonly observed. It is easily recognized by its jagged streaks of light (see Figure 30:21). Experts disagree, however, on the other two types. Some feel that sheet lightning is simply forked lightning that is located beyond the horizon and reflecting from nearby clouds. Others maintain that it is actually an electrical discharge taking place over a large area, but confess that they know very little about how it could function. Ball lightning is even more controversial because it is so rarely seen. According to the *Encyclopaedia Britannica*, it is a round, luminous mass of electrified air with an average diameter of only 20 centimeters (about 8 inches). The same source states that it is often red in color and travels toward the earth from a cloud until it explodes; it usually explodes in 3 to 5 seconds.

30:22 *One form of man-made lightning.*
(See chapter title page)

You have probably heard it said that lightning never strikes twice in the same place. This is untrue. The Washington Monument has been struck more than once during the same storm. The Empire State Building is struck 30 to 48 times a year. The taller a structure, the more often it is likely to be hit by lightning.

In the same year that he performed the kite experiment, Benjamin Franklin invented the **lightning rod** (see Figure 30:23). It is difficult to estimate how many lives have been saved by this simple but ingenious device. One or more pointed rods are fastened to the roof of a house or building. The lower ends of the rods are connected to the ground by means of wires.

30:23 *Lightning rods furnish a low-resistance electrical path around the outside of a house or building.*

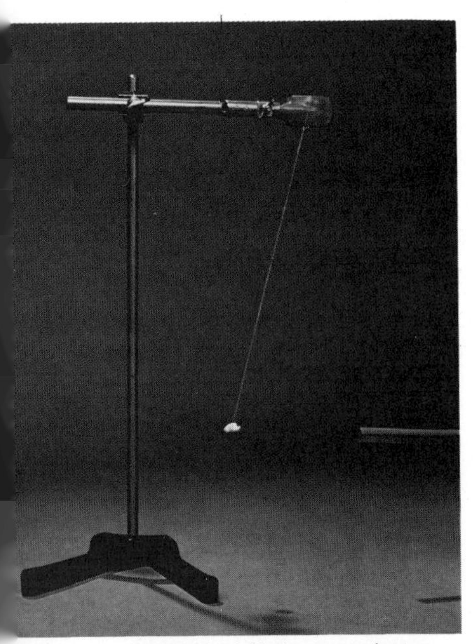

30:24 *A puffed-wheat electroscope as described in Experiment 30:2. Here the puffed-wheat with a positive charge is being repelled by a positively-charged lucite rod.*

Experiment 30:2 *Making a Puffed-Wheat "Electroscope"*

1. Thread a small needle with a one-foot length of sewing thread and tie a large knot in the end.
2. Draw the thread through a piece of puffed wheat.
3. Hang the thread from the screw of a clamp attached to a ring-stand, making sure that the puffed wheat is free to swing in all directions (see Figure 30:24). You are now ready to experiment with your "electroscope."
4. Rub a lucite or celluloid rod with wool to generate a positive charge.
5. Hold the rod near the puffed wheat and observe its attraction for the rod.
6. Now brush the rod against the puffed wheat and note how it repels the rod.
7. Repeat the operation using a negative charge generated by rubbing a hard rubber rod with wool.
8. While the negative charge is still on the puffed wheat, hold a positively charged rod near it and note the strong attraction.
9. Suggest a way that you could use the puffed-wheat "electroscope" to determine whether an unknown charge is positive or negative.

It is wise to keep a few basic rules of safety in mind concerning lightning:

1. The safest place to be during a thunderstorm is in a shielded enclosure, such as a car, or a building that is protected with lightning rods.

2. If you are caught out of doors when a thunderstorm occurs, stay away from isolated trees, high places, and bodies of water. If you are out in an open area, lie down.

3. Even indoors, you should still observe some precautions if a severe electrical storm is taking place in the immediate vicinity. Keep away from windows and outside doors. Do not touch electrical appliances or plumbing fixtures. (Lightning may strike a nearby power line, telephone pole, or water tower and induce a deadly amount of current into the surrounding wires and pipes.)

MICHAEL FARADAY *(1791-1867)*

English physicist

Michael Faraday is rated by science historians as the greatest of the experimental physicists. He is credited with the invention of the transformer, the electric motor, and the electric generator. It was Faraday who discovered the laws of electrolysis and added a new list of words to our scientific vocabulary—anode, cathode, ion, electrode, electrolyte, anion, cation, and electrolysis. Two basic units in physics are named in his honor—the **faraday**, a unit of electrical quantity, and the **farad**, a unit of electrical capacity. He also made many important contributions to our knowledge of diamagnetism, polarized light, and the liquefaction of gases. His discovery of benzene laid the foundation for aromatic organic chemistry. And, perhaps most noteworthy of all, he initiated one of the most profound developments in the history of physics—field theory. From his brilliant and imaginative mind came the terms "magnetic field" and "lines of force."

Faraday was an ardent student of the Scriptures. Without question, the knowledge and wisdom gained from his study of the Bible found their outworking in his life. A member of a small fundamental church in London, Faraday was faithful in his attendance at the all-day Sunday services and the midweek prayer meeting, even when the demands of research were heaviest. He was an exemplary Christian, both in the laboratory and in his testimony in the community. The fruit of the Spirit was abundantly evident in his life, and he professed his faith before his scientific colleagues, the noblemen of England, and members of the ruling family.

A month after his marriage at the age of 29, Faraday made a public profession of his faith in Christ. From this time on his spiritual growth was rapid. Before long, he was appointed to the office of elder in the church. In this capacity he was occasionally called upon to preach. Those who heard him were impressed with the depth of his thoughts and the wisdom of his exhortations. His Bible contained nearly three thousand carefully written notations in the margins—study aids, comments, and cross-references. But knowledge alone did not make him an effective spiritual leader; added to it was his tremendous enthusiasm, humble spirit, and kindly disposition. In the words of Bence Jones, one of Faraday's biographers, his was "a lifelong strife to seek and to say that which he thought was true, and to do that which he thought was kind." At the same time, this Christian man of science did not hesitate to condemn evil and defend the Faith against the attacks of unbelievers, cultists, and apostates.

For many years Faraday was director of the laboratory at the Royal Institution in London. His election as a Fellow of the Royal Society was one of the first of about 95 similar honors and distinctions conferred during his lifetime in recognition of his many outstanding achievements. The promise in Psalm 1:3 that "whatsoever he doeth shall prosper" was fulfilled in Michael Faraday's life in a most remarkable manner.

F. W. Boreham, in his book *A Handful of Stars*, relates that a reporter once asked Faraday about his speculations on the hereafter. "Speculations?" the physicist asked with astonishment. "I have none. I am resting on certainties. 'I *know* whom I have believed, and am persuaded that he is able to keep that which I have committed unto him against that day' " (II Timothy 1:12). Michael Faraday was truly a man who was ever ready to bear testimony of the hope that was within him. Although some of his scientific associates smiled scornfully at his beliefs, none of them could deny that his faith in Christ had produced a most virtuous life; and no one could say that Faraday's faith was the product of an inferior intellect.

List of Terms

ball lightning
Braun electroscope
capacitor
condenser
conductor
coulomb
Coulomb's law
current electricity
dosimeter
electrometer
electroscope
electrostatic precipitator
farad
faraday
forked lightning
insulator
inverse square law
Leyden jar
lightning rod
semiconductor
sheet lightning
static
static electricity
thunder
Van de Graaff generator
Wimshurst generator

Questions

1. (a) An ebonite comb takes on a negative charge as a person combs his hair. Suggest an explanation for what is happening. (b) Does the comb attract or repel the hair after it is charged? (c) Do the individual hairs attract or repel one another?
2. (a) Account for the fact that a moving car acquires an electrostatic charge if the weather conditions are right. (b) How could this be prevented?

3. You can discharge a charged electroscope by touching the knob or plate with your finger. What does this tell you about the conductivity of your body?
4. Does a fiberglas car offer its riders protection against lightning?
5. Graphite is a good conductor of electricity but its allotrope, diamond, is a poor conductor. Suggest a reason for the difference.
6. Name two disadvantages of the gold-leaf electroscope.
7. Why are the Wimshurst and Van de Graaff generators not dangerous in spite of the high voltage they produce?
8. How is the sound of thunder produced?
9. Pure water is a good insulator. Why is it dangerous to swim in a lake during a thunderstorm?
10. Why is it difficult to do research on ball lightning?
11. Why is a house located in a city reasonably safe from lightning, even though it is not equipped with lightning rods?
12. Explain the operation of a lightning rod.
13. According to Coulomb's law, what effect will moving two charged spheres twice as far apart have on the force between them?
14. Name some typical (a) conductors (b) insulators (c) semiconductors.
15. Name two ways (other than grounding) that an electroscope could be discharged.
16. What advantage, other than better fidelity, does FM radio have over AM?
17. (a) How is a Leyden jar constructed? (b) What two names are applied to this general type of electrical device?
18. A certain Wimshurst generator produces a spark 3 centimeters in length. Estimate the voltage between its terminals.
19. Even the best insulators will conduct electricity if they are subjected to a high enough voltage. Is this statement true or false?
20. Name three practical uses of static electricity.

Student Activities

1. Prepare a brief report on Faraday's work in electrostatics, including his famous "cage experiment."
2. Make a study of all the Scripture references to thunder and lightning, using a complete concordance.
3. Prepare a report on Charles Augustin de Coulomb or André Ampère.
4. Make a study of the Cottrell precipitator. Be prepared to explain its operation to the class.

Further Reading

Moore, A.D. *Electrostatics*. Garden City, N.Y.: Doubleday, 1968.
Sutton, A.M. *Nature on the Rampage*. Philadelphia: J.B. Lippincott, 1962 (chapter on lightning).

Photo next page
Substation in which voltage is stepped down from 230,000 to 115,000 volts by means of transformers.

UNIT X ELECTRICITY AND MAGNETISM

Chapter 31 Current Electricity

31:1 Harnessing the Electron Modern life would be quite different without electricity. Almost without thinking you switch on a light, a radio, or an air conditioner. You eat food that was kept in an electric refrigerator, wear clothes that were laundered in an electric washer, talk to your friends on the telephone, and glance at an electric clock to see what time it is. Electricity probably enters your life in more ways than you realize.

(a)

(b)

31:1 *Two of the ways in which electricity serves behind the scenes: (a) Electronic telephone switching equipment (b) Control console at a television station.*

As was stated in the last chapter, current electricity involves charges in motion. In this chapter we shall see how charges are put into motion to accomplish useful work. In most electrical devices electrons are caused to flow through a closed path called a **circuit**. If the electrons travel in only one direction through the circuit, the type of flow is called **direct current** (D.C.). If they travel back and forth, continuously reversing their direction, it is called **alternating current** (A.C.).

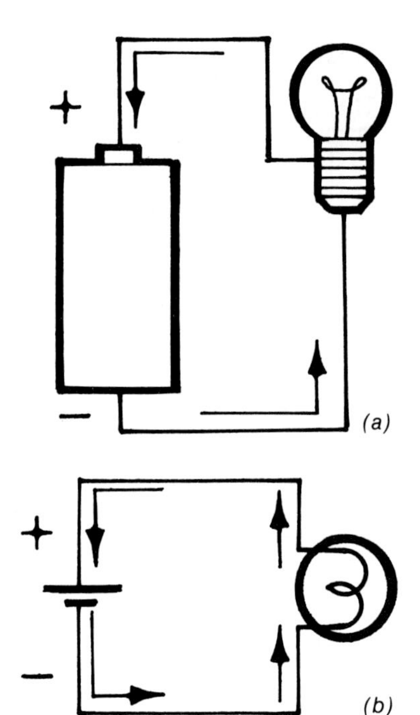

31:2 The simplest kind of circuit. (a) Pictorial diagram (b) Schematic diagram

In any circuit there must be a "pump" to supply the energy of motion to the electrons. This is usually a **battery** or a generator. The simplest kind of a circuit is shown in Figure 31:2a. It consists of a flashlight battery (dry cell) and a small bulb. Two wires are needed to complete the hookup—one from each end of the battery to each of the two metal contacts on the bulb. As soon as the second wire is connected, the bulb lights. Electrons are being pumped out of the lower end of the battery (negative terminal), through the bulb, and back to the upper end of the battery (positive terminal). After passing through the battery, electrons emerge at the negative terminal to start the trip again. Since there are always many electrons at every point in the circuit, the bulb lights continuously.

Figure 31:2a is an example of what is called a **pictorial diagram**. An actual picture of each circuit element is shown, with connecting wires between them. To save effort, those who work extensively with electrical circuits use the shorthand notation shown in Figure 31:2b, called a **schematic diagram**. Rather than a picture, a simple, easy-to-draw symbol is used for each circuit element. The symbol for the flashlight battery is two parallel lines—a long one and a short one. The short line is the negative terminal, the long one the positive terminal. The bulb is represented by a small coil enclosed in a circle. Some common symbols are shown in Table 31-1.

Table 31-1 *Common Circuit Symbols*

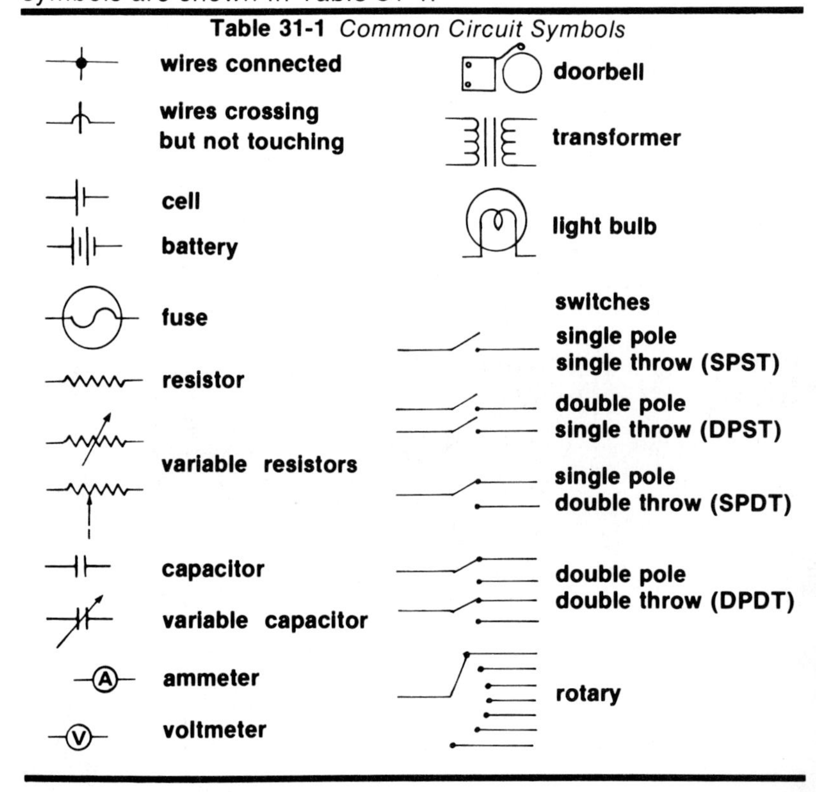

31:2 Series and Parallel Circuits Suppose we now attempt to connect *two* bulbs to our flashlight battery. There are two ways this can be done. We can arrange it so that all the electrons are forced to pass through each bulb in order (**series circuit**). Or we can allow the electrons to branch into two separate routes, some going through each bulb (**parallel circuit**). These two possibilities are diagrammed in Figures 31:3a and 31:3b.

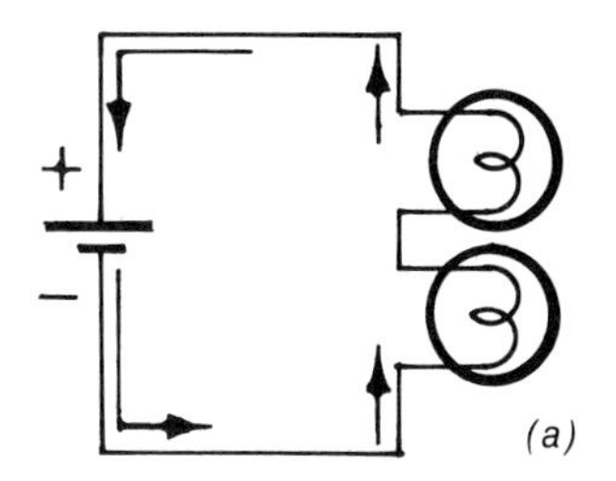

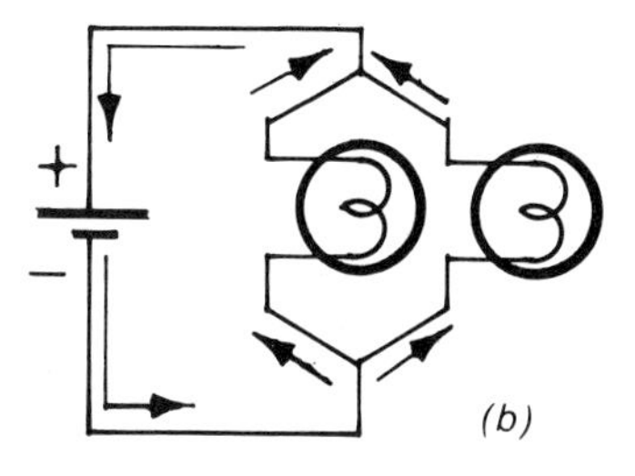

31:3 *A series circuit (a) and a parallel circuit (b).*

Experiment 31:1 *Series and Parallel Circuits*

1. Connect two $1\frac{1}{2}$-volt flashlight bulbs and a large dry cell as shown in Figure 31:3a. Observe how dimly the bulbs light. What happens when one bulb is unscrewed from its socket?
2. Reconnect the dry cell and bulbs as shown in Figure 31:3b. Note the brightness of the bulbs. What happens in this case when one of the bulbs is unscrewed from its socket?

An important feature of the parallel circuit is the independence of each branch. When one of the bulbs is disconnected, the other continues to burn. Not so with the series circuit. In the series circuit, if either bulb is disconnected, both will go out.

But we are not limited to only two bulbs. Three or more could be placed in series *or* parallel, provided, of course, that enough energy could be supplied by the battery. The appliances in our homes are wired in parallel, allowing each to be used independently of the others. Switches, however, are wired in series with the appliances they control. A **switch** is like a drawbridge that turns the electron flow on or off (see Figure 31:5). A switch consists of two conductors that are in contact with each other in the "on" position, but separated in the "off" position. Some typical switches are shown in the photograph. The simplest type of switch, called a single-pole single-throw (SPST) switch, makes and breaks the current in a single circuit. More elaborate switches, such as the channel selector on a television set, turn several different circuits off and on at the same time.

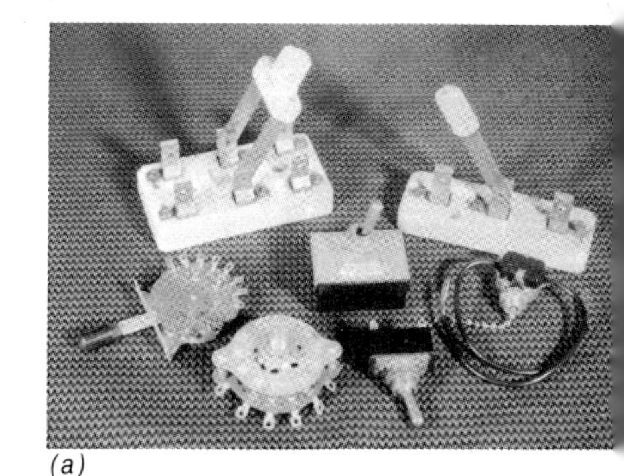

(a)

31:4 *(a) Some simple switches (b) More complex switches*

(b)

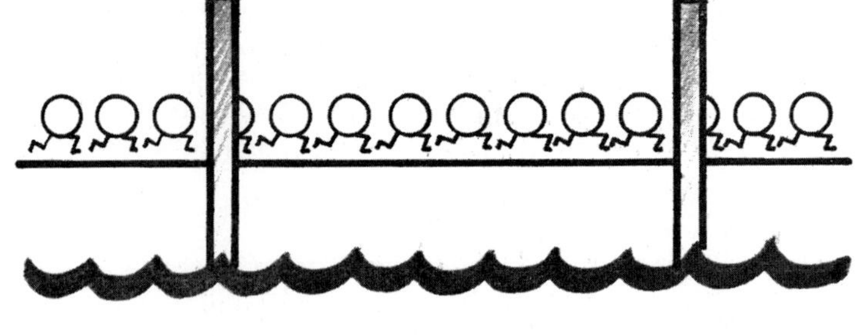

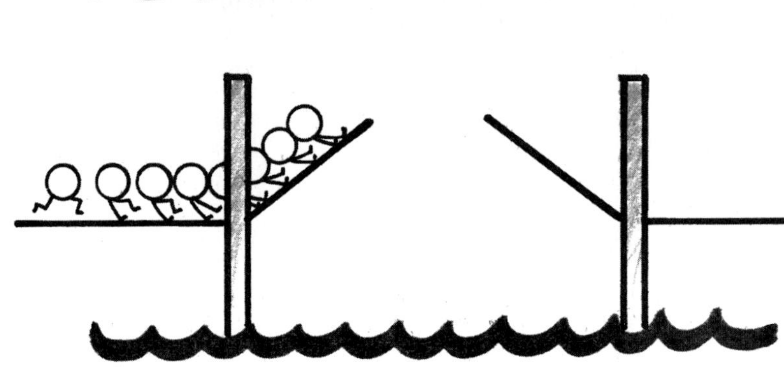

31:5 *A switch is like a drawbridge that turns the electron flow on or off.*

31:6 *A "blown" fuse.*

Fuses are also connected in series. A fuse is a current-sensitive device that protects a current from excessive electron flow. At the heart of a fuse is a low-melting metal alloy. If the current should rise to an unsafe level, the alloy melts and breaks the circuit. Once a fuse has "blown," there is no way to use it again. It should be replaced with another fuse having the same current rating. In some installations **circuit breakers** rather than fuses are used. These are electromagnetic devices that accomplish the same purpose as fuses, but the unit never has to be replaced. It is put back into operation simply by resetting a switch.

31:3 Potential Difference and Current Electricity in a circuit can be likened to water going over a waterfall. The height of the water corresponds to the **potential difference** or **electromotive force** (emf) pushing the electrons through the circuit. The thickness of the flow can be likened to the **current** in the circuit. These are by no means exact comparisons, but they give us at least an idea of what is happening.

The unit of potential difference is the **volt**. It is named after Alessandro Volta (1745-1827), the Italian physicist who built the first practical battery. Two points in a circuit have a potential difference of one volt if it takes one joule of work to move a coulomb of charge from the first point to the second. Because of the unit used, it has become commonplace to speak simply of **voltage** rather than potential difference or electromotive force.

The unit of current is the **ampere**. It is named in honor of the French physicist André M. Ampère (1775-1836). When a charge of one coulomb passes a given point in a circuit in one second, a current of one ampere is flowing. How can we visualize this? In a 100-watt light bulb operating under normal conditions the current is just about one ampere. Many of the small appliances in our homes operate at only a fraction of an ampere.

31:4 Ohm's Law In 1826 an important relationship was worked out by a German school teacher, Georg Simon Ohm (1787-1854). He recognized that in all electrical circuits there is some hindrance to the flow of current. He chose to call this hindrance **resistance**. His well-known law relates the voltage, the current, and the resistance of a circuit. Expressed as an equation, Ohm's law is

$$E = IR$$

$$\text{voltage} = (\text{current})\ (\text{resistance})$$

$$(\text{electromotive force}) = (\text{current intensity})\ (\text{resistance})$$

If the voltage is in *volts* and the current in *amperes*, the resistance will come out in **ohms**. The ohm is *the resistance needed to limit the current in a 1-volt circuit to one ampere.*

According to Ohm's law, in any circuit in which the resistance remains constant, the current is proportional to the voltage. Doubling the voltage causes twice the current to flow. Current and resistance, however, are *inversely* proportional. Therefore, in a circuit in which the voltage is constant but the resistance is variable, the current can be *increased* by *lowering* the resistance, or *decreased* by *raising* the resistance.

31:7 *Small components called resistors are included in electronic circuits to lower the voltage or limit the current.*

Problem: What voltage is needed to drive 7.50 amperes through a 15.0-ohm toaster?

Solution: We are given $I = 7.50$ amperes and $R = 15.0$ ohms. The unknown is E.

$$E = IR$$
$$E = (7.50)(15.0)$$
$$E = 112.5 \text{ volts}$$

Rounding to three significant figures, $E = 113$ volts.

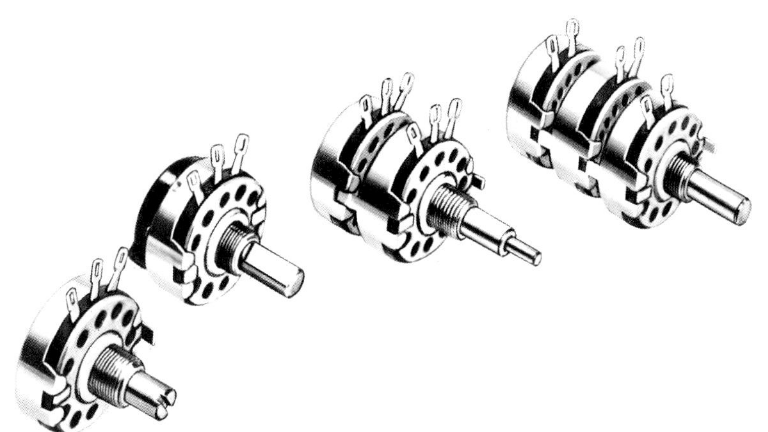

31:8 *Variable resistors such as those used to control volume and tone in a radio or television set.*

31:9 *Illustration of sample problem. Transistor radio operates from a 9-volt battery and has a resistance of 1800 ohms. How much current does it draw from the battery?*

Problem: What current flows in a transistor radio operating from a 9-volt battery if the resistance of the radio is 1800 ohms? (See Figure 31:9.)
Solution: The information given is: E = 9 volts; R = 1800 ohms. The unknown quantity is I.

$$E = IR$$
$$I = E/R$$
$$I = 9/1800 = 0.005 \text{ amperes.}$$

(The answer could also be expressed as 5 milliamperes.)

31:5 Short Circuits When electrical appliances malfunction, the difficulty can many times be traced to a **short circuit**, or "short." This means that the electrons are taking a shortcut which allows them to bypass some of their prescribed path. Electrons will always take the easiest route available to them. As soon as an easier alternate path opens up, they travel over it in preference to their intended path; and the work to which they were assigned is not accomplished. Moreover, the resistance of the shortcut is usually so low that excessive current is drawn from the house wiring, and a fuse is blown.

One example of a short circuit is shown in Figure 31:10. Let us assume that a lamp cord has frayed to the point where the wires are touching. As soon as the lamp is plugged in, there will be a spark at the plug and a fuse will blow. The lamp will not light even momentarily because the electrons bypass it rather than going through it. Wiring should be replaced long before it gets to this state.

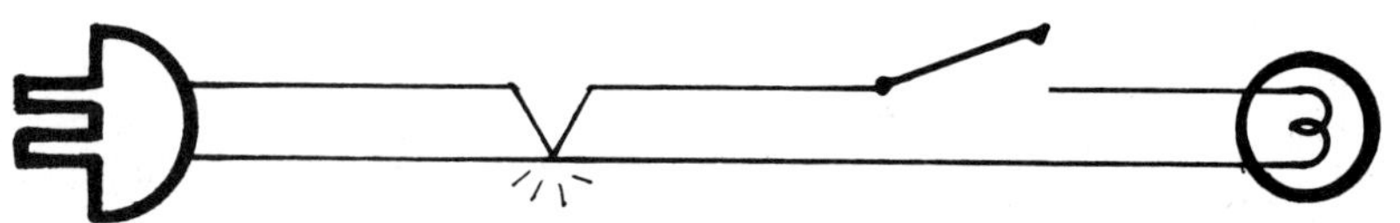

31:10 *Diagrammatic representation of a short circuit caused by a frayed wire.*

31:6 Shock Hazards Some electrical shocks are extremely dangerous; others are only mildly annoying. A knowledge of Ohm's law can help us to understand which is which. Research has shown that for a shock to be fatal, a certain minimum current must flow through the heart. Fortunately, there is a tendency for most of the current to travel along the surface of the body rather than at deeper levels. Also, a shock that does not include the heart in the circuit (such as one between two fingers of the same hand) will be far safer than one which includes the heart (such as one sustained from one hand to the other).

But, other things being equal, the problem comes down to an elementary Ohm's law calculation (E=IR). Between any two points on the body there will be a certain resistance (the actual amount varies from person to person and even with the same individual from one time to another). How much current will flow between these points depends on the applied voltage, provided, of course, that the power source is capable of delivering this amount of current. As we saw in the last chapter, there are devices such as the Wimshurst and Van de Graaff generators that are simply incapable of delivering lethal doses of current. But for a power source which can deliver the current, the higher the voltage the more current will flow.

In answer to the question, "Is it the voltage or the current that makes a shock dangerous?" we must say, "Both." It is the current that does the damage, but it is the voltage that provides the needed force to drive the current through the body. It is best to maintain a healthy respect for anything higher than 20 volts. Whether a given shock is potentially dangerous will depend, in addition to the voltage, on the amount of metal in contact with the skin, the body's resistance, and the portion of the body included in the circuit.

31:11 *It is extremely unwise to dig into a toaster with a metal object. It is especially foolhardy to do so in the vicinity of a grounded object such as a water faucet.*

31:7 Meters We can check up on how well our electrical circuits are performing by the use of meters. Voltage is measured with a **voltmeter**, current with an **ammeter** (a contraction of the words "ampere" and "meter"). Extremely tiny currents may be detected with a **galvanometer**. Some laboratory-type meters are shown in Figure 31:12. Each is equipped with binding posts for the connecting wires. Such a meter is "read" by observing the position of the pointer on the numbered scale.

31:12 *Typical laboratory meters.*

Figure 31:13 shows the use of an ammeter and voltmeter in a typical series circuit. In this hookup we have four 1.5-volt dry cells in series with a 6-volt, 0.25-ampere bulb. The ammeter, denoted by the symbol -Ⓐ-, is placed in *series* with the other circuit elements. It can be inserted at any point in the circuit, since the current is everywhere the same. The resistance of an ammeter is low enough that its inclusion in the hookup will not alter the circuit appreciably. Since the bulb is designed to operate at a current of 0.25 amperes when connected to a 6-volt source, the reading on the meter will be 0.25 amperes.

The voltmeter, denoted by the symbol -Ⓥ-, is connected across (in *parallel* with) points whose potential difference we desire to know. When it is placed across a single cell, as shown in Figure 31:13 it reads 1.5 volts. Across two adjacent cells it reads 3 volts. Three cells give a reading of 4.5 volts, while all four cells together register the full 6 volts. The same 6-volt reading is observed across the bulb. A voltmeter is designed in such a way

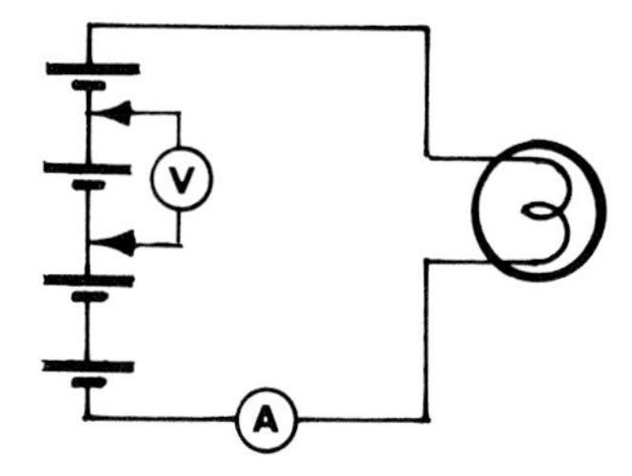

31:13 *How the ammeter and voltmeter are connected into a circuit—the ammeter in series and the voltmeter in parallel.*

31:14 *A multimeter combines the functions of voltmeter, ammeter, and ohmmeter.*

that it upsets the existing circuitry as little as possible. This is done by making its resistance high, so that it draws a minimum of current from the circuit.

A handy device for experimenters and electronics technicians is the **multimeter**, shown in Figure 31:14. By means of a rotary switch this instrument can be made to function not only as an ammeter and a voltmeter of various ranges, but also as an **ohmmeter**. An ohmmeter, in essence, is a sensitive ammeter in series with a battery. When it is connected to an unknown resistance, the meter gives an indication of how much current can be forced through the resistance at the voltage being used. The meter scale is calibrated directly in ohms.

Experiment 31:2 *Using the Voltmeter and Ammeter*

1. Connect the apparatus shown in Figure 31:13, omitting the voltmeter for the time being. Reverse the connections on the ammeter if it gives a backward reading.
2. Record the reading on the ammeter.
3. Using 2 wires connected to the voltmeter, read the voltage across each cell, then across each combination of 2, 3, and 4 cells.
4. Measure the voltage across the bulb and record this value.
5. Calculate the wattage of the bulb by multiplying the voltage across it by the current going through it.
6. Try connecting the ammeter at different points in the circuit to verify that the current is the same at all points.

31:8 Power in Electrical Circuits Several chapters ago we defined **power** as *the rate of doing work* or *work accomplished in a unit of time*. If one joule of work is accomplished in one second, one **watt** of power has been developed. While the watt is not exclusively electrical by its definition, it has come by common usage to be thought of primarily as an electrical unit. The watt is tied in with what we have been discussing in the following way: *A one-volt circuit in which the current is one ampere develops a power of one watt.* We can write an equation for finding the power in any circuit if the voltage and current are known:

$$P = EI$$

$$\text{power} = (\text{voltage})\ (\text{current})$$

Problem: A certain electric stove draws 12 amperes from a 220-volt line. Find its power in watts.

Solution: The information given is $E = 220$ volts and $I = 12$ amperes. P is the unknown quantity.

$$P = EI$$
$$P = (220)(12)$$
$$P = 2640 \text{ watts}$$

The answer could also be expressed as 2.64 kilowatts. (A **kilowatt** is 1000 watts.)

Problem: What current is drawn by a 40-watt porch light operating from a 110-volt circuit?
Solution: We are given P = 40 watts and E = 110 volts. The unknown quantity is I.

$$P = EI$$
$$I = P/E$$
$$I = 40/110 = 0.364 \text{ amperes}$$

31:15 *Watt-hour meter used to measure the amount of electrical energy used by the consumer.*

31:9 Electrical Energy How does an electric company figure a customer's bill at the end of the month? Two factors must enter in—how much *power* is used and for how long a *time*. Clearly, a customer who is away from home most of the month and uses little electricity should pay less than a customer who operates his appliances every day of the month. How can we include both the power and the time in the calculation?

Let us go back to the definition of power—*power is work per unit time:*

$$\text{power} = \text{work/time}$$

From the customer's standpoint a certain amount of *work* has been accomplished, but from the company's standpoint a certain amount of *energy* has been supplied. (Energy is the ability to do work.) The work and the energy are the same, and, as you may recall, they are measured in the same units. We could just as well write, therefore,

$$\text{power} = \text{energy/time}$$

Cross-multiplying, we obtain

$$\text{energy} = (\text{power})(\text{time})$$

The company can find out how much energy it has sold, then, by multiplying the power by the time over which the power was used. If a one-watt appliance is used for one hour, the energy consumed is one **watt-hour**. This is the basic unit of energy in which electricity is sold. A more practical unit is the **kilowatt-hour** (kwh), the amount of energy consumed by a 1000-watt appliance operating for one hour. The electric meter in your home measures the energy you use in kilowatt-hours. The meter is read each month, and you are billed at the rate of so many cents per kilowatt-hour. A graduated scale is generally used so that the more electricity a customer uses, the lower his average rate per kilowatt-hour.

Problem: At the rate of 2 cents per kilowatt-hour, how much would it cost to use an 1100-watt iron for 5 hours?
Solution: First we find the energy used.

$$\begin{aligned}\text{energy} &= (\text{power})(\text{time})\\ &= (1100 \text{ watts})(5 \text{ hours})\\ &= 5500 \text{ watt-hours or } 5.5 \text{ kilowatt-hours}\end{aligned}$$

At 2 cents per kilowatt-hour the cost would be (5.5)(2¢) = 11 cents.

31:16 *Substation in which voltage is stepped down from 230,000 to 115,000 volts by means of transformers.* *(See chapter title page)*

31:17 *Pole transformer which steps voltage down from 240 to 120 volts for use by the consumer.*

31:10 Power Distribution You are probably aware that alternating current is used in homes and buildings rather than direct current. The story behind this choice is interesting. In the late 1800's a controversy raged between American inventors Thomas A. Edison (1847-1931) and George Westinghouse (1846-1914). It had seemed natural to the early experimenters to take the simple D.C. circuits used in the laboratory and scale them up to larger proportions for home use. Edison was of this persuasion; he opposed the use of alternating current because of the difficulty of designing A.C. circuitry. Direct current seemed simple by comparison. Westinghouse, however, was convinced that this drawback was greatly outweighed by the flexibility of alternating current. The voltage could be stepped up or down at will by the use of transformers, making its transmission over large distances more economical. Eventually, Westinghouse gained enough support to win the argument. The choice of A.C. for general use has proved to be a good one.

31:18 *High tension lines used to transmit electricity from one city to another.*

31:11 The Transformer A doorbell operates on 10 volts A.C. If such a unit were plugged into a 110-volt outlet, it would quickly burn out or blow a fuse. To get the needed 10 volts for the doorbell from the 110-volt line, we could put some resistance in series with it. A light bulb of the proper wattage could serve as the resistor (see Figure 31:19a). However, this arrangement would waste energy. The energy used in lighting the bulb is not being used for ringing the bell. A far better way to reduce the 110 volts to the needed 10 volts is by means of a step-down transformer (see Figure 31:19b).

A **transformer** consists of two windings of insulated wire on a laminated iron core. ("Laminated" means simply *composed of many thin layers.*) Energy is put into a transformer on one winding(the **primary winding**) and taken out from the other (the **secondary winding**). If the secondary is wound with more turns of wire than the primary, it is a **step-up transformer**. By "step-up" we mean that *the voltage is increased*. Always, when the voltage is *increased* the current capabilities are *decreased*. This is a result of the law of conservation of energy.

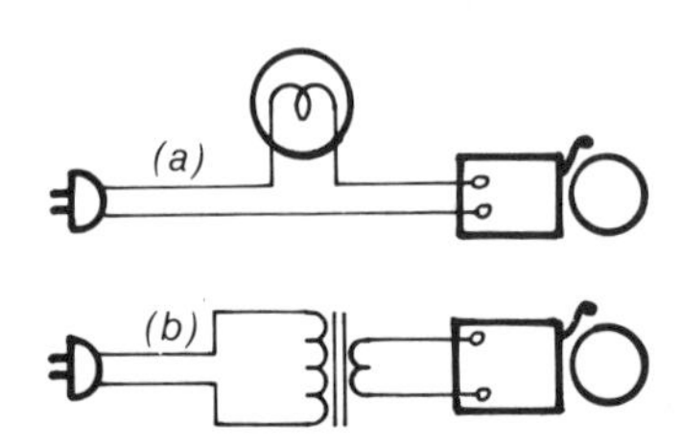

31:19 *Two ways of reducing the voltage for a doorbell. The second method is more efficient.*

The transformer in Figure 31:19b has *fewer* turns in the secondary than it does in the primary; hence, it is a **step-down transformer**. In contrast to the lightbulb method, very little energy is lost in the transformation process. Transformers are noted for their high efficiency. A well-designed unit can have an efficiency as high as 97 percent.

31:20 *Some small transformers used in electronics.*

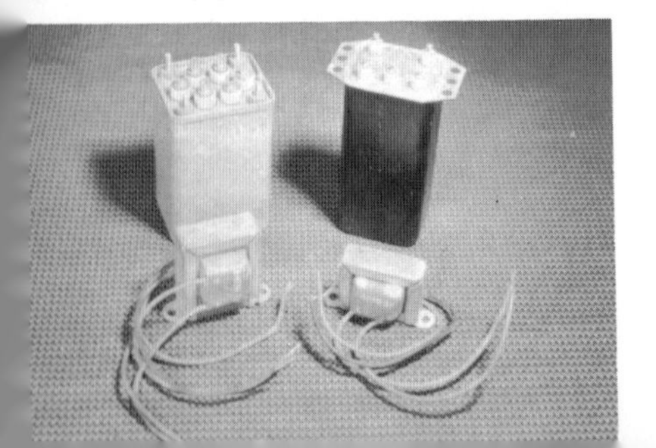

The transformer was invented by Michael Faraday in 1828. At the time, he was attempting to build a device that would convert mechanical to electrical energy. This objective was finally realized in 1831 when he constructed the first workable generator (see Faraday's biographical sketch, page 421). In experimenting with the transformer, Faraday soon came to realize that a current is induced into the secondary only when the current is *changing* in the primary. At the instant he connected a

D.C. source to the primary, there was a brief surge of current in the secondary. But when a steady direct current flowed in the primary, nothing happened in the secondary. Yet when the D.C. source was disconnected, another brief surge occurred in the secondary, this time in the opposite direction.

From these observations we may conclude the following: if the primary D.C. source is connected and disconnected at a fast enough rate, a continuous alternating current will flow in the secondary. This principle is used in the induction coil. A buzzer rapidly makes and breaks the current in the primary circuit. High voltage alternating current appears across the secondary. While interrupted or pulsating D.C. on the primary of a transformer will give satisfactory results for some applications, smoother operation can be obtained using A.C. (see Figure 31:21a). In this case the current is also changing continually, but there are never any abrupt changes as with the switched D.C. (see Figure 31:21b). This results in fewer maintenance problems and less interference to radio and television receivers. Fortunately, most transformers in common use *are* designed for A.C. operation.

31:21 *Alternating current (a) contrasted with pulsating direct current (b).*

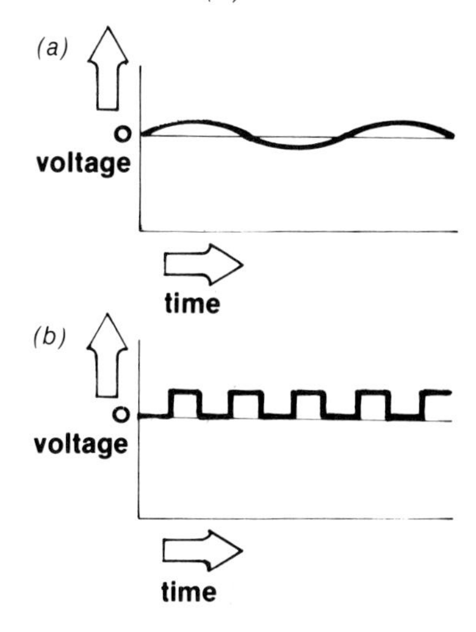

31:12 Heat from Electricity In Chapter 27 we discussed the production of electricity from heat. Now let us consider the generation of heat from electricity. This is a very natural and common energy transformation. Whenever an electric current flows through a wire, heat is produced. If copper wire is used, relatively little heat will be generated under normal conditions.

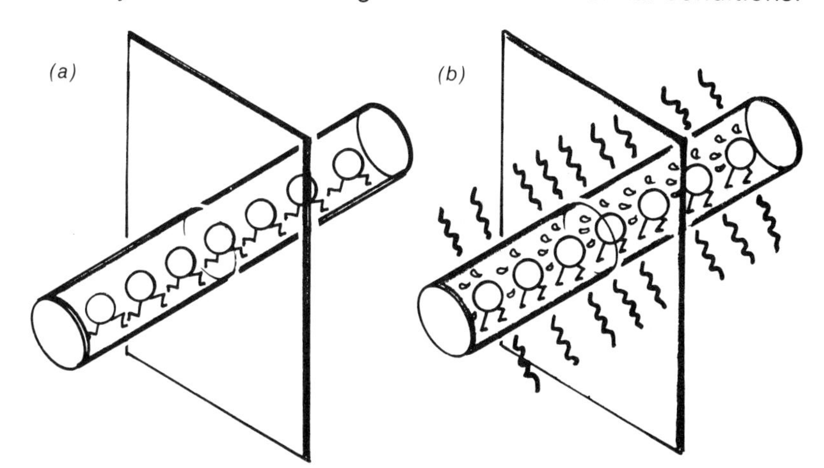

31:22 *Electrons flow with relative ease through copper wire (a) but encounter considerable resistance in nichrome wire (b). Heat is produced as the electrons are resisted.*

But higher resistance conductors such as **nichrome** (an alloy of nickel, iron, and chromium) become red-hot when current is passed through them. Electrons are hindered in their movement as they travel through these materials (see Figure 31:22). Such resistance to current flow causes a dissipation of energy in the form of heat. This effect can be thought of as "electrical friction." Appliances designed to produce heat (toasters, electric stoves, driers, irons, and the like) are equipped with heating elements made of nichrome or some similar material.

31:23 *A toaster uses a heating element made of nichrome or some similar high-resistance material.*

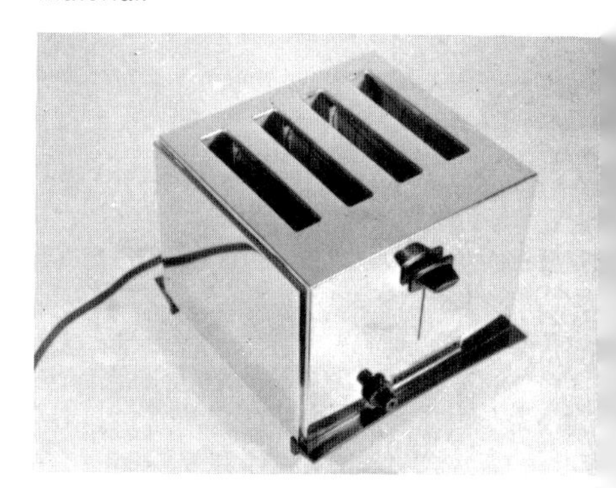

Another useful application of the heating effect of an electric current is the fuse, discussed previously in Section 31:2. Many conductors become hot as current flows through them—so hot, in fact, that they become a fire hazard. Too much current through a small conductor is particularly dangerous. To prevent this, a fuse is placed in series with the circuit to be protected. Constructed of a low-melting alloy, the fuse will "blow" and shut off the current before the temperature of the circuit rises to an unsafe level.

31:13 Light from Electricity The first electric lights were **arc lights**. British chemist Sir Humphrey Davy (1778-1829) found that a bright spark would flash between two carbon rods if they were connected to a battery of high enough voltage. Soon an arc light was developed based on this principle. Such lights were used for a time as street lights and floodlights, but they gave off too much glare for home use. Carbon arc lamps are still used today for motion picture projectors and other applications where extremely high light intensities are required.

But there are other ways to get light from electricity. When a current passes through a wire, light as well as heat can be given off if the wire becomes hot enough. The hotter the wire, the brighter it becomes. Many scientists in the last century recognized this as a possible method for home lighting. The major problem was that the heated wire, or **filament**, as it was called, burned up rapidly. Oxygen in the air combined with the metal to form an oxide. The oxide flaked off and the wire soon burned out. This problem could be solved by enclosing the filament in a glass bulb from which the air had been removed, but it was not until 1865 that a vacuum pump powerful enough to accomplish this was invented.

After much trial and error, Thomas Edison finally developed a workable **incandescent light bulb** in 1879. (*Incandescent* means simply *glowing with intense heat.*) At the heart of his bulb was a carbon filament made by baking a cotton thread. This filament gave enough light, but its life was short. Gradually light bulbs have been improved so that they work more efficiently and last much longer. The filaments today are made of tungsten, a heavy metal having a high resistance and a high melting point. The bulbs are filled with a mixture of argon and nitrogen rather than being left evacuated (see Section 21:4). The gas atmosphere greatly increases the life of the bulb by cutting down losses due to vaporization; and the low activity of the nitrogen-argon mixture assures that the filament will not be attacked chemically.

Still another approach to the problem of getting light from electricity is found in the *fluorescent light*, which will be discussed in Section 35:4. This method has the advantage of far less heat production than the incandescent light. Hence, less energy is wasted and a greater light intensity is obtained with a given amount of electricity.

31:14 Chemical Sources of Current Luigi Galvani (1737-1798), an Italian biologist, found that the leg muscle of a dissected frog twitched violently when it came into contact with two different metals. Galvani thought the convulsions were due to some kind of "animal electricity." The proper explanation was given by Italian physicist Alessandro Volta (1745-1827), who showed that the current was the result of the moist contact between two different metals. We have discussed this previously in the section on galvanic corrosion (see Section 22:11). The frog leg simply acted as the moist medium to conduct the current.

31:24 *Incandescent bulb used in slide projector. The coils of wire in the center are the tungsten filament.*

Experiment 31:3 *Current from a Chemical Reaction*

1. Connect the apparatus as shown in Figure 31:25, using 1 strip each of zinc and copper sheet metal for electrodes and alligator clips to make contact with them.
2. Fill the beaker about 2/3 full of dilute sulfuric acid and close the switch. Observe the current indicated on the ammeter. Note that gas bubbles are given off at both electrodes.
3. Open the switch and replace the ammeter with a flashlight bulb. Close the switch and observe that the bulb lights.
4. Measure the voltage across the bulb with a voltmeter. Note the degenerative changes taking place within the cell.

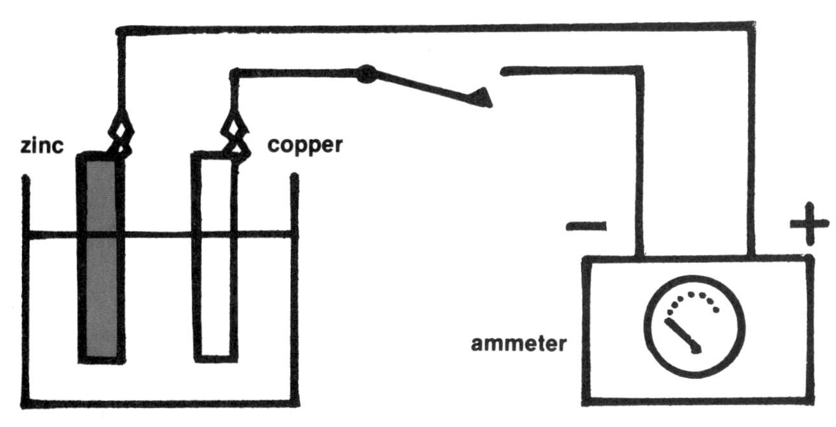

31:25 *Setup for Experiment 31:3.*

Volta eliminated the animal tissue and built what is known as a **voltaic pile**. The pile is a stack of alternate copper and zinc discs separated by pieces of leather or paper soaked in salt water or sodium hydroxide. One copper and zinc disc with the soaked separator between them made up a voltaic cell. The entire stack made up the pile or battery.

Electricity could be obtained from Volta's battery just as it could from electrostatic machines or a Leyden jar, but the battery had the advantage of supplying current continuously for some time before it had to be replenished. Also, the experimenter could control the amount of current by the number of cells in the stack. This forerunner of the modern battery dominated the field of electricity until the development of the generator. *Any device that changes chemical energy into electrical energy* is called a **voltaic cell.**

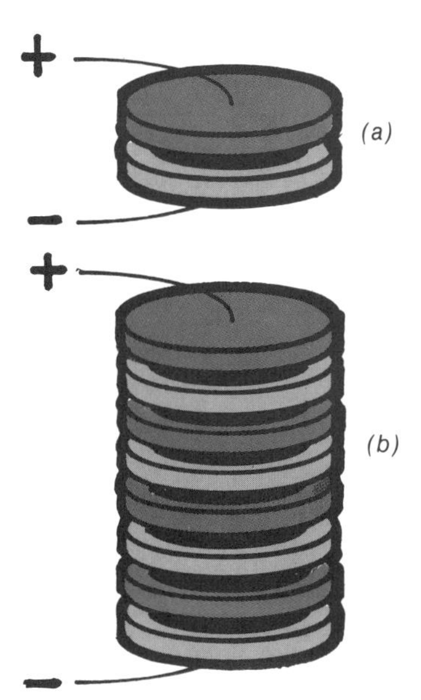

31:26 *(a) Voltaic cell (b) Voltaic pile.*

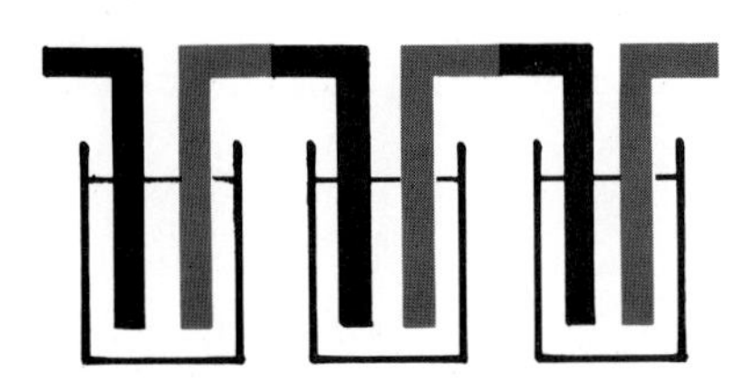

31:27 *Diagrammatic representation of a portion of Volta's crown of cups.*

Another type of battery developed by Volta is called a **crown of cups**. It consists of a series of vessels containing solutions of sodium chloride or sodium hydroxide joined by alternate strips of different metals (see Figure 31:27). Each unit is a cell, and the entire assembly of connected cells forms a battery. This is referred to as a **wet cell** because the current passes from one metal to the other by flowing through a liquid solution. A car battery contains several wet cells. A **dry cell**, on the other hand, uses an ionized paste in place of the liquid electrolyte. The most common dry cells in use today have zinc and carbon electrodes and a moist paste containing ammonium chloride, zinc chloride, and manganese dioxide. Other types of dry cells that have become popular in recent years because of their long life are mercury cells, alkaline cells, and nickel-cadmium cells.

In order to produce an electric current, chemical changes of a degenerative nature must take place in a voltaic cell. Let us examine the chemical reactions in a cell whose electrolyte is HCl in water, and whose electrodes are zinc and copper. In this cell the acid dissolves the zinc. Zinc ions go into solution, leaving their electrons on the zinc and giving it a negative charge.

$$Zn \longrightarrow Zn^{+2} + 2e^{-}\text{(on electrode)}$$

A few copper ions go into solution. Hydrogen ions pick up electrons at the copper electrode and go off as hydrogen gas. This leaves the copper electrode slightly more positive than the zinc electrode. A potential difference will then exist between the two. If the two electrodes are connected by a wire, electrons flow from the zinc through the wire to the copper anode. Here they neutralize more hydrogen ions and form more hydrogen gas. This will continue until all of the hydrogen ions have been neutralized. As long as chemical action continues in the electrolyte, the cell will yield current.

31:28 *Zinc-carbon dry cells and battery.*

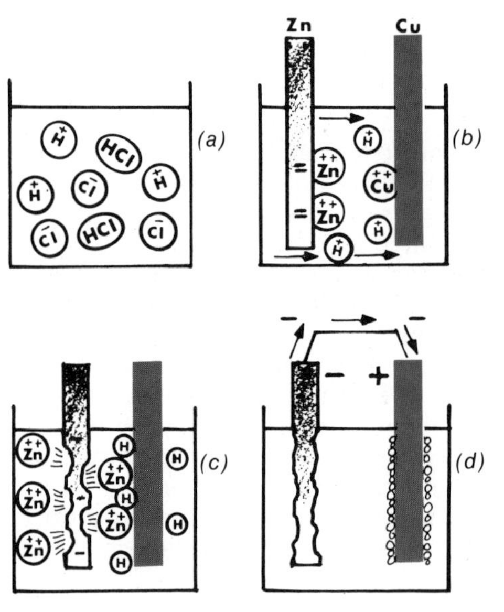

31:29 *(a) A copper-zinc voltaic cell using hydrochloric acid as the electrolyte. The acid already contains hydrogen and chloride ions. When zinc and copper electrodes are introduced (b), many zinc ions and a few copper ions go into solution. The zinc, being a more active metal, dissolves rapidly (c). Electrons flow in the external circuit from the zinc electrode to the copper electrode (d).*

The action of a wet-cell battery is similar to this. A lead storage cell contains a solution of sulfuric acid. The electrodes or plates are constructed of lead and lead peroxide. As the battery is used, both plates become converted to lead sulfate. This reaction is reversed when the battery is recharged. Recharging is carried out by passing a current through the battery in the direction opposite to that of the discharging current. This restores the plates and increases the chemical energy stored in the battery by raising the concentration of the sulfuric acid.

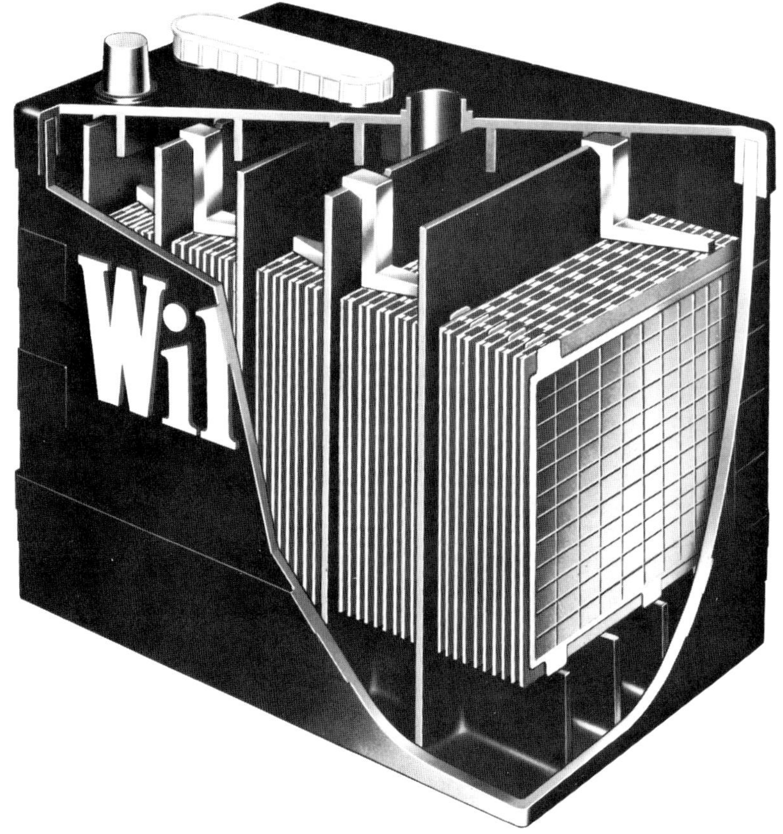

31:30 *Cutaway view of a car battery.*

31:15 Electrolysis Faraday coined several words to refer to the action occurring between the solution and the metals (or carbon) in the cell. The general process is called **electrolysis**, which means *the breaking down of a chemical compound in solution by the action of an electric current.* The solution is referred to as the **electrolyte** and the metals (or carbon) as **electrodes**. Ions (see Section 7:1) must be available in the solution for it to conduct current. If a voltage is applied to an electrolytic cell, the positive ions in solution drift toward the negative electrode (*cathode*), and the negative ions drift toward the positive electrode (*anode*). When an ion reaches an electrode, it gives up its charge and becomes neutral.

31:31 *Electrolysis of dilute hydrochloric acid.*

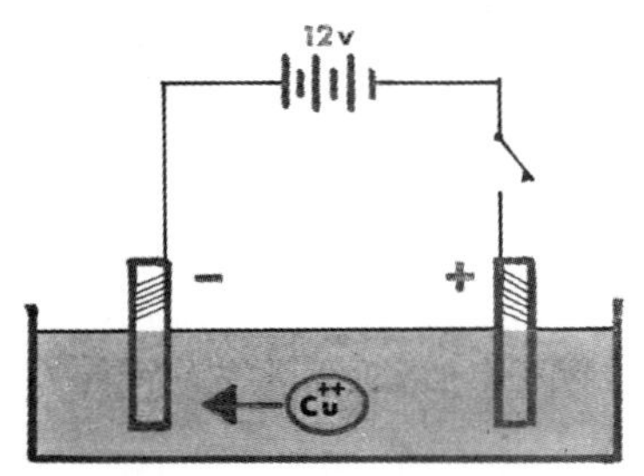

31:32 *Setup for Experiment 31:4. Electrolysis of copper sulfate.*

Experiment 31:4 *Electrolysis of Copper Sulfate*

1. Connect the apparatus as shown in Figure 31:32 using carbon rods from old dry cells for electrodes and copper sulfate solution for the electrolyte. You may use either a 12-volt battery or a 12-volt D.C. power supply as your current source. (A lower voltage may be used, but the reaction will be slower.)
2. Close the switch and notice the escape of gas bubbles from the anode. As the reaction proceeds, the cathode will become coated with copper and the solution will become less blue.

31:33 *Series of aluminum reduction cells in which bauxite is electrolyzed to free aluminum.*

If the solution is hydrochloric acid in water, the following reactions occur:

$$\text{Cathode: } 2H^{+} + 2e^{-} \longrightarrow H_2\uparrow$$

$$\text{Anode: } 2Cl^{-} \longrightarrow Cl_2\uparrow + 2e^{-}$$

Electrons given up at the anode follow around the external circuit to the cathode and neutralize the positive ions. The equation for the overall process is

$$2H^{+} + 2Cl^{-} \longrightarrow H_2\uparrow + Cl_2\uparrow$$

Hydrogen and chlorine gases are given off at the cathode and anode respectively.

If the positive ion in solution is a metal, it will "plate out" on the cathode, forming a coating. For example, the chemical equation for silver plating is

$$\text{Cathode: } Ag^{+} + e^{-} \longrightarrow Ag \text{ (coating)}$$

The article to be plated is made the cathode. A bar of pure silver is used for the anode. The plating bath (electrolyte) contains a soluble silver salt (commonly $AgNO_3$) and other chemicals. A limited amount of current is allowed to flow for some time until the article is fully coated. The current is kept low to insure an even coating that adheres to the surface. This process of metal coating by means of an electric current is called **electroplating**.

31:34 *Plant in which aluminum is electrolytically refined.*

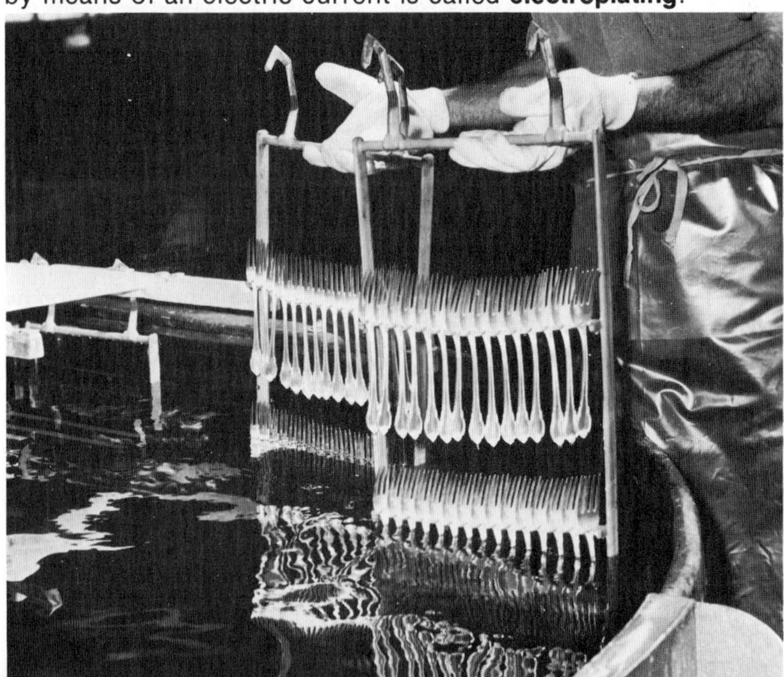

31:35 *Silverplating. Forks rotate in plating tank containing cyanide solution and bars of pure silver. Amount of electric current and length of time the pieces remain in the tank determine the amount of silver on each piece.*

31:16 Energy Transformations In this chapter we have seen the ability of electricity to be changed into heat, light, and chemical energy. In the next chapter we shall see how it can be transformed into magnetic, mechanical, and sound energy. We can now begin to appreciate how many different things electricity is able to do. Moreover, electricity can be produced in a variety of ways—from chemical energy, mechanical energy, sound, light, and heat, to mention a few. In all these energy changes the general principle of conservation of energy applies: *Energy can be changed from one form to another but can never be created or destroyed.* One of the greatest achievements of modern science and engineering has been the harnessing of such energy transformations for man's use. This is an example of man's subduing the earth (Genesis 1:28).

31:36 *(a) Power plant of the future? Model of a 300- to 500-megawatt fast breeder nuclear power plant scheduled for operation by 1980. Artist sketch (b) shows the projected size of the completed power plant.*

List of Terms

alternating current
ammeter
ampere
arc lights
battery
circuit
circuit breaker
crown of cups
current
direct current
dry cell
electrodes
electroplating
electrolysis
electrolyte
electromotive force
filament
fuse
galvanometer
incandescent light bulb
kilowatt
kilowatt-hour
multimeter
nichrome
ohm
ohmmeter
parallel circuit
pictorial diagram
potential difference
power
primary winding
resistance
schematic diagram
secondary winding
series circuit
short circuit
step-down transformer
step-up transformer
switch
transformer
volt
voltage
voltaic cell
voltaic pile
voltmeter
watt
watt-hour
wet cell

Questions

1. Distinguish between alternating current (A.C.) and direct current (D.C.).
2. What is the difference between a pictorial diagram and a schematic diagram?
3. Distinguish between static electricity and current electricity.
4. What are circuit breakers? What advantage do they have over fuses?
5. Explain the difference between series and parallel circuits.
6. How is the *volt* defined?
7. How is the *ampere* defined?
8. What do the letters "emf" stand for?
9. What is a short circuit?
10. How is the *ohm* defined?
11. Of what does an ohmmeter consist?
12. How is the *watt* defined? The *kilowatt*?
13. How is the *watt-hour* defined? The *kilowatt-hour*?
14. Name three ways in which light can be obtained from electricity.
15. Suggest a model to show how electric current produces heat.
16. How does an electric fuse work?
17. How are the ammeter and voltmeter connected to a circuit?
18. What two types of current can be used on the primary winding of a transformer?
19. What is nichrome? How does a nichrome wire differ from a copper wire of the same diameter and length?
20. What precautions are taken to insure that the filament of a light bulb will last as long as possible?
21. What are the products when hydrochloric acid is electrolyzed?

22. What happens if too high a current is used for electroplating?
23. Why is A.C. more practical for homes and buildings than D.C.?
24. Why would a shock between two fingers of the same hand be less dangerous than a shock sustained from one hand to the other?
25. What materials are used in the construction of the common dry cell?
26. Complete the following table of energy transformations:

Device	*Changes*	*Into*
dry cell	chemical energy	electrical energy
light bulb	____________	____________
toaster	____________	____________
transformer	____________	____________

Problems

1. What current will a 440-ohm light bulb draw from a 110-volt line? Express the answer in both amperes and milliamperes.
2. What voltage is needed to drive 1.3 amperes through a 50-ohm resistor?
3. What resistance is needed to limit the current in a 30-volt circuit to 0.15 amperes?
4. A person accidentally places himself in contact with a 110-volt line. If the portion of his body that is included in the circuit has a resistance of 18,500 ohms, what current will flow? Express the answer to the nearest milliampere.
5. An electric iron is rated at 1100 watts. What current flows through it when it is operated from a 110-volt line? What is its resistance?
6. An electric stove draws 15 amperes from a 220-volt line. Find its power in watts and kilowatts.
7. A 40-watt lamp draws 3 amperes. At what voltage is it operating?
8. At the rate of 2 cents per kilowatt-hour, how much would it cost to operate a 100-watt lamp for 24 hours? Round your answer to the nearest cent.
9. At the rate of 2 cents per kilowatt-hour, how much would it cost to operate a 2800-watt air conditioner for 30 days? Assume that the unit runs 50 percent of the time.
10. Find the cost of operating a 2-watt electric clock for 365 days at the rate of 2 cents per kilowatt-hour. Express your answer to the nearest cent.

Student Activities

1. Find out and report on the electrical rates per kwh in your area.
2. Make a lemon wet cell by inserting two strips of different metals into a lemon. Connect a galvanometer between the two pieces of metal to give a visible indication of the action in the cell.

★ 3. Sketch three separate graphs of Ohm's law as follows:
(a) current (ordinate) versus voltage (abscissa) at constant resistance
(b) resistance versus voltage at constant current
(c) current versus resistance at constant voltage

Further reading

Bender, Alfred. *Let's Explore with the Electron*. N.Y.: Sentinel Books, 1963.

Photo next page
Windmills, once a familiar part of the rural scenery in the United States, were used to convert mechanical energy from the air to electrical energy by means of a generator.

UNIT X ELECTRICITY AND MAGNETISM

Chapter 32 Magnetism

32:1 *Lodestone picking up paper clips.*

32:1 From Lodestone to Modern Technology Magnetism was recognized early in the history of mankind. The ancients discovered quite by accident that the mineral lodestone (now identified as magnetite, Fe_3O_4) would attract iron. Today, thousands of years later, after much painstaking research, we still do not have a satisfactory understanding of magnetism. But we have learned enough to put it to work for us. Many useful inventions depend upon magnetism for their operation—motors, generators, compasses, electromagnets, doorbells, buzzers, and electrical meters, to mention a few. In this chapter we shall try to gain some knowledge of the principles underlying these devices.

32:2 Attraction and Repulsion The principle for static electric charges—that like charges repel and unlike charges attract—applies also to magnets. We need only to change the wording slightly: like poles repel; unlike poles attract. You can verify this for yourself quite easily with a pair of bar magnets. If both are reasonably strong magnets, you will find invariably that the two south poles repel, the two north poles repel, and both north-south combinations attract (see Figures 32:2, 32:3).

32:2 *It is easy to verify for yourself that like poles of a magnet repel and unlike poles attract.*

32:3 *A "wobbly bar." One magnet "floats" above the other due to the repulsion of like poles.*

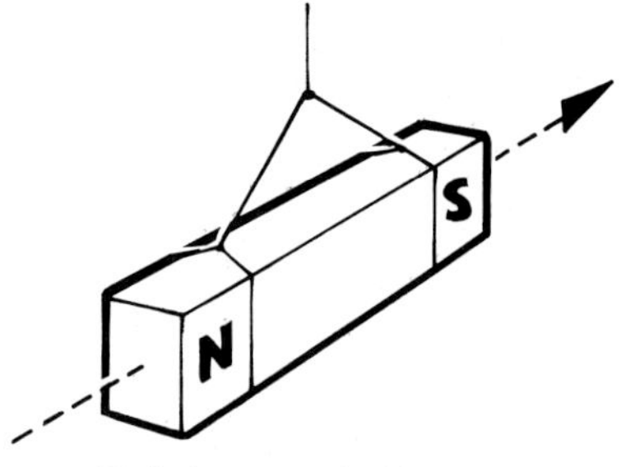

32:4 *A suspended bar magnet functions as a compass needle.*

If a bar magnet is suspended as shown in Figure 32:4, it will align itself approximately in a north-south direction. Such a suspended magnet is a **compass**. Sir William Gilbert (1540-1603), court physician to Queen Elizabeth, was the first to suggest that the earth is, in fact, a large magnet which acts upon and aligns a compass needle.

The end of a compass needle that is labeled "N" will point roughly toward the earth's north pole. However, there is usually a slight displacement to one side or the other because the earth's *magnetic* pole does not coincide with its *geographic* pole. In fact, they are approximately 1200 miles apart. The earth's magnetic poles are located as follows:

The earth's *south* magnetic pole (which attracts the "N" pole of a compass) is located near the Prince of Wales Island in northern Canada.

The earth's *north* magnetic pole (which attracts the "S" pole of a compass) is located near the coast of Antarctica. These two points are seen to be almost exactly opposite each other on a globe (see Figure 32:5).

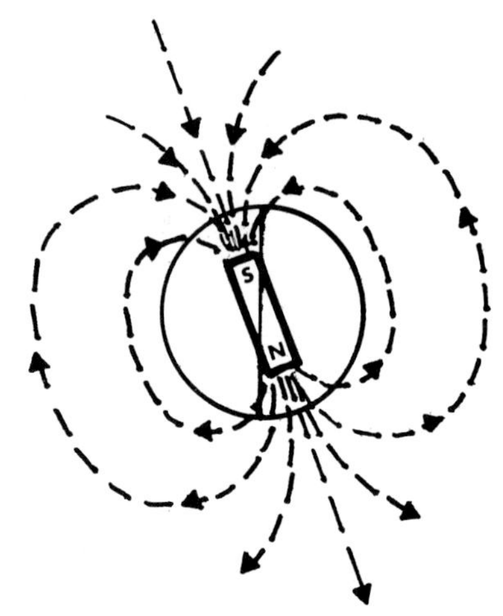

32:5 *The earth's magnetic poles are 1200 miles away from its geographic poles.*

A geomagnetic map, such as that shown in Figure 32:6, will help you to determine the correction for your compass at any given location. A correction of "10° W" means that the compass points 10 degrees west of true north. Those who are fortunate enough to live along the **agonic line** (line of zero correction, or zero declination, as it is called) can read true north directly from their compasses without need of a correction.

A **dipping needle** is similar to a compass, but it operates in a vertical plane rather than a horizontal plane (see Figure 32:7). When placed at either of the earth's magnetic poles, a dipping needle points straight up and down. Midway between the magnetic poles the needle is roughly horizontal. You can get some idea of your latitude by observing a dipping needle if suitable corrections are made.

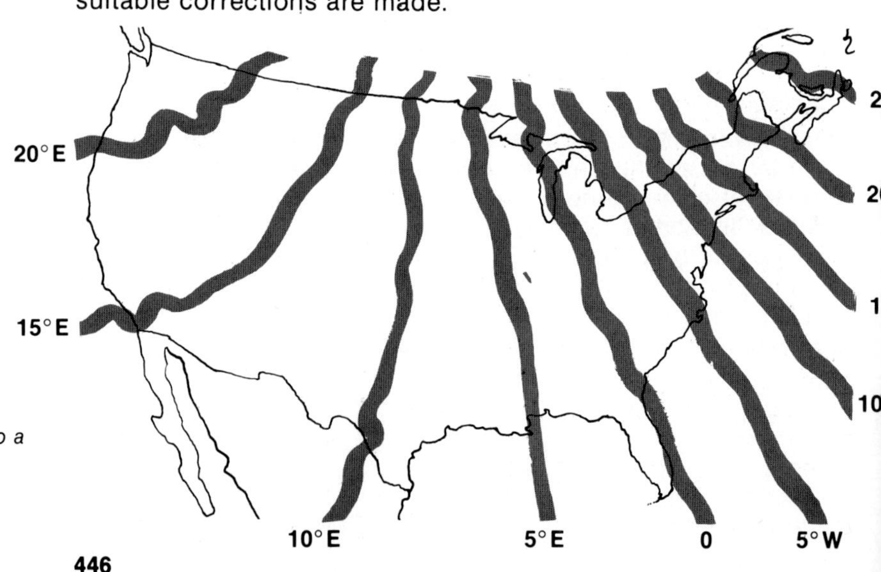

32:6 *A geomagnetic map of the United States. The declination lines show how much correction must be applied to a compass reading at a given place to locate true north.*

32:3 Magnetic Materials With enough patience, using the trial and error method, you could make a list of materials that are attracted by a magnet and another list of those that are not. Most materials—paper, glass, rubber, cloth, plastics, wood, stone, and even most metals—fail to respond to a magnet. The list of materials that are attracted is a very short one: *iron, cobalt, nickel, gadolinium*, and a few alloys; such alloys usually contain a certain percentage of one or more of these elements. For example, steel is an alloy of iron and carbon, with small percentages of various other metals sometimes added to give it special properties. The percentage of iron is high enough to make steel respond well to the attraction of a magnet. Another example of such an alloy is that used in a Canadian nickel. Its composition of 75 percent nickel-25 percent copper enables it to be picked up by a magnet, whereas the United States nickel, having a composition of 25 percent nickel-75 percent copper, does not respond even to a very powerful magnet.

32:7 *A dipping needle is a compass needle mounted so that it can swing in a vertical plane.*

32:4 The Field of a Magnet If a piece of cardboard is placed on top of a bar magnet and iron filings are sprinkled on the cardboard, the iron filings arrange themselves along curved **lines of force,** which extend from one pole to the other. The lines of force themselves are quite imaginary, really, but you can visualize the concept as follows. Pretend that the north pole can be removed from a tiny magnet and used separately. (Poles cannot be detached in real life, but this *unit north pole* is useful for

Experiment 32:1 *Magnetic Fields*

1. Place a horseshoe magnet on a table or desk and cover it with a piece of cardboard. Sprinkle a light layer of iron filings onto the cardboard. Tap the cardboard to make the pattern more clear. The iron filings align themselves along "lines of force," which extend from one pole to the other.
2. Repeat the procedure using a bar magnet in place of the horseshoe magnet. Notice that the lines of iron filings are longer this time because the poles are farther apart.
3. Now repeat the procedure using 2 bar magnets placed with like poles almost end to end, but with a slight gap between the magnets. Observe that the lines from one magnet do not extend to the other magnet.
4. Finally, repeat the last procedure using *unlike* poles. Note that the lines now extend continuously from the pole of one magnet to the pole of the other.

performing "thought experiments.") We now place our unit north pole at any point near the bar magnet. The unit pole will immediately experience a force, and the path along which it is pushed is called a "line of force." Because the lines are defined using a unit *north* pole, they are directed *away from* the north pole of the magnet and toward its south pole. We can similarly demonstrate the field of a horseshoe magnet, or combinations of two or more bar magnets. As a general rule, the lines extend continuously between unlike poles but refuse to join between like poles (see Experiment 32:1). Magnetic lines of force are shown in Figures 32:5 and 32:12.

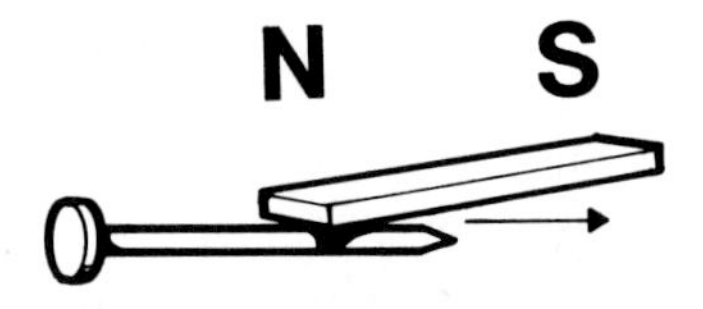

32:8 *One method of making a magnet. An unmagnetized piece of iron can be magnetized by stroking it repeatedly in the same direction with a strong magnet.*

32:5 Making a Magnet An iron nail can be magnetized by stroking it with a strong bar magnet. Every stroke should be made in the same direction to avoid any demagnetizing. As shown in Figure 32:8, if the nail is stroked with a north pole, the end of the nail toward which it is drawn will become a south pole.

A second method of making a magnet is by **induction**. When a strong magnet is used to pick up paper clips, it is often possible to pick up two or more in a row, as shown in Figure 32:9. This is called **magnetization by induction** or **induced magnetism**. We have actually made little magnets out of each of the paper clips merely by placing them in the strong field of the large magnet. Most of the magnetism will leave the paper clips when the large magnet is removed.

Still another method of making a magnet consists of wrapping the material to be magnetized with insulated wire and connecting the two ends of the wire to a direct current source. This will be covered under electromagnets.

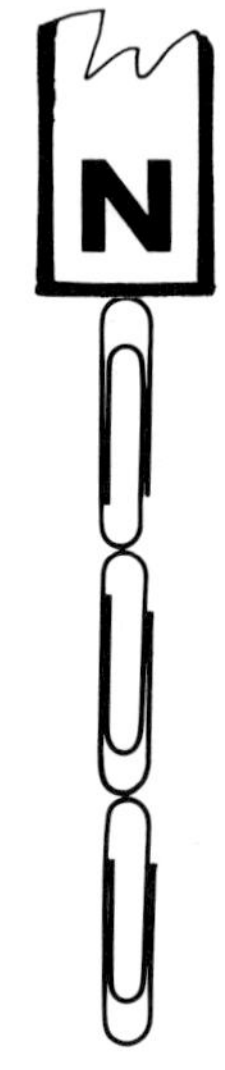

32:9 *Induced magnetism. If the magnet is strong enough, each paper clip also becomes a tiny magnet.*

32:6 The Domain Theory What happens inside a piece of iron that is being magnetized? The modern view is that the iron consists of separate regions called **domains**. Each domain contains many millions of elementary magnets. Within a given domain all the elementary magnets are aligned in the same direction. But in an unmagnetized piece of iron the domains themselves are aligned in various directions (see Figure 32:10). When a **magnetic field** is applied to the iron, those domains which are aligned in the direction of the field grow in size. Those which are aligned in some other direction decrease in size. If the magnetizing field is strong enough, some of the domains will rotate until they are lined up with the field. The magnetizing process is complete when all of the domains are oriented in the same direction.

Much experimental evidence indicates that the above view is correct. If the surface of a highly polished piece of iron is covered with a mixture of glycerin and powdered iron oxide, an examination with a microscope reveals that the particles of iron oxide have arranged themselves in definite lines. These lines are interpreted as boundaries between domains. When the iron is brought into a magnetic field, these lines are seen to move. The nature of their motion is consistent with the idea that some of the domains are growing in size while others are getting smaller.

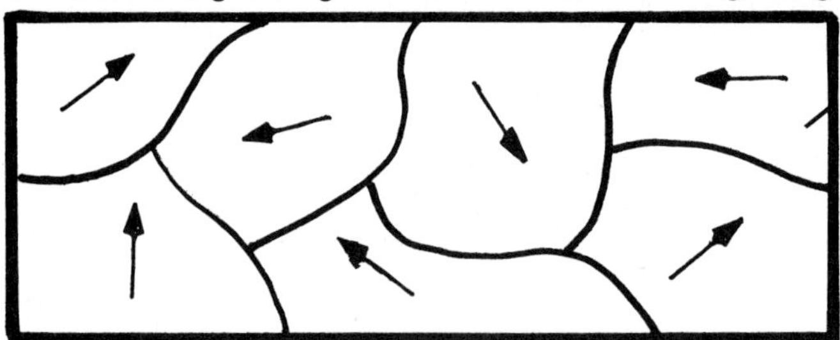

32:10 *There is experimental evidence that the domains in an unmagnetized piece of iron are randomly oriented.*

32:7 Oersted's Experiment In 1819 a Danish physicist, Hans Christian Oersted (1777-1851), discovered an important relationship between electricity and magnetism. Using a compass needle as a detector, he found that a wire carrying an electric current is surrounded by a magnetic field. The lines of force form circles centered on the wire and positioned at right angles to it. This effect is illustrated in Figure 32:11.

What this means is that a current-carrying wire is actually a magnet, although not a very strong one. Two nearby wires carrying currents in the same direction are slightly attracted to each other. If the current in either wire is reversed, there will be repulsion between the wires. A wire located in the field of a powerful magnet can experience a fairly strong deflection if the current in it is great enough. These are interesting observations, but do they have any practical significance?

Soon after Oersted's discovery, Ampère found that the magnetic effect of a wire could be made much greater by wrapping it into a coil. The type of coil called a **solenoid** can be made by wrapping insulated wire around a pencil, then removing the pencil (see Figure 32:12). When the solenoid is connected to a dry cell and tested with a compass, it is found to have a north pole at one end and a south pole at the other. The field around a solenoid is very similar to that of a bar magnet.

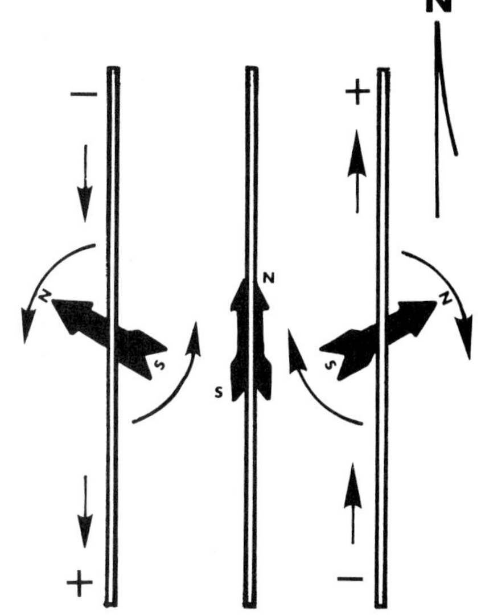

32:11 *Compass needle placed beneath a wire (center) is unaffected when there is no current passing through. At the left a current flowing down the wire deflects the needle to the left. At the right a current flowing up the wire deflects the needle to the right.*

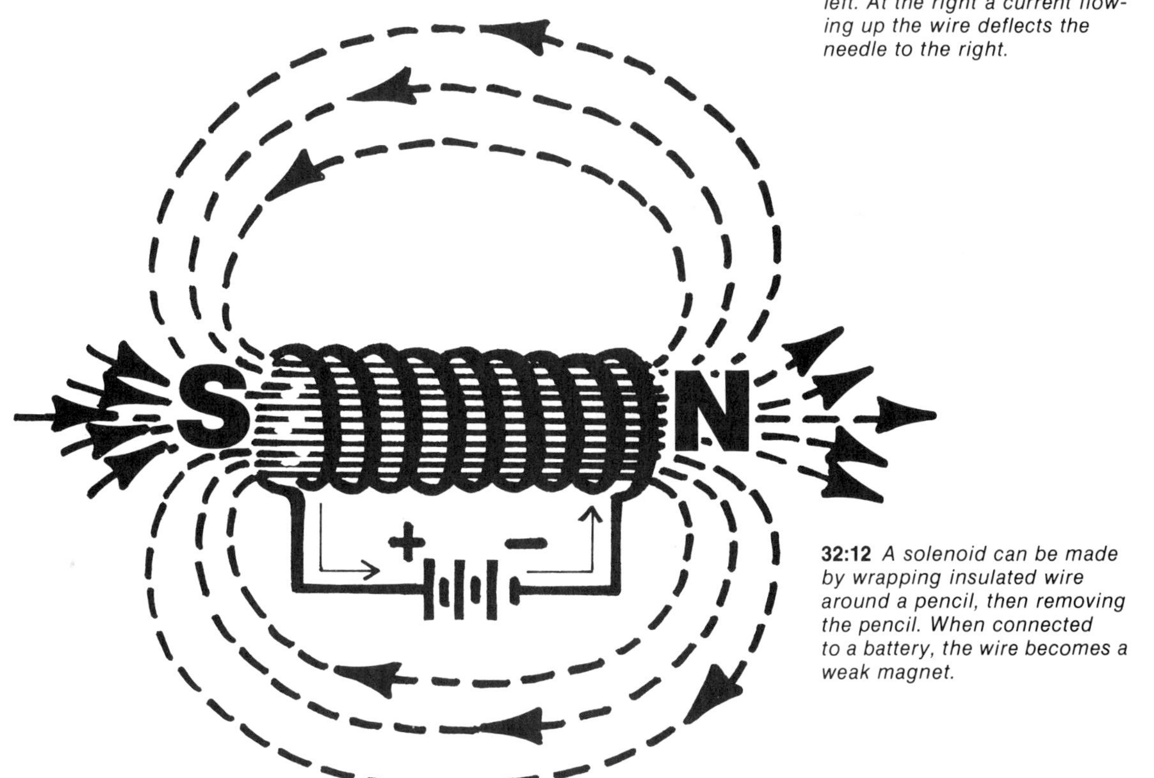

32:12 *A solenoid can be made by wrapping insulated wire around a pencil, then removing the pencil. When connected to a battery, the wire becomes a weak magnet.*

In 1823 an English experimenter named William Sturgeon (1783-1850) learned that the strength of a solenoid could be tremendously increased by placing a soft iron bar inside it. We explain this by saying that the iron gives the magnetic "lines of force" an easier path than they would have in air. Sturgeon thus made the first practical **electromagnet**. With the current turned on, his electromagnet would lift 20 times its own weight of iron. When the current was turned off, it returned essentially to its unmagnetized state. We can make a simple electromagnet by wrapping insulated wire on a nail. The more turns, the better it will work. The electromagnet is one of the most useful devices known to man. It is utilized in more ways than we could begin to enumerate. A few applications are illustrated in Figures 32:13-16.

32:13 *Electromagnets are used in this scrap iron hoist.*

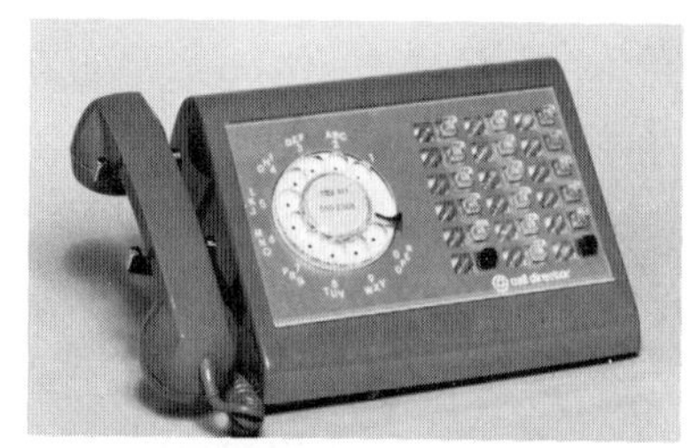

32:14 *Electromagnets are used inside the earphone of a telephone handset.*

32:16 *Electromagnets are used inside the generators of a diesel train.*

32:15 *Electromagnets are used inside an electric typewriter.*

Experiment 32:2 *Making an Electromagnet*

1. Wind 30-50 turns of insulated wire around a large nail, keeping the turns as close together as possible. It may be necessary to start a second layer of wire if your nail will not accommodate all the wire in one layer.
2. Strip the insulation from the two ends and connect them to the terminals of a large dry cell. Observe the effect of the electromagnet on a nearby compass needle.
3. Test the magnet's ability to pick up paper clips, thumb tacks, and small nails. How much induced magnetism are you able to observe?

32:8 The Doorbell Figure 32:17 shows the arrangement of parts used in a doorbell. When the circuit is completed by pressing the push button, two electromagnets in series attract the armature and clapper. As the clapper strikes the gong, the current is cut off by the "make-and-break," a normally closed switch attached to the armature. When the circuit is broken, the magnet stops attracting and releases the armature, allowing it to spring back and start the cycle over again. In most doorbells the clapper strikes the gong several times a second.

The buzzer is essentially the same as the doorbell but does not have the gong and clapper. Other related devices are the single-stroke door chime in which a solenoid causes a movable core to strike a metal bar and the two-tone door chime in which two different solenoids activate two metal bars having different pitches.

32:17 *A doorbell connected to a battery and switch.*

32:9 The Loudspeaker The means by which we hear music, speech, and other sounds electronically reproduced relies for its operation on electromagnetism. Figure 32:18 shows the working parts of a loudspeaker. A small coil of wire attached to the center of a paper cone is situated within the field of a strong permanent magnet. The coil is powered by alternating current whose frequency corresponds with that of the sound being reproduced. As the direction of the current in the coil changes back and forth, the coil and cone are alternately attracted to and repelled from the magnet. The rapid back-and-forth motion of the cone produces sound waves in the surrounding air.

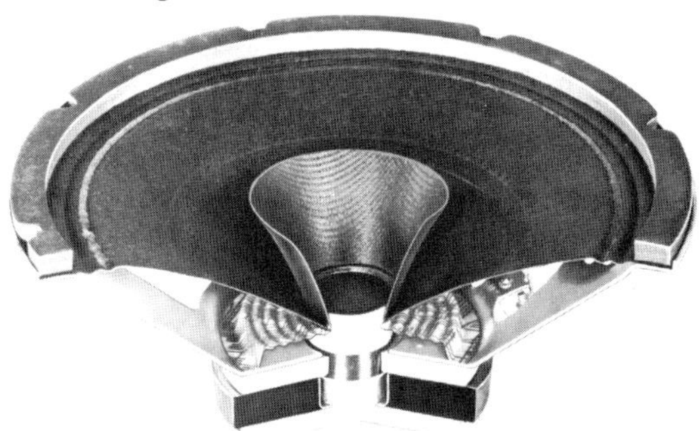

32:18 *Cutaway view showing the working parts of a loudspeaker.*

32:10 How Electrical Meters Work One very important application of the electromagnet is the **galvanometer**, the basic mechanism that forms the heart of most ammeters and voltmeters. It is named in honor of the Italian physician and physicist Luigi Galvani. An electromagnet is pivoted in such a way that it can turn between the poles of the horseshoe magnet. When a current flows through the electromagnet, it causes a repulsion between its poles and the poles of the fixed magnet. The stronger the current, the stronger the repulsion. A pointer needle attached to the electromagnet moves across a numbered scale to indicate the amount of current flowing. When the meter is disconnected from the current source, a hairspring returns the pointer to its original "zero" position (see Figure 32:19).

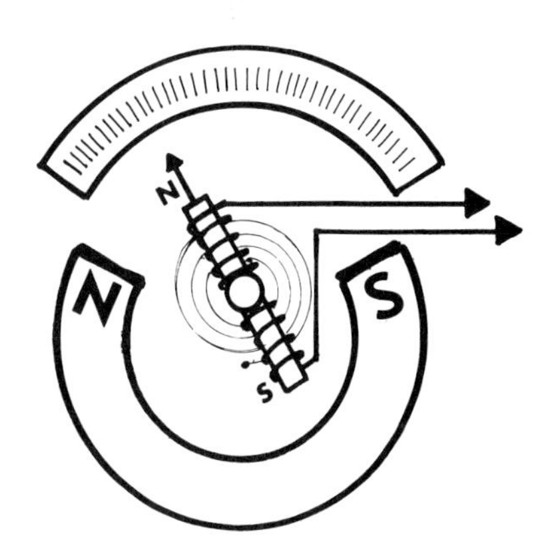

32:19 *A galvanometer works by magnetic repulsion.*

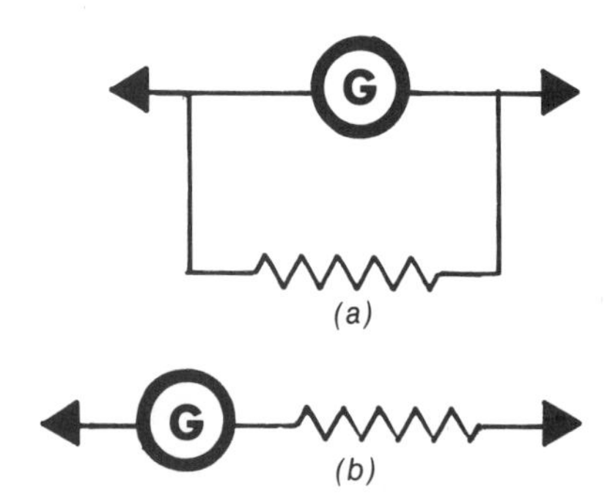

32:20 *Converting a galvanometer to an ammeter (a) by the addition of a parallel resistor, and to a voltmeter (b) by the addition of a series resistor.*

It is a fairly simple matter to convert a galvanometer to an ammeter or a voltmeter. Actually, the galvanometer is already an ammeter. For most applications, however, it is too sensitive and requires a **shunt**—a low resistance in parallel with it to carry some of the current. To make a voltmeter, a high resistance called a **multiplier** is placed in series with the galvanometer (see Figure 32:20). Various ranges of current and voltage can be accommodated by the use of different values of shunts and multipliers.

32:11 The Motor Suppose that the electromagnet in a galvanometer could keep turning continuously in the same direction. We would then have a **motor**. But there is one problem that must first be overcome. The poles of the magnets can repel only to a certain point. Beyond that point, attraction for the opposite poles sets in. The electromagnet would soon come to rest with north-south combinations locked together before it completed even one revolution.

To overcome this difficulty we make use of the fact that the poles of an electromagnet can be reversed by changing the direction of the current through the winding. If the polarity could be reversed every half revolution, the attraction and locking together could be avoided. And if the timing could be properly adjusted, the poles of the electromagnet would always be pushing away from the poles of the fixed magnet they have just passed. This is accomplished in a D.C. motor by two brass half-rings called a **split ring** or **commutator**.

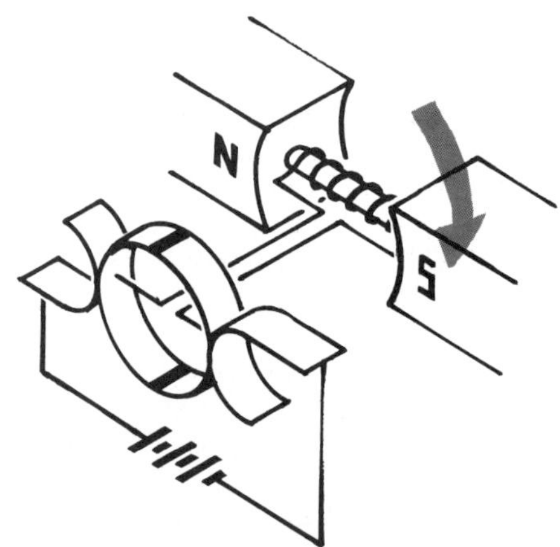

32:21 *A commutator (the split ring in the front) continually switches the incoming current to a D.C. motor so that the magnets are always repelling. The brushes are the two conductors which make contact with the commutator.*

32:22 *A St. Louis motor functions as a motor when connected to a battery, as a generator when connected to a galvanometer.*

The two wires from the electromagnet are connected to the commutator as shown. The commutator turns together on the same shaft with the electromagnet; the whole moving part of a motor is called the **armature**. Current is fed into the armature by means of **brushes**—conductors that are held against the commutator by spring action. The brushes on the St. Louis motor shown in Figure 32:22 are made of brass. Other more elaborate motors use graphite brushes held against the commutator by coil springs.

Motors designed for operation on alternating current use a different arrangement. Because the current is already changing direction many times a second, we do not have to switch it with a commutator. In one type of A.C. motor called the **conduction motor** two rings called **slip rings** are used as shown in Figure 32:23. These are not divided as in a D.C. motor. Another type called the **induction motor** uses neither brushes nor rings. Here the armature is caused to repel against the stationary electromagnet by induced magnetism. Some A.C. motors run at a certain prescribed speed, which is dictated by the frequency of the alternating current used (60 Hertz in most parts of the world). Electric clocks depend for their accuracy on the fact that power companies are able to supply very precisely regulated 60-Hertz A.C. to their customers.

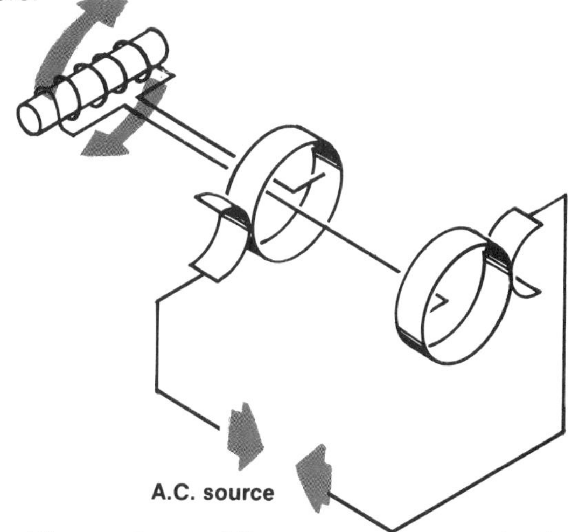

32:23 *One type of A.C. motor uses two complete brass rings called slip rings.*

32:12 The Generator The motor and the generator were both invented by the brilliant Christian physicist Michael Faraday. (See biographical sketch in Chapter 30.) Whereas a motor is a device which changes electrical energy to mechanical, a **generator** performs the reverse operation of changing mechanical energy to electrical. It is, therefore, an extremely useful invention. There is no easy way to transmit mechanical energy from place to place, but electricity can be conveniently sent long distances over wires.

The parts of a generator are substantially the same as those of a motor. In fact, a motor can be used as a generator without internal changes. A St. Louis motor, for example, can be connected to a galvanometer instead of a battery. When the armature is turned, a deflection is noted on the galvanometer. Reversing the direction of turning causes the meter to deflect in the opposite direction.

To generate electricity it is necessary to move a conductor in a magnetic field. Figure 32:24 shows one of the simplest possible arrangements, consisting of a wire, a horseshoe magnet, and a galvanometer. As the wire is moved up and down, the galvanometer swings first in one direction, then in the other.

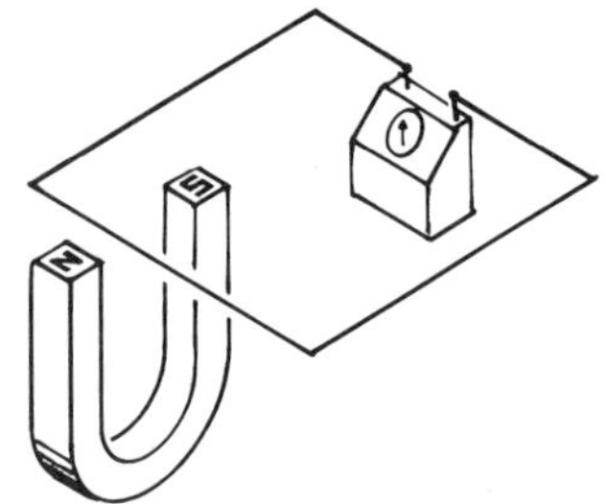

32:24 *A simple way to generate current. When the wire is moved up or down between the poles of the magnet, a current is indicated on the galvanometer.*

Better results can be obtained by using a coil of wire rather than a single wire, placing an iron core in the coil, using a rotary motion so that the speed can be increased, and increasing the strength of the magnet.

Generators supply over 99 percent of the electricity used in the United States. The mechanical energy to turn the generators comes from one of three sources:

1. *Fossil fuels* (usually coal) are burned to produce heat, which vaporizes water to form steam. The steam turns a turbine whose shaft is connected to the shaft of a generator.

2. *Hydroelectric power* utilizes a difference in height between two points along a river. Water is drawn off at a high point in the river and caused to fall against the blades of a water turbine which is connected to a generator. The water is channeled back into the river at a lower level than that from which it was removed. Thus, gravitational potential energy has been converted into electrical energy. Niagara Falls and Hoover Dam are the sites of two well-known hydroelectric plants.

3. *Atomic energy* derived from the fissioning (splitting) of uranium-235 is used to heat water. The steam produced powers a turbine, which in turn operates a generator.

32:25 *A hand crank generator. The armature turns between the poles of four U-shaped magnets.*

32:26 *Windmills, once a familiar part of the rural scenery in the United States, were used to convert mechanical energy from the air to electrical energy by means of a generator. (See chapter title page)*

32:27 *A row of generators in the Arizona wing of the Hoover Dam power plant.*

32:28 *Hoover Dam, part of a large hydroelectric plant.*

32:13 Decay of the Earth's Magnetic Field It is believed that the earth's magnetic field is produced by electric currents in its core. (This is exactly like Oersted's experiment, but on a far grander scale.) If this view is correct (and there has been no other workable theory to date), it is to be expected that these electric currents will gradually die out because of resistance in the core and the resulting loss of energy as heat. Consequently, the earth's magnetic field should decrease in strength as time goes on.

Interestingly enough, this is exactly what has been observed. In measurements made between 1835 and the present, a surprisingly rapid decline in strength has been noted. Dr. Thomas Barnes, physics professor at the University of Texas at El Paso, determined that the decay is exponential (similar to that of radioactive materials) and that its half-life is only 1400 years (See *Creation Research Society Quarterly*, June 1971, June 1972, and March 1973). This means that in 1400 years its intensity falls to half of what it was. In 2800 years it is reduced to one quarter of its former strength; in 4200 years it is down to one eighth, and so on.

Let us consider the significance of this for a moment. The earth's magnetic field serves a very important purpose. It protects the earth's inhabitants from cosmic rays. Potentially dangerous particles from outer space and occasional intense bursts of particles from solar flares are deflected by the field as they approach the earth. Instead of bombarding us, they are temporarily trapped in the **Van Allen belts** (see Figure 32:29). After bouncing between the earth's magnetic poles, some of the particles are hurled back into space. Others are leaked into the atmosphere, but at a slower, safer rate. This "spilling over" into the atmosphere gives rise to the **aurora borealis**, or northern lights, which are usually seen following a severe solar storm.

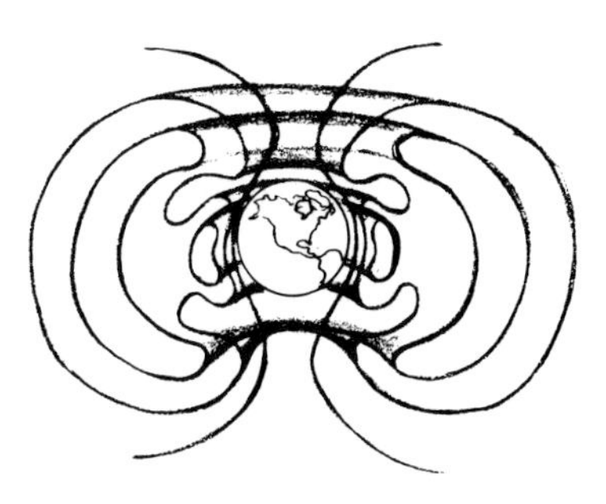

32:29 *Schematic representation of the earth's magnetosphere, showing the inner and outer Van Allen belts.*

The fact that we are rapidly losing our protection against cosmic rays is not a particularly appealing thought. One can imagine a time in the not-too-distant future when conditions on the earth will be unfit for living organisms. There are many who are unwilling to face the consequences of these findings and who claim that the earth's field periodically reverses its downward trend and regains its lost strength. But no increase in strength has ever been observed, and no likely model has ever been proposed that would allow this to happen. We are therefore forced to conclude that time is not as cheap a commodity as most people seem to think. Christians are exhorted to redeem (or buy up) the time, making the most of each opportunity (Ephesians 5:16, Colossians 4:5).

There are other important implications of Dr. Barnes' work. If the pattern of decay has been the same in the past as we now observe (and there is no good model for believing otherwise), a very strict limit is placed on the age of the earth. Based on these studies, Dr. Barnes believes that the age of the earth must be measured in thousands of years rather than millions or billions of years. Only 20,000 years ago, he calculates, the field strength would have been 18,000 gauss (compared to its present value of about half a gauss!), which would be stronger than the field between the pole pieces of the most powerful radar magnets. He questions whether the earth could have held together with the amount of heating in the core that would take place in producing such a high flux.

Dr. Barnes' studies offer one of many indications that the age of the earth has been grossly overestimated. You have probably not heard about such research being carried out by present-day Christian men of science. If you have not, it should indicate to you that somebody has been "managing the news." Very rarely are such evidences seriously considered in today's publications. One by one they are dismissed by those in authority so that their combined weight need never be faced. Evidence is being thrown out of court before the jury even becomes aware

32:30 *Aurora borealis (northern lights) caused by particles from the Van Allen belts "spilling over" into the atmosphere.*

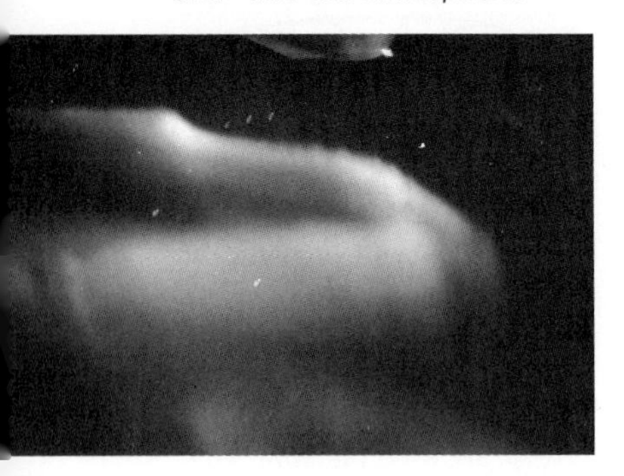

that it exists! Such is not the true spirit or methodology of science. It is the fervent hope and prayer of the authors that the students who use this textbook will learn to search out and consider *all* the available evidence.

SAMUEL F. B. MORSE (1791-1872)

American inventor

Almost every schoolboy is familiar with the quotation, "What hath God wrought!" These were the first words ever sent by intercity telegraph. But very few could state the source of the quotation—Numbers 23:23—or how it came to be selected. This and the events in the great Christian inventor's life which led up to it make a most fascinating story.

Samuel F. B. Morse was born in Charlestown, Massachusetts, in 1791. Though as the son of the town pastor he was reared in a Christian home, he did not accept Christ as his Saviour until early adulthood. He received his education at Philips Exeter Academy at Andover and Yale University, which were both still fundamental Christian schools at that time.

One of his most enjoyable classes at Yale was Professor Jeremiah Day's physics lecture. Morse was especially intrigued by the electrical demonstrations. The arcing of a spark gap was presented as evidence that the "electrical fluid" can be seen wherever a break occurs in the circuit. On one very memorable occasion the entire class joined hands in a circle while the good professor attached the students to the contacts of a high voltage generator. All feet left the floor in unison amidst a chaos of grunts and screams. The scientific purpose of the experiment is somewhat in doubt, but from that day forward Morse was duly impressed with the potentialities of electricity.

Many people are not aware that Morse was trained as an artist rather than an inventor. The year following his graduation from Yale, he went to England to study with Benjamin West, the most respected personage in the world of art at that time. (Those who have seen the magnificent West paintings in the Bob Jones University Chapel will have some appreciation for his work.) The training Morse received during his four-year sojourn abroad catapulted him into the ranks of America's foremost artists. He was later to paint such notables as President Monroe, Noah Webster (the portrait used as the frontispiece for Webster's dictionary), and General Lafayette. It was also during this period that he came under the influence of William Wilberforce and his zealous group of evangelical Christians known as the Clapham Sect. Morse was thoroughly impressed with Wilberforce and stated afterward, "What I saw of him in private gave me the most exalted opinion of him as a Christian." The witness of these Godly men soon brought him to trust Christ as his own Saviour. On returning to America he lost no time in coming forward in his own church and making a public profession of his faith.

While working in Concord, New Hampshire, he met his bride-to-be, Lucretia Pickering Walker. During their prolonged courtship his unrelenting efforts to win her to the Lord eventually bore fruit; by the time of their marriage she too had found Christ. Theirs was a happy home, marred only by his long absences and the financial problems that accompanied his chosen profession. But in spite of hard times he was able to say, "I feel satisfied that whilst engaged for God He will not suffer me to want."

It was on his second trip to Europe that the idea of the telegraph began to take shape. In Paris he saw the French semaphore system of communication. Signals were relayed from mountaintop to mountaintop across the land. How much better this was, he thought, than the mail system back home. During the return voyage aboard the packet ship *Sully*, Morse determined a specific plan of attack. When he disembarked in New York Harbor, his notebook contained sketches of the first crude electric telegraph.

Providentially, he received an appointment as professor of sculpture and painting at the University of the City of New York (later called New York University). This afforded him a place to experiment and served as a means of attracting those who were to become his assistants. After many months of labor, he was ready to demonstrate his collection of voltaic cells, coils, and wires for his colleagues at the university. A few showed interest in the invention, and he was soon able to enlist the help of a small group of friends. But only one, a fellow Christian named Alfred Vail, was to stick with him to the end.

One of the earliest public demonstrations took place in Vail's home town, Morristown, New Jersey. The sender and receiver were both at the same location, but the signals were traversing some two miles of coiled wire between the two units. The spectators were enthusiastic, and the Morristown *Jerseyman* extravagantly praised Professor Morse and his invention. In 1838 Morse and Vail gave a demonstration at the Franklin Institute in Philadelphia. Again the equipment functioned successfully and a favorable report was issued—the first acclaim to be received from representatives of the scientific community.

The next demonstration took place in Washington. It was their fond hope to secure funds for a trial line between two major cities some distance apart. Among those witnessing the demonstration were President Van Buren and several members of the cabinet. Unfortunately for Morse and his associates, our country was then in the midst of a severe depression. In 1841 a congressman informed him, "The treasury and the government are both bankrupt." But he kept praying and persevering. Late in 1842 he set up a demonstration line between the rooms of the House Committee on Commerce and the Senate Committee on Naval Affairs. Faithfully he remained on duty by his equipment, sending sample messages for the legislators and answering their questions. By the end of the year, the Committee on Commerce had submitted a favorable report. Representative Ferris, the chairman of the committee, recommended that $30,000 be appropriated to set up a trial intercity telegraph system.

The resolution came before the House on February 21, 1843. After several facetious remarks linking the telegraph with the occult sciences and religious fanaticism, the bill was carried 89 to 83, with 70 abstaining. The next hurdle was the Senate. A week and a half went by. When the last day of the session arrived, there were still 140 bills ahead of Morse's. His friends in the Senate became increasingly pessimistic and advised him to prepare for the worst. Evening came, and the lamps were lighted in the Senate chamber. Believing the situation to be utterly hopeless, Morse returned to his hotel room, had his devotions, and retired for the night.

The next morning at breakfast a waiter informed him that he had a visitor in the hotel parlor. Much to his surprise, it was the daughter of the Commissioner of Patents, Annie Ellsworth. She had come to bring the good news—his bill had passed around midnight without discussion and had been quickly signed by President Tyler. When he had recovered sufficiently from his mingled joy and disbelief, he made her a promise—when the lines were finally completed from Washington to Baltimore, she would be allowed to choose the text for the first message. Thus it was, on May 24, 1844, that the famous four words from the book of Numbers were transmitted from Baltimore to Washington and back again. Morse was delighted that appropriate words of Scripture had been chosen for the first intercity demonstration. In a letter to his brother he wrote, "That sentence of Annie Ellsworth's was divinely indited, for it is in my thoughts day and night, 'What hath God wrought!' It is *His* work, and He alone could have carried me thus far through all my trials and enabled me to triumph over the obstacles, physical and moral, which opposed me. 'Not unto us, not unto us, but to Thy name, O Lord, be all the praise.' "

List of Terms

- agonic line
- armature
- aurora borealis
- brushes
- commutator
- compass
- conduction motor
- dipping needle
- domain
- electromagnet
- galvanometer
- generator
- induced magnetism
- induction
- induction motor
- lines of force
- magnetic field
- magnetization by induction
- motor
- multiplier
- shunt
- slip rings
- solenoid
- split rings
- Van Allen belts

Questions

1. Of what does lodestone consist?
2. Who was the first to suggest that the earth is a large magnet?
3. What materials can be picked up by a magnet?
4. Why does a compass usually *not* point to "true north"?
5. What is an agonic line on a magnetic map?
6. Where are the earth's magnetic poles located?
7. What is a dipping needle?
8. Describe three ways a bar of iron can be made into a magnet.
9. Can magnetic poles be isolated? What do you suppose would happen if you attempted to saw off the south pole of a bar magnet?
10. What is a magnetic domain?
11. What important discovery did Oersted make?
12. What is a solenoid?
13. Name four devices that utilize electromagnets.
14. How can you convert a galvanometer into an ammeter? Into a voltmeter?
15. What is a commutator?
16. Why are electric clocks accurate?
17. Name three of Michael Faraday's inventions.
18. What kinds of energy can be utilized to generate electricity on a large scale?
19. What is the half-life of the decay of the earth's magnetic field? What does "half-life" mean in this case?
20. How is the earth's magnetic field thought to be generated?
21. What implication does the research on the decay of the earth's magnetic field have on our interpretation of earth history?
22. Complete the following table of energy transformations:

Device	*Changes*	*Into*
motor	electrical energy	mechanical energy
generator	______________	______________
loudspeaker	______________	______________
microphone	______________	______________
doorbell	______________	______________

Student Activities

1. Obtain a sample of lodestone and study it using a small compass, nails, iron filings, and similar materials. Attempt to locate its poles. Are there more than two poles?
2. Obtain samples of nickel, cobalt, and a Canadian nickel, and show that these materials are attracted by a magnet.
3. Prepare a brief report on William Gilbert or Hans Christian Oersted.
4. Make a more detailed study of the decay of the earth's magnetic field from the *Creation Research Society Quarterly* for June 1971, June 1972, and March 1973.

Further Reading

Bitter, Francis. *Magnets: the Education of a Physicist.* Garden City, N.Y.: Doubleday, 1959.

Nesbitt, E.A. *Ferromagnetic Domains.* Murray Hill, N.J.: Bell Telephone Laboratories, 1962.

Photo next page
The supertanker disturbs the water, making waves as it travels. Notice the various wave patterns it generates.

UNIT XI WAVE PHENOMENA

Chapter 33 Wave Theory

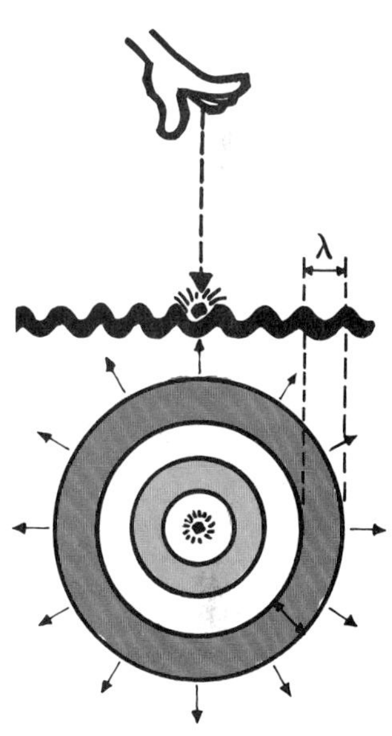

33:1 *A rock dropped into water generates circular waves on the surface of the water.*

33:1 Wave Motion When you throw a rock into a pond of calm water, a series of ringlets or waves moves out from the point where the rock went into the water. Because the behavior of water waves is typical of other types of waves—sound waves, light waves, and electrical waves—we can use them to illustrate **wave motion**.

If you could slice the water somehow and look along the slice, you could see the waves moving along with high and low spots. The high points are called the **crests** of the wave, and the low points are called **troughs** of the wave.

A combination of a crest and a trough makes up one **wavelength**. The distance from one crest to another or one trough to another is also a wavelength. The wavelength is represented by the Greek letter lambda (λ).

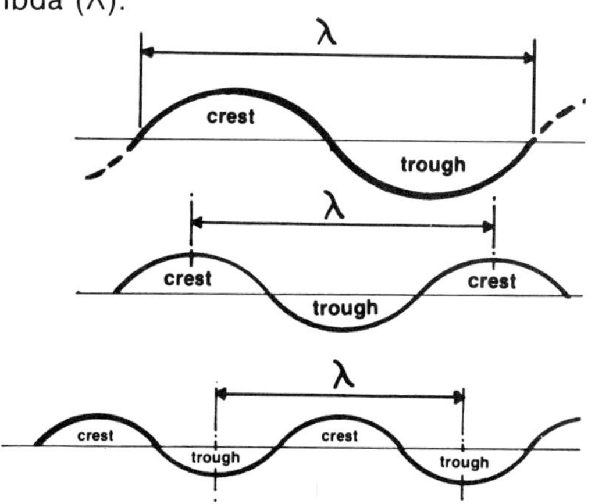

33:2 *Representation of the wavelength of various water waves.*

Again looking at our "slice" of water, we can find the **amplitude** of the wave. This is the maximum distance the crest rises above the former level of calm water, or the maximum distance the trough falls below the former level. Amplitude is represented by the letter *A*.

A
crest
former level of wat
trough
A

33:3 *Representation of the amplitude (A) of a water wave.*

We said that when the rock hit the water, waves moved out from the splash. If we could place a stationary marker somewhere in the water, we could count the waves as they pass the marker. We count the number of the crests passing the marker in a certain amount of time. Then we divide the number of crests by the time, and we have the **frequency** (*f*) of the wave. The frequency is the number of wavelengths passing a point in a unit of time.

Problem: 20 wave crests (or 20 wavelengths) pass a particular point in 10 seconds. Find the frequency of the wave.

Solution: f = 20 crests/10 seconds
f = 2/second
f = 2Hz or 2 wavelengths per second

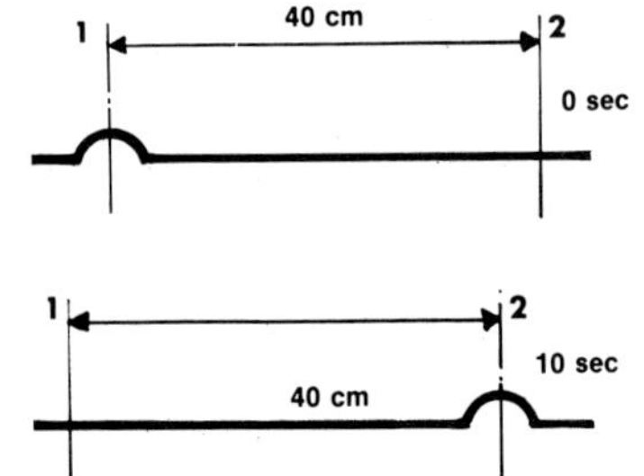

33:4 *Calculation of wave speed.*

The time it takes for one complete wave (a crest and a trough) to pass the marker is called the **period** of the wave and is denoted by T. Also,

$$T = 1/f$$

From the previous problem, the period of the wave is

$$T = 1/f = 1/2 \text{ sec}$$

It takes a half second for one complete wave to move past the stationary marker.

As the crests of waves move past a stationary marker, they have a certain speed. If we could follow one crest between two markers using a stopwatch, we could measure the wave **speed**.

Problem: It takes 10 seconds for the crest of a wave to move from marker one to marker two, which are 40 cm apart. Find the speed of the wave.
Solution: wave speed = 40 cm/10 sec = 4 cm/sec

If we know the direction of the movement of the wave, we can express the speed as a **velocity**. The wave velocity is 4 cm/sec from marker one to marker two. We have given the direction with the speed, and this makes it a velocity.

There is another way to calculate the speed of the wave.

Wave speed = (frequency) (wavelength) = $f\lambda$

Problem: The wavelength of a water wave is 10 cm and has a frequency of 2 per second. Find the speed of the wave.

Solution: Speed $= f\lambda$

$= (2/\text{sec})(10 \text{ cm})$

$= 20 \text{ cm/sec}$

If there is still confusion in your mind as to what a wave is, think of it as some kind of a disturbance. Picture again the calm pond of water. It is disturbed by someone throwing a rock into it with resultant waves emanating from the disturbance. A string on a violin gives off a sound wave when it is plucked by a finger or stroked by a bow. The string is disturbed. The slang expression "don't make any waves" indicates the nature of wave motion. It means, "do not disturb anything, and nothing will happen." If you disturb the "situation," waves result. A wave is a periodic traveling disturbance.

33:5 *The supertanker disturbs the water, making waves as it travels. Notice the various wave patterns it generates. (See chapter title page)*

Experiment 33:1 *Wave Motion*

1. Wave Traveling Down a Rope

Attach a long rope to a wall or a chair, or have someone hold the end of it securely. Stretch the rope out tightly; then jerk the loose end. A pulse (wave) will travel down the rope and back. Notice that a crest travels down the rope and a trough travels back. Measure the wave velocity and period.

2. Standing Waves on a Rope

This time rather than jerk the rope once, shake it up and down in as much of a rhythmic motion as possible. This sets up standing waves in the rope. The waves are referred to as standing waves since they do not move along the rope. The rope moves up and down, forming alternate crests and troughs. Certain points on the rope never move and are referred to as **nodes**. Measure the wavelength and amplitude of the waves. The same type of action can be demonstrated with a tightly-stretched rubber tube.

33:6 *Traveling wave pulse in a rope.*

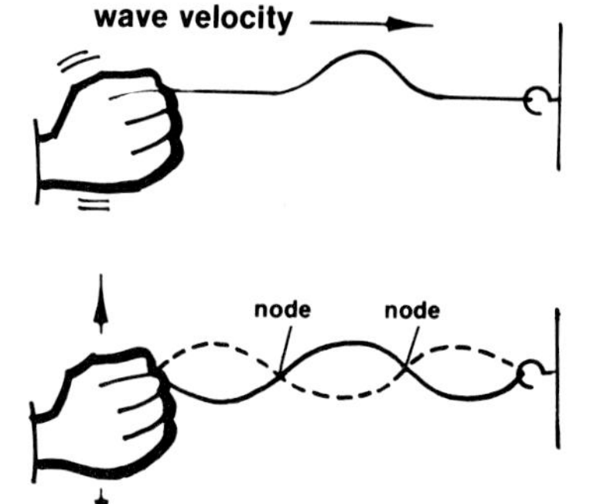

33:7 *Standing waves on a rope.*

3. Using a "Slinky" to Demonstrate Wave Motion

Stretch the Slinky out about five or six feet and attach one end to a chair, the wall, or have someone hold it securely. Jerk the loose end, and a pulse (wave) will move down the Slinky. Move the loose end back and forth rhythmically to produce a standing wave. Make measurements of wave velocity, period, wavelength, and amplitude whenever possible.

4. Using a Water Tank to Demonstrate Wave Motion

Construct a water tank approximately 2 by 3 feet with a glass bottom. Place a clear glass light bulb on the floor beneath the tank to cast wave shadows on the ceiling. Dip a finger in the water and observe the resultant water waves or ripples. Dip your finger in and out periodically. Measure wavelength, period, wave velocity. The waves produced by your finger are circular waves.

Straight waves can be produced by rocking a board back and forth on the surface of the water. Be sure the board is about the width of the tank. Measure wavelength, period, and wave velocity.

33:8 *Water tank used to demonstrate wave motion.*

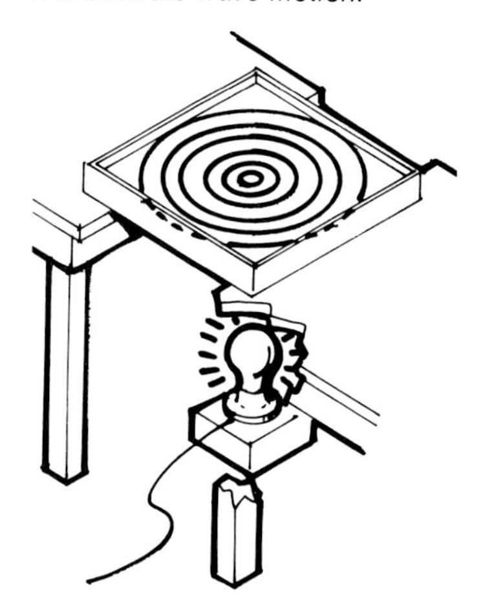

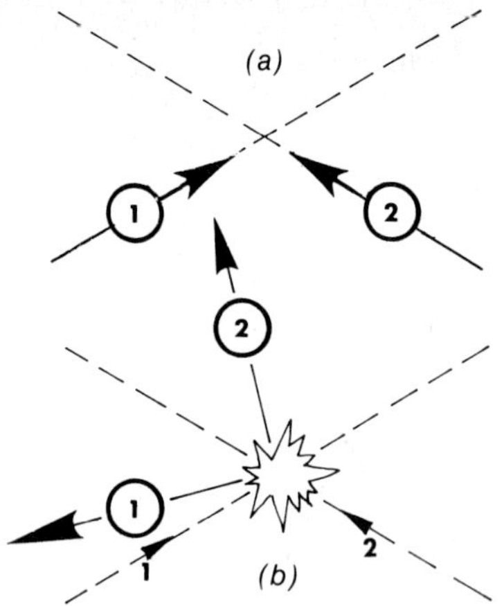

33:9 *Representation of the collision of two particles. Particles before collision. Particles after collision.*

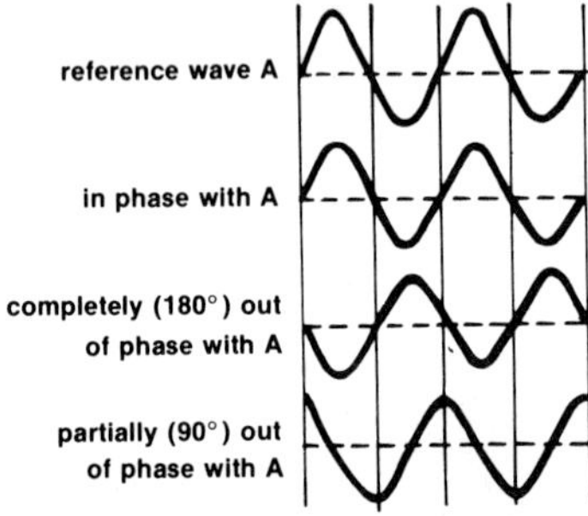

33:10 *Waves that are in phase and out of phase with each other.*

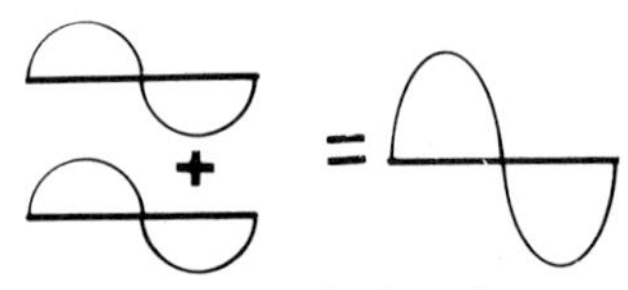

33:11 *Constructive interference.*

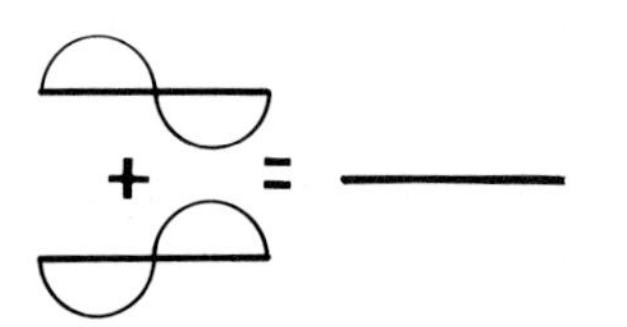

33:12 *Destructive interference.*

33:13 *The addition of two wave patterns to give a third wave pattern.*

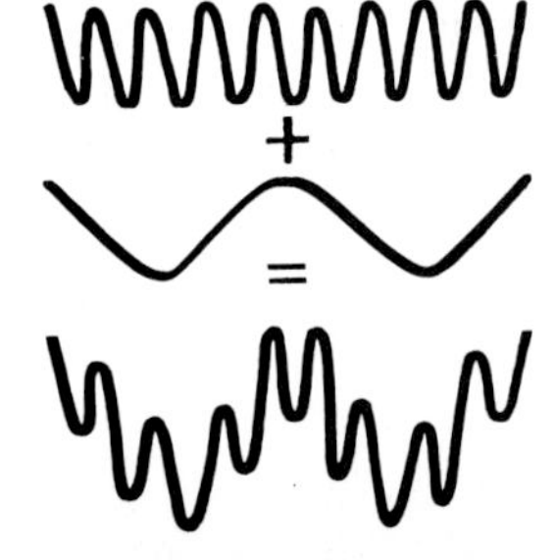

33:2 Interference of Waves When two particles collide, they fly off in different directions from each other. Waves do not collide like particles, but how do they act when they meet?

When waves from different sources cross each other's paths, the resulting effect is called **interference**. If the two waves are exactly in phase with each other, they will constructively interfere. The best way to illustrate whether waves are in phase or out of phase is by the use of a diagram. If two waves cross or come together in such a way that the crests and troughs of both waves fall together, they reinforce each other and form a wave with a larger crest and trough. This is **constructive interference.**

If two waves are exactly out of phase when they meet or cross, they will annihilate (destroy) each other. This is called **destructive interference**. Many times when waves interfere, both

Experiment 33:2 *Illustrations of Interference*

1. Drop 2 stones a slight distance away from each other into a large body of water, such as a pond. Notice how the waves interfere with each other until some distance away from the two "sources," a coherent wave front (two waves in phase and of one wavelength) is formed because of the "addition" of the two wave patterns.
2. Obtain a piece of exposed photographic film and a **monochromatic** source of light, such as a neon light or a sodium light. Make two parallel slits in the film as close together as possible without running into each other. Hold the film close to your eye and look through it into the light source. You should see a series of fringes produced by the interference of light coming through the slits.
3. Put a small amount of liquid soap or oil on a surface of water with an eyedropper. The oil or soap forms a very thin film on the surface of the water. Shine a light on the surface and notice the various color effects. This is a result of interference. As shown in Figure 33:15, some of the light is reflected back by the film surface and some by the water surface underneath the film. These two beams may interfere with each other, giving a rainbow effect.
4. Vibration of a tuning fork in an adjustable tube arrangement as shown in Figure 33:16 will produce either constructive or destructive interference of the sound waves depending upon the length of the two paths. As you slide the sections in and out, the intensity of the sound will increase or decrease.
5. Set one piece of plate glass on top of another. Separate the two plates of glass by placing a thin piece of paper between them at one edge. Shine a monochromatic beam of light onto the plates. If you place yourself so that you can see the reflected light (Figure 33:17), you will see interference bands rather than a continuous image.
6. If a plane-convex lens is placed on an optically flat plate of glass and illuminated from above with monochromatic light, as shown in Figure 33:18, a group of light and dark rings is produced. These are known as **Newton rings**.

constructive and destructive interference can occur, often yielding an interesting wave pattern. **Diffraction** of light and beats of sound waves are some interesting effects caused by interference of waves.

The Lord Jesus Christ wants to use Christians as His light bearers in this world. However, many times we simply interfere destructively with His will for our lives. Thus those around us get a rather poor picture of what a Christian is like. They cannot see Christ in us because our lives are sources of destructive interference.

33:14 *Interference wave patterns about a space capsule.*

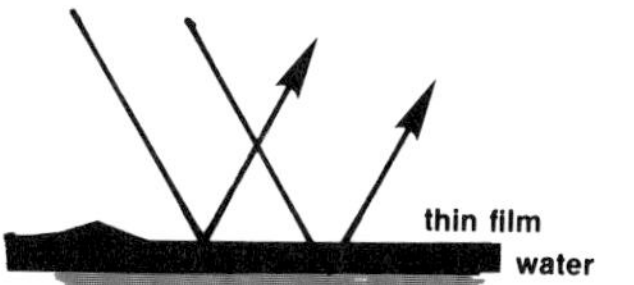

33:15 *Interference effects can be caused by the reflection of light from a thin film on water.*

List of Terms

amplitude	nodes
constructive interference	period
destructive interference	speed
diffraction	velocity
frequency	wave crest
interference (of waves)	wave motion
monochromatic	wave trough
Newton rings	wavelength

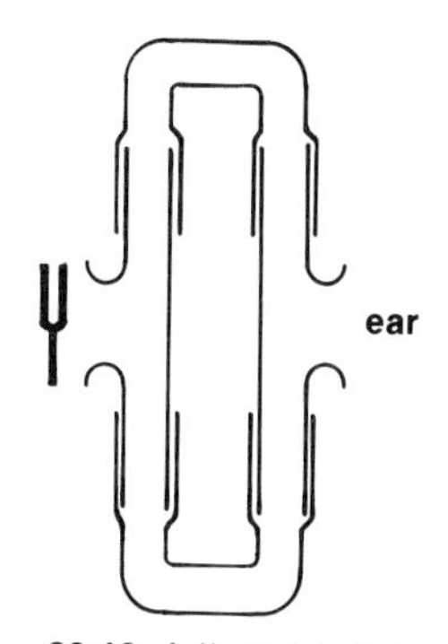

33:16 *Adjustable tube device used to produce wave interference.*

Questions

1. What is the crest of a water wave? Illustrate.
2. What is the trough of a water wave? Illustrate.
3. Define wavelength. What Greek letter represents wavelength?
4. What is the amplitude of a water wave?
5. Define wave frequency.
6. What is the frequency of the waves where 75 crests pass you in four seconds?
7. How many wavelengths come by a stationary point in 12 seconds if the frequency of the waves is 17/sec (or 17 Hz)?
8. What is the period of a wave? What are the units of a period?
9. If the frequency of a wave is 32/sec (or 32 Hz), what is its period?
10. The wavelength of a wave is two meters and its frequency is 30/sec (or 30 Hz).
    (a) What is its period?
    (b) What is its speed?
11. Define wave motion.
12. When waves meet, we do not refer to this as a collision. We call it ______________.
13. (a) Define constructive interference. Illustrate.
    (b) Define destructive interference. Illustrate.

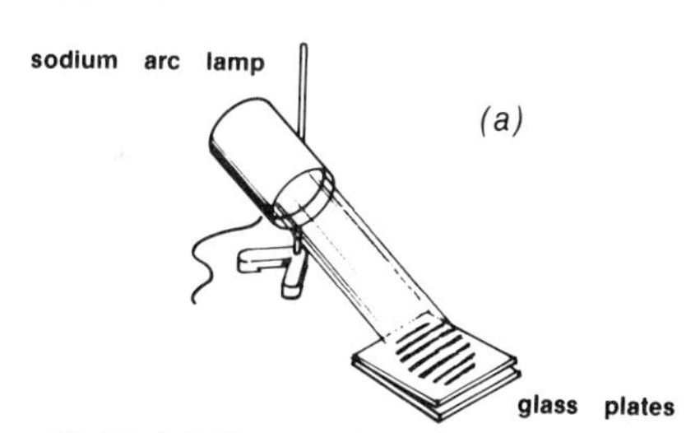

33:17 *(a) The production of interference fringes.*
(b) The path of the light rays to produce interference fringes using glass plates.

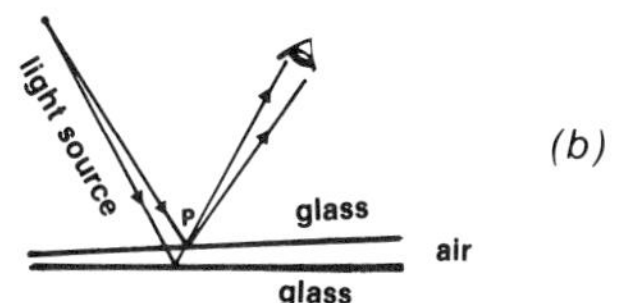

33:18 *The production of Newton rings.*

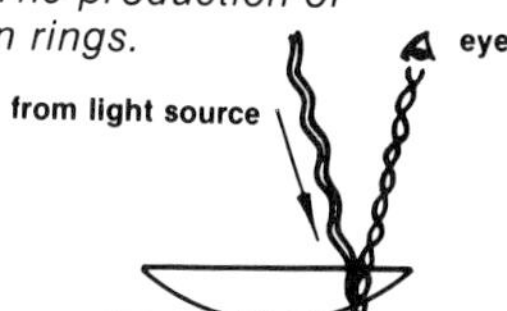

Student Activity

Make a scientific study of wave motion in water. Make a ripple tank and perform various experiments with the tank. Set various obstacles in the water and observe their interference effect. Write a paper discussing your results. Use illustrations as a method of describing your results.

70 60 50 40 30 20

UNIT XI WAVE PHENOMENA

Chapter 34 Visible Light

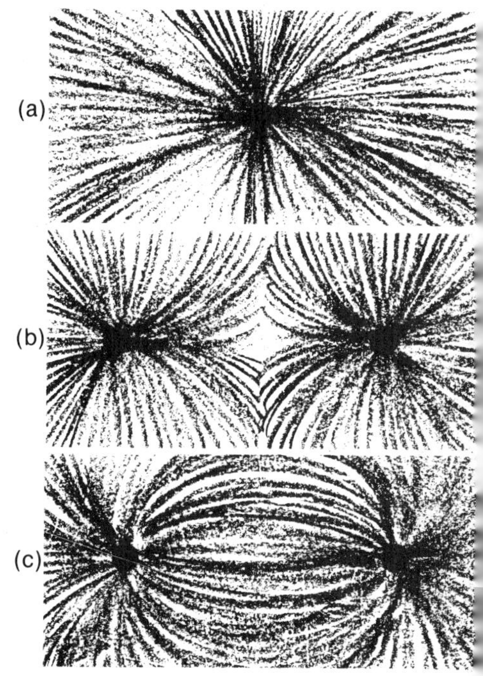

34:1 *Fields about charged particles. (a) a single charged particle and its field (b) repulsive field lines between two positive or two negative charges (c) attractive field between a positive and negative particle.*

34:1 What is Light? It is easy to describe the effects and properties of **light**. We will discuss such things as the reflection and refraction of light, but what *is* light? Visible light is a form of **electromagnetic radiation** to which the eye responds. But what is electromagnetic radiation?

The electromagnetic theory is one of the most sophisticated and abstract concepts in physics, and we can give you only extremely simple ideas. An electrically-charged particle has an electrical field associated with it. Imagine the field as its "sphere of influence." Many times you can detect a garbage can before you actually see it or stumble over it. Its odor influences the atmosphere for quite a distance. So a charged particle influences the space around it. This influence is called the field associated with the particle.

If we cause the electrical charge to move, a magnetic field develops around the charge. Thus, any moving charged particle has an **electrical** *and* **magnetic field** associated with it. A moving or accelerating electrical charge creates a disturbance known as an **electromagnetic wave**. All vibrating electrical charges radiate electromagnetic waves.

Remember from your reading and discussion on molecules in motion and atomic theory that matter is made up of atoms. These atoms are electrical in nature (electrons and protons are charged); atomic particles are constantly in motion. Hence, all physical objects in the universe—stars, planets, moons, rocks,

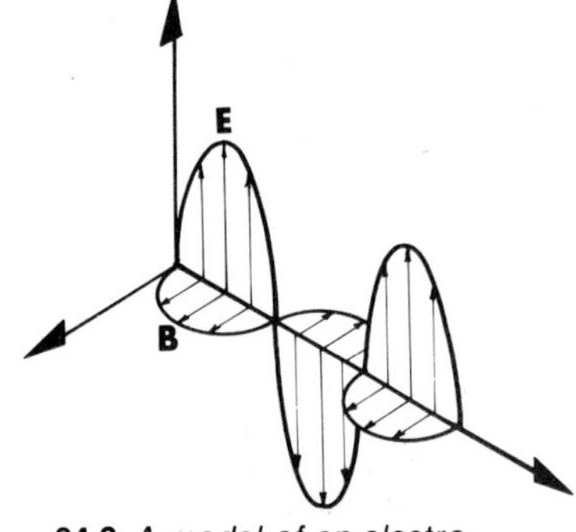

34:2 *A model of an electromagnetic wave showing the electric (E) and magnetic (B) parts of the wave.*

plants, and animal bodies—radiate electromagnetic waves. Since the material in matter vibrates at different frequencies, different wavelengths of electromagnetic radiation are emitted.

However, all electromagnetic waves move at the same speed in a vacuum, 300,000,000 (3×10^8) meters per second. This is the **speed of light**. Visible light has a frequency range of 4.3×10^{14} to 7×10^{14} Hz. Figure 34:3 shows the electromagnetic spectrum, which goes from very long wavelength radio waves to very short wavelength gamma rays. The shorter the wavelength of a wave, the more energetic and penetrating it is.

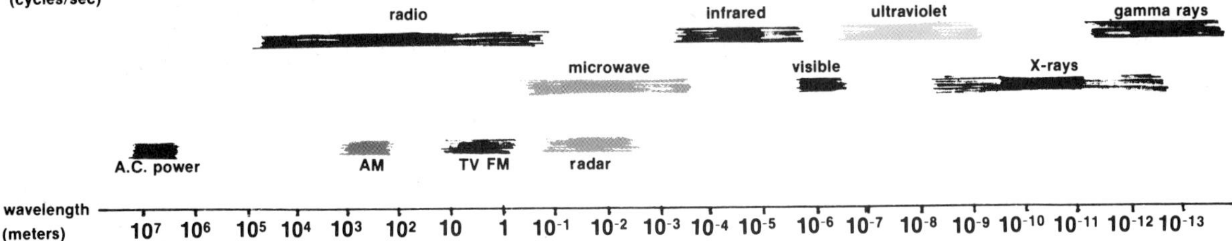

34:3 *The electromagnetic spectrum.*

Visible light occupies a very small band of wavelengths between 4×10^{-7} to 7×10^{-7} meters. Red light has the longest wavelength and violet light the shortest.

red - orange - yellow - green - blue - violet

← *increasing wavelength*

Table 34-1 *Color vs. Wavelength of Visible Light*

Approximate Wavelength (meters)	Color
4.2×10^{-7}	violet
4.7×10^{-7}	blue
5.3×10^{-7}	green
5.8×10^{-7}	yellow
6.3×10^{-7}	orange
6.8×10^{-7}	red

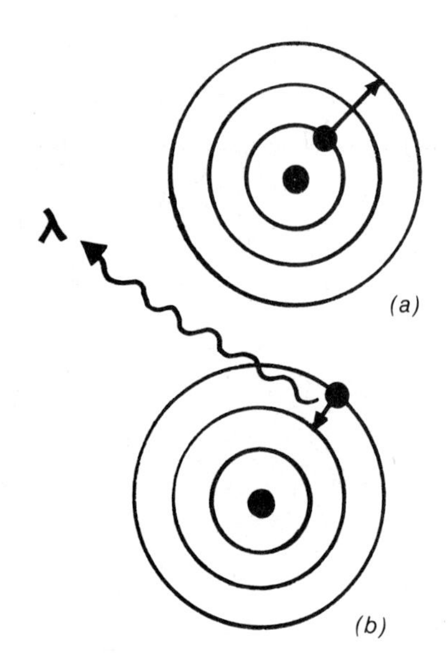

34:4 *Excitation of an electron and emission of a photon. (a) absorption of energy by orbiting electron (b) emission of a definite wavelength of electromagnetic radiation.*

34:2 Color and Light Emission How are light and different colors produced? We do not know, but we can use a model to help you visualize the imagined process. We will use the Bohr atom for our model. The electrons revolve about the nucleus in specific orbits. If an atom is disturbed by heat, other light, or electricity, it will absorb energy from these disturbances. The energy of the atom is increased, and some of this energy is absorbed by the orbiting electrons. The electrons that absorb energy are "excited" and are "promoted" to higher-energy orbits (farther away from the nucleus). When the electron returns to its normal lower-energy orbit, it releases the **excitation** energy in the form of a definite wavelength of light.

for page 468

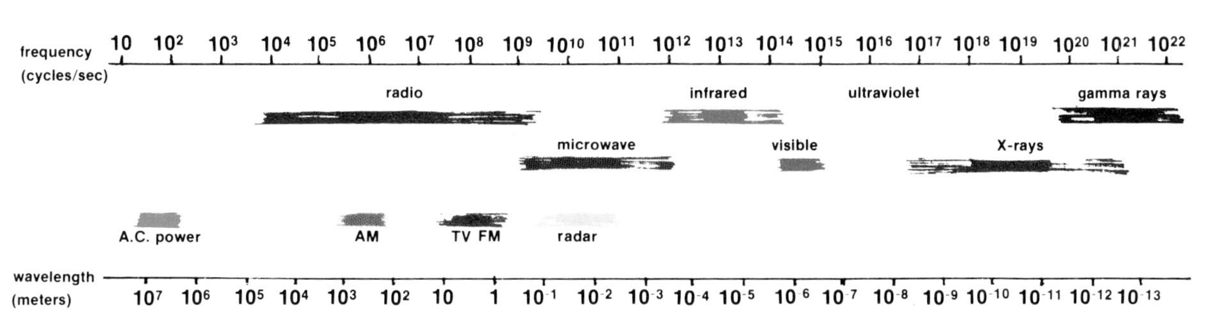

34:3 *The electromagnetic spectrum.*

for page 469

34:6 *Bright line spectrum of barium (representation).*

for page 476

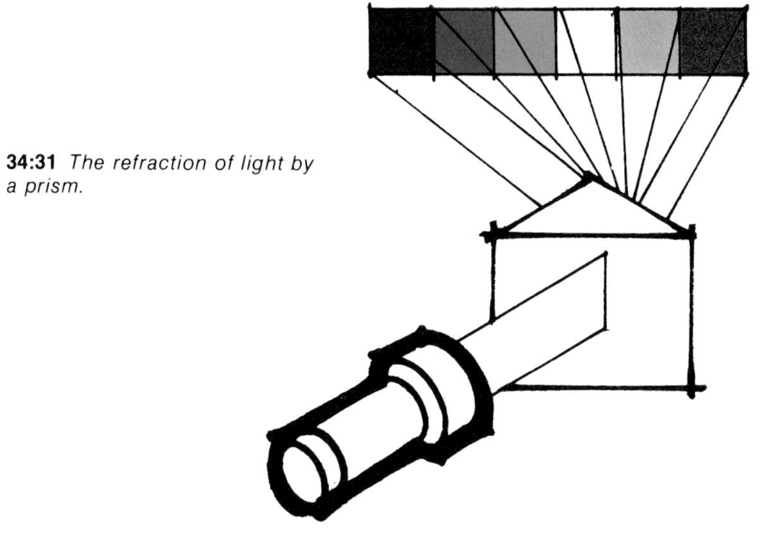

34:31 *The refraction of light by a prism.*

This process goes on in millions of atoms, millions of times in elemental matter. Many different wavelengths of light are being emitted. If an electron goes from orbit six to orbit five, a particular wavelength of light is given off. If an electron goes from orbit six to orbit four, a shorter wavelength of light is given off because more energy is released. Another entirely different wavelength of light is given off if an electron drops from orbit five to orbit four. Because atoms have many available orbits into which electrons can be excited, each element gives off a spectrum of wavelengths or colors. Since each element has a different number of electrons, the spectrum from each element is different; and each spectrum has many different colors in it.

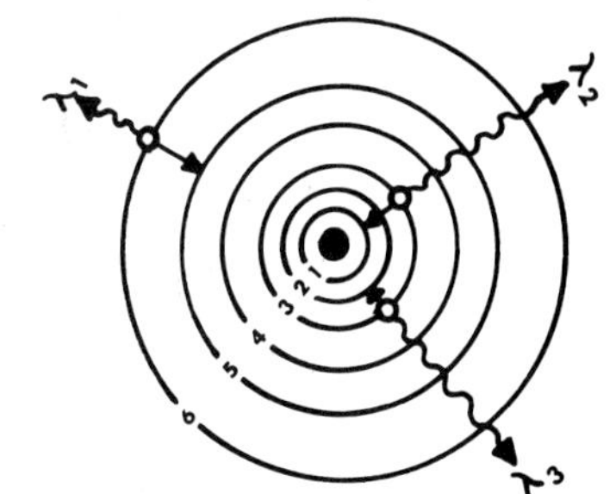

34:5 *A model showing the emission of three different wavelengths of electromagnetic radiation from an atom by electrons going from higher to lower energy orbits.*

34:6 *Bright line spectrum of barium (representation).*

All of the different wavelengths of light combine together to give each "excited" element a characteristic color. We discussed many of these colors in the flame test in the chemistry section. Neon gas used in advertising signs gives off an orange-red light. Mercury vapor street lights are very bright. This light has many blue and violet colors in it (watch as a street light comes on) and is a different "white" from a regular incandescent lamp.

34:3 Reflection of Light When we consider light moving from one place to another, we imagine that it travels in a straight line. The straight lines of light waves are referred to as rays of light.

If light rays hit an object through which they cannot pass, they bounce back. This is called **reflection**. If the surface of the object is rough, **diffuse reflection** occurs. Figure 34:8 is an illustration of this. The incoming rays are called **incident rays**, and the rays that "bounce back" are called **reflected rays**. When you see something such as a shirt, a car, or a tree, you are actually seeing reflected light rays from the object; and you are seeing diffusely scattered rays. Diffuse reflection makes an object visible to our eyes.

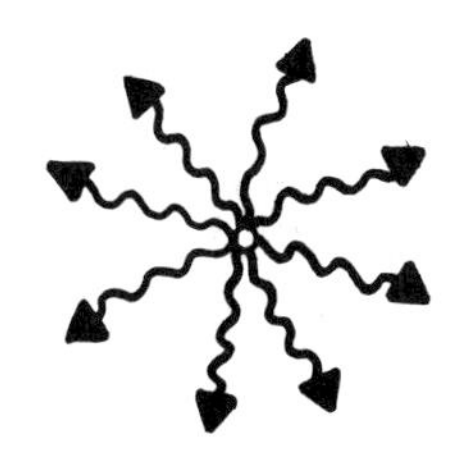

(a) wavy lines

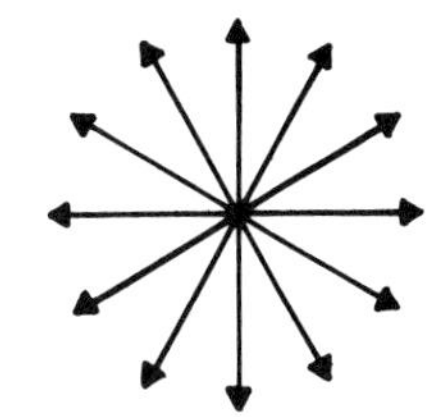

(b) rays

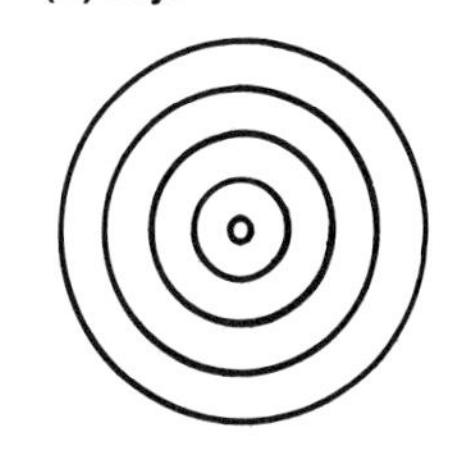

(c) concentric circles

34:7 *Various ways of showing waves traveling away from a source.*

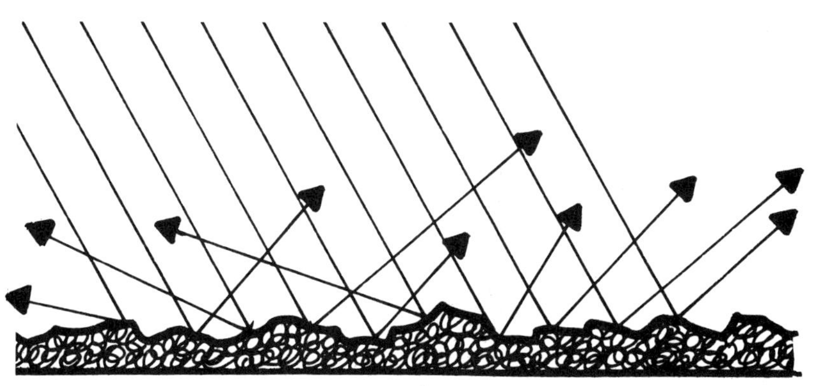

34:8 *Diffuse reflection of light rays from a surface.*

If the surface of an object is polished, rays leave the surface reflected in a regular manner. This behavior is characteristic of mirrors. The reflected rays leave a mirror at exactly the same angle as the incident rays strike it. In other words, the angle of reflection is equal to the angle of incidence. The angle usually measured is the angle that the incident and reflected rays make with a **normal** (perpendicular) to the mirror surface. We have pictured only one ray of incident and reflected light in Figure 34:9.

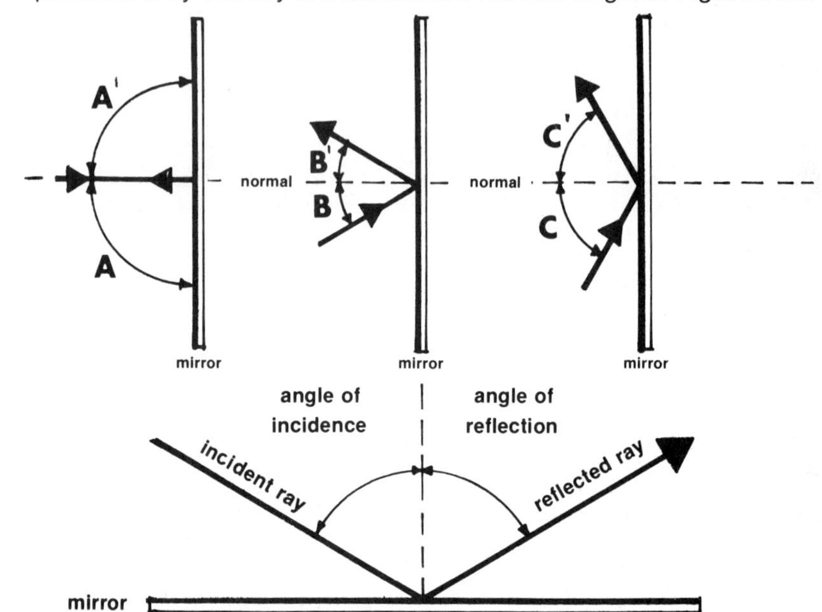

34:9 *Examples of regular reflection of light rays from a smooth surface. The angle of incidence equals the angle of reflection.*

Actually, a more proper way to visualize regular reflection is to imagine parallel rays of incident and reflected light. This is shown in Figure 34:10.

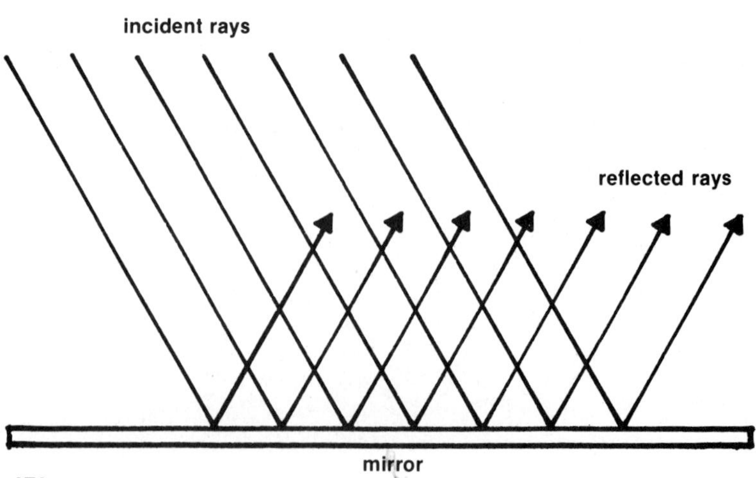

34:10 *All incident rays of light are parallel to one another, and all reflected rays of light are parallel to one another.*

Spiritually, the Lord Jesus Christ is the light of the world. We should be good reflectors so that others can see Christ in us.

34:4 Applications of Reflection A **concave** or converging **mirror** can be used to collect light rays from an object and focus them at a point or plane. The parallel incident rays hit the mirror at different places and the angle of curvature of the mirror causes them to be focused close to the mirror. This mirror can be used if you want to see an object that is a great distance away. A reflector **telescope** accomplishes this effectively. The curved mirror gathers the light and forms an image on a flat mirror. The light is then directed toward an eyepiece lens, where an image of the object can be viewed.

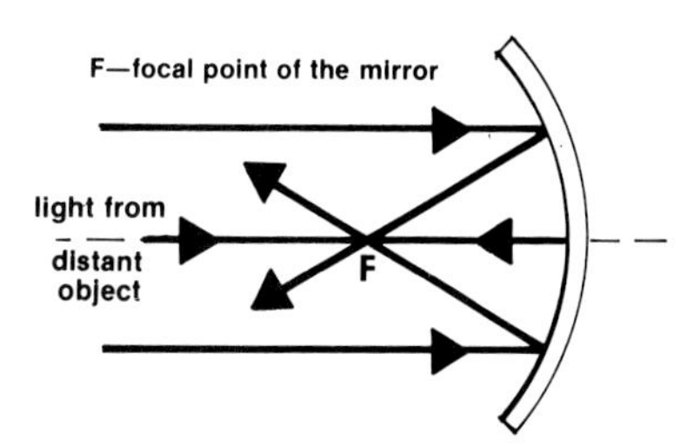

34:11 *Model of light rays being focused by a converging mirror.*

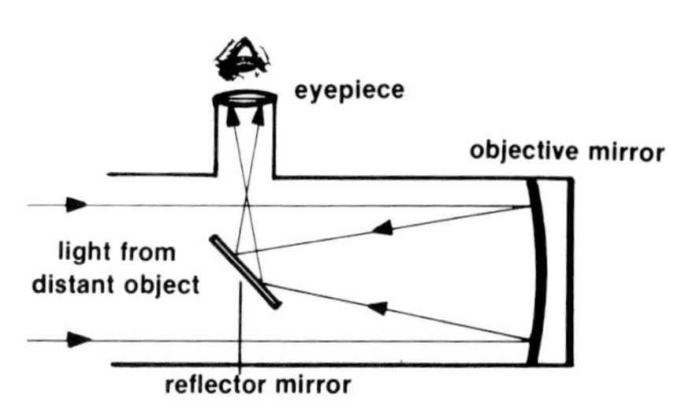

34:12 *Diagram of a reflector telescope.*

34:13 *A telescope mirror made from fused silica was the reflecting eye of a balloon observatory that ascended into the upper atmosphere.*

Experiment 34:1 *The Law of Reflection*

1. Stand a mirror upright on a piece of corrugated cardboard as shown by placing two pins behind it and one in front at the center. Using a protractor, draw a normal out from the pin in front of the mirror.
2. Place a pin anywhere to the right of the normal (pin A). Place another pin (pin B) to the left of the normal so that it lines up with the center pin and the image of pin A. Draw lines from A and B to the center pin. Using a protractor, measure the angles "i" and "r." These are the angles of incidence and reflection, respectively. Are they equal?
3. Remove pins A and B and repeat the experiment for several different positions of pin A. The law of reflection should hold true for all cases.

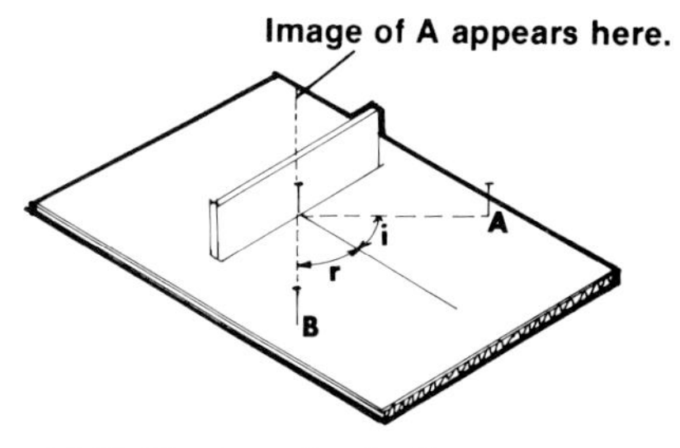

34:14 *The law of reflection apparatus.*

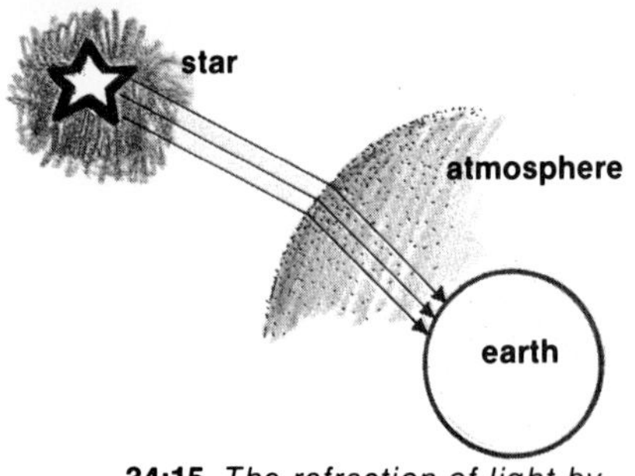

34:15 *The refraction of light by the atmosphere of the earth.*

34:5 Refraction of Light Imagine that light has just left a distant star and is approaching the earth. As long as the light does not hit anything or get interfered with in any way, it will continue traveling in parallel straight lines in the vacuum of outer space until it hits the atmosphere of the earth. Upon entering the atmosphere, it will be bent slightly as shown. If the light continues on until it passes through the surface of the water in a clear lake, it will be bent again. This bending of light when it passes from one medium to another is called **refraction**. Actually, part of the light will be reflected at the surface of the water, and the remainder of the light will be refracted as shown in Figure 34:17.

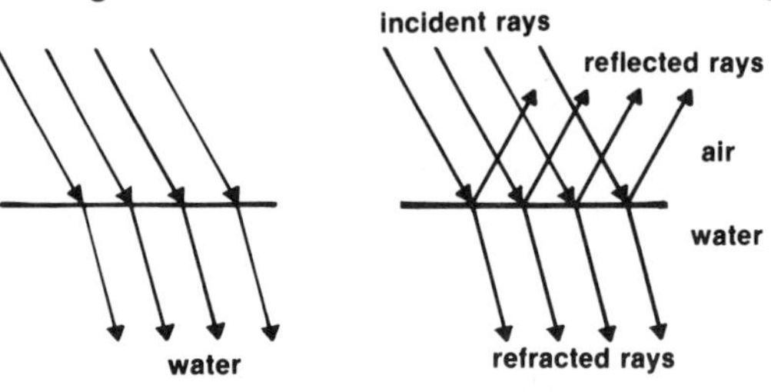

34:16 *The refraction of light in going from one medium to another.*

34:17 *The reflection and refraction of light at a surface of water.*

34:18 *Light rays enter and leave glass in the same direction and at the same speed. A is parallel to B.*

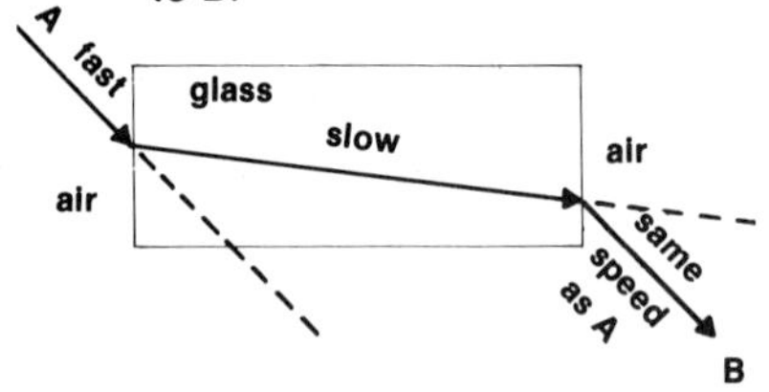

34:19 *The meterstick appears bent because of the refraction of light as the light leaves the water.*

Light speeds through a vacuum at 3×10^8 meters per second. As it enters a more dense medium, it slows down, causing a slight change in the path of the light ray. Interestingly, if the light re-enters the vacuum, it again assumes the speed of 3×10^8 meters per second. This observation is illustrated using air and glass in Figure 34:18. This refraction of light as it enters water can cause many illusions. A stick appears to be bent when it is dipped into water. Did you ever try to catch a fish that seems to be just under the surface of the water? You throw in your bait, but it lands far above the fish's head. Refraction of light makes the fish seem closer to the surface and fools you.

34:20 *Submerged objects appear to be nearer the surface than they actually are because of the refraction of light.*

On hot days, refraction of light makes illusions called *mirages*. If a layer of hot, less dense air is next to the ground, and cooler, more dense air is located above the hot layer, this density difference will cause a gradual bending of the light rays upward. People in the scorching heat of a desert often think they see pools of water ahead. Similarly, as you are driving on a hot day, the highway may appear wet in the distance. These are mirages due to refraction. Also, the wavy effect that appears over a hot highway is due to refraction. The same wavy pattern can be seen around an object placed on a hot stove.

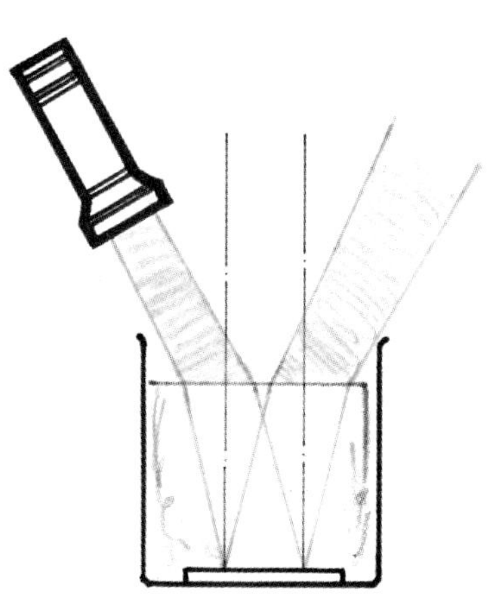

34:21 *Refraction of light upon entering and leaving a container of water.*

Experiment 34:2 *Refraction of Light*

1. Fill a cylindrical container (approximately 6 to 12 inches in diameter) with water and place a mirror at the bottom of the container.
2. Darken the room as much as possible.
3. Shine a flashlight on the mirror at an angle. Notice how the beam is bent when it enters the denser medium (water) and when it re-enters the less-dense medium (air).

34:22 *The focusing of parallel rays of light by a converging lens.*

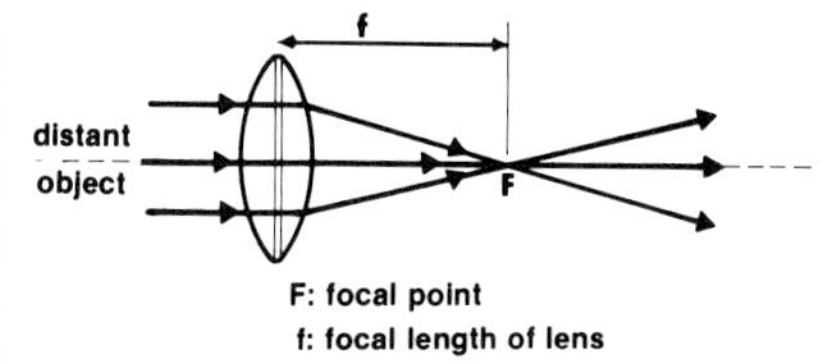

34:6 Use of Refraction If glass will cause light rays to bend (refraction), why not use this property to advantage? If we shape the glass in a correct way, we can take parallel rays and focus them at a point. The piece of glass that performs this function is called a **lens**. Lenses are used in eyeglasses to correct bad vision. They are used in cameras to focus the light on a photographic film. In telescopes and **microscopes**, lenses are used to magnify and focus objects for viewing.

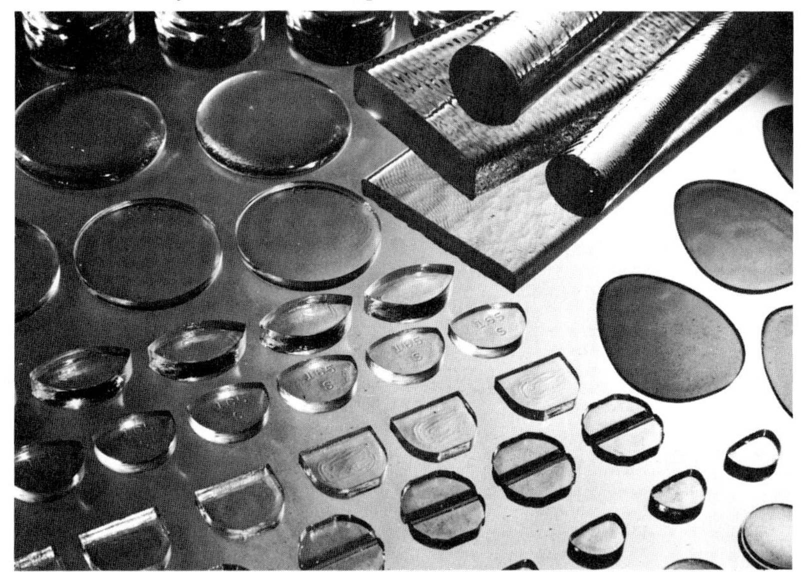

34:23 *Various sizes and shapes of lenses.*

34:24 *Good lenses must be used in high-quality cameras.*

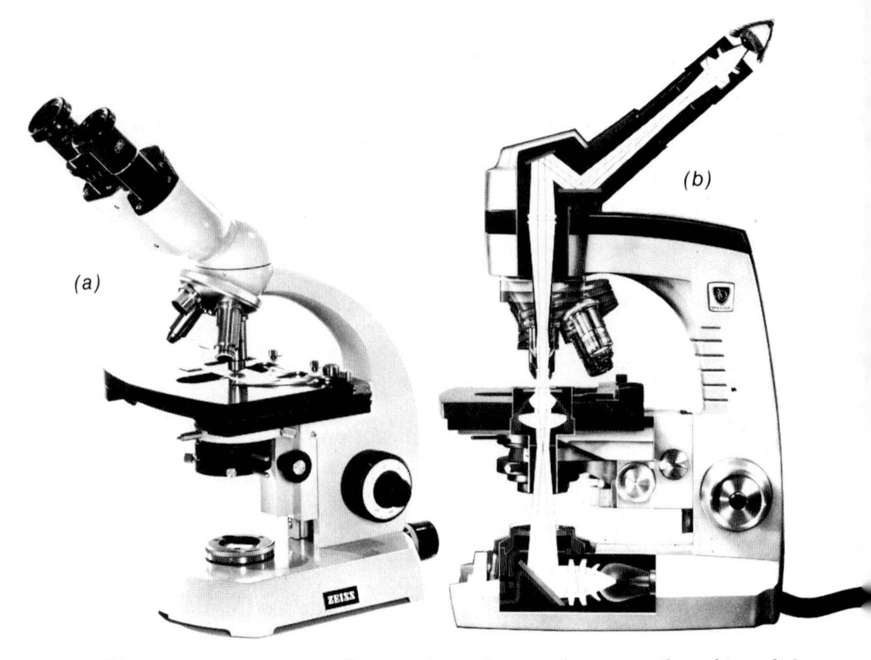

34:25 *(a) Many lenses are used in an optical microscope. (b) The optical path can be traced through a microscope. How many lenses do you see?*

The apparent size of an object depends upon the size of the image it makes on the retina of the eye. You can bring an object closer and closer to the eye to increase its apparent size. Think of seeing an airplane 10,000 feet in the air and then seeing the same plane 100 yards away on the runway. The apparent size of the airplane appears quite different at these distances from the eye.

(b)

34:26 *(a) A modern 24-inch refracting telescope at the United States Naval Observatory. (See chapter title page) (b) Galileo's first telescope.*

However, you cannot get certain objects too close to the eye, or they go out of focus. If a converging lens is placed in front of the object, it appears larger. A series of lenses in a microscope can give the instrument a magnification of 1000 times the actual size of the object. When large mirrors or lenses are used to collect the light in a telescope, other lenses can be used to magnify the object.

34:27 *Lighting ware. Which ones depend on the refraction of light for their usefulness?*

Experiment 34:3 *Focal Length of a Lens*

In this experiment we allow parallel rays to enter a lens and measure how far on the other side of the lens they come to a focus.

1. Place a flashlight having a fairly narrow beam at least 10 feet from the lens and aimed directly at the lens.
2. Place a piece of cardboard behind the lens and adjust it so the spot of light on it is as small and distinct as possible. Then measure the distance between the lens and the cardboard. This distance is called the **focal length** of the lens.

34:28 *Determining the focal length of a lens.*

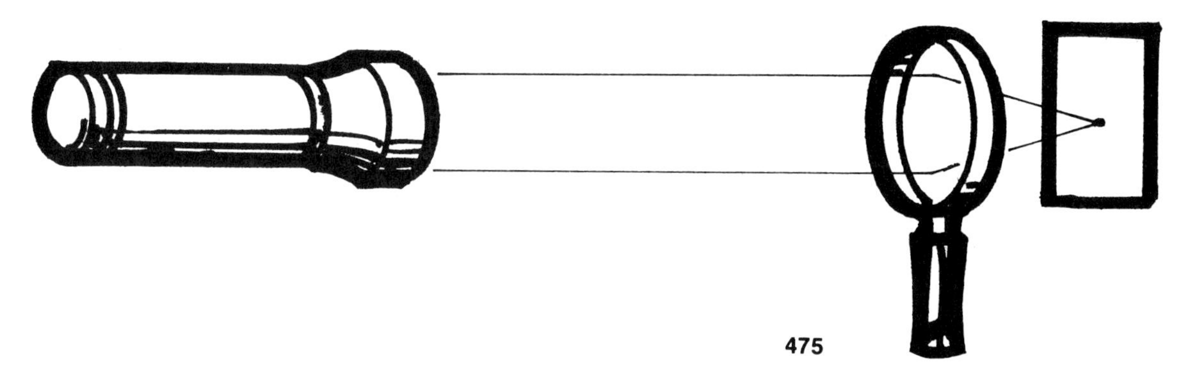

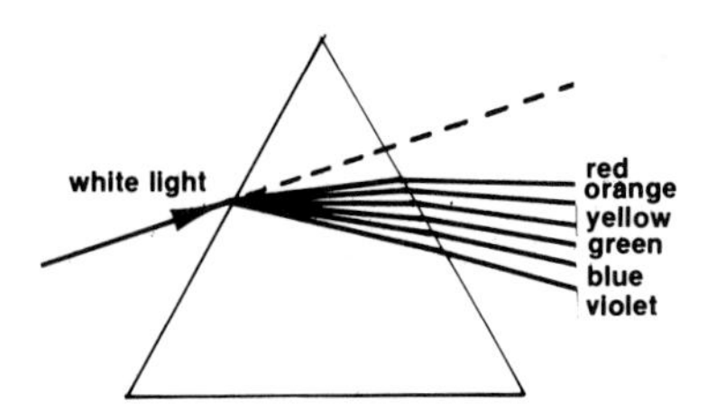

34:29 *Dispersion of light through a prism.*

34:7 Refraction in a Prism Light of one wavelength only is called **monochromatic light**. Light from the sun is made up of many different wavelengths, and this "mixture" is called **white light**. All of these different colors or wavelengths of light travel at the same speed in empty space. But upon entering a medium such as glass or water, these rays are reduced in speed, and the amount each is slowed up depends upon its wavelength. Red light travels fastest and violet slowest in a medium such as glass or water. Other colors assume speeds ranging between these extremes.

If white light is passed through a prism, the unequal speeds that result inside the glass cause the light to separate into its various colors. This separation effect is called **dispersion**.

34:30 *Prisms are employed in binoculars.*

Experiment 34:4 *Refraction of Light in a Prism*

Place a prism in a beam of light as shown in Figure 34:31. Adjust the screen behind the prism until the various colors of the light are seen dispersed on it.

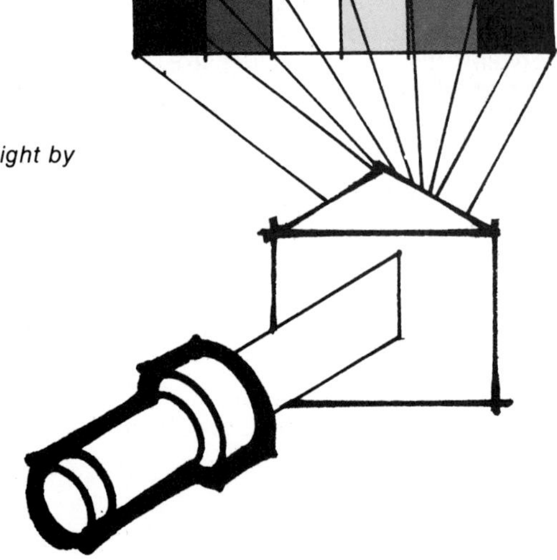

34:31 *The refraction of light by a prism.*

34:8 The Rainbow The **rainbow** is a divinely-ordained illustration of dispersion. Sunlight enters each raindrop and is dispersed into beautiful, separate colors. The raindrops act like prisms (see Figure 34:32). What you see in a rainbow depends upon where you are standing. No two people see the same rainbow. The dispersed rays come to each person from different raindrops.

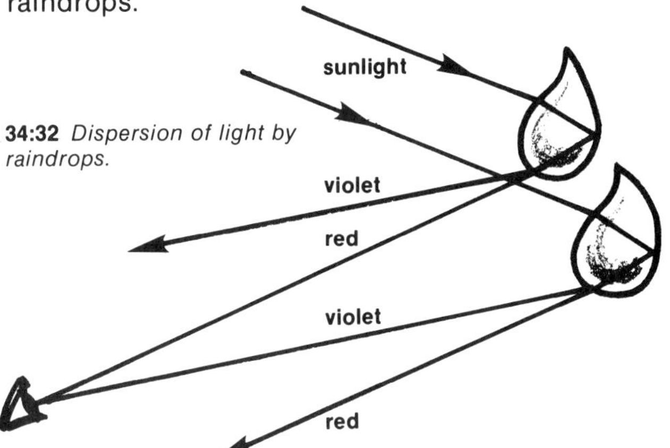

34:32 *Dispersion of light by raindrops.*

34:33 *An unusual photograph of two rainbows.*

God gave the rainbow in the sky as a sign to Noah and his descendants that He would never again destroy the earth with a worldwide flood (Genesis 9:11-17). Because it was given as a sign or token (something highly unusual or never before seen), it is logical to deduce that people living before the **Flood** had never witnessed a rainbow in the sky. Thus atmospheric raindrops would have been unknown until after the Flood. We are given a hint of the very different prediluvian water economy in Genesis 2:5-6. The ground, we are told, was watered by a mist from below, rather than by rain from above.

34:9 A Prediluvian Canopy and the Greenhouse Effect What features of the earth or its near surroundings could have prevented raindrops (and hence clouds) from forming before the Flood? Many Christian men of science are of the opinion that the pre-Flood earth was surrounded by a *canopy of water vapor* above the atmosphere. The canopy would have been invisible, allowing the sun, moon, and stars to be clearly seen. But its thermal properties would have been such as to prevent the formation of water droplets and clouds.

A water vapor canopy would have given the earth a universally mild climate. There is evidence from the fossil record that at one time the whole earth was covered with lush tropical vegetation. Of course, this is not the case today. How could this have been? Let us compare a water vapor canopy with a greenhouse. Both water vapor and glass are transparent. Visible light will pass through glass and water vapor. Plants in the greenhouse (or on the earth) absorb the rays. The plants reemit longer wavelengths of light by the process of radiation. Much of the reemitted light is in the form of infrared rays. Glass (or water vapor) will not transmit these infrared rays; they are reflected back into the greenhouse (or back onto the earth), producing a universal warming effect. This is the way a greenhouse will stay warm on a sunny day even in the winter. If the earth had been protected by a canopy, the canopy would have reflected the infrared light back onto the earth and kept the environment warm. At night the plants would have continued to emit infrared radiation. These rays would *not* have escaped through the canopy into space. Thus the nights would have been fairly warm also. Because of the protective canopy the seasonal changes would not have been marked by extremes in temperature. Large animals such as dinosaurs could have thrived in such an environment, feeding upon the luxuriant plant life. Men would have lived longer because harmful cosmic rays would have been excluded by the canopy.

We are not certain, of course, that there *was* in fact a water vapor canopy surrounding the earth before the Flood. (The idea is therefore sometimes referred to as the "canopy theory.") But the existence of such a canopy would answer a number of questions about prediluvian conditions.

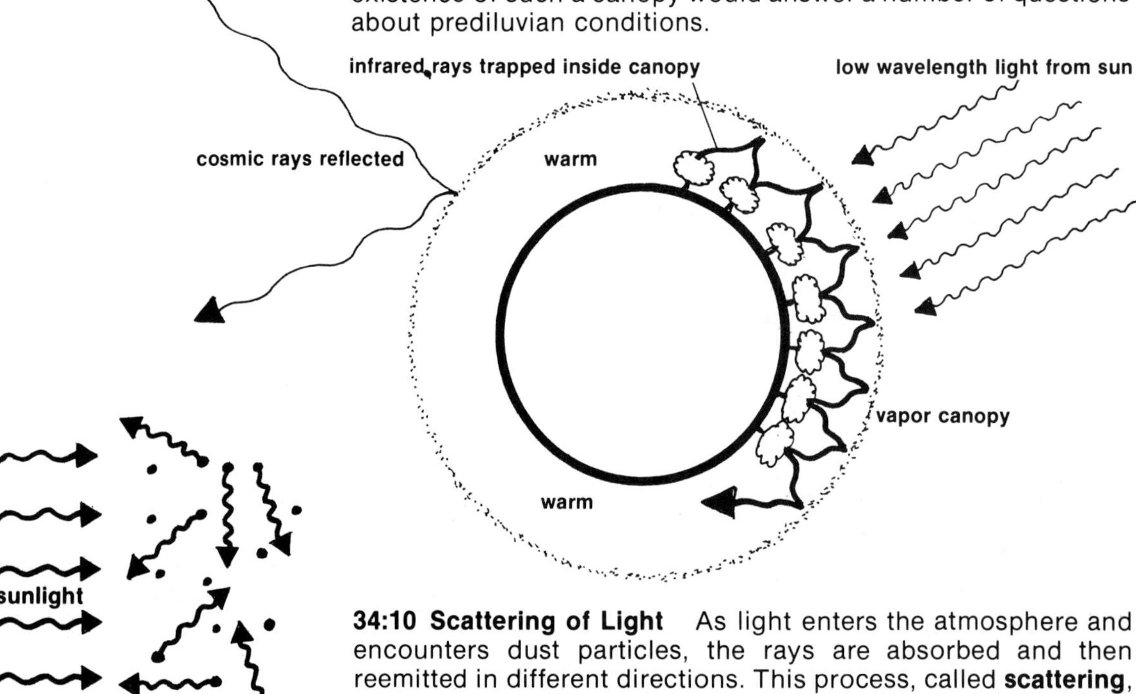

34:34 *Low wavelength radiation entering the earth's atmosphere is absorbed by objects on the earth. Higher wavelength infrared radiation is reemitted, but it cannot escape the canopy (greenhouse effect). It is reflected back to the earth, causing a universally mild climate over the whole earth.*

34:35 *Scattering of light by atmospheric dust particles.*

34:10 Scattering of Light As light enters the atmosphere and encounters dust particles, the rays are absorbed and then reemitted in different directions. This process, called **scattering**, is illustrated in Figure 34:35.

Scattering produces many interesting and beautiful effects. Blues and violets are scattered more than other colors when sunlight enters our atmosphere. Violet is scattered about 10 times as much as red. The longer the wavelength, the less the scattering. In the following list the colors are arranged in order according to the amount of scattering they undergo.

Colors More Easily Scattered
Violet
Blue
Green
Yellow
Orange
Red
Colors Less Easily Scattered

When all these scattered colors are combined, the resultant color is the blue color of the sky.

The red wavelengths are affected least by scattering. When the sun is rising or setting, its rays must travel farther through the atmosphere to reach the observer than at other times of the day. By the time the rays get to the observer, the shorter wavelengths have been scattered out, leaving only the longer wavelengths at the red end of the spectrum. Thus the rising or setting sun and any clouds located near it in the sky appear red.

The sun appears white outside our atmosphere. At noon on the earth the white light has the least distance to travel through our atmosphere, and it appears a yellow-white to an observer. As the sun moves to the west, the apparent color of the sun changes to orange; and finally, the setting sun is red.

Clouds are very bright and white because of the scattering of light caused by the water droplets in them. These droplets scatter so many different wavelengths of light that the combined effect produces a white color. Gray or black clouds are caused by shadows.

path of rays of sun at sunset (long)

sunlight

path of rays of sun at noon (short)

34:36 *Different path lengths of sunlight at different times of day.*

(a)

34:37 *(a) Sunrise. (b) Sunset.*

(b)

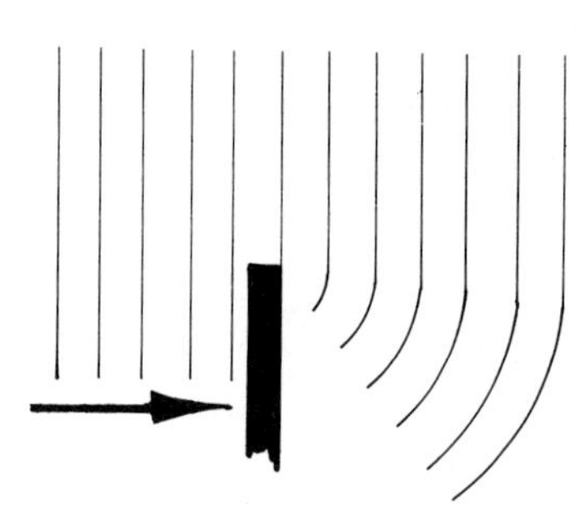

34:38 *Diffraction of waves around the edge of an obstacle.*

34:11 Diffraction of Light We have stated that light rays travel in straight lines. However, because of its wave nature, light under certain conditions can bend around the edge of an object. This effect is called **diffraction**. Diffraction can sometimes cause the edge of an object to appear blurred.

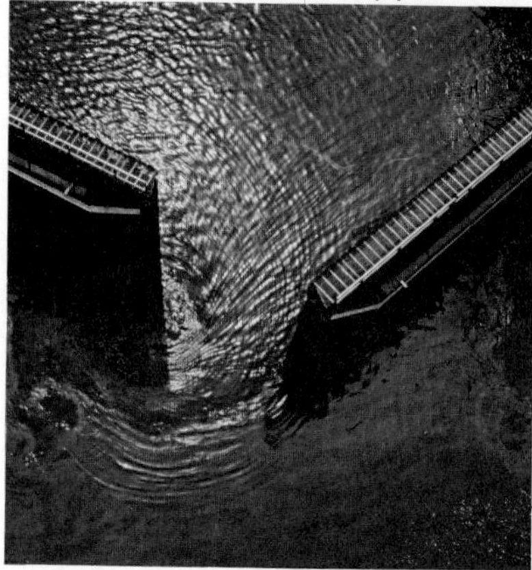

34:39 *The diffraction of water waves around a gate.*

34:40 *Diffraction of light through two slits.*

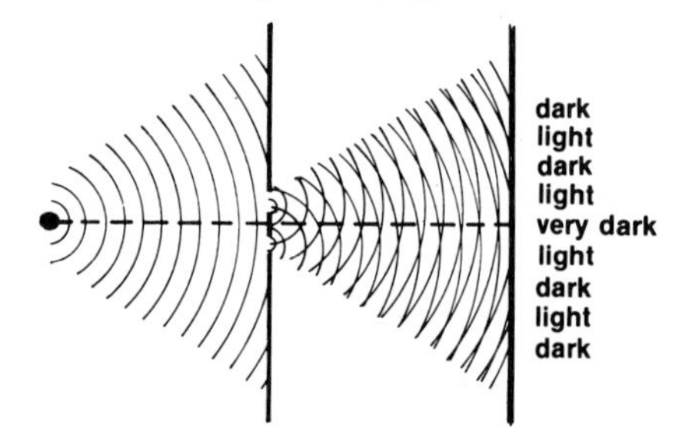

34:12 Light Energy To measure light energy, one must compare a beam of light to a given standard. Years ago light sources used to be compared to a specific kind of candle. The **intensity** of a standard candle at its source is one **candlepower**. Today the standard is the light produced when thorium oxide is heated to a temperature of 1775° C and viewed through a long tube with an opening of 1/60 cm^2 in area.

The rate at which light energy is carried from a source is measured in **lumens**. One candlepower is equal to approximately 12.5 lumens. A new 100-watt incandescent light bulb produces around 130 candlepower of light.

If you hold a piece of paper close to a light source, the paper is illuminated very brightly. As you move the paper away from the light source, the **illumination** decreases rapidly. The intensity of any light beam decreases inversely as the square of the distance from its source. What is the illumination on a piece of paper a distance of three meters away from the light?

34:41 *Interference pattern produced by diffraction of light by a wire.*

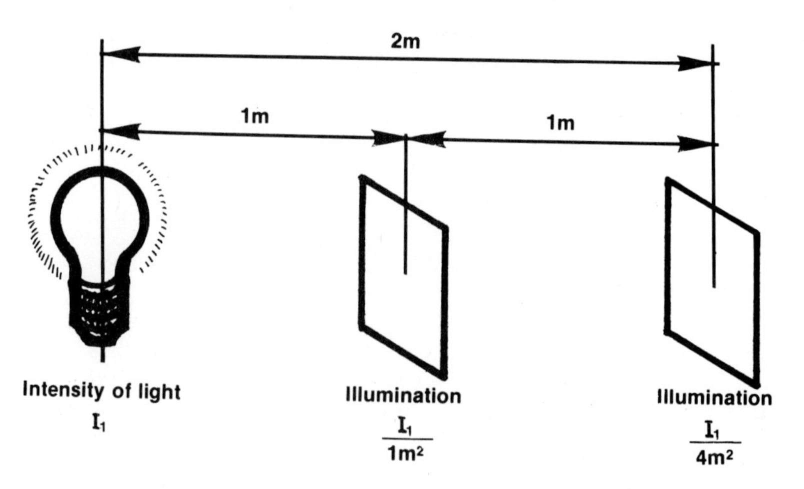

34:42 *The illumination of the light source on an object decreases as you move the object farther away from the source.*

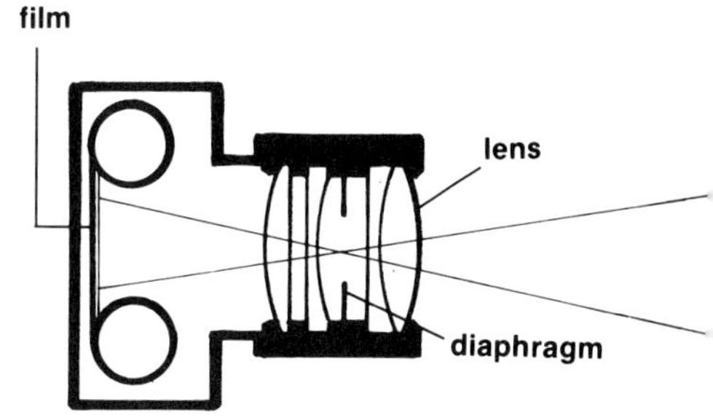

34:43 *Schematic diagram of a camera showing an adjustable diaphragm.*

34:13 Light Meters Since the human eye has an adjustable lens (the iris closes or opens to adjust to the intensity of light it receives), we have no need of light meters for ourselves. However, a camera has a diaphragm that has to be adjusted to control the ilumination received so that the film will not be underexposed or overexposed.

A light meter is used to measure the illumination directly. Inside the light meter is a photoelectric cell that changes the light energy into electric current. The brighter the illumination, the more electrical current will be produced in the meter. The current is used to deflect a needle over the scale on the light meter. The photographer can then adjust the meter to the type of film he is using and set his camera diaphragm to receive the proper amount of illumination.

34:44 *A light meter.*

34:14 The Eye "He that formed the eye, shall he not see?" (Psalm 94:9). Almost every waking moment we enjoy a continuous series of color pictures of the world around us. We are blessed with eyes that have automatic aiming, automatic focusing, and automatic aperture adjustment. The "film" on which they record is instant developing and self-renewing. Our eyes can function in a wide range of light intensities, from almost total darkness to bright sunlight. Resolution in the normal eye is so good that we can see clearly an object only 0.002 inches across—the diameter of a fine human hair. Our two eyes working together afford us stereoscopic vision—depth perception by triangulation from two different vantage points. On a typical day our eyes make some 100,000 separate motions, functioning up to 16 or more hours with seldom a complaint, then carry on their own maintenance work while we sleep.

So complex is the human eye that no one has yet constructed a working model of it. At the heart of the eye is a lens whose magnifying power can be varied by changing its thickness. No one has yet built such a lens. But this remarkable device, working in conjunction with the rest of the optical system, accounts for the eye's wondrous miniaturization. The eyeball is only about an inch in diameter—smaller than a ping-pong ball! Compare this with the dimensions of a high quality camera, which, in spite of its much greater size, has far less performance designed into it. A camera is focused by moving the lens in and out, varying its distance from the film. Think how inconvenient it would be if our eyes had to do this and how strange we would look when focused on a close object with our lenses all the way out! Yet this would be the closest modern science could come to making a working model of the eye.

Light entering the eye passes in course through the cornea, the aqueous humor, the lens, and the vitreous humor before striking the retina (see Figure 34:45). All of these are optically transparent and contribute to the formation of a sharp image on the retina by refracting each ray the proper amount. The retina corresponds to the film of a camera. It is here that the little-

34:45 *The optical system of the human eye is so sophisticated that scientists have been unable to build a working model of it. Light rays pass in turn through four different transparent materials—the cornea, the aqueous humor, the lens, and the vitreous humor—forming an inverted image on the retina. Aperture and focusing are automatically controlled by information which is fed back from the brain.*

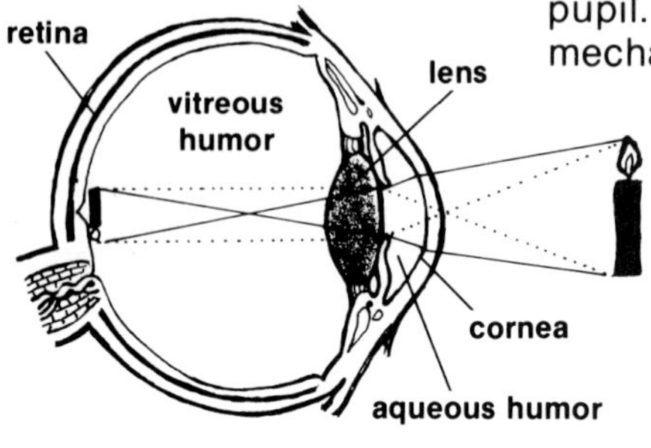

understood photoelectric conversion process takes place—light energy is changed to electrical impulses, which are carried by means of the optic nerve to the occipital lobe (back part) of the cerebrum. Exactly what takes place in the brain is even less well understood than the action in the retina.

Each eye is equipped with six powerful muscles which serve to aim it at the object being viewed (see Figure 34:46). These are synchronized with the six muscles of the other eye by "computer circuitry" in the brain. Another pair of muscles controls the opening and closing of each eye. Tiny paired sets of muscles inside the eyeball itself respond to electrical impulses fed back from the brain to adjust the focusing of the lens and the size of the pupil. The outside of the eye is provided with a "windshield wiper" mechanism that keeps the exposed surface moist and dust free.

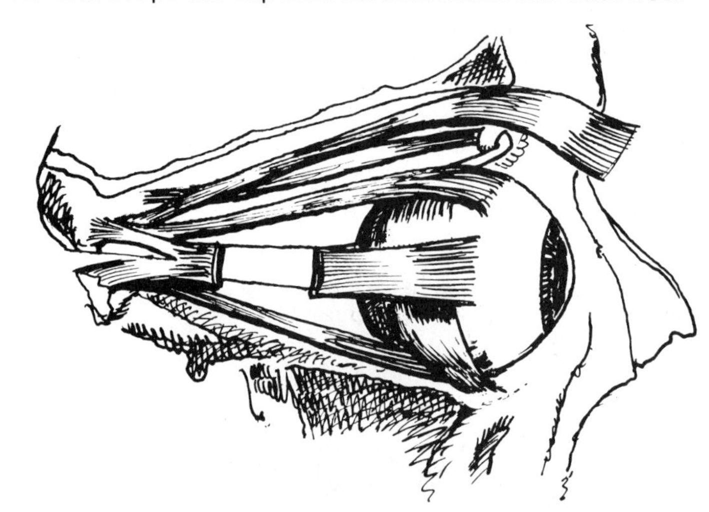

34:46 *Teleology. Intricate details of eye muscles testify to creation.*

In view of all these amazing mechanisms that work together in such a phenomenal way to produce human vision, it is difficult to understand how some people can believe that the eye came about through a trial-and-error process of evolution. The eye is generally regarded as an "adaptation" to our environment. Let us take the time to think this idea through, to see where the evolutionists go astray.

At one time, they tell us, no organism on earth possessed eyes. Sunlight flooded the earth, but the eyeless creatures swimming about in the "primitive oceans" were insensitive to these wavelengths. (Note carefully that no man was there to *observe* this imaginative picture.) How could such a blind world develop vision? Many of the evolutionary theorizers are strangely silent at this point. Light rays striking the skin of these marine organisms would in some way have to stimulate the formation of rudimentary eyes, if indeed the eye is an adaptive response to the environment. But how could this happen? Light rays striking skin never form eyes in any creature today. And even if such a thing *could* happen, the trait would not be passed on to the next generation. It is a well-established fact of biology that acquired characteristics, such as this would be, are not inheritable. In order to have a change passed on to the offspring, there must be a change *in the genes*. Yet the genes are *inside* the organism, totally oblivious to the light on the outside. The genes don't even "know" that the light is there, much less how to give instructions for the building of an eye.

Suppose, just for the sake of argument, that an imperfectly formed eye *is* able to make an appearance in some creature. Allegedly, an early eye would be a very crude device capable only of distinguishing between light and darkness. We are now asked to believe that it will keep bettering itself. Will it? Again, the genes do not even "know" in what ways the eye is defective, much less how to correct the problems that beset it. Some evolutionists place their faith in random mutations. But mutations have never been observed to carry out even one minor step of such a staggering undertaking. Evolution fails completely as an explanation for the eye. Creation is the only alternative.

"I will praise Thee; for I am fearfully and wonderfully made" (Psalm 139:14).

34:47 *Light spreads out rapidly from an incoherent source of light, such as a flashlight.*

34:15 The Laser The word **laser** means *L*ight *A*mplification by *S*timulated *E*mission of *R*adiation. It is a device that produces a coherent, concentrated, monochromatic beam of light. The light produced by an ordinary lamp is **incoherent**. This means that it consists of waves of many different frequencies. Light like this spreads out as it travels, and the beam becomes wider and wider and less intense with distance. A beam of coherent light will not spread out and become diffuse. Amazing things can be done with the intense beam of light that comes out of a laser.

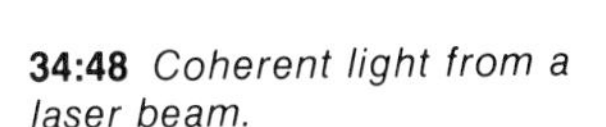

34:48 *Coherent light from a laser beam.*

34:49 *The helium-cadmium laser.*

The straightness of the beam makes it useful in surveying. Astronauts placed a laser reflector on the moon, and a laser beam was directed to it to measure the distance between the earth and the moon. Laser beams can be used to carry large amounts of information. Because they generate a tremendous amount of heat

(a)

(b)

34:50 *(a) A laser located 25 miles away from the camera. (b) The laser beam intercepted 25 miles away from its source is so intense that it has to be focused off-center from the camera so it will not overdevelop the film or injure the camera.*

in the area of the beam, they are used to weld and melt limited amounts of material. Since the beam can be used to transmit power, future applications seem limitless.

34:16 The Light-Year The speed of light has been measured by several different methods and found to be 186,000 miles per second or 3×10^8 meters per second in a vacuum. At this rate, it should travel about 11 million miles a minute. Because the sun is 93 million miles away, its light must take approximately 8.3 minutes to get to the earth. Or we could say that the sun is about 8.3 *light-minutes* away, defining the light-minute as the distance light travels in a minute. Sunlight requires more than five hours to reach the planet Pluto. We could, therefore, say Pluto is more than five *light-hours* from the sun.

In discussing distances to the stars, astronomers use a larger unit, the **light-year**. It is defined as the distance light travels in a year, 5.88 trillion miles. Often the astronomer rounds this number to 6 trillion miles for easy reckoning. The nearest stars, Alpha Centauri and its two companions, are about 4.3 light-years away, according to measurements made by triangulation. It should be borne in mind that the speed of light has only been determined within the solar system, the best measurements having been made by radar reflections from nearby planets (radio waves and light waves travel at the same speed). It seems like a logical assumption to many that light beyond the solar system should travel at the same speed as it does within the solar system; however, from a strictly scientific standpoint, there are no direct measurements to prove it.

It is a popular view that the universe must be billions of years old because the light from distant stars would require long expanses of time to reach the earth. As with so many pseudo-scientific arguments that sound attractive on the surface, the observations are lacking. No one has ever observed light in flight for such long periods of time. To do so would require separate measurements billions of years apart, quite beyond the realm of human capability.

The creationist view is that God set up a mature, fully-functioning universe at the beginning, with light from the distant stars already visible from the earth. This harmonizes with the fact that man, animals, plants, and the earth were all created in a mature state by rapid, miraculous acts of God rather than by slow, natural processes (Genesis 1, 2; Exodus 20:11; Psalm 33:6, 9). Many people are unwilling to ascribe this much power to God, yet 57 times the Bible describes Him as *Almighty* (all-powerful). The concept of a rapid Creation goes contrary to the natural tendency of the human mind to seek an evolutionary explanation for everything observed, but God's ways are not our ways. "For my thoughts are not your thoughts, neither are your ways my ways, saith the Lord. For as the heavens are higher than the earth, so are my ways higher than your ways, and my thoughts than your thoughts" (Isaiah 55:8-9).

James Clerk Maxwell (1831-1879)

Scottish physicist

James Clerk Maxwell was undoubtedly one of the greatest geniuses in the history of science. His mathematical equations predicted the existence of radio waves many years before their discovery and served to unite the separate sciences of optics, electricity, and magnetism within a single comprehensive framework. Albert Einstein called this achievement "the most profound and the most fruitful that physics has experienced since the time of Newton." At the same time, Maxwell was a humble and devout man of God. He would be called in today's terminology a "Fundamentalist," as would all the scientists whose biographical sketches are included in this book.

Maxwell was born in Edinburgh, Scotland, in 1831. His mother, a dedicated Christian, was in charge of his education until her death in 1839. She continually urged him to "look up through nature to nature's God." By the time he was eight years of age, he had memorized the entire 119th Psalm. He received his formal education at Edinburgh Academy, the University of Edinburgh, and Trinity College, Cambridge. At first he did not do outstanding work in his studies. Poor health often affected his attendance. But by the time he was graduated from the academy (high school), he ranked first in his class in mathematics and English. From this time on it became apparent to his instructors that something unusual was to be expected of him.

His brilliant career included teaching posts at three different institutions of higher learning. Two years after he had accepted his first teaching position at Marischal College, Aberdeen, Scotland, he married Katherine Mary Dewar, who proved to be a lifelong source of strength and encouragement, and who contributed in no small measure to his success. In 1860 he became professor of physics and astronomy at King's College in London, and in 1871 he took over the newly formed chair of experimental physics at Cambridge University. In connection with this position he was selected to supervise the planning and construction of the famous Cavendish Laboratory.

Maxwell's spiritual life was an example to all around him. Those who knew him best characterized him as humble, sincere, tender-hearted, devoted, and fearless in promoting that which he knew to be true and right. His beliefs were strictly orthodox, and his scientific research continually strengthened his conservative stand. He strongly disliked the type of preaching that was mere morality without the Gospel remedy. Through constant study he developed a remarkably thorough knowledge of the Scriptures, and he used this knowledge to good advantage. It was his custom to visit the sick and shut-in persons in the community and to read and pray with them if they desired. He was an elder in the Corsock Church near his home at Glenlair, Scotland, a church which had been founded and built through his spiritual leadership and financial assistance.

Maxwell's view of science as a God-ordained means of subduing the earth is shown in a prayer that was found among his notes: "Almighty God, Who hast created man in Thine own image, and made him a living soul that he might seek after Thee, and have dominion over Thy creatures, teach us to study the works of Thy hands, that we may subdue the earth to our use, and strengthen the reason for Thy service; so to receive Thy blessed Word, that we may believe on Him Whom Thou has sent, to give us the knowledge of salvation and the remission of our sins. All of which we ask in the name of the same Jesus Christ, our Lord."

An important work undertaken by Maxwell was the disproof of Laplace's nebular hypothesis. In 1796, the French atheist Laplace proposed a theory that the solar system had "evolved" from a large cloud, without need of a creator. The cloud, Laplace claimed, contracted over a period of millions of years and gradually produced the solar system as we know it today. Many who were of an anti-religious turn of mind accepted this idea without question. Maxwell, however, upon analyzing it mathematically, found two major flaws in the theory: (1) The material would never condense into planets; (2) There would be no way to slow the rapidly-spinning mass in the center to form our present slowly rotating sun. The theory was discarded, and to this day it has never been replaced with one that is truly workable.

Maxwell had no use for any theory of evolution—cosmic, chemical, biological, or otherwise. In a paper presented to the British Association at Bradford in 1873, he stated, "No theory of evolution can be formed to account for the similarity of molecules, for evolution necessarily implies continuous change. . . . The exact equality of each molecule to all others of the same kind gives it, as Sir John Herschel has well said, the essential character of a manufactured article, and precludes the idea of its being eternal and self-existent." In his discourse "On the Telephone," written in 1878, he took special pains to refute the evolutionary speculations of Herbert Spencer (1820-1903), an English philosopher who did much to popularize Darwin's theory of evolution.

Some of Maxwell's other contributions to science include experiments in color vision and optics, investigations of elastic solids, original extensions of pure geometry, mechanics, molecular physics, and a mathematical analysis of Saturn's rings. He extended Faraday's work, particularly the concept of "lines of force," placing it on a solid mathematical basis. He invented the optic bench and the opthalmoscope—the instrument used by doctors for looking inside the eye. The bell-shaped curve used in statistics is called a *Maxwellian distribution* in his honor. His name has also been memorialized in the physical unit, the "maxwell," a measure of magnetic flux.

Maxwell died of cancer in 1879 at the age of 48. During the following three years, two of his friends, Lewis Campbell and William Garnett, collected his letters and papers and penned his biography. Entitled *The Life of James Clerk Maxwell*, the massive volume was published by Macmillan of London in 1882. Several statements by colleagues who knew him intimately are included in the book. One of these was G. W. H. Tayler, Vicar of Trinity Church, Carlisle, who wrote as follows: "Maxwell has indeed left us a very bright memory and example. We, his contemporaries at college, have seen in him high powers of mind and great capacity and original views, conjoined with deep humility before his God, reverent submission to His will, and hearty belief in the love and atonement of that Divine Saviour Who was his portion and comforter in trouble and sickness, and his exceeding great reward" (page 174).

List of Terms

candlepower
concave mirror
diffraction
diffuse reflection
dispersion
electrical field
electromagnetic radiation
electromagnetic wave
emission
excitation
Flood
focal length
greenhouse effect
illumination
incident rays
incoherent
intensity
laser
lens
light
light-year
lumen
magnetic field
microscope
monochromatic light
normal
rainbow
reflected rays
reflection
refraction
scattering
speed of light
telescope
white light

Questions

1. What is visible light?
2. (a) Does an electrical charge have an electrical field associated with it?
   (b) What is the necessary condition for an electrical charge to have a magnetic field associated with it?
3. What is the speed of light in a vacuum?
4. If the frequency of a wave of visible light is 6×10^{14} Hz, find its wavelength.
5. If the wavelength of a wave of visible light is 5×10^{-7} m, find its frequency.
6. What is the color of the following wavelengths of visible light?
   (a) 4.31×10^{-7}m
   (b) 6.0×10^{-7}m
   (c) 5.4×10^{-7}m
7. (a) Is electromagnetic radiation with a wavelength of 6.7×10^{-7}m visible light? What is it?
   (b) Is electromagnetic radiation with a wavelength of 5×10^{-2}m visible light? What is it?
8. Using the Bohr model of the atom, explain how a particular color is produced. Illustrate.
9. How are different colors given off from the same atom? Use the Bohr model.
10. (a) What is a light ray?
    (b) Illustrate regular reflection of light from a surface.
    (c) Illustrate diffuse reflection of light from a surface.
11. How does a mirror "work"?
12. What is refraction of light?
13. Does light slow down or speed up when it passes into a denser medium?
14. (a) What is the purpose of a lens?
    (b) What effect does it employ?
15. (a) What is monochromatic light?
    (b) What is white light?
16. (a) The rainbow is caused physically by what?
    (b) Who ordained the rainbow? Give the reference.
17. (a) What conditions probably would have existed on the earth if there had been a vapor canopy over the earth?
    (b) How would a vapor canopy act?
18. (a) What is the scattering of light?
    (b) Name some effects caused by scattering.
19. What is the diffraction of light?
20. If the intensity of a light source is I_0, what is the illumination by this light on a piece of paper nine meters away?
21. What is the use of a light meter?
22. Name three automatic functions of the eye.
23. Why is it impossible to explain the origin of the eye by evolution?
24. (a) What does *laser* mean?
    (b) Compare the light coming from a table lamp and a laser.

Student Activities

1. Write a paper on one of the instruments discussed in this chapter.
2. Write a paper on light as a form of electromagnetic radiation.

Photo next page
Television antennas.

UNIT XI WAVE PHENOMENA

Chapter 35 Electromagnetic Spectrum

35:1 Infrared Radiation Look again at the chart of the electromagnetic spectrum in Section 34:1. As you move from the visible spectrum through the red colors, you move into the region of the invisible rays of electromagnetic radiation. Just below the visible red color is the infrared region. "Infra" means *below*.

Sir William Herschel (1738-1822) discovered **infrared radiation** and correctly labeled it "rays of heat." He refracted rays from the sun through a prism to separate the different colors. He then placed a thermometer into the dark region just below the red color (infrared); the temperature recorded was greater than in any other part of the spectrum. Thus, you can see why he called the region "rays of heat."

Infrared radiation has a wavelength range of about 0.0001 to 0.1 cm (10^{-6} to 10^{-3}m). What causes this kind of radiation? We do not know, but we can do some interesting experiments using infrared radiation that might suggest its origin. If we send infrared rays into matter in the form of compounds, we find that the molecules absorb certain wavelengths of the radiation. Each different compound will absorb different wavelengths of infrared. Therefore, we can turn this observation around and say that heated (thermally excited) molecules emit infrared radiation. This is how infrared originates.

Any substance, then, that is above absolute zero will give off infrared rays. Why do they give off only certain specific wavelengths? It is imagined that molecules vibrate and rotate as

What say, buddy?

Howdy, partner.

35:1 *Carbon and oxygen bonded together.*

I want to shake like this.

Not a chance.

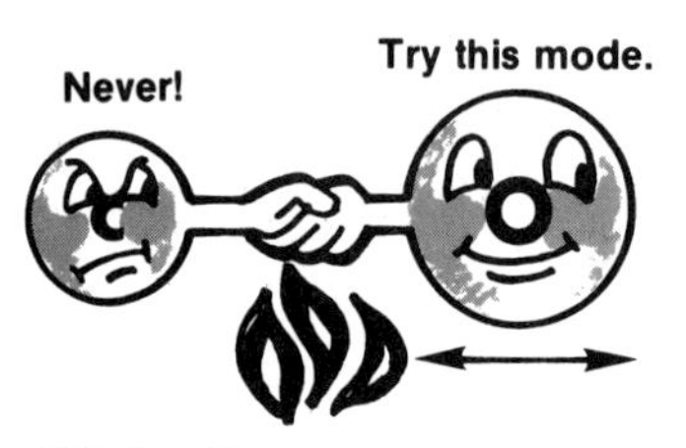

But if I join with hydrogen, we will do things differently.

We both agree on this rhythm.

35:2 *The carbon-oxygen bond tends to vibrate only in certain particular ways.*

they give off infrared energy. This can be illustrated by the use of models (not real atoms). Pretend that when a carbon atom combines with an oxygen atom, the molecule looks like a dumbbell. The oxygen end of the dumbbell is larger than the carbon end. The "handle" between the ends represents the molecular attraction forces, and it is flexible. As the molecule is heated, it can vibrate back and forth along the handle. Each different type of molecule will vibrate at certain particular frequencies because of the same atomic sizes and bonding forces. For instance, a C-O bond will always have the same vibrational frequency, but it will be different from that of the H-O bond. Each different bond type also has a certain rotational frequency. The molecule CH_4 has many vibrational and rotational modes (ways of vibrating and rotating).

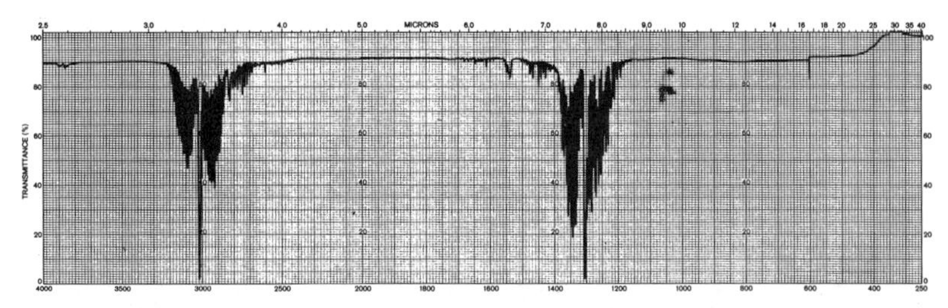

35:3 *Infrared spectrum of methane (CH_4).*

Since each substance has a different infrared spectrum, scientists can use this as a method of identifying unknown compounds. They can also determine how the atoms in a molecule are bonded together. The particular scientific technique is called **infrared spectroscopy**.

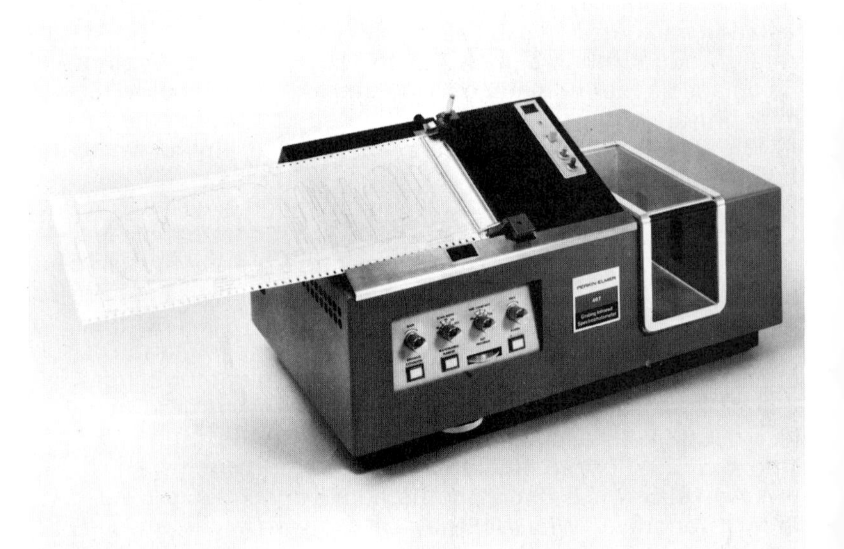

35:4 *Infrared spectrophotometer. A sample in a holder is placed into the chamber on the right-hand side of the spectrophotometer. An infrared beam passes through the sample, and its spectrum is recorded on the chart paper seen on the top left of the spectrophotometer.*

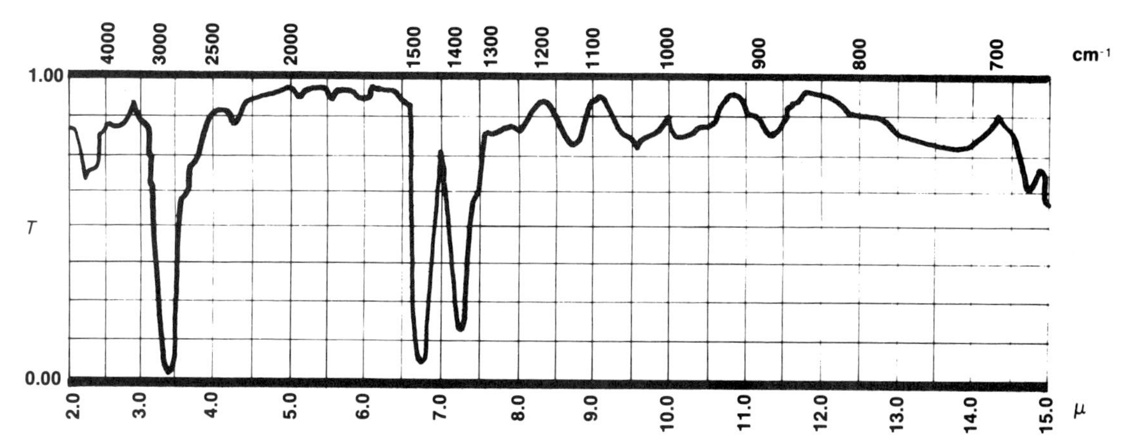

35:5 *Infrared spectrum of n-hexane ($CH_3CH_2CH_2CH_2CH_2CH_3$). All "upside-down peaks" on the spectrum are caused by the stretching, scissoring, and bending of various carbon-hydrogen bonds.*

Infrared radiation can penetrate through fog and haze and thus offers an excellent way to photograph many different things, particularly landscapes, at great distances. (Photographic paper sensitive to infrared can show details of the object, whereas ordinary photographic film would lose a substantial amount of detail.) Other practical uses of infrared sources are heat lamps for paint drying and physiotherapy.

(a)

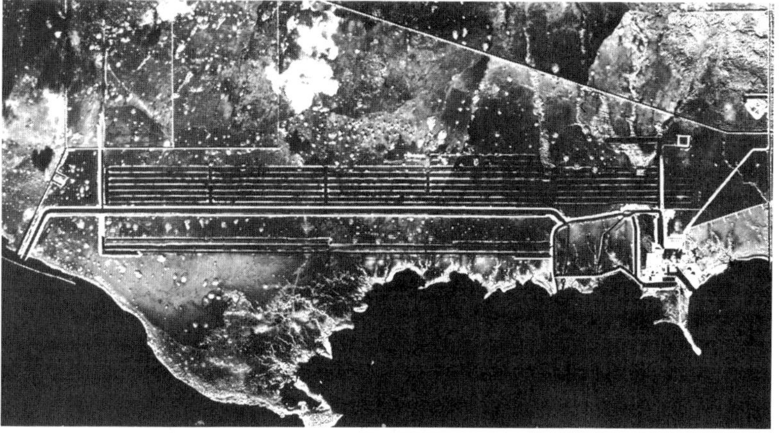

(b)

35:6 *(a) Standard color photograph. (b) Infrared photograph of the same area. Many times infrared photography can bring out more detail than standard methods.*

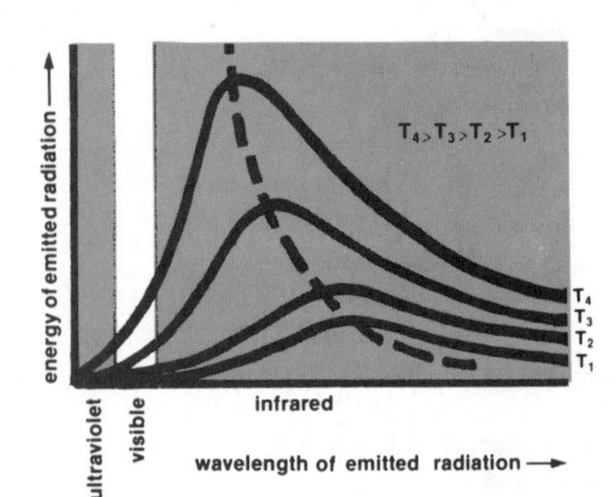

35:7 *An object can emit more high-energy infrared radiation as its temperature increases.*

Of course, the greatest source of infrared for the earth is the sun. It radiates energy corresponding to a body at 5800° K. The hotter any object gets, the more infrared it will emit. Figure 35:7 illustrates the energy given off by substances at different temperatures. Also, the hotter the source, the more energetic infrared radiation is given off. (The maximum of the curve shifts at higher temperatures to shorter wavelengths of radiation, which are more energetic.) The energy of any electromagnetic radiation is directly proportional to its frequency and inversely proportional to its wavelength.

35:2 Microwaves Moving to longer wavelength radiation in the electromagnetic spectrum, we come to the **microwave** region. It covers the wavelength range of 0.3 cm to 30 cm (3×10^{-3}m to 3×10^{-1}m). Microwaves behave similarly to visible light. They tend to travel in straight lines. They can be reflected from surfaces and can be focused into directed beams with curved metal mirrors.

Microwaves are used in **radar** (*r*adio *d*etection *a*nd *r*anging). They can be diffusely reflected by fairly small objects such as automobiles, airplanes, and ships, and large objects such as storm systems and land masses. Transmitters send out powerful microwave signals. The returning scattered radiation from an object is picked up by an antenna, and the time interval between transmission and reception is measured. This gives the distance from the transmitter to the object. Also, the signals can be transmitted in such a manner as to determine the speed of a car (again using the time interval between two distances).

35:8 *Microwave antenna.*

Problem: A state policeman has placed a radar unit on a highway to check for speeding cars. The first reflected signal reaches the receiver in 10^{-8} sec. The second reflected signal reaches the receiver in 9×10^{-9} sec. If the automobile has moved 10^{-7} feet in this time interval, how fast is it moving?
Solution:

$$\frac{10^{-7}\text{ ft}}{10^{-8} - (9 \times 10^{-9})\text{ sec}} = \frac{10^{-7}}{10^{-9}}\text{ ft/sec} = 100\text{ ft/sec} = 68.18\text{ mph}$$

Another application for microwave radiation is the microwave cooking oven. The microwaves set up a changing electric field (reversal of polarity) within the oven. This affects the water molecules in the food. If you remember from Chapter 11, water is a polar molecule with positive and negative ends. The changing electric field causes the water molecules to turn one way and then the other. As they turn, they collide with other molecules, giving off energy in the form of heat. Thus, the inside of the food is heated just about as fast as the surface. Cooking time is very quick because it is not necessary to wait for the heat to move from the surface to the interior of the food as in a standard oven.

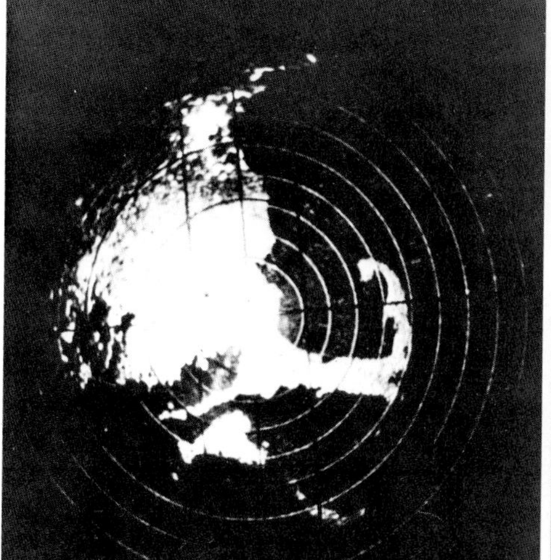

35:9 *Radar photograph of Cape Cod from 19,000 feet.*

35:10 *Airline radar weather and briefing room. Notice the radar scan in front of the man at the right.*

35:3 Radio and TV Waves If we continue moving down to longer wavelength electromagnetic radiation, we come to the radio and TV region. The majority of **radio** and **television waves** are normally transmitted at wavelengths *greater* than 30 cm (3×10^{-1}m).

Heinrich Hertz (1857-1894) was the first person to discover how to transmit electromagnetic waves, first called Hertzian waves. He used an induction coil as a source of alternating current. It was connected to leads which had a gap in them. When he turned the coil on, a spark would jump across the gap. Since he used alternating current, the polarity of the gap was continually changing; thus, the spark went from one conductor to the other and then in the reverse direction. He then placed a circular wire with a similar gap a short distance from the spark gap. He noted that a small spark was jumping the gap in the wire ring. He was actually transmitting electromagnetic radiation through air; the coil was his transmitter and the wire ring his receiver (antenna). The propagation of radio waves can be this simple.

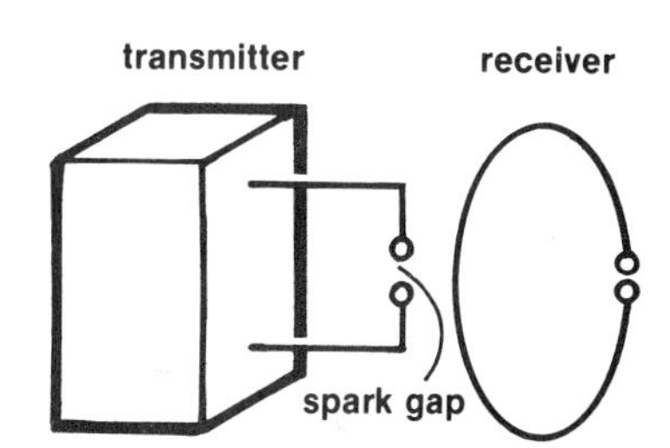

35:11 *An idealized drawing of the apparatus used by Hertz to transmit and receive radio waves.*

Radio signals can have wavelengths up to thousands of meters. With wavelengths this large they can be diffracted easily around trees or buildings, but large hills and mountains can interfere with transmission, causing "blacked-out" areas on the opposite side of the hills. However, because radio waves are reflected by the electrically charged layers in the upper atmosphere, it is possible for them to be transmitted for long distances and "over" mountains.

Local AM radio station transmission usually depends on surface-guided waves. For longer distance transmission, radio waves are reflected off the first layer (E-layer) of ionized air. For even greater distances and foreign broadcasts, shorter wavelengths are used and are reflected off a higher layer (F-layer) of ionized air.

FM and television transmission is done with very short wavelength radiation (around 1 meter). These waves travel in straight lines like microwaves, and the transmitters are usually placed on tall buildings or high ground. In most cases relay stations or coaxial cables must be used to transmit signals for distances greater than 50 miles.

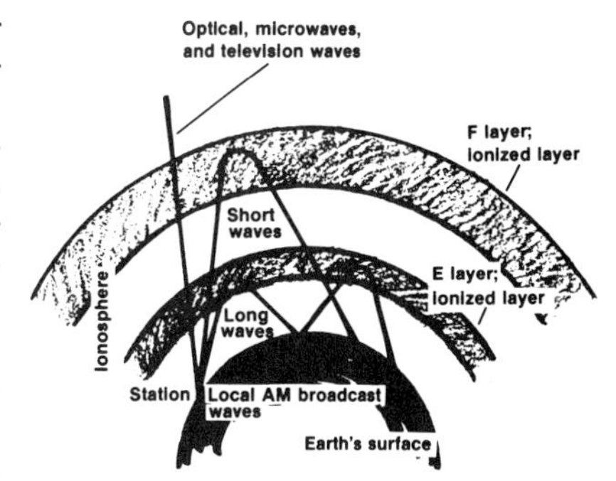

35:12 *Various paths taken by radio and television waves in the atmosphere.*

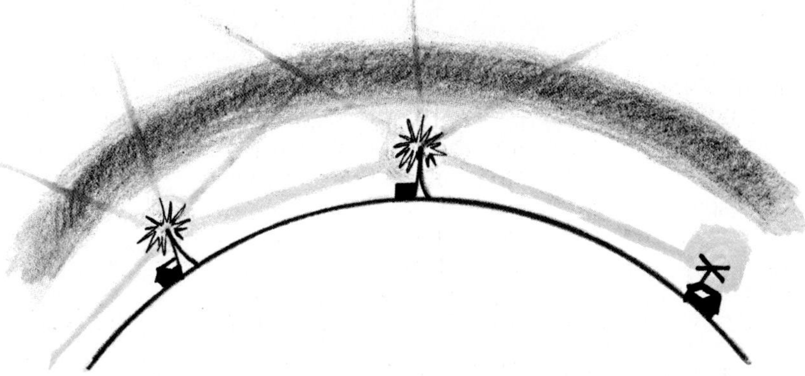

35:13 *Relay stations are needed to insure the proper transmission of television waves around the curvature of the earth.*

All of the various frequencies for broadcasting are shown in Figure 35:14.

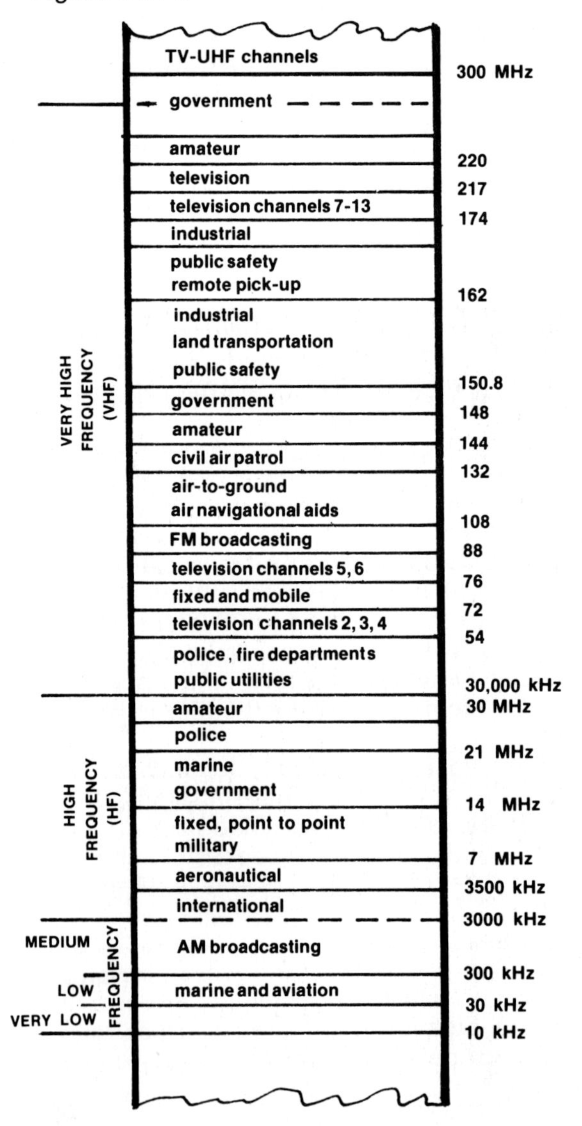

35:14 *Broadcasting frequency ranges.*

One type of radio transmitting antenna is shown in Figure 35:15. Each leg of the antenna is cut to 1/4 of the wavelength being transmitted. A source of high frequency electrical energy is connected at the center of the antenna. Electrons first race to the right-hand end of the antenna, making it negative. The electrons then return through the transmitter to the left-hand leg of the antenna, leaving the right-hand leg positive. This completes one cycle. There are hundreds of thousands or even millions of cycles each second, depending on the frequency of the transmitter. Electrons are rapidly accelerated back and forth during each cycle.

Figure 35:16 shows a vertical antenna in which the electrons are accelerated up and down. Any accelerating electromagnetic charge radiates energy, and the radiant energy from antennas such as these shown is a radio wave.

An antenna is needed to *receive* radio waves also. The first part of any radio or TV receiver is the antenna. Since the radio wave is electromagnetic in nature, it has a varying magnetic field associated with it. This is like the crests and troughs of a wave. This variable magnetic field induces an electrical signal (current) in the antenna. This electrical energy is carried to the radio or TV set and converted into sound and/or a picture by the devices in the set.

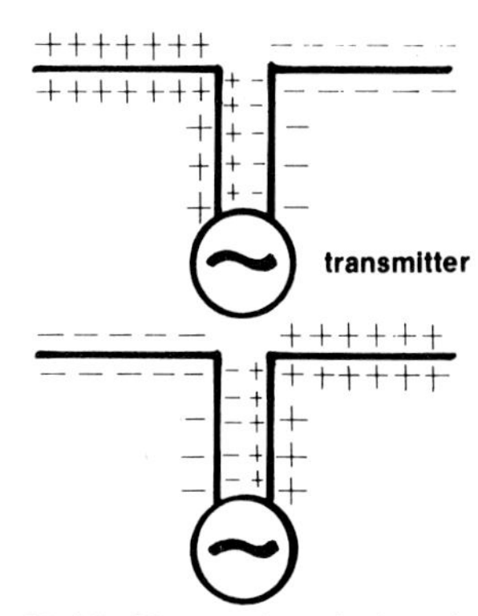

35:15 *Change in polarity of a two-wire antenna transmitter.*

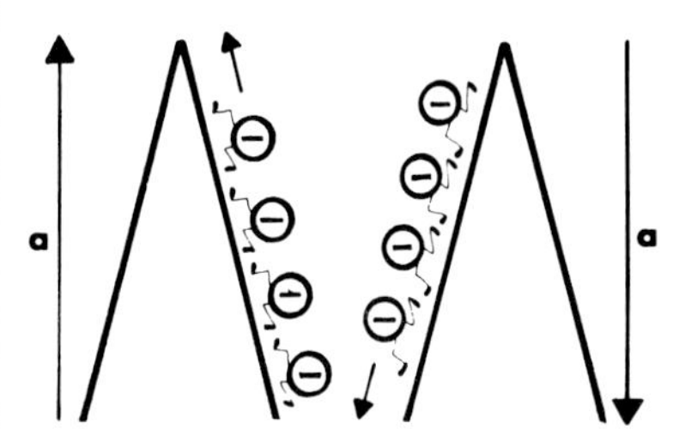

35:16 *Electrons accelerate up and down an antenna, causing the emission of radio waves.*

35:17 *Television antennas.* *(See chapter title page)*

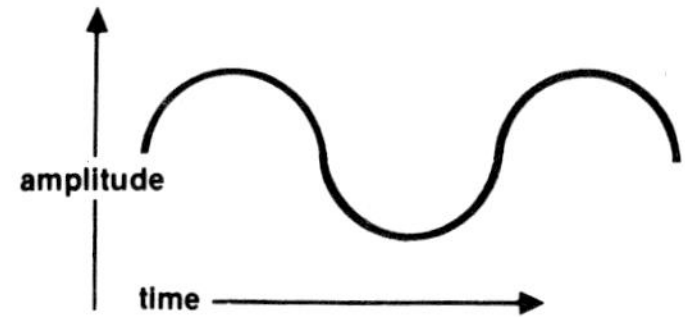

35:18 *Magnetic part of a wave. Its amplitude varies with time.*

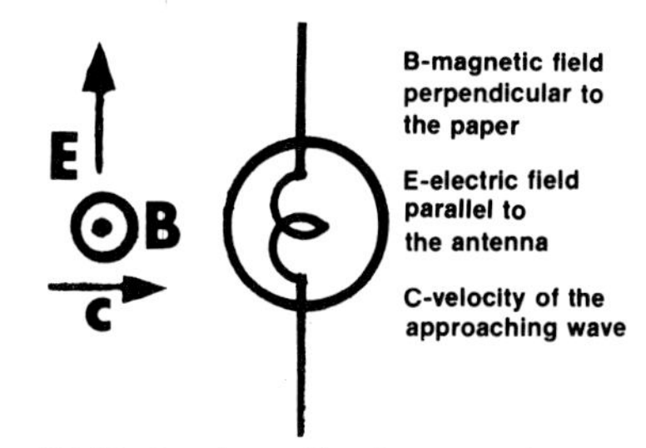

35:19 *A schematic diagram of an electromagnetic wave approaching an antenna.*

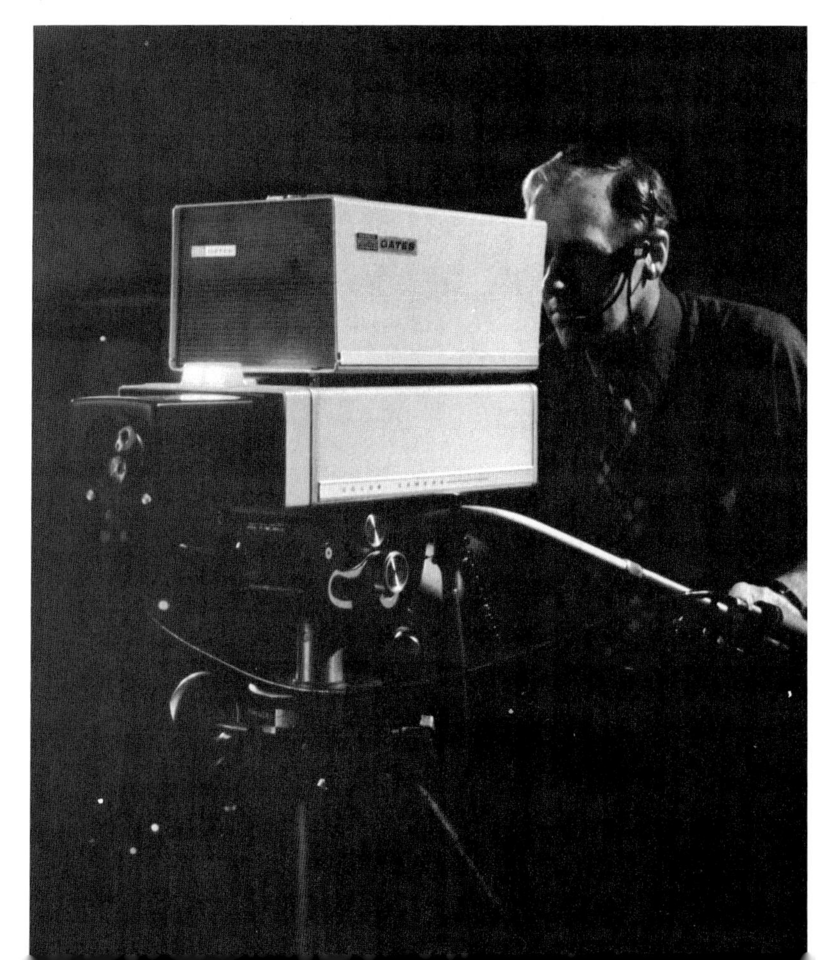

35:20 *A television camera.*

Radio waves are also received from astronomical bodies. A relatively new branch of science, **radio astronomy**, has been developed using radio telescopes to receive these waves. Observations on the earth are confined to wavelengths from 1 cm (1×10^{-2}m) to 50m, which can get through our atmosphere without being strongly absorbed. Radio telescopes have been placed on orbiting satellites also.

35:21 *The 50-foot radio telescope, atop the Administration Building of the United States Naval Research Laboratory is used in the radio astronomy program at frequencies ranging from about 200 to 35,000 megahertz.*

35:4 Ultraviolet Light Let us now go to the other side of the visible light region in the electromagnetic spectrum, just beyond the violet color to the ultraviolet region. "Ultra" means *beyond*. This region has a wavelength range of around 10^{-9}m to 4×10^{-7}m.

Ultraviolet radiation was discovered by J. W. Ritter (1776-1810) and W. H. Wollaston (1766-1828). A piece of paper dipped into silver nitrate solution was placed in the ultraviolet region of the spectrum (then unknown). The paper was darkened by the action of the radiation. This is one of the more important properties of ultraviolet—its actions on photographic substances. It causes many types of chemical changes.

Ultraviolet radiation is produced similarly to visible light. However, since the energies of ultraviolet rays are greater than those of visible light, the electrons close to the nucleus are involved in these transitions, whereas electrons farther away from the nucleus are involved in the formation of visible light.

Ultraviolet light (sometimes called black light) produces fluorescence in certain minerals. This makes ultraviolet light useful to prospectors.

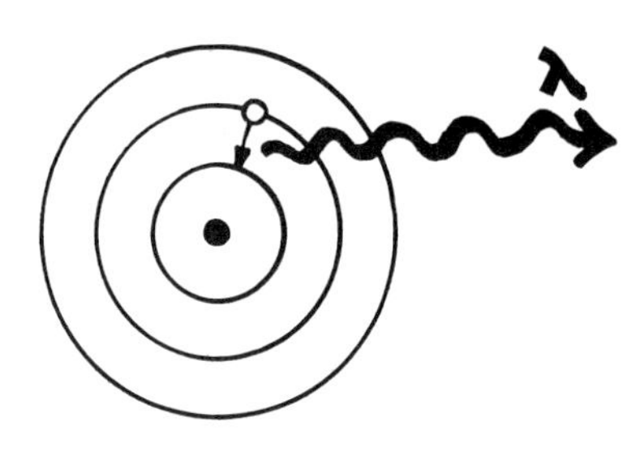

35:22 *Electron dropping from a higher to a lower energy orbit giving off ultraviolet radiation.*

35:23 *A unique photochromic glass that is sensitive to the ultraviolet in sunlight is used for a new automatic sunglass lens that adjusts its tint to the brightness of the day. Sunglasses are darkest when the sun is brightest but fade to lighter tints when less protection is needed.*

Mercury gives off ultraviolet radiation. Place mercury inside a glass tube which has been coated with a **fluorescent** material, add the filaments, and you have a fluorescent light. When the filaments are heated by the passage of electrical current through them, the mercury is vaporized. The hot vapor gives off ultraviolet rays which bombard the coating material, causing it to fluoresce (give off light).

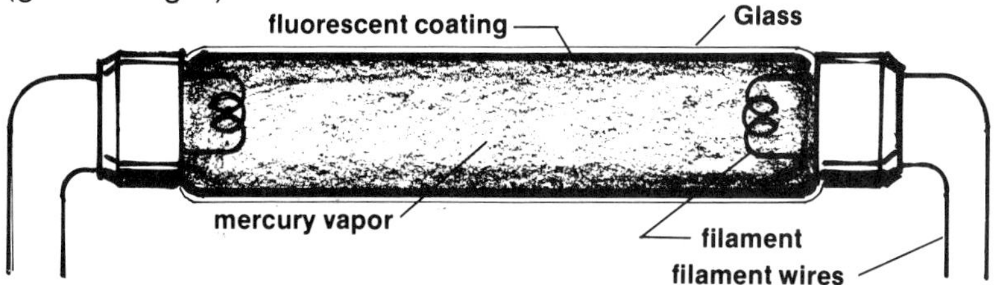

35:24 *A fluorescent light.*

Ultraviolet radiation has enough energy to remove electrons from the atoms and molecules it strikes. This action can cause serious damage in the cells of human bodies. Exposure to rays with wavelengths below 4×10^{-7}m may result in tanning, sunburn, freckles, nausea, and fever. Ultraviolet rays do not penetrate very deeply into the skin (1/10mm). Thus, a person can protect himself by applying large amounts of oils and creams to his skin. Ultraviolet light stimulates the manufacture of vitamin D from

35:25 *A fluorescent mineral shown in ultraviolet light.*

chemicals in the skin. This vitamin is needed for healthy bone structure, and certain foods such as milk are often exposed to ultraviolet rays to increase the vitamin D content.

Ultraviolet radiation kills bacteria; therefore, it is used to sterilize items and to aid in the preservation of foods. Ultraviolet light can be used to read maps and instrument panels in the dark cockpits of aircraft. Laundry marks written with invisible ink can be read in ultraviolet light. It is also used in fighting crime, revealing stains on objects that cannot be seen with visible light.

35:5 X-Rays W. K. Roentgen (1845-1923), the first winner of the Nobel prize, discovered **X-rays** in 1895 while studying cathode rays. X-rays range in wavelength from 10^{-12} to 10^{-8}m. They are formed when high energy electrons are rapidly decelerated by striking a solid target. Since they are not deflected by a magnetic field, X-rays have no charge. They affect photographic film the same way that visible light does. Once they pass through matter, a visible record of what they "see" can be obtained.

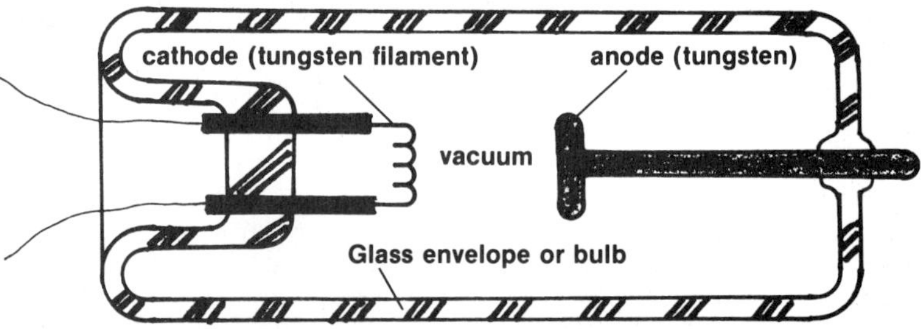

35:26 *A basic X-ray tube. A large potential difference between the anode and cathode accelerates the electrons from the heated filament towards the anode. When the electrons strike the anode they rapidly decelerate, releasing the energy in the form of X-rays.*

X-rays will ionize gases as they pass through them. When colliding with a valence electron in a substance, an X-ray will impart enough energy to the electron so that it can escape from the molecule. The best known property of X-rays is their ability to penetrate matter. This has made them very useful in medical work for viewing internal structures of the body. X-rays are used in industry to inspect large, thick, metal castings and forgings for internal defects. X-rays have also been used to determine whether a painting is genuine or forged.

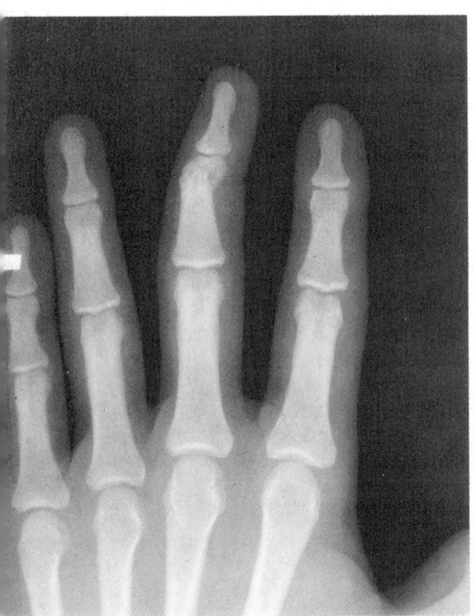

35:27 *An X-ray photograph of a hand with a broken finger bone.*

35:28 *Solid fuel rocket motors are checked for flaws by passing X-rays through them. The X-rays are generated by the betatron shown on the left.*

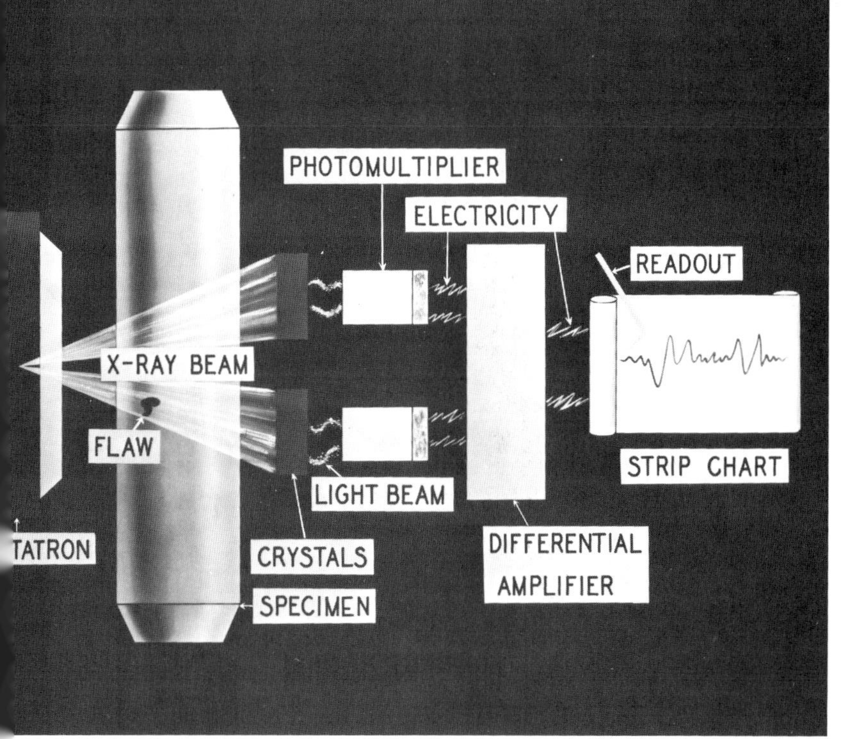

35:29 *A schematic diagram of how X-rays can be used to detect flaws in a metal part. X-rays, generated in the betatron, pass through the test piece. The exit rays are converted into light by the proper crystals. The light is then converted into an electrical impulse which goes through an amplifier and is recorded on a chart. A defect is noted as a peak on the chart.*

Prolonged exposure to X-rays can cause burns, cancer, and even death. X-rays are capable of destroying body cells and causing mutations in genetic material. It has been found that X-rays destroy unhealthy cells more easily than healthy cells; so they have been used in the treatment of certain types of cancer.

35:30 *"Open wide!" would be appropriate if this scene were in your dentist's office. However, this is a nondestructive testing laboratory, where a 97,000 pound part for a huge earthmoving machine is being X-rayed to determine the metal's structural excellence.*

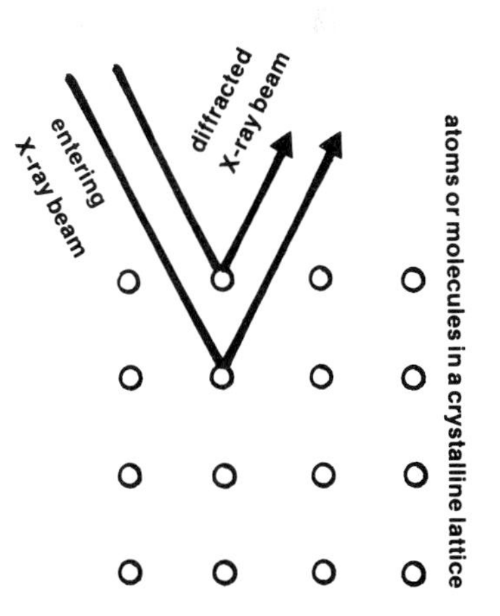

35:31 *Representation of diffraction of X-rays from crystals.*

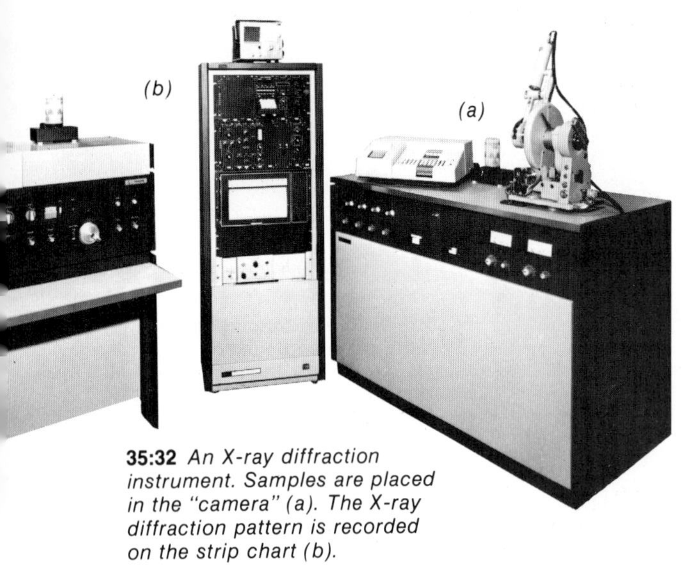

35:32 *An X-ray diffraction instrument. Samples are placed in the "camera" (a). The X-ray diffraction pattern is recorded on the strip chart (b).*

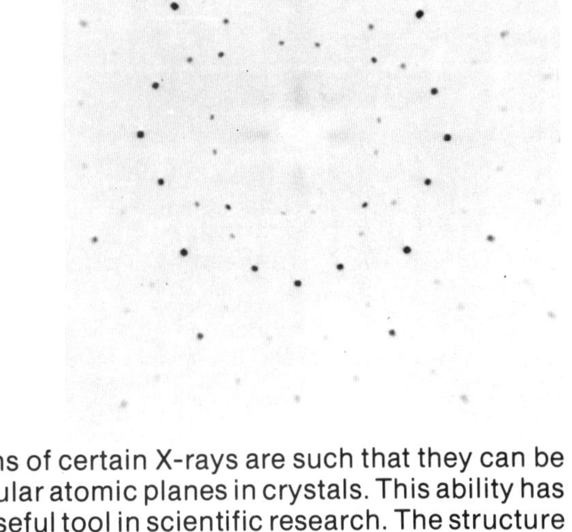

35:33 *A Laue transmission X-ray diffraction pattern of sodium chloride. Each dot represents a particular set of planes within the NaCl crystal.*

The wavelengths of certain X-rays are such that they can be diffracted by the regular atomic planes in crystals. This ability has proved to be a very useful tool in scientific research. The structure of crystals can be determined. The positions of atoms in complex molecules in crystals can be found by a proper analysis of X-ray diffraction patterns.

35:34 *A Debye-Scherer X-ray diffraction pattern of SiO_2. Each line represents a set of planes in the SiO_2 crystals.*

X-ray astronomy is a recently developed area of science. X-ray telescopes have been placed in satellites to gather information about X-ray emission from the sun and other astronomical bodies.

35:35 *Orbiting X-ray telescope. The 260-pound spacecraft, SOLARD 10, was mated to the Scout launch vehicle.*

35:6 Gamma Rays **Gamma (γ) rays** are very similar to X-rays, except that they have higher energies and thus have better penetrability of matter. The wavelength of γ rays is about 10^{-11}m downward. Gamma rays are given off by unstable nuclei. They have no charge, since they are not deflected by a magnetic field. The lack of charge in X-rays and γ rays enhances their ability to penetrate matter.

Gamma rays have many of the same uses as X-rays. In the medical field, γ radiation is used to treat cancer. In non-destructive testing, γ rays are sometimes used in place of X-rays on thicker objects; the exposure time can be lessened using γ rays rather than X-rays.

35:7 High Magnification Microscopes Extremely powerful microscopes have been developed using beams of electrons or helium ions rather than visible light. The **electron microscope** can magnify objects 400,000 times. Electrons are accelerated by a high voltage from a hot filament onto a specimen. A very high vacuum must be maintained inside the microscope to keep the electrons from being scattered by gas molecules. Magnetic coils are used for focusing the electrons rather than lenses, as in the case of visible light.

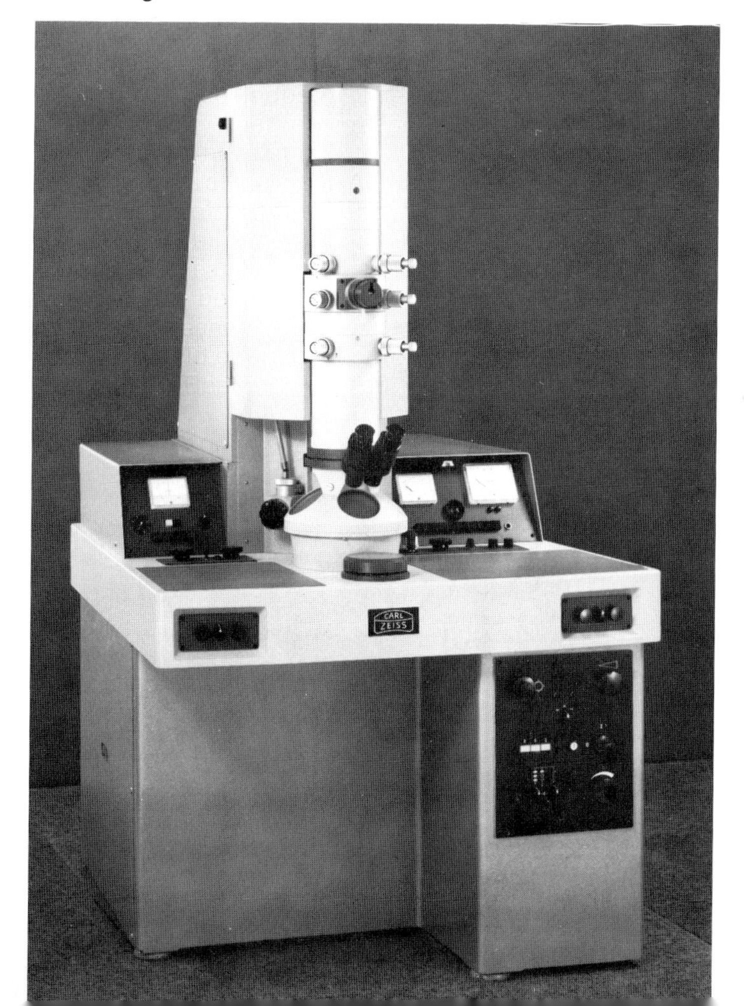

35:36 *An electron microscope.*

35:37 *Schematic diagram of an electron microscope. Electrons, accelerated from a heated tungsten filament by a high voltage at the anode, pass through a hole in the anode toward the specimen. They interact and pass through the specimen. The electrons are focused with a magnetic objective lens; and the image, further magnified with a magnetic projector lens, is observed on a fluorescent screen or recorded on a film in a camera. The gray areas indicate the path of the electron beam.*

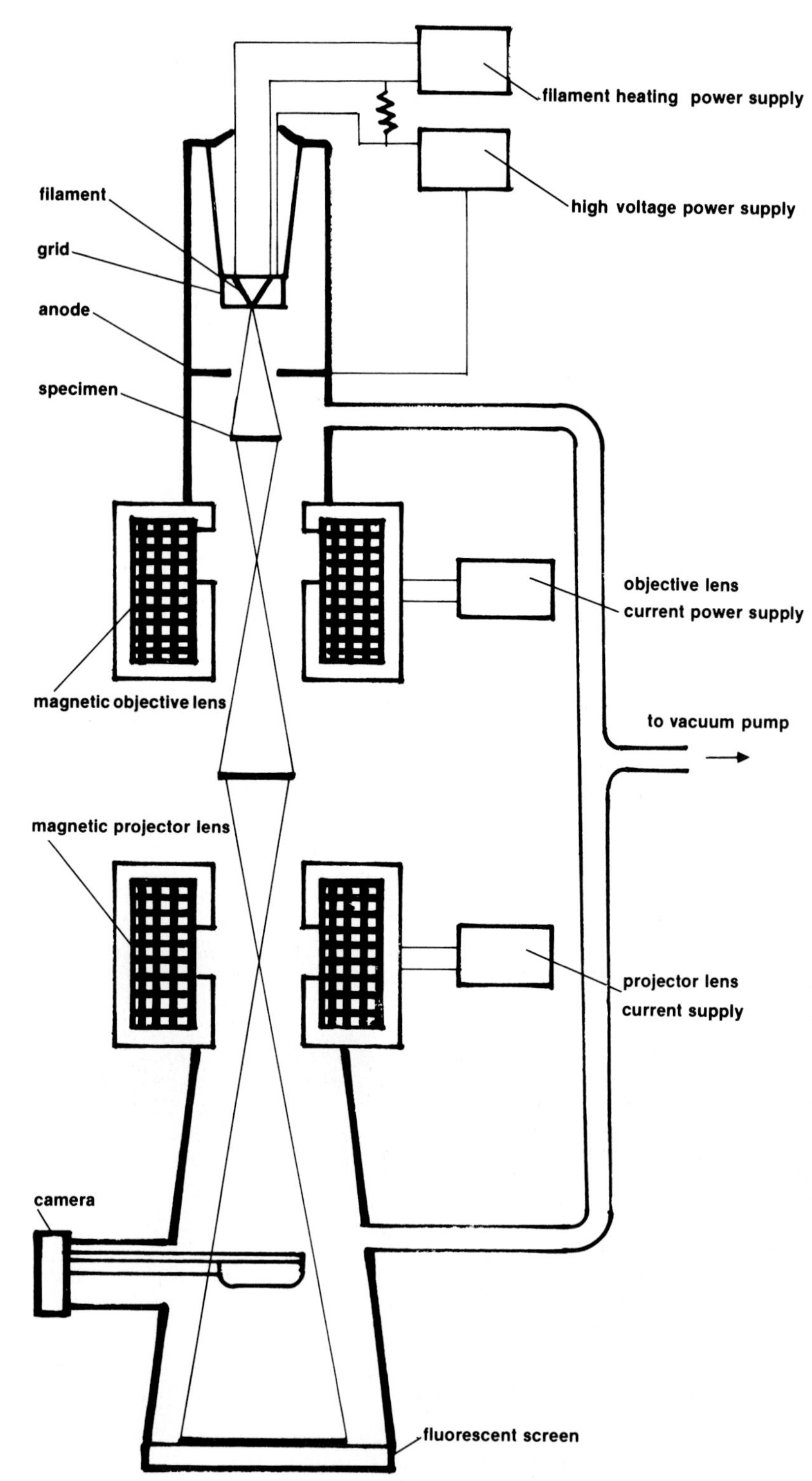

35:38 *Electrodeposited gold dendrites as seen at 10,000 magnifications in an electron microscope.*

The **field-ion microscope** employs helium ions. Helium gas is introduced into the unit. The specimen to be viewed is given a high positive charge. When the helium atom comes close to the specimen, electrons are taken from the gas, and helium ions are formed. The positively charged specimen repels the ion onto a screen. The spot on the screen corresponds to the atom doing the repulsion. This process repeated many times yields an image of the atomic structure. This technique is particularly useful in "seeing" vacant sites and dislocations in crystals. The field-ion microscope has the amazing power of 3 million magnifications.

35:39 *A field-ion microscope.*

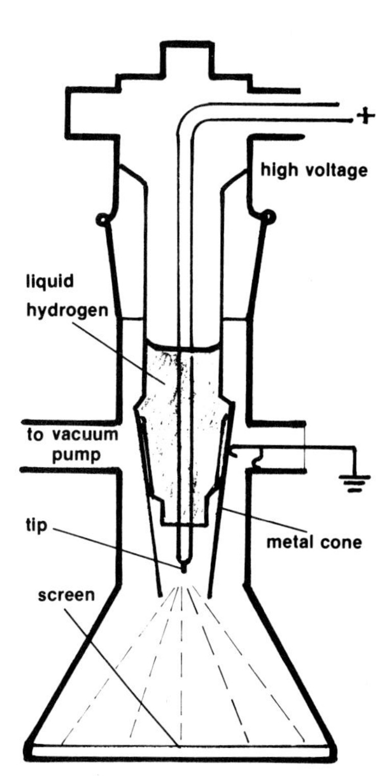

35:40 *Schematic diagram of a field-ion microscope. Polarized helium ions, attracted to the metal tip, are ionized on metal surface atoms. Upon ionization the helium ions are accelerated towards the screen. Many helium ions striking the screen form an image similar to Figure 35:41.*

35:41 *Photomicrograph of a tungsten crystal taken in a field-ion microscope. Original magnification on the screen was 1,300,000 diameters!*

35:8 Spectroscopy Sir Isaac Newton allowed light to pass through a narrow slit in a window shade; then the light was refracted through a prism to obtain a **spectrum** of colors. Almost 100 years after Newton's work, Thomas Melville (1726-1753) conducted experiments on the light emitted by hot gases. He noted that the spectra from the various gases consisted of definite lines, not a continuous, unbroken pattern of color such as is obtained from hot solids and liquids. This was the beginning of the important science of **spectroscopy**.

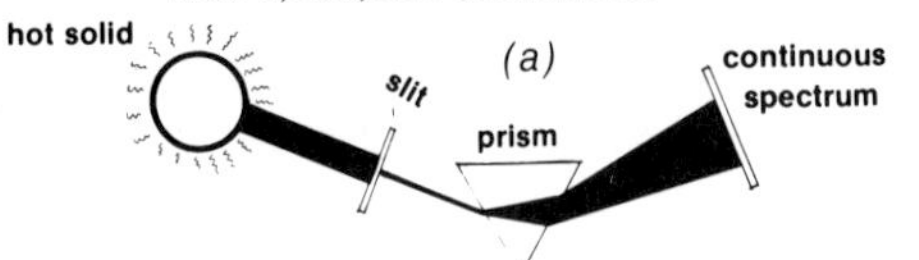

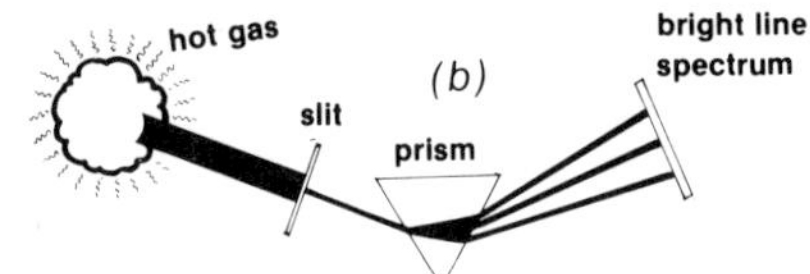

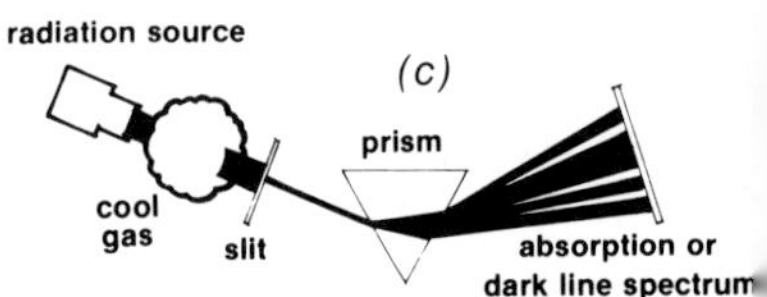

35:42 *(a) Schematic diagram of how a continuous spectrum is produced.*
(b) Schematic diagram of how a bright line spectrum is produced.
(c) Schematic diagram of how an absorption spectrum is produced.

As we said earlier, each element has a unique spectrum. Herschel suggested that each gas could be identified this way. As little as a millionth of a gram of a substance is sufficient to identify all of the elements in it by spectroscopic methods. We have already discussed the bright line spectrum of an element (see Figure 6:7). If white light is allowed to pass through a cool gas, as shown in Figure 35:42c, certain electron transitions in the gases absorb definite wavelengths of the white light, leaving dark areas (lines) on the film. This is called an absorption line spectrum.

The techniques of spectrochemical analysis are so good that a chemist can tell the percentage of a substance within certain limits by measuring the intensities of the lines on the spectrum. Spectrographic techniques have been extended even to the study of stars. The light from a star is collected in a telescope and passed through a **spectroscope**, which breaks the light up into its component wavelengths. Each star, including the sun, has a unique absorption spectrum which is determined by the elements it contains. This spectrum arises when light from the surface of the star passes through the cooler gases of the star's atmosphere, which absorb certain wavelengths. It is an interesting historical fact that the element helium was discovered from an unidentified set of absorption lines in the solar spectrum before it had ever been found on the earth (see Section 21:2).

35:43 *Absorption of electromagnetic energy by an electron in a Bohr model atom. The electron moves to a higher energy orbit.*

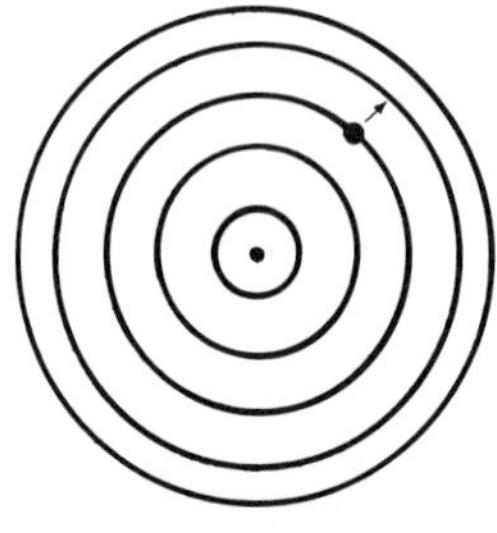

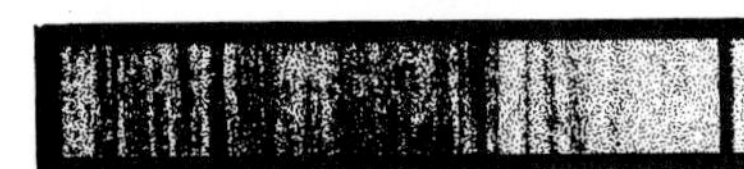

35:44 *Absorption spectrum of the sun (representation). Dark lines are the wavelengths that are absorbed.*

List of Terms

electron microscope
field-ion microscope
fluorescent
gamma rays
infrared radiation
infrared spectroscopy
microwaves
radar
radio and TV waves
radio astronomy
spectroscope
spectroscopy
spectrum
ultraviolet radiation
X-ray astronomy
X-rays

Questions

1. (a) Define "infra."
   (b) Who discovered infrared radiation?
2. (a) What may be the origin of infrared radiation?
   (b) Why will a particular type of molecule give off only certain frequencies of infrared radiation?
3. What is infrared spectroscopy?
4. List three possible uses of infrared radiation.
5. (a) The hotter any body becomes, the ________________ the *frequency* of infrared radiation it gives off.
   (b) The hotter any body becomes, the ________________ the *wavelength* of infrared radiation it gives off.
6. An automobile passes through a radar speed check. The first reflected signal reaches the receiver in 10^{-4} sec. The second reflected signal reaches the receiver in 7×10^{-5} sec. If the automobile has moved 10^{-3} ft. in this time interval, how fast is it moving?
7. How do microwaves heat food in a microwave oven?
8. (a) What two devices are necessary to send and pick up radio waves?
   (b) Who was the first scientist to discover how to transmit electromagnetic waves?
   (c) Radio waves have very short wavelengths. True or False?
9. FM transmission utilizes waves with shorter wavelengths than AM transmission. True or False?
10. How does a radio transmitting antenna work?
11. How does the receiving antenna in a radio operate?
12. Define "ultra."
13. How does the production of ultraviolet light by atoms differ from the production of visible light?
14. How does a mercury fluorescent light work?
15. (a) Name some harmful effects of ultraviolet radiation on the human body.
    (b) How does this happen?
    (c) Name some helpful uses of ultraviolet radiation.
16. How can elements be identified by spectrochemical analysis?
17. Each star has a unique absorption spectrum. True or False?
18. (a) Who discovered X-rays?
    (b) What is the wavelength range for X-rays? Are X-rays very energetic?
    (c) How are X-rays formed?
19. (a) What causes X-rays to diffract?
    (b) Why is this useful?
20. What are the two major differences in X-rays and gamma rays?
21. (a) Why must a vacuum be maintained inside an electron microscope?
    (b) What type of a microscope uses helium gas?
    (c) What magnification can be obtained with (b)?

Student Activity

Write a paper on one of the instruments or techniques discussed in this chapter covering its operation principles and related matters. Include drawings and illustrations.

Photo next page
Movable speakers in echoless rooms can be placed so that a man hears voices in the center of his head.

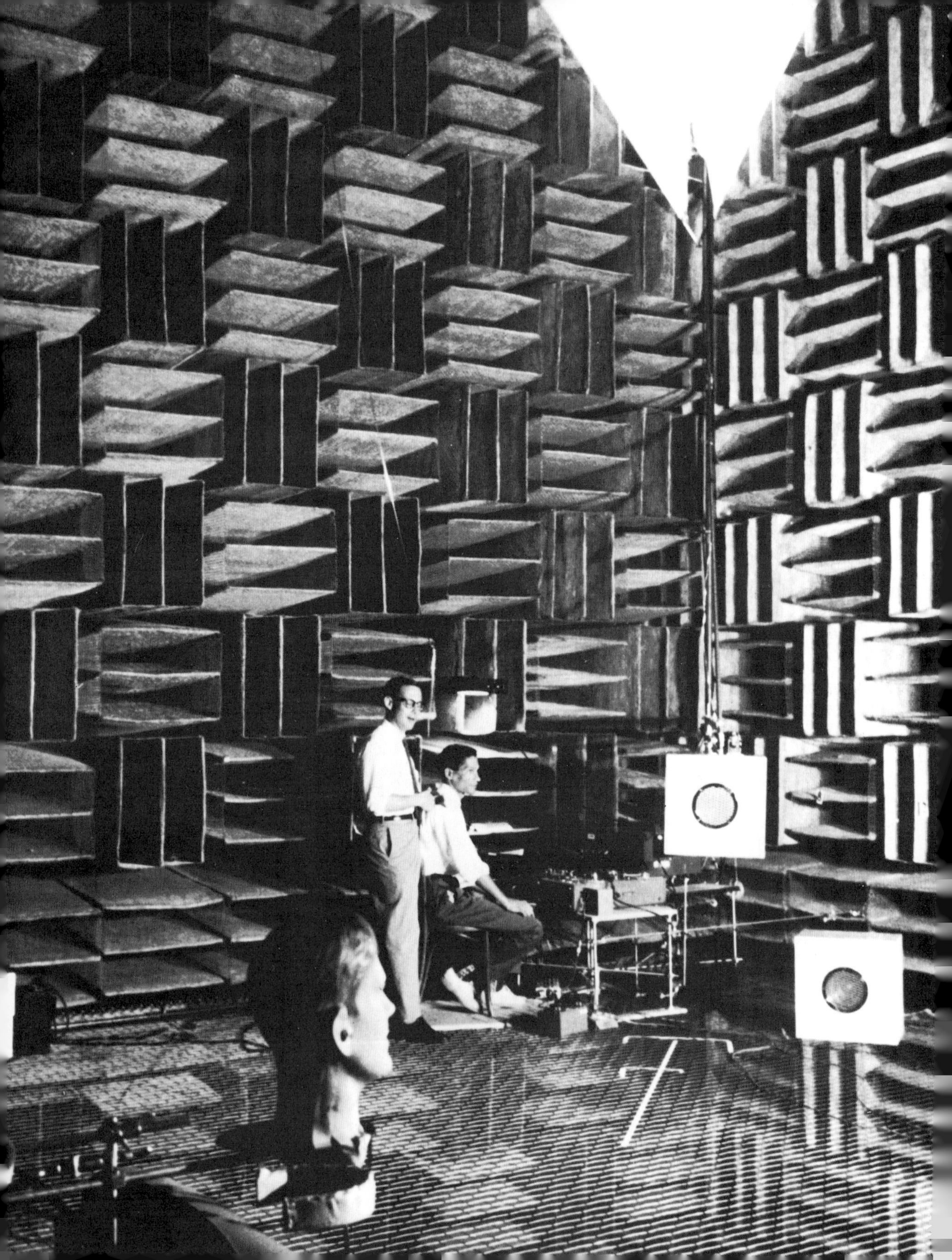

UNIT XI WAVE PHENOMENA

Chapter 36 Sound

36:1 Sound as Waves Waves in water, electromagnetic radiation, and **sound** have one thing in common—all three can be represented by a model of wave motion. Water waves and sound are different from electromagnetic radiation in that they need a medium in which to travel, but any form of electromagnetic energy can travel through a vacuum.

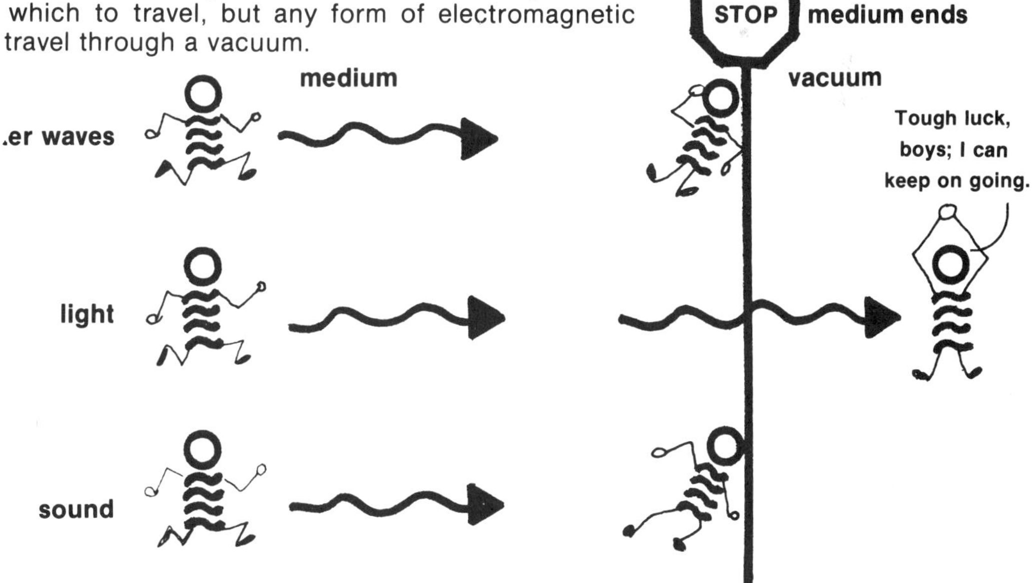

36:1 *Only electromagnetic radiation can travel through a vacuum.*

Experiment 36:1 *Sound Cannot Travel Through a Vacuum*

1. *Obtain either a doorbell and batteries or a mechanical alarm clock. Place it under a vacuum bell jar and hook to a vacuum pump as shown in Figure 36:2 or 36:3.*
2. *Pump out the air in the jar. Does the sound of the bell or alarm become louder or softer as the air is pumped out?*

bell jar
vacuum
cork or foam rubber
to vacuum pump

36:2 *Demonstration that sound does not pass through a vacuum.*

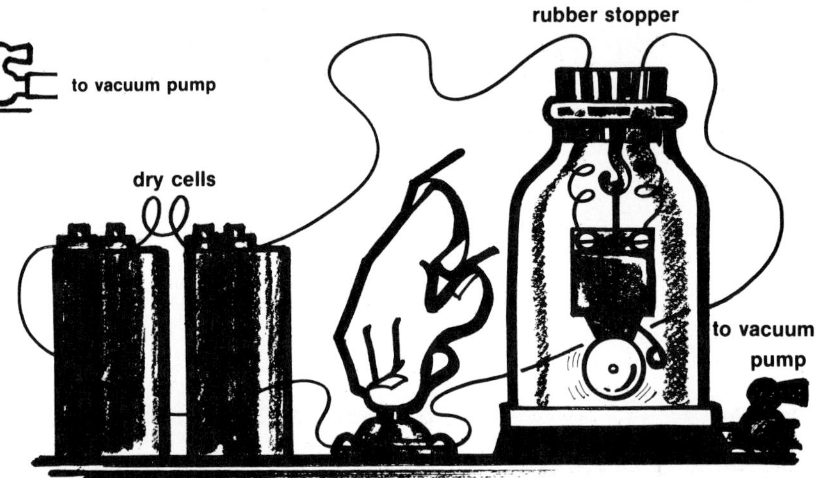

36:3 *Another way to demonstrate that sound does not travel through a vacuum.*

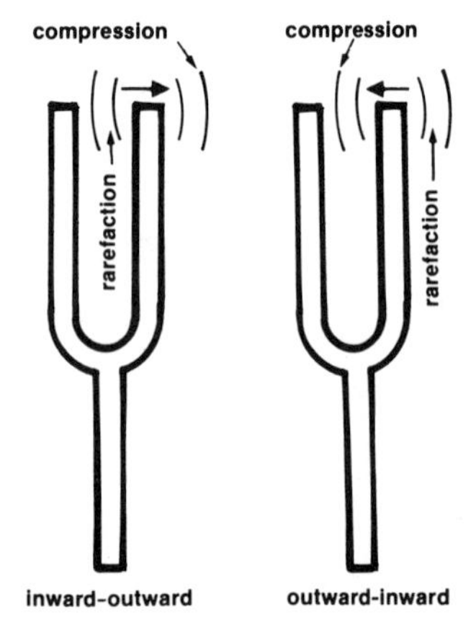

36:4 *A tuning fork compresses layers of air as it vibrates.*

36:5 *Representation of a sound wave.*

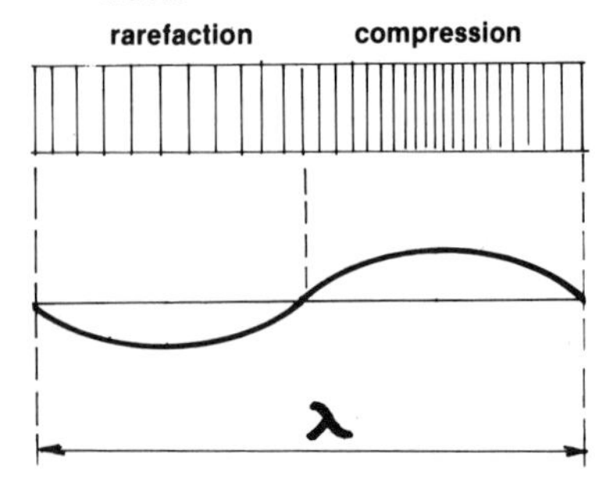

36:2 How Sound Originates Any vibration in a liquid, solid, or gas will produce a sound. Strike a tuning fork. It vibrates, and you can hear the sound. Imagine the molecules of air next to the vibrating end of the fork. As the end moves outward it "pressurizes" these adjacent "layers of air" and compresses the gas in this area. As the end of the fork moves in the opposite direction, it leaves an area of fewer molecules and lower density called a **rarefaction**. It generates another compressed area in the direction of its inward motion. When the end of the fork returns outward, an inward rarefaction follows that compression. The tuning fork continually produces these pressure waves followed by rarefactions. These pressure waves move on to new layers of air, causing the same compression, followed by an area of rarefaction. Thus, the model of the movement of sound in gases is that of alternate compression and rarefaction of the molecules of the gas.

As long as the tuning fork continues to vibrate, compressions and rarefactions will continue to move through the air. As the molecules collide with one another, some of the sound energy is converted into heat; and, eventually, after the sound waves have traveled for a good distance, the energy has completely dissipated. Any obstacle in the path of the moving waves will absorb some energy also.

36:6 *Sound waves spreading out from telephone receiver in circular condensations and rarefactions.*
36:7 *The pattern of sound waves, photographed by scientists at Bell Telephone Laboratories.*

Experiment 36:2 *Sound Waves Are Pressure Waves*

1. Support a pith ball by a thread as shown in Figure 36:8.
2. Strike a tuning fork so that it emits a sound.
3. Slide the tuning fork very slowly toward the ball. The ball will bounce away from the fork when they touch, and it will continue bouncing as long as the sound comes from the fork.

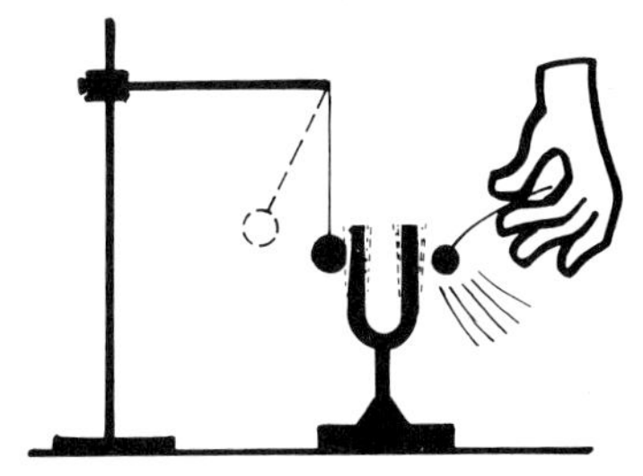

36:8 *Sound waves coming from the tuning fork will cause the pith ball to bounce.*

36:3 Media That Transmit Sound Any elastic substance can transmit sound. Solids are the best conductors of sound, liquids are next, and gases are last. Lay your head on the end of a table and have someone tap softly with a finger on the other end. The sound is very loud to you, whereas someone standing beside you may hear only a faint sound or nothing at all. Sounds travel readily through solids, as compared to gases. The next time you go swimming, duck your head under water and have someone strike two rocks or pieces of metal together. You will notice how loud the sound appears under the water. Liquids are good conductors of sound also.

A medium must be elastic and compressible for sound to go through it. It is difficult to imagine that a substance such as glass is compressible enough to transmit sound, but we must remember that the pressure caused by a sound wave is very slight indeed.

36:4 Audible Range Not all vibrations that reach the ear will be heard. The human ear responds to vibrations that have a frequency between 20 to 20,000 Hz (2×10^1 to 2×10^4). This range of hearing will vary from person to person. Some young people can hear vibrations up to 25,000 Hz. Certain people can hear vibrations at 10 Hz. This is one of the problems encountered when trying to analyze sound quantitatively. What is heard will vary, depending upon the hearer. Vibrations below the **audible range** are called **infrasonic**, whereas those above are called **ultrasonic**. Vibrations above 5×10^8 Hz are called **pretersonic**. Dogs can hear

36:9 *Sound travels through a solid (the table) much more easily than through air. The boy with his ear to the table can hear the tapping, whereas the boy standing to the side can hear nothing.*

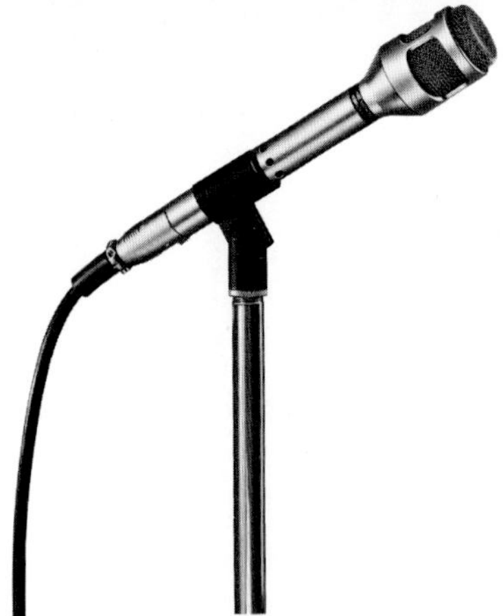

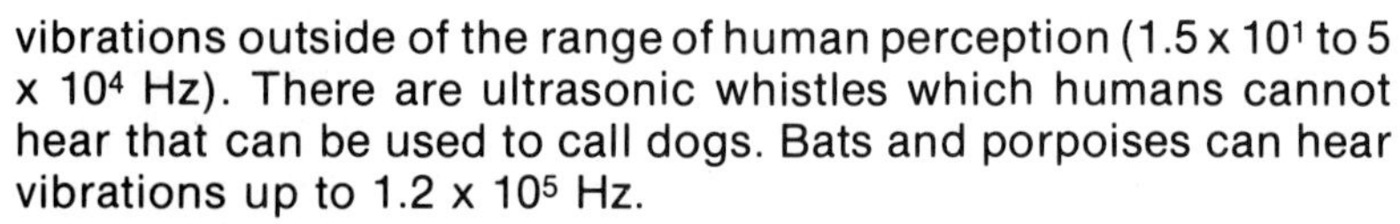

vibrations outside of the range of human perception (1.5×10^1 to 5×10^4 Hz). There are ultrasonic whistles which humans cannot hear that can be used to call dogs. Bats and porpoises can hear vibrations up to 1.2×10^5 Hz.

36:5 Speed of Sound Light moves at its maximum speed (3×10^8 m/sec) in a vacuum and slows down as it goes through denser media. Sound travels slowest in gases, faster in liquids, and fastest in solids, just about the reverse of light. The expression "keeping an ear to the ground" is an acknowledgment that sounds travel faster through the ground than through air. An early settler in the West could put his ear to the ground and hear the approach of riders on horses long before he could hear them coming by waves transmitted through the air. The vibrations caused by the hoofbeats reached the settler sooner through the ground than through the air. The approximate speed of sound in various media is shown in Table 36-1.

36:10 *Microphones are used to pick up sounds and transmit them farther. This particular microphone can be used for close talking.*

Table 36-1 *Speed of Sound in Various Media*

Substance	Temperature °C	Speed (m/sec)
glass	—	5500
iron	20	5230
brick	—	3630
water	—	1450
air	0	331
air	20	344
oxygen	0	317

36:11 *This microphone gives a wide-range frequency response to music and speech sounds.*

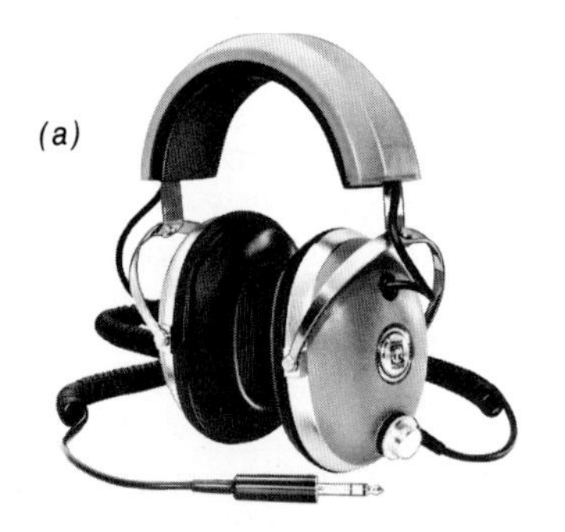

(a)

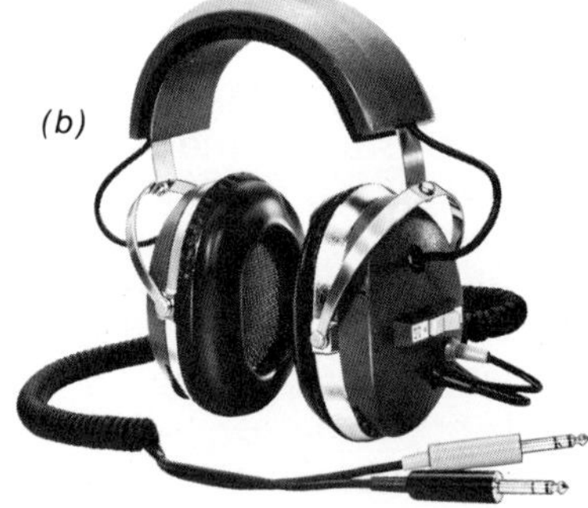

(b)

36:12 *Earphones used to emit transmitted sounds.* (a)

36:13 *Earphones capable of greater range of stereophonic response.* (b)

Notice that as the temperature of air goes up, sound travels through it faster. The hotter the air gets, the more elastic it becomes, and the faster vibrations can be transmitted through it.

Often, when discussing aircraft and rockets moving at speeds greater than the speed of sound, scientists use the term **supersonic**. Anything traveling at less than the speed of sound is moving at **subsonic** speeds. Travel at more than five times the speed of sound is called **hypersonic**.

36:6 Reflection of Sound Not only can sound be absorbed by obstacles; it can also be reflected by walls, buildings, and hills. An **echo** is an example of the reflection of sound. When sound is reflected off more than one surface, a series of echoes may be heard. If you yell into a canyon, you will hear repeated echoes.

Many years ago, sailors used the principle of the reflection of sound to determine how close they were to obstacles in a fog. They blew the ship's whistle and timed the echo; they could calculate the distance from the delay of the echo.

Experiment 36:3 *Reflection of Sound Waves*

1. Obtain 2 concave-shaped metal sheets, as shown in Figure 36:14. Mount them close to each other and place a slow-burning Bunsen burner between them close to one of the curved sheets.
2. Strike the sheet farthest from the burner. If the burner is placed at the "focal point" of the other sheet, the sound waves reflecting off this curved sheet will cause the flame to vibrate.

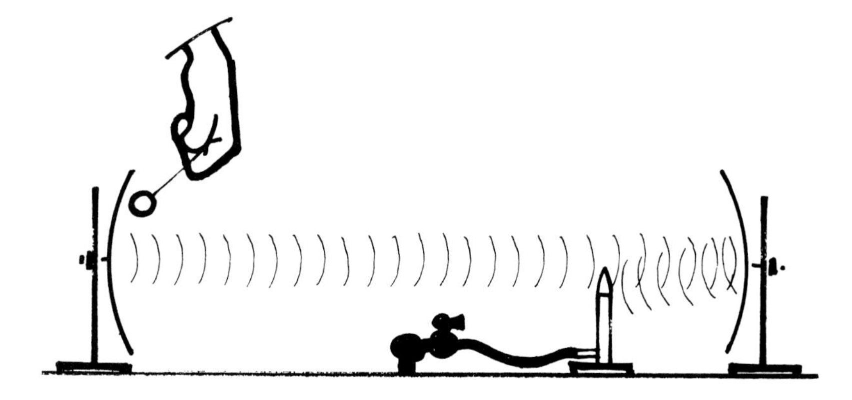

36:14 *Sound waves can be reflected.*

36:7 Loudness of Sound In a discussion of the **loudness** of sound, psychological factors enter in, making it difficult to get a satisfactory quantitative agreement between different people. What is loud to one person may not be loud to another. For instance, an unsaved person may sit by a radio listening to rock and roll music and not be bothered by its harsh loudness. A Christian who is used to Christ-honoring music could not stand the noise. So-called "sacred" rock would be too much for him also; its loudness level is about the same as the loudness level of "filthy" rock.

Basically, loudness is about the same thing as the intensity of a sound. Imagine that we hang some metal on a string and tap it lightly with a hammer. The sheet metal will vibrate slightly, and the resulting sound will not be loud. With slight vibrations, very little energy is imparted to the air in the form of pressure waves. Suppose we then hit the metal with all of our

36:15 *Two different reactions to loud sounds. A worldly person controlled by rock music starts gyrating when he hears the loud noise. The right kind of Christian cannot tolerate rock music.*

36:16 *(a) A light tap on the gong produces weak pressure waves and a soft sound. (b) A hard hit on the gong produces strong pressure waves and a loud sound.*

might. The sheet metal will vibrate violently, and the sound will be loud. A great deal more energy is imparted to the air. The pressure waves are "weak" in the first case and "strong" in the second.

Looking at this situation from a wave-model viewpoint, we can say that the intensity (I) of a sound is proportional to the **amplitude** (A) of the wave squared.

$$I \alpha A^2$$

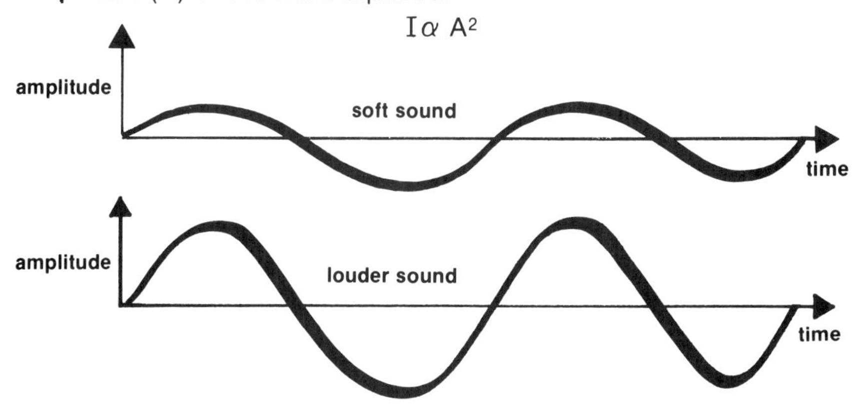

36:17 *The amplitudes of two different intensity sounds, one louder than the other.*

The greater energy a sound wave has, the larger is its amplitude. In the case mentioned earlier, the amplitude of the first wave was less than the amplitude of the wave generating the loud sound.

36:18 *A sound level meter that measures the sound intensity in decibels.*

Sound intensity is an actual physical quantity defined as *the amount of energy transported by the wave per unit time per unit area.* The energy input per unit time is **power**. Therefore, the units of intensity are watts per square centimeter. The lowest intensity in the audible range is 10^{-16}watt/cm^2, which corresponds roughly to the faintest sound which can be heard. This is used as the reference level for sound intensities. A sound that has an energy flow of 10^{-16}watt/cm^2 has an intensity level of 0.

36:19 *The loudest of all man-made sound, a space rocket at lift-off. The intensity of the sound is around 170 db.*

★ The intensity level of a sound is related by the logarithmic formula

$$\text{db} = 10 \log \frac{I}{I_0}$$

where I_0 is the reference intensity 10^{-16} watt/cm², I the intensity of the sound being investigated, and db the loudness of the sound above the reference level. When $I = 10^{-16}$ watt/cm², then

$$\text{db} = 10 \log \frac{10^{-16}}{10^{-16}} = 10 \log 1 = 0 \quad (\text{where } \log 1 = 0)$$

The intensity of noise level is expressed in **decibels** (db). Below is a chart of relative intensities versus intensity level of various familiar sounds. The maximum intensity which the human ear can tolerate is about 10^{-4}watt/cm², which corresponds to an intensity level of 120 db.

Table 36-2 *Loudness of Various Sounds*

	Intensity watts/cm²	Relative Intensity	Intensity of Noise Level (db)
threshold of hearing	10^{-16}	1	0
normal breathing or rustle of leaves	10^{-15}	10^{1}	10
average whisper	10^{-14}	10^{2}	20
library or quiet radio in home	10^{-12}	10^{4}	40
quiet automobile or quiet restaurant	10^{-11}	10^{5}	50
ordinary conversation	10^{-10}	10^{6}	60
busy traffic	10^{-9}	10^{7}	70
vacuum cleaner	10^{-8}	10^{8}	80
subway train	10^{-6}	10^{10}	100
threshold of pain	10^{-4}	10^{12}	120
small jet plane at takeoff	10^{-2}	10^{14}	140
space rocket at lift-off	10^{+1}	10^{17}	170

Problem: A sound has an intensity of 10^{-13} watt/cm². What is its noise level (db)?

Solution: $\text{db} = 10 \log (I/I_0) = 10 \log \frac{10^{-13}}{10^{-16}} = 10 \log 10^{3} = 30 \text{ db}$

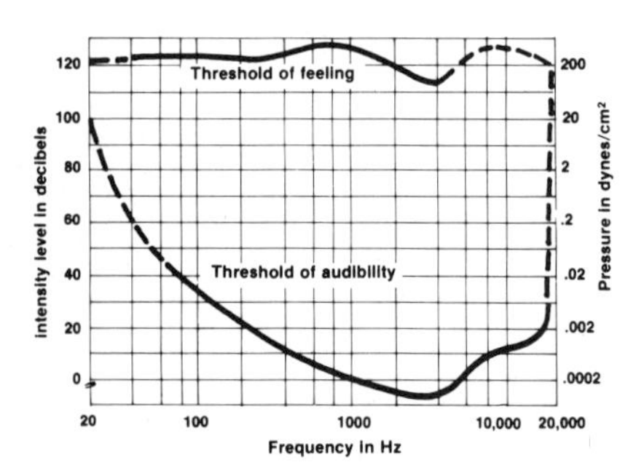

36:20 *Range of frequencies and their intensities which can be detected by the human ear.*

Our ability to hear a sound also depends upon its **frequency**. Normally, the higher the frequency of the sound, the easier it is to hear. This is best shown in Figure 36:20.

A sound wave with a frequency of 100 Hz normally cannot be heard unless it has an intensity level of about 30 db, whereas a sound with a frequency of 1000 Hz can be detected by the ear if it has an intensity level of only about 3 db.

Notice that if a sound gets too loud, it will hurt your ears; this is the threshold of pain.

36:8 Ultrasonic Vibrations A great many industrial uses have been made of ultrasonic vibrations. They are used in **nondestructive testing** to find internal flaws and cracks in metal parts. Suppose a flawless metal part is subjected to ultrasonic testing (Figure 36:21a). The reflections of the ultrasonic waves can be seen on an oscilloscope. The "blip" on the screen at M shows where the waves enter the metal part. The blip N shows where the waves are reflected from the lower surface of the part.

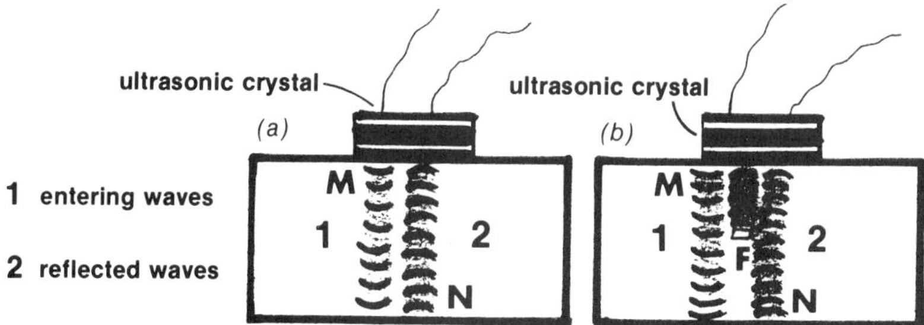

36:21 *Ultrasonic testing of metal parts without internal flaws (a) and with internal flaws (b).*

36:22 *(a) Ultrasonic vibrations cause a signal to appear on an oscilloscope screen corresponding to the upper (M) and lower (N) surfaces of the metal part. (b) Ultrasonic vibrations bounce off an internal flaw in the metal part, causing a new signal (F) to appear on the oscilloscope screen.*

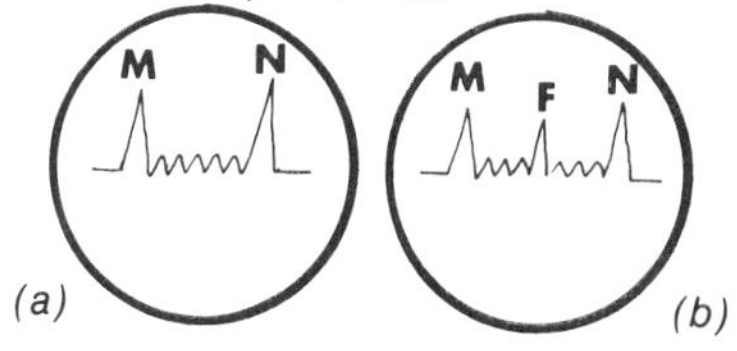

Suppose now that a metal part with a flaw is subjected to the same test (Figure 36:22b). The blips M and N represent the same surfaces as before; however, some of the ultrasonic waves were reflected off the flaw and cause a new blip (F) to appear between M and N. This blip indicates to the operator that a flaw is present in the metal. Present-day ultrasonic test equipment is often more automated than this example, but this will give you a general idea of how the method works.

36:23 *A high speed automated test installation in which material integrity is tested by the large flaw detection system at the left while material thickness is monitored by the ultrasonic digital thickness gauges mounted above. Three strip chart recorders provide a record of thickness.*

Ultrasonic vibrations are used to clean many items; that is, the item is dipped into a cleaning solution through which ultrasonic waves are generated. A great many objects can be cleaned quickly and thoroughly this way.

36:24 *A manually-operated ultrasonic digital thickness gauge. With it, readings may be made to accuracies of +0.0001 inch from one side of a workpiece.*

36:25 *Small ultrasonic cleaning units.*

36:26 *Large ultrasonic cleaning units.*

Sonar (*so*und *na*vigation and *r*anging) is an electronic method using ultrasonic vibrations for underwater detection (submarines, reefs, mines, wrecks, schools of fish), navigation, and communication. A beam is sent out through the water and reflected to the receiver. The time of return of the reflected beam is measured. This information can be converted into the distance to the object, if we know the speed at which the wave travels.

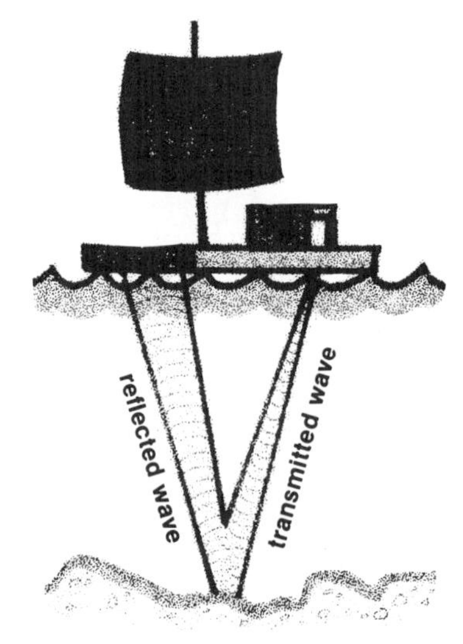

36:27 *Sonar units can be used to scan the ocean bottom.*

Ultrasonic vibrations can be used to do many other things, such as mix substances, emulsify oils, homogenize milk, and photograph bones. They could eventually replace X-rays as a means of revealing internal organs. An ultrasonic camera to do this job has been developed by the Stanford Research Center in California. It can "expose" more detail of internal structure than present X-ray cameras. An important advantage is that there is no danger of radiation in using ultrasonics as there is with X-rays.

36:28 *An underwater sonar pinger.*

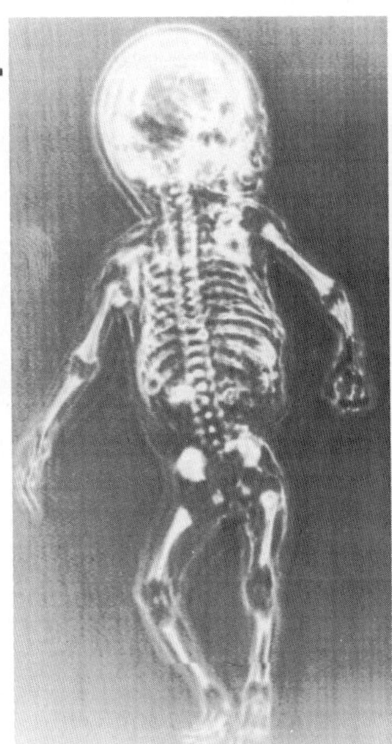

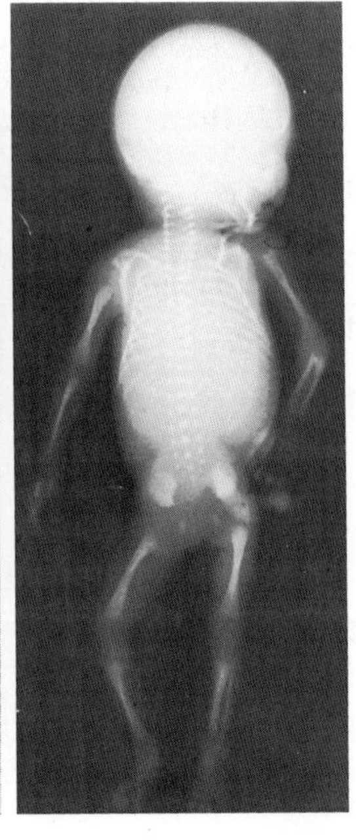

36:29 *Two pictures of a 17-week-old human fetus. The one of the right was obtained by conventional X-ray method, but the picture at the left was taken by an ultrasonic camera. Ultrasonic imaging, as it is known, can observe internal organs without the danger that accompanies X-rays.*

List of Terms

amplitude	power
audible range	pretersonic
decibel	rarefaction
echo	sonar
frequency	sound
hypersonic	sound intensity
infrasonic	subsonic
loudness	supersonic
nondestructive testing	ultrasonic

Questions

1. How do sound waves differ from electromagnetic waves?
2. How do sound waves originate and travel in air? Illustrate.
3. (a) What kind of media will transmit sound?
   (b) Arrange solids, liquids, and gases in order of sound-carrying capacity.
4. Humans have the same audible range as dogs. True or False?
5. If you were in the Old West, and you wanted to know if horses were galloping toward you, what would you do? Why?
6. Why does sound travel faster through a medium as the temperature rises?
7. Define supersonic and subsonic.
8. ★ Why cannot sound travel infinitely far in a medium?
9. What is an echo?
10. Relate loudness, intensity, and amplitude of sound.
11. Define sound intensity. In what units is it expressed?
12. ★ (a) A sound has an intensity of 10^{-3}watt/cm^2. What is its noise level in decibels?
    (b) A sound has an intensity of 10^{-12}watt/cm^2. What is its noise level in decibels?
13. Would you like to be standing close to an object making sound at a noise level of 170 db? Why?
14. (a) Can the human ear detect a 30 Hz sound of 20 db?
    (b) Can the human ear detect a 500 Hz sound of 20 db?
15. How can ultrasonic vibrations be used to detect internal flaws in metals?
16. Why can sound waves do work?

Student Activities

1. Write a paper on the use of ultrasonic vibrations in science and industry. Include the latest developments.
2. Write a paper on sound transmission by microphones, telephones, and the like. Discuss the principles of energy transfer in the particular instrument.

Photo next page
Console of a two-manual electronic organ.

UNIT XI WAVE PHENOMENA

Chapter 37 Musical Instruments and Acoustics

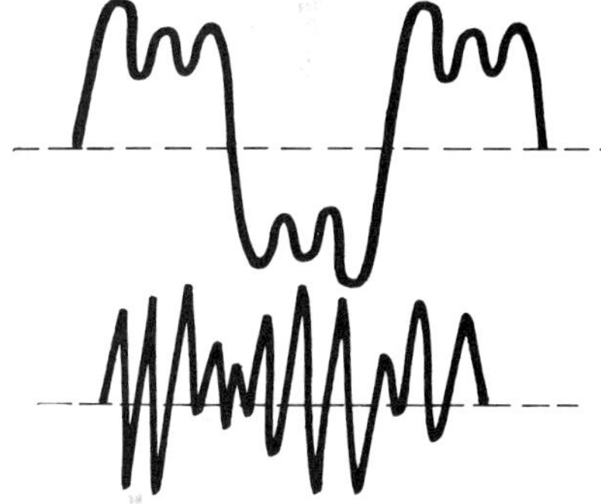

37:1 *A musical tone (upper drawing) has a regular wave pattern. Noise, shown in the lower drawing, has an irregular wave pattern.*

37:1 Pitch and Frequency We have already discussed the meaning of frequency. The frequency of a wave can be measured and is an objective quantity. The frequency of a sound determines its **pitch**. Generally, as the frequency of a sound increases, the pitch increases also. This is not a one-to-one relationship, however. Pitch is subjective and depends on the hearer. Pitch refers to how high or low a wave sounds to our ears.

37:2 Quality of a Musical Sound What makes music different from noise? Good music is pleasant to the ear, but noise is irritating and can produce fatigue. The difference lies in the wave pattern. Music has a *regular* wave pattern; noise has an *irregular* wave pattern (see Figure 37:1). Music differs from noise in the same way that order differs from disorder.

A tuning fork properly struck will give off a pure or simple **tone** (see Figure 37:2). A tone from a musical instrument, on the other hand, has a complex wave form. If a piano and a clarinet play the same note at the same loudness, you can distinguish between the two. Their complexity enables us to tell them apart. The wave forms of a piano note and a corresponding clarinet note are shown in Figure 37:3. The wavelength and amplitude are the same in each case, yet the shapes of the waves differ. The shape of the wave determines the **quality** of the tone produced.

How do the waves from musical instruments acquire the unusual shapes shown in Figure 37:3? Normally it is difficult, if not impossible, to obtain a pure tone from a musical instrument.

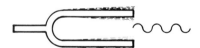

37:2 *A tuning fork properly struck emits a pure tone.*

37:3 *Waveforms of a note generated by (a) a piano, and (b) a clarinet. The frequency and amplitude are the same; only the quality differs.*

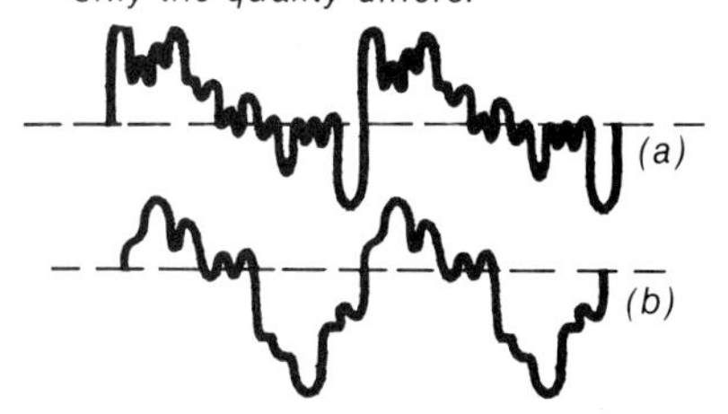

37:4 *Experimental arrangement capable of generating either musical sounds or noise.*

Experiment 37:1 *Musical Sounds and Noise*

1. Set up an apparatus as shown in Figure 37:4 (one set of holes equally spaced, the other randomly spaced). Use a pump having a compressed air output and a hose to fit.
2. With the disc spinning, direct the stream of compressed air to various points on the disc. Note that air blowing through equally spaced holes produces musical sounds, whereas air blowing through unequally spaced holes produces noise. From your observations, what can you say is the difference between music and noise?

When a person plays a note on an instrument, several wave forms combine to give a tone characteristic of that instrument. Let's use a violin string as an example. Suppose we pluck or stroke the string with a bow. The lowest frequency wave we can obtain is shown in Figure 37:5. This is called the **fundamental**.

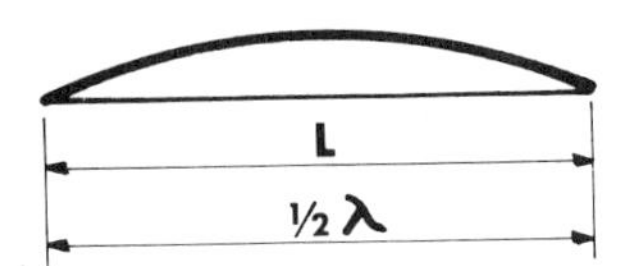

37:5 *The fundamental mode of vibration of a string. The length of the string is L. The wavelength emitted, λ, is equal to twice the length of the string. ($\lambda = 2L$)*

Yet other higher frequencies are possible, and many of these may develop on the string as well as the fundamental. These higher frequency vibrations are called **overtones**. Some of these are shown in Figure 37:6. Notice that there must always be a **node** at each end of the string, because the string is attached or touching at these points. A vibration of $\lambda = 1\frac{1}{4}$ L is impossible on this or any other string because there could only be a node at one end (see Figure 37:7). If the length of the string were changed, the frequencies of the fundamental and overtones would adjust themselves in such a way that there would still be a node at each end.

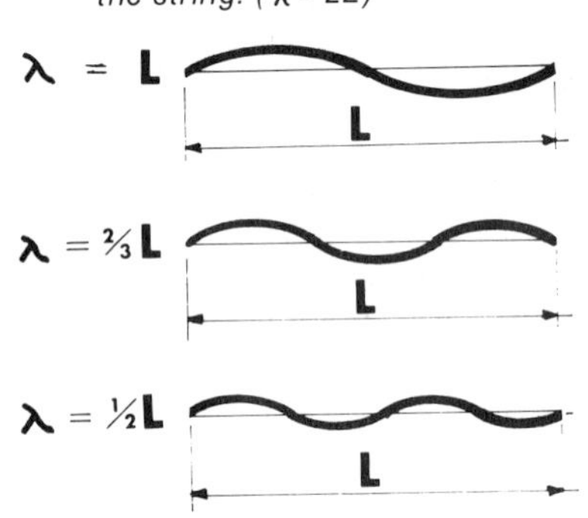

37:6 *The first overtone (second harmonic), second overtone (third harmonic), and third overtone (fourth harmonic) of a vibrating string. The string is said to be vibrating in segments.*

A musical tone, then, is a combination of the fundamental and several overtones to give the complex shape of the wave. The fundamental determines the pitch of the sound, and the number and intensities of the overtones present determine the quality of the tone. Each different instrument gives a characteristic sound because of its characteristic mixture of overtones.

Experiment 37:2 *Generating Overtones with a String*

1. Use the setup with the movable block that can be inserted, as shown in Figure 37:8, and a violin bow (if available). Bow or pluck the string with the movable block removed to generate the fundamental.
2. Place the block at the halfway point on the string, and bow or pluck either half. Observe that the frequency has doubled. This is the first overtone (also called the second harmonic).
3. Place the block at 1/3 the length of the string. Again bow or pluck either part. The frequency should now be three times the fundamental—the second overtone or third harmonic.
4. Repeat at 1/4 the length of the string, and keep repeating to see how many overtones you can detect.
5. Repeat the experiment using a larger weight on the string.

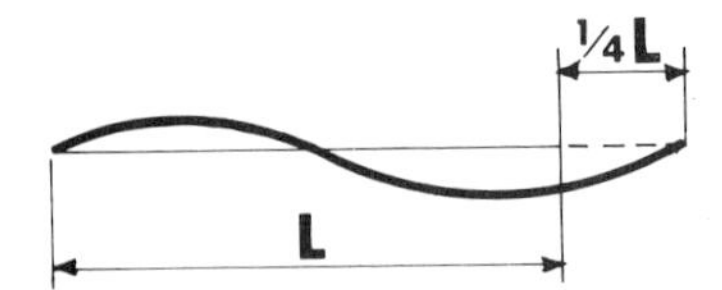

37:7 *An impossible situation. A string must have a node at both ends.*

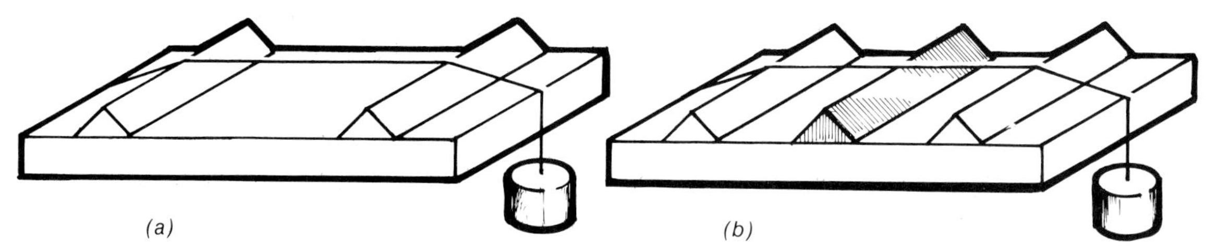

37:8 *(a,b) Setup for Experiment 37:2.*

Musical instruments are divided into four major categories: strings, brass, woodwinds, and percussion. We shall discuss them in this order.

37:3 Stringed Instruments There are probably more different stringed instruments in use in various parts of the world than could be covered in a large book on the subject. We will confine our attention to the bowed instruments used in a symphony orchestra. This family includes the *violin*, the *viola*, the *cello*, and the *string bass* or *double bass*. All of these use four strings stretched tightly between two wooden supports called the **nut** and the **bridge**. The bridge conducts vibrations to the front of the instrument, which forms one side of a hollow box. This box is an acoustical resonant chamber which greatly amplifies the sound. The lower the pitch, the larger the sounding box must be. The double bass has more than 100 times as much air space in its sounding box as the violin.

37:9 *Rare Italian violin, built by Nicolaus Amati in 1653.*

37:10 *Stringed instruments used in a symphony orchestra. (a) violin (b) viola (c) cello (d) double bass*

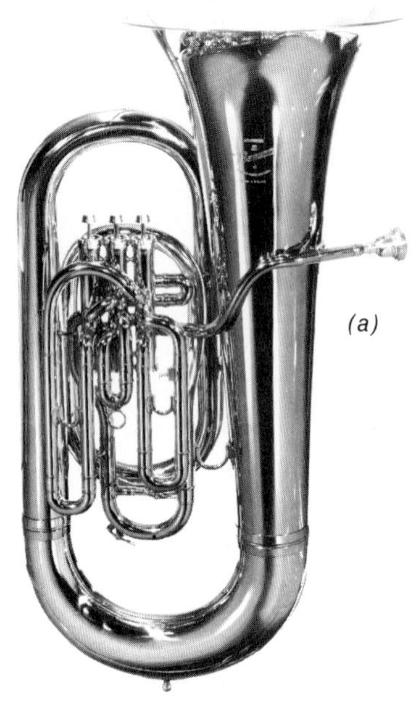

The pitch of a vibrating string depends upon three factors—its length, its tension, and its mass per unit length. The length is adjusted by the player as he fingers the strings with his left hand. The tension is regulated by the tuning pegs in the scroll (head) of the instrument. Differences in mass per unit length are designed into the strings themselves by the manufacturer. The lower strings on any given instrument are noticeably thicker and heavier than the upper strings.

The loudness of the tone produced depends primarily upon the amount of energy exerted by the player and his ability to utilize his energy efficiently. Quality is dependent upon the skill of the player, the point on the string where it is bowed or plucked, and the details of the construction of the instrument and bow.

37:4 Brass Instruments These instruments are named for the metal from which they are usually constructed. The brass family includes the *trumpet*, the *trombone*, the *baritone*, the *tuba*, the *French horn*, and several variations of these basic forms (*cornet, Flügelhorn, euphonium, sousaphone,* and others). All brass instruments utilize a vibrating air column whose length can be varied. Air columns have one thing in common with strings: the greater the length, the lower the pitch. In the trombone it is easy to see how the length is varied. When the slide is extended, the air column is lengthened, and the pitch is lowered. The other brass instruments have either valves or keys by which the player can switch in different lengths of air column. Usually there are three different lengths that can be added—singly, in pairs, or all together—to augment the length of the basic column.

37:11 *Brass instruments. (a) tuba (b) trumpet (c) trombone (d) sousaphone*

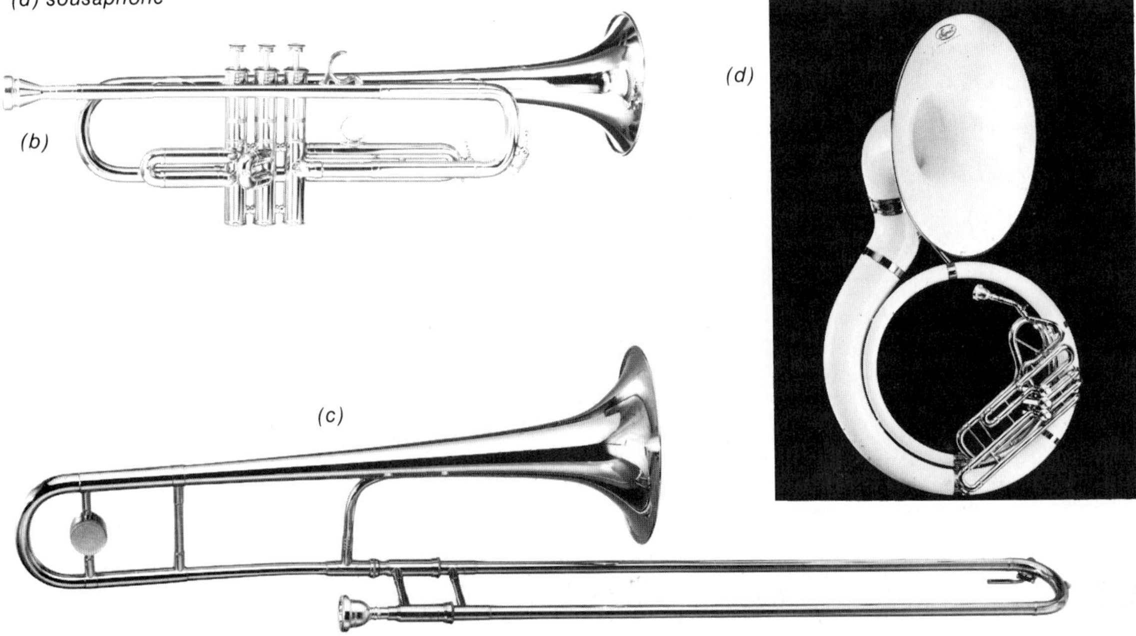

The air in the column is set into motion by the vibration of the player's lips. A cup-shaped device called the **mouthpiece** affords his lips the necessary flexibility to vibrate properly. By controlling the tension in his lips, he can regulate the mode of vibration in the column. Thus, he is able to select any of several pitches for each length of the air column. This can be understood by recalling that the different pitches produced by a *bugle* are all accomplished by the player's lips; there is no means of changing the length of the air column.

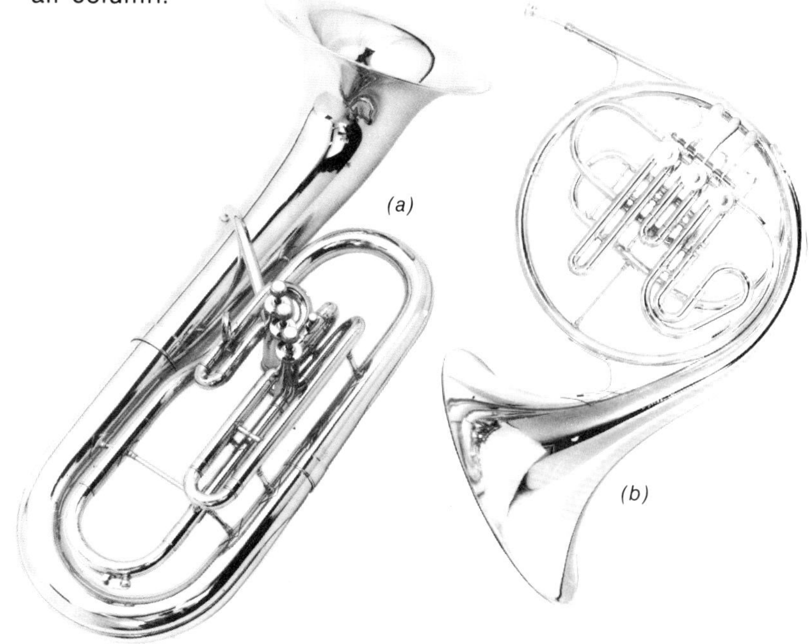

37:12 *More brass instruments. (a) baritone (b) French horn*

Brass instruments have many strong overtones, which combine to make them the loudest instruments in a band or orchestra. Their ease of audibility has made them popular among military leaders for signaling instructions to their battle troops.

37:5 Woodwind Instruments The woodwinds are wind instruments in which the pitch is varied by means of holes in the side of the air column. In this group are the *clarinet*, the *bass clarinet*, the *saxophone*, the *oboe*, the *English horn*, the *bassoon*, the *contrabassoon*, the *flute*, the *piccolo*, and many others which are not generally used in a regular band or orchestra, such as the *recorder*, the *ocarina*, and the *bagpipe*. The holes in the side of the air column are opened and closed either directly by the player's fingers or with flaps that are controlled by keys.

37:13 *Single-reed woodwind instruments. (a) saxophone (b) clarinet*

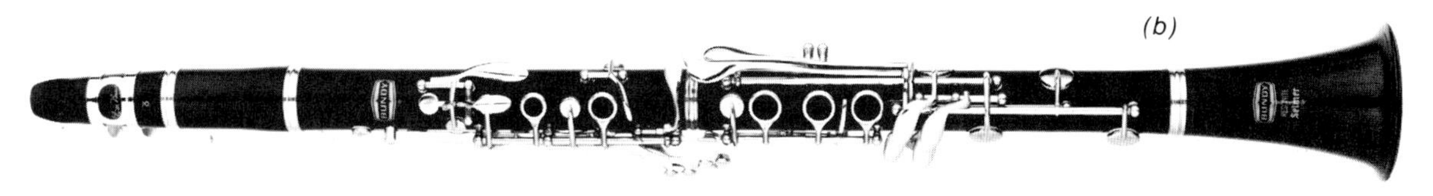

37:14 *Double-reed woodwind instruments. (a) oboe (b) English horn (c) bassoon*

37:15 *Woodwind instruments with no reed. (a) flute (b) piccolo*

37:16 *The Scottish bagpipe is a woodwind instrument. The player blows into an air column called the* chanter *using a double reed. The other three air columns called* drones *use single reeds. The bag furnishes a continuous supply of air.*

In the clarinet, the bass clarinet, and saxophone, the air column is excited by a single vibrating reed. A double reed is used in the oboe, English horn, bassoon, and contrabassoon. The Scottish bagpipe uses one double reed and three single reeds. The flute and piccolo are distinctive among orchestral instruments in that no reed is used; the player simply blows across one end of the air column.

37:6 Percussion Instruments These are instruments in which the tone is generated when some part is struck. Included in this group are the *drums, cymbals, tambourine, triangle, bells, chimes, xylophone, castanets, wood blocks,* and many others. The *piano* is usually also classified with the percussion instruments because its strings are struck with hammers.

37:17 *The piano is classified as a percussion instrument because its strings are struck with hammers.*

As with the other types of instruments, physical size is important in determining the pitch produced in a percussion instrument. The small snare drum has a much higher pitch than the large bass drum. The bars of a xylophone are graduated from the longest at the low end of the scale to the shortest at the high end.

37:7 The Harp and Organ These instruments are first mentioned in the fourth chapter of Genesis: "And his brother's name was Jubal: he was the father of all such as handle the harp and organ" (Genesis 4:21). Already in the eighth generation of man, music was highly developed. The harp was an instrument having several strings, each tuned to a different fixed pitch. Other ancient instruments related to the harp were the *lyre*, the *sackbut*, and the *psaltery*. The modern harp is a much larger instrument of 46

37:18 *Modern pipe organ.*

37:19 *Console of a three-manual pipe organ.*

strings played by plucking with the fingers of both hands.

The organ referred to in Scripture was a graduated series of pipes (similar to the well-known "pan-pipes") played by blowing over the tops of the air columns. The modern pipe organ, needless to say, is far more complex. There are many separate sets of graduated pipes activated by remote control (see Figures 37:18 and 37:19). The instrument is played by means of two or more keyboards similar to that of a piano, as well as a row of foot pedals. When a key or a pedal is depressed, an electrical circuit is completed, which admits air to the base of the corresponding pipe, causing it to vibrate and give off a sound. Any number of pipes can be activated simultaneously. Different sets of pipes having different qualities may be selected by the player using switches called **stops**. In a large organ the number of combinations of effects obtainable is almost unlimited. In recent years it has become possible to generate the sounds of an organ electronically. The console of an electronic organ is shown in Figure 37:20.

37:20 *Console of a two-manual electronic organ.*
(See chapter title page)

37:8 The Human Voice The human voice is used also as a musical instrument. It has a versatility that no man-made instrument can duplicate. Scientists have tried to design a machine that can "speak," but it requires many large computers just to generate the simplest "baby talk." The human voice can, of course, produce not only speech but words and music simultaneously. Scientists at the Bell Telephone Laboratories have studied the mechanism of the human voice in great detail. We cannot read the results of their research without being impressed that here again, as with so many other structures in the body, is clear evidence of design by a Higher Intelligence.

Again, physical size makes a difference. A woman's voice is pitched higher than a man's voice because of her shorter and less massive vocal cords. There are three general ranges recognized for women's voices. From highest to lowest these are **soprano**, **mezzo-soprano**, and **alto** (or **contralto**). Men's voices, from highest to lowest, are classified as **tenor**, **baritone**, and **bass**.

37:9 Resonance If two strings of a violin are tuned to the same pitch, bowing one of the strings will cause the other to vibrate. This will not happen when they are tuned to different pitches. (There is at least one exception to this statement—a resonance effect will be observed with strings that are tuned one octave apart.) The phenomenon illustrated here is called **resonance**. In general, resonance is *the condition existing when vibrators of the same natural frequency transfer energy one to the other.*

Resonance may be demonstrated very convincingly by the setup shown in Figure 37:21. Two identical tuning forks mounted on identical sounding boxes are placed several inches apart with the open ends of the boxes facing each other. When one of the tuning forks is struck with a rubber mallet and then suddenly damped by grasping it with the hand, the other tuning fork will still be heard. Energy has been transferred from one tuning fork to the other, the first functioning as the "transmitter," and the second as the "receiver." The process works equally well in the opposite direction.

37:21 *Resonance. Energy can be exchanged between tuning forks that are tuned to the same frequency.*

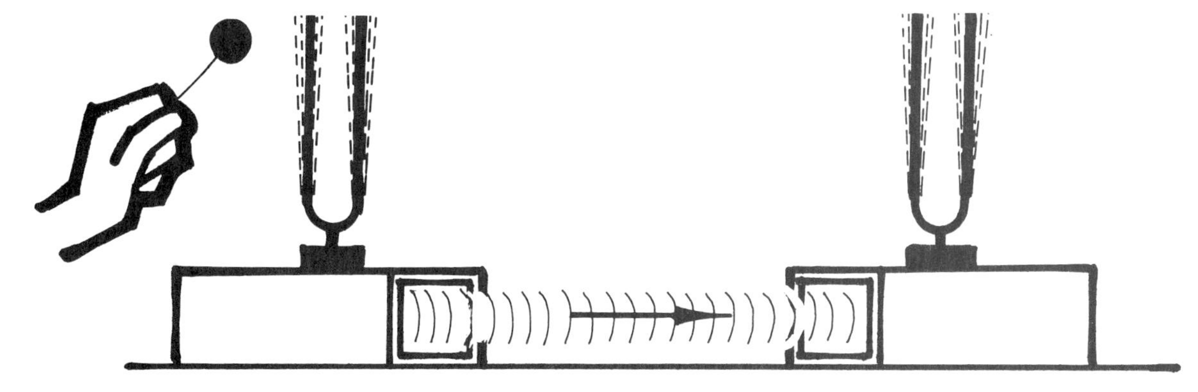

There are a number of important applications of the resonance principle. The sinuses (spaces) in the bones of the human skull act as resonant chambers to amplify the voice. The auditory canal of the ear is a broadly resonant air column which reinforces sounds in the mid-range of the audible spectrum. As we have seen, stringed instruments have sounding boxes that function as resonant chambers. Resonance also plays a role in electronics. A radio that is tuned to a certain station is in resonance with the transmitter of that station, and is therefore capable of receiving energy from it. All other stations are rejected because they differ in frequency from the dial setting of the radio.

Resonance effects have been known to cause severe damage. A number of years ago the Tacoma Bridge in Washington state was destroyed when a strong wind caused it to vibrate like a fiddlestring. The internal stresses were so great that the huge suspension bridge literally "fell apart." It is common practice for military leaders to order their troops to break step when crossing a bridge to prevent any possibility of damage to the bridge due to resonance.

Experiment 37:3 *Wavelength by the Resonance Method*

1. Take 2 paper cylinders (one to slide back and forth through the other—see Figure 37:22). Strike a 256-Hz tuning fork (middle C) near one end of the air column.
2. Move the outer cylinder back and forth slowly until the loudest sound that can be heard is obtained. Measure the total length of the air column. As this is a half-wave resonator (open at both ends), the wavelength for C-256 will be twice your measurement.
3. You can now calculate the speed at which the sound wave is traveling through the air using the relationship *Speed* = $f\lambda$. Use 256 for f and twice your measurement for λ. If your measurement is in meters, you will get the speed in meters per second; if it is in feet, the speed will come out in ft/sec.

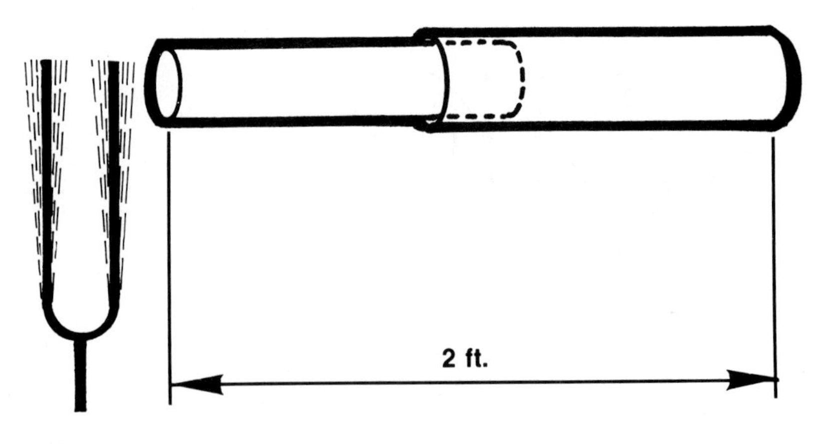

37:22 *Determining wavelength by the resonance method.*

37:10 Beats When two sources of sound having slightly different frequencies send out waves of about the same amplitude into the same place, they interfere with each other in such a manner that a regular variation in the intensity of the sound results. Such variations are called **beats**.

As shown in Figure 37:23, the two waves constructively interfere at certain times, producing a high amplitude and a loud sound. At other times the waves destructively interfere so that the amplitude is about zero. Thus, the sound "builds up" and "dies down" periodically. Each separate cycle from complete destructive interference through the maximum and back to complete destructive interference again is referred to as a beat.

Beats are sometimes used in tuning musical instruments. As two sounds are brought closer and closer together in frequency, the time between beats increases until the beats disappear. The instruments are then perfectly tuned.

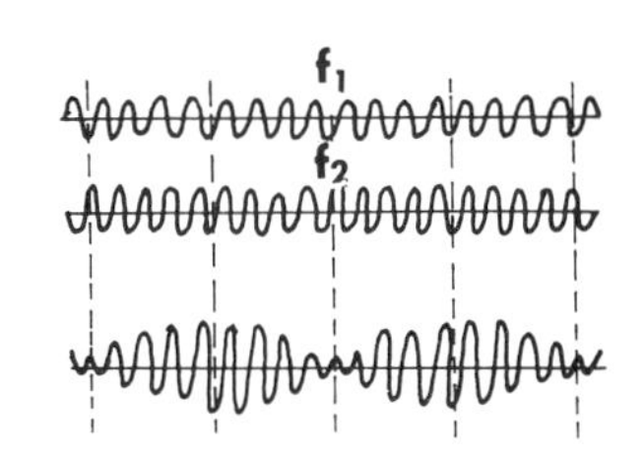

37:23 *Beats are generated when two waves interfere. The upper two waves have almost equal frequencies (f_1 and f_2). When these interfere they combine to give the waveform shown below.*

37:11 The Doppler Effect The pitch of a sound reaching your ear may vary, depending on whether the source of the sound is moving or stationary. If you are standing beside a railroad track and a train approaches you with its whistle blowing, you hear a certain pitch. As the train passes you, the pitch of the sound decreases. This is called the **Doppler effect** in honor of the Austrian physicist, Christian Doppler (1803-1853), who studied the phenomenon and offered a correct explanation for it in 1842.

The Doppler effect occurs because of the relative motion between the source and the observer. The train whistle has a frequency of f_1. If you and the train were both motionless and the whistle were blown, you would hear a pitch corresponding to f_1. As the train moves toward you, additional waves reach your ear each second because of the train's motion; and you hear a sound with an increased frequency f_2 ($f_2 > f_1$). As the train moves away from you, fewer waves reach your ear per second; and the sound you hear has a decreased frequency f_3 ($f_3 < f_1$). These two cases are illustrated in Figure 37:24.

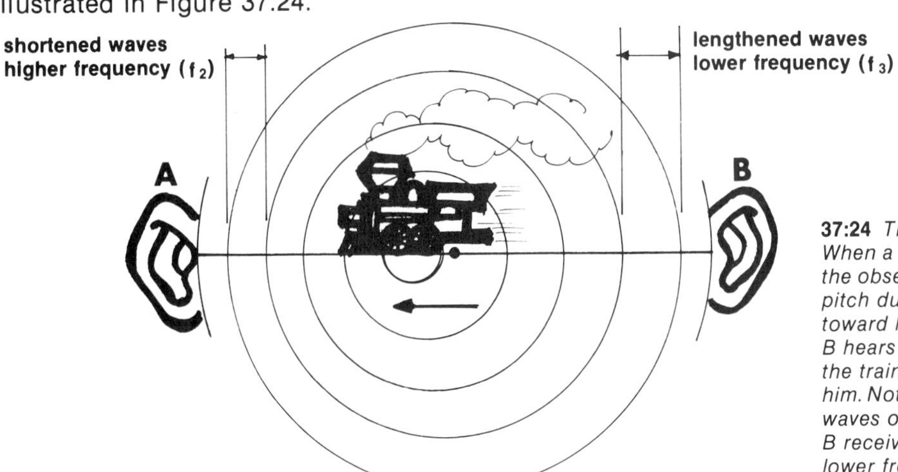

37:24 *The Doppler effect. When a train sounds its whistle, the observer at A hears a raised pitch due to the train's motion toward him. The observer at B hears a lowered pitch due to the train's motion away from him. Note that A receives shorter waves of higher frequency; B receives longer waves of lower frequency.*

37:25 *Slanted surfaces called splays are sometimes used at the sides of a stage to reflect sound outward to the audience.*

37:12 Acoustics When we speak of the **acoustics** of a building, we mean the "hearing qualities" of the building. Auditoriums must be designed properly to have good acoustics. Sound will be reflected in any type of enclosure. A person listening to a speaker or an orchestra in an improperly designed auditorium may be able to hear well in one seat but only very poorly in another. The sound may be too loud in the front seat of the auditorium and muffled in the rear. **Reverberations** (multiple echoes) may predominate somewhere else.

Years ago, before public address systems were available, curved reflecting surfaces were placed behind a public speaker so that the sound of his voice would carry into the audience. The design of the walls of an auditorium and stage can put the principle of sound reflection to good use. Notice in Figure 37:25 how the sides of the stage are equipped with slanted surfaces (called splays) to reflect the sound outward to the audience. Also, convex (curved outward) wall surfaces in the seating area reflect the sound to various points in order to achieve a more even distribution of sound (see Figure 37:26).

The materials of construction in an auditorium must be carefully chosen. If the sound is reflected back too strongly, reverberations can occur. This can be prevented by covering the walls and ceilings with a material that absorbs some of the sound energy from each wave which strikes it. A large number of small holes or openings in almost any material will serve to absorb sound effectively. Acoustical tile in particular has many such holes purposely designed into it to "trap" some of the sound energy.

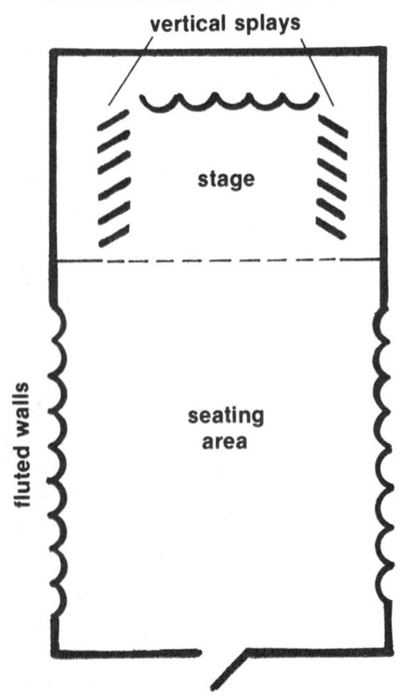

37:26 *A well-designed auditorium using both splays and fluted walls to achieve an even distribution of sound.*

37:27 *Various types of acoustical tile used to absorb sound energy. Note the rough porous surfaces.*

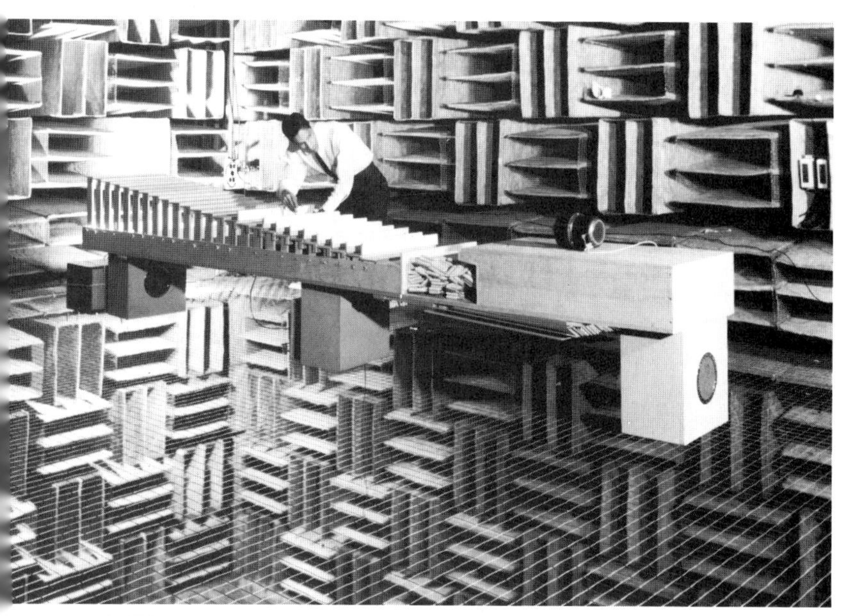

37:28 *An office where acoustical tile is used to advantage.*

37:29 *The ultimate in sound absorption. This specially designed anechoic chamber ("dead room") used for audio experiments eliminates 99.9 percent of reflected sound.*

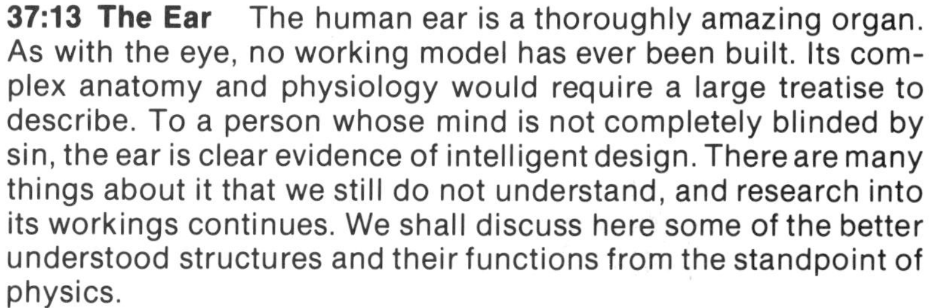

37:13 The Ear The human ear is a thoroughly amazing organ. As with the eye, no working model has ever been built. Its complex anatomy and physiology would require a large treatise to describe. To a person whose mind is not completely blinded by sin, the ear is clear evidence of intelligent design. There are many things about it that we still do not understand, and research into its workings continues. We shall discuss here some of the better understood structures and their functions from the standpoint of physics.

As you have probably learned in previous science courses, the ear consists of three parts—the **outer ear**, the **middle ear**, and the **inner ear** (see Figure 37:30). We shall consider briefly what happens in each part.

The outer ear consists of the *external ear* and the *external auditory canal*. The external ear aids us in determining the direction of a sound source by blocking sounds coming from behind. The external auditory canal is a quarter-wave air column (open at one end, closed at the other) having a length of about 2.7 centimeters and a diameter of 0.7 centimeters in an adult. This corresponds to a frequency of approximately 3200 Hz, but it is broadly resonant, enabling it to amplify sounds of many different frequencies throughout the mid-range of the audible spectrum.

The closed end of the canal is formed by the *tympanic membrane*, or *eardrum*, which separates the outer ear from the middle ear. A thin, flexible membrane, the eardrum is extremely sensitive to vibrations of the air in the column. It is reported that a normal eardrum can respond to pressure changes that move it as little as a billionth of a centimeter. This distance is only one tenth the diameter of a hydrogen atom! For ordinary sounds, the motion is about a millionth of a centimeter.

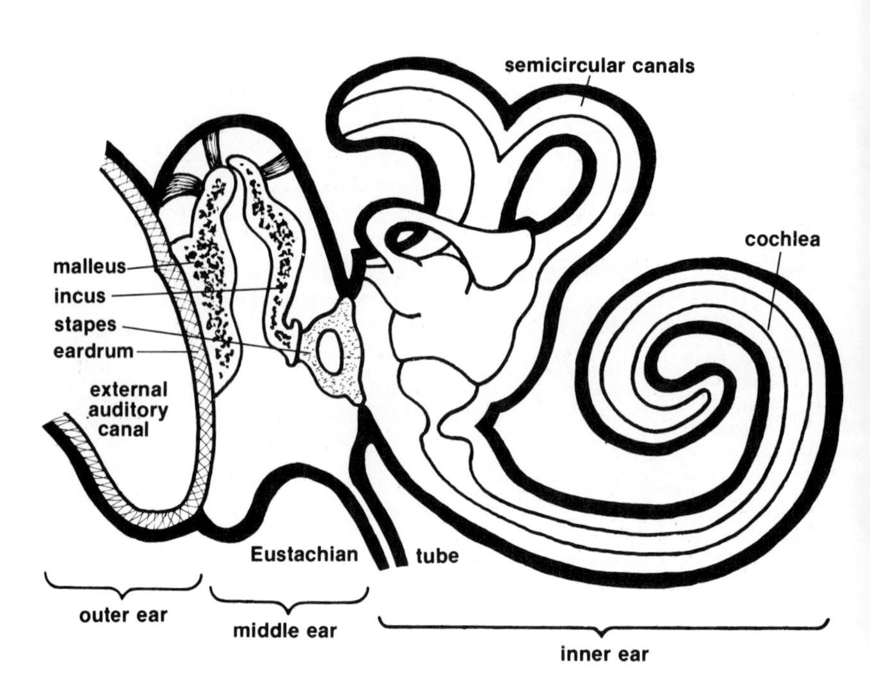

37:30 *Overall diagram of the ear.*

Before the sound impulses in the ear can be converted into nerve impulses and sent to the brain, it is necessary that the sound first enter a liquid, where contact can be made with the nerve endings. To transmit sound from air to a liquid efficiently, however, presents a most difficult and challenging problem. Suppose you are in a rowboat trying to converse with someone who is swimming under water. As you shout down at the water, most of the sound energy will simply be reflected back from the surface rather than entering the water. The efficiency of transfer from the one medium to the other is extremely poor, and you will be fortunate if you are heard at all. But the problem has been solved very neatly in the ear. An ingenious method is used to transmit sound from the air of the outer ear to the fluid of the inner ear. This is the function of the middle ear, which we shall now consider.

The middle ear consists of the eardrum and three bones or *auditory ossicles*, called the *malleus* (hammer), *incus* (anvil), and *stapes* (stirrup). Suspended at certain key points by ligaments, the bones function as little levers. The malleus is in direct contact with the eardrum, from which it transmits vibrations to the incus. The incus in turn passes the vibrations along to the stapes. The footplate of the stapes covers the oval window, the entrance to the fluid-filled cochlea of the inner ear. The fulcrums of the three bones are positioned in such a way that there is an increase in force of about 1.5 times from the eardrum to the oval window. But a much more substantial gain comes about by what is called **pressure amplification**. Because the eardrum has an area of approximately 25 times that of the oval window, the pressure is increased by a factor of 25 as vibrations are transmitted through the system. Coupled with the mechanical amplification of 1.5 due to the lever action, there is an overall increase of about 37½ in the middle ear, allowing us to hear sounds 1000 times as faint as we could if there were no amplification designed into the system.

There are two tiny muscles in the middle ear, the *tensor tympani* and the *stapedius*, whose function it is to restrict the vibration of the eardrum and the stapes, as a means of protection against uncomfortably loud sounds. These muscles are controlled by the brain and are analagous to the automatic diaphragm of the eye. Another middle ear structure is the *Eustachian tube*, a passageway connecting it with the throat. It allows the pressure to remain equal on both sides of the eardrum. Its action can be appreciated when you ascend rapidly in an elevator or an airplane. Usually the simple act of swallowing will relieve the discomfort in the eardrums by momentarily opening the Eustachian tubes.

The inner ear is the most intricate and least understood part of the ear. We cannot even name all of its structures in this brief description, much less discuss their workings. The inner ear contains the *semicircular canals* (the organ of balance) and the *cochlea*, the snail-shaped transducer that converts sound to electrical energy. Both parts of the inner ear are filled with a fluid called *perilymph*, whose viscosity is equal to about twice that of water.

For the sake of clearer understanding, let us pretend that the cochlea can be unrolled, as shown in Figure 37:31. Running almost the whole length of the unrolled cochlea is the *cochlear partition*, which divides the cochlea into roughly two halves, the *scala vestibuli* and the *scala tympani*. The division is not quite complete, however, and the two halves of the cochlea communicate with each other in the region at the end of the partition, called the *helicotrema*.

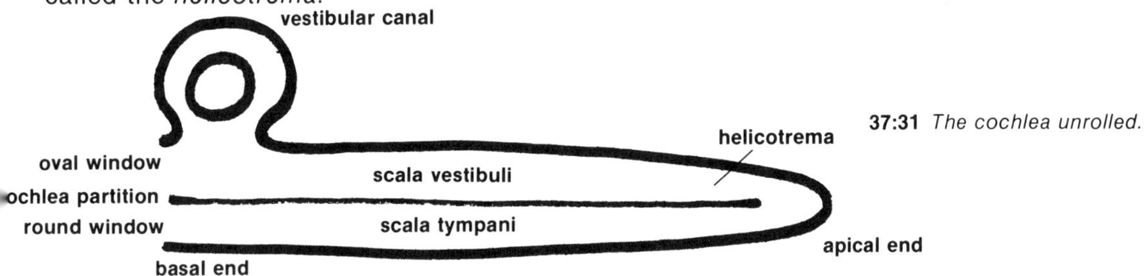

37:31 *The cochlea unrolled.*

The oval window is located at the basal end (large end) of the scala vestibuli. At the corresponding point of the scala tympani there is the membrane-covered *round window*, which also leads to the middle ear. The round window acts as an escape valve for slow pressure changes. If you descend rapidly in an elevator, for instance, the pressure on the outside of your ear will be greater than that on the inside. The increased pressure is transmitted through the eardrum and ossicles to the oval window. The inward push on the oval window is transmitted through the scala vestibuli, the helicotrema, and the scala tympani to the round window, where the pressure is released by pushing out the membrane into the middle ear cavity.

For rapid pressure fluctuations such as those encountered in sound waves, however, the perilymph does not have time to respond in this manner. Instead, the cochlear partition is caused to vibrate in a complex way in accordance with the incoming signals. Extending along the length of the cochlear partition is the *Organ of Corti*, into which project many thousands of endings of the auditory nerve. In some unknown way, about 30,000 sensory cells (called *hair cells*) in the Organ of Corti relay information to the auditory nerve endings corresponding to the pitch, quality, and loudness of the sounds being received. Many tens of thousands of separate signals are carried to the brain by way of several complicated relay stations—the *spiral ganglion*, the *cochlear nucleus*, the *superior olivary complex*, and the *medial geniculate body*, to mention a few. The signals arrive at the temporal lobes at the sides of the cerebrum, where they are interpreted. We do not even have a workable theory, though, as to just how this is accomplished.

In studying structures such as the eye and the ear, which we cannot fully fathom even with our most sophisticated science, the only attitude we can reasonably maintain is one of humility and awe. "Great things doeth he which we cannot comprehend" (Job 37:5).

List of Terms

acoustics
alto
baritone
bass
beat
brass instruments
bridge
contralto
Doppler effect
fundamental
inner ear
mezzo-soprano
middle ear
mouthpiece
nodal point
node
nut
overtone
outer ear
percussion instruments
pitch
pressure amplification
quality
resonance
reverberation
soprano
stops
stringed instruments
tenor
tone
woodwind instruments

Questions

1. What is meant by pitch?
2. What is the difference between a musical sound and noise?
3. Why does the same note played by two different musical instruments sound different?
4. What is meant by the term *fundamental*, as applied to vibrating strings or air columns?

5. What are overtones?
6. If the *fundamental* frequency of a vibrating string is 60 Hz, what are the first three overtones?
7. What are the four major categories of musical instruments?
8. Define resonance.
9. What wave effect is responsible for beats?
10. What is a beat?
11. What is the Doppler effect?
12. Name the three main parts of the ear. In what part of the ear is sound energy changed to electrical energy?
13. What takes place in the middle ear?
14. Why is it difficult to transfer sound from a gas to a liquid?
15. What are reverberations?
16. What purpose is served by vertical splays and fluted walls in auditoriums?
17. Where is music first mentioned in the Bible?
18. Name the stringed instruments normally used in a symphony orchestra.
19. What was the organ referred to in Scripture?
20. In general, what is the relationship between the physical size of a musical instrument and its pitch?
21. Name the different ranges of men's voices from highest to lowest. Give the three ranges for women's voices from highest to lowest.
22. What is a woodwind instrument?
23. What two woodwind instruments do not use reeds?
24. What is the function of the Eustachian tube?
25. How does the ear cope with sounds that are too loud for it?
26. In what part of the brain does hearing actually take place?
27. In what instrument is the pitch changed solely by adjustment of the player's lips?

Student Activities

1. Write a brief paper tracing the history of sound recording from the Edison wax cylinder to the cassette tape recorder.
2. Two identical tuning forks may be used to generate beats if one of them is altered slightly. Do this by sliding a safety pin of the proper size over each prong of one of the forks. Strike the two forks at the same time and place them close together against a table or desk top. Repeat using *two* safety pins on each prong.
3. Secure a loudspeaker from an old radio. Connect the two terminals of the speaker to the two terminals of a dry cell. Observe which way the cone moves. Now reverse the connections to the dry cell and observe that the cone moves in the opposite direction. This should help you to visualize how condensations and rarefactions are produced in the air in front of the speaker.

Further Reading

Baines, Anthony, ed. *Musical Instruments Through the Ages.* Baltimore, Md.: Penguin Books, 1961.

Benade, Arthur H. *Horns, Strings, and Harmony*. Garden City, N.Y.: Anchor Books, Doubleday, 1960.

Denes, P.B., and E.N. Pinson. *The Speech Chain*. Murray Hill, N.J.: Bell Telephone Laboratories, 1963.

Hill, W.H., F. Hill, and A.E. Hill. *Antonio Stradivari, His Life and Work*. N.Y. : Dover Publications, 1963.

Photo next page

The 80-inch liquid hydrogen bubble chamber contains 900 liters of liquid hydrogen at a temperature of -414° F. It is surrounded by a vacuum chamber and large magnet coils. As a pulse of highly energetic particles is guided magnetically into the chamber, the liquid hydrogen is superheated by a sudden reduction in pressure. The charged particles cause the superheated liquid hydrogen to boil, leaving a track of tiny bubbles to mark their paths. By measuring the curvature, length, and density of the tracks, scientists can determine the electric charge, momentum, mass, and other properties of the particles.

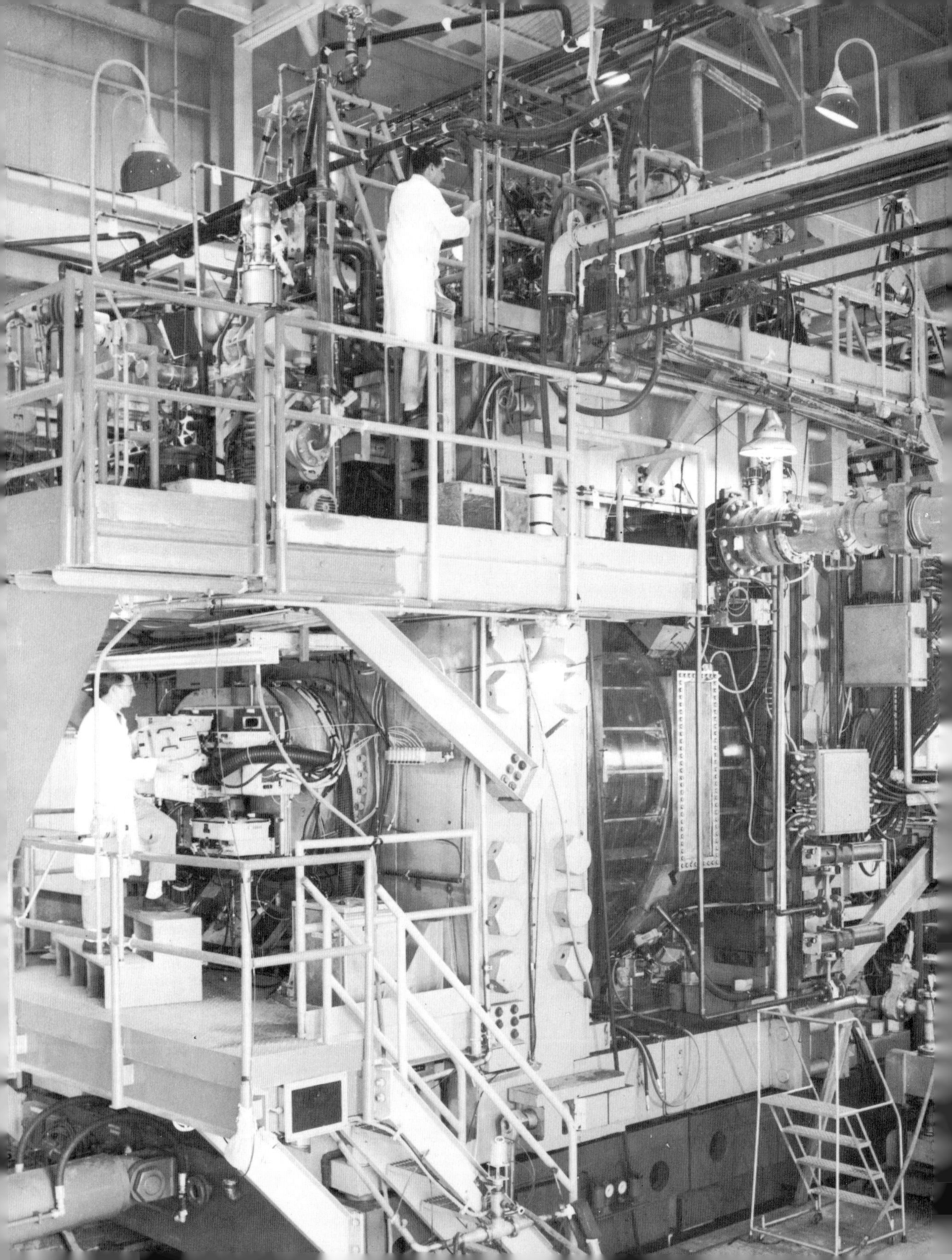

UNIT XII TWENTIETH CENTURY PHYSICS

Chapter 38 Natural Radioactivity

38:1 *Henri Becquerel (1852-1908).*

38:1 Discovery of Radioactivity Antoine Henri Becquerel (1852-1908), a French physicist, discovered **radioactivity** in 1896. He was working on the phenomena of fluorescence and phosphorescence when the discovery was made. Becquerel wrapped photographic film in thick black paper, placed a salt of uranium on top of the encased film, and exposed them to the sun (hoping the sun's rays would cause the salt to phosphoresce). He found that rays coming from the salt exposed the film. (The black paper prevented direct exposure by the sun.)

Later Becquerel planned to repeat the experiment, but the sun would not cooperate. He had placed the salt on another encased film, but the skies were too cloudy. He then placed the film with the salt on it back in a drawer to await a sunny day. Several days later, the Frenchman took the film and sample out of the drawer. He developed the film, expecting it to be only slightly exposed because of the lack of sunlight to encourage phosphorescence. However, to his amazement, Becquerel found that the film was more deeply exposed than if it had been in direct sunlight. He then correctly concluded that the rays coming from the sample did not have to be excited by sunlight.

The physicist investigated other compounds of uranium and the metal itself and found that all of them acted the same way. The more uranium present in the substance being tested, the darker the photographic film would be. He tried to excite the radiation with other light sources, but the rays given off did not depend on any external excitation. Becquerel found that this radiation from uranium could ionize the molecules in air, and it could discharge a positively or negatively charged body. These mysterious rays from uranium were called **Becquerel rays**.

38:2 *The Curies.*

38:2 Other Elements with Radioactive Properties Later, Marie Sklodowska Curie (1867-1934) and her husband, Pierre Curie (1859-1906), found that other elements would give off radiation similar to that given off by uranium. In their work with thorium and its compounds, they reached the conclusion that the emission of rays depended only on the presence of thorium atoms. The more of these atoms present in a sample, the more rays it gave off. Thus, in the case of both uranium and thorium, the emission of the rays was a property of the element.

The Curies used this conclusion to great advantage. In their studies of the radioactivity of pitchblende (an ore containing U_3O_8), they found that samples of the ores caused a greater ionizing effect of air than would normally be expected by the presence of uranium only. They concluded that there was another substance in the ore causing the extra radioactivity. The Curies chemically separated large amounts of the ore and isolated a new element, polonium, named for Marie Curie's native country, Poland. Later, they discovered another naturally radioactive element, radium. It was given that name because it was so highly radioactive.

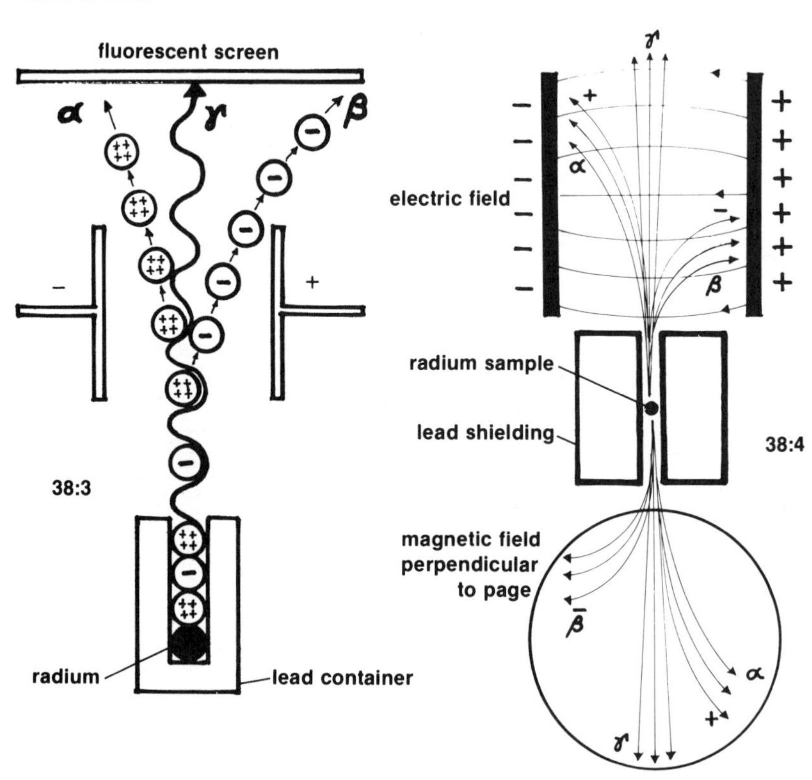

38:3 *Rutherford's experiment demonstrating that Becquerel rays were actually three different kinds of radiation.*

38:4 *The effect of electric and magnetic fields on Becquerel rays.*

38:3 The Nature of Becquerel Rays In 1889 Ernest Rutherford (1871-1937), investigating the Becquerel rays, discovered that they were actually three different kinds of rays. Rutherford's experiment consisted basically of placing a small piece of radium into the bottom of a hole in a lead block, and allowing the rays that escaped from the hole to pass through an electrostatic field. He then viewed the effect of the electrical field on the path of the rays by means of a zinc sulfide screen placed beyond the field. As the rays hit the screen, they produced a flash of light **(scintillation)**. He found that the rays separated into three parts—one part unaffected by the electrical field, another strongly attracted toward the positive side of the field, and the third slightly attracted toward the negative side of the field. The same effect can be observed using a magnetic field (see Figure 38:4). The ability of these three different rays to penetrate matter varies also. The rays easily absorbed by matter were called **alpha (α) rays** by Rutherford. The other more penetrating rays were called **beta (β) rays**. A French physicist, Paul Villard (1860-1934), identified the third type of radiation as **gamma (γ) rays**.

The α rays were deflected toward the negative side of the electric field, the β rays toward the positive side, and the γ rays passed through unaffected. In 1903, Rutherford listed the penetrating power of the three different types of radiation; his findings are summarized in Table 38-1.

Table 38-1 *Thickness of Aluminum Needed to Reduce the Radiation Intensity to One Half Its Initial Value*

Radiation	Thickness of Aluminum (cm)
α	0.0005
β	0.05
γ	8

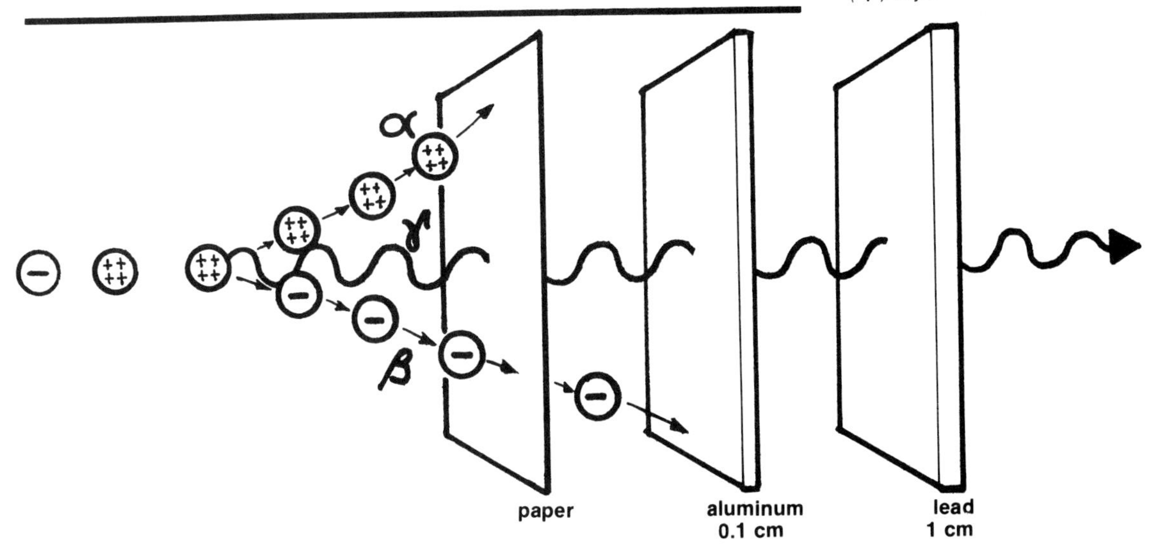

38:5 *Thin paper will stop alpha (α) rays; thin aluminum will stop beta rays (β). However, it takes fairly thick lead to stop gamma (γ) rays.*

38:4 Alpha Rays (Particles) Since alpha rays are attracted to the negative side of the electrical field, they are positively charged. Rutherford identified these positive rays to be actual particles, doubly ionized helium atoms, or simply helium nuclei.

Pierre Curie and Andre Debierne (1874-1949), a French chemist, had discovered that a gas (radon) was given off by radium. Radon was found to give off α rays. If radon was stored in a closed container, after some time helium gas could be found in the container also. Sir William Ramsay and English chemist Frederick Soddy (1877-1956) discovered that helium was also given off from radioactive radium bromide. This led Rutherford to suspect that alpha rays were helium nuclei, and he demonstrated this experimentally.

Since α particles are doubly charged, they have a high ionizing power. They like to "grab" electrons from surrounding matter. This ionizing power reduces their ability to penetrate matter (a layer of paper is enough to absorb α particles).

An interesting aspect in the identification of α particles is their origin. Where could they come from? The atom of an element such as radium or radon obviously is able to emit the nucleus of another element (helium)!

38:5 Beta Rays Since beta rays are attracted to the positive side of the electrical field, they are negatively charged. Becquerel experimentally determined that β rays were highly energetic electrons. The deflection of the rays in electric and magnetic fields was used to demonstrate this. β rays have some ionizing ability and can penetrate matter farther than α particles (it takes about one centimeter of aluminum to stop all β particles).

38:6 Gamma Rays Gamma rays are not affected by either electrical or magnetic fields. Thus, they possess no charge. γ rays have very little ionizing power, but their power of penetration is great. These rays can pass through many centimeters of lead or several feet of concrete before being absorbed. When anyone does experiments with γ rays, bulky and expensive shielding must be used.

Table 38-2 *Comparison of Radiation Properties*

	identity	mass	charge	relative ionizing power	relative penetrating power
alpha rays (α)	helium nucleus	7000 electron masses	2 electron charges	10,000	1
beta rays (β)	high-energy electron	mass of electron	electron charge	100	100
gamma rays (γ)	electromagnetic wave	none	none	1	10,000

38:7 Ionizing Power and Interaction We have been discussing the ionizing power of radiation. As any radiation passes through matter, it "collides with many molecules." These "collisions" may be nothing more than "near misses." The best term to use, since we cannot see what is happening on the atomic level, is to say the radiation interacts with the molecules of the substance through which it is passing. This **interaction** is nothing more than a transfer of some of its energy. During these interactions, enough of the kinetic energy of the ray may be transferred to the valence electrons of the affected molecule to allow these electrons to "escape from the atomic forces." After many, many of these interactions the ray will lose enough of its kinetic energy to be absorbed. The α particle "grabs" two freed electrons and forms a helium atom. The β ray is eventually captured by a positive ion and neutralized. A γ ray dissipates its energy differently from α and β particles; a discussion of this is beyond the scope of this book.

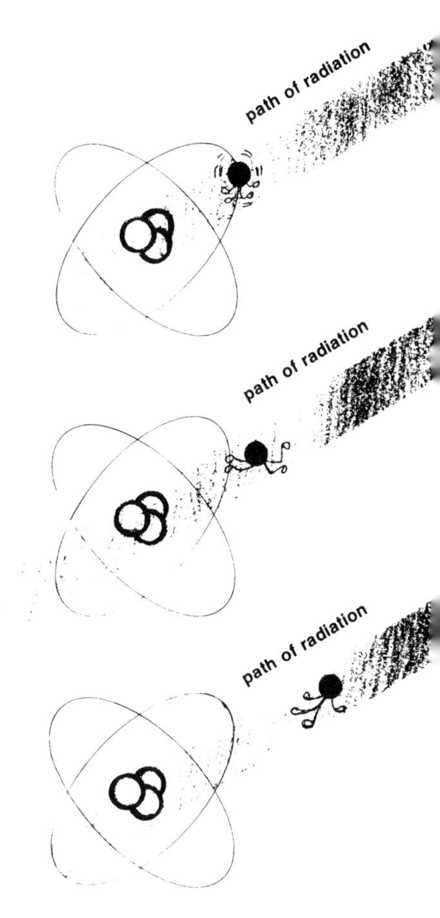

38:6 *Electrons bound to atoms are freed by picking up energy from passing radiation.*

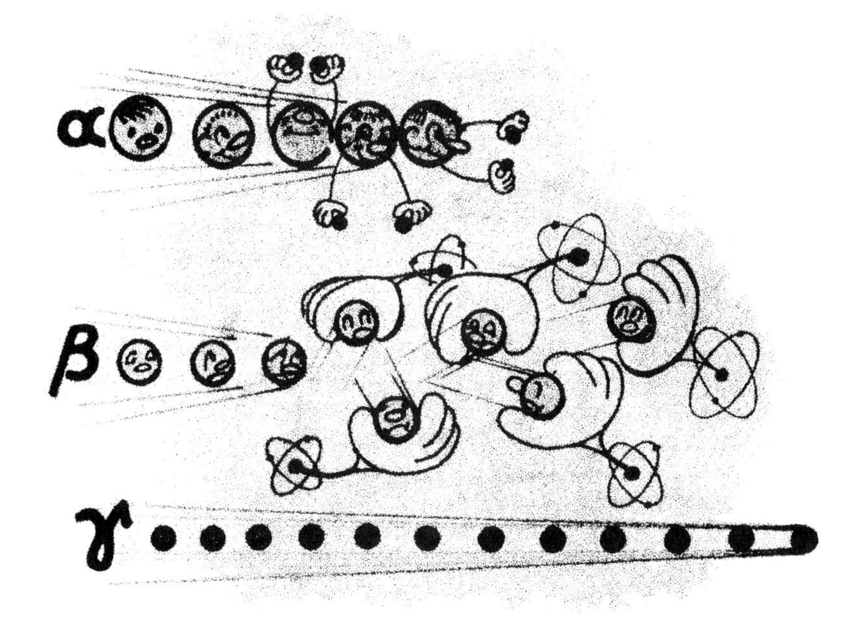

38:7 *Alpha rays "grab" electrons from the matter which they penetrate and have a high ionizing power but do not penetrate very far. Beta rays are "grabbed" by ions in the matter that they penetrate so they have a slight ionizing power but a greater power of penetration. Gamma rays are so energetic that they "zoom" right through most matter with little ionizing power.*

38:8 Detection of Radioactivity The ionization power of radiation can be used to detect its presence. Many of the pioneers in radioactive-elements research used electroscopes. We discussed these in Section 30:5. If a charged electroscope is placed within the vicinity of radiation, it will slowly discharge. The radiation passing through the air near the electroscope produces positive ions by removing electrons from many of the nitrogen and oxygen molecules. The freed electrons can be picked up by neutral molecules forming negative ions. If the electroscope is negatively charged, the positive ions neutralize this charge, causing the instrument to discharge. If the electroscope is positively charged, the negative ions will neutralize it again, causing discharge. The more powerful the source of radiation, the faster the discharge rate.

Another device utilizing the ionizing power of radiation for detection is the **Geiger-Müller tube counter**. The tube contains a small copper cylinder around a fine tungsten wire along the center. Some of the gas in the tube is pumped out, leaving a pressure less than atmospheric; then the glass container is sealed. The wire and tube are connected to a power supply, and a potential difference is developed between the two, with the wire being positive and tube negative. Radiation entering the glass chamber generates ions and freed electrons. The electrons move toward the positive wire under the influence of the potential difference, accelerating as they move. They ionize other molecules because of their energy. This multiplication of freed electrons causes an avalanche of electrons to move to the wire. This small current can be amplified electronically to cause the movement of a needle on a dial, or to operate a radio loud-speaker.

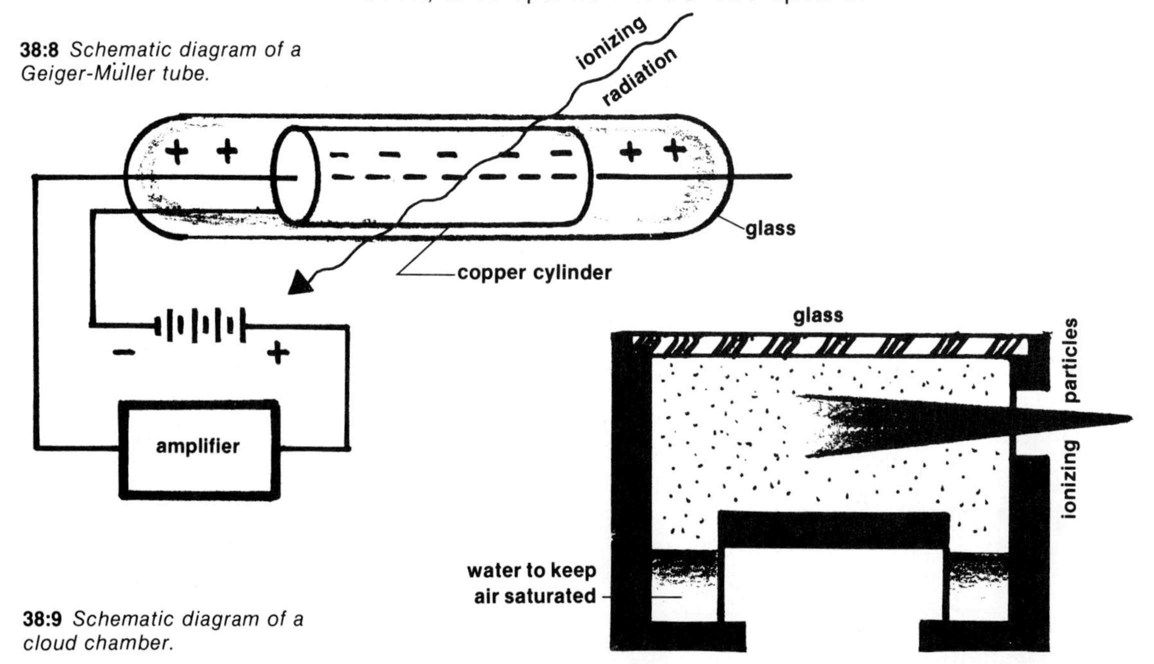

38:8 *Schematic diagram of a Geiger-Müller tube.*

38:9 *Schematic diagram of a cloud chamber.*

Charles T. R. Wilson (1869-1959), a Scottish physicist, developed the **cloud chamber**. The chamber must contain a gas supersaturated with a vapor. As radiation passes through the gas, it produces ions. The vapor condenses on the ions, leaving a visible path where the radiation has been. These streaks are similar to vapor trails left by high-flying aircraft. A magnet can be placed around the cloud chamber, causing the charged particles to curve. Examining the path of a particle in the magnetic field enables a scientist to determine the charge, speed, and energy of the particle.

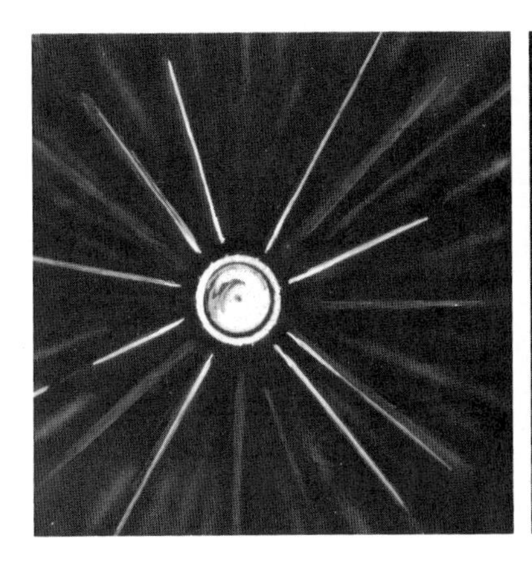

38:10 *Tracks of alpha particles from a radioactive sample in a small cloud chamber.*

38:11 *Tracks of beta particles from a radioactive sample in a small cloud chamber.*

38:12 *Diffusion cloud chamber for classroom use.*

Experiment 38:1 *Observation of Radiation in a Cloud Chamber*

1. Obtain a diffusion cloud chamber, such as the one pictured in Figure 38:12.
2. Place alcohol and dry ice in the bottle.
3. As the alcohol is vaporized, it will form droplets on any ions that are present, making the tracks from the radioactive sample visible.

Donald A. Glaser (1926-), an American physicist, developed the **bubble chamber**. A liquid is heated above its boiling point and prevented from boiling by a high pressure. Radiation enters the liquid through a window in the side of the chamber. Simultaneously, some of the pressure is released through a valve. A trail of bubbles is left by the radiation as it passes through the chamber. The ionization along its path provides nucleation sites for the boiling bubbles.

38:13 *The 80-inch liquid hydrogen bubble chamber contains 900 liters of liquid hydrogen at a temperature of -414° F. It is surrounded by a vacuum chamber and large magnet coils. As a pulse of highly energetic particles is guided magnetically into the chamber, the liquid hydrogen is superheated by a sudden reduction in pressure. The charged particles cause the superheated liquid hydrogen to boil, leaving a track of tiny bubbles to mark their paths. By measuring the curvature, length, and density of the tracks, scientists can determine the electric charge, momentum, mass, and other properties of the particles.*

(See chapter title page)

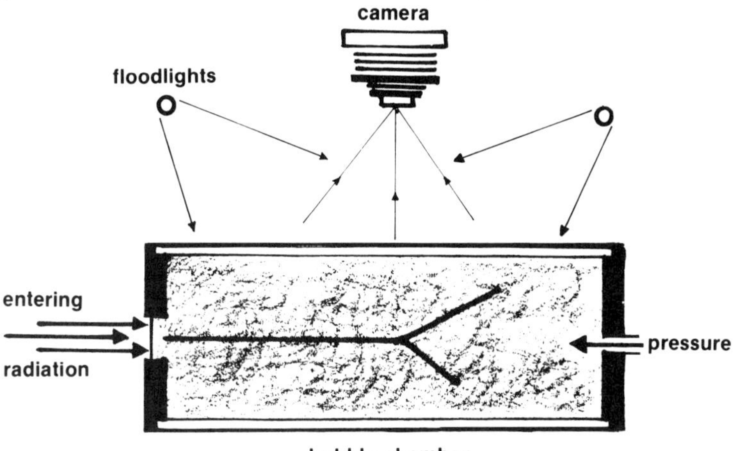

38:14 *Schematic diagram of a bubble chamber.*

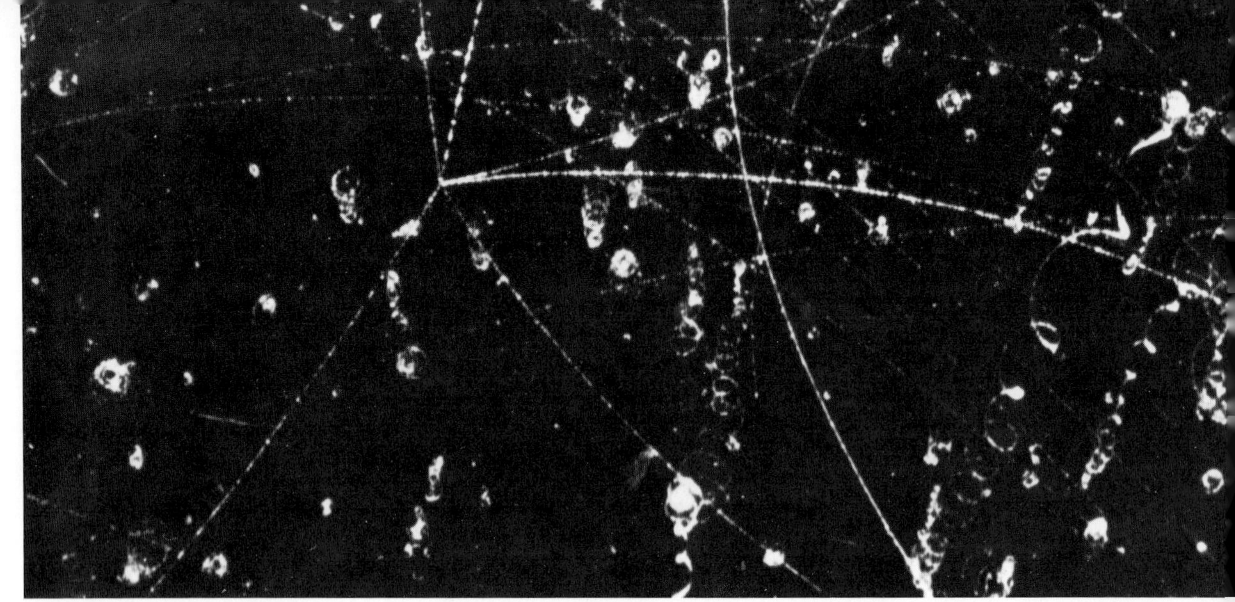

38:15 *This picture of nuclear particle tracks in the ten-inch diameter bubble chamber mounted inside a super-conducting magnet shows what happened to two negative K mesons that entered the bubble chamber. The mesons entered the bubble chamber from the right. Most prominent are the tracks of bubbles resulting from the interaction of a negative K meson and a helium nucleus.*

The scintillation method of detection used by Rutherford has already been discussed briefly in Section 38:3. Other detectors are **photographic emulsions** and **spark chambers**. When the emulsion is developed, grains of silver are deposited where the rays have passed. The chemical reaction is

$$Ag^{+} + e^{-} \text{ (freed by ionization)} \xrightarrow{\text{radiation}} Ag\downarrow$$

A series of oppositely charged plates with a gas between make up a spark chamber. As radiation passes through, ionization of the gas results. This in turn causes a series of sparks to pass from plate to plate following the path of the ray.

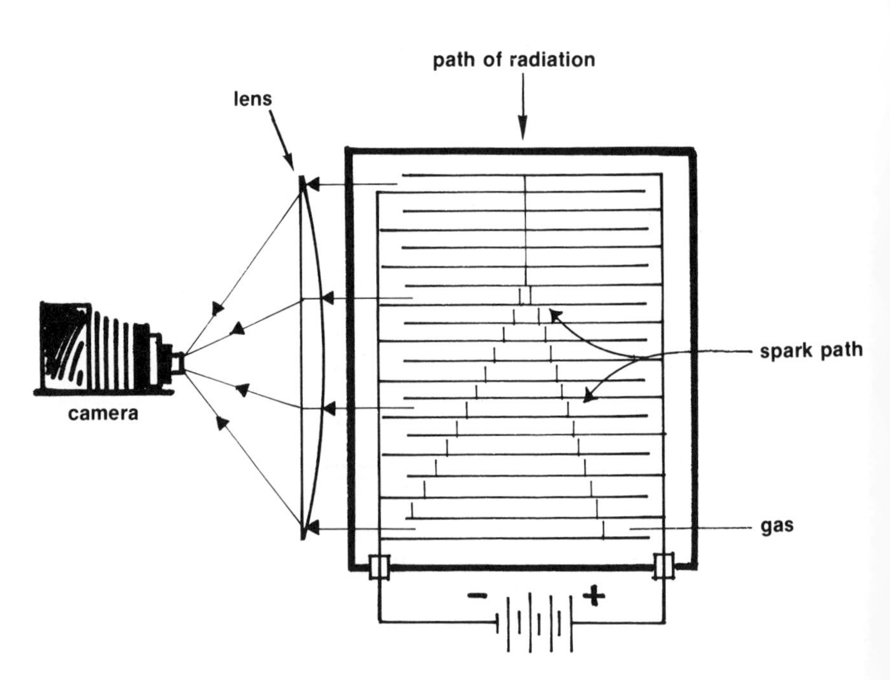

38:16 *Schematic diagram of a spark ohamber.*

38:9 Origin of Becquerel Rays Once α, β, and γ radiation had been discovered, scientists had to account for their origin. At the time of their discovery, most scientists believed in the atomic molecular structure of matter. However, it was thought that the atom was indestructible. Yet, if a radioactive atom emits something as massive as a helium nucleus, the parent atom cannot remain the same. Working with this information within the atomic model of matter, Rutherford and Soddy suggested a theory of radioactive transformation. They argued that when an α or β particle is emitted from an atom, the atom breaks into pieces! The emitted particle is the lighter fragment and the daughter atom left behind is the heavier fragment.

Radioactive transformations cannot be considered chemical reactions. For instance, radium can release energy for thousands of years without the need of adding other substances. No known chemical reaction can occur within the element itself. Normally, energy in chemical reactions is released when different elements or compounds react.

When Rutherford's idea of a central **nucleus**, where most of the mass of an atom exists, was accepted, *it* was considered the place from which the α and β rays were emitted.

When any emission occurs, the atom loses energy; thus radioactive transformation is referred to as radioactive decay. One way of visualizing radioactive decay is to imagine that the nuclei in elements such as uranium and radium are so massive that they are unstable.

38:17 *An easy way to designate the uranium-238 isotope.*

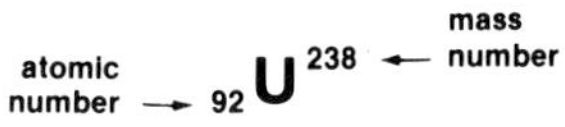

38:10 Radioactive Decay Series Uranium has an atomic number of 92. One of its isotopes has a mass number of 238 (92 protons and 146 neutrons in the nucleus). Figure 38:17 illustrates a simple way to designate the uranium-238 isotope. This decays by emitting an α particle, leaving an isotope of thorium. A **radioactive decay** equation similar to a chemical equation can be written for this transformation. Charge and mass on the left-hand side of the equation must equal charge and mass on the right-hand side. The parent nucleus is shown on the left; the daughter products are on the right side of the arrow.

$$_{92}U^{238} \longrightarrow {}_{2}He^{4} + {}_{90}Th^{234}$$

Th-234 will decay by emitting a beta particle and forming an isotope of protactinium.

$$_{90}Th^{234} \longrightarrow {}_{-1}\beta^{0} + {}_{91}Pa^{234}$$

The mass of the beta particle is considered so small that it is ignored. This radioactive disintegration series will continue until finally a stable isotope of lead (Pb-206) is formed. Fourteen successive decay steps take place before Pb-206 is formed.

38:11 The Life of Radioactive Atoms Since isotopes like U-238 are unstable, how long would it or any of its daughter products last? This is an impossible question to answer. Suppose we had a sample of 1,000,000 (10^6) atoms of uranium. Not all of the atoms would disintegrate at the same time, and no one can predict exactly when a given uranium atom will decay. One uranium atom may decay within a second; another may last more than 10^6 years. However, at the end of 4.51×10^9 (4.51 billion) years, 500,000 of the original uranium atoms will have decayed; and the other one half will still be U-238. At the end of 9.02×10^9 years, 750,000 of the original U-238 atoms will have decayed. Every 4.51×10^9 years, one half of the uranium atoms you started with will have decayed. If we let $T = 4.51 \times 10^9$ years, the following could be predicated:

Number of U-238 atoms	**Time**
1,000,000	0
500,000 (1/2 of original amount)	T
250,000 (1/4 of original amount)	2T
125,000 (1/8)	3T
62,500 (1/16)	4T
31,250 (1/32)	5T

The time T (4.51×10^9 years in the case of uranium) is called the **half-life** of the radioactive isotope. It is defined as *the length of time required for one half of the radioactive nuclei in a sample to decay.*

Problem: The half-life of radon-222 (Rn-222) is about 4 days. How much of a 64-gram sample will remain after 24 days?

Solution:

$$\frac{\text{Elapsed Time}}{\text{Half-life}} = \frac{24}{4} = 6 \text{ (no. of division)}$$

(1) $\frac{64}{2} = 32$ (3) $\frac{16}{2} = 8$, (5) $\frac{4}{2} = 2$

(2) $\frac{32}{2} = 16$ (4) $\frac{8}{2} = 4$ (6) $\frac{2}{2} = 1$ gram remaining

This information can be plotted on a graph (see Figure 38:18).

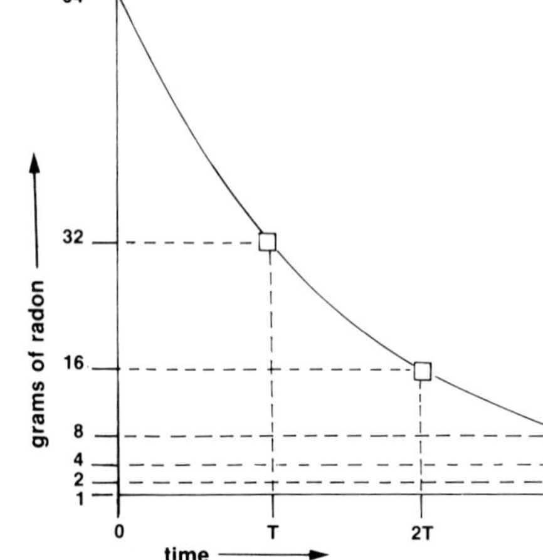

38:18 *Exponential decay of radon-222.*

We have been considering the disintegration of only one member of a decay series. A sample of pure U-238 would change composition in a complicated way. Th-234 would form, which would decay into Pa-234, and so forth. Eventually the sample would contain the parent element, U-238, and varying amounts of all of the daughter elements plus some helium gas from alpha decay.

38:12 Radioactive Dating Methods Radioactive isotopes appear to decay at a constant rate, regardless of their surroundings. Could we not, therefore, tell how long a given sample has been decaying by determining the relative amounts of end product and original material present? The answer is, "Yes, in some cases." There are, however, some strict limitations imposed by the nature of the assumptions underlying **radioactive dating**. Good results have been obtained on many samples less than 5000 years old using the **radiocarbon method**. But the results for objects older than this are *not* trustworthy, and there have even been a number of erroneous answers obtained on samples *less* than 5000 years old. Let us look briefly at some of the methods of age dating that have been attempted.

Radiocarbon: Carbon-14, a radioactive isotope of carbon having a half-life of 5730 years, is continually being formed from nitrogen by cosmic rays in the upper atmosphere. Combining with oxygen to form carbon dioxide, it finds its way into all living things, both plants and animals. Within the organism it slowly decays back to nitrogen, but that which decays is continually replaced by new carbon-14 brought in by way of food and carbon dioxide. Thus, there is a balance. A relatively constant amount of radioactivity per unit weight will remain in the organism.

When the organism dies, new carbon-14 is no longer brought in from the outside. That which is present continues to decay, resulting in a decrease in its radioactivity per unit weight. This quantity can be measured. The older the sample, the less radioactivity there should be. We can obtain a number of carbon-containing samples of known age, measure their radioactivity per unit weight, and construct a graph of radioactivity versus age. This is called a **calibration curve** (see Figure 38:18). Once we have such a graph, the age of a new sample can then hopefully be found by a simple two-step procedure: (1) Measure its radioactivity per unit weight; (2) Read its age from the graph.

Unfortunately, there are problems with the radiocarbon dating method. One difficulty is that the calibration curve cuts off at about 5000 years ago. The oldest sample of independently known age that could be found anywhere in the world was an Egyptian mummy from the First Dynasty, dating in the range of 2850-2950 B.C. To determine the age of anything older than this by radiocarbon requires *extrapolation*, which, you will recall, is extremely hazardous. Many workers, however, prefer to throw caution to the winds and extrapolate freely up to 40,000 or 50,000 years. The results from this procedure are, sad to say, represented to the public as though they were as securely established as the more recent dates.

Another problem is that the amount of carbon-14 in the atmosphere does not remain constant. Measurements have shown that it is continually increasing. A correction should be applied to the answers obtained to allow for this, but scientists disagree on how it should be done. Several members of the Creation Research Society who have made a study of this question are convinced that if the corrections are properly applied, all radiocarbon dates—even those obtained by extrapolation—will be telescoped

38:19 *Calibration curve for carbon-14 dating.*

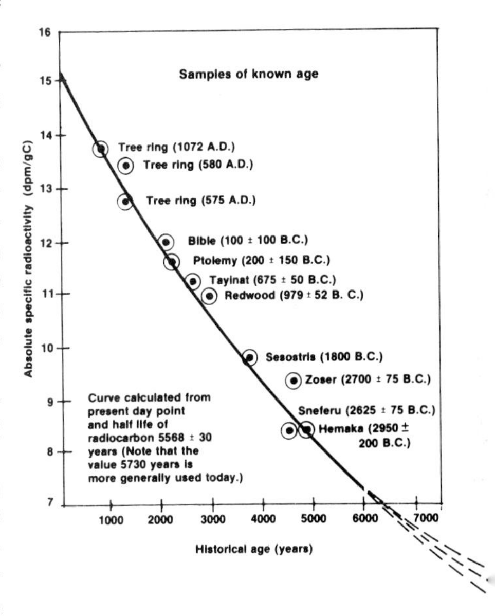

to within the last 10,000 years.

One of the most reliable dating methods known is tree-ring counting. It is, in fact, far more trustworthy than any radioactive method for the materials and time-spans to which it applies. What does this have to do with radiocarbon dating? Simply that, in some cases, there have been discrepancies between the dates obtained with the two methods, and in a few cases the discrepancies have been very great. In cases such as these, there is every reason to believe that the tree-ring date is right and the radiocarbon date is wrong. These are called **anomalous samples**. This further adds to the problems confronting the researchers. When new material is analyzed, there is always a slight possibility that it could be an anomalous sample. Usually there is no second method used to check the answers obtained, and the erroneous dates go undetected.

The "Long-Term" Methods: There are several dating methods which some scientists say are capable of providing ages running into millions and billions of years. The most popular of these are the **uranium-lead method**, the **potassium-argon method**, and the **rubidium-strontium method**. In each of these methods, an unstable original material is changing by radioactive decay to a stable end product. Uranium decays in a series of alpha and beta decay reactions to lead. Potassium decays to argon in a single step, and rubidium decays directly to strontium.

All of these so-called "long-term" dating methods require one major assumption which is highly questionable. The experimenter must assume that *none* of the end product was present in the sample at the beginning. Needless to say, no scientist was on the spot checking the sample at the beginning. Thus, the key ingredient of the scientific method—observation—is missing. Also, it is very presumptive for man to decree what the Creator could or could not have placed in the rocks at the beginning.

Recently it has become possible to check the validity of two of these methods on rocks of known age. *In every instance they have failed the test disastrously*. The dating methods are supposed to tell how many years have elapsed since the rock solidified. Volcanic rocks which were *observed* to solidify from lava flows *within the past 200 years* were tested with the uranium-lead and potassium-argon methods. Characteristically, they gave grossly exaggerated ages of millions and billions of years. Yet their true ages were less than 200 years! It is surprising that these methods are still taken seriously in the face of such findings.

There has been a general trend since the turn of the century to keep increasing the age estimates of the earth and of the universe. This has been referred to by some as "age inflation." Since the year 1900, the estimated age of the earth has been increasing at an average rate of 65,000,000 years per year, and the estimated age of the universe has been increasing exponentially.

It is obvious that truth cannot be this flexible. If there is any validity to the dates obtained by these methods, they would not have to be continually revised upward.

The next time you see some enormous age being given for a fossil or mineral sample, bear in mind that this "age" is based on extremely shaky assumptions that can in no way be verified. We may say in summary that many dates within the past 5000 years appear to be reasonably well established, but great caution should be exercised for anything older than this.

List of Terms

alpha rays
anomalous samples
Becquerel rays
beta rays
bubble chamber
calibration curve
cloud chamber
gamma rays
Geiger-Müller tube
half-life
interaction
nucleus
photographic emulsions
potassium-argon dating method
radioactive dating
radioactive decay
radioactivity
radiocarbon dating method
rubidium-strontium dating method
scintillation
spark chamber
uranium-lead dating method

Questions

1. (a) Who discovered radioactivity?
   (b) From where does this radiation come?
   (c) What are Becquerel rays ?
2. (a) Which of the rays are actually particles?
   (b) What is the electrical charge on each?
3. List the α, β, and γ rays in the order of penetrability of matter.
4. Why do α particles have such a high ionizing power?
5. What property of radiation is often used to detect it?
6. How does radiation produce a track in a cloud chamber?
7. If you start with 200 grams of U-238, plot a curve of mass of U-238 versus time for six half-lives.
8. Finish the following nuclear decay equations.

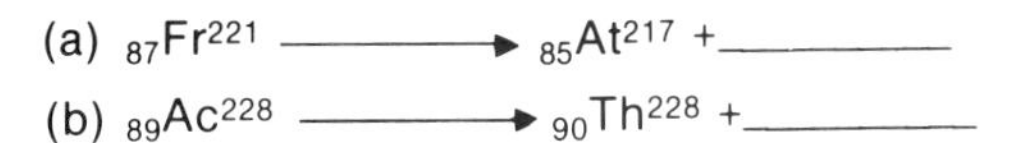

   (a) ${}_{87}Fr^{221} \longrightarrow {}_{85}At^{217} + \_\_\_\_\_\_$
   (b) ${}_{89}Ac^{228} \longrightarrow {}_{90}Th^{228} + \_\_\_\_\_\_$
9. How are Becquerel rays different from X-rays?
10. (a) What are the oldest samples that can be dated fairly reliably using radioactive dating methods?
    (b) What type of materials can be dated by the carbon-14 method?
    (c) Name some problems with the carbon-14 method of dating.
11. What is the big assumption in all of the long-term dating methods?
12. How does uranium decay to lead?
13. Are the long-term dating methods reliable? Why?

Student Activities

1. Write a paper on a method of the detection of radioactivity. Be sure to use charts and figures in your paper.
2. Write a paper on the radioactive dating methods as viewed from a creationist standpoint. Use material published by the Creation Research Society.

Photo next page

A Cockroft-Walton generator. The purpose of this machine is to provide the initial acceleration of 750,000 electron volts to the protons. Then they are injected into a 50 million-electron-volt linear accelerator before entering the orbit of an alternating gradient synchrotron.

UNIT XII TWENTIETH CENTURY PHYSICS

Chapter 39 Artificial Radioactivity

39:1 Transmutation of Elements Alchemists tried for hundreds of years to change one element into another. This operation is referred to as **transmutation.** The alchemists never succeeded, however, because they were using only chemical means. To transmute an element you must change its nucleus.

In 1919 Lord Rutherford was the first scientist to recognize that transmutation had occurred in one of his experiments. He was bombarding nitrogen gas with alpha particles from Bi-214 and observing what was happening through a scintillation screen. He noticed that particles were being produced which could travel further in the gas than the α particles. After much thought and more experimental work, Rutherford concluded that these particles were protons and that occasionally an α particle would collide with a nitrogen nucleus, causing it to break up and eject a proton (hydrogen nucleus). This artificial transmutation can be represented as

$$_7N^{14} + {_2He^4} \longrightarrow {_8O^{17}} + {_1H^1}$$

Since then many elements have been bombarded by high-energy particles and their nuclei artificially disintegrated.

One of the first problems scientists had to solve in dealing with nuclear disintegration is what happens when an α particle hits the nucleus. Does it simply hit the nucleus and knock out a proton, or is it actually captured by the nucleus, forming a brand new nucleus and quickly ejecting a proton? This was solved by

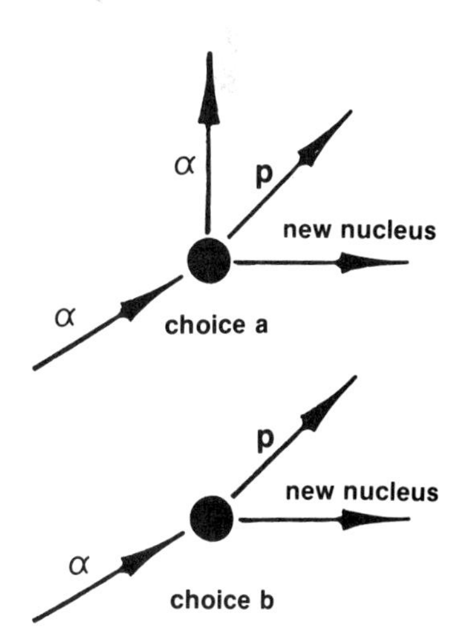

39:1 *Possibilities of what can happen when an alpha particle hits a nitrogen nucleus.*

P.M.S. Blackett in 1925. He looked at α bombardment of nitrogen in a cloud chamber. The only tracks which could be found were of the approaching α particle and, after the collision, the paths of the proton and the new nucleus. No tracks of the α particles could be found after the collision. Thus, it was concluded that the α particle is captured (absorbed) by the nitrogen nucleus.

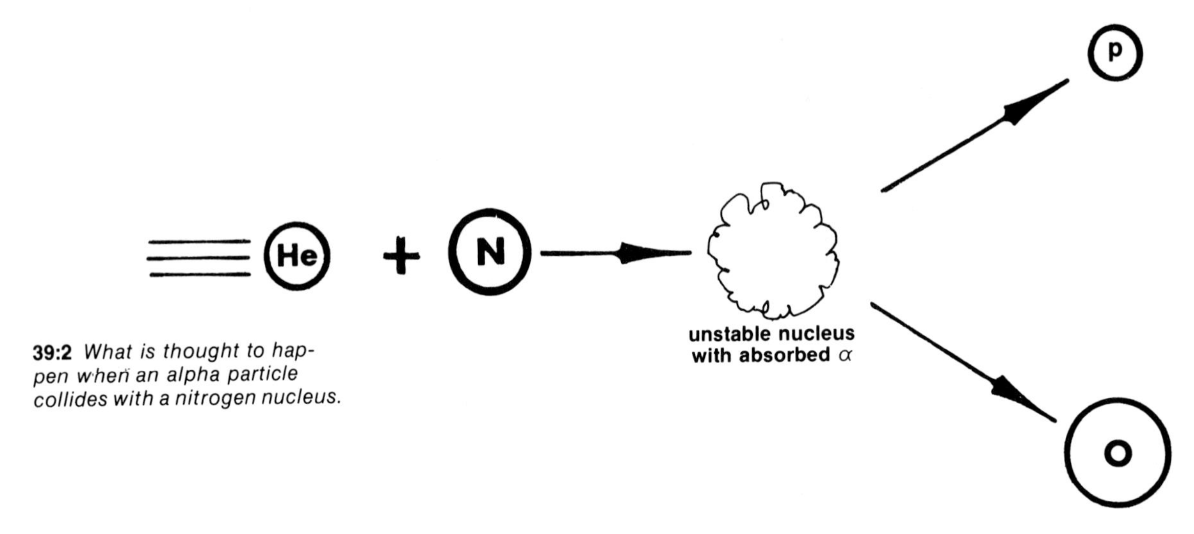

39:2 *What is thought to happen when an alpha particle collides with a nitrogen nucleus.*

39:2 Discovery of the Neutron Another problem facing physicists was that of explaining how negative particles (β) could come out of a positive nucleus if they accepted the Rutherford model of the atom. Rutherford himself suggested that a proton and an electron could be bound together in a nucleus to form a neutral particle. He named this yet unknown particle the **neutron**.

In 1930 two German physicists, Walther Bothe (1891-1957) and H. Becker, bombarded beryllium and boron with α particles. The radiation given off during this bombardment had no charge like γ rays, but its power of penetration was greater than that of γ rays. Using the principles of conservation of energy and momentum, Sir James Chadwick (1891-), an English physicist, deduced that this newly discovered radiation was actually a stream of neutrons. In the case of beryllium, the nuclear reaction proceeds as follows:

$${}_2He^4 + {}_4Be^9 \longrightarrow {}_6C^{12} + {}_0n^1$$

An interesting modern application of neutrons is in the nondestructive testing technique of **neutrography**. A beam of thermal (slow) neutrons is passed through a test object, then through a converter plate to produce ionizing radiation to affect the emulsion on a film. Such a developed film has a great deal of contrast in it. Low density as well as high density materials can be "seen" in the same assembly, as in the neutrograph of an HO scale locomotive in Figure 39:3.

39:3 *A model locomotive viewed using (a) X-rays and (b) neutrons. Notice the structural detail seen in the neutrograph.*

39:3 Particle Accelerators The work in nuclear physics was limited by the energy of the particles available to bombard the nucleus. The beams of α particles which had been used up to 1932 were of low intensity and kinetic energy. They were only energetic enough to bombard light element nuclei. The strong repulsive electrical forces of the heavy element nuclei prevented most of the low energy α particles from hitting them.

39:4 *A Cockroft-Walton generator. The purpose of this machine is to provide the initial acceleration of 750,000 electron volts to the protons. Then they are injected into a 50 million-electron-volt linear accelerator before entering the orbit of an alternating gradient synchrotron.*
(See chapter title page)

39:5 *An aerial view of the principal components of the National Accelerator Laboratory, Batavia, Illinois. The large circle is the main accelerator, or main ring. The medium circle is the booster accelerator. The smallest circle is an auditorium adjacent to the central laboratory. Experimental areas extend to the left. The accelerator is designed to produce 200-500 BeV protons.*

39:6 *The Stanford linear accelerator capable of producing 20 BeV electron beams.*

From the 1930's to the present, scientists and engineers have designed and built various kinds of **particle accelerators** to yield the high energy particles they need for nuclear research. One of the recently built accelerators is two miles long! A type of accelerator called a **synchrotron** that was constructed in Batavia, Illinois, can be used to accelerate protons to speeds of 99.999 percent that of light. Atomic reactors are excellent sources of high energy neutrons.

Physicists use accelerators to obtain high-energy particles to bombard nuclei. The nuclear reactions that occur because of the bombardment may yield information about nuclear structure and forces.

39:4 Radioactive Isotopes Many radioactive isotopes can be produced by bombarding elements with charged particles from accelerators and neutrons from atomic reactors. If Na-23 (a stable isotope) is bombarded with deuterons, Na-24 (an unstable isotope) is produced. The nuclear reaction is

$${}_{1}H^{2} + {}_{11}Na^{23} \longrightarrow {}_{11}Na^{24} + {}_{1}H^{1}$$

The unstable Na-24 breaks down to a stable isotope of magnesium Mg-24:

$${}_{11}Na^{24} \longrightarrow {}_{12}Mg^{24} + {}_{-1}\beta^{0} + \gamma + \nu$$

The ν in the above equation is the symbol for the neutrino, a particle that is always given off in beta decay.

More than 600 radioactive isotopes, called **radioisotopes**, have been produced so far. Many have been used in scientific and industrial applications. Na-24, called radiosodium, has a half-life of 15 hours and can be used in the human body to check for poor blood circulation. A small quantity of radiosodium chloride is injected into a vein and a Geiger counter is placed near another part of the body. If circulation is normal, the radiosodium is carried quickly through the body by the bloodstream and is detected by the counter. If circulation is poor, very little activity will be picked up by the counter after a reasonable time (a few seconds). The counter is moved from place to place on the body until a large amount of activity is detected. This indicates that the radiosodium has collected in this area and there is probably an obstruction in the bloodstream. Radioisotopes used this way are called **tracers**.

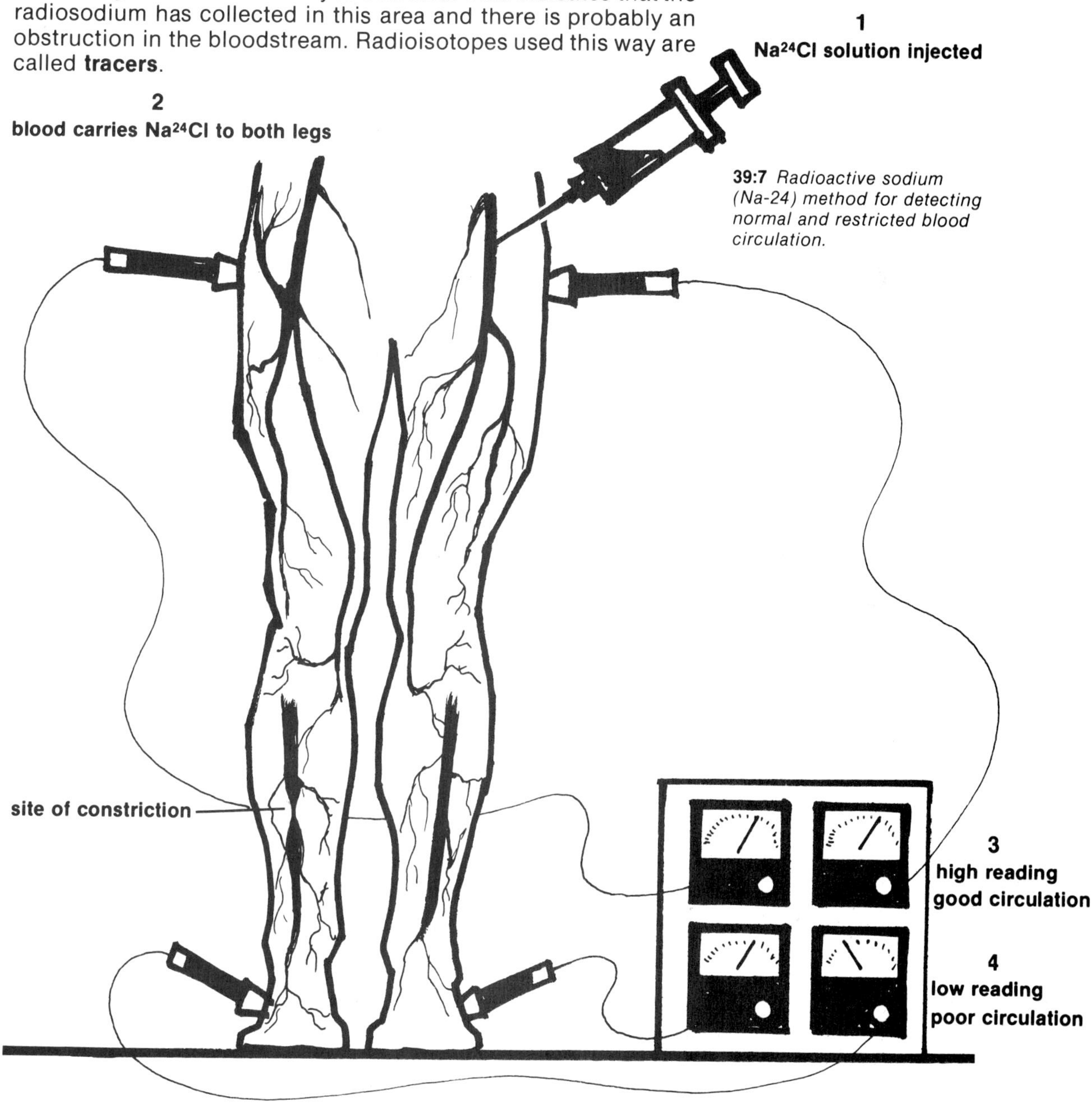

1
$Na^{24}Cl$ solution injected

2
blood carries $Na^{24}Cl$ to both legs

site of constriction

3
high reading
good circulation

4
low reading
poor circulation

39:7 *Radioactive sodium (Na-24) method for detecting normal and restricted blood circulation.*

Leaks in water pipes underneath floors can be discovered using radioactive iodine (I-131). This saves time and money because the leaky pipe can be located without tearing up the whole floor. The radioisotope is injected into the water system. A Geiger counter is moved down the length of pipe on the floor above. The radiation level drops when the counter is moved past the leak.

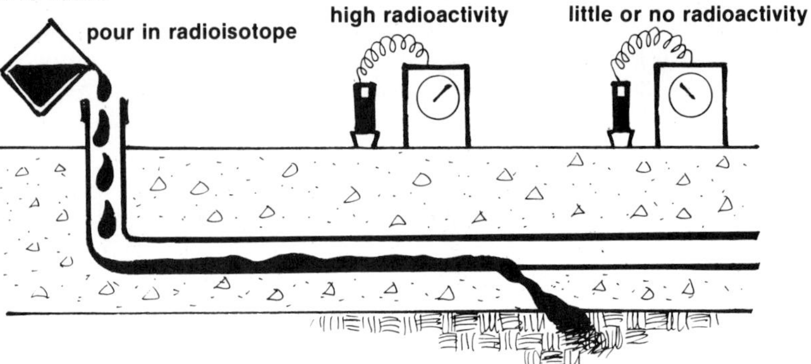

39:8 *Finding leaks in underground pipes.*

Radioisotopes can be used to check the wear of automobile pistons. The pistons are made radioactive and tested in an engine. As the pistons wear, minute particles of radioactive material collect in the engine oil. The oil is checked with a Geiger counter. The more wear, the greater the activity picked up by the detector. If the pistons are checked using different oils, it can be determined which oil cuts down on the piston wear the most. Any time it is necessary to check for small amounts of material in places that cannot be seen, it is likely that radioisotopes can be used.

39:5 Fission and Nuclear Energy Just as chemical reactions release or absorb energy, so do nuclear reactions. However, the difference is about a million times more energy per atom involved. The nucleus is a tremendous storehouse of energy. The discovery of **nuclear fission** which led to the utilization of nuclear energy was accidental, as were many scientific discoveries.

In the 1930's many scientists were bombarding various nuclei with neutrons, trying to create radioisotopes. Some interesting phenomena were observed in these studies. For example, when Al-27 is bombarded with neutrons, radioactive Al-28 is produced, which decays (half-life of 2.3 min.) by beta emission into a stable isotope of silicon (Si-28). The nuclear reactions are as follows.

$$_{0}n^{1} + {}_{13}Al^{27} \longrightarrow {}_{13}Al^{28} + \gamma$$

$$_{13}Al^{28} \longrightarrow {}_{14}Si^{28} + {}_{-1}\beta^{0} + \bar{\nu}$$

$\bar{\nu}$ is a symbol for the antineutrino, which is very similar to the neutrino. Thus, a heavier element is produced by bombarding a lighter element.

Uranium has the highest atomic number of the so-called naturally occurring elements (elements 1-92). Suppose uranium were bombarded with neutrons; would an element with atomic number 93 be produced? If so, it would be a "man-made element." Thus many scientists set out to do this. In 1939 Otto Hahn (1879-1968) and Fritz Strassman (1902-), two German chemists, found an isotope of barium (Ba-139) in a uranium mixture that had been bombarded with neutrons. Later an isotope of lanthanum (La-140) was found in a bombarded sample. These elements have atomic numbers of 56 and 57, lower than that of uranium. Smaller nuclei were being produced, not larger ones, as scientists had hoped. Lise Meitner (1878-1968) and Otto R. Frisch (1904-), Austrian physicists, suggested that the neutron collision causes a uranium nucleus to disintegrate into nuclei of roughly the same size. They named the process nuclear fission after a similar process of cell division in living organisms.

A typical fission reaction is shown below.

$$_{0}n^{1} + {}_{92}U^{235} \longrightarrow {}_{36}Kr^{91} + {}_{56}Ba^{142} + 3{}_{0}n^{1} + \text{energy}$$

The energy release of the reaction is about seven million times that of the explosion of a TNT molecule! Although the above equation is balanced, the total mass of the fission fragments is slightly less than that of the uranium nucleus. The mass loss is converted into energy. From Einstein's equation relating energy and mass

$$E = mc^2$$
E = energy
m = mass
c = speed of light

A small amount of mass will yield tremendous quantities of energy.

Problem: Suppose one α particle is converted completely into energy. How much energy would be generated?

Solution: m = mass of α particle = 6.65×10^{-27}kg
c = speed of light = 3×10^{8} m/sec
$E = mc^2 = 6.65 \times 10^{-27} (3 \times 10^{8})^2 \approx 6 \times 10^{-10}$j

It would take the explosion of about 10^8 molecules of TNT to yield the same amount of energy.

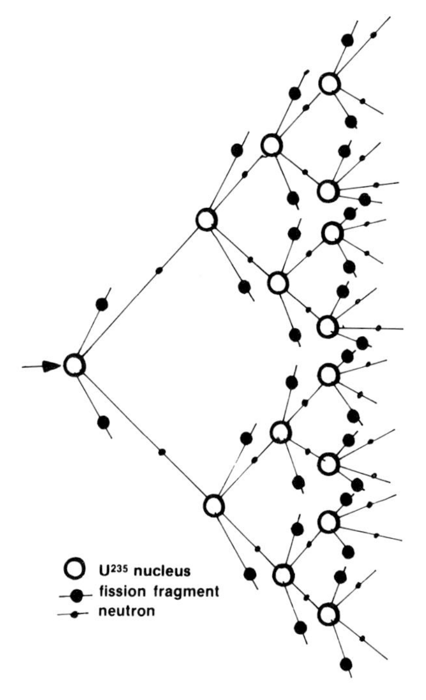

39:9 *A chain reaction.*

Notice that three neutrons are released in the fission of U-235. These neutrons can cause fissioning in three other U-235 atoms. If the same nuclear reaction occurred, nine more neutrons would be produced, then 27, and so on. This is called **chain reaction**; and once started, it can release enormous amounts of energy very rapidly.

Possibly you are wondering why chain reactions do not occur in uranium ores, destroying the metal before it can be removed from its ore. Naturally occurring uranium consists mainly of U-238, and this isotope does not fission. U-235, the isotope that is used in most fission processes, makes up less than one percent of the ore. It must be separated from the other isotopes of uranium before fission will occur. This separation is a difficult process which requires much engineering skill and technology.

39:10 *Aerial view of the Gaseous Diffusion Plant, Oak Ridge, Tennessee, where uranium isotopes are separated from each other.*

39:6 Atomic Bomb If an uncontrolled chain reaction is allowed to proceed, the result is a rapid release of tremendous quantities of energy that are capable of gross destruction. Such a device is referred to as an **atomic bomb**. Two isotopes, Pu-239 and U-235, can be used as "bomb fuel." Plutonium-239 is produced from the neutron bombardment of U-238. The nuclear reactions are given below.

$$ {}_0n^1 + {}_{92}U^{238} \longrightarrow {}_{92}U^{239} + \gamma $$

$$ {}_{92}U^{239} \longrightarrow {}_{93}Np^{239} + {}_{-1}\beta^0 + \nu + \gamma $$

$$ {}_{93}Np^{239} \longrightarrow {}_{94}Pu^{239} + {}_{-1}\beta^0 + \nu + \gamma $$

U-239 forms first, then decays into Np-239. Neptunium was the first man-made element and has an atomic number greater than 92. All elements with atomic numbers greater than 92 are referred to as *transuranic*, or *transuranium*, elements. Np-239 decays into the next transuranium element, Pu-239, which is capable of sustaining a chain reaction.

Two atomic bombs were used in World War II. A U-235 bomb was dropped on Hiroshima, Japan, on August 6, 1945; and a Pu-239 bomb was dropped on Nagasaki, Japan, three days later. Both cities were destroyed. The amount of uranium used in the Hiroshima bomb was about the size of a baseball. You can see the destruction potential in uncontrolled chain reactions! You may think that it was wrong for the United States to use such horrible weapons. World War II had been going on for over three years; Germany and Italy had been conquered; American fighting men were having to island-hop across the Pacific Ocean in amphibious operations to get to Japan. These dangerous operations were costing many lives. Also, to conquer Japan, a bloody amphibious landing on the islands of Japan would have been necessary. Many American and Japanese soldiers and civilians would have been slaughtered in these operations. Possibly the dropping of the atomic bombs was less costly. Certainly the war ended sooner.

Only when the Lord Jesus Christ comes back to this earth will wars end. But you must remember that just before He returns to reign, God will pour out His judgment on sinful men. God's wrath is much more destructive than any device man can manufacture. Sinful men have been killing since Cain, and God's destruction will purge the earth of wicked men and bring peace.

39:11 *God's wrath is more to be feared than man's destructive power. Men can destroy other men only physically. God can condemn a sinner eternally to the lake of fire.*

If you get some fissionable material and a source of neutrons, are you capable of starting a chain reaction? Not necessarily. There may not be enough fissionable material. Most of the neutrons would escape outside a small mass. Also, if the ratio of surface area to volume is too big for the chunk of material, too many neutrons will escape. To sustain a chain reaction, a certain **critical size**, **critical mass**, and **critical shape** of fuel are necessary. If all of these factors are correct, an atomic explosion can occur.

39:12 *Nuclear weapon of the "Fat Man" type, the kind detonated over Nagasaki, Japan, in World War II. The bomb is 60 inches in diameter and 128 inches long. The second nuclear weapon to be dropped weighed about 10,000 pounds and had a yield equivalent to approximately 20,000 tons of high explosives.*

A photograph of an atomic bomb is shown in Figure 39:12. It consists of a radioactive neutron source, subcritical masses of fissionable material, and a means of triggering the bomb. An explosive such as TNT can be detonated to drive the subcritical masses together close to the neutron source. The chain reaction takes over, and the atomic explosion occurs. Temperatures in the millions of degrees Celsius may be reached at the immediate site of the reaction. The surrounding air, expanded rapidly because of this heat, acts as a high shock wave. The tremendous heat and shock waves are capable of destroying cities.

39:13 *Nagasaki, Japan. Remains of the Nagasaki Medical College. Most of the buildings were of wood and were completely destroyed. Reinforced concrete buildings which remained suffered varying degrees of damage from blast and fire.*

39:7 Radioactive Fallout Besides the initial heat and shock wave of a nuclear explosion, there is still much harmful radioactivity in the immediate area. If a person is exposed to too much radiation, the consequences can be serious. If too much cell damage occurs in a living organism, it will die.

If there is a nuclear explosion in the atmosphere, radioactive dust may be circulated over the entire globe by currents in the stratosphere. The dust is then deposited over various places in the earth by different means such as rain and snow. As living organisms eat and breathe, fallout particles become lodged in their bodies. A list of typical radioactive elements found in fallout is given in Table 39-1.

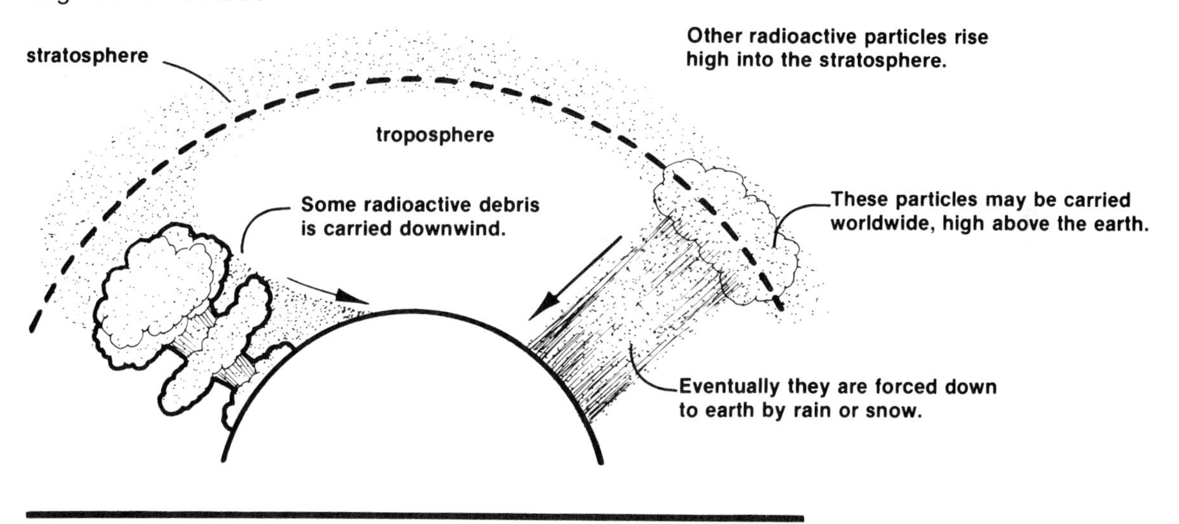

39:14 *Fallout distribution after a surface nuclear explosion.*

Table 39-1 *Some Radioactive Isotopes in Fallout*

Isotope	Half-Life
molybdenum-90	2.8 days
iodine-131	8.1 days
barium-140	12.8 days
strontium-89	53 days
strontium-90	28 years
cesium-137	30 years

Most of these decay fairly rapidly into harmless elements and are not too dangerous. However, the long half-life of Sr-90 and Cs-137 make them radioactive hazards. Sr-90 is similar chemically to calcium and tends to concentrate in dairy products, particularly milk. When taken internally, the Sr-90 collects in the bone structure of an individual and destroys material in the bone marrow. Radiation can also cause mutations in genetic material. The offspring of parents exposed to excessive radioactivity could be stillborn or malformed at birth.

39:8 Nuclear Reactors If the chain reaction process is controlled so that the heat from the nuclear reaction is given off at a much slower rate than an explosion, the energy can be utilized to generate electricity, just as burning coal and oil. The first **nuclear reactor** to achieve self-sustaining controlled release of energy was built by a team headed by Enrico Fermi (1901-1954), an Italian physicist, at the University of Chicago on December 2, 1942.

39:15 *In a pool of water (for shielding), 500,000 curies of cobalt-60 produce the glow known as Cerenkov radiation.*

The key to controlled fission is to control the rate of production of neutrons. Neutrons can be lost in a fissioning mass by the following processes:

1. Capture of neutrons by a fissionable atom with no resulting fission taking place.
2. Capture of neutrons by other material in the reactor.
3. Escape of neutrons from the reactor.

If 1, 2, or 3 happen excessively, the reactor will not operate properly. Proper design of the reactor, correcting this malfunction, comes under the field of nuclear engineering. The uranium fuel normally is placed in assemblies called **fuel elements**; these go in the core of the reactor where the fission process will occur. The fuel element placement and core are designed to utilize as many of the available neutrons as possible. A **moderator** and **reflector** "blanket" are placed around the core. The neutrons collide with molecules in the reflector and bounce back into the core. At the same time, the neutrons lose energy in these collisions. When they are reflected into the core, there is a greater chance that they will cause fission. More fission occurs

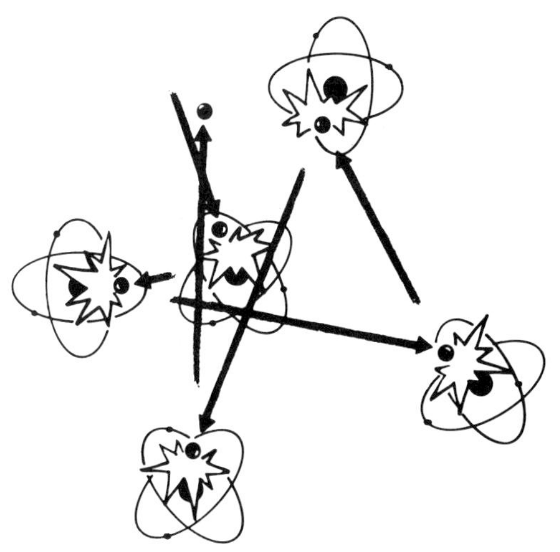

39:16 *Fast neutrons are moderated (slowed down) by collisions with certain types of atoms and then reflected back into the core as thermal neutrons (less energetic than fast neutrons).*

when U-235 is bombarded by slow neutrons (slowed down by collisions with moderator molecules) than by fast neutrons. Excellent moderator materials are carbon (graphite), heavy water, and beryllium.

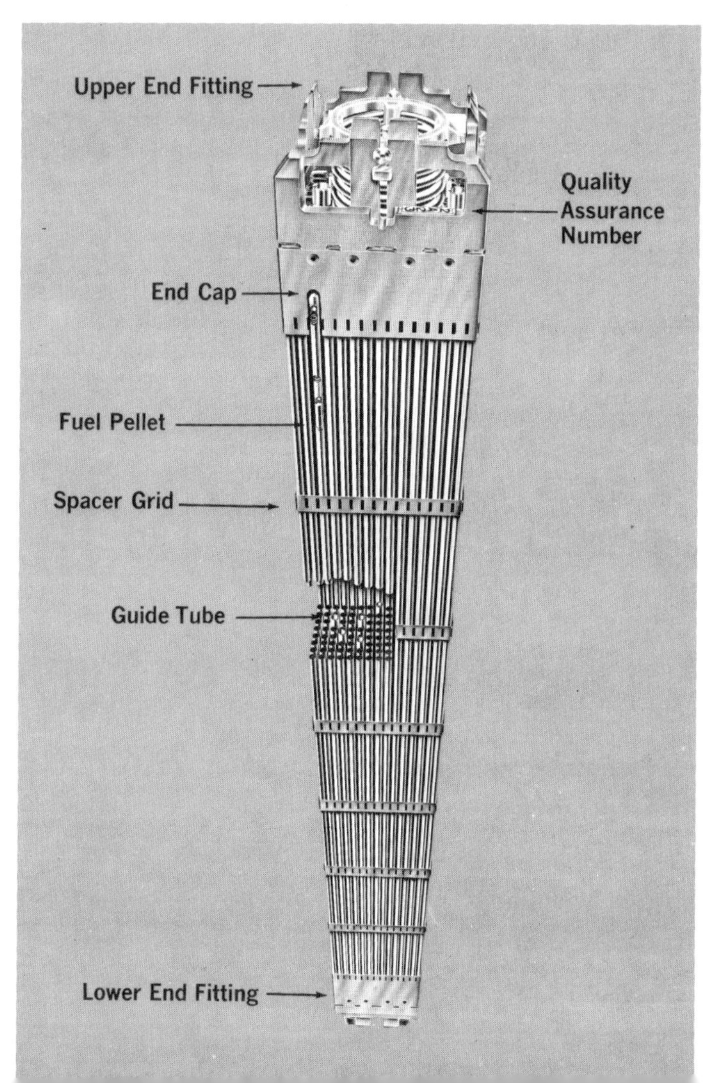

39:17 *Fuel element for a nuclear reactor.*

To keep the neutron production rate just right, **control rods** move up and down in the core. These are made from materials such as cadmium and boron, which absorb neutrons. If neutron production becomes too high (too fast a fission rate), the rods are moved further down into the core. When the neutron production rate becomes too slow, these rods are withdrawn a slight amount.

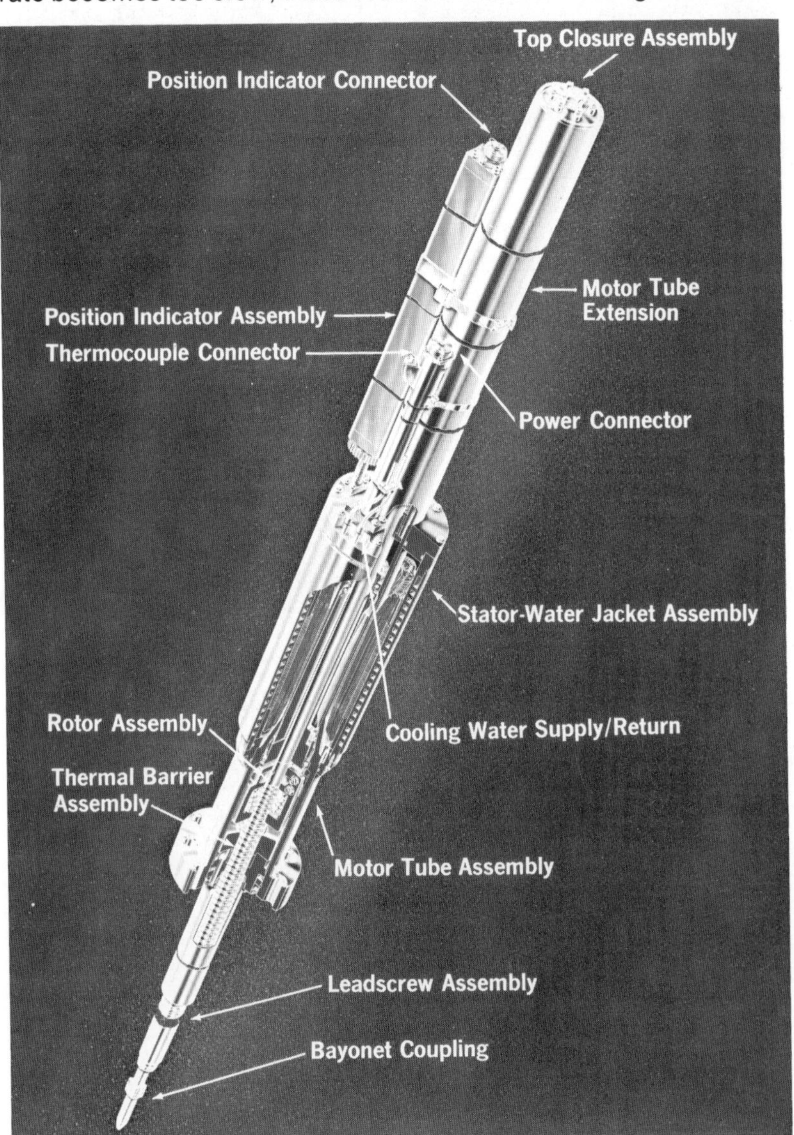

39:18 *Control rod drive for a nuclear reactor.*

If all of the rods are fully inserted in the core, no chain reaction is possible. These rods can be placed in the core quickly to prevent any sudden catastrophe. The reactor is enclosed in a special **reactor vessel** also, and this is surrounded by thick layers of concrete to prevent the escape of harmful radiation.

39:19 *A reactor vessel, weighing 425 tons, is carefully lowered to the barge that will carry it to a nuclear power station.*

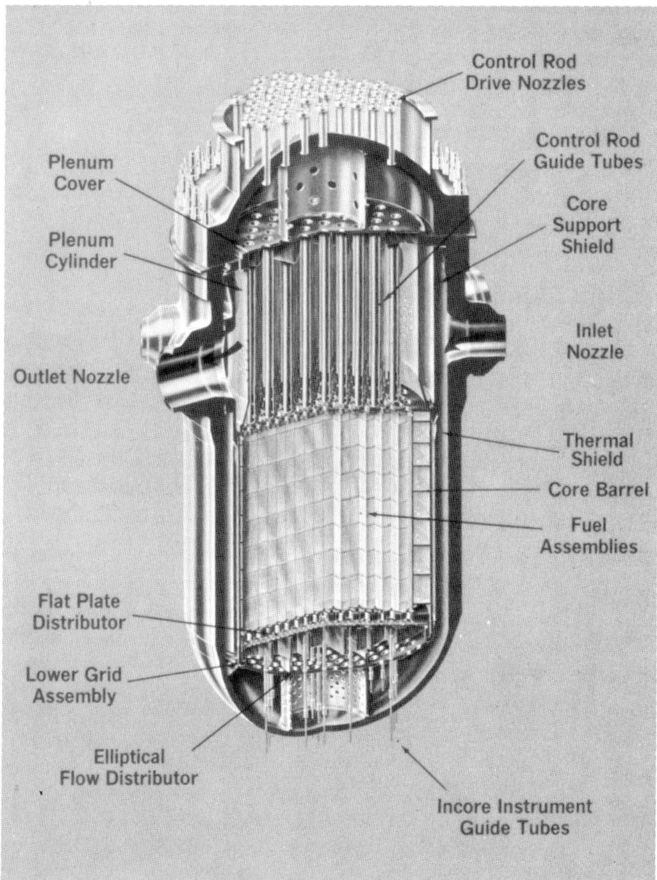

39:20 *A schematic diagram of a pressurized water reactor vessel.*

A fluid is circulated through the core to transfer the heat out of the reactor. This fluid normally goes to a heat exchanger, transfers its heat, and is pumped back through the core. The fluid on the other side of the heat exchanger is water. It is heated to form steam and generate electricity in a turbine. The primary reactor coolant fluid that has been used is water, but gases and liquid metals have been utilized also.

39:21 *A nuclear power station.*

Another type of nuclear reactor that offers promise is a **breeder reactor**. The elements U-238 or Th-232 are put in the fuel elements or in a breeder blanket around the core. Upon neutron bombardment, these isotopes change into fissionable Pu-239 and U-233 respectively. Thus the reactor produces more fissionable fuel than it uses.

Nuclear energy electrical power is now competitive with hydroelectric and fossil fuel generated power. More than half of the newly constructed power plants use nuclear energy. Nuclear power eliminates the air pollution associated with fossil fuel burning, though there is the problem with thermal pollution and low-level release of radioactivity. However, these problems are not serious, and thermal pollution can be overcome. One of the big problems with nuclear power generators is that the initial cost of the plant is high.

39:22 *"Hot cell." Scientists study nuclear fuel and components in this shielded laboratory.*

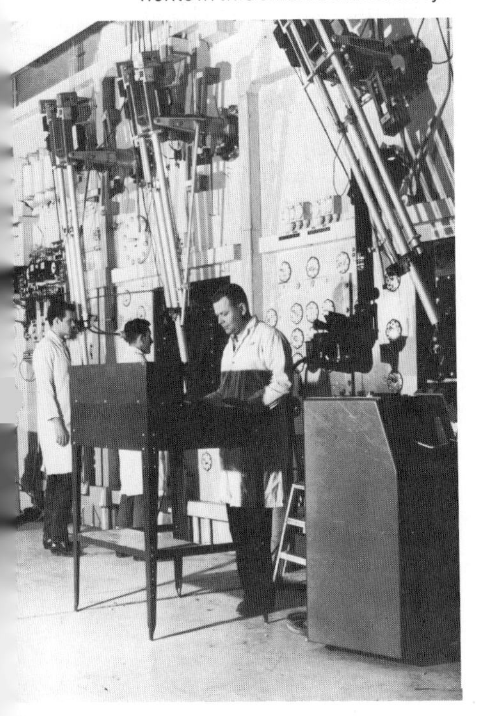

Radiation damage in reactors tends to embrittle structural materials of a reactor and reduce their resistance to corrosion. When fuel elements have to be replaced in a reactor, it may be advisable to recover the unused U-235 in the old elements. The fuel element has to be chemically processed in remote facilities so that workers are not exposed to the residual radiation. Any time a person is working with radioactive material, he must work behind proper shielding.

39:9 Uses of Nuclear Power We have discussed mainly the generation of electrical power. However, there are many present and potential uses for nuclear energy. Nuclear reactors are excellent sources of high energy neutron beams. Some reactors are devoted exclusively to scientific research. Others are used to prepare radioisotopes. Reactors are used to power ships and submarines. Much space can be saved by not having to carry fossil fuel on board. A nuclear submarine can remain submerged for months. Old oil-burning submarines had to surface often to take in fresh air.

39:23 *A research reactor. The experimental floor of the High Flux Beam Research Reactor (HFBR). This reactor has a maximum power of 40 million watts and a maximum neutron flux of 1.6 x* 10^{15} *neutrons per square centimeter per second.*

39:24 *Nuclear Task Force. The world's first all-nuclear surface task force, formed May 13, 1964, by the aircraft carrier* Enterprise, *the guided-missile cruiser* Long Beach, *and the guided-missile destroyer leader* Bainbridge *in the Mediterranean, was deployed with the Sixth Fleet. In the photo, the crew of the* Enterprise *spells Einstein's equation for the equivalence of energy and mass.*

39:25 *The nuclear powered submarine* Aspro *underway.*

39:26 *(a,b,c) The Subterrene rock-melting bit, operating at a temperature of about 1800 degrees Fahrenheit, bores a hole through a prepared rock specimen during a demonstration test and leaves a smooth glass lining. This method of excavating tunnels and boring deep holes promises considerable cost savings and faster work than present methods.*

Possible future uses of nuclear reactors are to provide energy to desalt sea water, convert atmospheric nitrogen into powdered fertilizers, and to manufacture fluid fuels from low-grade coal. Research work is being done on nuclear rocket engines. Since most of a rocket is taken up with fuel compartments, a great deal of space would be saved with nuclear power. The Atomic Energy Commission sponsored research on exploding nuclear devices to create lakes, harbors, and canals between oceans. Basically, nuclear energy is used to blow big holes in the ground quickly.

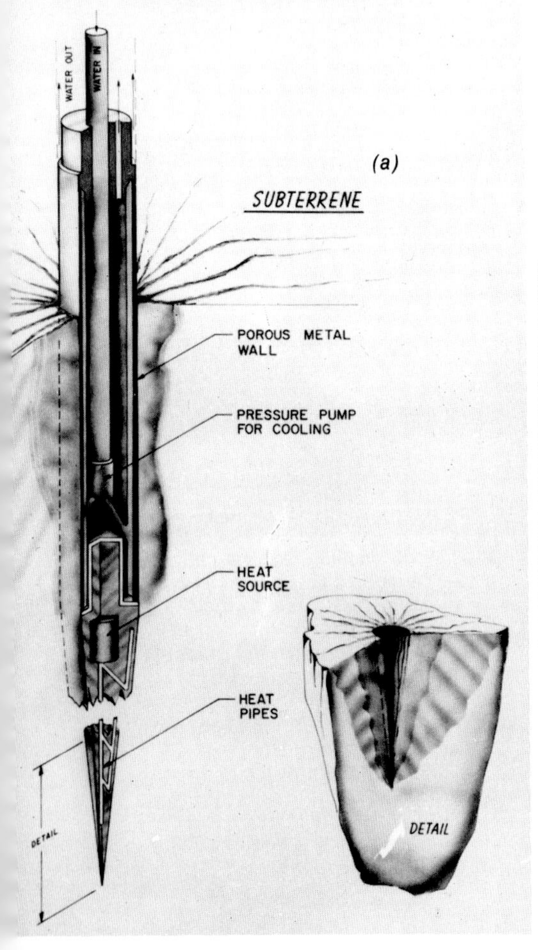

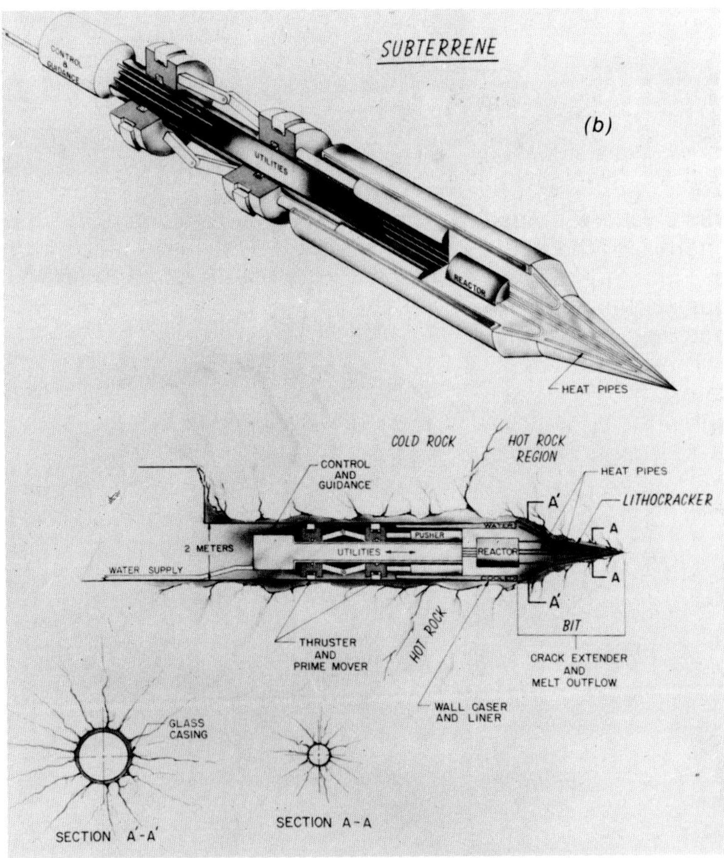

(c)

39:10 Nuclear Fusion Scientists have calculated the binding energy for the nuclei of each element. This information is shown in Figure 39:27. The element having the highest binding energy per nucleon is iron, and it is the most stable of all the naturally occurring elements. When an unstable uranium nucleus fissions, two smaller, more stable nuclei are formed; and energy is released. Chemically, physically, and biologically, any time a substance goes to a more stable state, it releases energy.

Looking below iron in Figure 39:27, you can see that the only way for nuclei to become more stable (and release energy) is to

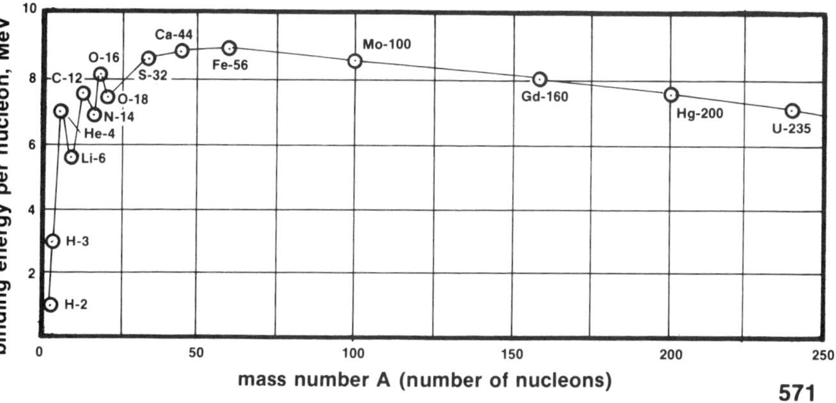

39:27 *Nuclear binding energy curve.*

combine or fuse; that is, nuclear fusion. Nuclear fusion reactions have been accomplished in the laboratory by bombarding light isotopes with other light high-energy particles such as deuterons. Typical fusion reactions are shown below.

$$ {}_1H^2 + {}_1H^2 \longrightarrow {}_1H^3 + {}_1H^1 + \text{energy} \qquad (1) $$
$$ {}_1H^2 + {}_1H^3 \longrightarrow {}_2He^4 + {}_0n^1 + \text{energy} \qquad (2) $$
$$ {}_1H^2 + {}_2He^3 \longrightarrow {}_2He^4 + {}_1H^1 + \text{energy} \qquad (3) $$

${}_1H^3$ is tritium, an isotope of hydrogen. In these typical fusion reactions about 10^6 (1,000,000) times more energy is released than in the explosion of a TNT molecule. This is not as much as in the fissioning of a uranium nucleus; but, gram for gram, more energy is released in the fusion process because there are more hydrogen isotope nuclei than uranium nuclei per gram.

39:11 Fusion Hydrogen Bomb The light nuclei must collide at a very high speed for fusion to occur. These speeds correspond to the high temperatures found in stars. If an atomic bomb is exploded, the temperatures necessary to start fusion are available. Thus an atomic bomb can be used to trigger a **hydrogen bomb**. Critical mass considerations limit the size of an atomic bomb, but there are no such restrictions on hydrogen bombs. A hydrogen bomb can be made as big as necessary. There is a limitation if a bomb is to be transported by airplane or rocket. It cannot exceed the carrying capacity. Since an atomic bomb is needed as a fuse, it is obvious that a hydrogen bomb has far more destructive potential.

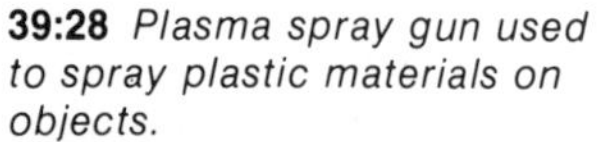

39:28 *Plasma spray gun used to spray plastic materials on objects.*

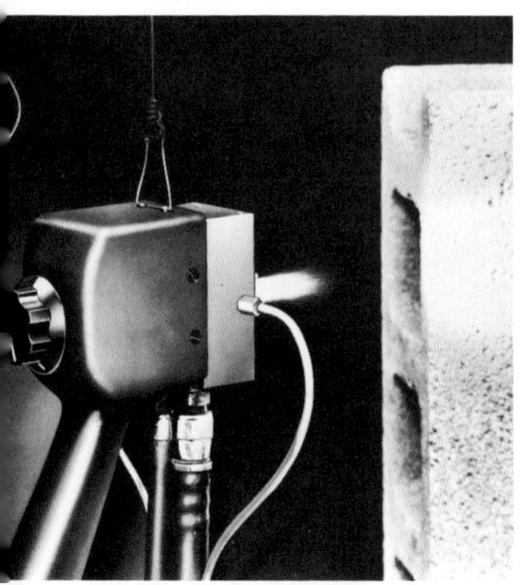

39:12 Controlled Fusion The fusion of deuterium and tritium (reaction 2) could be used to generate electric power. Deuterium is found in water; eight gallons of water contain about one gram of this isotope. Thus the oceans offer an almost unlimited supply. If deuterium could be used in controlled fusion, it would be a tremendous source of energy.

There are problems that have not been solved as yet to make controlled fusion possible. Since the positive nuclei must collide to fuse, the repulsive forces of like charges must be overcome. This means the nuclei must travel at high speeds and be confined in a region where they can collide without escaping or being absorbed by any container walls. There must be many collisions in a short space of time so that energy can be obtained from the process. All of these circumstances require that the nuclei must be at temperatures around 100 million degrees. At these temperatures all of the electrons have escaped from the nuclei and the positive and negative particles form a *plasma*. There is no known container material that can hold hot plasma. The container walls would simply melt or vaporize. Theoretically, the plasma can be held with a magnetic field and the charges accelerated with an electric field. However, sustained controlled fusion to produce electric power is not yet commercially feasible. If you are interested in making a career in some field of science, this one would be very challenging.

39:29 *An experimental apparatus for controlled fusion reactions.*

39:30 *A stellarator for performing fusion research.*

39:13 Reactions in the Sun and Stars Present theory holds that the sun and stars derive their energy from fusion reactions. The main reaction in the sun is believed to be the stepwise combination of four hydrogen nuclei to form a helium nucleus, two positrons and two neutrinos. The overall reaction for all the steps is

$$4\,{}_1H^1 \longrightarrow {}_2He^4 + 2\,{}_1\beta^0 + 2\nu + 3\,\gamma$$

39:31 *Globular Cluster M13, NGC-6205.*

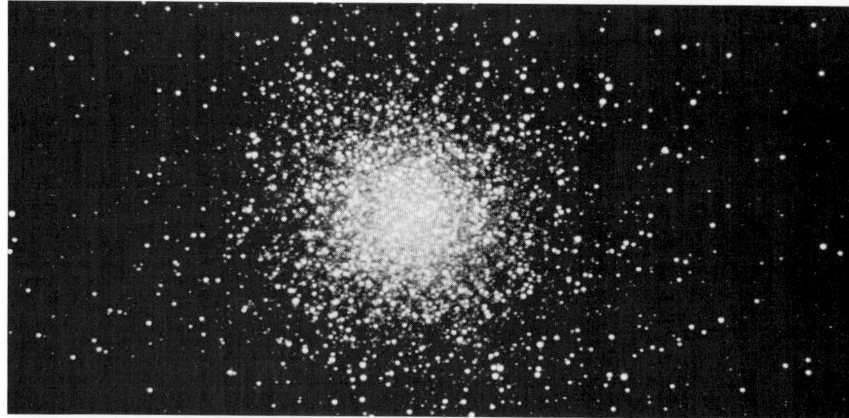

39:32 *Our sun.*

It has been estimated that the sun is consuming its hydrogen at the astonishing rate of 4½ million tons per second to maintain its present level of energy output. The energy produced by the sun is radiated out into space in all directions, never to be recovered. We have here a good example of the degeneration processes that are so prevalent throughout the universe. Nature is pursuing an inexorable downhill trend in accordance with the second law of thermodynamics.

Stars having temperatures higher than that of the sun can carry out additional fusion reactions, such as the combination of two helium nuclei to form a beryllium nucleus, a beryllium and a helium nucleus to form a carbon nucleus, and so on, forming even-numbered elements of increasing atomic weight until iron is reached. Elements formed in this way can be thought of as "ashes" of the fusion process. Eventually, if the Lord permits the present order to continue long enough, all usable energy-producing reactions will be exhausted. The sun and stars will run down completely and burn out. Science has found that the heavens are indeed growing old and wearing out, exactly as the Bible indicates (see Psalm 102:25-26; Hebrews 1:10-11).

List of Terms

atomic bomb
breeder reactor
chain reaction
control rods
critical mass
critical shape
critical size
fuel element
hydrogen bomb
moderator
neutrography
neutron
nuclear fission
nuclear fusion
nuclear reactor
particle accelerator
radiation damage
radioactive fallout
radioisotopes
reactor vessel
reflector
synchrotron
tracers
transmutation

Questions

1. Define transmutation.
2. (a) When an alpha particle interacts with a nitrogen nucleus (1) does it hit the nucleus and knock a proton out of it or (2) is the alpha particle absorbed and the proton emitted?
   (b) Offer proof for your answer to part (a).
3. (a) Why is a converter plate used in neutrography?
   (b) Why is this necessary?
4. Why are particle accelerators used in physics?
5. What is the primary use of high-energy particles?
6. (a) What are radioactive isotopes?
   (b) What is a stable isotope?
7. Name 3 uses of radioisotopes.
8. Suggest some evidence that the nucleus is involved in radioactive decay and not the electrons as in chemical reactions.
9. What is nuclear fission?
10. Calculate the energy release if three electrons are converted completely into energy. Mass of an electron = 9.11×10^{-31}kg.
11. Calculate the energy release if 1 gram is converted completely into energy.
12. What is a chain reaction?
13. Why are uranium ores not destroyed by natural fissioning?
14. What is an atomic bomb?
15. When is radioactive fallout dangerous?
16. (a) Who was the first physicist to achieve a nuclear chain reaction?
   (b) What is the value of controlled fission?
17. (a) What assemblies in a nuclear reactor contain the fissionable material?
   (b) What assemblies in a nuclear reactor insure safe operation?
18. What is a breeder reactor?
19. What is nuclear fusion?
20. Which process offers the most energy release per unit mass, fusion or fission?
21. What are the problems involved in controlled fusion?
22. How do scientists think that stars get their energy?

Student Activities

1. Write a paper on particle accelerators. Be sure to include a description of each type.
2. Write a paper on the peaceful uses of atomic energy. Write to the Atomic Energy Commission, Washington, D.C. 20545, for helpful information.

Photo next page

The Howell Memorial Science Building at Bob Jones University.

HOWELL MEMORIAL
SCIENCE BUILDING

UNIT XII TWENTIETH CENTURY PHYSICS

Chapter 40 Modern Physics

40:1 Waves and Particles If we analyze the motion of a planet about the sun, we treat the planet as if it were a **particle**. If we investigate the reflection of sound, we treat the sound as if it were a **wave**. The two scientific models are mutually exclusive. It would seem strange to talk about a moving planet as wave motion, or sound traveling as particles. If something is a wave, it is not a particle; and if it is a particle, it is not a wave. This seems to make sense; but does it? Remember we are dealing only with scientific models, not reality. A planet is a planet, and sound is sound, but does science really describe either correctly? No. The models are good, but they are far from perfect. They are the best models we have now; so we will continue to use them until something better is developed.

40:2 Photoelectric Effect In using the wave and particle concepts, we encounter some problems in modern physics. For many years light was considered to be wavelike. It could be reflected and refracted, and these abilities are properties of waves. Then came the problem of the **photoelectric effect**. In 1887, the German physicist Heinrich Hertz (1857-1894) noticed that when light of a short wavelength shines on a metal surface, electric charges are emitted. Since both light and electricity are involved, it was named the photoelectric effect. The charges were later found to be electrons.

In studying the photoelectric effect, scientists have found that the current coming off any metal surface exposed to light is

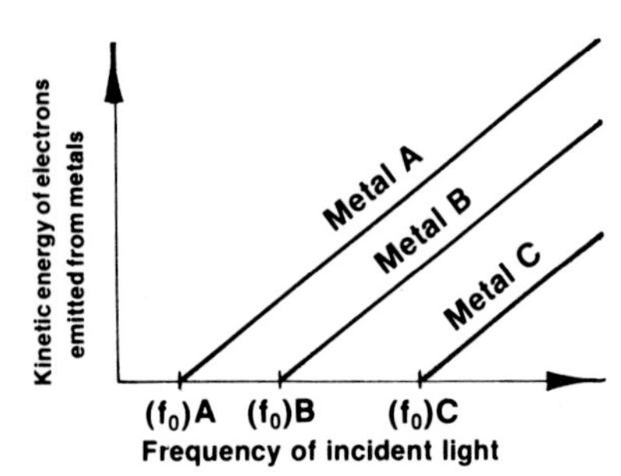

40:1 *Photoelectric effect. Each metal gives off photoelectrons after a certain minimum frequency (f_0) of light is directed onto its surface. This threshold frequency is different for each metal.*

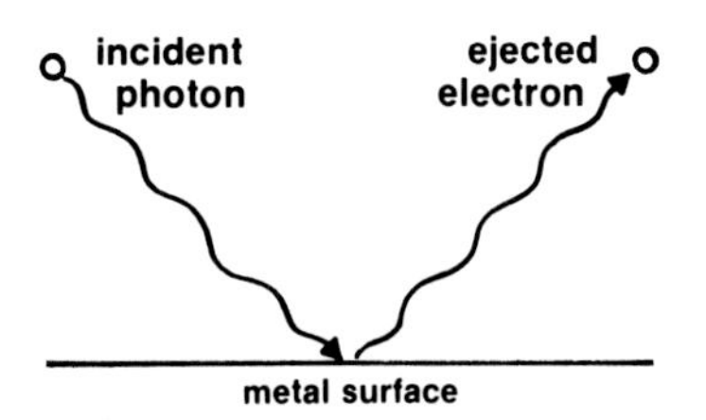

40:2 *Photoelectric effect—one photon can eject one electron from a metal surface.*

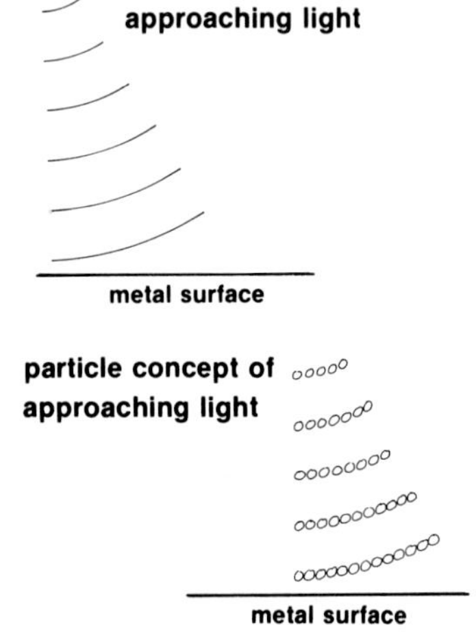

40:3 *Suggested models for particulate and wave nature of light.*

dependent on the frequency of the light. If the frequency of the light is too low, no electrons will be ejected, no matter how long you shine the light on the metal surface. If you increase the intensity of the light beam but do not change the frequency of the light, still no electrons will be emitted. This just doesn't make good sense according to the wave theory. The more intense any light wave is, the more energetic it is, according to the wave theory. If you want to knock an electron off the metal surface, you just "hit it more often" with light of the same frequency. But this does not work.

However, once the light gets up to a certain frequency where electrons are emitted, if the intensity of the beam is increased, more photoelectrons are emitted. Also, the kinetic energy of the photoelectron is dependent upon the frequency of the incoming light and not on the intensity of the beam. Again according to the wave theory, if the intensity of the beam increases, the kinetic energy of the photoelectron should increase. Thus, the wave theory of light fails to explain the photoelectric effect.

Albert Einstein (1879-1955), a German physicist, developed an "explanation" for the photoelectric effect that was later proven experimentally by Robert A. Millikan (1868-1953), an American physicist. Einstein proposed the amazing idea that light arrived at the metal surface in small bundles, or **quanta**! These individual bundles are now called **photons**.

Instead of a broad wave front of light, the energy of the beam is divided into small chunks. If the light is of a single frequency, all of the chunks or quanta have exactly the same energy. The energy of a single quantum is proportional to its frequency (f) where h is a very small proportionality constant (Planck's constant).

$$E \propto f$$
$$E = hf$$

It takes a certain amount of energy (E_o) to knock an electron out of a metal surface. If the energy (E_p) of the photon is less than E_o, an electron will not be ejected. If the energy of the photon is greater than E_o, an electron will be ejected. You can imagine that the individual photons collide with individual electrons and impart their energy to the electrons in this way. No matter how many photons with energy below E_o hit an individual electron, it will not leave the metal surface. In other words, when $E_o > E_p$, the intensity of the light beam does not have any effect. When the light is of such a frequency that $E_p > E_o$, more photons hitting the metal surface produce more photoelectrons. In other words, when $E_p > E_o$, the intensity of the light beam does have an effect (see Figure 40:4). If you had a milk bottle filled with concrete, and you threw bean bags at it, you could never tip the bottle over, no matter how many bean bags you threw. However, if you threw a heavy ball at a bottle, you could tip it over easily if you hit it. Low frequency photons are like bean bags, and high frequency photons are like the balls. The high frequency photons have enough energy to remove an electron.

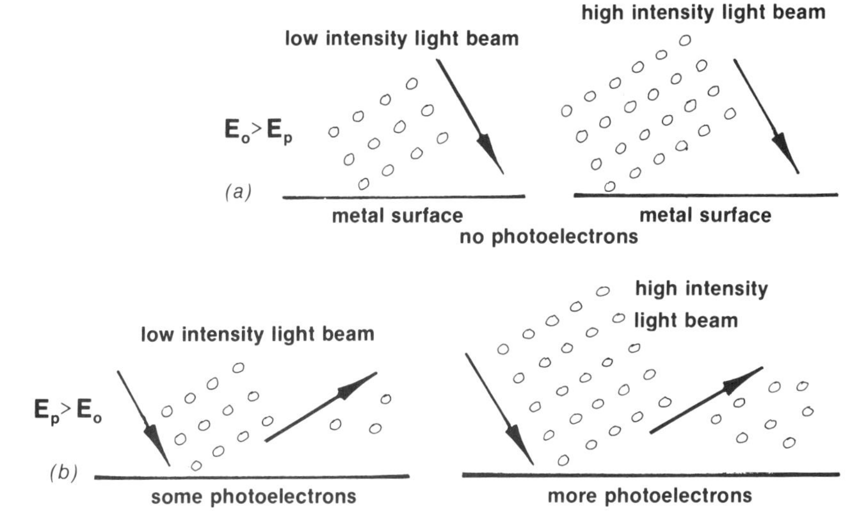

40:4 *(a) If the energy of the photons is less than the threshold energy, no photoelectrons will be emitted regardless of the intensity of the beam. (b) If the energy of the incident photons is greater than the threshold energy, photoelectrons will be emitted in an amount dependent upon the intensity of the beam.*

40:3 What is Light? Well, is light particulate or wavelike? Scientists still use wave theory to "explain" refraction and reflection of light, and particle (quantum) theory to "explain" the photoelectric effect. Light is neither a wave nor a particle, but it has properties of both. Light acts like a wave sometimes and a particle at others. What is wrong with light? Can't it make up its mind? There is nothing wrong with light. The problem is with our scientific models. They are insufficient and incomplete. But that cannot be helped. Scientists try to describe phenomena in nature the best way they can with the tools they have available. As we have said before, scientists are humans, and humans are limited.

To be blunt, we do not know what light is. We can operate only within the framework of our models. It may help you to realize that when we can localize something (know its position), we treat it as a particle. When we cannot localize something, we treat it as a wave. Actually, the best models of light, electrons, and so forth are mathematical models; but the trouble with them is that they give you nothing to visualize. You can best understand the submicroscopic phenomena (light, electrons, protons, and so forth) of physics mathematically. The mathematics you need is beyond the scope of this book.

40:4 Uncertainty Suppose you want to localize an electron so that you can see it. Let us say that you are going to use some form of electromagnetic radiation to view the object. By some means, you have isolated an electron. You direct the radiation to where you think the electron is. When it hits the electron, it will be reflected backwards through an optical system, and you will see the image of the electron! Unfortunately, no. As you direct the radiation onto the electron, let us say it is at rest. It is completely localized. The lenses in the optical system of your "instrument" have a certain field of view. You focus them toward the electron, but you see nothing. The radiation hitting the electron imparted kinetic energy to it, and it sped from the field of view.

This imaginary, impossible experiment serves to illustrate a point. Any radiation used to view submicroscopic particles will give the object such a large amount of kinetic energy that it will speed away before you can see it. Thus in any measurement of a quantity involved with submicroscopic particles there will always be an **uncertainty** in the measurement. The problem with our experiment was that we claimed to be able to localize an electron exactly and know its exact speed (zero). When two simultaneous measurements are made in science there is a limit to how accurate they will be.

40:5 *An electron cannot be localized long enough to see it.*

Look at this problem another way. Suppose you want to measure the speed and position of a moving automobile. You place two boards in the street and with a stop watch you see how much time it took for the car to travel from one board to the other. You know the position of the car at all times, but your speed is only an average (distance/time). Suppose you decide to obtain the exact, instantaneous speed of the automobile at an exact location. You move the boards closer and closer together, you use electronic devices rather than the stop watch, then you use radar, and so on. Finally, with the boards almost touching, you will get the exact position and speed. No, again. With the boards so close together, there is no instrument that can record the small time interval taken between the boards as the automobile moves over them. You have the exact position of the car, but you have no idea of the speed.

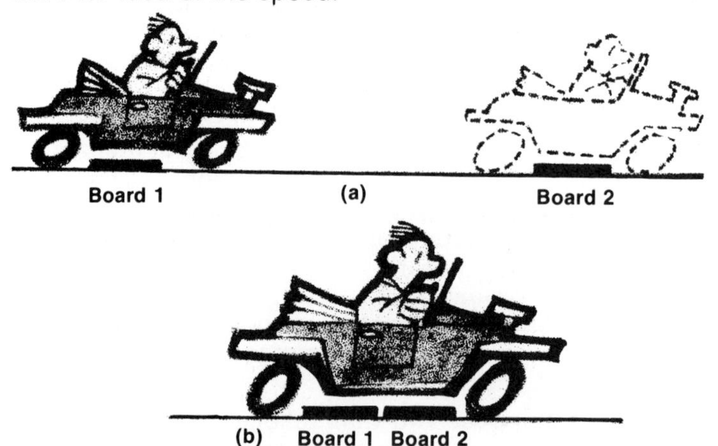

40:6 *(a) The average speed of the car can be measured between boards 1 and 2. (b) Although you know the exact position of the car, the boards are too close to get a speed at that exact location.*

This is the same situation that existed with our supposed isolated electron. If you are willing to accept a little error in the measurement of the position and speed of an automobile, you can get the two readings nicely—not exactly, but with an error so small as to be insignificant. However, the electron is so small that the uncertainty in the same measurements is quite large.

40:5 Probability, not Position From the previous discussion you should understand, then, that we never can know where something so small as an electron is. Rather than try, scientists work with atomic and subatomic problems assuming this uncertainty. A scientist may say an electron is probably within a certain region of space 95 percent of the time. The smaller the region, the lower the **probability.** Such a region is called an "electron cloud." A *1s* cloud is the "place" where we will most likely find a *1s* electron in an atom. Refer to our discussion in Chapter 6. Remember, if the electron "acts" like a wave, it has no definite position. So in quantum mechanics we deal with wave functions instead of exact quantities. Wave functions are mathematical statements that represent electrons, protons, and the like. The invisible atomic world is a world of wave functions, probability amplitudes, and mathematical mazes. It may be complicated, but it works!

40:7 *An electron cloud.*

40:6 Christian Interpretations The success of the scientific method has led many people to believe that results obtained from the field of science are absolute and unquestionable. The views developed from quantum physics have tended to tame this rash idea. The uncertainty and probabilistic nature of science should convince anyone that caution should be exercised in scientific interpretations. There is no exactness in science.

Quantum theory has deflated the balloon of absoluteness in science. The proper emphasis is placed on uncertainty and possibilities. Many times no visual model is possible; only the solutions of wave equations are available.

Do not base your soul's eternal destiny on man's ideas and vague conceptions. Put your trust in something reliable and unchangeable. Only the Lord can offer you fixed values and truth. We do not condemn true science; we only want you to see its proper worth in your life. You cannot expect more out of science than it is capable of offering. If there is an area where God's Word and men's pronouncements conflict, obviously the humanistic concept is wrong.

40:7 Aether For centuries many scientists felt that space in the universe was not a vacuum but was actually filled with a fluid called **aether** (sometimes spelled "ether"). This was the medium in which light was thought to travel; it had been suggested by Christian Huygens (1629-1695), a Dutch mathematician. The aether was an unusual fluid—the planets could travel through it without being slowed down, and it was invisible. The aether was absolutely stationary; hence, it would make a good reference frame for moving objects. Two American physicists, Albert Michelson (1852-1931) and Edward Morley (1838-1923), designed an elaborate apparatus to check the motion of the earth relative to the aether. They concluded, from the results of the experiment, that aether did not exist. Also, they learned that the speed of light is not affected by the motion of an observer. Light coming to the earth from star (1) or (2), as illustrated

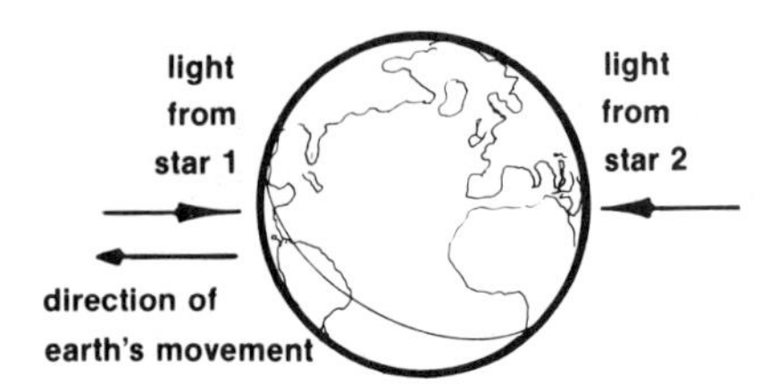

40:8 *According to the Michelson-Morley experiment, the speed of light coming from star (1) and from star (2) are exactly the same if measured on the moving earth.*

in Figure 40:8, had the same speed. The motion of the earth did not cause the speed of light from (1) to be faster or the speed of light (2) to be slower. According to the old ideas of relative speed, the speed of light from (1) to an observer on the earth should be $v_1 + v_e$ since the earth is moving toward (1) where

v_1 = speed of light in a vacuum, and
v_e = speed of earth in orbit.

Also, the speed of light from (2) to an observer on the earth should be $v_1 - v_e$, since the earth is moving away from (2). Actually, the speed of the light in both cases is 3.00×10^8 m/sec.

40:8 Special Relativity Einstein suggested the **special theory of relativity** in 1905, in which he assumed the nonexistence of aether and the consistent speed of light in the same medium. There are some serious consequences of this last assumption. The one physical absolute in the universe is assumed to be the speed of light. It is a maximum in a vacuum (3×10^8 m/sec), and it never changes, regardless of the speed of its source or of the observer.

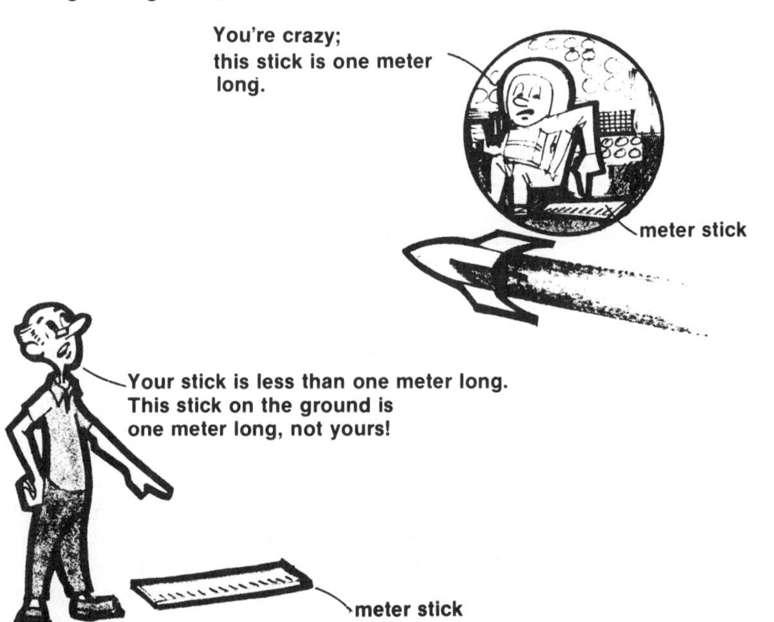

40:9 *Length contraction.*

If this theory is accepted, the usual definitions of length and time have to be revised. For instance, suppose you want to measure the length of a meterstick on the earth. You measure and find that it is the expected one meter. You then send the meterstick off into space on a rocket at 10^3 miles per hour. Pretend you can measure the length of the meterstick on the rocket from your position on earth. The stick measures less than one meter! If you get on the rocket with the stick and measure the length, it is one meter. If you get on the rocket, leave the meterstick on the earth, and measure its length from the moving rocket, it is less than one meter! The length of the stick changes as far as you are concerned, depending upon your frame of reference. As long as you and the meterstick are in the same reference frame (on earth together, or on the rocket together), the stick will measure a length of one meter. Interestingly, the same is true of time. If you put a clock on a rocket and kept a watch with you on the earth, the clock in the moving rocket would appear to move more slowly to you than your watch. If you were on the rocket with the clock, it would appear to run normally. Again, what you observe depends upon your frame of reference.

Another interesting consequence of the special theory of relativity is that an object can have a variable mass depending upon its speed. The faster an object moves, the more massive it becomes! Electrons have been accelerated to speeds of 0.999999996 times the speed of light, and their mass increases over 10,000 times their rest mass (9.11×10^{-31} kg). Things we have discussed, such as **length contraction**, **time dilation**, and **relativistic mass increase**, only become important at very high speeds, that is, those approaching the speed of light.

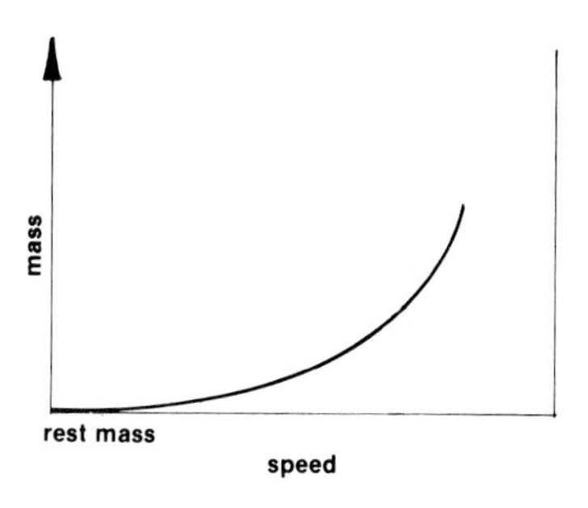

40:10 *The increase of the mass of a particle with speed.*

40:9 Energy and Mass We mentioned earlier the equation $E = mc^2$. One interpretation from this equation is that mass is just one form of energy. The laws of **conservation of mass** and **conservation of energy** are combined with this equation. When mass is conserved, energy is conserved. Thus, there is a mass equivalence of energy.

In this course we have discussed our physical world from two standpoints—chemistry, the study of matter, and physics, the study of energy. Yet, in the present chapter we have seen that matter and energy are really both manifestations of the same thing. How intriguing it is to contemplate that something which occupies space and has mass can also at the same time be the ability to do work! At present we are unable to comprehend this.

40:10 The Grand Perspective In looking back over the material we have covered in this book, we find that certain key concepts stand out. One of these is the *orderliness of nature*. We do not see the chaos and confusion that would reign if the world had originated by chance. The universe operates in accordance with definite laws ordained by the Creator. Man has attempted to formulate and codify the laws and principles of nature. We invariably fall short of our ideal of "thinking God's thoughts after Him," but the God-given order in the things we investigate is unmistakable at every level of scientific inquiry. We see it under the microscope. We see it in the behavior of heat, light, sound, electricity, and magnetism. We see it in the motions of the stars and planets.

Another concept of supreme importance is the *conservation of energy* (the first law of thermodynamics). This is a sweeping generalization encompassing all of the observed universe. Under the present economy—the laws by which the universe has operated since the end of the creation week—energy can be neither created nor destroyed. Creation is finished (Genesis 2:1-3). No new material or energy is coming into being. That which was formed at the beginning is being preserved until the time God chooses to dissolve the present order and create a new heaven and a new earth (Revelation 21:1).

Finally, we have seen an overall principle of *deterioration and decay* **(degeneration)** in the world about us (the second law of thermodynamics). Although the total amount of energy in the universe is remaining constant, its availability for accomplishing useful work is decreasing. Rather than remaining in a compact or concentrated form, energy tends to spread out. As it does so, much of it is lost. The sun and stars are continually dissipating their energy into space. The earth is slowing down in its rotation. Its magnetic field is decaying. Continents are eroding into the oceans. Truly, the heavens and the earth are waxing old like a garment (Psalm 102:25-26; Hebrews 1:10-11). The Christian researcher recognizes that many of the principles outlined in Scripture have been borne out by scientific observations in a most striking way.

40:11 A Challenge In studying about scientists and their work, you have probably come to realize that science is nothing more than man's feeble effort to understand God's creation and to harness some of its forces. Our success in these undertakings has been less than complete. But in retrospect we see progress. We can do things today that we could never do before, go places we could never go before. More of the universe "makes sense" to us than it used to. But while some of the cosmos has become intelligible, much of it remains shrouded in mystery. Each question that is answered by science raises several new ones which await future study.

Perhaps you have felt challenged to pursue some avenue of scientific research. More Christian people are needed in the sciences today than ever before. Too often, competent men of courage and character have been missing when false assumptions and ungodly theories have needed to be contested. Many areas of study cry out for further investigation. Current ideas need to be scrutinized and reassessed. This is a vital ministry. It is not beyond the realm of possibility that someone who reads this book might find himself called to this fascinating work.

List of Terms

aether
conservation of energy
conservation of mass
degeneration
length contraction
particle
photoelectric effect
photons
probability
quanta
quantum theory
relativistic mass increase
special theory of relativity
time dilation
uncertainty
wave

Questions

1. In classical physics when we discuss motion, two different models of natural situations are mentioned as capable of motion. What are these extremely different models?
2. Are scientific descriptions correct? Why?
3. (a) In the photoelectric effect does light appear to act as a wave or a particle?
   (b) In reflection and refraction, does light appear to act as a wave or a particle?
   (c) Is light wavelike or particulate?
4. Light is shone onto a metal surface. The energy required to release a photoelectron is E_o.
   (a) If the energy of the beam of light is E_{L_1} and $E_o > E_{L_1}$, will any photoelectrons be released?
   (b) Suppose we allow much more light of energy E_{L_1} to flood the metal surface; will any photoelectrons be released?
   (c) If we increase the energy of the beam of light to E_{L_2} and $E_{L_2} > E_o$, will any photoelectrons be released?
5. Has anyone ever seen an electron? Why?
6. The *exact* speed and position of an automobile can be measured simultaneously. True or False?
7. A fairly accurate measurement of the speed and position of an automobile can be measured simultaneously. True or False?
8. What is an electron cloud?
9. Is science exact? Why?
10. What was the aether supposed to be?
11. (a) What is the speed of light in a vacuum?
    (b) If an observer is moving at 1.85×10^5 m/sec in the same direction as the light is moving, how fast does the light appear to be traveling to him?
12. You are standing on the earth as a meterstick zooms by your head at 3×10^3 m/sec. Does the meterstick appear longer or shorter than one meter (according to the special theory of relativity)?
13. Mass is one form of energy. True or False?
14. Is nature orderly or disorderly? Why?
15. (a) Can man create or destroy energy?
    (b) Is the universe deteriorating?
16. What is a quantum?
17. There is a need for Christian men of science to reinterpret ungodly theories and ideas. True or False?

Student Activities

1. Write a paper on the Michelson-Morley experiment showing how they disproved the existence of aether.
2. Write a paper on quantum physics explaining the wave-particle paradox.
3. Write a paper on the unusual considerations of speed, mass, time, and length when one accepts the special theory of relativity.

Appendix

APPENDIX I

Scientific Notation—*A Shorthand Method of Writing Numbers*

I. Numbers which are multiples of 10 are represented as follows:

1,000,000	=	10^6
100,000	=	10^5
10,000	=	10^4
1,000	=	10^3
100	=	10^2
10	=	10^1
1	=	10^0
0.1	=	10^{-1}
0.01	=	10^{-2}
0.001	=	10^{-3}
0.0001	=	10^{-4}
0.00001	=	10^{-5}
0.000001	=	10^{-6}

II. Numbers which are *not* multiples of 10 are represented as a product of two numbers. For example, the number 186,000 would be written as 1.86×10^5, which is equivalent to 1.86 times 100,000. The number 93,000,000 would be written 9.3×10^7. There is a convention which states that the number should be written in such a way that the part which comes before the times sign lies between 1 and 10.

Numbers less than one are handled in a similar way, but with negative exponents. For example, the number 0.005 is written as 5×10^{-3}, which is equivalent to 5 times 0.001.

The great time-saving value of scientific notation can best be appreciated by considering the comparative ease of writing very large or very small numbers. For example, the mass of the sun is 2×10^{30} kilograms. To write this out longhand would require 30 zeros. The mass of the electron is 9.1×10^{-28} grams, which would require 27 zeros when written as an ordinary number.

APPENDIX II

The Elements—*Their Symbols, Atomic Numbers, and Atomic Weights*

Atomic Number	Atomic Weight	Element	Symbol
1	1.00797	hydrogen	H
2	4.0026	helium	He
3	6.939	lithium	Li
4	9.0122	beryllium	Be
5	10.811	boron	B
6	12.01115	carbon	C
7	14.0067	nitrogen	N
8	15.9994	oxygen	O
9	18.9984	fluorine	F
10	20.183	neon	Ne
11	22.9898	sodium	Na
12	24.312	magnesium	Mg
13	26.9815	aluminum	Al
14	28.086	silicon	Si
15	30.9738	phosphorus	P
16	32.064	sulfur	S
17	35.453	chlorine	Cl
18	39.948	argon	Ar
19	39.102	potassium	K
20	40.08	calcium	Ca
21	44.956	scandium	Sc
22	47.90	titanium	Ti
23	50.942	vanadium	V
24	51.996	chromium	Cr
25	54.938	manganese	Mn
26	55.847	iron	Fe
27	58.933	cobalt	Co
28	58.71	nickel	Ni
29	63.54	copper	Cu
30	65.37	zinc	Zn
31	69.72	gallium	Ga
32	72.59	germanium	Ge
33	74.922	arsenic	As
34	78.96	selenium	Se
35	79.909	bromine	Br
36	83.80	krypton	Kr
37	85.47	rubidium	Rb
38	87.62	strontium	Sr
39	88.905	yttrium	Y
40	91.22	zirconium	Zr
41	92.906	niobium	Nb
42	95.94	molybdenum	Mo
43	(97)	technetium	Tc
44	101.07	ruthenium	Ru
45	102.905	rhodium	Rh
46	106.4	palladium	Pd
47	107.870	silver	Ag

48	112.40	cadmium	Cd
49	114.82	indium	In
50	118.69	tin	Sn
51	121.75	antimony	Sb
52	127.60	tellurium	Te
53	126.904	iodine	I
54	131.30	xenon	Xe
55	132.905	cesium	Cs
56	137.34	barium	Ba
57	138.91	lanthanum	La
58	140.12	cerium	Ce
59	140.907	praseodymium	Pr
60	144.24	neodymium	Nd
61	(145)	promethium	Pm
62	150.35	samarium	Sm
63	151.96	europium	Eu
64	157.25	gadolinium	Gd
65	158.924	terbium	Tb
66	162.50	dysprosium	Dy
67	164.930	holmium	Ho
68	167.26	erbium	Er
69	168.934	thulium	Tm
70	173.04	ytterbium	Yb
71	174.97	lutetium	Lu
72	178.49	hafnium	Hf
73	180.948	tantalum	Ta
74	183.85	tungsten	W
75	186.2	rhenium	Re
76	190.2	osmium	Os
77	192.2	iridium	Ir
78	195.09	platinum	Pt
79	196.967	gold	Au
80	200.59	mercury	Hg
81	204.37	thallium	Tl
82	207.19	lead	Pb
83	208.980	bismuth	Bi
84	(210)	polonium	Po
85	(210)	astatine	At
86	(222)	radon	Rn
87	(223)	francium	Fr
88	(226)	radium	Ra
89	(227)	actinium	Ac
90	232.038	thorium	Th
91	(231)	protactinium	Pa
92	238.03	uranium	U
93	(237)	neptunium	Np
94	(244)	plutonium	Pu
95	(243)	americium	Am
96	(247)	curium	Cm
97	(247)	berkelium	Bk
98	(251)	californium	Cf
99	(254)	einsteinium	Es
100	(253)	fermium	Fm
101	(256)	mendelevium	Md
102	(254)	nobelium	No

103	(257)	lawrencium	Lr
104	(257)	kurchatovium	Ku
105	(260)	hahnium	Ha

In cases where no stable isotope exists, the atomic weight of the *most* stable isotope is given in parentheses.

APPENDIX III

Activity Series

Metals and Hydrogen	Nonmetals
potassium	fluorine
sodium	chlorine
barium	bromine
strontium	iodine
calcium	
magnesium	
aluminum	
manganese	
zinc	
chromium	
cadmium	
iron	
cobalt	
nickel	
tin	
lead	
hydrogen	
copper	
arsenic	
bismuth	
antimony	
mercury	
silver	
platinum	
gold	

How to Use the Activity Series:

The elements in the Activity Series are arranged in order of their chemical activity from the most active at the top of the listing to the least active at the bottom. For the single replacement reaction

$$Zn + CuSO_4 \longrightarrow ZnSO_4 + Cu$$

it is necessary for the zinc to be more active than the copper for a reaction to take place. As can be seen from the Activity Series, the zinc *is* in fact more active than the copper, and the reaction *will* therefore take place. The following reaction would *not* take place because the copper is not active enough to replace the zinc:

$$Cu + ZnSO_4 \longrightarrow 2NaCl + Zn$$

Nonmetals are handled in a similar way. Consider the reaction

$$Cl_2 + 2NaBr \longrightarrow 2NaCl + Br_2$$

Chlorine is seen from the Activity Series to be more active than bromine. It will therefore replace bromine from its compounds. The reverse reaction

$$Br_2 + 2NaCl \longrightarrow 2NaBr + Cl_2$$

will *not* take place.

Glossary

absolute temperature scale See Kelvin temperature scale.

absolute zero That temperature at which, theoretically, an ideal gas would have no volume (it would exert no pressure), (-273° C. 0° K); also a body at 0° K would contain no thermal energy.

absorb To take into solution.

acceleration Change in velocity for a given period of time. Acceleration is a vector quantity, possessing both a magnitude (size) and a direction.

acceleration due to gravity The change in velocity per unit of time, undergone by a body in free fall. It is equal to approximately 32 ft/sec^2 or 9.8 m/sec^2.

accurate As exact as possible; carefully measured.

acetic acid $C_2H_4O_2$ or CH_3COOH—the chief acid in vinegar.

acetylene C_2H_2—the simplest alkyne; a colorless, gaseous hydrocarbon.

acid A substance that can donate a proton to another substance.

acid anhydride a nonmetal oxide such as SO_2 which forms an acid when added to water.

acoustics The science of sound; the qualities of a room that have to do with how clearly sounds can be heard or transmitted in it.

actinide elements The elements whose atomic numbers are 89 to 103 inclusive.

activated charcoal Charcoal that has been heated in steam; it is used to adsorb substances.

active metals Metals that are chemically active such as lithium (Group IA and IIA elements).

activity series Substances arranged according to their chemical activity.

addition reaction A reaction in which two molecules combine to yield a single molecule of product.

adhesive A substance used to glue objects together.

adsorb A substance holds another substance on its surface.

aerosol Colloidal suspension of liquid drops in a gas.

aether The imagined medium in the universe in which light traveled.

agonic line A line on the earth's surface along which no correction is required for a compass reading.

air resistance The retarding force of air on an object moving through it.

albumen The white of an egg.

alcohol A type of organic compound in which the OH group replaces hydrogen in a saturated hydrocarbon—general structural formula is ROH.

algae Any plant of a subdivision (Algae) of the Thallophyta consisting of chlorophyll-bearing plants widely distributed in fresh and salt water and in moist land locations—includes seaweeds, kelps, diatoms, pond scums, etc.

alkali metals The extremely active metals of Group IA of the periodic table.

alkaline earth metals The elements of Group IIA of the periodic table.

alkane Hydrocarbons characterized by having carbon-carbon single bonds.

alkene A hydrocarbon characterized by having carbon-carbon double bonds.

alkyne A hydrocarbon characterized by having carbon-carbon triple bonds.

allotropes (allotropic forms) Forms of an element having different arrangements of the atoms, such as ordinary oxygen (O_2) and ozone (O_3).

alloy A substance that has metallic properties and is composed of two or more chemical elements of which at least one is a metal.

alpha particle (α) A helium nucleus consisting of two protons and two neutrons.

alpha ray A stream of alpha particles.

alternating current Electrical current in which the electrons reverse their direction many times a second.

alto Same as contralto.

amino acid An organic acid formed by the hydrolysis of proteins having the general formula

$$R-\underset{NH_2}{\overset{H}{\underset{|}{\overset{|}{C}}}}-COOH.$$

amino group NH_2–.

ammeter electrical instrument used to determine the intensity of the electron flow in a circuit.

ammonia NH_3.

ammonium ion NH_4^+.

ammonium salt A compound in which a nonmetal or a negative radical is combined with NH_4^+.

amorphous Without structure.

ampere The electrical unit defined as the amount of current flowing when a coulomb of charge passes a given point in one second.

amplitude The maximum height of a crest or trough or the maximum distance an object vibrates from its rest position.

aneroid barometer An instrument which uses changes in a partially evacuated metal box to determine atmospheric pressure.

angular momentum A physical quantity that expresses in a mathematical way how much rotation or revolution is taking place. The angular momentum of a point mass in orbit around a central body is calculated by multiplying its mass times its velocity times the radius of the orbit through which it is traveling.

anhydrous Without water.

anion A negative ion.

anneal To "soften" metals and glasses by heating to high temperatures. The material annealed remains in the "solid" state. Internal stresses within the material are relieved.

anode The positive electrode of an electrolytic cell.

anthracite Hard coal; contains more carbon than bituminous coal.

antimatter Every type of particle in nature has a sister particle with exact opposite properties. For every type of particle (matter) there is an antiparticle (antimatter).

antineutrino See neutrino.

antiseptic Substance used to kill and prevent the growth of pathogenic and putrefactive bacteria.

aqua regia A mixture of hydrochloric and nitric acid.

area A measure of two-dimensional surface. Area is calculated by multiplying length times width and is expressed in square units such as in^2 or m^2.

armature The part that turns in an electric motor or generator. The armature consists of a laminated iron core wound with insulated wire.

aromatic hydrocarbons Ring compounds of carbon and hydrogen (possessing a stable ring).

asbestos A non-combustible, non-conducting, chemically resistant mineral.

atom The smallest part of an element that can enter into chemical combination with other elements.

atomic bomb Uncontrolled chain reaction releasing a tremendous amount of energy.

atomic energy Heat energy released when atoms fission.

atomic mass the mass of a given atom expressed in atomic mass units.

atomic number The number of protons in the nucleus of an atom.

atomic oxygen Oxygen in which the atoms are separate rather than combined into O_2 or O_3 molecules.

atomic weight The relative weight of an atom as compared to the weight of one atom of carbon-12.

audible range Range of sound frequencies that can be detected by the ear.

aurora borealis Luminous bands or streamers of light sometimes appearing in the night sky of the Northern Hemisphere; also called northern lights.

B

ball lightning A rarely seen form of lightning said to consist of small round luminous masses of electrified air.

base A substance that will accept a proton from an acid.

basic anhydride A metal oxide such as CaO which forms a base when added to water.

basic oxygen furnace A furnace for refining pig iron to steel where gaseous oxygen is bubbled through the molten charge to oxidize the impurities.

baritone A male voice with a range between tenor and bass; a brass instrument having a similar range.

barometer An instrument for measuring atmospheric pressure.

bass The lowest male singing voice.

battery A group of two or more cells connected in series or parallel; it has also become common usage to refer to a single cell as a battery.

bauxite $Al_2O_3 \cdot XH_2O$, $1 \leq x \leq 3$.

beat (beat frequency) A third tone which appears when two tones close in frequency are sounded simultaneously.

Becquerel rays Alpha, beta, and gamma rays emitted from radioactive materials.

ben zene An aromatic hydrocarbon—C_6H_6.

beta decay A radioactive decay process in which an electron is given off.

beta rays Stream of energetic electrons.

bimetal Containing two different metals or alloys.

biochemistry The chemistry of living organisms.

bituminous coal Soft coal.

blast furnace A furnace in which the reduction of iron ore to pig iron is accomplished.

bleach A substance used to remove the color from an object—to whiten.

block and tackle An arrangement of pulleys and ropes or cables for pulling or hoisting large heavy objects.

boiling point The temperature at which the upward vapor pressure of a liquid equals the downward pressure of the atmosphere.

bond To join together.

borax $Na_2B_4O_7 \cdot 10H_2O$.

Boyle's law The volume of a dry gas is inversely proportional to the applied pressure if its temperature is held constant. $PV=C$

brass A copper-zinc alloy.

brass instruments The family of musical instruments often constructed of brass, including the trumpet, trombone, baritone, tuba, and French horn.

breeder reactor A nuclear reactor that produces more fissionable material than it burns up.

brimstone Biblical term for sulfur.

bronze A copper-tin alloy.

Brownian movement (or motion) The constant zigzag movement of colloidal dispersions in a fluid medium.

brush (electrical) A piece of graphite, bronze, or other material used as a conductor between an outside circuit and the armature of a motor or generator.

bubble chamber Device used to detect radiation by a bubble trail left in the chamber fluid by moving particle.

buffer A substance capable of reacting with an acid or base used to control the concentration of H_3O^+ ions of a solution.

bull's-eye diagram A two-dimensional representation of an atom in which the nucleus and shells are drawn as concentric circles.

Bunsen burner A type of gas-burner in which a mixture of gas and air is burned at the top of a short metal tube.

buret A graduated glass tube from which a small quantity of a solution can be drawn off at a time.

C

caloric theory The erroneous belief that heat is a form of matter.

calorie Quantity of energy that will raise the temperature of one gram of water 1°C.

candlepower The intensity of a standard candle at its source.

capacitor An electrical device consisting of two conductors separated by an insulator (also called a condenser).

capillary Fine, slender, glass tube having a hairlike bore (inside diameter).

carbohydrates A class of compounds containing carbon, hydrogen, and oxygen. The hydrogen and oxygen in the compounds are in the same ratio as found in water (two hydrogens to one oxygen).

carbon black A black substance resulting from the incomplete combustion of natural gas.

catalyst A substance used to speed up or slow down a chemical reaction.

catalytic decomposition Decomposition of a compound with the aid of a catalyst.

category A class of things; a division in a scheme of classification.

cathode The negative electrode of an electrolytic cell.

cathode ray A stream of electrons.

cation A positive ion.

cellulose The carbohydrate that makes up the skeletal material in all plants.

Celsius temperature scale The temperature scale on which water under normal pressure freezes at 0° and boils at 100°.

center of gravity A central point at which all the mass of a body may be imagined to be concentrated.

Centigrade temperature scale Same as Celsius temperature scale.

centimeter (cm) The metric length unit equal to 1/100 of a meter.

ceramics The science of producing clay and refractory products.

cgs system A subdivision of the metric system utilizing the *centimeter*, the *gram*, and the *second* as the basic units of measurement.

chain reaction A self-sustaining reaction which, once started, steadily provides the energy and matter necessary to continue the reaction.

charcoal A form of carbon prepared by heating wood or bones in the absence of air.

charge The quantity of electricity an object contains.

Charles' law The volume of a dry gas is directly proportional to its absolute temperature, provided the pressure remains constant. $V \alpha T$

chemical change A change in which a substance loses its characteristics and changes into one or more new substances.

chemical evolution The imagined formation of complex compounds from simpler compounds and elements by purely natural processes before any life existed on the earth.

chemical metallurgy Science that deals with the extraction of metals from their ores.

chemistry The study of matter and the changes it undergoes.

chloride Compound formed by the reaction of chlorine with an electropositive element.

circuit Closed path traversed by electrons.

circuit breaker Electromagnetic device which automatically switches off an overloaded circuit.

classification An arrangement of things according to some system.

clay A mixture of hydrous aluminum silicates with other substances.

cloud chamber Device used to detect ionizing radiation by using vapor condensation on the ions generated in the path of the moving radiation.

cloud, electron See electron cloud.

coal An impure form of carbon found in the forms of lignite, bituminous, and anthracite.

coefficient of linear expansion The ratio of increase in length of a material per unit of length, per degree of temperature increase.

coherent For wave motion—waves in phase with one other.

coin gold A gold alloy containing 10 percent copper.

coin silver A silver alloy containing 10 percent copper.

coke A form of carbon prepared by the high temperature distillation of coal in the absence of air.

cold-work Deforming a metal while it is below its recrystallization temperature.

colloid Particles of size 1×10^{-7} to 1×10^{-5} cm.

colloidal suspension A mixture of colloidal particles in a medium.

combined gas law A relationship which combines Boyle's law and Charles' law into one equation: $\frac{P_1V_1}{T_1} = \frac{P_2V_2}{T_2}$.

combustion The rapid combination of other substances with oxygen, usually considered the state of being on fire.

commutator Device used in D.C. motors and generators for changing the direction of the current.

compass A device consisting of a magnetic needle swinging freely on a pivot, capable of aligning itself with the earth's magnetic lines of force.

composition (chemical) What a substance is made up of.

compound A substance made up of two or more elements chemically united.

concave Curved like the inside of a circle or sphere.

concentration (ore) The process used to separate and concentrate a desired mineral from the undesirable substances that were mined with it.

condenser (chemical) A device used to cool vapor to the liquid state.

condenser (electrical) An electrical device consisting of two conductors separated by an insulator (see capacitor).

conduction A process by which heat flows from a region of high temperature to a region of low temperature within a medium or between different mediums in direct physical contact—energy is transmitted by direct molecular communication.

conductor Material capable of carrying heat (in the case of a thermal conductor) or electrons (in the case of an electrical conductor).

conservation Preservation.

constant A quantity which remains the same throughout a given experiment or calculation.

constructive interference Waves reinforce each other when they interfere.

contralto A female voice of the lowest range.

control rods Devices made of high neutron absorption materials used to control the rate of the fission (splitting apart) process in a nuclear reactor.

convection A process of energy transport by the combined actions of heat conduction, energy storage, and mixing motion—most important mechanism of energy transfer between a solid surface and a liquid or gas.

converge To come together.

cornea The transparent front of the outer coat of the eyeball.

corrosion Gradual chemical attack on a metal by various substances.

Coulomb's law Two electrically charged bodies attract or repel each other with a force that is directly proportional to the product of their charges and inversely proportional to the distance between them squared.

covalent bond A chemical bond where the atoms share valence electrons.

creationism The belief that the universe and the things in it were spoken into existence in substantially their present form by miraculous acts of God as described in Genesis chapters 1 and 2.

creationist One who adheres to the doctrine of creationism.

critical mass The minimum mass of fissionable material needed to sustain a chain reaction.

critical shape The shape of fissionable material needed to sustain a chain reaction.

critical size The minimum size of fissionable material needed to sustain a chain reaction.

crucible A ceramic container made of a refractory (heat resistant) material designed to hold hot substances.

cryogenic Pertaining to extremely low temperatures.

cryolite Na_3AlF_6.

crystal A solid with a definite internal structure.

cube A solid with six equal, square sides.

cubic centimeter (cm^3) The unit of volume equal to the volume of a cube one centimeter on a side (also called a milliliter).

cubic meter A large measure of volume equal to 1000 liters; a cube measuring one meter on each side.

current The intensity of flow in an electrical circuit, measured in amperes.

current electricity Charges in motion, usually traveling in a complete path, or circuit.

D

data (pl. of datum) Facts or figures gathered during the course of an experiment.

DDT Dichloro-diphenyl-trichloroethane—a powerful insecticide.

decibel Unit used to express the intensity of a sound.

declination The correction in degrees that must be applied to a compass reading to find the direction of true north.

decompose To separate into parts.

decomposition reaction To break a substance into parts in a chemical reaction.

deformation The process used to change the shape of an object.

deformation, elastic See elastic deformation.

dehydration The removal of water from a substance.

deliquescence Having the capacity to absorb moisture from the atmosphere and dissolve in it.

density The quantity expressing how massive or heavy an object is for its size. In the metric system it is customary to use mass per unit of volume for density; in the English system weight per unit of volume is preferred.

destructive interference Waves cancel each other when they interfere.

detergent A non-soap cleaning agent.

deuterium The isotope of hydrogen having one neutron in its nucleus.

dextrins Sticky substances formed when starch decomposes in hot water.

diamond A very beautiful, hard allotropic form of carbon.

diatomic Containing two atoms per molecule.

diffraction The bending of waves around the edge of an object.

diffusion The spontaneous mixing of two or more substances by molecular motion.

dipole An object having a negative and positive pole.

dipping needle (dip needle) A vertically suspended magnetic needle used to indicate the angle at which the lines of force of the earth's field enter (or leave) the ground.

direct current Electric current in which the electrons move in one direction only. Abbreviation: D. C.

direct proportion A situation in which two quantities vary in the same way. Both increase or decrease together.

disperse To separate.

displacement Distance.

distance principle The general rule that if force is gained, distance must be sacrificed; conversely, if speed or distance is gained, force must be sacrificed.

distillation The separation of parts of a substance by heating. The more volatile (quickly evaporating) parts go off and the less volatile parts remain.

diverge To extend in different directions from a point.

DNA A nucleoprotein which assures that living organisms will reproduce after their kind.

domain Region of a magnetic material in which all the elementary magnets are aligned in the same direction.

Doppler effect a) The apparent change in the frequency of a sound caused by relative motion between the sound source and the observer; b) A similar effect for light waves; light from an approaching source is shifted to the blue end of the visible spectrum; light from a receding source is shifted toward the red end of the spectrum.

dosimeter Pocket-size electrometer to monitor radiation exposure.

double bond Two electron-pairs form a bond between atoms.

double replacement reaction Chemical reaction in which two compounds exchange ions to form two new compounds.

doubtful figure The last significant figure of a properly reported measurement.

drawing An operation used in forming metal parts where the metal is pulled through a pattern that has the desired shape of the final product.

dross Same as slag.

dry cell A voltaic cell which utilizes an ionized paste in place of a liquid electrolyte.

dry ice Solid CO_2.

ductile Capable of being drawn out; pliable.

dye (verb) To fix a color in a substance. (noun) The substance used to color an object.

dynamid A close combination of a negative and a positive charge in the Lenard model of the atom.

dyne The amount of force needed to impart an acceleration of one cm/sec^2 to a mass of one gram.

E

echo The reflection of sound.

ecology The totality or pattern of relations between organisms and their environment.

effervescence Bubbling or foaming.

effort Input force.

elastic deformation Temporary changes in the dimensions of an object caused by stress. The object returns to its original dimensions after removal of the stress.

elastic (elasticity) That property of matter that enables it to return to its original dimensions after a stress has been removed from it.

electric field The region of influence surrounding an electric charge.

electrochemical Electricity is used to effect a chemical change.

electrocleaning bath A cleaning solution that is electrified to aid in the cleaning process (the part to be cleaned is usually one of the electrodes).

electrode Either of two terminals of an electric source, an anode or cathode.

electrolysis Decomposition of a compound by means of an electric current.

electrolyte The solution in an electrochemical cell; any substance capable of conducting an electric current when dissolved or melted.

electrolytic process A process involving the use of electricity—usually to decompose a substance.

electromagnet Magnet consisting of an iron core surrounded by a coil of wire. The core becomes magnetized when current flows through the wire.

electromagnetic radiation All wave phenomena that can move from one body to another although there is space between the bodies.

electrometer A calibrated electroscope.

electromotive force Electron-moving force, more commonly called potential difference or voltage.

electron A negatively charged particle having a mass of 1/1837 of that of a proton.

electron affinity The amount of energy released when an electron is added to an atom.

electron cloud The place where an electron is supposed to be.

electron microscopy The viewing of minute details of objects using electron beams.

electron orbitals The place where electrons can be found in atoms—similar to electron clouds.

electron shell A particular electron arrangement in atoms. Electrons in the same shell have approximately the same energy.

electronegativity The tendency to attract shared electrons.

electroplating Metal coating by means of an electric current.

electropositive Capable of becoming positively charged.

electroscope Device for detecting static electrical charges.

element A substance that cannot be decomposed by ordinary chemical means.

emission The energy released from a substance.

emulsion Colloidal suspension of a liquid in a liquid.

energy The ability or capacity to do work.

energy, thermal Heat.

English system The system of measurements that uses inches, feet, yards, miles; quarts, gallons, pounds, tons; seconds, minutes and hours.

enzyme A protein that acts as a catalyst to speed up chemical reactions in living cells.

epsom salt $MgSO_4 \cdot 7H_2O$.

erg A metric unit used to measure energy; defined as the amount of work done when a force of one dyne acts through a distance of one centimeter.

escape velocity The velocity needed to escape the gravitational pull of an astronomical body.

etch To engrave or cut by chemical means normally using a corrosive liquid.

evaporating dish A ceramic or glass dish in which mixtures can be poured so that volatile parts may be allowed to escape from the top of the dish.

evolution An imaginary process by which nature is said to continually improve itself through gradual development. It has become fashionable today to apply evolutionary thinking to the entire universe. Stars, galaxies, planets, plants, animals, and people are all said to "evolve" from simpler states of matter.

evolution, chemical See chemical evolution.

evolutionist One who attempts to explain the origin of things by theories of gradual development.

excitation Increasing the energy of a particle.

exothermic Giving off heat.

exponent A small number written above and to the right of a quantity. It denotes the power to which the quantity is to be raised (how often the quantity is to be multiplied by itself).

extrusion An operation used in forming metal parts where the metal is pushed through a pattern that has the desired shape of the final product (like squeezing toothpaste out of a tube).

F

Fahrenheit temperature scale The temperature scale on which water at normal atmospheric pressure freezes at 32° and boils at 212°.

family A vertical column of the periodic table which forms part of a group.

fat Compounds called glycerides containing carbon, hydrogen, and oxygen found in living organisms.

fertilizer A substance used to feed plant life; provides the necessary nutrients for plant growth.

field-ion microscopy The viewing of minute details of objects using helium ions.

filament The fine metal wire in a light bulb or vacuum tube which becomes heated by an electric current.

filter To strain one medium out of another—usually a solid out of a liquid.

first-class lever A lever in which the fulcrum is located between the effort and the resistance.

first law of thermodynamics Energy can neither be created nor destroyed; it is conserved in any process.

fixer Bath used to dissolve the unreduced AgBr on photographic film.

Flood The Noachian Deluge as recorded in Genesis 6-9.

flotation The concentration of minerals from ores by agitation of the ground ore with suitable liquids.

fluorescence That property of certain substances enabling them to absorb a certain wavelength or radiation and reemit a longer wavelength. Fluorescence occurs only as long as the substance is subjected to radiation.

fluorspar CaF_2.

flux A material added to molten metal to remove impurities from the metal.

foam Colloidal suspension of gas bubbles in a liquid.

focal length The distance from the optical center of a lens or mirror to the point when the rays converge.

fog Colloidal suspension of liquid drops in a gas.

force A push or a pull.

force field A space in which forces are exerted due to the presence of a charge, particle, etc.

forging An operation used in forming metal parts where hammer-like strokes are applied to a metal on an anvil until it conforms to the desired shape.

forked lightning The most commonly observed type of lightning, characterized by jagged streaks of light. (also called streak, zigzag, or chain lightning).

fossil fuels Coal, oil, gas.

fractional distillation Separation of a mixture into components by distillation.

Frasch process Industrial process for removing native sulfur from salt domes.

free electron theory of metals The valence electrons in a metal are free to roam throughout the metal structure.

free fall Descent under the influence of gravity with no retarding force.

freezing point The temperature at which a liquid cooled under normal pressure changes to a solid (also called the melting point).

frequency The number of vibrations or cycles per unit of time.

friction Retarding force due to surface contact acting on a moving object.

fuel element Assembly containing the fissionable material in a nuclear reactor.

fulcrum The pivot point of a lever.

fumigant Any substance whose vapors are capable of destroying vermin and insects in an enclosed space—disinfectant.

function A mathematical quantity dependent on other quantities.

fundamental The lowest frequency obtainable from a vibrating string, air column, etc.

fungicide A fungus killer.

fuse Safety device used in an electric circuit. The fuse melts and breaks the circuit if the current rises above a safe level.

G

galvanic corrosion A localized corrosion that results when two dissimilar metals are in contact.

galvanizing Coating steel with zinc—the steel is passed through a bath of molten zinc.

galvanometer Instrument for detecting very small electrical currents.

gamma rays Electromagnetic radiation with a wavelength of less than 10^{-11}m given off by certain radioactive substances.

gas The state of matter having neither a definite shape nor a definite volume.

Gay-Lussac's law The pressure of a confined gas is proportional to its absolute temperature, provided its volume is held constant.

Geiger-Müller tube Device used to detect ionizing radiation by using a highly charged wire inside a gas tube.

gel Colloidal suspension of a liquid in a solid.

gelatin A colloidal protein produced from bones, white connective tissue, and skins of animals—forms a gel with water.

generator A device which changes mechanical energy to electrical energy.

glass A supercooled liquid, usually a hard and brittle, transparent substance.

glaze (ceramic) A glass coating on an object.

glucose A sugar with the chemical formula $C_6H_{12}O_6$.

graduate cylinder (graduated cylinder) A laboratory vessel marked for measuring volume units.

gram A metric unit of mass and weight; mass equal to one cubic centimeter of water at 4° C.

granite An igneous rock composed of quartz, feldspar, and mica.

graphite A form of carbon with a hexagonal crystal structure.

gravitation The force by which every particle in the universe is attracted by every other particle.

gravity The force that tends to draw all bodies in the earth's sphere toward the center of the earth.

greenhouse effect Short waves of light entering an enclosure, absorbed by objects, and reemitted at higher wavelengths so that they cannot escape from the enclosure.

group A vertical column of the periodic table, containing two families, A and B.

gypsum $CaSO_4 \cdot 2H_2O$.

gypsy·moth Imported moth whose larvae can consume a square foot of tree leaves daily.

H

Haber process The industrial process where nitrogen and hydrogen are reacted to form ammonia.

half-life The time required for one half of the radioactive nuclei in a sample to decay.

Hall process The industrial operation used to remove aluminum from bauxite.

halide A fluoride, chloride, bromide, or iodide.

halogens Salt reformers; the elements of Group VIIA of the periodic table.

hardness The property of matter that enables it to resist yielding and scratching.

hard water Natural water containing large amounts of dissolved calcium and magnesium.

heat Thermal energy.

heat capacity Same as specific heat.

heat exchanger A device used to transfer heat from one place or substance to another.

heavy water D_2O Variety of water in which the regular hydrogen atoms are replaced with deuterium atoms.

hematite Fe_2O_3.

hemoglobin Protein that carries oxygen in the blood.

herbicide A weed killer.

Hertz (Hz) The unit of frequency equal to one cycle per second.

heterogeneous Consisting of dissimilar parts.

homogeneous Of the same or similar parts.

hormone A protein that controls processes in living organisms.

horsepower The English unit of power equal to 550 foot-pounds per second.

hot-work Deforming a metal above its recrystallization temperature.

humus Decayed organic matter in soil.

hydride A compound formed by the union of hydrogen with one other element.

hydrocarbon A compound containing only hydrogen and carbon.

hydrochloric acid HCl.

hydrofluoric acid HF.

hydrogen bomb Uncontrolled fusion processes releasing a tremendous amount of energy.

hydrogen bond A weak chemical bond formed between a hydrogen atom in one molecule and a highly electronegative atom in an adjacent molecule.

hydrogen ion Proton.

hydrometer An instrument for determining the specific gravity of a liquid by flotation, most generally used to assess the condition of the electrolyte in a storage battery.

hydronium ion H_3O^+

hydroxide ion OH^-.

hygrometer An instrument used for measuring humidity, composed of a wet and dry bulb thermometer.

hypo Photographer's term for water solution of sodium thiosulfate.

hypothesis A guess or an idea not backed by observed facts.

I

ideal gas An imaginary gas which obeys Boyle's and Charles' laws exactly.

ignite To set on fire.

illuminate To supply with light.

image A visible representation of something.

incandescent Glowing with intense heat.

inclined plane A slanted surface on which heavy objects can be raised more easily than they could be lifted.

induced magnetism The process by which a piece of unmagnetized iron or other material becomes magnetized by its nearness to a strong magnet.

inert Chemically inactive.

inertia The property of matter which causes it to resist a change in its state of motion.

infrared radiation The electromagnetic radiation emitted from objects with temperatures above 0° K—the wavelength range is 10^{-6} to 10^{-3}m.

infrared spectroscopy Forming and analyzing the infrared spectra emitted by substances.

infrasonic Below the audible range—less than 20 Hz.

inhibitor Agent which retards a chemical reaction; a negative catalyst.

insecticide An insect killer.

insulator A material which does not conduct. At least two types should be distinguished—*thermal insulators*, which hinder the flow of heat, and *electrical insulators,* which hinder the flow of electrons.

intensity Relative strength or brightness; relative degree of energy content.

interaction The effect one substance has on another.

interference of waves The resulting effect when waves from two different sources cross each other's path—the effect can be destructive, constructive, or a combination of both.

interpretation A scientist's opinion of why something happened.

inverse proportion A situation in which one quantity decreases as another increases. When two quantities are inversely proportional, their product is equal to a constant.

inverse square law A relationship commonly encountered in physics in which the intensity falls off as the square of the distance. Gravitation, electrostatic forces, magnetic forces, electromagnetic radiation, and sound all follow inverse square laws.

ion An atom or group of atoms with a charge.

ion-exchange A replacement reaction in which the ions in one substance replace the ions in another substance.

ionic bond The chemical bonding between a positive and negative ion.

ionic character A measure of the extent to which a shared electron pair is dominated by one atom or the other in a covalent bond.

isomer Different compounds that have the same molecular formula.

isotopes Forms of the same element differing only in the number of neutrons in the nucleus.

J

jackscrew Combination of the screw and the wheel-and-axle.

joule The amount of work done when a force of one newton acts over a distance of one meter.

K

Kekulé model A model for the structure of benzene.

Kelvin temperature scale The temperature scale on which water at normal atmospheric pressure freezes at 273° and boils at 373°. The zero of the Kelvin scale coincides with absolute zero.

kilogram (kg) Unit of mass equal to 1000 grams. The kilogram is the mass of 1 liter of water at 4° C.

kiloHertz (kHz) The unit of frequency equal to 1000 Hertz.

kilometer (km) Metric unit of length equal to 1000 meters.

kilowatt (kw) A unit of power equal to 1000 watts.

kinetic energy The energy of a body due to its motion.

kitchen match A match containing phosphorus in its head—can be lit by friction of the head on a surface.

L

laminated Composed of many thin layers, as the core of a transformer.

lanthanide elements The series whose atomic numbers are 57-71 inclusive.

laser Light amplification by stimulated emission and radiation.

laughing gas N_2O.

law A statement of a naturally occurring, uniform pattern of behavior. Some laws are mathematical relationships expressed as equations; others are simply verbal descriptions of what has been observed to happen.

law of charges Like charges repel; unlike charges attract.

law of conservation of angular momentum The angular momentum of any rotating or revolving system remains constant during any process involving an interaction within the system.

law of conservation of mass-energy The total quantity *matter plus energy* in the universe does not change.

law of conservation of momentum (linear) The momentum of a system is conserved in any process involving an interaction within the system.

law of magnetic poles Like poles repel; unlike poles attract.

leguminous plants Plants of the family Leguminosae—nitrogen-fixing bacteria–grow on the roots of these plants.

length contraction The apparent shrinking of a moving body—a consequence of the special theory of relativity.

lens Any device for concentrating or dispersing radiation by refraction.

lever A rigid bar capable of turning about a fixed point called the fulcrum.

Leyden jar Simple condenser formed by coating a jar both inside and out with tin or lead foil.

light (visible) A form of electromagnetic radiation with a wavelength range of 4×10^{-7} to 7×10^{-7}m—thought to be produced by electronic transitions in matter.

lightning rod A pointed rod placed at the top of a house or building to protect it against lightning.

light-year The distance light travels in a year.

lignite A form of coal called brown coal. It has a high moisture content.

lime CaO.

limestone $CaCO_3$—A water-deposited sedimentary material.

limewater $Ca(OH)_2$ + water.

lines of force Imaginary lines radiating from or directed toward a magnetic pole. The lines represent paths that would be followed by a tiny unit north pole as it is attracted or repelled by the magnetic pole. Similar lines are imagined surrounding electrostatic and gravitational centers.

lipid See fat.

liquid The state of matter which has a definite volume but no definite shape.

liter (ℓ) Metric unit of volume (equal to the volume of a cube 10 centimeters on a side).

litmus paper Paper containing a deep blue dye that will turn red in an acid solution and remains blue in a basic solution. Red litmus paper will turn blue in a basic solution.

lodestone (loadstone) A magnetic rock consisting of the mineral magetite, Fe_3O_4.

long form table The form of the periodic table having 18 vertical columns. This expanded arrangement serves to separate the A families from the B families.

loudness Subjective measure of the intensity of a sound.

lumen Unit of light-energy flowing away from a source.

luminous Emitting light.

luster, metallic See metallic luster.

lye Sodium hydroxide.

M

machining Cutting a metal part to a desired shape by use of a sharp tool—the part is usually rotated rapidly against the tool.

magnetic field The region of influence surrounding a magnet or a conductor carrying an electric current.

magnetite A black, strongly magnetic oxide of iron having the formula Fe_3O_4.

magnetosphere The region surrounding the earth in which particles are trapped by the earth's magnetic field, including the two Van Allen belts and the space between them.

main-group elements The 44 elements which are not transition elements.

malaria A disease caused by any of certain animal parasites (genus Plasmodium) which are introduced by the bite of the infected anopheles mosquito and invade the red corpuscles of the blood.

malleability Property that determines the ease of deforming a material.

mass A measure of the quantity of matter in a body. The mass of an object can be measured by its resistance to acceleration, or by its gravitational behavior.

mass spectrograph A large research instrument used to identify elements, isotopes, and free radicals in an unknown sample. The identification is based on the masses of the fragments.

match (kitchen) See kitchen match.

matter Anything which occupies space and has mass.

Maxwellian distribution A normal (bell-shaped) curve, exemplified by the velocities of gas molecules.

mechanical advantage The factor by which the force is multiplied in a machine.

mechanical energy The form of energy studied in mechanics. Mechanical energy includes *kinetic energy* (energy of motion) and *potential energy* (stored mechanical energy).

mechanical work Work done by a machine, such as the compression of a gas.

mechanism Natural forces that cause certain events to occur.

median The middle value in a set of numbers or measurements.

megaHertz (MHz) The unit of frequency equal to 1,000,000 Hertz.

melting point The temperature at which a solid heated under normal pressure changes to a liquid (also called the freezing point).

meniscus The curved upper surface of a column of liquid.

mercurial barometer A device in which the height of a column of mercury is used to determine the pressure of the atmosphere.

metal A hard, heavy, lustrous, malleable, tough substance which is a good conductor of heat and electricity—chemically, a substance able to donate electrons in a chemical reaction.

metallic bond The chemical bond between metal atoms where valence electrons are not localized.

metallic character A measure of the ease with which an element loses electrons and forms positive ions.

metallic luster The shiny appearance of freshly cut or polished metal surfaces.

metalloid A substance that has chemical properties of both metals and nonmetals.

metallurgical microscope A microscope that uses light reflected off the surface of a metal rather than light transmitted through the specimen as with a biological microscope.

metallurgy The science that deals with the extraction of metals from their ores and their adaption for commercial use.

metallurgy, chemical See chemical metallurgy.

metallurgy, physical See physical metallurgy.

meter The metric unit of length defined as 1,650,763.73 wavelengths of the orange light emitted from krypton-86.

methane The simplest known hydrocarbon—CH_4.

metric system A decimal system of weights and measures based on the meter.

mezzo-soprano A female voice with a range between soprano and contralto.

mica A mineral silicate that readily separates into very thin leaves.

micron 10^{-4}cm or 10^{-6}m.

microscope An optical instrument for assisting the eye in observing minute objects and minute features of objects.

microscope, metallurgical See metallurgical microscope.

microstructure (metal) The structure of polished and etched metal and alloy specimens as revealed by the microscope.

microwaves Electromagnetic radiation with a wavelength range of 3×10^{-3} to 0.3 m.

milk of magnesia A suspension of $Mg(OH)_2$ in water.

milligram (mg) Metric unit of mass equal to 1/1000 of a gram.

milliliter (ml) Metric unit of volume equal to 1/1000 of a liter.

millimeter (mm) Metric unit of length equal to 1/1000 of a meter.

mixture A material containing two or more different substances.

mks system A subdivision of the metric system utilizing the *meter*, the *kilogram*, and the *second* as the basic units of measurement.

model A simplified picture used by a scientist to represent the phenomenon he is studying. Two kinds of models are geometrical and mathematical.

moderator Material used in a nuclear reactor to thermalize (slow down) the fast neutrons emitted in a nuclear reaction.

molecular orbital An electron cloud in a molecule—representation of electron position in a bond.

molecular weight The sum of the atomic weights of the atoms that make up the molecule.

molecule The smallest particle of an element or a compound that can exist in the free state and still retain the characteristics of the substance.

moment Torque or turning tendency calculated by multiplying the applied force by the lever arm through which it acts.

momentum The vector quantity defined as mass times velocity.

monochromatic Single wavelength.

monoclinic A crystal system having two oblique axes and a third perpendicular to both.

monosaccharide A sugar which cannot be hydrolyzed by water.

mortar A construction material made by mixing calcium hydroxide, sand, and water. It solidifies and hardens with age.

mortar and pestle A hard bowl and mixing rod used for grinding or pounding chemical substances to a powder.

motor Device which converts electrical energy to mechanical energy.

multimeter An instrument which can be switched to operate alternatively as a voltmeter, an ammeter, or an ohmmeter.

multiplier Resistor placed in series with a galvanometer to convert it to a voltmeter.

mutarotation An internal reaction in some sugars to form ring compounds.

mutation A degenerative variation in some inheritable characteristic of a plant or animal.

N

native metal A metal found in its pure state in nature.

natural gas A gaseous hydrocarbon consisting mainly of methane generated naturally in underground oil deposits.

nebular hypothesis An attempt to explain the origin of the solar system by speculating that it began with the contraction of an interstellar cloud; this theory has been found to be mathematically untenable.

negative Characterized by an excess of electrons. The negative terminal of a battery supplies electrons to the external circuit.

neutralization reaction When an acid reacts with a base and the two neutralize each other forming a salt plus water.

neutrino A massless, uncharged particle that is emitted along with an electron during beta decay—it possesses energy, linear, and angular momentum.

neutrography A nondestructive testing method where materials are subjected to a stream of neutrons and an image of the interior of the part is obtained on photographic film.

neutron A nuclear particle with no charge and approximately the same mass as a proton.

newton (N) The force needed to impart an acceleration of one m/sec^2 to a mass of one kilogram.

Newton's first law Objects at rest tend to remain at rest unless acted upon by an unbalanced force. Objects in motion tend to remain in motion in the same direction and at the same speed unless acted upon by an unbalanced force.

Newton's law of gravitation Any two bodies attract each other with a force proportional to the product of their masses and inversely proportional to the square of the distance between them.

Newton's second law Force equals mass times acceleration. (F=ma).

Newton's third law For every action there is an equal and opposite reaction.

nichrome An alloy of nickel, iron, and chromium used in electrical heating devices.

nitrate A compound composed of a cation and the NO_3^-.

nitrate ion NO_3^-.

nitric acid HNO_3.

nitride A compound formed when a metal combines with nitrogen only.

nitrogen cycle The sequence of physical and chemical processes by which atmospheric nitrogen is taken into the soil, utilized by plants and animals, and eventually returned to the atmosphere.

noble gases The elements of Group VIIIA of the periodic table.

noble metals Metals that do not react easily; very passive chemically.

nodal point See node.

node Stationary point in a vibrating string or air column.

nondestructive testing Checking materials for internal flaws without destroying the part being tested.

nonmetal A substance that accepts electrons in a chemical reaction.

northern lights See aurora borealis.

NTA Nitrolotriacetic acid.

nuclear fission The breaking up of large atoms into smaller atoms and other fragments.

nuclear fusion Two small atoms combine to form a larger atom.

nuclear power Energy available from the heat taken from a nuclear reactor.

nuclear reactor An assembly in which controlled nuclear fission occurs.

nucleon Any proton or neutron that exists inside the atomic nucleus.

nucleoprotein A protein found in the reproductive system of living organisms-responsible for the transmission of hereditary characteristics.

nucleus The central core of some atom models containing all of the positive charge.

O

observation The act of observing or noticing something; the basis of the scientific method.

octet Complete shell of eight electrons.

ohm(Ω) The unit of resistance named in honor of Georg Simon Ohm. One ohm is the resistance needed to limit the current in a one-volt circuit to one ampere.

Ohm's law Voltage equals current times resistance (E=IR).

ohmmeter An instrument for measuring resistance directly. The ohmmeter consists of a sensitive ammeter in series with a battery.

oil Liquid fat.

open hearth furnace A furnace for melting metal in which the bath is heated by the convection of hot gases over the surface of the metal and by radiation from the roof—normally a furnace where pig iron is refined into steel.

optic Pertaining to the eye or to vision.

optics The science of light, vision, and sight.

orbit The path of a body revolving around another body.

ore A mineral from which a metal can be extracted profitably.

ore reduction A process by which a metal is removed from its ore.

organic Derived from living organisms.

organic chemistry The chemistry of compounds that contain carbon.

oscillate To vibrate or swing back and forth.

Ostwald process The commercial process where nitric acid is produced from ammonia.

overflow can A device to measure volume of an irregular object by the water displacement method.

overtones Higher frequencies produced by a vibrating string, air column, etc. The frequencies of the overtones are multiples of the fundamental frequency.

oxidation a)The process whereby oxygen combines with some other substance. b) Raising the valence of an element.

oxide A binary compound of oxygen with an element.

oxidizing agent a) A chemical capable of supplying oxygen to another substance. b) A chemical, or an element contained in the chemical, which raises the valence of another element.

oxidizing atmosphere An atmosphere containing a substantial amount of oxygen so that any chemical reactions would normally be of substances with oxygen.

ozone The allotropic form of oxygen having three atoms per molecule. Ozone is represented as O_3.

ozone layer The ozone-containing portion of the stratosphere which absorbs harmful ultraviolet radiation from the sun.

P

paper A mat of short cellulose fibers.

parabola The graph obtained when one quantity varies as the square of another.

parallel circuit An electrical circuit in which electrons branch into alternate routes, and later come back together.

parathion An organic phosphate used as a nerve gas and insecticide.

particle A phenomenon exhibiting mass and a definite position.

particle accelerator A machine used to speed up particles so that they can be used to bombard the nuclei of substances.

particulate pollution (air) One- to ten-micron particles in the air.

passive metals Metals that are relatively inactive chemically, such as platinum.

pendulum A body hung so that it can swing freely back and forth.

perception The awareness of physical objects through the senses.

percussion instruments Musical instruments in which the tone is generated when some part is struck.

period (chemical) A horizontal row of the periodic table (also called a *series*).

periodic law The properties of the elements are a periodic function of their atomic numbers; that is, when the elements are arranged in order by atomic number, their properties repeat according to a definite pattern.

period of a pendulum The length of time required for one complete cycle of a pendulum, swinging back and forth.

period (wave motion) Time required for a wavelength to pass a point or the time required for one complete vibration of an object.

peroxide An oxide in which there is more oxygen in the compound than in the normal oxide—chemically, there is an oxygen-oxygen bond in the compound.

petroleum An oily, liquid mixture of many hydrocarbons.

pH A scale used to classify the strength of acid and basic solutions. (It is the negative logarithm of the hydrogen ion concentration in moles per liter in a solution.)

phenomenon (pl. phenomena) An event or action that can be observed.

phenophthalein A crystalline compound ($C_{20}H_{14}O_4$) which is colorless in an acid solution and pink in a basic solution.

phlogiston theory The erroneous belief that combustible materials contain a substance called phlogiston, said to be released as the substance burns.

phosphorescence The emission of radiation from a substance exposed to a light source after the light has been turned off.

phosphoric acid H_3PO_4.

photoelectric Electric effects due to the action of light.

photoelectric effect The emission of electrons from a metal surface when light is shined on it.

photon A particle of light.

physical change A change in a substance which alters its properties without changing it into a new substance.

physical metallurgy Science that deals with the adaption of metals for commercial use.

physical science The branch of science that includes chemistry and physics.

physicist One who pursues physics as a vocation.

physics The study of energy and the changes it undergoes.

pickling An industrial term for the use of HCl to remove rust and scale from steel.

pig iron The impure form of iron taken from a blast furnace.

pigment A coloring agent.

pitch a) The distance between two adjacent threads of a screw or bolt. b) The effect of the frequency of sound waves on the ears.

plasma Extremely hot, ionized gas.

plaster of Paris $(CaSO_4)_2 \cdot H_2O$ formed by heating gypsum.

plastics A large class of organic substances usually consisting of long-chain molecules which can be molded into a variety of shapes.

plastic deformation Permanent change in dimensions of an object under action of a stress.

plasticity That property of matter that enables it to assume a given shape under a load and remain in that shape after the load is removed.

pneumatic trough Large open container for water, used in collecting gases by the water displacement method.

pneumonitis A lung disease.

point mass An imaginary simplification in which the entire mass of an object is considered to reside in a single point.

polar Term used to describe compounds which possess positive and negative poles in their molecules.

pollution Contamination: physically, a substance contains undesirable and obnoxious impurities.

polymer A long-chain molecule consisting of many repeated chemical units.

polysulfides Polymer sulfides.

Portland cement A cement made by heating limestone with clays.

positive Characterized by a deficiency of electrons. The positive terminal of a battery takes up electrons from the external circuit.

potassium-argon dating method Determining the age of minerals containing radioactive potassium and a stable argon end-product, assuming that no argon was present in the mineral when it solidified.

potential difference Voltage or electromotive force.

potential energy Stored energy possessed by virtue of position or condition.

power The rate of doing work; work per unit of time.

prebiotic Before life appeared.

precipitate A solid substance that comes out of a liquid solution because of physical change or chemical reaction.

predictability The ability to anticipate what will happen in an experimental situation under a certain prescribed set of conditions.

prediction A foretelling of what will happen under certain conditions.

prejudice Previous learning or conditioning on a subject; bias.

pressure Force per unit area.

pretersonic Sound waves having a frequency greater than 5×10^8 Hz.

primary standard (for the kilogram) A platinum-iridium cylinder located at Sèvres, France.

primary winding The winding of a transformer which is connected to the source of electricity (the "input" of the device).

prism A solid whose bases or ends are any similar, equal, and parallel plane figures and whose lateral faces are parallelograms.

probability The likelihood of an event—prediction based on several factors when certainty is impossible.

process A sequence of operations or a sequence of chemical reactions.

protein A complex nitrogen-containing organic compound found in all living organisms.

proton A positively charged particle; a hydrogen atom nucleus.

pulley A wheel with a grooved rim.

pulsating direct current Direct current which is rapidly switched off and on.

pure substance Uncontaminated matter with a definite, fixed chemical composition, containing no other substance.

Pyrex glass Commercial name for a particular type of heat-resistant glass.

Q

qualitative Descriptive information in science not involving numbers.

quality The characteristics of a sound, determined by the mixture of overtones, that allow us to distinguish it from other sounds.

quanta Plural of quantum—discrete or separate packages or units of energy.

quantitative Using numbers or amounts to record scientific information.

quantum theory The theory that energy exists in discrete quantities rather than as a continuous entity.

quartz A transparent crystalline form of SiO_2.

quench To remove the heat rapidly from an object by dipping it quickly into a cold medium.

R

radar *Ra*(dio) *d*(etection) *a*(nd) *r*(anging).

radiation A process by which heat flows from a high temperature body to a body at lower temperature when the bodies are separated in space; also a term applied to all types of electromagnetic-wave phenomena.

radiation damage The embrittlement of structural materials by radiation, particularly neutron bombardment.

radical An ion made up of two or more elements.

radioactive dating Finding the age of samples based on the amount of radioactive substance present in the material.

radioactive decay The loss of energy from matter by radioactive emission.

radioactive fallout The descent to the earth of radioactive particles following a nuclear explosion.

radioactive isotope A particular isotope of an element that is radioactive.

radioactivity The emission of radiant energy, usually the spontaneous emission by certain elements of alpha, beta, and gamma rays.

radio astronomy The reception and analysis of radio spectra from astronomical bodies.

radiocarbon dating method Determining the age of once-living organisms using the radiocarbon (C-14) remaining in the substance.

radiometer A device used to detect radiant energy.

radio waves Electromagnetic radiation with a wavelength range of greater than 20 cm produced by electrical impulses.

rainbow An arch of light exhibiting the colors of the spectrum. It is formed opposite the sun during or after the close of a shower and is caused by refraction, reflection, and dispersion of light in drops of water.

react The act of producing other substances chemically.

reactor core The area in a nuclear reactor where the nuclear fission process occurs.

reactor vessel Large steel vessel used to contain the nuclear reactor.

reducing agent a) A chemical agent which removes oxygen from another compound. b) A chemical, or element thereof, which lowers the valence of another element.

reducing atmosphere An atmosphere containing substances that would prevent oxidation reactions. Oxides would be reduced or decomposed in a reducing atmosphere.

reduction a) Removal of oxygen from a compound. b) Lowering the valence of an element.

reduction, ore See ore reduction.

refine To purify.

reflect To turn back.

reflector, reactor Material used to reflect thermal neutrons back into the core of the nuclear reactor.

refraction The bending of rays as they pass from one medium to another—a wave phenomenon.

refractory material A material that is resistant to the action of extreme heat.

regular solid A solid of a standard geometric shape, such as a cube.

relativistic mass increase The faster a body travels the more massive it becomes —a consequence of the special theory of relativity.

replacement reaction An element replacing another element in a compound (same as "single replacement reaction").

resistance a) The load opposing the force when work is accomplished. b) Hindrance to current flow in an electrical circuit.

resonance The condition existing when vibrating systems of the same natural frequency transfer energy from one system to the other.

retina The inner membrane of the eyeball containing the light-sensitive rods and cones which receive the optical image.

reverberation Multiple echoes.

reversible reaction A reaction capable of reversing itself under the proper conditions. It can go forward or backward.

rhombic Having the shape of a rhombus (an equilateral parallelogram usually having oblique angles).

RNA A nucleoprotein.

rolling An operation used in forming metal sheet. A metal piece is passed between cylindrical rolls and squeezed into a sheet.

S

safety match A match that can be lit only by using a special striking surface containing phosphorus.

salt A substance formed when an acid reacts with a base.

saturated hydrocarbons Hydrocarbons containing only single bonds between carbons.

scalar A quantity having a magnitude but no direction. Examples: mass, density, volume.

scatter To disperse in different directions.

science See Section 1:1 (pages 3 and 4)

scientific journal A publication in which scientists report their research findings. Journals can be weekly, monthly, or quarterly.

scintillation A flash of light.

screening effect The weakening of the pull of a nucleus on the outer electrons because of the electron shells in between.

screw The simple machine which consists of an inclined plane wrapped around a cylinder or cone.

second (sec) The basic unit of time in both the metric and English systems. It is defined as 9,192,631,770 vibrations of an atom of cesium-133.

secondary standards Standards measured from the primary standard for the kilogram. In the United States designated as Kg number 20, kept at the Bureau of Standards in Washington, D.C.

secondary winding The winding of a transformer from which energy is taken (the "output" of the device).

second-class lever A lever in which the resistance is located between the fulcrum and the effort.

second law of thermodynamics Heat spontaneously goes from a hot object to a cold object—the law of science dealing with the direction of natural processes.

semiconductor A substance that is an electrical conductor at high temperatures but acts as an insulator at low temperatures. Examples: silicon and germanium.

series A horizontal row of the periodic table.

series circuit An electrical circuit in which the electrons are required to travel through each circuit element in order.

sewage Waste from domestic, commercial, and industrial establishments.

sheet lightning Controversial form of lightning believed by some to be a highly diffused electrical discharge occurring simultaneously over a vast area.

shell (electron) See electron shell.

short circuit A short cut through which electrons in an electrical circuit bypass a portion of their prescribed path.

short form table Form of the periodic table which has eight groups plus the rare gases. The A and B families are not separated as in the long form table.

shunt A resistor placed in parallel with a galvanometer to convert it to an ammeter.

significant figures Reliable figures in a number.

silica SiO_2, sand.

silicate The salt formed when a basic substance reacts with silicon dioxide (SiO_2).

silicone Molecular chains of silicon, oxygen, carbon, and hydrogen—inorganic rubbers.

silt Small soil and rock particles.

single bond Electron-pair bond between atoms.

single replacement reaction Same as "replacement reaction."

slag The product of the reaction of the undesirable parts of an ore or mineral with another substance.

slaked lime $Ca(OH)_2$.

slip rings Brass rings used to make contact with the brushes in A. C. motors and generators.

slug The English unit of mass weighing 32 lbs. at the earth's surface; that mass which would be given an acceleration of one ft/sec^2 by a force of one pound.

smelting A thermal process of ore reduction.

smog (photochemical) Air pollution held in an area by a temperature inversion at high altitudes. Light initiates many of the reactions necessary to produce the smog.

soap Sodium or potassium salts of long-chain fatty acids.

sodium hypochlorite NaClO—used as a bleach.

sol Colloidal suspension of a solid in a liquid.

solar cell A cell powered by energy taken from sunlight.

solenoid A coil of wire carrying an electric current and having the properties of a magnet.

solid A state of matter that has a definite shape and volume.

soluble Able to be dissolved.

solute A material that dissolves in another.

solution A homogeneous mixture composed of a solute and a solvent.

solvent A substance that dissolves another substance.

sonar *So*(und) *na*(vigation) and *r*(anging) using ultrasonic vibrations.

soprano A female voice of the highest range.

sound The sensation of hearing—physical cause is wave motion (vibration) through an elastic medium.

sound intensity Amount of energy transported by a sound wave per unit of time per unit of area.

spark chamber A device used to detect ionizing radiation by the ionization of a gas between electrically-charged plates.

special theory of relativity All the laws of physics must be the same for all inertial frames moving relative to each with uniform translational motion (a constant velocity).

specific heat The heat energy absorbed per unit of mass of a substance, per degree of rise in temperature.

specimen A piece or part of a class of substances that is representative of the complete class—used for testing.

spectroscope An optical instrument for forming and analyzing spectra emitted by substances.

spectroscopy The forming and analyzing of spectra emitted by substances.

spectrum An image formed by radiation directed through a spectroscope and brought to a focus in which each wavelength corresponds to a characteristic emission of the substance being tested.

speculation The act of theorizing or imagining.

speed A measure of how fast an object is moving. Speed differs from velocity in that speed is a scalar quantity, having a magnitude but no direction.

speed of light 3×10^8 m/sec in a vacuum.

split rings See commutator.

spontaneous Happening naturally without an external motivation

square centimeter A unit of area equal to the area of a square, one centimeter on a side.

square meter A unit of area equal to the area of a square, one meter on a side.

stable Having atoms so strongly bonded together that only extremely high temperatures can separate them; not radioactive.

stainless steel Alloy steels with high chromium-nickel content.

stalactite An icicle-like formation of calcium carbonate on the roof of a cave.

stalagmite A cylindrical or conical formation of calcium carbonate on the floor of a cave.

starch A class of carbohydrates—a polymer formed from the union of many sugar molecules.

state Particular physical condition or position of a substance.

state function (electron) Similar to electron clouds.

static Electrical discharges in the atmosphere that interfere with radio reception. The discharges are observed as noise in an AM radio.

static electricity The type of electricity characterized by charges at rest.

steel An iron-carbon alloy—a large class of iron-carbon alloys of variable composition.

step-up transformer A transformer which increases the voltage. Such a transformer is constructed with more turns of wire in the secondary than in the primary winding.

sterling silver A silver alloy containing 7.5 percent copper.

storage battery An electrical battery that can be repeatedly recharged.

S. T. P. Standard temperature and pressure, defined as 0°C and 760 mm of mercury respectively.

stringed instruments (string instruments) Musical instruments in which strings are set into vibration by bowing or plucking.

sublimation A solid changing directly into a gas without going through the liquid state.

subscript A small number or symbol written below and to the right of a quantity, as the 2 in T_2.

substance Anything physical that can be perceived by the senses.

sugar A class of crystalline carbohydrates with a sweet taste.

sulfate Any compound where a substance has combined with SO_4^{-2}.

sulfate ion SO_4^{-2}.

sulfide A compound where a metal has combined with sulfur.

sulfuric acid H_2SO_4.

suspension A uniform mixture of fine particles of one substance in another.

switch A device for turning an electric current off and on.

symmetry of time Time has no effect on the result of the same reproducible event (positive time will work just as well as negative time in physics equations).

synthesis Putting together; the formation of a compound by the combining of two or more simpler compounds, elements, or radicals.

synthesis reaction Two or more substances react to form a new substance.

T

talc A soft mineral consisting of a basic magnesium silicate.

Teflon Commercial name for a tetrafluorethylene polymer.

teleology Purposiveness or design in nature as an explanation of natural phenomena.

telescope An optical instrument for enlarging the image of a distant object.

temperature That property that is a measure of the thermal energy of the body.

terminal velocity The velocity of a falling object at which the upward force due to air resistance equals the downward force of gravity.

tetrahedron A geometric form found often in nature as the internal structure of compounds—a solid bounded by four plane triangular faces.

theistic evolution The attempted harmonization of evolution with a belief in God.

theory An idea that appears to be backed by some observed facts.

thermal energy Heat.

thermal expansion The increase in the dimensions of a material as its temperature increases.

thermal pollution The heating of natural water that has been used as a coolant in an industrial operation.

thermodynamics That branch of science which deals with the conversion of heat into work and energy transformations in general.

thermometer A device used to measure temperature.

thermonuclear power Energy available from the heat given off when atoms fuse.

thermos bottle See vacuum bottle.

thermostat A device for the automatic regulation of temperature.

third-class lever A lever in which the effort is located between the fulcrum and the resistance.

thistle tube Long tube with a funnel-type top to allow for the pouring of liquid in stoppered containers.

time dilation The apparent slowing down of time of a moving body—a consequence of the special theory of relativity.

transformation A change.

transformer An electrical device used to increase or decrease the voltage available from an alternating current source.

transistor An electronic device that controls electrical current flow by the use of semiconductors.

transition elements (transition metals) Those metallic elements whose atomic numbers are 21-30, 39-48, 57-80, and 89-105 inclusive. Although there are some exceptions, the transition elements are usually defined as elements which place electrons in outer shells before inner shells are filled up.

transmutation The process of changing one substance into another.

triatomic Consisting of three atoms.

tribulation period The seven-year judgment God brings upon the earth (read the book of Revelation) because of man's sin.

triple bond Three electron-pairs form a bond between atoms.

tritium The radioactive isotope of hydrogen having two neutrons in its nucleus.

tsetse fly A small blood-sucking fly of southern Africa whose bite can transmit the parasite that causes sleeping sickness.

turbine An engine of one or more rotary units (like a fan) mounted on a shaft—will turn when acted on by a fluid under pressure.

Tyndall effect The scattering of light due to its passage through a medium of suspended particles.

U

ultramicroscope A microscope with a strong light to view colloidal particles.

ultrasonic Above the audible range—greater than 2×10^4 Hz.

ultraviolet radiation Electromagnetic radiation with a wavelength range of 10^{-9} to 4×10^{-7} m—thought to be produced by electronic transitions in matter.

unbalanced force Any force on a body which is not opposed by an equal force in the opposite direction.

uncertainty In science, the impossibility of measuring exactly two quantities at the same time.

uniformitarianism The belief that the present is the key to the past.

units A label, such as "inches" or "feet," used to specify the terms in which a measurement is being reported.

universal gravitational constant The proportionality constant in Newton's law of gravitation, represented by G.

universal negative A blanket statement of denial applied to an extensive region of space or time.

unsaturated hydrocarbons Hydrocarbons containing carbon-carbon double or triple bonds.

uranium-lead dating method Determining the age of minerals containing radioactive uranium and a stable lead end-product, assuming that no lead was present in the mineral when it solidified.

V

vacuum bottle (thermos bottle) A thermos bottle used to contain liquids which slows down heat transfer by conduction, convection, and radiation.

vacuum distillation The process of boiling and recondensing a liquid at reduced pressure.

valence The number of electrons gained or lost by an atom in a chemical reaction.

Van Allen belts Two regions of the earth's near-surround in which charged particles from space become temporarily trapped by the earth's magnetic field.

Van de Graaff generator Machine used to generate extremely high voltages at very low currents. Charges are transported to or from a metal dome by a motor-driven belt.

van der Waals' force Weak bonding forces—electron distribution in an atom causes an instantaneous dipole to form—the opposite charged dipoles attract each other briefly.

vapor The gaseous state of a substance.

variable A quantity which can change during the course of an experiment or calculation.

vector A directed line segment; a quantity having both a magnitude and a direction. Examples: velocity, acceleration, and momentum.

velocity Change in location per unit time. Velocity is a vector quantity, having both a magnitude and a direction.

viral encephalitis Sleeping sickness caused by a virus transmitted through an insect bite—inflammation of the brain (brain fever).

vital processes Processes necessary to maintain life in an organism.

vitamins Naturally occurring organic substances that are necessary for life processes to operate.

volatile Easily evaporated at a relatively low temperature.

volt (v) The unit of potential difference named in honor of Alessandro Volta. (Two points in a circuit have a potential difference of one volt if it takes one joule of work to move a coulomb of charge from one point to the other.)

voltage Potential difference or electromotive force expressed in volts.

voltaic cell Any device that changes chemical energy into electrical energy.

voltmeter An instrument for measuring potential difference or electromotive force in volts.

volume A measure of the amount of space occupied by a three-dimensional object. Volume is expressed in cubic units such as ft^3 or m^3.

volume expansion Usually three times the linear expansion of any material.

vulcanization The process of treating crude rubber with sulfur at high temperatures—the resultant rubber is stronger and more elastic.

W

water displacement method The technique of collecting a gas by allowing it to bubble into an inverted bottle of water.

water glass A solution of sodium silicate in water.

water (hard) See hard water.

water of hydration Water incorporated into the crystal structure of a compound (also called water of crystallization).

water vapor Water in the gaseous state.

watt (w) The power needed to accomplish one joule of work in one second.

wave A phenomenon exhibiting wave motion.

wave crest The "high" portion of a wave.

wavelength The combination of a crest and a trough of a wave.

wave mechanics The branch of physics which treats atomic particles as wave phenomena.

wave motion A periodic traveling disturbance.

wave trough The "low" portion of a wave.

wedge The simple machine formed by combining two inclined planes.

weight The downward force on an object due to the earth's gravity.

weightlessness A condition of zero gravitational pull, as in an earth satellite.

wet cell A voltaic cell in which the electrolyte is a liquid.

wheel-and-axle The simple machine consisting of two pulleys of different sizes fastened to the same shaft.

white light Light from the sun or light containing a mixture of wavelengths.

Wimshurst generator Rotary action static electric machine used to develop high voltages at extremely low currents.

windlass Special type of wheel-and-axle in which a crank takes the place of the wheel.

woodwind instruments Wind instruments having openings in the side of the air column. The openings can be opened or closed in various combinations to change the pitch of the instrument.

work The physical quantity defined as the product of force and distance.

workability Ability of an experiment or piece of apparatus to produce the desired result; useful, practical, possible.

working hypothesis A temporary explanation based on limited knowledge.

X

X-ray astronomy The reception and analysis of X-ray spectra from astronomical bodies.

X-rays Electromagnetic radiation with a wavelength range of 10^{-12} to 10^{-8}m—produced by the rapid deceleration of electrons.

Index

Credits

Cover Photo Union Carbide Nuclear Corporation; **Page XII** Copyright by the California Institute of Technology and the Carnegie Institution of Washington; **Page 20** Parke, Davis & Company; **Page 64** Standard Oil of California; **Page 92** Cryogenic Technology; **Page 112** Environmental Science Services Administration; **Page 226** United States Navy Photo; **Page 248** Douglas Whiteside; **Page 382** Unusual Films; **Page 508** Bell Telephone Laboratories.

Figure 1:5 United States Atomic Energy Commission; **Figure 1:12** United States Naval Observatory; **Figure 2:2** National Aeronautics and Space Administration; **Figure 2:8** Copyright by the California Institute of Technology and The Carnegie Institution of Washington; **Figure 3:9** (a) Photo by Evan Hansen (b) Creation Research Society; **Figure 3:14** (b) National Transportation Safety Board; **Figure 3:15** Questar Corporation; **Figure 3:16** Pfizer Incorporated.

Figure 4:1 Firestone Tire and Rubber Company; **Figure 4:2** W. R. Grace and Company; **Figure 4:3** Parke, Davis and Company; **Figure 4:4** Pfizer Incorporated; **Figure 4:5** General Motors Corporation; **Figure 4:6** National Aeronautics and Space Administration; **Figure 4:8** Corning Glass Works; **Figure 4:9** Linde Division, Union Carbide; **Figure 4:11** Wilson Instrument Division, American Chain & Cable Company; **Figure 4:14** United States Steel Corporation; **Figure 4:15** (a) Pfizer Incorporated (b) Chamber of Mines of South Africa; **Figure 5:6** Duke Power Company; **Figure 5:9** Westinghouse Photo; **Figure 5:10** Cryogenic Technology; **Figure 5:16** National Aeronautics and Space Administration; **Figure 6:1** Fisher Scientific Company; **Figure 7:7** Corning Glass Works; **Figure 7:8** EM Laboratories; **Figure 7:9** Courtesy of Hach Chemical Company.

Figure 9:1 The Stanley Works; **Figure 9:3** National Bureau of Standards; **Figure 9:9** National Bureau of Standards; **Figure 9:10** Ohaus Scale Corporation; **Figure 9:11** Ohaus Scale Corporation; **Figure 9:12** Ohaus Scale Corporation; **Figure 9:14** National Bureau of Standards; **Figure 9:15** National Bureau of Standards; **Figure 9:16** National Bureau of Standards; **Figure 9:17** United States Navy Photo; **Figure 9:19** Pickett Industries; **Figure 9:20** Keuffel & Esser Company; **Figure 10:3** (a and b) Central Scientific Company; **Figure 10:5** Kollsman Instrument Company; **Figure 10:6** United States Navy Photo; **Figure 10:8** (a and b) Taylor Instruments; **Figure 10:9** Sikorsky Aircraft; **Figure 10:10** United States Air Force Photo; **Figure 10:11** United States Department of Commerce; **Figure 10:13** Gas Equipment Department, Air Products and Chemicals; **Figure 10:17** Taylor Instruments; **Figure 10:18** Mettler Instrument Corporation.

Figure 11:1 Photo by Roy E. Waite; **Figure 11:2** Photo by Evert Fruitman; **Figure 11:3** Geological Survey of Canada; **Figure 11:5** Union Pacific Railroad Color Photo; **Figure 11:6** Bendix Environmental Science Division; **Figure 12:4** Linde Division, Union Carbide Corporation; **Figure 12:5** Arthur D. Little, Incorporated; **Figure 12:6** United States Navy Photo; **Figure 12:7** Matheson Gas Products; **Figure 12:10** Airco Welding Products; **Figure 12:12** Photo by Evert Fruitman; **Figure 12:13** Matheson Gas Products; **Figure 12:17** United States Department of Commerce; **Figure 12:18** Union Carbide Nuclear.

Figure 13:3 Utah Department of Fish and Game; **Figure 13:4** Morton-Norwich Products; **Figure 13:5** Morton-Norwich Products; **Figure 13:6** Church and Dwight Company; **Figure 13:7** Atomics International Division of North American Rockwell; **Figure 13:8** Israeli Embassy; **Figure 13:9** Union Carbide Corporation, Nuclear Division; **Figure 13:10** Virginia Polytechnic Institute and State University; **Figure 13:11** United States Gypsum; **Figure 13:12** Portland Cement Association; **Figure 13:13** United States Bureau of Reclamation; **Figure 13:15** Luray Caverns, Virginia; **Figure 13:16** Creation Research Society, Photo by Tex Helm; **Figure 13:18** Calgon Corporation; **Figure 13:19** Crystal Research Laboratories Incorporated; **Figure 13:21** Dow Chemical Corporation; **Figure 13:22** Brush Wellman, Incorporated; **Figure 14:1** South African Information Service; **Figure 14:6** Pickards Mather and Company; **Figure 14:7** Borg-Warner Corporation; **Figure 14:8** Cities Service Corporation; **Figure 14:9** United States Bureau of Reclamation; **Figure 14:10** United States Atomic Energy Commission; **Figure 14:11** United States Atomic Energy Commission; **Figure 14:12** International Rectifier Corporation; **Figure 14:19** Brockway Glass Company; **Figure 14:20** PPG Industries; **Figure 14:21** Brockway Glass Company; **Figure 14:22** Corning Glass Works; **Figure 14:23** Corning Glass Works; **Figure 14:24** General Shale Products Corporation; **Figure 14:25** Rockwell International; **Figure 14:26** Dow Corning Corporation; **Figure 14:27** Ralph Keen; **Figure 14:28** Motorola Corporation; **Figure 14:29** Motorola Corporation; **Figure 15:4** Universal Oil Products Company; **Figure 15:6** Mobile Oil Corporation; **Figure 15:13** United States Department of Agriculture Soil Conservation Service; **Figure 15:15** Eastman Kodak Company; **Figure 15:16** Crompton and Knowles Corporation; **Figure 15:17** Firestone Tire and Rubber Company; **Figure 15:18** E. I. DuPont de Nemours and Company; **Figure 15:19** Goodyear Tire and Rubber Company; **Figure 16:1** Pfizer Incorporated; **Figure 16:2** Savannah Sugar Refining Corporation; **Figure 16:3** Philips Electronic Instruments; **Figure 16:5** Champion International; **Figure 16:6** Huyck Corporation; **Figure 16:7** Georgia Pacific Corporation; **Figure 16:8** Pet Incorporated; **Figure 16:9** Consolidated Natural Gas Company; **Figure 16:13** The Ealing Corporation; **Figure 16:14** Corning Glass Works; **Figure 17:3** Westinghouse Photo; **Figure 17:4** Allis-Chalmers.

Figure 18:1 Matheson Gas Products; **Figure 18:3** John Deere; **Figure 18:4** United States Department of Agriculture Soil Conservation Service; **Figure 18:5** United States Department of Agriculture Soil Conservation Service; **Figure 18:7** W. R. Grace and Company; **Figure 18:8** W. R. Grace and Company; **Figure 18:11** United States Navy; **Figure 18:12** United States Navy Photo; **Figure 18:14** Edgewood Arsenal, Department of the Army; **Figure 18:17** Diamond Match Company; **Figure 18:18** W. R. Grace and Company; **Figure 18:19** Viking Corporation; **Figure 19:1** Freeport Minerals Company; **Figure 19:6** Argonne National Laboratory; **Figure 19:13** Firestone Tire and Rubber Company; **Figure 20:4** PPG Industries; **Figure 20:6** Inland Steel Company; **Figure 20:7** Wear-Ever Aluminum Incorporated; **Figure 20:8** Unusual Films; **Figure 20:9** Unusual Films; **Figure 20:10** Eastman Kodak Company; **Figure 20:11** Eastman Kodak Company; **Figure 21:2** Goodyear Tire and Rubber Company; **Figure 21:3** United States Navy Photo; **Figure 21:4** Arthur D. Little; **Figure 21:5** Arthur D. Little; **Figure 21:7** Photo by Evert Fruitman; **Figure 21:9** Brookhaven National Laboratory; **Figure 21:10** GTE Sylvania; **Figure 21:12** Central Scientific Company; **Figure 21:13** Photo by Evert Fruitman; **Figure 21:14** Argonne National Laboratory; **Figure 21:15** Argonne National Laboratory; **Figure 21:16** Argonne National Laboratory.

Figure 22:1 Bausch and Lomb; **Figure 22:2** Specimens prepared and photomicrographs taken by Carl Scumaker, Continental Telephone Laboratories; **Figure 22:5** Homestake Mining Company; **Figure 22:6** St. Joe Minerals Corporation; **Figure 22:7** Anaconda Company; **Figure 22:8** (a) Republic Steel Corporation (b) Bethlehem Steel Corporation; **Figure 22:10** (a) Republic Steel Corporation (b) Bethlehem Steel Corporation; **Figure 22:11** American Iron and Steel Institute; **Figure 22:12** Bethlehem Steel Corporation; **Figure 22:14** (a) Bethlehem

Steel Corporation (b) Jones and Laughlin Steel Corporation (c) Cincinnati Milacron Incorporated; **Figure 22:15** Bethlehem Steel Corporation; **Figure 22:16** Allegheny Ludlum Steel Corporation; **Figure 22:18** Kaiser Aluminum Corporation; **Figure 22:20** Reynolds Metals Company; **Figure 22:21** Reynolds Metals Company; **Figure 22:22** Anaconda Company; **Figure 23:1** (a) Redevelopment Authority of the City of Philadelphia (b) United States Department of Agriculture Soil Conservation Service; **Figure 23:2** (a and b) Photo by United States Forest Service (c) United States Department of Agriculture Soil Conservation Service; **Figure 23:4** Union Carbide Corporation; **Figure 23:5** Montana Power Company; **Figure 23:7** Environmental Protection Agency; **Figure 23:8** United States Department of Agriculture Soil Conservation Service; **Figure 23:9** Lab Con Company; **Figure 23:10** Joy Manufacturing Company.

Figure 24:1 Bell Labs Photo; **Figure 24:12** (a) The Stanley Works (b) Bucyrus-Erie Company; **Figure 24:14** Bucyrus-Erie Company; **Figure 24:17** Schwinn Bicycle Company; **Figure 24:20** The Stanley Works; **Figure 24:23** United States Department of Transportation; **Figure 24:25** (a, b, and c) The Stanley Works; **Figure 24:27** Eastman Kodak Company; **Figure 24:28** Toce Brothers, Manufacturing Limited; **Figure 24:30** Caterpillar Tractor Company; **Figure 25:9** United States Air Force Photo; **Figure 25:13** United States Air Force Photo; **Figure 25:17** Young & Rubicam; **Figure 25:18** Brunswick Corporation; **Figure 25:22** United States Naval Observatory; **Figure 25:25** Carson Astronomical Instruments Incorporated; **Figure 25:26** The Stanley Works; **Figure 25:27** United States Bureau of Reclamation; **Figure 26:7** National Aeronautics and Space Administration; **Figure 26:8** National Aeronautics and Space Administration; **Figure 26:13** Photo by Evert Fruitman; **Figure 26:15** The Torsion Balance Company; **Figure 26:16** National Aeronautics and Space Administration; **Figure 26:17** United States Navy Photo; **Figure 26:18** Goodyear Tire and Rubber Company; **Figure 26:20** United States Navy Photo; **Figure 26:21** United States Navy Photo.

Figure 27:3 Cincinnati Milacron; **Figure 27:17** Exxon Corporation; **Figure 27:28** Carolina Power and Light; **Figure 27:29** Westinghouse Electric Company; **Figure 28:11** York Division, Borg-Warner Corporation; **Figure 28:19** Bucyrus-Erie Company; **Figure 28:20** Cities Service; **Figure 28:21** Cities Service; **Figure 28:22** Cities Service.

Figure 30:16 Photo by Evert Fruitman; **Figure 30:18** Xerox Corporation; **Figure 30:20** Photo by Roy E. Waite; **Figure 30:21** Photo by Evert Fruitman; **Figure 30:22** General Electric; **Figure 31:1** (a) Western Electric (b) Photo by Evert Fruitman; **Figure 31:4** DW "Multi-Switch," Switches Switchcraft; **Figure 31:7** Allen Bradley Company; **Figure 31:8** Allen Bradley Company; **Figure 31:9** Photo by Evert Fruitman; **Figure 31:14** Simpson Electric Company; **Figure 31:16** Georgia Power Company; **Figure 31:17** Georgia Power Company; **Figure 31:18** United States Department of Interior; **Figure 31:28** Burgess Division, Gould; **Figure 31:30** ESB Corporation; **Figure 31:33** International Silver Company; **Figure 31:34** Kaiser Aluminum & Chemical Corporation; **Figure 31:35** Reynolds Metals Company; **Figure 31:36** United States Atomic Energy Commission; **Figure 32:1** Ward's Natural Science Establishment; **Figure 32:13** General Electric; **Figure 32:14** Western Electric; **Figure 32:15** Western Electric; **Figure 32:16** Union Pacific Railroad Colorphoto; **Figure 32:18** Utah Electronics; **Figure 32:26** Photo by Evert Fruitman; **Figure 32:28** Bureau of Reclamation; **Figure 32:30** United States Department of Commerce.

Figure 33:5 IU International Corporation; **Figure 33:14** United States Navy Photo; **Figure 34:13** Corning Glass Works; **Figure 34:23** Corning Glass Works; **Figure 34:24** Honeywell Corporation; **Figure 34:25** (a) Carl Zeiss, Incorporated (b) American Optical Company; **Figure 34:26** (a) United States Naval Observatory (b) Bausch and Lomb; **Figure 34:27** Corning Glass Works; **Figure 34:30** Bausch and Lomb; **Figure 34:33** Photo by Mrs. Earl Nutz; **Figure 34:37** (a) Photo by Evert Fruitman (b) United States Navy Photo; **Figure 34:39** Dravo Corporation; **Figure 34:44** Weston Instruments; **Figure 34:48** Bell Laboratories; **Figure 34:50** Spectra-Physics; **Figure 35:3** Perkin-Elmer Corporation; **Figure 35:4** Perkin-Elmer Corporation; **Figure 35:6** Florida Power and Light Company; **Figure 35:8** Photo by Evert Fruitman; **Figure 35:9** Bell Telephone Laboratories; **Figure 35:10** Bell Telephone Laboratories; **Figure 35:17** Harris-Intertype Corporation; **Figure 35:20** Gates Division, Harris-Intertype Corporation; **Figure 35:21** United States Naval Research Laboratory; **Figure 35:23** Corning Glass Works; **Figure 35:26** Philips Electronic Instruments; **Figure 35:27** Barge Memorial Hospital; **Figure 35:28** Allis-Chalmers; **Figure 35:30** Allis-Chalmers; **Figure 35:32** Philips Electronic Instruments; **Figure 35:33** Philips Electronic Instruments; **Figure 35:34** Philips Electronic Instruments; **Figure 35:35** National Aeronautics and Space Administration; **Figure 35:36** Carl Zeiss, Incorporated; **Figure 35:37** JOEL (U. S. A.); **Figure 35:38** JOEL (U. S. A.); **Figure 35:41** Central Scientific Company; **Figure 36:6** Bell Telephone Laboratories; **Figure 36:7** Bell Telephone Laboratories; **Figure 36:10** Shure Brothers; **Figure 36:11** Shure Brothers; **Figure 36:12** Koss Stereophones; **Figure 36:13** Koss Stereophones; **Figure 36:18** General Radio Company; **Figure 36:19** National Aeronautics and Space Administration; **Figure 36:20** From H. Fletcher reviews of *Modern Physics*, January 1940; **Figure 36:23** Krautkramer-Branson; **Figure 36:24** Krautkramer-Branson; **Figure 36:25** Bendix Corporation; **Figure 36:26** Bendix Corporation; **Figure 36:28** United States Navy Photo; **Figure 36:29** Stanford Research Institute; **Figure 37:10** William Lewis & Son; **Figure 37:11** Selmer Division of the Magnavox Company; **Figure 37:12** King Musical Instruments; **Figure 37:13** Selmer Division of the Magnavox Company; **Figure 37:14** Selmer Division of the Magnavox Company; **Figure 37:18** M. P. Moller, Incorporated; **Figure 37:19** M. P. Moller, Incorporated; **Figure 37:20** Allen Organ Company; **Figure 37:27** Johns-Manville Corporation; **Figure 37:28** Johns-Manville Corporation; **Figure 37:29** Bell Telephone Laboratories.

Figure 38:10 Sargent-Welch Scientific Company; **Figure 38:11** Sargent-Welch Scientific Company; **Figure 38:12** Sargent-Welch Scientific Company; **Figure 38:14** Brookhaven National Laboratory; **Figure 38:15** Argonne National Laboratory; **Figure 39:3** General Electric Corporation; **Figure 39:4** Brookhaven National Laboratory; **Figure 39:5** National Accelerator Laboratory; **Figure 39:6** Stanford Linear Accelerator Center; **Figure 39:10** Nuclear Division, Union Carbide Corporation; **Figure 39:12** Los Alamos Scientific Laboratory; **Figure 39:13** United States Atomic Energy Commission; **Figure 39:15** Brookhaven National Laboratory; **Figure 39:16** Atomic Energy Division, Babcock & Wilcox Company; **Figure 39:18** Atomic Energy Division, Babcock & Wilcox Company; **Figure 39:19** Babcock & Wilcox Company; **Figure 39:20** Atomic Energy Division, Babcock & Wilcox Company; **Figure 39:21** Commonwealth Edison Company; **Figure 39:22** Atomics International; **Figure 39:23** Brookhaven National Laboratory; **Figure 39:24** United States Navy Photo; **Figure 39:25** United States Navy Photo; **Figure 39:26** Los Alamos Scientific Laboratory; **Figure 39:28** Sealectro Corporation; **Figure 39:29** Allis-Chalmers; **Figure 39:30** Allis-Chalmers; **Figure 39:31** United States Naval Observatory; **Figure 39:32** Questar Corporation; **Figure 40:1** Unusual Films.

About the Authors

Emmett L. Williams is an acknowledged authority in the related fields of chemistry and physics and a well-known spokesman for the creationist position. He holds bachelor of science and master of science degrees from Virginia Polytechnic Institute and the doctor of philosophy degree from Clemson University.

After serving as metallurgical engineering professor at VPI, he did research in the ceramic engineering department at Clemson in 1964. Dr. Williams joined the Bob Jones University science faculty in 1966. In 1972 he was named head of the department of physics.

Author Williams has been a board member of the Creation Research Society for a number of years and has served as its vice president. He is a member of the American Association of Physics Teachers and the American Institute of Physics; and he has been chosen for Alpha Sigma Mu (metallurgical engineering honorary), Sigma Gamma Epsilon (earth science honorary), and Sigma Xi (research honorary).

Dr. Williams has served with the Atomic Energy Division of Babcock and Wilcox, Union Carbide Nuclear Corporation, Inland Motor Corporation, Electrotec Corporation, and Superior Continental Corporation.

He has authored many magazine and technical articles, including "This Old World Isn't Old," "Entropy and the Solid State," "Is the Universe a Thermodynamic System," and "Faith in Fact or Fairy Tale?"

George Mulfinger, Jr., is uniquely qualified for authorship of a textbook. He has a broad background in the fields of education, research, and scientific journalism.

Mr. Mulfinger was graduated *summa cum laude* in chemistry from Syracuse University with awards in physical chemistry and organic chemistry and membership in Phi Beta Kappa. He also received the master of science degree in physics from Syracuse. He has completed most of the work required for the doctor of philosophy degree, with additional graduate study at Harvard University and the University of Georgia. The physics specialist has received research grants from the National Science Foundation and the Creation Research Society.

Prior to joining the science faculty at Bob Jones University in 1965, author Mulfinger taught science on the high school level for 10 years. He was head of the mathematics and science departments of Elbridge (New York) Central School and later chairman of the science department at Syracuse Central Technical High School.

Mr. Mulfinger is a member of several scientific societies and a regular contributor to magazines and scientific journals. He has authored a number of booklets on science and the Bible, including *The Flood and the Fossils* and *How Did the Earth Get Here?*